APPLIED AND COMPUTATIONAL COMPLEX ANALYSIS

VOLUME 1

Power Series—Integration—Conformal Mapping —Location of Zeros

PETER HENRICI
Professor of Mathematics
Eidgenössische Technische Hochschule, Zürich

A WILEY-INTERSCIENCE PUBLICATION

JOHN WILEY & SONS, New York . London . Sydney . Toronto

Copyright © 1974, by John Wiley & Sons, Inc.

All rights reserved. Published simultaneously in Canada.

No part of this book may be reproduced by any means, nor transmitted, nor translated into a machine language without the written permission of the publisher.

Library of Congress Cataloging in Publication Data:
Henrici, Peter, 1923–
  Applied and computational complex analysis.

  (Pure and applied mathematics)
  "A Wiley-Interscience publication."
  Bibliography: v. 1, p.
  1. Analytic functions. 2. Functions of complex variables. 3. Mathematical analysis. I. Title.

QA331 .H453   1974   515′.9   73-19723

ISBN 0-471-37244-7

Printed in the United States of America

10 9 8 7 6 5 4 3 2 1

Dedicated to the Memory of

HEINZ RUTISHAUSER

PREFACE

This book constitutes the first installment of a projected three-volume work that will present applications as well as the basic theory of analytic functions of one or several complex variables. Applications are made to other branches of mathematics, to science and engineering, and to computation. The algorithmic attitude toward mathematics—not to consider a problem solved unless an algorithm for constructing the solution has been found—prevails not only in the sections devoted to computation but forms one of the work's unifying themes.

A short overview of the three volumes is in order. The first volume, after laying the necessary foundations in the theory of power series and of complex integration, discusses applications and basic theory (without the Riemann mapping theorem) of conformal mapping and the solution of algebraic and transcendental equations. The second volume will cover topics that are broadly connected with ordinary differential equations: special functions, integral transforms, asymptotics, and continued fractions. The third volume will center similarly around partial differential equations and will feature harmonic functions, the construction of conformal maps, the Bergman-Vekua theory of elliptic partial differential equations with analytic coefficients, and analytical techniques for solving three-dimensional potential problems.

In collecting all these topics under one cover, I have been guided by the idea that for today's applied mathematician it is not enough to specialize, however deeply, in any single narrowly restricted area. He should also be made aware as forcefully as possible of the light that radiates from the basic theories of mathematics into the neighboring fields of science. What I have tried to do here for complex analysis, should as part of an applied mathematics curriculum also be done, for instance, in real analysis and linear algebra.

The prerequisites for Volume I include advanced calculus, matrix calculus, and a smattering of modern algebra. I have not attempted to hold the level of presentation to a uniform level; however, even in those places in which advanced notions are introduced (e.g., §2.1, 3.5, 4.6, 5.9, and 6.12) the reader will find that the presentation is self-contained and, hopefully, well motivated. It is felt that, with some omissions perhaps, this book could form the basis of a well-rounded full-year course at the senior/first-year graduate level of a modern, computation-oriented curriculum in applied mathematics.

A word about the layout of the contents of the present volume may be necessary. It is well known that there are two essentially different approaches to complex variable theory: Riemann's approach, based on complex differentiability, and the approach of Weierstrass, based on power series. Although the Weierstrassian example has been followed in a venerable series of classical texts (Whittaker and Watson [1927], first ed. [1903], Hurwitz and Courant [1929], Dienes [1931]), the "geometric" method of Riemann, brilliantly exposed, for example, by Ahlfors [1953], seems to have been preferred by most analysts to the "computational" method of Weierstrass, at least until recently, when under the influence of Bourbaki (Cartan [1961]) power series came back into fashion.

Since this book is deliberately oriented toward computation, it was almost a foregone conclusion that the theory be based on power series. A computing machine cannot really deal with functions defined on a continuum, but it can readily handle sequences of power series coefficients and perform all kinds of algebraic operations on them. The elementary functions, and many of the important transcendental functions of mathematical physics, are most easily defined by power series. In my treatment of power series I make a sharp distinction between the purely algebraic aspects (Chapter 1) and the analytic aspects of the theory. In treating the analytic aspects I distinguish between properties of functions analytic at a point (Chapter 2) and functions analytic in a region (Chapter 3).

The emphasis on power series has several didactic advantages. First, it provides the student with the opportunity to renew his acquaintance with the basic notions of algebra (group, ring, field, isomorphism) which are illustrated here by interesting examples that are too elaborate to be treated in detail in the usual algebra course. Second, it teaches him to differentiate between the algebraic-computational and the analytic-functional content of the theory and to appreciate the increase in depth when proceeding to the latter. Third and most important, the early treatment of the basic facts on analytic continuation made possible by power series will impress the student at an early stage with the fundamental inner coherence of analytic functions which so completely sets them apart from more general types of function.

After having described the overall trend of the book, I now wish to comment briefly on the treatment of certain details. The hypergeometric

PREFACE

identities developed in §1.6 are not immediately required for the theory that follows. They are inserted here as a source of illustrations and as an example of how the "formal" apparatus of power series can lead to concrete results. The formal treatment of the Lagrange-Bürmann expansion in Section 1.9, on the other hand, is indispensable to the theory of the inverse function in §2.4.

Chapter 2 discusses power series with values in a Banach algebra, not only because this discussion can be carried through with absolutely no increase in complexity over the usual treatment of power series but also because it provides an excellent opportunity to introduce certain matrix-valued functions that are required in the theory of ordinary differential equations (Chapter 9). By using some simple notions from functional analysis we can once again relate complex analysis to another important branch of theoretical mathematics.

In Chapter 3 the discussion of analytic continuation along an arc calls for the concept of homotopy, which is also required for a version of the Cauchy integral theorem. A nonstandard topic in Chapter 3 is the constructive treatment of Weierstrassian analytic continuation (§3.6).

In the treatment of integration (Chapter 4) the local version of Cauchy's theorem becomes a triviality if analyticity is defined by power series. The fact that Goursat's generalization finds no place here does no irreparable damage because this result, although of great historical and intellectual interest, is seldom used in applied mathematics. In regard to nonlocal versions of Cauchy's theorem I have heeded the advice of my colleague and former teacher Albert Pfluger to present a homotopy version of the theorem before entering into homology. This suffices for a treatment of the Laurent series, which includes some applications to Bessel functions and Fourier series. In place of a formal treatment of homology theory, whose full power again is seldom required in applied mathematics, I use winding numbers to prove the Jordan curve theorem (for piecewise smooth curves). Thus in the theory of residues the concept of the "interior" of a Jordan curve with all its great intuitive appeal can be applied without embarrassment. In addition to the applications of residues to integration, I discuss the summation of infinite series by summatory functions, including the Plana summation formula.

Chapter 5, on conformal mapping, after a brief discussion of elementary maps, features a careful study of the Joukowski map, a thorough understanding of which is essential to many applications. The treatment of Moebius transformations emphasizes symmetry rather than the cross-ratio. Added to technically motivated applications of conformal mapping to electrostatics and fluid dynamics is a treatment of Poisson's equation which includes a proof of Saint-Venant's isoperimetric inequality for torsional rigidity. This chapter concludes with an extensive discussion of the Schwarz-Christoffel mapping function, including a systematic treatment of the rounding of corners in Schwarz-Christoffel maps.

In Chapter 6 we enter virgin territory for most complex analysts. The goal here is to discuss complex variable methods for determining the zeros of any given polynomial (and, in some cases, of arbitrary analytic functions) to arbitrary precision. Our survey contains algorithms based on the ideas of search (D. H. Lehmer), exclusion (H. Weyl), iteration (Newton-Schröder-Laguerre), and descent (K. Nickel and H. Rutishauser), preceded by classical material on the geometry of zeros, including the algorithms of Sturm, Routh, and Schur-Cohn for determining the number of zeros in an interval, a half-plane, and a disk. A new feature here is the use of circular arithmetic, which reveals the well-known theorem of Laguerre as a special case of a much more general result.

Chapter 7, on partial fractions, begins with a section on the actual construction of the partial fraction decomposition, a topic that is frequently neglected. This is followed by applications of partial fractions to the summation of series, to interpolation by sums of exponentials, to combinatorial analysis, and to difference equations. The main part of the chapter is then taken up by a presentation of the quotient-difference algorithm due to Rutishauser. Contrary to earlier treatments, we discuss fully a number of exceptional cases (e.g., poles of higher order, sets of equimodular poles). It is hoped that our presentation will convince the reader that classical analysis indeed has something to offer to computation.

No author should be so dictatorial that he expects his book to be read in strict order, from cover to cover. If, as in this volume, the size is considerable, it seems almost mandatory to keep the chapters reasonably self-contained by limiting their mutual dependence to a few clearly referenced key facts. In my case, I was more or less forced to adopt this attitude by presenting most of the chapters as separate short courses at the Swiss Federal Institute of Technology in Zurich (ETHZ) and some of them at the University of California at Los Angeles (UCLA). No power series are required to understand Chapter 5, and integration is essential only late in this chapter. Conversely, a careful study of §5.4 will enable the reader to understand most of Chapter 4 with only cursory references to the first three chapters. (Here we even sketch a proof of Cauchy's theorem based on differentiability, using the Stokes formula.) Chapter 6 (polynomials) has few prerequisites except Moebius transformations, the Schwarz lemma, and the principle of the argument. Chapter 7 (partial fractions) is independent of Chapter 6 and presupposes little more than the Cauchy coefficient estimate and the theory of isolated singularities. From Chapter 4 on many sections can be regarded as self-contained small essays and as such

PREFACE

xi

are suitable for seminar presentation. A number of them also invite numerical experimentation. Some possibilities for experimentation are formulated at the end of each chapter as "seminar assignments" most of which have been tried by students at the ETHZ, either in the form of term papers or in connection with seminar presentations.

Some of my work on Chapter 6 was supported by the Office of Naval Research during the summer of 1969 at UCLA. I also wish to express my gratitude to numerous individuals who helped me to shape this treatise. In the first place I should mention those anonymous undergraduate students at ETHZ who by their reactions to my lectures often encouraged me to seek better ways to organize my material. Sincere thanks are due to my assistants A. Friedli, M. Gutknecht, and M.-L. Kaufmann for reading parts of the manuscript and suggesting many improvements. Among my former Ph.D. students and other collaborators, K. M. Brown, W. B. Gragg, Rolf Jeltsch, P. Pfluger, and notably I. Gargantini have substantially contributed to the matters dealt with in this volume. Colleagues at ETHZ —A. Huber and C. Blatter—and at UCLA—E. A. Coddington and E. G. Straus in casual conversations often provided me with flashes of understanding. Finally, I wish to record my debt to my teachers who introduced me to the beautiful subject of complex analysis, both those whom I met as a student—F. Oberhettinger, A. Ostrowski, A. Pfluger, E. Stiefel, and M. J. O. Strutt—and those whom I came to know mainly through their works, notably L. Ahlfors and the giants of the past—R. Courant, G. N. Watson, and A. Hurwitz.

It remains for me to express my appreciation to the staff of John Wiley & Sons for their expert handling of all problems of editing and production that arose in connection with my manuscript.

I dedicate this volume to the memory of Heinz Rutishauser, my former colleague and one of the few true geniuses of modern numerical mathematics. Until his untimely death in 1970 his alert mind was a source of insight and inspiration whose originality was perennially refreshing. I can only hope that Rutishauser would have approved of the form in which some of his work is presented here. As a small token of my indebtedness, I have adopted the Rutishauser symbol $:=$ (meaning "is defined by"), which he introduced in his fundamental 1949 paper on automatic programming and which in recent years has come into increasing use in areas outside numerical analysis and programming.

PETER HENRICI

Fort Collins, Colorado
March 1973

CONTENTS

1 Formal Power Series 1

    1.1. Algebraic Preliminaries: Complex Numbers, 1
    1.2. Definition and Algebraic Properties of Formal Power Series, 9
    1.3. A Matrix Representation of Formal Power Series, 14
    1.4. Differentiation of Formal Power Series, 18
    1.5. Formal Hypergeometric Series and Finite Hypergeometric Sums, 27
    1.6. The Composition of a Formal Power Series with a Nonunit, 35
    1.7. The Group of Almost Units Under Composition, 45
    1.8. Formal Laurent Series: Residues, 51
    1.9. The Lagrange-Bürmann Theorem, 55
    Seminar Assignments, 64
    Notes, 65

2 Functions Analytic at a Point 66

    2.1. Banach Algebras: Functions, 66
    2.2. Convergent Power Series, 75
    2.3. Functions Analytic at a Point, 87
    2.4. Composition and Inversion of Analytic Functions, 94
    2.5. Elementary Transcendental Functions, 106
    2.6. Matrix-Valued Functions, 122
    2.7. Sequences of Functions Analytic at a Point, 133
    Seminar Assignments, 137
    Notes, 138

3 Analytic Continuation. 139

    3.1. Rearrangement of Power Series; Derivatives, 139
    3.2. Analytic Extension and Continuation, 144
    3.3. New Determination of the Radius of Convergence, 155
    3.4. Sequences of Functions Analytic in a Region, 159

- 3.5. Analytic Continuation Along an Arc: Monodromy Theorem, 165
- 3.6. Numerical Analytic Continuation Along an Arc, 170
  - Seminar Assignments, 179
  - Notes, 179

4 Complex Integration 180

- 4.1. Complex Functions of a Real Variable, 180
- 4.2. The Integral of a Function Along an Arc, 187
- 4.3. Integrals of Analytic Functions, 195
- 4.4. The Laurent Series: Isolated Singularities, 206
- 4.5. Applications of the Laurent Series: Bessel Functions, Fourier Series, 219
- 4.6. Continuous Argument, Winding Number, Jordan Curve Theorem, 230
- 4.7. Residue Theorem: Cauchy Integral Formula, 241
- 4.8. Applications of the Residue Theorem: Evaluation of Definite Integrals, 249
- 4.9. Applications of the Residue Theorem: Summation of Infinite Series, 265
- 4.10. The Principle of the Argument, 276
  - Seminar Assignments, 285
  - Notes, 285

5 Conformal Mapping 286

- 5.1. Geometric Interpretation of Complex Functions, 286
- 5.2. Moebius Transformations: Algebraic Theory, 299
- 5.3. Moebius Transformations: The Riemann Sphere, 307
- 5.4. Moebius Transformations: Symmetry, 316
- 5.5. Holomorphic Functions and Conformal Maps, 323
- 5.6. Conformal Transplants, 333
- 5.7. Problems of Plane Electrostatics, 342
- 5.8. Two-Dimensional Ideal Flows, 356
- 5.9. Poisson's Equation, 371
- 5.10. General Results on Conformal Maps, 377
- 5.11. Symmetry, 387
- 5.12. The Schwarz-Christoffel Mapping Function, 396
- 5.13. Mapping the Rectangle. An Elliptic Integral, 416
- 5.14. Rounding Corners in Schwarz-Christoffel Maps, 422
  - Seminar Assignment, 432
  - Notes, 432

CONTENTS

6 Polynomials 433

- 6.1. The Horner Algorithm, 433
- 6.2. Sign Changes. The Rule of Descartes, 439
- 6.3. Cauchy Indices: The Number of Zeros of a Real Polynomial in a Real Interval, 443
- 6.4. Disks Containing a Specified Number of Zeros, 450
- 6.5. Geometry of Zeros (Theorems of Gauss-Lucas, Laguerre, and Grace), 463
- 6.6. Circular Arithmetic, 475
- 6.7. The Number of Zeros in a Half-Plane, 485
- 6.8. The Number of Zeros in a Disk, 491
- 6.9. Methods for Determining Zeros: A Survey, 497
- 6.10. Methods of Search for a Single Zero, 500
- 6.11. Methods of Search and Exclusion for the Simultaneous Determination of All Zeros, 513
- 6.12. Fixed Points of Analytic Functions: Iteration, 523
- 6.13. Newton's Method for Polynomials, 534
- 6.14. Methods of Descent, 541
  Seminar Assignments, 550
  Notes, 551

7 Partial Fractions 553

- 7.1. Construction of the Partial Fraction Representation of a Rational Function, 553
- 7.2. Partial Fractions: Miscellaneous Applications 562
- 7.3. Some Applications to Combinatorial Analysis, 578
- 7.4. Difference Equations, 584
- 7.5. Hankel Determinants, 591
- 7.6. The Quotient-Difference Algorithm, 608
- 7.7. Hadamard Polynomials, 622
- 7.8. Matrix Interpretation, 634
- 7.9. Poles with Equal Moduli, 641
- 7.10. Partial Fraction Expansions of Meromorphic Functions, 655
  Seminar Assignments, 662
  Notes, 662

Bibliography 663

Index 673

APPLIED AND COMPUTATIONAL COMPLEX ANALYSIS

Volume I

1

FORMAL POWER SERIES

§1.1. ALGEBRAIC PRELIMINARIES: COMPLEX NUMBERS

Let us recall some basic algebraic concepts and terminology.

A set $\mathcal{G}$ of mathematical objects (such as numbers, mappings, transformations) is called a **group** if it satisfies the following conditions:

1. A **rule of composition** is defined in $\mathcal{G}$ in the following sense: with every ordered pair (a,b) of elements of $\mathcal{G}$ there is associated a unique element of $\mathcal{G}$, called the **product** of a and b. The product of a and b may be denoted by a symbol such as $a*b$. It is not required that $b*a = a*b$.

2. The product is associative: for any three elements a, b, c of $\mathcal{G}$ there holds

$$(a*b)*c = a*(b*c).$$

3. There exists a left **identity element** e in $\mathcal{G}$ with the property that

$$e*a = a \quad \text{for all } a \text{ in } \mathcal{G}.$$

4. For each a in $\mathcal{G}$ there exists a **left inverse** $\bar{a}$ in $\mathcal{G}$ such that

$$\bar{a}*a = e.$$

It can be proved from these axioms that a left inverse $\bar{a}$ of a is also a right inverse (i.e., that $a*\bar{a} = e$) and that a left identity element is also a right identity element. Furthermore, it can be shown that the identity element is unique and that each a in $\mathcal{G}$ has precisely one inverse element.

Examples of groups:

1. The integers where $*$ denotes addition (verify the axioms; what is e?).
2. The positive rational numbers if $*$ denotes multiplication.
3. The rotations of a rigid body around a fixed point in space.

4. The square matrices of a given order with real elements if $*$ denotes matrix addition. How about matrix multiplication?

If, in a group $\mathcal{G}$, the **commutative law** holds for the operation $*$, i.e., if

$$a*b = b*a$$

for all a, b in $\mathcal{G}$, then $\mathcal{G}$ is called an **abelian group**. Among the four examples given, **1, 2, 4** are abelian groups, but the remaining group is not.

A mathematical system in which only the axioms 1 and 2 hold and the existence of the identity and inverse is not required is called a **semigroup**. If the commutative law holds, the semigroup may be called *abelian* or *commutative*.

Examples of semigroups:

5. All integers under the operation of multiplication form an abelian semigroup.
6. The square matrices of a given order with real elements under the operation of multiplication form a (nonabelian) semigroup.

Groups have *one* rule of composition. Many algebraic systems of interest have two. For us the two most important algebraic systems with two rules of composition are the *integral domain* and the *field*. For convenience we identify the two rules of composition by the familiar symbols for addition and multiplication. In fact, these designations are used in most applications.

A system $\mathfrak{D}$ of mathematical objects is called an **integral domain** if addition and multiplication, the two rules of composition, are defined in it and satisfy the following laws:

1. With respect to $+$, $\mathfrak{D}$ forms an abelian group.
2. With respect to $\cdot$, $\mathfrak{D}$ forms an abelian semigroup with an identity element. The identity elements for the operations $+$ and $\cdot$ are distinct.
3. In addition to the commutative and associative laws already postulated, the **distributive law**

$$a(b+c) = ab + ac$$

holds for any three elements a, b, c of $\mathfrak{D}$.

4. The cancellation law for multiplication holds: If 0 denotes the identity element for the operation $+$ ($a+0=a$ for all a) and

$$ac = 0, \qquad a \neq 0,$$

then $c=0$. (There are no **divisors of zero**.)

Examples of integral domains:

7. All integers under the ordinary interpretation of addition and multiplication.

ALGEBRAIC PRELIMINARIES 3

8. All polynomials with real coefficients (for instance) under the ordinary rules of algebra. What are the identity elements?

9. All numbers of the form $a+b\sqrt{3}$, where a and b are integers.

A set $\mathfrak{F}$ of mathematical objects is called a **field** if it is an integral domain such that the elements $\neq 0$ (the identity element with respect to $+$) not only form a semigroup with identity but actually a group. Thus the following axioms hold in a field $\mathfrak{F}$:

1. With respect to $+$ the elements of $\mathfrak{F}$ form an abelian group with identity element 0.
2. With respect to $\cdot$ the elements $\neq 0$ form an abelian group and all elements form an abelian semigroup.
3. The distributive law holds.

It follows from these axioms that a field cannot have any divisors of zero, for if
$$ac=0, \qquad a\neq 0,$$
then a^{-1} (the inverse of a with respect to multiplication) exists and
$$a^{-1}(ac)=(a^{-1}a)c=c=a^{-1}0=0;$$
hence $c=0$.

Examples of fields:

10. The field of real numbers.
11. The field of rational numbers.
12. The residue classes modulo p, where p is a prime number. This is an example of a field with a finite number of elements.

For us the most important field is the **field of complex numbers**, frequently denoted by $\mathbb{C}$. Some intuitive knowledge of complex numbers is assumed. Let us recall that from a logical point of view the complex numbers are most easily introduced as ordered pairs (α,β) of two real numbers α and β. Addition is defined by

$$(\alpha,\beta)+(\gamma,\delta)=(\alpha+\gamma,\beta+\delta).$$

It is obvious that the associative law holds. The identity element with respect to addition is the pair $(0,0)$, where 0 denotes the real number zero. The inverse of (α,β) with respect to addition is $(-\alpha,-\beta)$.

Multiplication of two ordered pairs is defined by

$$(\alpha,\beta)\cdot(\gamma,\delta)=(\alpha\gamma-\beta\delta,\alpha\delta+\beta\gamma).$$

It can be verified that the associative law holds. The commutative and distributive laws are also easily seen to hold. The identity element with respect to multiplication is the ordered pair $(1,0)$. Let us verify the existence of the inverse element with respect to multiplication. Given a pair $(\alpha,\beta)\neq(0,0)$, we wish to find (ξ,η) such that

$$(\alpha\xi-\beta\eta, \alpha\eta+\beta\xi)=(1,0).$$

This is equivalent to the system of two equations with two unknowns:

$$\alpha\xi-\beta\eta=1,$$
$$\beta\xi+\alpha\eta=0.$$

The determinant is $\alpha^2+\beta^2\neq 0$. Thus the system has a unique solution given by

$$\xi=\frac{\alpha}{\alpha^2+\beta^2}, \quad \eta=-\frac{\beta}{\alpha^2+\beta^2}.$$

The inverse with respect to multiplication of the complex number $(\alpha,\beta)\neq(0,0)$ thus is the complex number

$$\left(\frac{\alpha}{\alpha^2+\beta^2}, -\frac{\beta}{\alpha^2+\beta^2}\right).$$

The two rules for addition and multiplication of complex numbers can be combined if we introduce the **imaginary unit** i. Write the complex number (α,β) in the form $\alpha+i\beta$ and proceed as if i were a real number with the additional property that $i^2=-1$. Obviously it would not be satisfactory to define complex numbers in this manner because the nature of the symbol i would remain unclear. From our point of view i is merely an abbreviation for the ordered pair $(0,1)$, and the notation $\alpha+i\beta$ is to be regarded as an abbreviation for

$$\alpha(1,0)+\beta(0,1),$$

in which the multiplication of a complex number by a real number is performed componentwise, as in vector algebra.

For complex numbers of the special form $(\alpha,0)$ the above rules of operation yield

$$(\alpha,0)+(\beta,0)=(\alpha+\beta,0),$$
$$(\alpha,0)\cdot(\beta,0)=(\alpha\beta,0),$$

showing that sums and products of complex numbers $(\alpha,0)$ are formed in the same manner as sums and products of real numbers. The field of complex numbers thus contains a subfield that is "isomorphic" to the field

ALGEBRAIC PRELIMINARIES

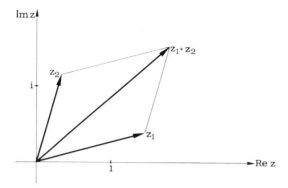

Fig. 1.1a. Addition of complex numbers.

of real numbers. In the following no distinction is made between real numbers α and complex numbers $(\alpha, 0)$.

It is natural to associate with the complex number (α, β) the point in the euclidean plane with the coordinates α and β. The addition of two complex numbers $z_1 = (\alpha_1, \beta_1)$ and $z_2 = (\alpha_2, \beta_2)$ then corresponds to the ordinary addition of the position vectors of these points (see Fig. 1.1a). Multiplication of a complex number by a real number likewise has the familiar geometrical meaning found in vector algebra.

Multiplication of two arbitrary complex numbers also has a simple geometric interpretation. If $z = (\alpha, \beta) = \alpha + i\beta$ is a complex number, we define (consistent with the usage for real numbers) the **absolute value** or **modulus** of z by

$$|z| = (\alpha^2 + \beta^2)^{1/2}.$$

Furthermore, if $z = (\alpha, \beta) \neq (0, 0) = 0$, we denote by arg z and call **argument** of z any real number ϕ satisfying the two relations

$$\alpha = |z|\cos\phi, \qquad \beta = |z|\sin\phi.$$

Geometrically, $\phi = \arg z$ is one of the angles formed by the vector $z = (\alpha, \beta)$ with the positive real axis (Fig. 1.1b). Evidently the number ϕ is determined only up to integral multiples of 2π. Elementary relations between trigonometric functions now show that for any two complex numbers z_1 and z_2

$$|z_1 z_2| = |z_1||z_2|,$$

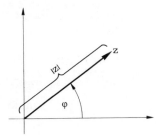

Fig. 1.1b. Absolute value and argument of complex number.

and, if $z_1 z_2 \neq 0$, for an arbitrary choice of the arguments,

$$\arg z_1 z_2 = \arg z_1 + \arg z_2 \quad (\text{mod } 2\pi).$$

Therefore, if complex numbers are multiplied, their absolute values are multiplied but the arguments are *added* (Fig. 1.1c). Similarly, it can be shown that if $z_2 \neq 0$

$$\left| \frac{z_1}{z_2} \right| = \frac{|z_1|}{|z_2|},$$

and also if $z_1 z_2 \neq 0$

$$\arg \frac{z_1}{z_2} = \arg z_1 - \arg z_2 \quad (\text{mod } 2\pi).$$

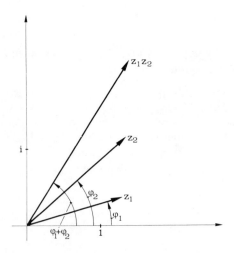

Fig. 1.1c. Multiplication of complex numbers.

ALGEBRAIC PRELIMINARIES

More notation is required. If $z = (\alpha, \beta) = \alpha + i\beta$ is a complex number, we write

$$\alpha = \operatorname{Re} z, \quad \text{called the } \textbf{real part} \text{ of } z,$$

$$\beta = \operatorname{Im} z, \quad \text{called the } \textbf{imaginary part} \text{ of } z.$$

The number $(\alpha, -\beta) = \alpha - i\beta$ is the (complex) **conjugate** of $z = (\alpha, \beta)$ and is denoted by $\bar{z}$. We take note of the trivial relations

$$z\bar{z} = |z|^2, \quad \frac{z + \bar{z}}{2} = \operatorname{Re} z, \quad \frac{z - \bar{z}}{2i} = \operatorname{Im} z.$$

It is easily shown that for any two complex numbers z_1, z_2

$$\overline{z_1 + z_2} = \overline{z_1} + \overline{z_2}, \quad \overline{z_1 - z_2} = \overline{z_1} - \overline{z_2},$$

$$\overline{z_1 z_2} = \overline{z_1}\, \overline{z_2},$$

and if $z_2 \neq 0$

$$\overline{\left(\frac{z_1}{z_2}\right)} = \frac{\overline{z_1}}{\overline{z_2}}.$$

Thus the operations of taking complex conjugates and of forming sums, differences, products, or quotients may be interchanged.

From the very definition of Re and Im we have

$$|\operatorname{Re} z| \leq |z|, \quad |\operatorname{Im} z| \leq |z|.$$

We use some of the above rules to obtain an algebraic proof of the **triangle inequality**

$$|z_1 + z_2| \leq |z_1| + |z_2|,$$

valid for arbitrary complex z_1 and z_2. Indeed

$$|z_1 + z_2|^2 = (z_1 + z_2)\overline{(z_1 + z_2)}$$

$$= (z_1 + z_2)(\overline{z_1} + \overline{z_2})$$

$$= z_1 \overline{z_1} + z_1 \overline{z_2} + z_2 \overline{z_1} + z_2 \overline{z_2}$$

$$= |z_1|^2 + 2\operatorname{Re} z_1 \overline{z_2} + |z_2|^2$$

$$\leq |z_1|^2 + 2|z_1||z_2| + |z_2|^2$$

$$= (|z_1| + |z_2|)^2,$$

and the desired result follows because absolute values are non-negative.

We return for a moment to general fields to introduce one more concept. Let e be the identity element of the multiplicative group of a field $\mathfrak{F}$. Because under the operation $+\mathfrak{F}$ forms an additive group, the elements

$$e+e=:2e,$$

$$e+e+e=:3e,\ldots,$$

belong to $\mathfrak{F}$. It may happen that one of these sums ne (n is a positive integer) equals zero, the identity element of the additive group. (It *must* happen if the field is finite.) If p is the smallest integer for which this happens, the field is said to have **characteristic** p. If it never happens, the field is said to have characteristic zero. The field of rationals clearly has characteristic zero and so have all fields that contain the field of rationals as a subfield, such as the field of real or complex numbers. In all that follows "field" will always mean a field of characteristic zero.

PROBLEMS

1. Show that the matrices of the form
$$\begin{pmatrix} a & b \\ c & d \end{pmatrix},$$
where a, b, c, and d are integers such that $ad-bc=1$, become a nonabelian group under the operation of matrix multiplication.

2. Show that the matrices of the form
$$\begin{pmatrix} a & b \\ 0 & a \end{pmatrix}$$
(a,b real) become an abelian group under addition and an abelian semigroup with identity under matrix multiplication. Do they form an integral domain?

3. Show that the numbers $p+q\sqrt{2}$, where p and q are rational, form a field.

4. Show that the matrices
$$\begin{pmatrix} a & b \\ -\bar{b} & \bar{a} \end{pmatrix}$$
(a,b complex) form a "skew field," i.e., a field in which the commutative law of multiplication does not hold.

5. Deduce from the group axioms that any left inverse is also a right inverse and that a left identity is a right identity. [Consider the relation $a^{-1}aa^{-1}=ea^{-1}=a^{-1}$ and multiply from the left by an inverse of a^{-1}.]

6. If $z=\rho(\cos\phi+i\sin\phi)$, show that, for all integers n, $z^n=\rho^n(\cos n\phi+i\sin n\phi)$.

7. Determine the absolute value and all arguments of the complex number $z=(1+i\sqrt{3})^{-1}$. For which integers is z^n real?

8. Write $(\sqrt{3}+i)^{12}$ in the form $\alpha+i\beta$.

DEFINITION AND ALGEBRAIC PROPERTIES

9. In the identity $\cos n\phi + i \sin n\phi = (\cos \phi + i \sin \phi)^n$ (n is a positive integer) expand the right member by the binomial formula and compare real and imaginary parts, thus proving **Moivre's formula** representing $\cos n\phi$ and $\sin n\phi$ as polynomials in $\cos \phi$ and $\sin \phi$.

10. Characterize (in your own words or in a sketch) the sets of those points z in the complex plane that satisfy the following conditions:
    (a) $|z| \leq 2$;   (b) $|z| > 1$;   (c) $\operatorname{Re} z \geq \frac{1}{2}$;
    (d) $0 \leq \operatorname{Im}(-iz) < \pi$;   (e) $\operatorname{Re} z^2 = 1$;   (f) $|z^2 + 1| = 1$.

11. For arbitrary complex z and positive integers n let
$$s_n(z) := 1 + z + z^2 + \ldots + z^n.$$
    (a) Determine graphically by vector addition in the complex plane the position of the points $s_n[(1+i)/2]$, $n = 1, 2, \ldots, 8$.
    (b) Prove that for $z \neq 1$ and for $n = 1, 2, \ldots$, the following formula holds:
$$s_n(z) = \frac{1 - z^{n+1}}{1 - z}.$$

[Subtract $z s_n(z)$ from $s_n(z)$.]

§1.2. DEFINITION AND ALGEBRAIC PROPERTIES OF FORMAL POWER SERIES

From the abstract point of view a **formal power series** (fps) over a field $\mathfrak{F}$ is an infinite sequence $\{a_0, a_1, a_2, \ldots\}$ of elements of $\mathfrak{F}$. Equivalently, it is a function from the set of non-negative integers to $\mathfrak{F}$:

$$\{0, 1, 2, \ldots\} \to \mathfrak{F}.$$

For practical reasons, however, the fps $\{a_0, a_1 a_2, \ldots\}$ is written in the form

$$a_0 + a_1 x + a_2 x^2 + \cdots + a_n x^n + \cdots.$$

Although this looks like an ordinary infinite series, it should not be so considered. No value is assigned to the symbol x. Since no norm has been imposed on $\mathfrak{F}$, we could not speak of convergence even if we replaced x with an element of $\mathfrak{F}$. If a norm had been defined, it would still be uncertain that we had convergence. The symbol x merely serves as a placekeeper (or "indeterminate"); it and the $+$ signs are introduced because they are strongly suggestive of certain algebraic operations to be defined.

A formal power series is usually denoted by a capital letter such as $P, Q, R, \ldots$. The element a_n of the fps

$$P = a_0 + a_1 x + a_2 x^2 + \ldots$$

is called the *n*th **coefficient** of P. The field $\mathfrak{F}$ from which the coefficients are taken is an arbitrary field of characteristic zero; in later applications it is usually the field of complex numbers.

Examples of formal power series:

1. $1 + \frac{1}{1!}x + \frac{1}{2!}x^2 + \cdots$;

2. $1 + 1!x + 2!x^2 + \cdots$;

3. Every *polynomial* with coefficients in $\mathfrak{F}$, in particular the series

$$I := 1 + 0x + 0x^2 + \cdots,$$

$$X := 0 + 1x + 0x^2 + 0x^3 + \cdots.$$

We now define addition and multiplication of a fps. It is shown later that with these operations the fps form an *integral domain*. If

$$P := a_0 + a_1 x + a_2 x^2 + \cdots,$$

$$Q := b_0 + b_1 x + b_2 x^2 + \cdots$$

are two formal power series, their *sum* is defined by

$$P + Q := (a_0 + b_0) + (a_1 + b_1)x + (a_2 + b_2)x^2 + \cdots.$$

It is clear that with respect to this sum the formal power series form an abelian group. The reader will have little difficulty in identifying the identity element of this group and the inverse of the fps P.

The *product* of the two formal power series P and Q is defined in the manner suggested by polynomials, namely $PQ := a_0 b_0 + (a_0 b_1 + a_1 b_0)x + (a_0 b_2 + a_1 b_1 + a_2 b_0)x^2 + \cdots$ or $PQ = c_0 + c_1 x + c_2 x^2 + \cdots$, where

$$c_n := \sum_{k=0}^{n} a_k b_{n-k} = \sum_{k=0}^{n} b_k a_{n-k}, \qquad n = 0, 1, 2, \ldots.$$

This is also called the **Cauchy product** of two formal power series to distinguish it from other products that can be formed. It is clear the the Cauchy product is always well defined. To obtain the *n*th coefficient of the product a finite number of algebraic operations with elements from $\mathfrak{F}$ must be performed. Whether the coefficients c_n can be expressed by a simple formula (given that the a_n and b_n can be so expressed), is, of course, a different question.

EXAMPLE

4. To find the product of the fps

$$P := 1 + 1x + 1x^2 + 1x^3 + \cdots,$$

$$Q := 1 + 2x + 3x^2 + 4x^3 + \cdots.$$

DEFINITION AND ALGEBRAIC PROPERTIES

According to the general rule, $PQ = c_0 + c_1 x + c_2 x^2 + \cdots$, where

$$c_n = 1 + 2 + \cdots + (n+1) = \frac{(n+1)(n+2)}{2}.$$

Thus

$$PQ = 1 + 3x + 6x^2 + 10x^3 + 15x^4 + \cdots.$$

It is clear that the multiplication of fps thus defined is commutative and a little computation shows that it is also associative. There exists an identity element, to be denoted by I:

$$I := 1 + 0x + 0x^2 + \cdots.$$

Thus the formal power series under the operation of Cauchy multiplication form an abelian semigroup with identity. Do they form a group? No, because the fps $0 + 1x + 0x^2 + \cdots$ (for instance) possesses no reciprocal (although it differs from the identity element $0 := 0 + 0x + 0x^2 + \cdots$ of the additive group).

As already asserted, however, the formal power series form an integral domain. It remains to be shown that

(a) the distributive law holds,
(b) there are no divisors of zero.

To prove (a) let $P := a_0 + a_1 x + \cdots$, $Q := b_0 + b_1 x + \cdots$, $R := c_0 + c_1 x + \cdots$ be three formal power series. The nth coefficient of PQ is $a_0 b_n + a_1 b_{n-1} + \cdots + a_n b_0$; of PR it is $a_0 c_n + a_1 c_{n-1} + \cdots + a_n c_0$. Thus the nth coefficient of $PQ + PR$ is $a_0(b_n + c_n) + a_1(b_{n-1} + c_{n-1}) + \cdots + a_n(b_0 + c_0)$, but because the kth coefficient of $Q + R$ is $b_k + c_k$ the last expression also equals the nth coefficient of $P(Q + R)$. Thus it follows that $PQ + PR = P(Q + R)$.

To show the absence of divisors of zero we prove that $P \neq 0$ and $Q \neq 0$ imply that $PQ \neq 0$. (Here 0 denotes the identity element with respect to addition, i.e., the fps with all coefficients zero.) Let a_n be the first nonzero coefficient of P and let b_m be the first nonzero coefficient of Q. We shall show that the coefficient c_{n+m} of the product PQ is $\neq 0$. This follows from

$$c_{n+m} = a_0 b_{n+m} + \cdots + a_{n-1} b_{m+1} \quad (= 0 \text{ because } a_i = 0)$$
$$+ a_n b_m \quad (\neq 0 \text{ because } a_n \neq 0 \text{ and } b_m \neq 0,$$
$$\text{and } \mathfrak{F} \text{ has no divisors of zero})$$
$$+ a_{n+1} b_{m-1} + \cdots + a_{n+m} b_3 \quad (= 0 \text{ because } b_j = 0).$$

We have completed the proof of Theorem 1.2a.

THEOREM 1.2a

The formal power series over a field $\mathfrak{F}$ form an integral domain.

We have already seen that the multiplicative semigroup of a fps is not a group; i.e., not every nonzero element has an inverse. This is not to say that *no* element (except the identity element I) has an inverse. In any integral domain $\mathfrak{D}$ the elements that have inverses with respect to multiplication are called the **units** of $\mathfrak{D}$. (Thus a field may be described as an integral domain in which all nonzero elements are units.)

THEOREM 1.2b

The units of the integral domain of formal power series $P = a_0 + a_1 x + \cdots$ over a field $\mathfrak{F}$ are precisely the series with $a_0 \neq 0$.

Proof. If $P = a_0 + a_1 x + a_2 x^2 + \cdots$, we wish to show that there exists a fps $Q = b_0 + b_1 x + b_2 x^2 + \cdots$ such that $PQ = I$ if and only if $a_0 \neq 0$. By the definition of multiplication $PQ = I$ if and only if the following infinite number of relations hold:

$$a_0 b_0 = 1$$
$$a_0 b_1 + a_1 b_0 = 0 \qquad (1.2\text{-}1)$$
$$a_0 b_2 + a_1 b_1 + a_2 b_0 = 0$$
$$\cdots$$

The first relation is possible if and only if $a_0 \neq 0$. Thus the condition $a_0 \neq 0$ is *necessary* in order that P may be a unit. To show that it is also *sufficient*, suppose $b_0, b_1, \ldots, b_{n-1}$ have been determined so that the first n relations (1.2-1) hold. The $(n+1)$st relation may then be written as

$$a_0 b_n = -a_1 b_{n-1} - a_2 b_{n-2} - \cdots - a_n b_0, \qquad (1.2\text{-}2)$$

which determines b_n in view of $a_0 \neq 0$ and determines it uniquely. Thus, if $a_0 \neq 0$, the series P possesses an inverse Q, which is unique. (The fact that the inverse of a unit is unique actually is a consequence of the axioms of an integral domain.) ∎

To avoid confusion in terminology we call the inverse with respect to multiplication of a fps P the **reciprocal** of P, denoted by P^{-1}. (The inverse with respect to addition is the negative of P, denoted by $-P$.)

The formal power series in which $a_0 = 0$ and which according to Theorem 1.2b are not units, are called **nonunits**.

ALGEBRAIC PRELIMINARIES

PROBLEMS

1. Find the reciprocal of the fps $1-x$.
2. If Q is an arbitrary fps and $P=1-x$, show that the nth coefficient of $P^{-1}Q$ is the sum of the first n coefficients of Q.
3. The Fibonacci numbers a_n are defined by $a_0=1$, $a_1=1$, $a_{n+1}=a_n+a_{n-1}$ ($n=1,2,\ldots$). If $P=\sum a_n x^n$, determine P^{-1}.
4. The fps

$$P := \frac{1}{1!} + \frac{1}{2!}x + \frac{1}{3!}x^2 + \cdots$$

is a unit, hence has a reciprocal. The **Bernoulli numbers** B_n are defined by

$$P^{-1} = \sum_{n=0}^{\infty} \frac{B_n}{n!} x^n.$$

(a) Show that $B_0=1$, $B_1=-\frac{1}{2}$, $B_2=\frac{1}{6}$, $B_3=0$, $B_4=\frac{1}{30}$.
(b) Establish the recurrence relation

$$B_n = \sum_{k=0}^{n} \binom{n}{k} B_k, \qquad n=2,3,\ldots.$$

5. The formal power series over a field $\mathcal{F}$ could be molded into an algebraic system by defining the sum of the two series $P=a_0+a_1x+\cdots$ and $Q=b_0+b_1x+\cdots$ in the usual way and the product by

$$P \times Q := a_0 b_0 + a_1 b_1 x + a_2 b_2 x^2 + \cdots$$

(**Hadamard Product**). Show that with respect to this multiplication the formal power series form an abelian semigroup with identity. Do they form an integral domain?

6. Let $P := a_0 + a_1 x + a_2 x^2 + \cdots$ be a formal power series over the field of real numbers such that $a_n > 0$, $a_{n+2} a_n - a_{n+1}^2 > 0$, $n=0,1,2,\ldots$. If $P^{-1} = b_0 + b_1 x + b_2 x^2 + \ldots$, show that $b_n < 0$, $n=1,2,\ldots$. (Kaluza [1928]).

7. Let $s_1, s_2, \ldots$ be complex numbers. Show that the reciprocal of the series

$$P := 1 - s_1 \frac{x}{1!} + \begin{vmatrix} s_1 & 1 \\ s_2 & s_1 \end{vmatrix} \frac{x^2}{2!} - \begin{vmatrix} s_1 & 1 & 0 \\ s_2 & s_1 & 2 \\ s_3 & s_2 & s_1 \end{vmatrix} \frac{x^3}{3!} + \cdots$$

is given by

$$Q := 1 + s_1 \frac{x}{1!} + \begin{vmatrix} s_1 & -1 \\ s_2 & s_1 \end{vmatrix} \frac{x^2}{2!} + \begin{vmatrix} s_1 & -1 & 0 \\ s_2 & s_1 & -2 \\ s_3 & s_2 & s_1 \end{vmatrix} \frac{x^3}{3!} + \cdots.$$

[*Elemente der Math.* **16**, 87]

§1.3. A MATRIX REPRESENTATION OF FORMAL POWER SERIES

In §1.2 we found the recurrence relation (1.2-2) for the coefficients of the reciprocal series of a formal power series. We can derive an explicit formula and further elucidate the algebraic properties of formal power series by associating with

$$P := a_0 + a_1 x + a_2 x^2 + \cdots \tag{1.3-1}$$

the infinite triangular matrix

$$\mathbf{A} = \begin{bmatrix} a_0 & a_1 & a_2 & a_3 & \cdots \\ & a_0 & a_1 & a_2 & \cdots \\ & & a_0 & a_1 & \cdots \\ & & & a_0 & \\ & \cdot & \cdot & \cdot & \end{bmatrix}. \tag{1.3-2}$$

Matrices of this special form ($a_{ij} = a_{j-i}$ for $j \geq i$, $a_{ij} = 0$ for $j < i$) are called **semicirculant matrices**. (The name *circulant matrices* is usually given to matrices of the form

$$\begin{bmatrix} a_1 & a_2 & a_3 & \cdots & a_n \\ a_n & a_1 & a_2 & \cdots & a_{n-1} \\ \cdot & \cdot & \cdot & & \\ a_2 & a_3 & a_4 & \cdots & a_1 \end{bmatrix}.)$$

Semicirculant matrices are a special case of **upper triangular matrices**, i.e., of matrices $\mathbf{A} = (a_{ij})$, where $a_{ij} = 0$ for $j < i$. It is easy to see that the product of two infinite upper triangular matrices always exists and is again upper triangular. Moreover, if we define the nth **section** $\mathbf{A}_n$ of an infinite matrix $\mathbf{A}$ as the finite submatrix composed of the first n rows and columns of $\mathbf{A}$, the product of the nth sections of two upper triangular matrices $\mathbf{A}$ and $\mathbf{B}$ will equal the nth section of the product:

$$(\mathbf{AB})_n = \mathbf{A}_n \mathbf{B}_n, \qquad n = 1, 2, \ldots. \tag{1.3-3}$$

In particular this holds for semicirculants.

A MATRIX REPRESENTATION

We describe the correspondence between the fps (1.3-1) and the semicirculant matrix (1.3-2) as

$$P \to A,$$

and we call A the *associated* semicirculant (matrix) of P. Conversely, if a semicirculant matrix A is given, we call the fps whose coefficients are the elements in the first row of A the associated fps of A.

It is trivial that the associated semicirculant of a sum of two formal power series is the sum of the associated semicirculants:

$$\text{if } P \to A, \quad Q \to B, \quad \text{then} \quad P + Q \to A + B. \tag{1.3-4}$$

Surprisingly, the same is true for the product:

$$\text{if } P \to A, \quad Q \to B, \quad \text{then} \quad PQ \to AB. \tag{1.3-5}$$

Indeed, if $A = (a_{ij})$, $B = (b_{ij})$, where $a_{ij} = a_{j-i}$, $b_{ij} = b_{j-i}$ for $j \geq i$, then

$$C := AB = (c_{ij}),$$

where

$$c_{ij} = \sum_{k=0}^{(\infty)} a_{ik} b_{kj} = \sum_{k=i}^{j} a_{ik} b_{kj}$$

$$= \sum_{k=i}^{j} a_{k-i} b_{j-k} = \sum_{m=0}^{j-i} a_m b_{j-i-m}$$

$$= c_{j-i},$$

and

$$PQ =: c_0 + c_1 x + c_2 x^2 + \cdots.$$

Because $PQ = QP$, we obtain the side result that semicirculant matrices commute.

The relations (1.3-4) and (1.3-5) express the fact that the mapping $\to$ of the set of formal power series onto the set of semicirculants is an **isomorphism**.

Let P be a unit in the integral domain of formal power series; i.e., let $a_0 \neq 0$. The reciprocal series P^{-1} then exists, and if

$$P \to A, \quad P^{-1} \to B,$$

according to (1.3-5),

$$AB = BA = I,$$

where **I** denotes the (semicirculant) unit matrix,

$$\mathbf{I} = \begin{bmatrix} 1 & & & 0 \\ & 1 & & \\ & & 1 & \\ 0 & & & \cdots \end{bmatrix}.$$

The matrix **B** thus is the inverse of $\mathbf{A}, \mathbf{B} = \mathbf{A}^{-1}$. Furthermore, from (1.3-3)

$$\mathbf{B}_n = (\mathbf{A}_n)^{-1}.$$

Let

$$P =: a_0 + a_1 x + a_2 x^2 + \cdots,$$

$$P^{-1} =: b_0 + b_1 x + b_2 x^2 + \cdots,$$

and

$$\mathbf{A}_{n+1} = \begin{bmatrix} a_0 & a_1 & \cdots & a_n \\ & a_0 & \cdots & a_{n-1} \\ & & \cdots & \\ 0 & & & a_0 \end{bmatrix},$$

$$(\mathbf{A}_{n+1})^{-1} = \begin{bmatrix} b_0 & b_1 & \cdots & b_n \\ & b_0 & \cdots & b_{n-1} \\ & & \cdots & \\ 0 & & & b_0 \end{bmatrix}.$$

If $\mathbf{C} := (c_{ij})$ is any nonsingular matrix, the elements of $\mathbf{D} := (d_{ij}) := \mathbf{C}^{-1}$ are calculated by the formula

$$d_{ij} = \frac{\text{algebraic complement of } c_{ji}}{\text{determinant of } \mathbf{C}}.$$

(The algebraic complement of c_{ji} is the determinant of the matrix obtained from **C** by crossing out the row and the column containing c_{ji}, multiplied by $(-1)^{i+j}$. See §7.5 for a review of the properties of determinants used

A MATRIX REPRESENTATION

here.) We apply this formula to obtain $b_n = b_{1,n+1}$. The algebraic complement of $a_{1,n+1}$ is

$$(-1)^n \begin{vmatrix} a_1 & a_2 & \cdots & a_n \\ a_0 & a_1 & \cdots & a_{n-1} \\ 0 & a_0 & \cdots & a_{n-2} \\ & & \cdots & \\ 0 & 0 & a_0 & a_1 \end{vmatrix}.$$

Obviously, the determinant of $\mathbf{A}_{n+1}$ equals

$$\det \mathbf{A}_{n+1} = a_0^{n+1}.$$

This implies

THEOREM 1.3 (Wronski formula)

If $P := a_0 + a_1 x + a_2 x^2 + \cdots$ is a formal power series, $a_0 \neq 0$, the coefficients of the reciprocal series $P^{-1} =: b_0 + b_1 x + b_2 x^2 + \cdots$ are given by

$$b_n = \frac{(-1)^n}{a_0^{n+1}} \begin{vmatrix} a_1 & a_2 & \cdots & a_n \\ a_0 & a_1 & \cdots & a_{n-1} \\ 0 & a_0 & \cdots & a_{n-2} \\ & & \cdots & \\ 0 & 0 & a_0 & a_1 \end{vmatrix}, \quad n = 1, 2, \ldots. \quad (1.3\text{-}6)$$

This formula is used in our discussion of the quotient-difference algorithm (see Chapter 7). The first few coefficients, written out explicitly, are

$$b_0 = \frac{1}{a_0}, \quad b_1 = -\frac{a_1}{a_0^2}, \quad b_2 = \frac{1}{a_0^3} \begin{vmatrix} a_1 & a_2 \\ a_0 & a_1 \end{vmatrix},$$

$$b_3 = -\frac{1}{a_0^4} \begin{vmatrix} a_1 & a_2 & a_3 \\ a_0 & a_1 & a_2 \\ 0 & a_0 & a_1 \end{vmatrix}.$$

PROBLEMS

1. Verify the Wronski formula for the fps

   (a) $P = 1 + x + x^2 + \cdots$,

   (b) $P = 1 + 2x + 3x^2 + \cdots$.

2. Let $P := a_0 + a_1 x + a_2 x^2 + \cdots$ be a fps over the field of complex numbers, $a_0 \neq 0$. Show: (a) There are exactly two formal power series $Q =: b_0 + b_1 x + b_2 x^2 + \cdots$ such that $Q^2 = P$. The two possible values of b_0 are the two solutions z of $z^2 = a_0$. (b) For every positive integer n there are n series $Q =: b_0 + b_1 x + b_2 x^2 + \cdots$ such that $Q^n = P$. The n possible values of b_0 are the n solutions of the equation $z^n = a_0$.

§1.4. DIFFERENTIATION OF FORMAL POWER SERIES

Differentiation in the integral domain of formal power series, viewed as sequences of elements of the underlying field $\mathfrak{F}$, is defined purely algebraically as the operator that, acting on a sequence $P = \{a_0, a_1, a_2, \ldots\}$, produces the sequence $\delta P := \{a_1, 2a_2, 3a_3, \ldots\}$. Because this operator has many of the properties of differentiation, familiar from analysis, it is customary to write $\delta P =: P'$ and to call P' the **derivative** of P. Thus, if

$$P := a_0 + a_1 x + a_2 x^2 + \cdots,$$

then

$$P' := a_1 + 2a_2 x + 3a_3 x^2 + \cdots.$$

The integers $2, 3, \ldots$ that appear as factors in the definition of P' are to be viewed as abbreviations of $e + e, e + e + e, \ldots$, where e is the multiplicative identity in the field $\mathfrak{F}$.

EXAMPLES

1. $(x + 1! x^2 + 2! x^3 + \cdots)' = 1 + 2! x + 3! x^2 + \cdots$.

2. $I' = 0$.

The last example has the following converse.

THEOREM 1.4a

If P is a formal power series (over a field $\mathfrak{F}$ of characteristic zero) such that $P' = 0$, then P is a constant series.

By a *constant series* we mean, of course, a series of the form $P = a_0 + 0x + 0x^2 + \cdots$.

DIFFERENTIATION OF FORMAL POWER SERIES

Proof. Let $P =: a_0 + a_1 x + a_2 x^2 + \cdots$. The statement that $P' = 0$ is equivalent to $(n+1)a_{n+1} = 0, n = 0, 1, 2, \cdots$. Because $\mathfrak{F}$ has characteristic zero, $n + 1 \neq 0, n = 0, 1, 2, \ldots$ follows, and because $\mathfrak{F}$ has no divisors of zero we conclude that $a_{n+1} = 0, n = 0, 1, 2, \ldots$; that is, only a_0 can be different from zero. ∎

It is clear that the differentiation operator δ is a linear operator: if P and Q are any two formal power series and a and b are any two elements of $\mathfrak{F}$, then

$$\delta(aP + bQ) = a\delta P + b\delta Q,$$

or, expressed differently,

$$(aP + bQ)' = aP' + bQ'.$$

We shall now prove the **product rule**

$$(PQ)' = P'Q + PQ'. \tag{1.4-1}$$

Let $P =: a_0 + a_1 x + \cdots, Q =: b_0 + b_1 x + \cdots$. If $(PQ)' =: c_0 + c_1 x + \cdots$, we have, by definition,

$$c_n = (n+1)(a_0 b_{n+1} + a_1 b_n + \cdots + a_{n+1} b_0), \quad n = 0, 1, 2, \ldots.$$

On the other hand, the nth coefficient of $P'Q$ is given by

$$c_n^{(1)} = a_1 b_n + 2a_2 b_{n-1} + \cdots + (n+1) a_{n+1} b_0$$

and that of PQ' by

$$c_n^{(2)} = a_0 (n+1) b_{n+1} + a_1 n b_n + \cdots + a_n b_1.$$

Evidently $c_n^{(1)} + c_n^{(2)} = c_n$, which demonstrates the desired result. ∎

Other elementary differentiation rules follow from the product rule in the usual manner, e.g.,

$$(P^2)' = P'P + PP' = 2PP',$$

and by induction,

$$(P^n)' = nP^{n-1}P', \quad n = 1, 2, \ldots. \tag{1.4-2}$$

Furthermore, let P be a unit and set $Q := P^{-1}$. Differentiating the identity $PQ = I$ we obtain

$$P'Q + PQ' = 0,$$

hence

$$PQ' = -P'Q$$

or, multiplying by Q,

$$Q' = -P'Q^2 = -P'P^{-2}.$$

Since multiplication is commutative, we may use the fraction bar to denote reciprocals. The result then appears in the suggestive form

$$\left(\frac{1}{P}\right)' = -\frac{P'}{P^2}. \tag{1.4-3}$$

We are now ready to consider simple **formal differential equations** (fde), which are equations involving a formal power series and one or several of its derivatives. The object usually is to find one or more formal power series that satisfy the given fde. We consider two examples.

PROBLEM 1

If a is a given element of $\mathfrak{F}$, determine all fps P such that

$$P' = aP. \tag{1.4-4}$$

If $P = : b_0 + b_1 x + b_2 x^2 + \cdots$, (1.4-4) is equivalent to

$$b_1 + 2b_2 x + 3b_3 x^2 + \cdots = a(b_0 + b_1 x + b_2 x^2 + \cdots).$$

Comparing coefficients, we find

$$nb_n = ab_{n-1}, \qquad n = 1, 2, \ldots,$$

from which there follows (again making use of the fact that $\mathfrak{F}$ has characteristic zero)

$$b_1 = ab_0, \qquad b_2 = \frac{a^2}{2} b_1,$$

and generally by induction

$$b_n = \frac{a^n}{n!} b_0, \qquad n = 1, 2, \ldots.$$

Thus

$$P = b_0 \left(1 + \frac{a}{1!} x + \frac{a^2}{2!} + \cdots \right)$$

or $P = b_0 E_a$, where E_a denotes the **exponential series** (with subscript a)

$$E_a := 1 + \frac{a}{1!} x + \frac{a^2}{2!} x^2 + \cdots. \tag{1.4-5}$$

DIFFERENTIATION OF FORMAL POWER SERIES

Thus it turns out that there exists a solution series P of (1.4-4) and that it is fully determined by its zeroth coefficient b_0, which may be chosen arbitrarily.

Concerning the exponential series defined by (1.4-5) we assert the following theorem.

THEOREM 1.4b (Addition theorem for the exponential series)

For arbitrary a and b in $\mathfrak{F}$

$$E_a E_b = E_{a+b}. \qquad (1.4\text{-}6)$$

Proof. For the proof of this theorem we calculate the derivative of the fps $P := E_a E_b - E_{a+b}$. Using the product rule and the differential equation satisfied by E_a, we obtain

$$P' = E'_a E_b + E_a E'_b - E'_{a+b}$$

$$= aE_a E_b + bE_a E_b - (a+b)E_{a+b}$$

$$= (a+b)(E_a E_b - E_{a+b})$$

$$= (a+b)P.$$

There follows $P = c_0 E_{a+b}$, where c_0 is the zeroth coefficient of P. This coefficient is readily seen to be 0. Thus $P = 0$, as asserted. ∎

As a result of the addition theorem we have for any $a \in \mathfrak{F}$

$$E_a^{-1} = E_{-a}. \qquad (1.4\text{-}7)$$

Another consequence, admittedly minor, is as follows. We can also calculate the product $E_a E_b$ explicitly. The nth coefficient is

$$\frac{a^n}{n!} \frac{b^0}{0!} + \frac{a^{n-1}}{(n-1)!} \frac{b^1}{1!} + \cdots + \frac{a^1}{1!} \frac{b^{n-1}}{(n-1)!} + \frac{a^0}{0!} \frac{b^n}{n!}$$

or, using the familiar notation for the **binomial coefficient**,

$$\binom{n}{k} := \frac{n!}{k!(n-k)!},$$

$$\frac{1}{n!} \left[\binom{n}{0} a^n + \binom{n}{1} a^{n-1} b + \cdots + \binom{n}{n-1} ab^{n-1} + \binom{n}{n} b^n \right].$$

In view of (1.4-6) this equals

$$\frac{1}{n!}(a+b)^n.$$

We have proved the *binomial theorem*. The result is trivial, but the method is capable of generalization.

PROBLEM 2

Let a be an element of $\mathfrak{F}$; find all fps P such that

$$P' = \frac{a}{1+x}P. \qquad (1.4\text{-}8)$$

Again let $P = :b_0 + b_1 x + \cdots$. The given functional relation is then equivalent to

$$(1+x)(b_1 + 2b_2 x + 3b_3 x^2 + \cdots) = a(b_0 + b_1 x + \cdots)$$

or, forming the Cauchy product on the left,

$$b_1 + (2b_2 + b_1)x + (3b_3 + 2b_2)x^2 + \cdots = ab_0 + ab_1 x + \cdots.$$

Comparing coefficients, this means the same as

$$nb_n + (n-1)b_{n-1} = ab_{n-1}, \qquad n = 1, 2, \ldots,$$

or

$$b_n = \frac{a-n+1}{n} b_{n-1}.$$

There follows

$$b_n = \binom{a}{n} b_0,$$

where $\binom{a}{n}$ denotes the (generalized) binomial coefficient,

$$\binom{a}{0} := 1, \qquad \binom{a}{n} := \frac{a(a-1)\cdots(a-n+1)}{n!}, \qquad n = 1, 2, \ldots.$$

Thus we find that P satisfies (1.4-8) if and only if

$$P = b_0 B_a,$$

DIFFERENTIATION OF FORMAL POWER SERIES

where b_0 is the zeroth coefficient of P and

$$B_a := 1 + \binom{a}{1} x + \binom{a}{2} x^2 + \cdots$$

is called the **binomial series** with subscript a.

Again an interesting functional relation may be proved.

THEOREM 1.4c

For arbitrary a and b in $\mathfrak{F}$

$$B_a B_b = B_{a+b}. \tag{1.4-9}$$

Proof. The method of proof is the same. We differentiate

$$P := B_a B_b - B_{a+b}$$

and use the differential equation (1.4-8) to obtain

$$P' = B_a' B_b + B_a B_b' - B_{a+b}'$$

$$= \frac{1}{1+x} [a B_a B_b + b B_a B_b - (a+b) B_{a+b}]$$

$$= \frac{1}{1+x} (B_a B_b - B_{a+b})$$

$$= \frac{a+b}{1+x} P.$$

By the above P is proportional to B_{a+b}, but, because the zeroth coefficient of P equals $1 \cdot 1 - 1 = 0$, $P = 0$ follows. ∎

The method previously used to verify the binomial theorem now yields something more interesting. Calculating the nth coefficient of $B_a B_b$ explicitly, we get

$$\binom{a}{0}\binom{b}{n} + \binom{a}{1}\binom{b}{n-1} + \cdots + \binom{a}{n}\binom{b}{0}.$$

By the functional relation just established this equals

$$\binom{a+b}{n}.$$

Thus we have established **Vandermonde's theorem**,

$$\sum_{k=0}^{n}\binom{a}{k}\binom{b}{n-k}=\binom{a+b}{n}, \qquad (1.4\text{-}10)$$

a classical formula with numerous applications in combinatorics and in special function theory.

For many applications it is preferable to write Vandermonde's theorem in a different form, using factorial notation. For any $a \in \mathcal{F}$ and any non-negative integer n we define the **generalized factorial** or **Pochhammer symbol** by

$$(a)_n := \begin{cases} 1, & \text{if } n=0, \\ \underbrace{a(a+1)(a+2)\cdots(a+n-1)}_{n \text{ factors}}, & \text{if } n=1,2,\ldots. \end{cases}$$

The binomial coefficient can now be expressed as

$$\binom{a}{n} = \frac{a(a-1)\cdots(a-n+1)}{n!}$$

$$= (-1)^n \frac{(-a)(-a+1)\cdots(-a+n-1)}{n!}$$

$$= (-1)^n \frac{(-a)_n}{n!}.$$

Similarly, assuming that $b \neq 0, -1, -2, \ldots, -n+1$,

$$\binom{b}{n-k} = \binom{b}{n} \frac{n(n-1)\cdots(n-k+1)}{(b-n+1)\cdots(b-n+k)}$$

$$= (-1)^n \frac{(-b)_n}{n!} (-1)^k \frac{(-n)_k}{(b-n+1)_k}.$$

Thus (1.4-10) may be written

$$\sum_{k=0}^{n} \frac{(-a)_k (-n)_k}{k!(b-n+1)_k} = \frac{(-a-b)_n}{(-b)_n}.$$

DIFFERENTIATION OF FORMAL POWER SERIES

We simplify by replacing $-a$ with a and $b-n+1$ with c. Then $(-a-b)_n$ turns into $(a-c-n+1)_n = (-1)^n(c-a)_n$, $(-b)_n = (-c-n+1)_n = (-1)^n \times (c)_n$ (see Problem 5), and Vandermonde's theorem assumes the form

$$\sum_{k=0}^{n} \frac{(a)_k(-n)_k}{(c)_k k!} = \frac{(c-a)_n}{(c)_n}. \qquad (1.4\text{-}11)$$

A generalization of this formula due to Gauss, in which the finite sum is replaced by an infinite series, is established in §8.6.

Vandermonde's theorem can frequently be used to express sums involving binomial coefficients in closed form. As a very simple example we mention the case $a=b=n$ of (1.4-10) [or $a=-n$, $c=1$ of (1.4-11)] which yields

$$\sum_{k=0}^{n} \binom{n}{k}^2 = \binom{2n}{n}. \qquad (1.4\text{-}12)$$

PROBLEMS

1. Let m and n be integers $\neq 0$. Determine all solutions $Q = 1 + b_1 x + \cdots$ of the equation

   $$Q^m = (1+x)^n.$$

   (Compare Problem 2, §1.3.)

2. Let Q and R be given fps. Prove that the linear fde of order 1,

   $$P' = QP + R,$$

   always posses solutions and determine the structure of the general solution.

3. Show that the fps

   $$P := 0! - 1!x + 2!x^2 - 3!x^3 + \cdots$$

   formally satisfies the differential equation

   $$x^2 P' = x - P.$$

   Which "sum" would it be reasonable to assign to the series P? (This problem is to be solved heuristically; a theory is not given until Chapter 11.)

4. Find a formal solution of the differential equation

   $$xP'' + (c-x)P' - aP = 0$$

   $(a, c \in \mathcal{F}, c \neq 0, -1, -2, \ldots)$.

5. Prove the following relations for the Pochhammer symbol ($a \in \mathfrak{F}, n, m = 0, 1, 2, \ldots$):

   (a) $(a)_n = (-a-n+1)_n (-1)^n$,

   (b) $(a)_{n+m} = (a)_n (a+n)_m = (a)_m (a+m)_n$,

   (c) $(2a)_{2n} = (a)_n (a+\tfrac{1}{2})_n 2^{2n}$.

6. If n is a positive integer, show that
$$(n+1)(n+2) \cdots (2n) = 2^n 1 \cdot 3 \cdots (2n-1).$$

7. For non-negative integers r and u show that
$$\sum_{s=0}^{r} (-1)^s \binom{r}{s} \binom{2r-s}{u-s} = \binom{r}{u}.$$

[S. Lynn].

8. For non-negative integers s, t, r show that
$$\sum_{p=0}^{r} (-1)^p \binom{p+s}{t} \binom{r}{p} = \begin{cases} 0, & t < r, \\ (-1)^t, & t = r, \\ (-1)^r (t-r)! (s-t+1+r)_{t-r}, & t > r \end{cases}$$

[A. Huber].

9. If m and r are integers, $m \geqslant r > 0$, show that
$$\frac{1}{m}\binom{r}{0} - \frac{1}{m-1}\binom{r}{1} + \frac{1}{m-2}\binom{r}{2} - \cdots + (-1)^r \frac{1}{m-r}\binom{r}{r}$$
$$= (-1)^r \frac{r!(m-r-1)!}{m!}.$$

10. Let the polynomial p be defined by $p(x) = \binom{x}{n}^2$. Prove that
$$p(x) = \binom{x}{n}\binom{n}{0}\binom{n}{0} + \binom{x}{n+1}\binom{n}{1}\binom{n+1}{1} + \cdots + \binom{x}{2n}\binom{n}{n}\binom{2n}{n}.$$

11. If $n > m > 0$, prove that
$$\sum_{k=m}^{n} \binom{k}{k-m}\binom{n}{n-k} = \binom{n}{m} 2^{n-m}.$$

§1.5. FORMAL HYPERGEOMETRIC SERIES AND FINITE HYPERGEOMETRIC SUMS.

Let p, q be non-negative integers and let $a_1, a_2, \ldots, a_p; b_1, b_2, \ldots, b_q$ be elements of $\mathfrak{F}$, subject to the condition that for all i and all non-negative integers n, $b_i + n \neq 0$. The formal power series

$$F := \sum_{n=0}^{\infty} \frac{(a_1)_n (a_2)_n \cdots (a_p)_n}{(b_1)_n (b_2)_n \cdots (b_q)_n n!} x^n \qquad (1.5\text{-}1)$$

is called a **generalized hypergeometric series**. The following two notations are commonly used for F and appear indiscriminately in this book:

$$F = {}_pF_q \begin{bmatrix} a_1, \ldots, a_p; x \\ b_1, \ldots, b_q \end{bmatrix}; \qquad (1.5\text{-}1a)$$

$$F = {}_pF_q(a_1, a_2, \ldots, a_p; b_1, b_2, \ldots, b_q; x). \qquad (1.5\text{-}1b)$$

The special case $p = 2$, $q = 1$ is called the **classical** (or **Gaussian**) **hypergeometric series** and is frequently denoted by

$$F(a, b; c; x).$$

The generalized hypergeometric series (ghs) are of great importance in the theory of special functions of mathematical physics. Many such functions are defined as special cases of these series. Besides the Gaussian series, the case $p = q = 1$, called the confluent hypergeometric series, occurs frequently and is denoted by

$${}_1F_1(a; c; x) \quad \text{or} \quad \Phi(a, c, x).$$

Although not required by logical development, some formal aspects of the generalized hypergeometric series are discussed at this early stage to enable us to illustrate the theory of formal power series with some nontrivial examples.

The parameters a_i in (1.5-1) are called *numerator parameters*, the b_i, *denominator parameters*. Obviously the order in which these parameters appear is unimportant. Also, if a numerator parameter equals a denominator parameter, it may be canceled; for instance,

$${}_2F_2(a, b; a, c; x) = {}_1F_1(b; c; x).$$

By the series $_pF_q(\ldots;\ cx)$ we mean the series (1.5-1) with the factor c^n inserted in the nth coefficient.

Examples of generalized hypergeometric series:

1. $_1F_0(1;\ x) = 1 + x + x^2 + \ldots$, the geometric series. (This explains the name given to the more general series.)

2. $E_a(x) = 1 + \dfrac{ax}{1!} + \dfrac{(ax)^2}{2!} + \cdots = {}_0F_0(ax).$

3. $B_a(x) = 1 + \dbinom{a}{1} x + \dbinom{a}{2} x^2 + \cdots = {}_1F_0(-a;\ -x).$

4. A series such as $1 + \tfrac{1}{2}x + \tfrac{1}{3}x^2 + \cdots$ may be written in hypergeometric form by noting that

$$\frac{1}{n+1} = \frac{(1)_n}{(2)_n}, \qquad n = 0, 1, \ldots.$$

Thus

$$1 + \tfrac{1}{2}x + \tfrac{1}{3}x^2 + \cdots = {}_2F_1(1, 1;\ 2;\ x).$$

5. $1 + 1!x + 2!x^2 + \cdots = {}_2F_0(1, 1;\ x).$

6. The Bessel function of order zero has the fps expansion

$$J_0(x) = 1 - \frac{x^2}{2^2(1!)^2} + \frac{x^4}{2^4(2!)^2} - \frac{x^6}{2^6(3!)^2} + \cdots$$

$$= {}_0F_1\left(1;\ -\frac{x^2}{4}\right).$$

If in a generalized hypergeometric series one of the numerator parameters is a negative integer, say $-n$, then because

$$(-n)_m = 0 \quad \text{for} \quad m > n$$

all coefficients after the nth are zero. The series thus "terminates" and the insertion of an element of $\mathcal{F}$ for the indeterminate x defines a value of the series in a purely algebraic way. In a terminating series of type $_{p+1}F_p$ the value of the series for $x = 1$ is of frequent interest. One example is *Vandermonde's theorem*, which may be written in the form

$$F(a, -n;\ c;\ 1) = \frac{(c-a)_n}{(c)_n}. \tag{1.5-2}$$

As an example of a functional relation between hypergeometric series,

HYPERGEOMETRIC SERIES AND SUMS

we shall now prove **Kummer's first identity**,

$$_0F_0(x)\,_1F_1(a;c;-x) = {}_1F_1(c-a;c;x). \qquad (1.5\text{-}3)$$

(The first factor, in ordinary notation, is just E_1 or e^x.) The coefficient of x^n in the Cauchy product of the two formal power series on the left is

$$\sum_{k=0}^{n} \frac{(a)_k(-1)^k}{(c)_k k!} \cdot \frac{1}{(n-k)!} = \frac{1}{n!}\sum_{k=0}^{n}\frac{(a)_k(-n)_k}{(c)_k n!}$$

$$= \frac{1}{n!} F(a,-n;c;1) = \frac{(c-a)_n}{(c)_n n!},$$

by Vandermonde's formula. The result is the nth coefficient of the series on the right in (1.5-3).

A Differential Equation Satisfied by $_pF_q$. The differential operator δ defined in §1.4 maps x^n onto nx^{n-1}. For formal purposes it is often preferable to work with the differential operator $\theta := x\delta$ in place of δ. We then have

$$\theta x^n = nx^n, \qquad n = 0,1,2,\ldots.$$

Products, sums, and scalar multiples of θ are defined in the obvious manner. It is easy to see that any differential operator

$$\Theta := a_0\theta^m + a_1\theta^{m-1} + \cdots + a_m \qquad (1.5\text{-}4)$$

$(a_i \in \mathfrak{F}, i=1,2,\ldots,m)$ can always be written in the form

$$\Theta = b_0 x^m \delta^m + b_1 x^{m-1}\delta^{m-1} + \cdots + b_m, \qquad (1.5\text{-}5)$$

where the b_i are suitable elements of $\mathfrak{F}$, $b_0 = a_0$, $b_m = a_m$. Conversely, every operator of the form (1.5-5) can be converted into an operator of the form (1.5-4).

EXAMPLE 7

The operator $\theta^2 - 3\theta + 1$ transforms x^n into $(n^2 - 3n + 1)x^n$ and we have $n^2 - 3n + 1 = n(n-1) - 2n + 1$. Thus the same result is achieved by the operator $x^2\theta^2 - 2x\delta + 1$.

Let P_n denote the nth term of the generalized hypergeometric series (1.5-1):

$$P_n := \frac{(a_1)_n(a_2)_n\cdots(a_p)_n}{(b_1)_n(b_2)_n\cdots(b_q)_n n!} x^n.$$

Operating on P_n with $\theta + c$ ($c \in \mathfrak{F}$) adds the factor $n+c$ to the numerator. In particular, for $c = b_1 - 1$ we obtain

$$(\theta + b_1 - 1) P_n = \frac{(a_1)_n \cdots (a_p)_n}{(b_1)_{n-1}(b_2)_n \cdots (b_q)_n n!} x^n.$$

Repeating the process with $c = b_2 - 1, \ldots, b_q - 1, 0$, we get (if $n \geq 1$)

$$\theta(\theta + b_1 - 1) \cdots (\theta + b_q - 1) P_n$$

$$= \frac{(a_1)_n \cdots (a_p)_n}{(b_1)_{n-1} \cdots (b_q)_{n-1}(n-1)!} x^n$$

$$= x(\theta + a_1) \cdots (\theta + a_p) P_{n-1}. \tag{1.5-6}$$

We now easily obtain:

THEOREM 1.5

Let $a_1, a_2 \ldots, a_p, b_1, \ldots, b_q$ be arbitrary elements of $\mathfrak{F}$ such that no b_i is zero or a negative integer. Then the generalized hypergeometric series (1.5-1) is the only solution of the formal differential equation

$$\theta(\theta + b_1 - 1) \cdots (\theta + b_q - 1) P = x(\theta + a_1) \cdots (\theta + a_p) P \tag{1.5-7}$$

with zeroth coefficient 1.

Proof. That $P = F$ is a solution is seen by comparing the terms involving x^n on both sides of (1.5-7) and using (1.5-6). That F is the only solution may be seen as follows: let $P := c_0 + c_1 x + c_2 x^2 + \cdots$ be a solution and assume that

$$c_m = \frac{(a_1)_m \cdots (a_p)_m}{(b_1)_m \cdots (b_q)_m m!}$$

is true for $m = n$. (It is certainly true for $m = 0$.) Then the coefficient of x^{n+1} on the right of (1.5-7) is

$$(a_1 + n) \cdots (a_p + n) c_n;$$

on the left it is

$$(n+1)(b_1 + n) \cdots (b_q + n) c_{n+1}.$$

HYPERGEOMETRIC SERIES AND SUMS

Because none of the factors $n+1, b_1+n, \ldots, b_q+n$ is zero,

$$c_{n+1} = \frac{(a_1+n)\cdots(a_p+n)}{(b_1+n)\cdots(b_q+n)(n+1)} c_n$$

follows. The result is by induction with respect to n. ∎

EXAMPLES

8. The series $_0F_0(x)$ satisfies $\theta P = xP$, i.e. $P' = P$, as shown before.

9. The series $_1F_0(a; x)$ satisfies $\theta P = x(\theta+a)P$, i.e., $(1-x)P' = aP$, again in agreement with a previous result.

By expressing θ in terms of δ the differential equation (1.5-7) assumes the form

$$\left(x^{q+1}\delta^{q+1} + s_q x^q \delta^q + \cdots + s_1 x\delta \right) P$$
$$= x\left(x^p \delta^p + r_{p-1} x^{p-1} \delta^{p-1} + \cdots + r_0 \right) P, \quad (1.5\text{-}8)$$

where $r_0, \ldots, r_{p-1}$ and $s_1, \ldots, s_q$ are certain constants, depending on the a_i and the b_j, respectively. One may ask whether every differential equation of the form of (1.5-8) can be solved formally by a suitable series (1.5-1). In the first place this requires writing the differential operators appearing in parentheses in (1.5-8) as polynomials in θ. This can always be done. In order to achieve the standard form (1.5-7), one must then write these polynomials as products of linear factors $\theta + b_j - 1$ and $\theta + a_i$, respectively. This in general is possible only if $\mathscr{F}$ is the field of complex numbers. It finally may happen that a b_j is zero or a negative integer, in which case there is no solution of the form (1.5-1).

The fact remains, however, that in the field of complex numbers the differential equation (1.5-8) can always be reduced to the standard form (1.5-7). Differential equations of the form (1.5-8) are easily recognized: if we define the *weight* of the term $x^n \delta^m$ by $n-m$, the weights of all terms in (1.5-8) are either 0 or 1.

As an example we consider the equation

$$xP'' + (c-x)P' - aP = 0.$$

Here the weights are 0 and -1; they become 1 and 0 if the equation is multiplied by x. It may then be written

$$(x^2 \delta^2 + cx\delta) P = x(x\delta + a) P.$$

Because $x^2\delta^2 = \theta^2 - \theta$, this is the same as

$$[\theta^2 + (c-1)\theta]P = x(\theta + a)P,$$

which is in the form of (1.5-7), where $p = q = 1, a_1 = a, b_1 = c$. Thus, if $c \neq 0, -1, -2, \ldots$, the equation has the formal solution

$$P = {}_1F_1(a; c; x),$$

the only solution with zeroth coefficient 1.

As a second example we determine the equation satisfied by the Gaussian hypergeometric series $F(a,b; c; x)$. By Theorem 1.5 this series is the only formal solution with initial coefficient 1 of the equation

$$\theta(\theta + c - 1)P = x(\theta + a)(\theta + b)P. \tag{1.5-9}$$

In conventional notation it appears as

$$x(1-x)P'' + [c - (a+b+1)x]P' - abP = 0.$$

A somewhat lengthy computation, which we shall not reproduce, shows that the same differential equation is also satisfied by the formal power series

$$B_{c-a-b}(-x)F(c-a, c-b; c; x).$$

Because it also has zeroth coefficient one, there follows (on replacing $c-a$ with a and $c-b$ with b) the formula

$${}_1F_0(c-a-b; x)\,{}_2F_1(a,b; c; x) = {}_2F_1(c-a, c-b; c; x), \tag{1.5-10}$$

called **Euler's first identity** in the theory of hypergeometric series.

Let us compare coefficients in (1.5-10). On the left the coefficient of x^n is

$$\sum_{k=0}^{n} \frac{(a)_k(b)_k}{(c)_k k!} \frac{(c-a-b)_{n-k}}{(n-k)!} = \frac{(c-a-b)_n}{n!} \sum_{k=0}^{n} \frac{(a)_k(b)_k(-n)_k}{(c)_k(a+b-c-n+1)_k k!}$$

$$= \frac{(c-a-b)_n}{n!} {}_3F_2\left[\begin{matrix} a, b, -n; \\ c, a+b-c-n+1 \end{matrix}\ 1\right].$$

HYPERGEOMETRIC SERIES AND SUMS

Since the nth coefficient on the right obviously equals

$$\frac{(c-a)_n(c-b)_n}{(c)_n n!},$$

we obtain the identity

$${}_3F_2\left[\begin{array}{c} a,b,-n;\ 1 \\ c,a+b-c-n+1 \end{array}\right] = \frac{(c-a)_n(c-b)_n}{(c)_n(c-a-b)_n}, \qquad (1.5\text{-}11)$$

known as the **Saalschütz formula**.

NOTE

A terminating ${}_1F_0$ with unit argument can always be "summed" (i.e., expressed as products and quotients of Pochhammer symbols) and (according to the binomial theorem) has the value 0:

$${}_1F_0(-n;\ 1) = 0, \qquad n = 1, 2, \ldots.$$

A terminating ${}_2F_1$ with unit argument can always be summed by Vandermonde's theorem:

$${}_2F_1\left[\begin{array}{c} -n,a;\ 1 \\ c \end{array}\right] = \frac{(c-a)_n}{(c)_n}, \qquad n = 0, 1, 2, \ldots.$$

A terminating ${}_3F_2$ with unit argument can be summed by the Saalschütz formula if the sum of the denominator parameters exceed the sum of the numerator parameters by 1:

$${}_3F_2\left[\begin{array}{c} -n,a,b;\ 1 \\ c,a+b-c-n+1 \end{array}\right] = \frac{(c-a)_n(c-b)_n}{(c)_n(c-a-b)_n}, \qquad n = 0, 1, 2, \ldots.$$

Another case in which a terminating ${}_3F_2$ can be summed is given by Dixon's formula (see Problem 10, §1.6).

PROBLEMS

1. Prove the identity

$$_1F_0(a; x) \,_1F_0(b; x) = \,_1F_0(a+b; x).$$

2. Prove the relation (valid for $a, b, a+b-1 \neq 0, -1, -2, \ldots$)

$$_0F_1(a; x)\,_0F_1(b; x) = \,_2F_3\left[\begin{array}{c} \frac{1}{2}(a+b-1), \frac{1}{2}(a+b); 4x \\ a, b, a+b-1 \end{array}\right].$$

Note the special cases $a=b, a=2-b$.

3. Prove the identity (1.5-3) by a differential equation method.

4. Which homogeneous differential equation of order 2 is satisfied by the fps

$$P := 0! + 1!x + 2!x^2 + \cdots ?$$

5. For $n = 0, 1, 2, \ldots$, if c is not zero or a negative integer,

$$_2F_1\left[\begin{array}{c} -\frac{n}{2}, -\frac{n}{2} - \frac{1}{2}; 1 \\ c \end{array}\right] = \frac{2^n(c+\frac{1}{2})_n}{(2c)_n}.$$

6. Let n be a non-negative integer. Show that for arbitrary a, c such that the sum is defined

$$_3F_2\left[\begin{array}{c} -\frac{n}{2}, -\frac{n}{2} - \frac{1}{2}, a; 1 \\ c, a-c-n+\frac{1}{2} \end{array}\right] = \frac{(2c-2a)_n(c+\frac{1}{2})_n}{(2c)_n(c+\frac{1}{2}-a)_n}.$$

[Distinguish between even and odd values of n and apply the Saalschütz formula.]

7. For $n = 0, 1, 2, \ldots$ and $c \neq 0, -1, -2, \ldots$,

$$_3F_2\left[\begin{array}{c} -\frac{n}{3}, -\frac{n}{3} - \frac{1}{3}, -\frac{n}{3} - \frac{2}{3}; 1 \\ c, -c-n \end{array}\right] = (\tfrac{4}{3})^n \frac{(\frac{3}{2}c+\frac{1}{2})_n(\frac{3}{2}c+1)_n}{(3c)_n(c+1)_n}.$$

COMPOSITION WITH A NONUNIT

§1.6. THE COMPOSITION OF A FORMAL POWER SERIES WITH A NONUNIT

The problem considered here is the "substitution of one power series into another." In general, this cannot be done. Let

$$P := a_0 + a_1 x + a_2 x^2 + \cdots \tag{1.6-1}$$

be a fps and let

$$Q := b_0 + b_1 x + b_2 x^2 + \cdots \tag{1.6-2}$$

be another. We wish to substitute Q for x in P. The series Q^n exist for $n = 0, 1, 2, \ldots$; we write

$$Q^n = (b_0 + b_1 x + b_2 x^2 + \cdots)^n$$
$$=: b_0^{(n)} + b_1^{(n)} x + b_2^{(n)} x^2 + \cdots. \tag{1.6-3}$$

The coefficients $b_k^{(n)}$ can be found recursively by multiplying Q^{n-1} by Q. (A more efficient way to calculate the $b_k^{(n)}$ is given by the J. C. P. Miller formula; see below.) If we substitute these series for x^n into P, we get

$$a_0 + a_1 \left(b_0^{(1)} + b_1^{(1)} x + b_2^{(1)} x^2 + \cdots \right)$$
$$+ a_2 \left(b_0^{(2)} + b_1^{(2)} x + b_2^{(2)} x^2 + \cdots \right)$$
$$+ \cdots$$

Collecting coefficients of like powers, we find for the coefficient of x^n ($n = 0, 1, \ldots$)

$$c_n = a_0 b_n^{(0)} + a_1 b_n^{(1)} + a_2 b_n^{(2)} + \cdots. \tag{1.6-4}$$

This, in general, has no meaning; we have not defined convergence, and if we had we would not know whether it would take place here.

Assume, however, that Q is a nonunit, i.e., that $b_0 = 0$. Then, for $n = 1, 2, \ldots$, the series Q^n begins at the earliest with the coefficient of x^n, i.e.,

$$b_k^{(n)} = 0, \quad n > k.$$

Consequently, the sums (1.6-4) terminate after the nth term and the fps obtained by formally substituting Q into P exists. We denote it by $P \circ Q$ or by $P(Q)$ and call it the **composition** of P with Q. If

$$P \circ Q =: c_0 + c_1 x + c_2 x^2 + \cdots,$$

we have, using the notation in (1.6-3), $c_0 = a_0$ and

$$c_n = a_1 b_n^{(1)} + a_2 b_n^{(2)} + \cdots + a_n b_n^{(n)}, \qquad n = 1, 2, \ldots. \qquad (1.6\text{-}5)$$

Of course, as in a simple fps, we do not touch the convergence problem; no numerical value is assigned to the composed series for any value of x. Also, although the composition of a fps and a nonunit always exists, there may not always be a simple formula for its coefficients. One of our goals is to establish simple recurrence relations in certain cases.

EXAMPLE 1

To calculate the coefficients $c_0, c_1, c_2, \ldots,$ in the expansion

$$E_1 \circ Q =: c_0 + c_1 x + c_2 x^2 + \cdots,$$

where $Q := yx - \frac{1}{2}x^2$ and y is a complex constant, we have

$$Q^n = x^n (y - \tfrac{1}{2} x)^n$$

$$= y^n x^n - \frac{1}{2}\binom{n}{1} y^{n-1} x^{n+1} + \frac{1}{2^2}\binom{n}{2} y^{n-2} x^{n+2}$$

$$+ \cdots + (-1)^n \frac{1}{2^n}\binom{n}{n} x^{2n}.$$

Thus

$$b_k^{(n)} = \begin{cases} 0, & k < n, \\ (-1)^{k-n} \dfrac{1}{2^{k-n}} \binom{n}{k-n} y^{2n-k}, & n \leq k \leq 2n, \\ 0, & k > 2n, \end{cases}$$

and we find

$$c_k = \sum_{n=0}^{k} a_n b_k^{(n)}$$

$$= \sum_{n=0}^{k} \frac{(-1)^{k-n}}{n!} \frac{1}{2^{k-n}} \binom{n}{k-n} y^{2n-k}.$$

COMPOSITION WITH A NONUNIT

Letting $k-n=p$, we get for $y \neq 0$

$$c_k = y^k \sum_{p=0}^{n} \frac{1}{p!(k-2p)!} \left(-\frac{1}{2y^2}\right)^p$$

$$= \frac{y^k}{k!} {}_2F_0\left(-\frac{k}{2}, -\frac{k}{2}+\frac{1}{2}; -\frac{2}{y^2}\right).$$

The c_k are certain polynomials of degree k in y known as **Hermite polynomials**.

The Distributive Laws of Composition. Let A, B be arbitrary formal power series

$$A =: a_0 + a_1 x + a_2 x^2 + \cdots,$$

$$B =: b_0 + b_1 x + b_2 x^2 + \cdots,$$

and let Q, R be nonunits,

$$Q =: q_1 x + q_2 x^2 + \cdots,$$

$$R =: r_1 x + r_2 x^2 + \cdots.$$

We shall continue to use the notation

$$Q^k = \sum q_n^{(k)} x^n, \qquad R^k = \sum r_n^{(k)} x^n, \qquad k=1,2,\ldots.$$

Our goal here is to calculate the coefficients $c_0, c_1, c_2, \ldots$, in the product

$$C := (A \circ Q) \cdot (B \circ R),$$

where the compositions are to be performed first. By (1.6-5) these coefficients are

$$c_n = \sum_{i+j=n} \sum_{k=0}^{i} a_k q_i^{(k)} \sum_{m=0}^{j} b_m r_j^{(m)}. \tag{1.6-6}$$

We claim that these coefficients can also be formed in a different manner. For $k, m = 0, 1, 2, \ldots$, let

$$Q^k R^m =: \sum p_n^{(k,m)} x^n.$$

THEOREM 1.6a

With the above notation

$$c_n = \sum_{k+m \leq n} a_k b_m p_n^{(k,m)}. \tag{1.6-7}$$

Intuitively this means that to calculate the product

$$C = (a_0 + a_1 Q + a_2 Q^2 + \cdots)(b_0 + b_1 R + b_2 R^2 + \cdots)$$

we must first form all possible cross products,

$$C = a_0 b_0 + a_1 b_0 Q + a_0 b_1 R + a_2 b_0 Q^2 + a_1 b_1 QR + \cdots,$$

expand each term separately, and collect coefficients of like powers. Since Q and R are nonunits, only the cross products $Q^k R^m$ where $k + m \leq n$ contribute to the coefficient of x^n.

Proof. We show that (1.6-7) reduces to (1.6-6):

$$p_n^{(k,m)} = \sum_{i+j=n} q_i^{(k)} r_j^{(m)};$$

hence

$$\sum_{k+m \leq n} a_k b_m p_n^{(k,m)} = \sum_{k+m \leq n} a_k b_m \sum_{i+j=n} q_i^{(k)} r_j^{(m)}$$

$$= \sum_{i+j=n} \sum_{k+m \leq n} a_k q_i^{(k)} b_m r_j^{(m)}$$

$$= \sum_{i+j=n} \sum_{k=0}^{i} a_k q_i^{(k)} \sum_{m=0}^{j} b_m r_j^{(m)}. \blacksquare$$

We note some special cases.

1. If $Q = R$, then $p_n^{(k,m)} = q_n^{(k+m)}$. Thus, summing with respect to $s := k + m$, (1.6-7) becomes

$$c_n = \sum_{s=0}^{n} q_n^{(s)} \sum_{k+m=s} a_k b_m. \tag{1.6-8}$$

This however, is precisely the nth coefficient in the composition of the Cauchy product AB with Q and we have proved the **right distributive law for composition**,

$$(A \circ Q) \cdot (B \circ Q) = (A \cdot B) \circ Q. \tag{1.6-9}$$

COMPOSITION WITH A NONUNIT

2. We may also use (1.6-7) to calculate the product $(A \circ Q) \cdot B$. Here $R = X = 0 + 1x + 0x^2 + \cdots$, and

$$p_n^{(k,m)} = q_{n-m}^{(k)}.$$

Hence (1.6-7) yields

$$c_n = \sum_{k+m \leqslant n} a_k b_m q_{n-m}^{(k)}$$

or

$$c_n = \sum_{k=0}^{n} a_k \sum_{m=0}^{n-k} b_m q_{n-m}^{(k)}. \quad (1.6\text{-}10)$$

This means that to calculate the product

$$C := (a_0 + a_1 Q + a_2 Q^2 + \cdots) B$$

we may multiply out

$$C = a_0 B + a_1 Q B + a_2 Q^2 B + \cdots,$$

expand each term individually, and collect like powers. Thus a sort of distributive law also holds and it obviously does not matter whether the factor B is placed on the left or right.

EXAMPLE 2

Let

$$A := {}_2F_1(a,b;\, c;\, x),$$

$$B := {}_1F_0(a;\, x),$$

$$Q := -\frac{x}{1-x} = -x - x^2 - x^3 - \cdots.$$

Because $Q = -x\, {}_1F_0(1; x)$, we have by Problem 1 of §1.5

$$Q^k = (-x)^k\, {}_1F_0(k; x),$$

$$Q^k B = (-x)^k\, {}_1F_0(a+k; x),$$

with the nth coefficient

$$(-1)^k \frac{(a+k)_{n-k}}{(n-k)!} = \frac{(-n)_k (a)_n}{n!\, (a)_k}.$$

Thus by Vandermonde's theorem

$$c_n = \sum_{k=0}^{n} \frac{(a)_k (b)_k}{(c)_k k!} \frac{(-n)_k (a)_n}{n!(a)_k}$$

$$= \frac{(a)_n}{n!} \sum_{k=0}^{n} \frac{(-n)_k (b)_k}{(c)_k k!}$$

$$= \frac{(a)_n (c-b)_n}{(c)_n n!},$$

which equals the nth coefficient in $_2F_1(a, c-b; c; x)$. Thus we have proved the formula

$$_1F_0(a; x) \,_2F_1\left(a, b; c; -\frac{x}{1-x}\right) = \,_2F_1(a, c-b; c; x), \quad (1.6\text{-}11)$$

called **Euler's second identity**. Equivalent forms are

$$_2F_1(a, b; c; x) = \,_1F_0(a; x) \,_2F_1\left(a, c-b; c; -\frac{x}{1-x}\right)$$

$$= \,_1F_0(b; x) \,_2F_1\left(c-a, b; c; -\frac{x}{1-x}\right).$$

Differentiation of Composed Formal Power Series. Here we wish to establish the **chain rule** of differentiation.

THEOREM 1.6b

If A is a formal power series and Q a nonunit,

$$(A \circ Q)' = (A' \circ Q) \cdot Q'. \quad (1.6\text{-}12)$$

Proof. In the earlier notation of the coefficients of powers the coefficient of x^n in $(A \circ Q)'$ is

$$c_n := (n+1)\left(a_1 q_{n+1}^{(1)} + a_2 q_{n+1}^{(2)} + \cdots + a_{n+1} q_{n+1}^{(n+1)}\right).$$

The nth coefficient in $(A' \circ Q) Q'$ can be calculated by using the distributive law (Theorem 1.6a):

$$(a_1 + 2a_2 Q + 3a_3 Q^2 + \cdots) Q' = a_1 Q' + 2a_2 QQ' + 3a_3 Q^2 Q' + \cdots. \quad (1.6\text{-}13)$$

COMPOSITION WITH A NONUNIT

According to §1.4,

$$kQ^{k-1}Q' = (Q^k)'.$$

Thus the coefficient of x^n in $kQ^{k-1}Q'$ is $(n+1)q_{n+1}^{(k)}$, and for the coefficient of x^n in (1.6-13) we find

$$a_1(n+1)q_{n+1}^{(1)} + a_2(n+1)q_{n+1}^{(2)} + \cdots + a_{n+1}(n+1)q_{n+1}^{(n+1)}$$

in agreement with what we had before. ∎

As an application, we shall prove a useful formula for raising a power series (with zeroth coefficient 1) to an arbitrary power. Let $a \in \mathfrak{F}$ be arbitrary, let B_a denote the binomial series, and let

$$Q := b_1 x + b_2 x^2 + \cdots \tag{1.6-14}$$

be a nonunit. We wish to find recurrence relations for the coefficients c_i in the series

$$B_a \circ Q =: c_0 + c_1 x + c_2 x^2 + \cdots. \tag{1.6-15}$$

Evidently $c_0 = 1$. Differentiating, we get

$$(B_a \circ Q)' = c_1 + 2c_2 x + 3c_3 x^2 + \cdots.$$

By the chain rule this equals $(B_a' \circ Q)Q'$. Thus by the differential equation satisfied by the binomial series $B_a' = aB_{-1}B_a = aB_{a-1}$

$$(B_a \circ Q)' = a(B_{a-1} \circ Q)Q'.$$

Multiplying by $B_1 \circ Q$, using the distributive law, and observing $B_1 B_{a-1} = B_a$, there follows

$$(B_1 \circ Q)(B_a \circ Q)' = a(B_a \circ Q)Q' \tag{1.6-16}$$

or, written out in full,

$$(1 + b_1 x + b_2 x^2 + \cdots)(c_1 + 2c_2 x + 3c_3 x^2 + \cdots)$$
$$= a(c_0 + c_1 x + c_2 x^2 + \cdots)(b_1 + 2b_2 x + 3b_3 x^2 + \cdots).$$

Forming the Cauchy products and comparing the coefficients of x^{n-1}, we have (if $b_0 := 1$)

$$n_{n-1}c_1 + 2b_{n-2}c_2 + \cdots + nb_0 c_n = a(c_{n-1}b_1 + 2c_{n-2}b_2 + \cdots + nc_0 b_n).$$

Solving for c_n, we get the following

THEOREM 1.6c

If $Q := b_1 x + b_2 x^2 + \cdots$ and B_a denotes the binomial series, then $B_a \circ Q = c_0 + c_1 x + c_2 x^2 + \cdots$, where $c_0 = 1$,

$$c_n = \frac{1}{n} \sum_{k=0}^{n-1} [a(n-k) - k] c_k b_{n-k}$$

$$= \frac{1}{n} \sum_{k=1}^{n} [(a+1)k - n] c_{n-k} b_k, \qquad n = 1, 2, \ldots . \qquad (1.6\text{-}17)$$

The recurrence relation (1.6-17) is called the **J. C. P. Miller formula**. If programmed with care, it permits the computation of the coefficients $c_1, c_2, \ldots, c_n$ in n^2 multiplications and $n - 1$ divisions.

EXAMPLE 3

We compute the coefficients c_i in the expansion

$$B_{-1/2} \circ (-2x - 3x^2) =: c_0 + c_1 x + c_2 x^2 + \cdots .$$

Here, in the notation of Theorem 1.6c, $a = -\frac{1}{2}$, $b_1 = -2$, $b_2 = -3$, and $b_k = 0$ for $k > 2$; (1.6-17) thus reduces to the three-term recurrence

$$c_n = \frac{1}{n} [(2n-1)c_{n-1} + 3(n-1)c_{n-2}], \qquad n = 1, 2, \ldots .$$

We readily obtain

n	1	2	3	4	5	6
c_n	1	3	7	19	51	141

The c_n turn out to be integers (compare Example 3, §1.9).

PROBLEMS

1. Form the composition $B_{-1/2} \circ Q$, where B is the binomial series and $Q := -2yx + x^2$, and express the coefficients as hypergeometric sums with the argument y. (These coefficients are known as **Legendre polynomials**.)
2. Prove the formal identity

$$_1F_0(a; x) \, _2F_1\left(\frac{a}{2}, \frac{1+a}{2} - b; 1 + a - b; -\frac{4x}{(1-x)^2}\right) = {_2F_1}(a, b; 1 + a - b; x).$$

[Apply the Saalschütz formula.]

COMPOSITION WITH A NONUNIT

3. Prove
$$_2F_1(a,b; a+b+\tfrac{1}{2}; 4x(1-x)) = {}_2F_1(2a,2b; a+b+\tfrac{1}{2}; x).$$
[Distinguish between coefficients of even and of odd powers.]

4. Prove **Whipple's formula**
$$_1F_0(a; x) \, _3F_2\left[\begin{array}{c} \frac{a}{2}, \frac{a}{2}+\frac{1}{2}, 1+a-b-c; \; -\frac{4x}{(1-x)^2} \\ 1+a-b, 1+a-c \end{array}\right]$$
$$= {}_3F_2\left[\begin{array}{c} a,b,c; \; x \\ 1+a-b, 1+a-c \end{array}\right].$$

5. The **Gegenbauer polynomials** $C_n^\nu(y)$ are defined as the coefficients c_n in the expansion
$$B_{-\nu} \circ Q =: c_0 + c_1 x + c_2 x^2 + \cdots,$$
where $Q := -2yx + x^2$. Show that the J. C. P. Miller formula yields a three-term recurrence relation for the $C_n^\nu(y)$.

6. Devise a formula similar to the J. C. P. Miller formula for the coefficients c_n in
$$E_a \circ Q =: c_0 + c_1 x + c_2 x^2 + \cdots,$$
where E_a is the exponential series and Q is a given nonunit. Use the formula to form the composition of E_1 with the series $x - \tfrac{1}{2}x^2 + \tfrac{1}{3}x^3 - \cdots$. (For an application to Stirling's series see §11.3.)

7. Find a three-term recurrence relation for the Hermite polynomials.

8. Verify that for $a = -1$ the J. C. P. Miller formula reduces to the recurrence relation for the coefficients of the reciprocal series; see §1.2.

9. If P is a unit and Q is a nonunit, show that $P^{-1} \circ Q = (P \circ Q)^{-1}$.

10. Setting $a = -n$ in Whipple's formula (see Problem 4) and replacing x with 1, prove **Dixon's formula**
$$_3F_2\left[\begin{array}{c} -n,a,b; \; 1 \\ 1-n-a, 1-n-b \end{array}\right] = \begin{cases} \dfrac{(2m)!}{m!} \dfrac{(a+b+m)_m}{(a+m)_m (b+m)_m}, \\ \text{if } n \text{ is even, } n = 2m, \\ 0, \text{ if } n \text{ is odd.} \end{cases}$$

11. For any positive integer n find a closed expression for the sum
$$1 - \binom{n}{1}^3 + \binom{n}{2}^3 - \cdots + (-1)^n \binom{n}{n}^3.$$

[Use Dixon's formula.]

12. Using Dixon's formula, express the following products by single hypergeometric series in x^2:

   (a) $\quad {}_2F_0(a,b; x)\, {}_2F_0(a,b; -x)$,

   (b) $\quad {}_1F_1(a; b; x)\, {}_1F_1(a; b; -x)$,

   (c) $\quad {}_0F_2(a,b; x)\, {}_0F_2(a,b; -x)$.

13. Using formal differential equations, prove the identity

$$(B_a \circ X)(B_a \circ (-X)) = B_a \circ (-X^2)$$

and apply it to show that

$$
{}_2F_1\left[\begin{array}{c} -n, b; \; -1 \\ 1-n-b \end{array}\right] = \begin{cases} \dfrac{(m+1)_m}{(b+m)_m}, & \text{if } n \text{ is even, } n = 2m; \\ 0, & \text{if } n \text{ is odd.} \end{cases}
$$

14. Prove

$$
\sum_{k=0}^{n} (-1)^k \binom{n}{k}^2 = \begin{cases} (-1)^m \dfrac{(2m)!}{(m!)^2}, & \text{if } n = 2m, \\ 0, & \text{if } n \text{ is odd.} \end{cases}
$$

15. Using Euler's first transformation, show that

$$
{}_2F_1\left[\begin{array}{c} a, -n; \; -1 \\ 1+a+n \end{array}\right]
$$

equals the coefficient of x^n in the series

$$
(1-x^2)^n {}_2F_1\left[\begin{array}{c} 1+n, a+n; \; x \\ 1+a+n \end{array}\right].
$$

Then prove the formula

$$
{}_2F_1\left[\begin{array}{c} a, -n; \; -1 \\ 1+a+n \end{array}\right] = \frac{(1+a)_n}{(1+a/2)_n}.
$$

16. Express by single hypergeometric series

   (a) $\quad {}_0F_1(a; x)\, {}_0F_1(a; -x)$,

   (b) $\quad {}_0F_1(1+a; x)\, {}_0F_1(1-a; -x)$.

17. Let

$$F := 1 + \tfrac{1}{3}x^2 + \tfrac{1}{5}x^4 + \cdots,$$
$$P := 2(x - x^3 + x^5 - x^7 + \cdots).$$

Prove

$$F \circ P = (1 + x^2) F.$$

§1.7. THE GROUP OF ALMOST UNITS UNDER COMPOSITION

In this section we are concerned with nonunits in the integral domain of formal power series with a nonzero first coefficient, $P = a_1 x + a_2 x^2 + \cdots$, where $a_1 \neq 0$. For brevity we call them **almost units** (in the integral domain of fps). It follows from the discussion in §1.6 that the composition of two almost units again yields an almost unit. Thus the set of almost units is closed under composition. Also, there is an identity element, $X = 1x + 0x^2 + 0x^3 + \cdots$, such that $X \circ P = P \circ X = P$ for any almost unit P.

THEOREM 1.7a

Under the operation of composition the almost units in the integral domain of formal power series over a field $\mathfrak{F}$ form a group.

Proof. For the proof it remains to be shown that the associative law holds and that every almost unit P has an inverse under composition. This is accomplished most easily by matrix methods.

If $P := a_1 x + a_2 x^2 + \cdots$, we continue to use the notation

$$P^k = \sum_n a_n^{(k)} x^n, \quad k = 1, 2, \ldots.$$

With any nonunit P we now associate the matrix $\mathbf{A} = (a_{ij})$ with the elements $a_{ij} := a_j^{(i)}$, $i, j = 1, 2, \ldots$. In §1.6 we saw that $a_{ij} = 0$ for $j < i$. Thus the matrix $\mathbf{A}$ is triangular:

$$\mathbf{A} = \begin{bmatrix} a_1^{(1)} & a_2^{(1)} & a_3^{(1)} & a_4^{(1)} & \cdots \\ & a_2^{(2)} & a_3^{(2)} & a_4^{(2)} & \\ & & a_3^{(3)} & a_4^{(3)} & \\ & & & a_4^{(4)} & \\ 0 & & & & \ddots \end{bmatrix}.$$

The product of any two such matrices exists and is again triangular; see §1.3.

The above correspondence is denoted by the symbol $P \to \mathbf{A}$. No matrix $\mathbf{A}$ can correspond to two different P because the first row of $\mathbf{A}$ contains the coefficients of P.

LEMMA 1.7b

Let P and Q be nonunits and let $P \to \mathbf{A}$, $Q \to \mathbf{B}$. Then

$$P \circ Q \to \mathbf{AB}. \tag{1.7-1}$$

Indeed, if $P =: a_1 x + a_2 x^2 + \cdots$, $Q =: b_1 x + b_2 x^2 + \cdots$, and $\mathbf{AB} =: (c_{ij})$, then for $n \geq m$

$$c_{mn} = a_m^{(m)} b_n^{(m)} + a_{m+1}^{(m)} b_n^{(m+1)} + \cdots + a_n^{(m)} b_n^{(n)}.$$

The expression on the right also arises by forming the nth coefficient in the composition $P^m \circ Q$. By virtue of the distributive law of composition, $P^m \circ Q = (P \circ Q)^m$. Thus c_{mn} equals the nth coefficient in $(P \circ Q)^m$, as required. ∎

The associative law for composition is now easily proved. Let P, Q, R be nonunits and let $P \to \mathbf{A}$, $Q \to \mathbf{B}$, $R \to \mathbf{C}$. By the lemma $P \circ Q \to \mathbf{AB}$, hence $(P \circ Q) \circ R \to (\mathbf{AB})\mathbf{C}$. In particular, the coefficients of $(P \circ Q) \circ R$ appear in the first row of the matrix $(\mathbf{AB})\mathbf{C}$, but because matrix multiplication is associative the same coefficients appear also by forming $\mathbf{A}(\mathbf{BC})$, which arises from the composition $P \circ (Q \circ R)$. Thus $(P \circ Q) \circ R = P \circ (Q \circ R)$, as required.

For the proof of Theorem 1.7a it remains to be shown that every almost unit $P = a_1 x + a_2 x^2 + \cdots$ has an inverse. By general group theory it suffices to show that it has a *left* inverse. $Q = b_1 x + b_2 x^2 + \cdots$ is a left inverse of P if and only if its coefficients satisfy

$$b_1 a_1^{(1)} = 1,$$

$$b_1 a_n^{(1)} + b_2 a_n^{(2)} + \cdots + b_n a_n^{(n)} = 0, \qquad n = 2, 3, \ldots. \tag{1.7-2}$$

Because $a_n^{(n)} = a_1^n \neq 0$ ($n = 1, 2, 3, \ldots$), these recurrence relations have a unique solution $\{b_1, b_2, \ldots\}$ such that $b_1 \neq 0$. This completes the proof of Theorem 1.7a. ∎

In order to distinguish the inverse with respect to composition from the inverse with respect to Cauchy multiplication, we denote the former by

GROUP OF ALMOST UNITS UNDER COMPOSITION

$P^{[-1]}$ and also call it the **reversion** of the almost unit P. In a similar sense we write

$$P^{[2]} := P \circ P, \ldots.$$

By virtue of the associative law no ambiguity can arise by writing $P^{[n]}$ for the n-fold composition of P with itself, no matter how formed.

It is easy to determine the matrix associated with $P^{[-1]}$. Let $P \to \mathbf{A}$ and $P^{[-1]} \to \mathbf{B}$. Because the series $X := x + 0x^2 + 0x^3 + \cdots$ corresponds to the identity matrix $\mathbf{I}$, it follows from Lemma 1.7b that $\mathbf{AB} = \mathbf{I}$.

THEOREM 1.7c

If P is an almost unit, and $P \to \mathbf{A}$, then $P^{[-1]} \to \mathbf{A}^{-1}$.

If (as in §1.3) we denote the nth section of the triangular matrix $\mathbf{A}$ by $\mathbf{A}_n$, we have $(\mathbf{A}^{-1})_n = (\mathbf{A}_n)^{-1}$. Theorem 1.7c thus reduces the problem of reverting a power series to one of inverting a finite triangular matrix.

EXAMPLE 1

Let $a \in \mathcal{F}$ be arbitrary. We wish to revert the series

$$P := x \,_1F_0(a; x) = x B_{-a} \circ (-X)$$

$$= x + \frac{(a)_1}{1!} x^2 + \frac{(a)_2}{2!} x^3 + \cdots.$$

Because

$$P^k = x^k \,_1F_0(ka; x) = x^k B_{-ka} \circ (-X),$$

the matrix corresponding to P is $A := (a_{mn})$, where

$$a_{mn} := \frac{(ma)_{n-m}}{(n-m)!}, \qquad n \geq m, \tag{1.7-3}$$

i.e.,

$$\mathbf{A} = \begin{bmatrix} 1 & \frac{(a)_1}{1!} & \frac{(a)_2}{2!} & \frac{(a)_3}{3!} & \cdots \\ & 1 & \frac{(2a)_1}{1!} & \frac{(2a)_2}{2!} & \\ & & 1 & \frac{(3a)_1}{1!} & \\ & & & 1 & \\ 0 & & & & \ddots \end{bmatrix}.$$

FORMAL POWER SERIES

We assert that the elements of the inverse matrix $\mathbf{B} := (b_{mn}) := \mathbf{A}^{-1}$ are given by

$$b_{mn} = \frac{m}{n} \frac{(-na)_{n-m}}{(n-m)!}, \qquad n \geq m, \qquad (1.7\text{-}4)$$

i.e.,

$$\mathbf{A}^{-1} = \begin{bmatrix} 1 & \frac{1}{2}\frac{(-2a)_1}{1!} & \frac{1}{3}\frac{(-3a)_2}{2!} & \frac{1}{4}\frac{(-4a)_3}{3!} & \cdots \\ & 1 & \frac{2}{3}\frac{(-3a)_1}{1!} & \frac{2}{4}\frac{(-4a)_2}{2!} & \\ & & 1 & \frac{3}{4}\frac{(-4a)_1}{1!} & \\ & & & 1 & \\ 0 & & & & \ddots \end{bmatrix}.$$

The verification may be based on Vandermonde's theorem. It is clear that $a_{mm}b_{mm} = 1$, whereas for $n > m$

$$\sum_{k=m}^{n} a_k^{(m)} b_n^{(k)} = \sum_{k=m}^{n} \frac{(ma)_{k-m}}{(k-m)!} \frac{k}{n} \frac{(-na)_{n-k}}{(n-k)!}$$

$$= \sum_{p=0}^{n-m} \frac{(ma)_p}{p!} \frac{m+p}{n} \frac{(-na)_{n-m-p}}{(n-m-p)!}$$

$$= c' + c'',$$

where

$$c' := \frac{m}{n} \frac{(-na)_{n-m}}{(n-m)!} \,{}_2F_1\!\left[\begin{array}{c} ma, -n+m; 1 \\ na-n+m+1 \end{array}\right] = \frac{m}{n} \frac{(ma-na)_{n-m}}{(n-m)!},$$

$$c'' := \frac{1}{n} \sum_{p=1}^{n-m} \frac{(ma)_p}{(p-1)!} \frac{(-na)_{n-m-p}}{(n-m-p)!}$$

$$= \frac{ma}{n} \frac{(-na)_{n-m-1}}{(n-m-1)!} \,{}_2F_1\!\left[\begin{array}{c} 1+ma, -n+m+1; 1 \\ na-n+m+2 \end{array}\right]$$

$$= \frac{ma}{n} \frac{(ma-na+1)_{n-m-1}}{(n-m-1)!},$$

and $c' + c'' = 0$ follows readily. Thus we have

$$P^{[-1]} = x + \frac{(-2a)_1}{2!} x^2 + \frac{(-3a)_2}{3!} x^3 + \frac{(-4a)_3}{4!} x^4 + \cdots.$$

GROUP OF ALMOST UNITS UNDER COMPOSITION

We require a formula for the derivative of the reversion of an almost unit P. Let $Q := P^{[-1]}$; then by definition $Q \circ P = X$. By differentiating both sides and applying the chain rule, we get

$$(Q' \circ P) P' = I. \tag{1.7-5}$$

Because P is an almost unit, P' is a unit whose reciprocal exists; hence

$$Q' \circ P = (P')^{-1}. \tag{1.7-6}$$

Application of $\circ Q$ from the right and use of the fact that $Q = P^{[-1]}$ yields $Q' = (P')^{-1} \circ Q$. Because $A^{-1} \circ Q = (A \circ Q)^{-1}$ (see Problem 9, §1.6), we finally get the following:

THEOREM 1.7d

If P is an almost unit and $Q := P^{[-1]}$, then

$$Q' = (P' \circ Q)^{-1}. \tag{1.7-7}$$

This formula can sometimes be used to find Q itself. An important example of this is the following:

EXAMPLE 2

To find the reversion of

$$P := E_a - I = \frac{a}{1!} x + \frac{a^2}{2!} x^2 + \cdots, \quad (a \neq 0).$$

Let $Q := P^{[-1]}$; then by (1.7-7)

$$Q' = [(E_a - I)' \circ Q]^{-1}.$$

But $(E_a - I)' = a E_a$; hence

$$(E_a - I)' \circ Q = a E_a \circ Q = a(E_a - I) \circ Q + a I \circ Q = a(X + I),$$

and

$$Q' = a^{-1}(I + X)^{-1} = a^{-1} B_{-1} = \frac{1}{a}(1 - x + x^2 - x^3 + \cdots).$$

Since the zeroth coefficient of Q is zero, the last relation implies $Q = (1/a)L$, where

$$L := x - \tfrac{1}{2} x^2 + \tfrac{1}{3} x^3 - \tfrac{1}{4} x^4 + \cdots \tag{1.7-8}$$

is called the **logarithmic series**.

As an application of Example **2** we prove the following:

THEOREM 1.7e

Let E_a, B_a, and L denote the exponential, binomial, and logarithmic series, respectively. Then

$$E_a \circ L = B_a. \qquad (1.7\text{-}9)$$

Proof. Let $E_a \circ L =: P$. Then P has zeroth coefficient 1. Moreover, $P' = (E_a' \circ L) L' = aPL'$, and because $L' = 1 - x + x^2 - \cdots = (1+x)^{-1}$ we find that P satisfies the differential equation $(1+x)P' = aP$. It follows from the definition of B_a that $P = B_a$. ∎

PROBLEMS

1. Show by means of an example that, in general, $P \circ Q \ne Q \circ P$.
2. Give an alternate proof of Equation (1.7-7) by differentiating the identity $P \circ Q = X$.
3. If $a \ne 0$ and

$$P := \frac{E_{2a} - I}{E_{2a} + I}, \qquad Q := P^{[-1]},$$

show that $aQ = x + \frac{1}{3}x^2 + \frac{1}{5}x^5 + \cdots$. $[P' = 4aE_{2a}/(E_{2a}+I)^2 = a(I+P)(I-P) = a(I-P^2)$; hence $P' \circ Q = a(I-X^2)$, $Q' = (P' \circ Q)^{-1} = (1/a)(I-X^2)^{-1}$.]

4. The exponential series $E := E_1$ satisfies the relation $E^{-1} = E \circ (-X)$. Prove that the fps P satisfies $P^{-1} = P \circ (-X)$ if and only if $P = G + H$, where H is a nonunit and $G = \pm B_{1/2} \circ H^2$.

5. For $n, k = 1, 2, \ldots$, let the coefficients $\delta_{k,n}$ be defined by

$$B_k L = \sum_{n=1}^{\infty} (-1)^n \delta_{k,n} x^n.$$

Show that

$$\delta_{k,n} = \begin{cases} (-k)_n \left(\dfrac{1}{k} + \dfrac{1}{k-1} + \cdots + \dfrac{1}{k-n+1} \right), & n \leqslant k, \\ -\dfrac{k!}{(-n)_{k+1}}, & n > k. \end{cases}$$

6. Let $b \ne 0$, $P := (1/b)(B_b - I)$. Show that $P^{[-1]} = B_{1/b} \circ (bX) - I$. [Apply the method in Example 2.]

7. Let $P := \alpha_1 x + \alpha_2 x^2 + \cdots$ be a formal power series over the field of real numbers and let $\alpha_1 > 0$. Show that there are two different fps S such that $S \circ S = P$ (square roots under composition) and find a recurrence relation for their coefficients. Apply the result to calculate the first few coefficients of S for the series

   (a) $P := XB_{-1}$,

   (b) $P := E_1 - I$.

8. Let P be a nonunit. Show that the only solution $Q = 1 + b_1 x + b_2 x^2 + \cdots$ of the equation $Q^2 = 1 + P$ (see Problem 2, §1.3) is given by $Q := B_{1/2} \circ P$.
9. This problem illustrates the application of formal power series to combinatorial problems. Let q_n be the number of ways of associating a product of n factors $a_1 a_2 \cdots a_n$ in a nonassociative algebra; for example, $q_3 = 2$ because $a_1(a_2 a_3)$ and $(a_1 a_2) a_3$ are the only possibilities. Similarly $q_4 = 5$ because of the five cases $a_1(a_2(a_3 a_4))$, $a_1((a_2 a_3) a_4)$, $(a_1 a_2)(a_3 a_4)$, $(a_1(a_2 a_3)) a_4$, $((a_1 a_2) a_3) a_4$. We define $q_1 = 1$.
   (a) Show that for $n \geq 2$
   $$q_n = \sum_{k=1}^{n-1} q_k q_{n-k}.$$
   (b) If $P := q_1 x + q_2 x^2 + \cdots$, show that $P^2 - P = X$.
   (c) Show that $P = \frac{1}{2}[1 - B_{1/2} \circ (-4X)]$, hence that $q_n = 4^{n-1} (\frac{1}{2})_{n-1} / n!$, $n = 1, 2, \ldots$. [I. Niven]

§1.8. FORMAL LAURENT SERIES; RESIDUES

Let $\mathcal{P}$ denote the totality of formal power series over a field $\mathcal{F}$. We have seen in §1.2 that $\mathcal{P}$ is an integral domain. It is not a field, because it has elements—the nonunits—that do not possess a reciprocal in $\mathcal{P}$.

The question may be asked whether it is possible to "embed" $\mathcal{P}$ in a field; i.e., whether there is a field $\mathcal{Q}$ containing $\mathcal{P}$ whose rules of composition for elements $P \in \mathcal{P}$ agree with those already defined in $\mathcal{P}$. If such a field $\mathcal{Q}$ exists, even nonunits $P \neq 0$ in $\mathcal{P}$ would have a reciprocal P^{-1} in $\mathcal{Q}$ such that $PP^{-1} = I$, the identity element in $\mathcal{P}$.

The problem of embedding a given integral domain in a field is studied in abstract algebra, in which it is shown that every integral domain $\mathcal{P}$ can be embedded in a field $\mathcal{Q}$ in the above sense. In all fields containing $\mathcal{P}$ there is a uniquely determined (up to isomorphism) *minimal* field $\mathcal{Q}_0$. This is the **quotient field** of $\mathcal{P}$; it can be defined abstractly as the set of equivalence classes of pairs (a, b) of elements of $\mathcal{P}$, where $b \neq 0$ and two pairs (a, b) and (c, d) are considered equivalent if $ad = bc$.

EXAMPLE 1

The quotient field of the integral domain of integers is the field of rational numbers.

When the above construction is carried through for the integral domain $\mathcal{P}$ of formal power series, it yields a field that is isomorphic to the *field of formal Laurent series*. Rather than giving the details of construction, we present the set of formal Laurent series as a new entity and verify *a posteriori* that it has the required properties.

A **formal Laurent series** (fLs) over a field $\mathcal{F}$ is defined abstractly as a sequence of elements of $\mathcal{F}$ whose indices run through the set of *all* integers, subject to the condition that only a finite number of elements with negative indices are different from zero. In line with the usage employed for formal power series, however, we view the formal Laurent series $L := \{\ldots, 0, 0, a_k, a_{k+1}, \ldots\}$ (where k may be a negative integer) as a series of powers of an indeterminate x which contains at most a finite number of powers with negative exponents:

$$L = a_k x^k + a_{k+1} x^{k+1} + \cdots.$$

[This use of the term "Laurent series" differs from that in analytical theory (see §4.4), in which an infinite number of powers with negative exponents may occur.] A formal *power* series evidently may be regarded as a formal *Laurent* series in which no powers with negative exponents occur.

The *sum* of two formal Laurent series is defined in the obvious manner by adding the coefficients of like powers. The identity element with respect to addition is the fLs 0 whose coefficients are all zero. The *product* of two formal Laurent series is defined in the manner suggested by Cauchy multiplication: if

$$L := \sum a_l x^l, \qquad M := \sum b_m x^m$$

are two formal Laurent series, then

$$LM := \sum c_n x^n,$$

where

$$c_n := \sum_{l+m=n} a_l b_m. \tag{1.8-1}$$

These sums are finite for every integer n because only a finite number of nonzero a_l and b_m have negative indices. Moreover, for sufficiently small (negative) n we have $c_n = 0$ for the same reason. Thus the product of two formal Laurent series is always a formal Laurent series.

EXAMPLE 2

If k is any integer (positive, negative, or zero), then

$$\left(\sum_{n=k}^{\infty} x^n\right)^2 = \sum_{n=2k}^{\infty} (n - 2k + 1) x^n.$$

We denote the totality of a formal Laurent series by $\mathcal{L}$. It is clear that for elements in $\mathcal{L}$ that are also in $\mathcal{P}$ the operations of addition and multiplication defined above agree with the corresponding operations defined in $\mathcal{P}$.

FORMAL LAURENT SERIES RESIDUES

They are obviously associative and commutative also in $\mathcal{L}$ and the distributive law holds. The identity element with respect to multiplication in $\mathcal{P}$,

$$I := 1 + 0x + 0x^2 + \cdots,$$

remains the identity element in $\mathcal{L}$. It may be shown, as in $\mathcal{P}$, that $\mathcal{L}$ has no divisors of zero. Hence $\mathcal{L}$ is an integral domain.

THEOREM 1.8a

The formal Laurent series over a field $\mathcal{F}$ themselves form a field.

Proof. It remains to be shown that every series $L \in \mathcal{L}$ that is not the zero series possesses a reciprocal. It is clear that the fLs $X := 1 \cdot x + 0 \cdot x^2 + \cdots$ has the reciprocal $X^{-1} := 1 \cdot x^{-1} + 0 \cdot x^0 + \cdots$; more generally, X^k has the reciprocal $X^{-k} := 1 \cdot x^{-k} + 0 \cdot x^{-k+1} + \cdots$ for any integer k. Now let $L := a_k x^k + a_{k+1} x^{k+1} + \cdots$ be a nonzero element of $\mathcal{L}$, $a_k \neq 0$ (k can be any integer, positive, negative, or zero). We then have $L = X^k P$, where $P := a_k + a_{k+1} x + \cdots$ is a unit in $\mathcal{P}$ and consequently has a reciprocal P^{-1}. We assert that the reciprocal of L is $P^{-1} X^{-k}$, and indeed $P^{-1} X^{-k} L = P^{-1} X^{-k} X^k P = I$. ∎

We note that the reciprocal of a formal Laurent series with first nonvanishing power x^k is a formal Laurent series with first nonvanishing power x^{-k}.

If

$$L := \sum a_k x^k$$

is a formal Laurent series, its **derivative** L' is defined by

$$L' := \sum (k+1) a_{k+1} x^k. \tag{1.8-2}$$

This again is a formal Laurent series and for those series that are in $\mathcal{P}$ the definition agrees with the earlier one. The usual rules for the derivative of sums and products are easily seen to remain valid.

The coefficient a_{-1} of x^{-1} in $L := \sum a_k x^k$, called the **residue** of L, is especially important, and we write $a_{-1} = \operatorname{res}(L)$.

EXAMPLE 3

Let $P := x - x^2$ and find the residue of P^{-k} for any positive integer k. It is evident that $P = x(1-x)$. Hence, by the proof of Theorem 1.8a and by Theorem 1.4c,

$$P^{-k} = x^{-k}(1-x)^{-k} = x^{-k} \sum_{n=0}^{\infty} \frac{(k)_n}{n!} x^n = \sum_{n=0}^{\infty} \frac{(k)_n}{n!} x^{n-k}.$$

There follows

$$\mathrm{res}(P^{-k}) = \frac{(k)_{k-1}}{(k-1)!},$$

the $(k-1)$st coefficient in the unit $(1-x)^{-k}$.

THEOREM 1.8b

A formal Laurent series is a derivative if and only if its residue is zero.

Proof. If L is a formal Laurent series, the coefficient of x^{-1} in the series L' by definition equals $0 \cdot a_0 = 0$. Conversely, let

$$M := \sum c_k x^k$$

be a formal Laurent series such that $c_{-1} = 0$. If we define

$$a_k = \frac{1}{k} c_{k-1}, \quad k \neq 0$$

and let a_0 be arbitrary, then clearly $M = L'$, where $L := \sum a_k x^k$. ∎

PROBLEMS

1. Admitting doubly infinite matrices (the indices of the elements run from $-\infty$ to ∞), show how to extend the matrix representation of formal power series (see §1.3) to formal Laurent series and prove the corresponding theorems.
2. Show that the composition of a formal Laurent series with a nonunit in the integral domain of formal power series may be defined and establish a formula for its coefficients.
3. Extend the matrix representation of a formal power series with respect to composition (see §1.7) to the composition of a formal Laurent series with a nonunit in the integral domain of a formal power series.
4. Let L be a formal Laurent series and let P be a nonunit in $\mathcal{P}$:

$$P := a_p x^p + a_{p+1} x^{p+1} + \cdots, \quad p > 0, \quad a_p \neq 0.$$

   Show that

$$\mathrm{res}[(L \circ P) P'] = p \, \mathrm{res}(L).$$

5. Let $L := \sum a_n x^n$ be a formal Laurent series with lowest nonzero coefficient a_q and let k be any integer. If

$$L^k =: \sum b_n x^n,$$

show that $b_n = 0$, $n < kq$, $b_{kq} = a_q^k$, and

$$b_n = \frac{1}{(n-kq)a_q} \sum_{m=1}^{n-kq} [(k+1)m - (n-kq)] a_{q+m} b_{n-m}, \quad n > kq.$$

[Extension of the J. C. P. Miller formula.]

6. If M and N are arbitrary formal Laurent series, show that

$$\text{res}(M'N) = -\text{res}(MN').$$

§1.9. THE LAGRANGE-BÜRMANN THEOREM

Let $\mathcal{P}$ and $\mathcal{L}$ denote, respectively, the integral domain of formal power series and the field of formal Laurent series over a basic field $\mathcal{F}$. Here we return to the problem of constructing the *reversion* (i.e., inverse with respect to composition) of the almost unit

$$P := a_1 x + a_2 x^2 + \cdots, \quad a_1 \neq 0$$

in $\mathcal{P}$. It was shown in §1.7 that this reversion can be found by finding the inverse of the matrix

$$\mathbf{A} := \begin{bmatrix} a_1^{(1)} & a_2^{(1)} & a_3^{(1)} & \cdots \\ & a_2^{(2)} & a_3^{(2)} & \\ & & a_3^{(3)} & \\ 0 & & & \ddots \end{bmatrix}, \quad (1.9\text{-}1)$$

where $P^k =: \sum a_n^{(k)} x^n$. The coefficients of $Q := P^{[-1]}$ are just the elements in the first row of the matrix $\mathbf{B} := \mathbf{A}^{-1}$. More generally, the coefficients of Q^k are the elements in the kth row of $\mathbf{B}$.

We now show that the coefficients $b_n^{(k)}$ of Q^k can also be calculated very simply in a completely different way.

THEOREM 1.9a   (Schur-Jabotinski theorem)

Let $P := a_1 x + a_2 x^2 + \cdots$ be an almost unit in $\mathcal{P}$ ($a_1 \neq 0$) and let

$$P^k =: \sum_{n=k}^{\infty} a_n^{(k)} x^n \quad (1.9\text{-}2)$$

for $k = \pm 1, \pm 2, \ldots$. If $Q := P^{[-1]}$, then for all positive integers m

$$Q^m = \sum_{n=m}^{\infty} b_n^{(m)} x^n,$$

where

$$b_n^{(m)} := \frac{m}{n} a_{-m}^{(-n)}, \qquad m \geqslant n. \tag{I}$$

Proof. Let **A** be defined by (1.9-1) and let $\mathbf{B} = (b_{mn})$, where $b_{mn} := 0$ for $n < m$ and

$$b_{mn} := \frac{m}{n} a_{-m}^{(-n)}, \qquad n \geqslant m.$$

It is asserted that $\mathbf{B} = \mathbf{A}^{-1}$. It suffices to show that **B** is a *right* inverse of **A**.

Let $\mathbf{C} := (c_{rs}) := \mathbf{AB}$. It is clear that $c_{rs} = 0$ for $s < r$ because the product of two upper triangular matrices is upper triangular. Furthermore,

$$c_{rr} = a_{-r}^{(-r)} a_r^{(r)} = a_1^{-r} a_1^r = 1.$$

It remains to be shown that $c_{rs} = 0$ for $s > r$, in which case

$$c_{rs} = \sum_{k=r}^{s} a_{rk} b_{ks} = \frac{1}{s} \sum_{k=r}^{s} k a_k^{(r)} a_{-k}^{(-s)}.$$

The last sum equals the coefficient of x^{-1} in the Cauchy product of the derivative of

$$P^r = a_r^{(r)} x^r + a_{r+1}^{(r)} x^{r+1} + \cdots$$

and of the formal Laurent series

$$P^{-s} = a_{-s}^{(-s)} x^{-s} + a_{-s+1}^{(-s)} x^{-s+1} + \cdots.$$

Consequently

$$sc_{rs} = \operatorname{res}[(P^r)' P^{-s}],$$

but

$$(P^r)' P^{-s} = r P^{r-1} P' P^{-s} = r P^{r-s-1} P' = \frac{r}{r-s} (P^{r-s})'.$$

Thus sc_{rs} equals a constant times the residue of a derivative, hence is zero

LAGRANGE-BÜRMANN THEOREM

by Theorem 1.8b; $c_{rs} = 0$ ($s > r$) follows and

$$AB = I$$

as desired. ∎

Formula I reduces the problem of calculating the coefficients of the series $P^{[-1]}$ and of its powers to the problem of calculating the series P^k for negative integers k. This calculation may be accomplished by a modification of the J. C. P. Miller algorithm; see Problem 5, §1.8.

In some cases Formula I can even be used for the explicit determination of the coefficients of $P^{[-1]}$.

EXAMPLES

1. We return to the problem considered in example 1 of §1.7, namely that of finding the reversion $Q =: b_1 x + b_2 x^2 + \cdots$ of the series

$$P := x\,{}_1F_0(a; x) = x B_{-a}(-x).$$

Because

$$P^k = x^k\,{}_1F_0(ka; x)$$

for all integers k, we have

$$a_n^{(k)} = \frac{(ka)_{n-k}}{(n-k)!}, \qquad n \geq k;$$

hence

$$b_n^{(m)} = \frac{m}{n} a_{-m}^{(-n)} = \frac{m}{n} \frac{(-na)_{n-m}}{(n-m)!},$$

in agreement with (1.7-4). In particular,

$$b_n = b_n^{(1)} = \frac{(-na)_{n-1}}{n!}, \qquad n = 1, 2, \ldots.$$

2. To find the reversion of the series

$$P := XE_a = x + \frac{a}{1!} x^2 + \frac{a^2}{2!} x^3 + \cdots,$$

we have

$$P^k = X^k E_{ka}$$

for all integers k; hence

$$a_n^{(k)} = \frac{(ka)^{n-k}}{(n-k)!}, \qquad n \geq k. \tag{1.9-3}$$

There follows

$$P^{[-1]} = \frac{1}{1}a_1^{(-1)}x + \frac{1}{2}a_1^{(-2)}x^2 + \frac{1}{3}a_1^{(-3)}x^3 + \cdots$$

$$= x + \frac{(-2a)^1}{2!}x^2 + \frac{(-3a)^2}{3!}x^3 + \cdots.$$

Next we establish a more sophisticated version of Theorem 1.9a. Let P be an almost unit in $\mathcal{P}$, let $Q := P^{[-1]}$, and let $R := c_0 + c_1 x + c_2 x^2 + \cdots$ be an arbitrary element of $\mathcal{P}$. We wish to express the coefficients of

$$R \circ Q = : d_0 + d_1 x + d_2 x^2 + \cdots$$

in terms of the coefficients of R and P. Evidently $d_0 = c_0$ and

$$d_n = \sum_{k=1}^{n} c_k b_n^{(k)}, \qquad n = 1, 2, \ldots.$$

Using the result of Theorem 1.9a, we have

$$d_n = \frac{1}{n} \sum_{k=1}^{n} c_k k a_{-k}^{(-n)}.$$

The sum is readily recognized as the residue of $R' P^{-n}$. Hence

$$d_n = \frac{1}{n} \mathrm{res}(R' P^{-n}), \qquad n = 1, 2, \ldots.$$

The Lagrange-Bürmann expansion follows:

THEOREM 1.9b (Lagrange-Bürmann expansion)

Let P be an almost unit in $\mathcal{P}$, $Q := P^{[-1]}$ and let $R := c_0 + c_1 x + \cdots$ be an arbitrary element of $\mathcal{P}$. Then

$$R \circ Q = c_0 + \sum_{n=1}^{\infty} \frac{1}{n} \mathrm{res}(R' P^{-n}) x^n. \tag{II}$$

Another form of the expansion which is frequently useful is obtained by forming the derivative of (II). By the chain rule (Theorem 1.6b) we get on the left

$$(R \circ Q)' = (R' \circ Q) Q'.$$

LAGRANGE-BÜRMANN THEOREM

On the right differentiation yields the series

$$\sum_{n=0}^{\infty} \operatorname{res}(R'P^{-n-1})x^n.$$

Now R was an arbitrary series in $\mathcal{P}$; therefore $S := R'$ is arbitrary.

COROLLARY 1.9c

For any $S \in \mathcal{P}$ and any almost unit $P \in \mathcal{P}$, if $P^{[-1]} =: Q$, there holds

$$(S \circ Q)Q' = \sum_{n=0}^{\infty} \operatorname{res}(SP^{-n-1})x^n. \qquad (\text{III})$$

The full power of the Lagrange-Bürmann theorem and its corollary will become apparent only when we discuss *convergent* power series (see §2.4). Nevertheless, the identities (II) and (III) can also be useful for formal purposes. They can be read from left to right or from right to left and the latter method can occasionally be used to "sum" certain infinite series. By this we mean the identification of certain formal power series whose coefficients appear to be defined in a complicated way; for instance, if (III) is to be applied, the method consists in identifying the general term of the series as a residue of a formal Laurent series of special form SP^{-n-1}.

EXAMPLES

3. Let the successive powers of the trinomial $1+x+x^2$ be written in a symmetrical triangular array:

$$(1+x+x^2)^0 = \quad \mathbf{1}$$
$$(1+x+x^2)^1 = \quad 1+\mathbf{x}+x^2$$
$$(1+x+x^2)^2 = \quad 1+2x+\mathbf{3x^2}+2x^3+x^4$$
$$(1+x+x^2)^3 = 1+3x+6x^2+\mathbf{7x^3}+6x^4+3x^5+x^6$$

$\cdot \ \cdot \ \cdot \qquad\qquad \cdot \ \ \cdot \ \ \cdot$

What is the sum of the boldface terms in the array's center column?

Solution. The coefficient of the center term in the nth row evidently equals

$$\operatorname{res}[x^{-n-1}(1+x+x^2)^n] = \operatorname{res}(SP^{-n-1}),$$

where

$$P := \frac{x}{1+x+x^2}, \quad S := \frac{1}{1+x+x^2}.$$

By (III) the sum in question equals $(S \circ Q)Q'$, where $Q := P^{[-1]}$, i.e.,

$$x = \frac{Q}{1+Q+Q^2}.$$

Solving a quadratic yields

$$Q = \frac{2x}{1-x+W}, \qquad (1.9\text{-}4)$$

where

$$W := (1-2x-3x^2)^{1/2} := B_{1/2} \circ (-2x-3x^2).$$

Thus

$$S \circ Q = \frac{1}{1+Q+Q^2} = xQ^{-1} = \tfrac{1}{2}(1-x+W).$$

Differentiating (1.9-4) yields

$$Q' = \frac{2}{(1-x+W)W}.$$

There follows

$$(S \circ Q)Q' = \frac{1}{W} = (1-2x-3x^2)^{-1/2}.$$

4. Determine the generating function[1] of the **Legendre polynomials** P_n defined by

$$P_n(h) := \frac{1}{2^n n!} \frac{d^n}{dh^n} (h^2-1)^n, \quad h \in \mathbb{C},$$

the derivative being understood in the algebraic sense.

Solution. The difficulty lies in interpreting $P_n(h)$ as a residue. However, if we define a formal power series over $\mathbb{C}$ by

$$F := 2^{-n}\left[(h+x)^2 - 1\right]^n,$$

then by Taylor's theorem for polynomials $P_n(h)$ equals the nth coefficient of F. Thus it is the zeroth coefficient in the expansion of P^{-n}, where

$$P := \frac{2x}{(h+x)^2 - 1},$$

---

[1] By the **generating function** of a sequence $\{a_0, a_1, a_2, \ldots\}$ we mean the formal power series $\sum a_n x^n$.

LAGRANGE-BÜRMANN THEOREM

or the residue of SP^{-n-1}, where

$$S := \frac{2}{(h+x)^2 - 1}.$$

(In order to classify P and S as formal power series we have to assume that $h^2 \neq 1$; the cases $h = \pm 1$ are trivial). By (III) the generating function equals $(S \circ Q)Q'$, where $Q := P^{[-1]}$; Q by definition satisfies

$$x = \frac{2Q}{(h+Q)^2 - 1},$$

which by solving a quadratic yields

$$Q = \frac{(h^2 - 1)x}{1 - hx + W}, \qquad (1.9\text{-}5)$$

where

$$W := (1 - 2hx + x^2)^{1/2} := B_{1/2} \circ (-2hx + x^2).$$

There follows

$$S \circ Q = \frac{2}{(h+Q)^2 - 1} = xQ^{-1} = \frac{1 - hx + W}{h^2 - 1}.$$

From (1.9-5)

$$Q' = \frac{h^2 - 1}{W(1 - hx + W)}$$

and there follows

$$(S \circ Q)Q' = \frac{1}{W}$$

or

$$\sum_{n=0}^{\infty} P_n(h) x^2 = (1 - 2hx + x^2)^{-1/2}. \qquad (1.9\text{-}6)$$

PROBLEMS

1. Prove

$$\operatorname{res}\left[\frac{E_2 + I}{E_2 - I}\right]^n = \begin{cases} 1, & n \text{ odd}, \\ 0, & n \text{ even}. \end{cases}$$

[Compare Problem 3, §1.7.]

2. Let $F := x - \frac{1}{3}x^3 + \frac{1}{5}x^5 - \dots$,

$$P := \frac{2x}{1 - x^2}, \qquad Q := P^{[-1]}.$$

Prove the functional relation
$$F \circ Q = \tfrac{1}{2} F.$$

3. If
$$P := \frac{x + x^9 + x^{25} + x^{49} + \cdots}{1 + 2x^4 + 2x^{16} + 2x^{36} + \cdots},$$

show numerically that
$$P^{[-1]} = x + 2x^5 + 15x^9 + 150x^{13} + \cdots.$$

[Weierstrass]

4. If $a \neq -1, -2, \ldots$, there holds
$$\left(\frac{1 + \sqrt{1-x}}{2} \right)^{-a} = {}_2F_1\left(\frac{a}{2}, \frac{a}{2} + \frac{1}{2}; a+1; x \right).$$

[The expression on the left is the composition of $B_{-a}(-X) = {}_1F_0(a; x)$ with
$$Q := \frac{1 - \sqrt{1-x}}{2}.]$$

5. For $m = 1, 2, \ldots,$
$$\left(\frac{1 + \sqrt{1-x}}{2} \right)^m = {}_2F_1\left(-\frac{m}{2}, -\frac{m}{2} + \frac{1}{2}; -m+1; x \right)$$
$$- \left(\frac{x}{4} \right)^m {}_2F_1\left(\frac{m}{2}, \frac{m}{2} + \frac{1}{2}; m+1; x \right).$$

[See Problem 4.]

6. Assuming the formal identity
$$F(a, b; a+b+\tfrac{1}{2}; x) = F(2a, 2b; a+b+\tfrac{1}{2}; y),$$

where $y := \tfrac{1}{2}(1 - \sqrt{1-x})$, prove **Watson's formula**
$${}_3F_2\left[\begin{array}{c} 2a, 2b, -n; 1 \\ a+b+\tfrac{1}{2}, -2n \end{array} \right] = \frac{(a+\tfrac{1}{2})_n (b+\tfrac{1}{2})_n}{(a+b+\tfrac{1}{2})_n (\tfrac{1}{2})_n}.$$

7. Show that for $a \neq 0$
$$\left[\frac{d^{n-1}}{dt^{n-1}} \left(\frac{t-1}{t^a - 1} \right)^n \right]_{t=1} = (-1)^n \left(-\frac{1}{a} \right)_n.$$

LAGRANGE-BÜRMANN THEOREM

[By Taylor's theorem the expression on the left equals $(n-1)!$ times the coefficient of x^{n-1} in the expansion of

$$\left[\frac{x}{(1+x)^a - 1}\right]^n,$$

i.e., $(n-1)!\operatorname{res} P^{-n}$, where $P := (1+x)^a - 1$.]

8. *Jacobi's transformation.* Let $0 \leqslant \alpha \leqslant \pi/2$, $\mu := \cos\alpha$. Then

$$\frac{d^{n-1}}{d\mu^{n-1}}\left[(\sin\alpha)^{2n-1}\right] = (-1)^{n-1}\frac{1\cdot 3\cdots(2n-1)}{n}\sin n\alpha. \qquad (1.9\text{-}6)$$

[Denote the expression on the left by $c_n(\mu)$. Then

$$c_n(\mu) = \frac{d^{n-1}}{d\mu^{n-1}}\left[(1-\mu^2)^{n-1/2}\right]$$

$= (n-1)!$ times coefficient of x^{n-1} in the expansion of $\left[1-(\mu+x)^2\right]^{n-1/2}$

$$= (n-1)!\operatorname{res}\frac{\left[1-(\mu+x)^2\right]^{n-1/2}}{x^n}$$

$$= (-1)^n 2^n (n-1)!\operatorname{res}(SP^{-n}),$$

where

$$S := \left[1-(\mu+x)^2\right]^{-1/2}, \qquad P := -\frac{2x}{1-(\mu+x)^2}.$$

By (III) the residue equals the coefficient of x^{n-1} in $(S \circ Q)Q'$, where $Q := P^{[-1]}$. By letting $\omega := e^{i\alpha}$,

$$W := (1 - 2\mu x + x^2)^{1/2} = (1-\omega x)^{1/2}(1-\bar{\omega}x)^{1/2},$$

the computation yields

$$Q + \mu = \frac{1-W}{x}, \qquad Q' = \frac{1-W-\mu x}{x^2 W}$$

and

$$(S \circ Q)^{-2} = 1 - (\mu + Q)^2 = -2\frac{1-W-\mu x}{x^2}.$$

Using the identity

$$2\frac{1-W-\mu x}{x^2} = \left[\frac{(1-\bar{\omega}x)^{1/2}-(1-\omega x)^{1/2}}{x}\right]^2$$

$$= \left[\frac{\omega-\bar{\omega}}{(1-\bar{\omega}x)^{1/2}+(1-\omega x)^{1/2}}\right]^2,$$

we find

$$(S\circ Q)Q' = i\frac{\omega-\bar{\omega}}{(1-\bar{\omega}x)^{1/2}+(1-\omega x)^{1/2}} \cdot \frac{1}{2W}$$

$$= i\frac{(1-\bar{\omega}x)^{1/2}-(1-\omega x)^{1/2}}{x} \cdot \frac{1}{2W}$$

$$= i\frac{(1-\omega x)^{-1/2}-(1-\bar{\omega}x)^{-1/2}}{2x},$$

and the result follows by the binomial theorem.]

SEMINAR ASSIGNMENTS

1. Given an almost unit $P := a_1 x + a_2 x^2 + \cdots$, the series $P^{[-1]} =: b_1 x + b_2 x^2 + \cdots$ can be found by a variety of methods:

   (a) from the recurrence relation (1.7-2);
   (b) by matrix inversion (Theorem 1.7c);
   (c) by the Lagrange-Bürmann formula (Theorem 1.9b).

   Make a study of the relative efficiency of these methods. Also compare the computation of the coefficients $(1/n)a_{-1}^{(-n)}$ required in the Lagrange-Bürmann formula
   (a) individually by the J. C. P. Miller formula;
   (b) recursively by computing P^{-1} once and for all and using $P^{-n} = P^{-1}P^{-(n-1)}$.

2. Write a subroutine package for computations with formal power series and use it to solve the computational problems in Chapter 1 and the combinatorial problems in §7.3.

3. Make a study of the order of magnitude of the Bernoulli numbers by letting $b_n := B_{2n}/(2n)!$ and determine experimentally

   (a) $$\lim_{n\to\infty} (b_n)^{1/n},$$

NOTES

(b) $$\lim_{n\to\infty} \frac{b_{n+1}}{b_n},$$

using any available numerical device of convergence acceleration.

4. Show in detail that the quotient field of the integral domain of formal power series is isomorphic to the field of formal Laurent series.

5. Show that versions (II) and (III) of the Lagrange-Bürmann theorem can be extended to the case in which R and S are formal Laurent series (see Whittaker & Watson [1927], p. 131, for an analytical version of this result).

NOTES

The following contain classical expositions of analytic function theory based on power series: Whittaker & Watson [1927] (1st ed. 1902), Hurwitz & Courant [1929] (1st ed., 1922), Dienes [1931], H. Cartan [1961], and A. Dinghas [1968]. Among them only Cartan deals explicitly with *formal* power series. An authoritative treatment of the algebraic theory of formal power series is to be found in Bourbaki [1950] and Niven [1969], who also presents some interesting applications, offers a very readable introduction. Computational aspects of formal series are treated by Knuth [1969].

§1.1. For a broadminded introduction to modern algebra Birkhoff & MacLane [1941] is still an excellent reference.

§1.3. See Wronski [1811]. Interesting inequalities for the coefficients of the reciprocal series were revealed by Kaluza [1928] and generalized by Dahlquist [1959].

§1.5. For concise accounts of the generalized hypergeometric series see Bailey [1935] or Erdelyi [1953]. Many different notations for special types of hypergeometric series are in use. The Clebsch-Gordan coefficients occurring in quantum mechanics (see Edmonds [1957]) are expressible in terms of terminating hypergeometric series.

§1.6. For Theorem 1.6c I am indebted to Professor J. C. P. Miller (verbal communication). It was apparently known to Euler; see Gould [1974].

§1.7. For Theorem 1.7c see Schur [1947] or (in a nonformal context) Jabotinski [1953]. The name "almost unit" was suggested by Professor I. Kaplansky (verbal communication).

§1.9. For the formal proof of the Lagrange-Bürmann theorem given here, see Henrici [1964]. A different formal proof, for which combinatorial analysis was used, is given by Raney [1960]. Many nonformal proofs (for convergent power series) are known. Whittaker & Watson [1927] list Bürmann's theorem (p. 129) and Lagrange's theorem (p. 133) as two separate results. Except for matters of notation, however, they are identical. Hurwitz & Courant [1929, p. 138] call the series (II) the Bürmann-Lagrange series. So do Polya & Szegö [1925], p. 125, from which Examples 3 and 4 are taken. Good [1960] generalizes the Lagrange-Bürmann theorem (for convergent series) to functions of several variables, and several proofs of Jacobi's identity (Problem 8) are given by Watson [1944], pp. 27–28.

2
FUNCTIONS ANALYTIC AT A POINT

§2.1. BANACH ALGEBRAS: FUNCTIONS

Chapter 1 was wholly concerned with *formal* series $\sum a_n x^n$ with coefficients from an arbitrary field $\mathfrak{F}$. No value was assigned to the symbol x; this symbol and its powers were used merely as a notational device to suggest certain algebraic operations.

From this chapter on the coefficient field $\mathfrak{F}$ is always, even without explicit mention of this assumption, the *field of complex numbers*.

From this chapter on also we shall replace x with some mathematical object $\mathbf{Z}$ which may be substituted in the power series in place of the "indeterminate" x and which enables us to define (in favorable cases) the *value* of the series as a limit of the sequence of its partial sums. In order to make this possible the space of mathematical objects $\mathbf{Z}$ which can be substituted should have the following properties:

1. If $\mathbf{Z}$ belongs to the space, then the powers $\mathbf{Z}^n$, $n = 0, 1, 2, \ldots$ should be defined and should belong to the space,

2. Multiplication by a complex scalar should be defined as an element in the space in order to define the individual terms of the series.

3. Finite sums of elements of the space should be defined as elements of the space in order that partial sums of the series may be defined.

4. A metric should be defined in the space in order that Cauchy sequences may be defined and identified.

5. With the metric so defined the space should be complete to allow Cauchy sequences to converge to an element of the space.

It is clear that the space of complex numbers (with the distance between the numbers z_1 and z_2 defined by $|z_1 - z_2|$) has all the above properties.

BANACH ALGEBRAS: FUNCTIONS

However, this space is too special for some purposes; for instance in the theory of ordinary differential equations (see Chapter 9) it is frequently desirable to study formal power series in which the indeterminate is replaced by a matrix.

Fortunately there is a well-developed mathematical concept that is broad enough to include the space of matrices of any order, yet at the same time enjoys the properties (1) through (5) listed above. This is the concept of a *Banach algebra*. Some preliminary definitions are required to introduce it.

I. A set $\mathcal{L}$ of mathematical objects $x, y, \ldots$, is called a **linear space** if the elements of $\mathcal{L}$ form an additive group and if, moreover, multiplication by complex scalars is defined as follows: if a is a complex number and x is any element of $\mathcal{L}$, then $ax = xa$ is a well-defined element of $\mathcal{L}$. For arbitrary scalars $a, b, c, \ldots$, and arbitrary $x, y, \ldots$, in $\mathcal{L}$ the following relations hold:

$$(a+b)x = ax + bx, \tag{2.1-1a}$$

$$a(x+y) = ax + ay, \tag{2.1-1b}$$

$$a(bx) = (ab)x, \tag{2.1-1c}$$

$$1x = x. \tag{2.1-1d}$$

It can be deduced from these axioms and from the group axioms that $0x = \mathbf{0}$ for all x, where 0 denotes the complex number zero and $\mathbf{0}$ the identity element of the additive group $\mathcal{L}$.

The most familiar example of a linear space is $\mathbb{C}^n$, the complex Euclidean space of dimension n, whose elements (or "points") are n-tuples of complex numbers. The sum of two n-tuples

$$\mathbf{x} = \begin{bmatrix} x_1 \\ x_2 \\ \vdots \\ x_n \end{bmatrix}, \quad \mathbf{y} = \begin{bmatrix} y_1 \\ y_2 \\ \vdots \\ y_n \end{bmatrix}$$

is defined by

$$\mathbf{x} + \mathbf{y} := \begin{bmatrix} x_1 + y_1 \\ x_2 + y_2 \\ \vdots \\ x_n + y_n \end{bmatrix}$$

and scalar multiplication by

$$a\mathbf{x} := \begin{bmatrix} ax_1 \\ ax_2 \\ \vdots \\ ax_n \end{bmatrix}.$$

II. A **linear space** $\mathcal{L}$ is called **normed** if with every element $\mathbf{x} \in \mathcal{L}$ there is associated a real number $\|\mathbf{x}\|$, called the **norm** of $\mathbf{x}$, which satisfies the following conditions:

$$\|\mathbf{x}\| \geq 0 \quad \text{for all } \mathbf{x} \in \mathcal{L}, \tag{2.1-2a}$$

$$\|\mathbf{x}\| = 0 \quad \text{if and only if } \mathbf{x} = \mathbf{0}, \tag{2.1-2b}$$

$$\|a\mathbf{x}\| = |a|\, \|\mathbf{x}\| \quad \text{for all scalars } a \text{ and all } \mathbf{x} \in \mathcal{L}, \tag{2.1-2c}$$

$$\|\mathbf{x} + \mathbf{y}\| \leq \|\mathbf{x}\| + \|\mathbf{y}\| \quad \text{for all } \mathbf{x}, \mathbf{y} \in \mathcal{L}. \tag{2.1-2d}$$

By way of an example the space $\mathbf{C}^n$ can be endowed with a norm in several different ways; for instance, the norm of an element

$$\mathbf{x} := \begin{bmatrix} x_1 \\ x_2 \\ \vdots \\ x_n \end{bmatrix}$$

of $\mathbf{C}^n$ may be defined by

$$\|\mathbf{x}\|_1 := \sum_{k=1}^{n} |x_k|, \tag{2.1-3}$$

by

$$\|\mathbf{x}\|_2 := \left[\sum_{k=1}^{n} |x_k|^2 \right]^{1/2} \tag{2.1-4}$$

(this is called the **Euclidean norm**), or by

$$\|\mathbf{x}\|_\infty := \max_{1 \leq k \leq n} |x_k|. \tag{2.1-5}$$

BANACH ALGEBRAS: FUNCTIONS

It is easily seen that all these expressions satisfy the norm postulates (2.1-2).

Let $\{x_n\}$ be a sequence of elements of a normed linear space $\mathcal{L}$; $\{x_n\}$ is called a **Cauchy sequence** if, given any $\epsilon > 0$, there exists an integer $N = N(\epsilon)$ such that

$$\|x_n - x_m\| < \epsilon \quad \text{for all } n > N \text{ and all } m > N.$$

The sequence $\{x_n\}$ is called **convergent** if there exists an element $x \in \mathcal{L}$ such that

$$\lim_{n \to \infty} \|x_n - x\| = 0.$$

It is clear that every convergent sequence is a Cauchy sequence. The converse, however, need not hold.

III. A normed linear space $\mathcal{L}$ is called **complete** if every Cauchy sequence of elements of $\mathcal{L}$ converges. A complete normed linear space is also called a **Banach space**.

The space $\mathbb{C}^n$ is a Banach space under any of the norms defined above, for if the sequence of points

$$\mathbf{x}_k := \begin{bmatrix} x_1^{(k)} \\ x_2^{(k)} \\ \vdots \\ x_n^{(k)} \end{bmatrix}, \quad k = 0, 1, 2, \ldots,$$

is a Cauchy sequence the n sequences of complex numbers $\{x_m^{(k)}\}$, $m = 1, 2, \ldots, n$, are Cauchy sequences and therefore have limits $x_1, \ldots, x_n$ by virtue of the completeness of $\mathbb{C}$. It is then easily seen that the sequence $\{\mathbf{x}_k\}$ has the limit

$$\mathbf{x} := \begin{bmatrix} x_1 \\ x_2 \\ \vdots \\ x_n \end{bmatrix} \in \mathbb{C}^n.$$

These axioms evidently take care of the postulates (2) through (5) stated at the beginning of this section. However, we still cannot multiply. For multiplication, an analogous set of axioms is required.

IV. A set $\mathcal{A}$ of mathematical objects $\mathbf{A}, \mathbf{B}, \ldots$, is called an **algebra** if it is a linear space and if, in addition to addition and scalar multiplication, every ordered pair $(\mathbf{A}, \mathbf{B})$ of elements of $\mathcal{A}$ has a unique product $\mathbf{AB}$ such that, for arbitrary $\mathbf{A}, \mathbf{B}, \mathbf{C}, \ldots$, of $\mathcal{A}$

$$(\mathbf{AB})\mathbf{C} = \mathbf{A}(\mathbf{BC}), \qquad (2.1\text{-}6a)$$

$$\mathbf{A}(\mathbf{B}+\mathbf{C}) = \mathbf{AB} + \mathbf{AC}, \qquad (2.1\text{-}6b)$$

$$(\mathbf{A}+\mathbf{B})\mathbf{C} = \mathbf{AC} + \mathbf{BC}. \qquad (2.1\text{-}6c)$$

Multiplication and scalar multiplication commute,

$$a\mathbf{A}b\mathbf{B} = ab\mathbf{AB}, \qquad (2.1\text{-}7)$$

and we require the existence of an identity element $\mathbf{I}$ such that

$$\mathbf{IA} = \mathbf{AI} = \mathbf{A} \qquad (2.1\text{-}8)$$

for all $\mathbf{A} \in \mathcal{A}$. It is *not* required that multiplication be commutative and we permit the existence of divisors of zero.

A simple nontrivial example is the algebra $\mathbb{C}^{n,n}$ of square matrices of a fixed order n with complex elements, in which addition, multiplication, and scalar multiplication are defined by the usual rules of matrix algebra.

V. An **algebra** $\mathcal{A}$ is called **normed** if as a linear space it is normed and if the norm, in addition to (2.1-2), satisfies the relation

$$\|\mathbf{AB}\| \leq \|\mathbf{A}\| \, \|\mathbf{B}\| \qquad (2.1\text{-}9)$$

for all $\mathbf{A}, \mathbf{B} \in \mathcal{A}$. We also require that the norm of the multiplicative identity element be one:

$$\|\mathbf{I}\| = 1. \qquad (2.1\text{-}10)$$

The algebra $\mathbb{C}^{n,n}$ of square matrices of order n can be endowed with a norm in the following manner: let the space $\mathbb{C}^n$ be endowed with some norm $\|\mathbf{x}\|$; then for each $\mathbf{A} \in \mathbb{C}^{n,n}$ and each $\mathbf{x} \in \mathbb{C}^n$, if the product of a matrix and a column vector is defined in the usual manner, $\mathbf{Ax}$ is again an element of $\mathbb{C}^n$. We now define the norm of $\mathbf{A}$ by

$$\|\mathbf{A}\| := \sup_{\mathbf{x} \neq 0} \frac{\|\mathbf{Ax}\|}{\|\mathbf{x}\|}. \qquad (2.1\text{-}11)$$

We leave it to the reader to verify that this definition of a norm satisfies (2.1-9), (2.1-10), and the equivalents of (2.1-2).

BANACH ALGEBRAS: FUNCTIONS

The norm (2.1-11) is called the matrix norm **induced** by the vector norm $\|\mathbf{x}\|$. If $\mathbf{A} := (a_{ij})$, it can be shown that the special vector norms $\|\mathbf{x}\|_1$, $\|\mathbf{x}\|_2$, and $\|\mathbf{x}\|_\infty$ defined above induce the following special matrix norms:

$$\|\mathbf{A}\|_1 := \max_{1 \leqslant j \leqslant n} \sum_{i=1}^{n} |a_{ij}| \quad \text{(column sum norm)}, \tag{2.1-12}$$

$$\|\mathbf{A}\|_2 := \text{square root of greatest eigenvalue of } \mathbf{A}^*\mathbf{A} \quad \text{(spectral norm)}, \tag{2.1-13}$$

$$\|\mathbf{A}\|_\infty := \max_{1 \leqslant i \leqslant n} \sum_{j=1}^{n} |a_{ij}| \quad \text{(row sum norm)}: \tag{2.1-14}$$

Here $\mathbf{A}^*$ denotes the conjugate transpose of $\mathbf{A}$.

VI. Finally, a **normed algebra** $\mathfrak{A}$ is called **complete** if it is complete as a normed linear space. A complete normed algebra is also called a **Banach algebra**.

With any of the norms introduced above the space $\mathbb{C}^{n,n}$ is easily seen to be complete by virtue of the completeness of the complex number system. Thus with any of these norms *the square matrices of a fixed order n with complex elements form a Banach algebra*.

Some other examples of Banach algebras follow:

1. $\mathbb{C}$, the algebra of complex numbers, where $\|z\| = |z|$. (This is the special case $\mathbb{C}^{1,1}$ of the algebra of matrices.)

The next examples are of great mathematical interest. Because they are not needed in the context of this book, they are not discussed extensively.

2. The bounded linear operators in a Banach space. (This is a generalization of $\mathbb{C}^{n,n}$.)

3. The continuous complex-valued functions f defined, say, on a fixed closed bounded interval I, where addition, multiplication, and scalar multiplication are defined pointwise and where the norm is defined by

$$\|f\| := \sup_{x \in I} |f(x)|.$$

Completeness of this algebra follows from the fact that the uniform limit of a sequence of continuous functions is continuous.

4. The set of absolutely convergent Fourier series, with the norm

$$\left\| \sum_{n=-\infty}^{\infty} c_n e^{int} \right\| := \sum_{n=-\infty}^{\infty} |c_n|.$$

The Banach algebras mentioned in Examples 3 and 4 are commutative.

In every algebra $\mathcal{C}$ there are elements **A** with the property that for some $\mathbf{B} \in \mathcal{C}$, $\mathbf{AB} = \mathbf{BA} = \mathbf{I}$. Any such **B** is called an **inverse** of **A** and is denoted by $\mathbf{A}^{-1}$; for instance, the identity element **I** possesses the inverse **I**. An element of an algebra $\mathcal{C}$ which possesses an inverse is called a **regular element** of $\mathcal{C}$; in the algebra $\mathbf{C}^{n,n}$ the set of regular elements consists of *nonsingular* matrices. One of the most striking results of the advanced theory of Banach algebras, due to N. Wiener, states that in the Banach algebra of absolutely convergent Fourier series (see Example 4) every element $f(t) := \sum_{n=-\infty}^{\infty} c_n e^{int}$ such that $f(t) \neq 0$ for all t is regular.

If $P := a_0 + a_1 x + a_2 x^2 + \cdots$ is a formal power series with complex coefficients and $\mathcal{C}$ is a Banach algebra, it is possible, by virtue of (IV), to define the partial sums

$$P_n(\mathbf{Z}) := a_0 \mathbf{I} + a_1 \mathbf{Z} + a_2 \mathbf{Z}^2 + \cdots + a_n \mathbf{Z}^n, \qquad n = 0, 1, 2, \ldots, \quad (2.1\text{-}15)$$

for any element $\mathbf{Z} \in \mathcal{C}$. By virtue of (V) it makes sense to ask whether $\{P_n(\mathbf{Z})\}$ is a Cauchy sequence. If it is, then by (VI) its limit is an element of $\mathcal{C}$. This limit is called the *value of the series* P at the point $\mathbf{Z}$ and is denoted by

$$P(\mathbf{Z}) = a_0 \mathbf{I} + a_1 \mathbf{Z} + a_2 \mathbf{Z}^2 + \cdots.$$

On the set of those $\mathbf{Z} \in \mathcal{C}$ for which P converges it defines a function with values in $\mathcal{C}$. It is with these functions that we are primarily concerned in this chapter. Although we expect the reader to possess a certain familiarity with the general concept of a function, it is convenient at this point to recall some basic notions. (Fig. 2.1a).

Let A and B be two sets of mathematical objects. A **function** f from A to B is a set of ordered pairs (x, y), in which $x \in A$ and $y \in B$, subject to the condition that any $x_0 \in A$ occurs at most once as a first element of an ordered pair $(x, y) \in f$. Thus, if $(x, y) \in f$, y is uniquely determined by x; we write $y = f(x)$ and call y the **value** of f at x. The subset of all x such that $(x, y) \in f$ is called the **domain of definition**, or briefly the **domain**, of the function f and may be denoted by $D(f)$. The subset of B consisting of all $y \in B$ such that $(x, y) \in f$ is called the **domain of values**, or the **range**, of f and may be denoted by $W(f)$.

Fundamental concepts in the general theory of functions are those of a *limit* and of *continuity*. Intuitively, if f is a function from A to B and $a \in A$, f has the limit $b \in B$ at a if the values of f at points sufficiently close to a are close to b; f is said to be continuous at a if its limit at a exists and equals $f(a)$. In order to formulate these intuitive notions precisely it is

BANACH ALGEBRAS: FUNCTIONS 73

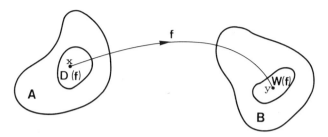

Fig. 2.1a. The function concept.

necessary to give a mathematical meaning to "closeness." This is accomplished in the widest possible context in a branch of mathematics known as the theory of topological spaces. For our purposes, however, it is sufficient to consider functions defined on subsets of normed linear spaces.

Let A be a normed linear space (such as $\mathbb{C}^n$ or $\mathbb{C}^{n,n}$) and let x_0 be a point of A. If ϵ is any positive number, the set of all $x \in A$ such that $\|x - x_0\| < \epsilon$ is called a ϵ-**neighborhood** of x_0 and is denoted by $N(x_0, \epsilon)$. We shall use the symbol $N(x_0)$ in a generic way to denote any ϵ-neighborhood of x_0.

If A is a normed linear space and $D \subset A$, a point $x_0 \in A$ is called a **point of accumulation** of D if every $N(x_0)$ contains points $x \neq x_0$ of D. The **closure** of D by definition is the set D augmented by all points of accumulation of D.

If f is a function from A to B and both A and B are normed linear spaces, the f is said to possess the **limit** y_0 at x_0 if x_0 is a point of accumulation of $D(f)$ and if for every neighborhood $N(y_0)$ there exists a neighborhood $N(x_0)$ such that for all $x \in D(f) \cap N(x_0)$ satisfying $x \neq x_0$, $f(x) \in N(y_0)$. That f has the limit y_0 at x_0 is expressed by the notation

$$\lim_{x \to x_0} f(x) = y_0.$$

The function f is said to be **continuous** at a point $x_0 \in D(f)$ if x_0 is a point of accumulation of $D(f)$ and

$$\lim_{x \to x_0} f(x) = f(x_0).$$

If f and g are functions from a set A to a linear space B such that $D(f) = D(g) =: D$, then for arbitrary $a, b \in \mathbb{C}$ we denote by $af + bg$ the function whose value at $x \in D$ is $af(x) + bg(x)$. Similarly, if B is an algebra, fg denotes the function whose value at x is $f(x)g(x)$. It is easily seen that if A and B are normed linear spaces and f and g are continuous at $x_0 \in A$ the function $af + bg$ is continuous at x_0. Similarly, if B is a normed algebra, the function fg is continuous at $\mathbf{x}_0$. It now easily follows that if $\mathcal{C}$ is a normed algebra, any function defined for all $\mathbf{Z} \in \mathcal{C}$ by

$$f(\mathbf{Z}) := a_0\mathbf{I} + a_1\mathbf{Z} + \cdots + a_n\mathbf{Z}^n$$

is continuous at all points $\mathbf{Z}_0 \in \mathcal{C}$. Thus, in particular, the partial sums of a series $P(\mathbf{Z})$ are continuous everywhere, whether or not the series converges.

We finally consider sequences $\{f_n\}$ of functions from a normed linear space A to a Banach space B. It is assumed that all f_n have the same domain of definition $D \subset A$. Such a sequence is said to **converge** on a set $D_0 \subset D$ if for every $\mathbf{x} \in D_0$

$$f(\mathbf{x}) := \lim_{n \to \infty} f_n(\mathbf{x})$$

exists. The sequence is said to **converge uniformly** on D_0 if it converges on D_0 and if the sequence of real numbers

$$\sigma_n := \sup_{\mathbf{x} \in D_0} \|f_n(\mathbf{x}) - f(\mathbf{x})\|, \quad n = 0, 1, 2, \ldots, \qquad (2.1\text{-}16)$$

tends to zero for $n \to \infty$.

THEOREM 2.1.

Let A be a normed linear space, let B be a Banach space, and let $\{f_n\}$ be a sequence of functions from a set $D \subset A$ to B. If the functions f_n are all continuous at a point $\mathbf{x}_0 \in D$ and the sequence $\{f_n\}$ converges uniformly on the intersection of a neighborhood $N(\mathbf{x}_0)$ with D, the limit function $f := \lim f_n$ is continuous at $\mathbf{x}_0$.

In brief, the limit function of a uniformly convergent sequence of continuous functions is continuous.

Proof. Let $D_0 := D \cap N(\mathbf{x}_0)$ and let $\epsilon > 0$ be given. For all n and all $\mathbf{x} \in D_0$ we have

$$f(\mathbf{x}) - f(\mathbf{x}_0) = [f(\mathbf{x}) - f_n(\mathbf{x})] + [f_n(\mathbf{x}) - f_n(\mathbf{x}_0)] + [f_n(\mathbf{x}_0) - f(\mathbf{x}_0)];$$

hence by the triangle inequality (Axiom 2.1-2d)

$$\|f(\mathbf{x}) - f(\mathbf{x}_0)\| \leq 2\sigma_n + \|f_n(\mathbf{x}) - f_n(\mathbf{x}_0)\|,$$

where σ_n is defined by (2.1-16). Fix n so that $\sigma_n < \epsilon/3$ (this is possible because $\sigma_n \to 0$) and then pick $\delta > 0$ so that

$$\|f_n(\mathbf{x}) - f_n(\mathbf{x}_0)\| < \frac{\epsilon}{3}$$

CONVERGENT POWER SERIES

for all $x \in D_0$ such that $\|x - x_0\| < \delta$ (this is possible because f_n is continuous at x_0). Then for all such x

$$\|f(x) - f(x_0)\| < \epsilon,$$

which implies that f is continuous at x_0. ∎

PROBLEMS

1. Let $\{x_n\}$ be a sequence of elements of a Banach space B. The infinite series

$$x_0 + x_1 + x_2 + \cdots = \sum_{k=0}^{\infty} x_k \qquad (2.1\text{-}17)$$

is said to be **convergent** if the sequence of its partial sums

$$s_n := x_0 + x_1 + \cdots + x_n, \quad n = 0, 1, 2, \ldots,$$

is convergent. Prove that the series (2.1-17) is convergent if and only if, given any $\epsilon > 0$, there exists $n_0 = n_0(\epsilon)$ such that for all $n > n_0$ and all $k > 0$

$$\|x_n + x_{n+1} + \cdots + x_{n+k}\| < \epsilon.$$

2. Let $\{f_n\}$ be a sequence of functions from a set D to a Banach space B. The infinite series

$$f_0 + f_1 + f_2 + \cdots = \sum_{k=0}^{\infty} f_k \qquad (2.1\text{-}18)$$

is said to **converge** on D if the series $\sum_{k=0}^{\infty} f_k(x)$ converges for every $x \in D$. The series (2.1-18) is said to **converge uniformly** on D if the sequence of partial sums

$$s_n(x) := f_0(x) + f_1(x) + \cdots + f_n(x)$$

converges uniformly on D. Prove that the series (2.1-18) converges uniformly on D if there exists a sequence of real numbers $\{\mu_n\}$ such that

(a) $\sup_{x \in D} \|f_n(x)\| \leq \mu_n, \quad n = 0, 1, 2, \ldots;$

(b) $\sum_{n=0}^{\infty} \mu_n$ is convergent.

[**Weierstrass M-test** for uniform convergence.]

§2.2. CONVERGENT POWER SERIES

The reader is presumed to be familiar with the rudiments of the theory of limits of sequences of real numbers. We recall that not every such sequence has a limit; on the other hand, every monotone bounded sequence has a finite limit. Monotone unbounded sequences have the "improper" limits $+\infty$ or $-\infty$, depending on whether they are monotone increasing or monotone decreasing.

We require the notion of the *limes superior* (upper limit) of a sequence of real numbers. Let $\{\alpha_n\}$ be a sequence of real numbers. For $n=0,1,2,\ldots$, we consider the set $\{\alpha_n, \alpha_{n+1}, \alpha_{n+2}, \ldots\}$. If this set is bounded from above, we denote by σ_n its supremum (least upper bound). If it is unbounded, we set $\sigma_n = \infty$. If $\sigma_n < \infty$ for some n, then $\sigma_n < \infty$ for all n and the sequence $\{\sigma_n\}$ is monotonically decreasing. Thus it has a finite limit or it tends to $-\infty$. In either case we write

$$\sigma := \lim_{n \to \infty} \sigma_n. \tag{2.2-1}$$

If $\sigma_n = \infty$ for some, hence for all n, we put

$$\sigma := \infty. \tag{2.2-2}$$

The (extended) real number σ defined by (2.2-1) or (2.2-2) is called the **limes superior** of the given sequence $\{\alpha_n\}$; we write

$$\sigma =: \limsup_{n \to \infty} \alpha_n.$$

The definition of the limes superior may be summarized by the formula

$$\sigma = \lim_{n \to \infty} \left(\sup_{m \geq n} \alpha_m \right).$$

If the limes superior is finite, it may also be characterized as the unique number σ such that for every $\epsilon > 0$

$\alpha_n > \sigma + \epsilon$ at most for finitely many n,

$\alpha_n > \sigma - \epsilon$ for infinitely many n.

In a completely analogous manner the **limes inferior** of a sequence $\{\alpha_n\}$ of real numbers is defined by

$$\omega := \lim_{n \to \infty} \left(\inf_{m \geq n} \alpha_m \right).$$

If $\omega \neq \pm \infty$, this is the unique number such that for every $\epsilon > 0$

$\alpha_n < \omega - \epsilon$ at most for finitely many n,

$\alpha_n < \omega + \epsilon$ for infinitely many n.

CONVERGENT POWER SERIES

If the (proper or improper) limit

$$\lambda := \lim_{n\to\infty} \alpha_n$$

exists, then

$$\limsup_{n\to\infty} \alpha_n = \liminf_{n\to\infty} \alpha_n = \lambda;$$

conversely, if the limes superior and inferior agree, the ordinary limit exists and equals their common value.

Let

$$P := a_0 + a_1 x + a_2 x^2 + \cdots$$

be a formal power series over the field of complex numbers and let $\mathfrak{B}$ be a Banach algebra. If we substitute an element $\mathbf{Z} \in \mathfrak{B}$ for the indeterminate x, we obtain an infinite series

$$P(\mathbf{Z}) := a_0 \mathbf{I} + a_1 \mathbf{Z} + a_2 \mathbf{Z}^2 + \cdots \tag{2.2-3}$$

of elements of $\mathfrak{B}$ and it is now meaningful to ask whether this series converges. We recall that by convergence of (2.2-3) we mean the convergence of the sequence of partial sums

$$P_n(\mathbf{Z}) := a_0 \mathbf{I} + a_1 \mathbf{Z} + \cdots + a_n \mathbf{Z}^n, \qquad n = 0, 1, 2, \ldots,$$

i.e., the existence of an element of $\mathfrak{B}$, which we again call $P(\mathbf{Z})$, such that

$$\lim_{n\to\infty} \|P_n(\mathbf{Z}) - P(\mathbf{Z})\| = 0.$$

For any formal power series P we call **radius of convergence** and denote by $\rho(P)$ the supremum of all numbers ρ such that the series (2.2-3) converges for all $\mathbf{Z}$ satisfying $\|\mathbf{Z}\| \leq \rho$. Thus, by definition, if $0 < \rho(P) < \infty$, $P(\mathbf{Z})$ converges for all $\mathbf{Z}$ such that $\|\mathbf{Z}\| < \rho(P)$ and diverges for some $\mathbf{Z}$ such that $\|\mathbf{Z}\| > \rho(P)$. It may or may not converge if $\|\mathbf{Z}\| = \rho(P)$.

THEOREM 2.2a (Cauchy-Hadamard formula)

If

$$P := a_0 + a_1 x + a_2 x^2 + \cdots \tag{2.2-4}$$

is any formal power series, its radius of convergence is given by the formula

$$\rho(P) = \left[\limsup_{n\to\infty} |a_n|^{1/n} \right]^{-1}. \tag{2.2-5}$$

Proof. We show that the number

$$\sigma := \left[\limsup_{n \to \infty} |a_n|^{1/n} \right]^{-1}$$

satisfies $\sigma \leqslant \rho(P)$ and $\sigma \geqslant \rho(P)$.

1. Let $\sigma > 0$. We show that $P(\mathbf{Z})$ converges for every $\mathbf{Z}$ such that $\|\mathbf{Z}\| < \sigma$. Let $\|\mathbf{Z}\| < \tau < \sigma$; then

$$\tau^{-1} > \limsup_{n \to \infty} |a_n|^{1/n},$$

hence $|a_n|^{1/n} > \tau^{-1}$ for at most a finite number of n, and $|a_n| \leqslant \tau^{-n}$ say, for $n > N$. To show that the partial sums $P_n(\mathbf{Z})$ form a Cauchy sequence let $n > m > N$. Then

$$\|P_n(\mathbf{Z}) - P_m(\mathbf{Z})\| = \|a_{m+1}\mathbf{Z}^{m+1} + \cdots + a_n\mathbf{Z}^n\|$$

$$\leqslant |a_{m+1}| \|\mathbf{Z}\|^{m+1} + \cdots + |a_n| \|\mathbf{Z}\|^n$$

$$\leqslant \tau^{-m-1} \|\mathbf{Z}\|^{m+1} + \cdots + \tau^{-n} \|\mathbf{Z}\|^n$$

$$< \left(\frac{\|\mathbf{Z}\|}{\tau} \right)^m \frac{\|\mathbf{Z}\|}{\tau - \|\mathbf{Z}\|}$$

because $\|\mathbf{Z}\| < \tau$. The last term can be made $< \epsilon$ by choosing m large enough. It follows that $\{P_n(\mathbf{Z})\}$ is a Cauchy sequence and that the series $P(\mathbf{Z})$ converges.

2. Let $\sigma < \infty$. Choose $\tau > \sigma$. Then $|a_n|^{1/n} > \tau^{-1}$ for an infinite number of n and $|a_n| > \tau^{-n}$ for an infinite number of n. Letting $\mathbf{Z} = \tau \mathbf{I}$, where $\mathbf{I}$ is the identity element of $\mathfrak{B}$, we have

$$\|P_n(\mathbf{Z}) - P_{n-1}(\mathbf{Z})\| = \|a_n \mathbf{Z}^n\| = |a_n| \|\tau^n \mathbf{I}\| > \tau^{-n} \tau^n = 1$$

for an infinite number of n, which shows that $\{P_n(\mathbf{Z})\}$ is *not* a Cauchy sequence. Hence $P(\mathbf{Z})$ diverges for $\mathbf{Z} = \tau \mathbf{I}$, which shows that $\rho(P) \leqslant \tau$ and, because $\tau > \sigma$ was arbitrary, $\rho(P) \leqslant \sigma$. ∎

Going beyond the statement of the theorem, the proof shows that if $\rho(P) < \infty$ the series actually diverges for all $\mathbf{Z} \in \mathfrak{B}$ such that $\|\mathbf{Z}\| > \rho(P)$ and

$$\|\mathbf{Z}^n\| = \|\mathbf{Z}\|^n, \quad n = 2, 3, \ldots. \tag{2.2-6}$$

If $\mathfrak{B}$ is the Banach algebra of complex numbers, (2.2-6) holds for all elements of $\mathfrak{B}$. In general, however, (2.2-6) need not hold; for instance, if,

CONVERGENT POWER SERIES

in the algebra of 2×2 matrices endowed with the spectral norm,

$$\mathbf{A} := \begin{bmatrix} 0 & a \\ 0 & 0 \end{bmatrix},$$

we have $\|\mathbf{A}\| = |a|$ but $\|\mathbf{A}^n\| = 0$ for $n > 1$.

An element $\mathbf{Z}$ of an algebra $\mathfrak{B}$ such that $\mathbf{Z}^n = \mathbf{0}$ for some positive integer n is called **nilpotent**. It is clear that any power series P (even if $\rho(P) = 0$) converges for all nilpotent elements.

In some cases the radius of convergence of a power series may be computed from a formula simpler than (2.2-5). From the definition of limes superior it is clear that

$$\rho(P) = \left[\lim_{n \to \infty} |a_n|^{1/n} \right]^{-1}$$

whenever the limit on the right exists.

THEOREM 2.2b

Let $P := a_0 + a_1 x + a_2 x^2 + \cdots$ *be a formal power series such that* $a_n \neq 0$ *for all sufficiently large n. Then*

$$\rho(P) = \lim_{n \to \infty} \left| \frac{a_n}{a_{n+1}} \right| \qquad (2.2\text{-}7)$$

whenever the limit on the right exists.

Proof. Let

$$\gamma := \lim_{n \to \infty} \left| \frac{a_n}{a_{n+1}} \right|.$$

Evidently $\gamma \geq 0$. If $\gamma < \infty$, then for any $\theta > \gamma$ there exists m such that $n \geq m$ implies

$$\left| \frac{a_n}{a_{n+1}} \right| < \theta;$$

hence $|a_n| \geq |a_m| \theta^m \theta^{-n}$ and

$$\limsup_{n \to \infty} |a_n|^{1/n} \geq \theta^{-1}.$$

Similarly, if $\gamma > 0$, we can show that for every θ such that $0 < \theta < \gamma$
$$\limsup_{n \to \infty} |a_n|^{1/n} \leq \theta^{-1}.$$
The above implies
$$\rho(P) = \left[\limsup_{n \to \infty} |a_n|^{1/n} \right]^{-1} = \gamma. \quad \blacksquare$$

We apply formula (2.2-7) to determine the radius of convergence of the generalized hypergeometric series
$$_pF_q \left[\begin{array}{c} \alpha_1, \ldots, \alpha_p; x \\ \beta_1, \ldots, \beta_q \end{array} \right],$$
in which none of the denominator parameters $\beta_1, \ldots, \beta_q$ is zero or a negative integer. We may also assume that none of the numerator parameters is zero or a negative integer, for otherwise the series terminates and the radius of convergence is clearly ∞. From
$$a_n = \frac{(\alpha_1)_n \cdots (\alpha_p)_n}{(\beta_1)_n \cdots (\beta_q)_n n!}$$
it then follows that
$$\frac{a_n}{a_{n+1}} = \frac{(\beta_1 + n) \cdots (\beta_q + n)(n+1)}{(\alpha_1 + n) \cdots (\alpha_p + n)}$$
and the limit of $|a_n/a_{n+1}|$ is clearly ∞, 1, or 0, depending on whether $p \leq q$, $p = q+1$, or $p > q+1$. Thus, if $_pF_q$ is a nonterminating series,
$$\rho(_pF_q) = \infty \quad \text{if } p \leq q,$$
$$\rho(_pF_q) = 1 \quad \text{if } p = q+1,$$
$$\rho(_pF_q) = 0 \quad \text{if } p > q+1;$$
for instance, the exponential series
$$E_t(\mathbf{Z}) := {}_0F_0(t\mathbf{Z}) = 1 + \frac{t}{1!}\mathbf{Z} + \frac{t^2}{2!}\mathbf{Z}^2 + \cdots$$
converges for all $\mathbf{Z}$, whereas the binomial series
$$B_{-a}(-\mathbf{Z}) := {}_1F_0(a; \mathbf{Z}) = 1 + \frac{a}{1!}\mathbf{Z} + \frac{a(a+1)}{2!}\mathbf{Z}^2 + \cdots$$

CONVERGENT POWER SERIES

$(a \neq 0, -1,...)$ converges for all $\mathbf{Z}$ such that $\|\mathbf{Z}\| < 1$. If neither a nor b is a negative integer, a series such as $_2F_0(a,b;\, z)$ converges for no complex number $z \neq 0$. As we shall see in Chapter 11, however, such series can still have an analytic meaning as *asymptotic expansions*.

We now return to general questions of convergence. With regard to convergence of a series $P(\mathbf{Z})$ for elements $\mathbf{Z}$ such that $\|\mathbf{Z}\| < \rho(P)$, the following strengthening of Theorem 2.2a holds.

THEOREM 2.2c

Let the series $P := a_0 + a_1 x + a_2 x^2 + \cdots$ *have radius of convergence* $\rho(P) > 0$ *and let* $0 < \tau < \rho(P)$. *If $\mathfrak{B}$ is any Banach algebra, the series $P(\mathbf{Z})$ converges* **uniformly** *for all* $\mathbf{Z} \in \mathfrak{B}$ *such that* $\|\mathbf{Z}\| \leq \tau$.

Proof. By Theorem 2.2a the series converges for $\mathbf{Z} = \gamma \mathbf{I}$, where $\gamma := \frac{1}{2}(\tau + \rho(P))$. Hence

$$\lim_{n \to \infty} \gamma^n a_n = 0,$$

which implies that

$$\gamma^n |a_n| < \mu, \qquad n = 0, 1, 2 \ldots \qquad (2.2\text{-}8)$$

for a suitable constant μ. Denoting the sum of the series (2.2-3) by $P(\mathbf{Z})$ and its nth partial sum by $P_n(\mathbf{Z})$, we let

$$\sigma_n := \sup_{\|\mathbf{Z}\| \leq \tau} \|P(\mathbf{Z}) - P_n(\mathbf{Z})\|.$$

We have, by (2.2-8),

$$\sigma_n = \sup_{\|\mathbf{Z}\| \leq \tau} \left\| \sum_{k=n+1}^{\infty} a_k \mathbf{Z}^k \right\|$$

$$\leq \sum_{k=n+1}^{\infty} |a_k| \tau^k \leq \mu \sum_{k=n+1}^{\infty} \left(\frac{\tau}{\gamma}\right)^k$$

$$= \frac{\mu}{1 - \tau/\gamma} \left(\frac{\tau}{\gamma}\right)^{n+1}$$

and because $0 < \tau < \gamma$ this clearly tends to 0 as $n \to \infty$, proving uniform convergence. ■

Because the partial sums $P_n(\mathbf{Z})$ are continuous, Theorem 2.1 immediately yields the following:

THEOREM 2.2d

If the series P has radius of convergence $\rho(P) > 0$, the function defined by its sum $P(\mathbf{Z})$ is continuous at all points $\mathbf{Z}$ such that $\|\mathbf{Z}\| < \rho(P)$.

Suppose now that $P(\mathbf{Z})$ also converges for some $\mathbf{Z}_0$ such that $\|\mathbf{Z}_0\| = \rho(P)$. Is the function $P(\mathbf{Z})$ still continuous at $\mathbf{Z}_0$? In other words, does

$$\lim_{\mathbf{Z} \to \mathbf{Z}_0} P(\mathbf{Z}) = P(\mathbf{Z}_0) \tag{2.2-9}$$

hold? Naturally, the approach of $\mathbf{Z}$ to $\mathbf{Z}_0$ has to be restricted to those $\mathbf{Z}$ for which the series converges [which could include some $\mathbf{Z} \neq \mathbf{Z}_0$ such that $\|\mathbf{Z}\| = \rho(P)$]. Disappointingly, there are examples that show that (2.2-9) does not always hold, not even if the approach of $\mathbf{Z}$ to $\mathbf{Z}_0$ is restricted to $\mathbf{Z}$ such that $\|\mathbf{Z}\| < \rho(P)$. The following weaker result can be established, however:

THEOREM 2.2e (Abel's limit theorem)

Let P be a power series with radius of convergence $\rho(P) > 0$. Then for every $\mathbf{Z}_0$ such that $P(\mathbf{Z}_0)$ converges[1]

$$\lim_{\tau \to 1-} P(\tau \mathbf{Z}_0) = P(\mathbf{Z}_0). \tag{2.2-10}$$

Proof. For $n = 0, 1, 2, \ldots$, we let

$$P_n(\mathbf{Z}) := \sum_{k=0}^{n} a_k \mathbf{Z}^k.$$

We shall show that $P_n(\tau \mathbf{Z}_0) \to P(\tau \mathbf{Z}_0)$ *uniformly* for $0 \leq \tau \leq 1$. The limit relation (2.2-10) then follows from Theorem 2.1.

Putting $\mathbf{A}_n := P(\mathbf{Z}_0) - P_n(\mathbf{Z}_0)$, the hypothesis implies $\|\mathbf{A}_n\| \to 0$. Setting

$$\sigma_n := \sup_{k \geq n} \|\mathbf{A}_k\|, \tag{2.2-11}$$

there follows

$$\lim_{n \to \infty} \sigma_n = 0. \tag{2.2-12}$$

---

[1] The notation $\tau \to 1-$ means that τ approaches 1 through real values < 1.

Because $a_n Z_0^n = A_{n-1} - A_n$, we have for $0 \leq \tau < 1$

$$P(\tau Z_0) - P_n(\tau Z_0) = \sum_{k=n+1}^{\infty} a_k \tau^k Z_0^k$$

$$= \sum_{k=n+1}^{\infty} (A_{k-1} - A_k) \tau^k$$

$$= \sum_{k=n}^{\infty} A_k \tau^{k+1} - \sum_{k=n+1}^{\infty} A_k \tau^k$$

$$= A_n \tau^{n+1} - (1-\tau) \sum_{k=n+1}^{\infty} A_k \tau^k;$$

hence, using (2.2-11),

$$\|P(\tau Z_0) - P_n(\tau Z_0)\| \leq \|A_n\| \tau^{n+1} + (1-\tau) \sigma_{n+1} \frac{\tau^{n+1}}{1-\tau} \leq 2\sigma_n.$$

This evidently also holds for $\tau = 1$, and the uniform convergence of the sequence $\{P_n(\tau Z_0)\}$ for $0 \leq \tau \leq 1$ therefore follows from (2.2-12). ∎

Aside from theoretical applications, Abel's limit theorem is frequently useful for the summation of special series. Bypassing the systematic development, we offer two examples based on special results to be established in §§2.5 and 3.3, although to most readers they will be familiar from calculus.

1 It is easily shown that for $-1 < \tau < 1$

$$\tau - \tfrac{1}{2}\tau^2 + \tfrac{1}{3}\tau^3 - \tfrac{1}{4}\tau^4 + \cdots = \log(1+\tau).$$

The power series on the left has radius of convergence 1, and converges for $\tau = 1$; hence

$$1 - \tfrac{1}{2} + \tfrac{1}{3} - \tfrac{1}{4} + \cdots = \lim_{\tau \to 1^-} \left(\tau - \tfrac{1}{2}\tau^2 + \tfrac{1}{3}\tau^3 - \cdots \right)$$

$$= \lim_{\tau \to 1^-} \log(1+\tau)$$

$$= \log 2.$$

2 From

$$\tau - \tfrac{1}{3}\tau^3 + \tfrac{1}{5}\tau^5 - \tfrac{1}{7}\tau^7 + \cdots = \arctan \tau \qquad (-1 < \tau < 1)$$

we may similarly conclude that

$$1 - \tfrac{1}{3} + \tfrac{1}{5} - \tfrac{1}{7} + \cdots = \frac{\pi}{4}.$$

Our next result produces an estimate for the size of the coefficients of a convergent power series. Suppose the formal power series

$$P := a_0 + a_1 x + a_2 x^2 + \cdots$$

has radius of convergence $\rho(P) > 0$. Then, if $\|\mathbf{Z}\| = \rho < \rho(P)$, the series $a_0 \mathbf{I} + a_1 \mathbf{Z} + a_2 \mathbf{Z}^2 + \cdots$ converges and its terms thus tend to zero. If $\mathbf{Z}$ is such that $\|\mathbf{Z}^n\| = \|\mathbf{Z}\|^n$ (e.g., if $\mathbf{Z} = z\mathbf{I}$), it follows that the sequence with the nth term $\|a_n \mathbf{Z}^n\| = |a_n| \rho^n$ tends to zero, hence is bounded. Thus for some (unspecified) constant μ,

$$|a_n| \leq \frac{\mu}{\rho^n}, \qquad n = 0, 1, 2, \ldots. \tag{2.2-13}$$

The above estimate for $|a_n|$ is not explicit enough for some purposes. It turns out, however, that the value of μ can be specified more precisely. For complex numbers z such that $|z| < \rho(P)$ consider the **scalar power series**

$$P(z) := a_0 + a_1 z + a_2 z^2 + \cdots.$$

In the Banach algebra of complex numbers this defines a function that is continuous on every disk $|z| \leq \rho$ where $\rho < \rho(P)$. Thus the quantity

$$\mu(\rho) := \sup_{|z| = \rho} |P(z)|$$

is finite for $0 \leq \rho < \rho(P)$.

THEOREM 2.2f (Cauchy's estimate)

The inequality (2.2-13) holds with $\mu = \mu(\rho)$; i.e., for $0 < \rho < \rho(P)$ and $m = 0, 1, 2, \ldots$,

$$|a_m| \leq \frac{\mu(\rho)}{\rho^m}. \tag{2.2-14}$$

Cauchy's estimate for the coefficients of a power series is usually proved by Cauchy's integral formula (see §4.7) and that proof is very short. The proof given below is longer but it does not require the concept of the integral. As a preparation we require the following:

CONVERGENT POWER SERIES

LEMMA 2.2g

Let n be a positive integer and let $w := \exp(2\pi i/n)$. Then for all integers k such that $0 < |k| < n$

$$1 + w^k + w^{2k} + \cdots + w^{(n-1)k} = 0.$$

Proof. Because $w^k \neq 1$, we have by the well-known summation formula for the truncated geometric series

$$1 + w^k + w^{2k} + \cdots + w^{(n-1)k} = \frac{w^{nk} - 1}{w^k - 1} = 0$$

since $w^{nk} = \exp(2\pi i k) = 1$. ∎

Proof of Theorem 2.2f. This proof is carried out first for the case in which P is a polynomial,

$$P(z) := a_0 + a_1 z + \cdots + a_{n-1} z^{n-1}.$$

Let m be an integer, $0 \leq m \leq n-1$, and for $|z| = \rho > 0$ let

$$g(z) := z^{-m} P(z) = a_0 z^{-m} + a_1 z^{-m+1} + \cdots + a_{n-1} z^{-m+n-1}.$$

If w is defined as in Lemma 2.2g,

$$g(w^k \rho) = a_0 \rho^{-m} w^{-mk} + a_1 \rho^{-m+1} w^{(-m+1)k} + \cdots$$

$$+ a_m + \cdots + a_{n-1} \rho^{n-1-m} w^{(n-1-m)k};$$

hence by the lemma

$$\frac{1}{n} \sum_{k=0}^{n-1} g(w^k \rho) = a_m.$$

On the other hand, because $|g(z)| = \rho^{-m} |P(z)|$,

$$|g(w^k \rho)| \leq \rho^{-m} \mu(\rho)$$

and

$$|a_m| \leq \frac{1}{n} \sum_{k=0}^{n-1} |g(w^k \rho)| \leq \frac{\mu(\rho)}{\rho^m}, \qquad m = 0, \ldots, n-1.$$

Thus the assertion of Theorem 2.2f is proved for polynomials.

To prove the theorem for a nonterminating power series P with radius of convergence $\rho(P)>0$ let $\epsilon>0$ be given. Depending on $\rho<\rho(P)$ and on the index m for which (2.2-14) is to be proved, choose $n \geqslant m$ such that

$$|P(z)-(a_0+a_1z+\cdots+a_nz^n)| \leqslant \epsilon \qquad (2.2\text{-}15)$$

for all z such that $|z| \leqslant \rho$ (this is possible because the series converges uniformly for $|z| \leqslant \rho$). It follows from the above that

$$|a_m| \leqslant \frac{\mu_n(\rho)}{\rho^m},$$

where

$$\mu_n(\rho):= \sup_{|z|=\rho} |a_0+a_1z+\cdots+a_nz^n| \leqslant \mu(\rho)+\epsilon,$$

by virtue of (2.2-15). Hence

$$|a_m| \leqslant \frac{\mu(\rho)+\epsilon}{\rho^m},$$

and the assertion follows from the fact that $\epsilon>0$ was arbitrary. ∎

As a corollary of Cauchy's estimate, we obtain a weak form of what is known as the **principle of the maximum**.

COROLLARY 2.2h

If P is a power series with positive radius of convergence, the function $\|P(\mathbf{Z})\|$ cannot have an isolated maximum at $\mathbf{Z}=\mathbf{0}$.

Proof. Assuming the contrary would mean the existence of $\rho>0$ such that

$$\|P(\mathbf{Z})\| < \|P(\mathbf{0})\| = |a_0|$$

for all $\mathbf{Z}$ such that $0<\|\mathbf{Z}\| \leqslant \rho$ and a fortiori that

$$|P(z)| = \|P(z\mathbf{I})\| < |a_0|$$

for all z such that $|z|=\rho$. Because the set of these z is bounded and closed, $\mu(\rho)<|a_0|$ would follow, thus contradicting Cauchy's estimate for $m=0$. ∎

PROBLEMS

1. If $P:=a_0+a_1x+a_2x^2+\cdots$ and $Q:=a_0+qa_1x+q^2a_2x^2+\cdots$, where $q \neq 0$, then $\rho(Q)=q^{-1}\rho(P)$.

2. If the series $F := a_0 + a_1 x + a_2 x^2 + \cdots$ has radius of convergence σ, prove that for every integer m the series $G := a_m + a_{m+1} x + a_{m+2} x^2 + \cdots$ likewise has radius of convergence σ. (If $m < 0$, the coefficients $a_{-1}, a_{-2}, \ldots, a_m$ may be defined arbitrarily.)
3. Prove that the series P and its formal derivatives $P', P'', \ldots$ all have the same radius of convergence.

§2.3. FUNCTIONS ANALYTIC AT A POINT

Let f be a function defined on a subset D of some Banach algebra $\mathcal{B}$, with values in $\mathcal{B}$.

DEFINITION

The function f is said to be **analytic** *at some point $\mathbf{Z}_0 \in D$ if there exist $\rho > 0$ and a formal power series $F = a_0 + a_1 x + a_2 x^2 + \cdots$ with radius of convergence $\geq \rho$ such that*
  (a) *the neighborhood $N(\mathbf{Z}_0, \rho)$ belongs to D;*
  (b) *for all $\mathbf{Z} \in N(\mathbf{Z}_0, \rho)$, if $\mathbf{H} := \mathbf{Z} - \mathbf{Z}_0$,*

$$f(\mathbf{Z}) = f(\mathbf{Z}_0 + \mathbf{H}) = F(\mathbf{H}). \tag{2.3-1}$$

If f is analytic at $\mathbf{Z}_0$ and if (2.3-1) holds, we shall say that f is **represented** by F in $N(\mathbf{Z}_0, \rho)$.

The simplest case of an analytic function arises if f is *defined* by the power series F; for instance, the function defined for all $\mathbf{Z}$ by the *exponential series*

$$f(\mathbf{Z}) := E_1(\mathbf{Z}) := \mathbf{I} + \frac{1}{1!} \mathbf{Z} + \frac{1}{2!} \mathbf{Z}^2 + \cdots,$$

called the **exponential function**, evidently is analytic at $\mathbf{0}$.

On the other hand, it is not required for analyticity at $\mathbf{Z}_0$ that the representation (2.3-1) hold for all $\mathbf{H}$ such that the series will converge, not even for all $\mathbf{H}$ satisfying $\|\mathbf{H}\| < \rho(F)$. All that matters is that there exists $\rho > 0$ (which naturally must satisfy $\rho \leq \rho(F)$) such that the representation holds for all $\mathbf{H}$ such that $\|\mathbf{H}\| < \rho$.

By introducing $\mathbf{Z} - \mathbf{Z}_0$ as a new variable we can frequently assume that the point at which f is analytic is the origin.

Let $f(\mathbf{Z}_0 + \mathbf{H})$ for $\|\mathbf{H}\| < \rho$ be represented by the series F, as in (2.3-1). By Theorem 2.2d the sum of the series is continuous for all such $\mathbf{H}$.

THEOREM 2.3a

If the function f is analytic at Z_0, it is continuous in a neighborhood of the point Z_0.

Can a function that is analytic at Z_0 be represented by two different power series in $H = Z - Z_0$? The answer is no. In fact, the following stronger result holds.

THEOREM 2.3b (Identity principle for analytic functions)

Let the functions f and g be analytic at Z_0 and let $f(Z) = g(Z)$ for all Z of a set whose intersection Q with the set of all points $Z_0 + hI$ (h complex) has Z_0 as a point of accumulation. Then the power series representing f and g near Z_0 are identical, and consequently $f(Z) = g(Z)$ in a full neighborhood of the point Z_0.

Proof. Without loss of generality we may assume that $Z_0 = 0$. If the assertion were untrue, then by subtracting the power series representing f and g we would find that the identity

$$0 = a_n Z^n + a_{n+1} Z^{n+1} + \cdots$$

would hold for all $Z \in Q, \|Z\| < \rho$, where $a_n \neq 0$. Because Q consists of elements of the form hI, h complex, there follows

$$0 = (a_n h^n + a_{n+1} h^{n+1} + \cdots)I = h^n(a_n + a_{n+1}h + \cdots)I.$$

Because $cI = 0$ implies $c = 0$ by (2.1-2), there follows for $0 < |h| < \rho$, $hI \in Q$ that $k(h) := a_n + a_{n+1}h + \cdots = 0$, and because Q has the point of accumulation 0 we conclude that

$$\lim_{h \to 0} k(h) = 0,$$

yet by Theorem 2.3a this limit equals $a_n \neq 0$. This contradiction establishes Theorem 2.3b. ∎

The following example shows that the conclusion of Theorem 2.3b is false if we merely assume that $f(Z) = g(Z)$ on a set with point of accumulation Z_0. Let $\mathcal{B} = \mathbb{C}^{2,2}$, let $Z_0 = 0$, and consider any power series of the form

$$f(Z) = a_2 Z^2 + a_3 Z^3 + \cdots, \qquad a_2 \neq 0.$$

Then, by Theorem 2.3b (where $g = 0$), f does not vanish identically, yet

FUNCTIONS ANALYTIC AT A POINT

$f(\mathbf{Z}) = \mathbf{0}$ on the set of all 2×2 matrices of the special form

$$\mathbf{Z} := \begin{bmatrix} 0 & z \\ 0 & 0 \end{bmatrix},$$

which has the point of accumulation $\mathbf{0}$.

If $\mathcal{B} = \mathbb{C}$, the algebra of complex numbers, then a function f analytic at some $z_0 \in \mathbb{C}$ is called a **scalar analytic function**. Because in $\mathbb{C}$ all elements are of the form $z\mathbf{I}$, it follows from Theorem 2.3b that $f(z) = g(z)$ on *any* set having point of accumulation z_0 implies that $f(z) = g(z)$ in a full neighborhood of z_0.

An important conclusion may be drawn on the *zeros* of scalar analytic functions. A **zero** of a complex-valued analytic function is any point z such that $f(z) = 0$. Applying the above to the function $g = 0$, we obtain the following:

THEOREM 2.3c

If f is scalar analytic at z_0 and z_0 is a point of accumulation of zeros of f, then f vanishes identically in a neighborhood of z_0.

In other words, if f does not vanish identically but $f(z_0) = 0$, there exists a neighborhood $N(z_0)$ such that $f(z) \neq 0$ in $N(z_0)$ except at z_0. The power series representing f near z_0 then has the form

$$f(z) = a_n h^n + a_{n+1} h^{n+1} + \cdots ;$$

$h := z - z_0$, where a_n is the first nonvanishing coefficient. Its index n is called the **order of the zero** z_0.

Returning to functions defined on an arbitrary Banach algebra $\mathcal{B}$, we now state the central result of this section.

THEOREM 2.3d

In any Banach algebra $\mathcal{B}$ the totality of functions analytic at a fixed point $\mathbf{Z}_0 \in \mathcal{B}$ forms an integral domain. The units are the functions f with $f(\mathbf{Z}_0) \neq \mathbf{0}$. The power series representing sums, products, and reciprocals of analytic functions are the formal sums, products, and reciprocals of the series representing these functions.

The proof is found by establishing quantitative versions of the several statements of the theorem. Without loss of generality we may assume that $\mathbf{Z}_0 = \mathbf{0}$.

THEOREM 2.3e

In a Banach algebra $\mathfrak{B}$ let the functions f and g be analytic at $\mathbf{0}$. Let F and G be the series representing these functions and let $\rho_0 > 0$ be such that

$$f(\mathbf{Z}) = F(\mathbf{Z}), \qquad g(\mathbf{Z}) = G(\mathbf{Z})$$

for all $\mathbf{Z} \in N(\mathbf{0}, \rho_0)$. Then for the same $\mathbf{Z}$

$$(f+g)(\mathbf{Z}) = (F+G)(\mathbf{Z}), \qquad (2.3\text{-}2)$$

$$(fg)(\mathbf{Z}) = (FG)(\mathbf{Z}). \qquad (2.3\text{-}3)$$

Proof. The relation (2.3-2) is trivial because convergent series may be added term by term. To prove (2.3-3) let $F =: a_0 + a_1 x + \cdots$, $G =: b_0 + b_1 x + \cdots$, and denote by

$$P_n := a_0 b_0 + (a_0 b_1 + a_1 b_0) x + \cdots + (a_0 b_n + \cdots + a_n b_0) x^n$$

the nth partial sum of the Cauchy product FG. We wish to show that for $\mathbf{Z} \in N(\mathbf{0}, \rho_0)$

$$\|P_n(\mathbf{Z}) - f(\mathbf{Z}) g(\mathbf{Z})\| \to 0 \qquad (2.3\text{-}4)$$

as $n \to \infty$. Let F_n and G_n denote the nth partial sums of the series representing f and g. Because $F_n(\mathbf{Z}) \to f(\mathbf{Z})$, $G_n(\mathbf{Z}) \to g(\mathbf{Z})$ for $\mathbf{Z} \in N(\mathbf{0}, \rho_0)$, (2.3-4) [hence (2.3-3)] will be established if

$$\|P_n(\mathbf{Z}) - F_n(\mathbf{Z}) G_n(\mathbf{Z})\| \to 0. \qquad (2.3\text{-}5)$$

We have

$$F_n(\mathbf{Z}) G_n(\mathbf{Z}) - P_n(\mathbf{Z}) = (a_n b_1 + a_{n-1} b_2 + \cdots + a_1 b_n) \mathbf{Z}^{n+1}$$

$$+ (a_n b_2 + a_{n-1} b_3 + \cdots + a_2 b_n) \mathbf{Z}^{n+2} + \cdots + a_n b_n \mathbf{Z}^{2n}.$$

Let $\|\mathbf{Z}\| < \rho < \rho_0$. Then by Theorem 2.2f (Cauchy's estimate) there exists μ such that

$$|a_n| \leqslant \frac{\mu}{\rho^n}, \qquad |b_n| \leqslant \frac{\mu}{\rho^n}$$

FUNCTIONS ANALYTIC AT A POINT

for $n=0,1,2,\ldots$. Letting $\gamma := \|\mathbf{Z}\|\rho^{-1}$, we have, in view of $|\gamma|<1$,

$$\|F_n(\mathbf{Z})G_n(\mathbf{Z}) - P_n(\mathbf{Z})\|$$
$$\leq (n\gamma^{n+1} + (n-1)\gamma^{n+2} + \cdots + \gamma^{2n})\mu^2 \leq \mu^2 n^2 \gamma^{n+1},$$

and (2.3-5) follows, for $n^2\gamma^n \to 0$ for $n\to\infty$. ∎

Theorem 2.3e shows that the functions analytic at **0** form an algebra. It is clear that this algebra is commutative and it is easily seen that there are no divisors of zero, for, if f and g are analytic at **0** but not identically equal to zero in any neighborhood of **0**, neither F nor G can be the zero series. Hence by Theorem 1.2a FG is not the zero series and thus by Theorem 2.3b cannot represent the function zero.

We thus have shown that the functions analytic at **0** form an integral domain. Now we shall identify its units.

LEMMA 2.3f

Let $F := a_0 + a_1 x + a_2 x^2 + \cdots$ be a formal power series with radius of convergence $\rho_0 > 0$, $a_0 \neq 0$. The reciprocal series F^{-1} then has a positive radius of convergence. More precisely, if for $0 < \rho < \rho_0$

$$\mu(\rho) := \sup_{|z|=\rho} |F(z)|$$

the radius of convergence of F^{-1} is not less than

$$\sigma := |a_0| \sup_{0 < \rho < \rho_0} \frac{\rho}{|a_0| + \mu(\rho)} > 0. \qquad (2.3\text{-}6)$$

Proof. By Cauchy's estimate, if $0 < \rho < \rho_0$,

$$|a_k| \leq \frac{\mu(\rho)}{\rho^k}, \qquad k = 0,1,2,\ldots. \qquad (2.3\text{-}7)$$

Let $F^{-1} =: c_0 + c_1 x + c_2 x^2 + \cdots$. We assert that for $0 < \rho < \rho_0$

$$|c_k| \leq \frac{\mu(\rho)}{|a_0|^2}\left[\frac{|a_0| + \mu(\rho)}{|a_0|\rho}\right]^{k-1}, \qquad k=1,2,\ldots. \qquad (2.3\text{-}8)$$

Because $c_1 = -a_1 a_0^{-2}$ and $|a_1| \leq \mu(\rho)\rho^{-1}$, this holds for $k=1$. Assuming the truth of (2.3-8) for $k=1,2,\ldots,n-1$, we get from the formula

$$c_n = -\frac{1}{a_0}(a_n c_0 + a_{n-1} c_1 + \cdots + a_1 c_{n-1}), \qquad n \geq 1,$$

using $c_0 = a_0^{-1}$ and estimating the a_k by (2.3-7),

$$|c_n| \leqslant \frac{\mu(\rho)}{\rho^n |a_0|^3} \left\{ |a_0| + \mu(\rho) \sum_{k=0}^{n-2} \left[\frac{|a_0| + \mu(\rho)}{|a_0|} \right]^k \right\}$$

$$= \frac{\mu(\rho)}{\rho^n |a_0|^2} \left\{ 1 + \left[\left(\frac{|a_0| + \mu(\rho)}{|a_0|} \right)^{n-1} - 1 \right] \right\},$$

thus establishing (2.3-8) for $k = n$. The estimate (2.3-8) now implies that

$$\limsup_{k \to \infty} |c_k|^{1/k} \leqslant \frac{|a_0| + \mu(\rho)}{|a_0| \rho}$$

for every ρ such that $0 < \rho < \rho_0$. It follows that the radius of convergence of F^{-1} is at least σ, as defined by (2.3-6). ∎

THEOREM 2.3g

In a Banach algebra $\mathcal{B}$ let f be analytic at $\mathbf{0}$, $f(\mathbf{0}) = a_0 \mathbf{I} \neq \mathbf{0}$, and let $\rho_0 > 0$ be such that $f(\mathbf{Z}) = F(\mathbf{Z})$ for $\|\mathbf{Z}\| < \rho_0$. If σ is defined by (2.3-6) and $\rho_1 := \min(\rho_0, \sigma)$, then, for $\|\mathbf{Z}\| < \rho_1$, $f(\mathbf{Z})$ is a regular element of $\mathcal{B}$. The reciprocal function f^{-1}, whose value at $\mathbf{Z}$ is $[f(\mathbf{Z})]^{-1}$, is represented by

$$f^{-1}(\mathbf{Z}) = F^{-1}(\mathbf{Z}), \qquad \|\mathbf{Z}\| < \rho_1, \tag{2.3-9}$$

hence is analytic at $\mathbf{0}$.

Proof. For $\|\mathbf{Z}\| < \rho_1$ define a function g, analytic at $\mathbf{0}$, by $g(\mathbf{Z}) := F^{-1}(\mathbf{Z})$. By Theorem 2.3e, if $\|\mathbf{Z}\| < \rho_1$,

$$f(\mathbf{Z}) g(\mathbf{Z}) = g(\mathbf{Z}) f(\mathbf{Z}) = (FF^{-1})(\mathbf{Z}) = \mathbf{I}.$$

Hence $f(\mathbf{Z})$ is regular and $f^{-1}(\mathbf{Z}) = g(\mathbf{Z})$. We have already seen that g is analytic at $\mathbf{0}$. ∎

The statements of Theorem 2.3d are now fully proved. These results give a deeper meaning to many of the formal identities proved in Chapter 1; for instance, the formal identity

$$_1F_0(a; x) \,_1F_0(b; x) = \,_1F_0(a+b; x)$$

(equivalent to Vandermonde's formula) can now be interpreted as follows: if a and b are complex numbers and $\|\mathbf{Z}\| < 1$, the three series

$$_1F_0(a; \mathbf{Z}), \,_1F_0(b; \mathbf{Z}), \,_1F_0(a+b; \mathbf{Z})$$

FUNCTIONS ANALYTIC AT A POINT

all converge and the product of the values of the first two equals the value of the third. In particular, if $b = -a$, we have

$$_1F_0(a; \mathbf{Z})\,_1F_0(-a; \mathbf{Z}) = \mathbf{I};$$

hence all elements of $\mathfrak{B}$ representable in the form $_1F_0(a; \mathbf{Z})$, where $\|\mathbf{Z}\| < 1$, have an inverse

$$[_1F_0(a; \mathbf{Z})]^{-1} = {_1F_0}(-a; \mathbf{Z}).$$

We conclude, in particular, that all elements $\mathbf{Z} \in \mathfrak{B}$ such that $\|\mathbf{Z} - \mathbf{I}\| < 1$ have an inverse, for if $\mathbf{H} := \mathbf{I} - \mathbf{Z}$ then $\mathbf{Z} = \mathbf{I} - \mathbf{H} = {_1F_0}(-1; \mathbf{H})$, where $\|\mathbf{H}\| < 1$, and

$$\mathbf{Z}^{-1} = {_1F_0}(1; \mathbf{H}) = \mathbf{I} + \mathbf{H} + \mathbf{H}^2 + \cdots.$$

THEOREM 2.3h

If $\mathfrak{B}$ is any Banach algebra, the set of those elements of $\mathfrak{B}$ that have an inverse is open.

Proof. Let $\mathbf{Z}_0 \in \mathfrak{B}$ have an inverse $\mathbf{Z}_0^{-1}$. We assert that all $\mathbf{Z} \in N(\mathbf{Z}_0, \|\mathbf{Z}_0^{-1}\|^{-1})$ have an inverse. Indeed, if $\mathbf{D} := \mathbf{Z}_0 - \mathbf{Z}$, then $\mathbf{Z} = \mathbf{Z}_0[\mathbf{I} - \mathbf{Z}_0^{-1}\mathbf{D}]$. The second factor is of the form $\mathbf{I} - \mathbf{H}$, where, by the definition of $N(\mathbf{Z}_0, \|\mathbf{Z}_0^{-1}\|^{-1})$, $\|\mathbf{H}\| < 1$, hence has the inverse $_1F_0(1; \mathbf{H})$. The first factor has an inverse by hypothesis; hence [since $(\mathbf{AB})^{-1} = \mathbf{B}^{-1}\mathbf{A}^{-1}$]

$$\mathbf{Z}^{-1} = {_1F_0}(1; \mathbf{Z}_0^{-1}(\mathbf{Z}_0 - \mathbf{Z}))\mathbf{Z}_0^{-1}. \quad \blacksquare$$

PROBLEMS

1. Give examples showing that the radius of convergence of the series $F + G$ and FG can be larger than $\min(\rho(F), \rho(G))$.

2. A function f, if $D(f)$ is symmetric about $\mathbf{0}$, is called **even** if
$$f(-\mathbf{Z}) = f(\mathbf{Z})$$
for all $\mathbf{Z} \in D(f)$ and **odd** if
$$f(-\mathbf{Z}) = -f(\mathbf{Z})$$
for all $\mathbf{Z} \in D(f)$. If an even (odd) function is analytic at $\mathbf{0}$, show that the power series representing it near $\mathbf{0}$ contains only even (odd) powers of $\mathbf{Z}$. [Consider the functions $g(\mathbf{Z}) := f(\mathbf{Z}) \mp f(-\mathbf{Z})$.]

3. From the fact that
$$(1-z)^{-1} = 1 + z + z^2 + \cdots$$

FUNCTIONS ANALYTIC AT A POINT

for $|z| < 1$ deduce by Theorem 2.3e that

$$(1-z)^{-2} = 1 + 2z + 3z^2 + \cdots, \quad |z| < 1,$$

and more generally that for $k = 1, 2, \ldots$,

$$(1-z)^{-k} = \sum_{n=0}^{\infty} \frac{(k)_n}{n!} z^n, \quad |z| < 1. \tag{2.3-10}$$

4. Deduce the result (2.3-10) from Theorem 2.3g.

§2.4. COMPOSITION AND INVERSION OF ANALYTIC FUNCTIONS

In this section we establish results that give analytic meaning to the theory of composition and inversion of formal power series developed in the latter part of Chapter 1. Only theoretical aspects are discussed; the study of special examples is deferred to §2.5. We begin by recalling some concepts from general function theory.

I. Composition and Inversion of Functions

Let A, B, C, be sets and let f be a function from A to B and g a function from B to C, subject to the condition that between the subsets $W(f)$ and $D(g)$ of B there holds the relation $W(f) \subset D(g)$. Then for every $(x, y) \in f$ there exists a uniquely determined element $(y, z) \in g$, and we may define a function from A to C as the set of all pairs (x, z), where $z := g(y)$ and $y := f(x)$. This function is called the **composition** of g with f and is denoted by $g \circ f$ (see Fig. 2.4a). (Note that $f \circ g$ need not be defined and, if defined, need not agree with $g \circ f$.)

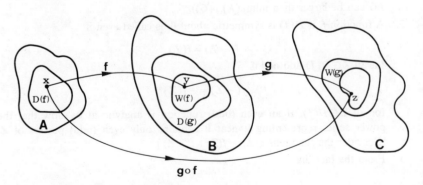

Fig. 2.4a. Composition of functions.

COMPOSITION AND INVERSION

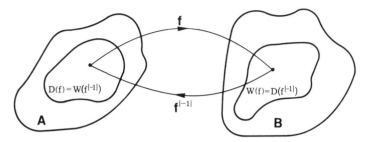

Fig. 2.4b. The inverse function.

Let f again be a function from A to B. We now assume not only that every $x_0 \in A$ occurs at most once as a first element of an ordered pair $(x,y) \in f$ but also that every $y_0 \in B$ occurs at most once as a second element. The function f is then called **one-to-one**. If f is one-to-one from A to B, it is possible to define a function $f^{[-1]}$ from B to A as the set of pairs (y,x), where $(x,y) \in f$. This is called the **inverse function** of the function f (see Fig. 2.4b). It is clear that $D(f^{[-1]}) = W(f)$, $W(f^{[-1]}) = D(f)$. The function $f^{[-1]} \circ f$ is from A to A; it is **trivial** in the sense that it consists of all pairs (x,x), where $x \in D(f)$. Similarly, $f \circ f^{[-1]}$ is trivial in $W(f)$.

II. Composition of Analytic Functions

Here the above concepts are applied to the case in which the sets $A, B, \ldots$, are all identical with the same Banach algebra $\mathfrak{B}$ and in which the functions $f, g, \ldots$, are analytic at certain points of $\mathfrak{B}$. Without loss of generality we may assume that the point of analyticity is the origin; it seems unnecessary to encumber the formulation of the theorems with the general case. We first show that the composition of two analytic functions is analytic.

THEOREM 2.4a

In a Banach algebra $\mathfrak{B}$ let the functions f and g be analytic at $\mathbf{0}$, $f(\mathbf{0}) = \mathbf{0}$. Let F and G be the series representing f and g and let $\rho_0 > 0$ and $\sigma_0 > 0$ be such that

$$g(\mathbf{W}) = G(\mathbf{W}) \quad \text{for} \quad \|\mathbf{W}\| < \sigma_0 \qquad (2.4\text{-}1)$$

and

$$f(\mathbf{Z}) = F(\mathbf{Z}), \quad \|f(\mathbf{Z})\| < \sigma_0 \quad \text{for} \quad \|\mathbf{Z}\| < \rho_0. \qquad (2.4\text{-}2)$$

FUNCTIONS ANALYTIC AT A POINT

The function $g \circ f$ is then analytic at $\mathbf{0}$, and

$$g \circ f(\mathbf{Z}) = G \circ F(\mathbf{Z}) \quad \text{for} \quad \|\mathbf{Z}\| < \rho_0. \tag{2.4-3}$$

Proof. Let $F =: a_1 x + a_2 x^2 + \cdots$, $G =: b_0 + b_1 x + \cdots$, $C := G \circ F =: c_0 + c_1 x + \cdots$. Writing

$$C_n(\mathbf{Z}) := c_0 \mathbf{I} + c_1 \mathbf{Z} + \cdots + c_n \mathbf{Z}^n,$$

we wish to show that for every $\mathbf{Z}$ such that $\|\mathbf{Z}\| < \rho_0$

$$\|C_n(\mathbf{Z}) - g(f(\mathbf{Z}))\| \to 0, \tag{2.4-4}$$

as $n \to \infty$. Let $\|\mathbf{Z}\| < \rho_0$. Then, since $\|f(\mathbf{Z})\| < \sigma_0$, if $G_n(\mathbf{W}) := b_0 \mathbf{I} + b_1 \mathbf{W} + \cdots + b_n \mathbf{W}^n$,

$$G_n(f(\mathbf{Z})) \to g(f(\mathbf{Z})),$$

and (2.4-4) will be established if it is shown that

$$\|C_n(\mathbf{Z}) - G_n(f(\mathbf{Z}))\| \to 0. \tag{2.4-5}$$

By Cauchy's estimate

$$|b_n| \leq \frac{\mu}{\sigma^n}, \quad n = 0, 1, 2, \ldots, \tag{2.4-6}$$

for every $\sigma < \sigma_0$, where

$$\mu := \mu(\sigma) := \sup_{|w| = \sigma} |G(w)|.$$

Likewise by Cauchy's estimate for every $\rho < \rho_0$

$$|a_n| \leq \frac{\sigma_0}{\rho^n}, \quad n = 1, 2, \ldots;$$

more generally, if the coefficients $a_n^{(k)}$ are (as in §1.6) defined by

$$F^k =: a_k^{(k)} x^k + a_{k+1}^{(k)} x^{k+1} + \cdots,$$

then (since, for $|z| < \rho_0$, $|F^k(z)| = |F(z)|^k < \sigma_0^k$)

$$|a_n^{(k)}| \leq \frac{\sigma_0^k}{\rho^n}, \quad n \geq k \geq 0. \tag{2.4-7}$$

We recall that $c_0 = b_0$:

$$c_m = b_1 a_m^{(1)} + b_2 a_m^{(2)} + \cdots + b_m a_m^{(m)}, \quad m = 1, 2, \ldots.$$

There follows

$$G_n(f(\mathbf{Z})) = C_n(\mathbf{Z}) + \sum_{m=n+1}^{\infty} (b_1 a_m^{(1)} + \cdots + b_n a_m^{(n)})\mathbf{Z}^m,$$

and to prove (2.4-5) we have to show that the sum tends to zero as $n \to \infty$. Its mth term by (2.4-6) and (2.4-7) is bounded by

$$\mu \rho^{-m} \left\{ \frac{\sigma_0}{\sigma} + \left(\frac{\sigma_0}{\sigma}\right)^2 + \cdots + \left(\frac{\sigma_0}{\sigma}\right)^n \right\} \|\mathbf{Z}\|^m = n\mu \left(\frac{\sigma_0}{\sigma}\right)^n \left(\frac{\|\mathbf{Z}\|}{\rho}\right)^m.$$

By selecting ρ such that $\|\mathbf{Z}\| < \rho < \rho_0$ the whole sum is bounded by

$$n\mu \left(\frac{\sigma_0 \|\mathbf{Z}\|}{\sigma \rho}\right)^n \frac{\|\mathbf{Z}\|}{\rho - \|\mathbf{Z}\|},$$

which tends to zero as $n \to \infty$ if σ is chosen to make

$$\frac{\sigma_0}{\sigma} < \frac{\rho}{\|\mathbf{Z}\|}.$$

This completes the proof of (2.4-5), hence of Theorem 2.4a. ∎

III. Inversion of Analytic Functions

The problem of determining the function that is the inverse of a given analytic function is solved in Theorem 2.4c, the proof of which is reduced to Theorem 2.4a by the following:

THEOREM 2.4b

Let the formal power series $F := a_1 x + a_2 x^2 + \cdots, a_1 \neq 0$, *have a positive radius of convergence. The reversion* $F^{[-1]}$ *of* F *then also has a positive radius of convergence.*

Proof. Let F have radius of convergence $\rho > 0$. We shall produce a lower bound for the radius of convergence of the series $F^{[-1]}$ in terms of a quantity $\nu(\sigma)$, defined for $0 \leqslant \sigma < \rho$ by

$$\nu(\sigma) := \inf_{|z|=\sigma} |F(z)|,$$

where $F(z) = a_1 z + a_2 z^2 + \cdots$. It is clear that $\nu(\sigma)$ is continuous and that for σ_0 sufficiently small $\nu(\sigma) > 0$ for $0 < \sigma < \sigma_0$. We denote by ρ_0 the supremum of all such numbers $\sigma_0 \leqslant \rho$.

By a version of the Lagrange-Bürmann formula (Theorem 1.9a) the coefficients of the reversion

$$F^{[-1]} =: b_1 x + b_2 x^2 + \cdots$$

are given by

$$b_n = \frac{1}{n} a_{-1}^{(-n)}, \qquad n = 1, 2, \ldots,$$

where for arbitrary integers m we have put

$$F^m = \sum a_k^{(m)} x^k.$$

In order to estimate the $a_k^{(m)}$ we consider the series

$$P := a_1 + a_2 x + a_3 x^2 + \cdots.$$

Because $a_1 \neq 0$, the reciprocal series

$$P^{-1} =: c_0 + c_1 x + c_2 x^2 + \cdots$$

exists, and because the radius of convergence of P equals that of F, hence is positive, P^{-1} by Lemma 2.3f likewise has a radius of convergence $\rho_1 > 0$. By Cauchy's estimate, the coefficients c_k satisfy for $0 < \sigma < \rho_1$

$$|c_k| \leq \frac{\mu(\sigma)}{\sigma^k}, \qquad k = 0, 1, 2, \ldots,$$

where

$$\mu(\sigma) := \max_{|z| = \sigma} |P^{-1}(z)|.$$

More generally, if for $n = 1, 2, \ldots,$

$$P^{-n} =: c_0^{(n)} + c_1^{(n)} x + c_2^{(n)} x^2 + \cdots,$$

then

$$|c_k^{(n)}| \leq \frac{[\mu(\sigma)]^n}{\sigma^k}, \qquad k = 0, 1, 2, \ldots. \tag{2.4-8}$$

Now, if $\sigma < \rho$ and $\sigma < \rho_1$,

$$\mu(\sigma) = \max_{|z| = \sigma} |P^{-1}(z)| = \left[\min_{|z| = \sigma} |P(z)| \right]^{-1}$$

COMPOSITION AND INVERSION

and

$$\min_{|z|=\sigma} |P(z)| = \min_{|z|=\sigma} \left| \frac{F(z)}{z} \right| = \frac{1}{\sigma} \nu(\sigma),$$

and (2.4-8) yields for $0 < \sigma < \sigma_1 := \min(\rho, \rho_1)$

$$|c_k^{(n)}| \leq \frac{\sigma^{n-k}}{[\nu(\sigma)]^n}, \quad n, k = 0, 1, 2, \ldots. \tag{2.4-9}$$

Since $F = XP$, $a_{-1}^{(-n)} = c_{k+n}^{(n)}$, and thus in particular

$$b_n = \frac{1}{n} c_{n-1}^{(n)}.$$

Thus (2.4-9) implies for $0 < \sigma < \sigma_1$

$$|b_n| \leq \frac{1}{n} \frac{\sigma}{[\nu(\sigma)]^n}, \quad n = 1, 2, \ldots,$$

and for every $\sigma \in (0, \sigma_0)$ we get

$$\rho(F^{[-1]}) = \liminf_{n \to \infty} |b_n|^{-1/n} \geq \nu(\sigma);$$

hence

$$\rho(F^{[-1]}) \geq \max_{0 < \sigma < \sigma_1} \nu(\sigma) > 0. \quad \blacksquare \tag{2.4-10}$$

It is shown in §3.3 that ρ_1, the radius of convergence of P^{-1}, satisfies $\rho_1 \geq \rho_0$. Because $\rho_0 \leq \rho$, it follows that $\sigma_1 = \rho_0$, hence that

$$\rho(F^{[-1]}) \geq \max_{0 < \sigma < \rho_0} \nu(\sigma).$$

THEOREM 2.4c

In a Banach algebra $\mathfrak{B}$ let f be analytic at $\mathbf{0}$, $f(\mathbf{0}) = \mathbf{0}$, and let the series F representing f in a neighborhood of $\mathbf{0}$ be an almost unit in the integral domain of formal power series. Then for every sufficiently small $\rho_1 > 0$ there exists $\sigma_1 > 0$ such that

(a) *f is one-to-one on the set $S := N(\mathbf{0}, \rho_1)$;*
(b) *$f(S)$ contains the set $T := N(\mathbf{0}, \sigma_1)$;*
(c) *$f^{[-1]}(\mathbf{W}) = F^{[-1]}(\mathbf{W})$ for all $\mathbf{W} \in T$.*

Proof. Let σ_0 be the radius of convergence of $F^{[-1]}$. By Theorem 2.4b $\sigma_0 > 0$. We define g by

$$g(\mathbf{W}) := F^{[-1]}(\mathbf{W}), \qquad \|\mathbf{W}\| < \sigma_0.$$

If $\rho_1 > 0$ is chosen such that

$$f(\mathbf{Z}) = F(\mathbf{Z}) \quad \text{and} \quad \|f(\mathbf{Z})\| < \sigma_0 \quad \text{for } \mathbf{Z} \in S := N(\mathbf{0}, \rho_1),$$

then by Theorem 2.4a

$$g(f(\mathbf{Z})) = F^{[-1]} \circ F(\mathbf{Z}) = \mathbf{Z}, \qquad \mathbf{Z} \in S.$$

Thus, if $\mathbf{Z}_1 \in S$ and $\mathbf{Z}_2 \in S$, $f(\mathbf{Z}_1) = f(\mathbf{Z}_2)$, there follows

$$\mathbf{Z}_1 = g(f(\mathbf{Z}_1)) = g(f(\mathbf{Z}_2)) = \mathbf{Z}_2,$$

proving (a). Now choose $\sigma_1 \leqslant \sigma_0$ such that (see Fig. 2.4c)

$$\|g(\mathbf{W})\| < \rho_1 \quad (\text{i.e., } g(\mathbf{W}) \in S) \quad \text{for } \mathbf{W} \in T.$$

Then again by Theorem 2.4a

$$f(g(\mathbf{W})) = F \circ F^{[-1]}(\mathbf{W}) = \mathbf{W}, \qquad \mathbf{W} \in T, \qquad (2.4\text{-}11)$$

showing that every $\mathbf{W} \in T$ is assumed by f at some point $\mathbf{Z} \in S$, namely at point $\mathbf{Z} := g(\mathbf{W})$. This proves (b). Because f is one-to-one on S, the inverse function $f^{[-1]}$ is defined on $f(S)$ and thus in particular on T. As shown by (2.4-11) on T the inverse function agrees with g, proving (c). ∎

IV. Lagrange-Bürmann Expansions

Let H be a formal power series, F an almost unit in the integral domain of formal power series, and $G := F^{[-1]}$. We recall the following formal identi-

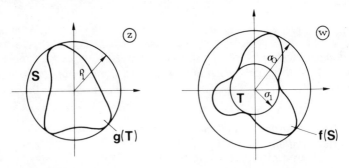

Fig. 2.4c. Notations of Theorem 2.4c.

COMPOSITION AND INVERSION

ties, stated in a slightly different notation as Theorem 1.9b (Lagrange-Bürmann expansion) and Corollary 1.9c:

$$H \circ F = H(0) + \sum_{n=1}^{\infty} \frac{1}{n} \text{res}(H'G^{-n}) x^n,$$

$$(H \circ F)F' = \sum_{n=0}^{\infty} \text{res}(HG^{-n-1}) x^n.$$

As a result of Theorems 2.4a and 2.4c we are now able to give some analytical content to these formulas.

THEOREM 2.4d

In a Banach algebra $\mathcal{B}$ let f and h be analytic at 0, $f(0) = 0$. Let F and H be the series representing f and h and let $\rho > 0$ and $\tau > 0$ be such that

$$h(\mathbf{W}) = H(\mathbf{W}) \quad \text{for} \quad \|\mathbf{W}\| < \tau$$

and

$$f(\mathbf{Z}) = F(\mathbf{Z}), \quad \|f(\mathbf{Z})\| < \tau \quad \text{for} \quad \|\mathbf{Z}\| < \rho.$$

Let f' be the analytic function defined by $f'(\mathbf{Z}) := F'(\mathbf{Z})$, $\|\mathbf{Z}\| < \rho$, where F' is the formal derivative of F. Let $f'(0) \neq 0$ and let $G := F^{[-1]}$. Then for $\|\mathbf{Z}\| < \rho$ the following expansions hold:

$$h(f(\mathbf{Z})) = h(0) + \sum_{n=1}^{\infty} \frac{1}{n} \text{res}(H'G^{-n}) \mathbf{Z}^n, \tag{2.4-12}$$

$$h(f(\mathbf{Z}))f'(\mathbf{Z}) = \sum_{n=0}^{\infty} \text{res}(HG^{-n-1}) \mathbf{Z}^n. \tag{2.4-13}$$

Either formula can be used from left to the right, i.e., to expand the functions involved, or right to the left, i.e., to sum the series indicated. In both cases the series G can be obtained by direct inversion of F or by expanding $f^{[-1]}$. Several examples are given in the problems on §2.5 after the elementary transcendental functions have been formally introduced.

V. Analytic Equations

Theorem 2.4c answers the questions of existence, uniqueness, and construction of solutions of the equation

$$f(\mathbf{Z}) = \mathbf{W}, \tag{2.4-14}$$

if $\|\mathbf{W}\|$ is sufficiently small and f is represented by a power series

$$F(\mathbf{Z}) = a_1 \mathbf{Z} + a_2 \mathbf{Z}^2 + \cdots,$$

where $a_1 \neq 0$. Indeed, the Theorem affirms the existence of numbers $\rho_1 > 0$ and $\sigma_1 > 0$ such that if $\|\mathbf{W}\| < \sigma_1$ (2.4-14) will possess exactly one solution $\mathbf{Z}$ such that $\|\mathbf{Z}\| < \rho_1$ and that this solution will be given by

$$\mathbf{Z} = F^{[-1]}(\mathbf{W}).$$

More explicitly, the series representing $\mathbf{Z}$ may be written

$$\mathbf{Z} = \sum_{n=1}^{\infty} \frac{1}{n} \operatorname{res}(F^{-n}) \mathbf{W}^n. \tag{2.4-15}$$

We now consider the problem of solving (2.4-14) in the case in which the series representing f has the form

$$F(\mathbf{Z}) = a_n \mathbf{Z}^n + a_{n+1} \mathbf{Z}^{n+1} + \cdots, \qquad n > 1, a_n \neq 0. \tag{2.4-16}$$

In a formal sense this equation can be reduced to the case considered above. Let

$$Q := \frac{a_{n+1}}{a_n} x + \frac{a_{n+2}}{a_n} x^2 + \cdots; \tag{2.4-17}$$

then

$$F = a_n X^n (I + Q) = a_n X^n (B_1 \circ Q),$$

where $I := 1 + 0x + 0x^2 + \cdots$ and B_a denotes the binomial series. Let c denote any number such that $c^n = a_n$. Then, since $B_a B_b = B_{a+b}$ (see Theorem 1.4c), the series

$$H := c X (B_{1/n} \circ Q) \tag{2.4-18}$$

satisfies $H^n = c^n X^n (B_1 \circ Q)$; hence

$$H^n = F. \tag{2.4-19}$$

H is an almost unit; therefore it has a reversion $H^{[-1]}$. If F has a positive radius of convergence, so has Q, hence H. Thus by Theorem 2.4c there exist numbers $\rho_1 > 0$ and $\sigma_1 > 0$ such that $H(\mathbf{Z})$ defines a function that is one-to-one on the set $S := N(\mathbf{0}, \rho_1)$ and assumes every value $\mathbf{V}$ in the set $T := N(\mathbf{0}, \sigma_1)$. Moreover, we may assume that ρ_1 is chosen such that $H^n = F$

represents $f(\mathbf{Z})$ for $\|\mathbf{Z}\| < \rho_1$.

LEMMA 2.4e

Equation 2.4-14 [*in which f is represented by* (2.4-16)] *can have a solution* $\mathbf{Z} \in S$ *only if there exists* $\mathbf{V} \in \mathfrak{B}$ *such that* $\mathbf{V}^n = \mathbf{W}$. *The existence of a solution* $\mathbf{Z} \in S$ *is guaranteed if* $\mathbf{V} \in T$.

Proof. (a) Let $f(\mathbf{Z}) = \mathbf{W}$, $\mathbf{Z} \in S$. Define $\mathbf{V} := H(\mathbf{Z})$; then

$$\mathbf{W} = F(\mathbf{Z}) = H^n(\mathbf{Z}) = \mathbf{V}^n,$$

thus $\mathbf{W}$ has the form $\mathbf{V}^n$. (b) Let $\mathbf{W} = \mathbf{V}^n$, where $\mathbf{V} \in T$. Define $\mathbf{Z} := H^{-1}(\mathbf{V})$. Then $H(\mathbf{Z}) = \mathbf{V}$ and $\mathbf{Z} \in S$;

$$F(\mathbf{Z}) = H^n(\mathbf{Z}) = \mathbf{V}^n = \mathbf{W}.$$

Thus (2.4-14) has a solution $\mathbf{Z}$ in S. ∎

In (b), if $\mathbf{V} \neq \mathbf{0}$, we can immediately assert the existence of n different solutions, for if $\mathbf{V}^n = \mathbf{W}$ then

$$(r^k \mathbf{V})^n = \mathbf{W}, \qquad k = 0, 1, \ldots, n-1,$$

where $r := e^{2\pi i / n}$; hence all n elements $\mathbf{Z}_k := H^{[-1]}(r^k \mathbf{V})$ are solutions. All are distinct because $H^{[-1]}$ is one-to-one on T.

In general not every element $\mathbf{W}$ of a Banach algebra $\mathfrak{B}$ is an nth power; for instance the 2×2 matrix over the field of complex numbers

$$\mathbf{A} := \begin{bmatrix} 0 & a \\ 0 & 0 \end{bmatrix}$$

is for $a \neq 0$ not representable as the square of any matrix. Thus, if $n > 1$, we cannot always assert the existence of solutions of (2.4-14) for $\mathbf{W} \neq \mathbf{0}$, no matter how small the $\|\mathbf{W}\|$. The situation is different, however, if $\mathfrak{B}$ is the algebra of complex numbers, for every complex number w can be represented in the form $w = v^n$ and $|w| = |v|^n$. Thus Lemma 2.4e yields the following:

THEOREM 2.4f

Let f be a function in the complex plane analytic at 0, *and for* $|z| < \rho_0$ *let*

$$f(z) = F(z) := a_n z^n + a_{n+1} z^{n+1} + \cdots,$$

where $a_n \neq 0$ and $n \geq 1$. Then for every sufficiently small $\rho_1 > 0$ there exists $\sigma_1 > 0$ such that for every $w \in N(0, \sigma_1)$, $w \neq 0$, the equation

$$f(z) = w \qquad (2.4\text{-}20)$$

has precisely n solutions $z \in N(0, \rho_1)$. The n solutions are

$$z_k := H^{[-1]}(r^k v), \qquad k = 1, 2, \ldots, n,$$

where H is given by (2.4-18), $r := e^{2\pi i/n}$, and v is any solution of $v^n = w$.

The series $H^{[-1]}$ can be expressed directly in terms of F. By the Lagrange-Bürmann formula $H^{[-1]} = b_1 x + b_2 x^2 + \cdots$, where

$$b_m = \frac{1}{m} \operatorname{res} H^{-m};$$

but $H = cX(B_{1/n} \circ Q)$, where c is an nth root of a_n and Q is defined by (2.4-17). Thus

$$b_m = \frac{1}{mc^m} \operatorname{res}[X^{-m}(B_{-m/n} \circ Q)].$$

COROLLARY 2.4g

The numbers z_k are given by

$$z_k = \sum_{m=1}^{\infty} \frac{1}{m} \operatorname{res}[X^{-m}(B_{-m/n} \circ Q)] \left(\frac{r^k v}{c} \right)^m, \qquad k = 1, 2, \ldots, n, \quad (2.4\text{-}21)$$

where c denotes any number satisfying $c^n = a_n$.

The number $\operatorname{res}[X^{-m}(B_{-m/n} \circ Q)]$ equals the $(m-1)$st coefficient in $B_{-m/n} \circ Q$, and thus is readily computed by the J. C. P. Miller formula (Theorem 1.6c).

VI. The Maximum Principle for Scalar Analytic Functions

Theorem 2.4f enables us to prove for scalar analytic functions a strengthened form of the **maximum principle** given earlier as a corollary of Cauchy's estimate (Corollary 2.2h).

THEOREM 2.4h.

Let the complex-valued function f be represented in a neighborhood of the point z_0 by a nonconstant power series in $z-z_0$. Then in every neighborhood of z_0 there exist points z such that $|f(z)|>|f(z_0)|$; i.e., the function $|f|$ cannot have a maximum (isolated or nonisolated) at z_0.

Proof. Without loss of generality we may assume that $z_0=0$. Since there exists $\rho>0$ such that, for $|z|<\rho$, $h(z):=f(z)-f(0)=a_n z^n+a_{n+1}z^{n+1}+\cdots$, where $a_n\neq 0$ and $n\geqslant 1$, there exists, by Theorem 2.4f, for every sufficiently small $\rho_1<\rho$ a number σ_1 such that the equation $h(z)=w$ has solutions z satisfying $|z|<\rho_1$ for every w satisfying $|w|<\sigma_1$. If w is chosen such that $|f(0)+w|>|f(0)|$ and z is a corresponding solution, then

$$|f(z)|=|h(z)+f(0)|=|w+f(0)|>|f(0)|. \quad \blacksquare$$

VII. Polynomials

Let p be a polynomial of degree $n>0$ with complex coefficients

$$p(z)=a_n z^n+a_{n-1}z^{n-1}+\cdots+a_0, \quad a_n\neq 0.$$

By letting $z=z_0+h$, expanding $z^k=(z_0+h)^k$ by the binomial theorem, and rearranging, we see immediately that p is representable by a (terminating) power series in h, hence analytic at the point z_0. Because z_0 was arbitrary, it follows that p is *analytic at every point of the complex plane*. The celebrated theorem due to Gauss on the existence of zeros of complex polynomials can be proved as an easy corollary of Theorem 2.4h.

THEOREM 2.4i (Fundamental theorem of algebra)

Let p be a nonconstant polynomial with complex coefficients. Then p assumes the value zero.

Proof. The proof is by contradiction and therefore not constructive. Assume that $p(z)\neq 0$ for all complex numbers z. Then, since p is analytic everywhere, the function f defined by $f(z):=[p(z)]^{-1}$ by Theorem 2.3d is analytic everywhere. Hence $|f(z)|$ is a continuous real function everywhere of $z=x+iy$. Simple estimates prove the existence of a number $\rho>0$ such that $|f(z)|<|f(0)|$ for $|z|>\rho$. Let S denote the compact set $|z|\leqslant\rho$ and let μ be the supremum of the values taken by $|f(z)|$ on S. Then $\mu<\infty$, and by a theorem of Weierstrass there exists z_0 such that $|f(z_0)|=\mu$. The inequality $|f(z)|\leqslant|f(z_0)|$ now holds for $z\in S$ and, since $z\notin S$ implies $|f(z)|<|f(0)|$ $\leqslant|f(z_0)|$, for $z\notin S$. Because f is not constant, this contradicts the principle of the maximum. $\blacksquare$

PROBLEMS

1. Show that in the algebra of 2×2 matrices over the field of complex numbers the matrix

$$\mathbf{W} := \begin{bmatrix} 0 & w \\ 0 & 0 \end{bmatrix}$$

has no square root.

2. For $n = 1, 2, \ldots$, let

$$s_n(z) := z + z^2 + \cdots + z^n.$$

Show that for $|z|$ sufficiently small

$$s_n(z) = \frac{z}{1-z} + \sum_{m=n+1}^{\infty} (-1)^{m-n} \binom{m-2}{n-1} \left(\frac{z}{1-z}\right)^m.$$

3. Let the series F and G have radius of convergence infinity, let F be a nonunit, and let

$$f(\mathbf{Z}) := F(\mathbf{Z}), \qquad g(\mathbf{Z}) := G(\mathbf{Z})$$

for all $\mathbf{Z}$. Show that

$$g \circ f(\mathbf{Z}) = G \circ F(\mathbf{Z})$$

for all $\mathbf{Z}$.

§2.5. ELEMENTARY TRANSCENDENTAL FUNCTIONS

Here we apply the results of §§2.3 and 2.4 to the study of certain special functions well known in real analysis: the *exponential function*, the *logarithm*, and the *general power function*. Throughout this section the algebra ⓑ is the algebra of complex numbers. Thus the functions to be considered are defined on certain subsets of the complex plane, and are complex valued.

I. The Exponential Function

We saw in §2.2 that the exponential series has radius of convergence infinity. Thus we may define for all complex z a function "exp" by

$$\exp(z) := E_1(z) = 1 + \frac{z}{1!} + \frac{z^2}{2!} + \cdots . \qquad (2.5\text{-}1)$$

ELEMENTARY TRANSCENDENTAL FUNCTIONS

This is called the **exponential function** and is also denoted by e^z, for reasons to become apparent presently. By definition, the function exp is analytic at $z=0$; moreover, by Theorem 2.2d it is continuous at all points z.

Evidently $E_1(z) = E_z(1)$. Thus, since $E_a E_b = E_{a+b}$, Theorem 2.3d yields $E_1(a) E_1(b) = E_1(a+b)$, hence

$$\exp(a)\exp(b) = \exp(a+b) \qquad (2.5\text{-}2)$$

for arbitrary complex a and b. This is known as the **addition theorem** of the exponential function.

Letting $a = z$, $b = -z$, we find

$$\exp(z)\exp(-z) = \exp(0) = 1;$$

hence $\exp(z)$ *never vanishes* and

$$\exp(-z) = [\exp(z)]^{-1} \qquad (2.5\text{-}3)$$

for all complex z.

An induction on (2.5-2) yields

$$\exp(z_1)\exp(z_2)\cdots\exp(z_n) = \exp(z_1 + z_2 + \cdots + z_n)$$

for arbitrary $z_1, z_2, \ldots, z_n$. Setting $z_1 = z_2 = \cdots = z_n = a$, we get

$$[\exp(a)]^n = \exp(na), \qquad n = 1, 2, \ldots, \qquad (2.5\text{-}4)$$

for arbitrary complex a. We choose $a = 1/n$, and there follows

$$\left[\exp\left(\frac{1}{n}\right)\right]^n = e, \qquad (2.5\text{-}5)$$

where

$$e := \exp(1) = 1 + \frac{1}{1!} + \frac{1}{2!} + \cdots = 2.71828\ldots$$

is the transcendental number known as the **basis of natural logarithms**. Since $\exp(1/n) > 0$, (2.5-5) means the same as

$$\exp\left(\frac{1}{n}\right) = e^{1/n}, \qquad n = 1, 2, \ldots,$$

where, in the established parlance of elementary algebra, $y^{1/n}$ for $y > 0$

denotes the unique solution $x>0$ of $x^n=y$. Now using (2.5-4), where $a=1/m$ (m is a positive integer), we have

$$\exp\left(\frac{n}{m}\right) = \left[\exp\left(\frac{1}{m}\right)\right]^n = (e^{1/m})^n = e^{n/m}.$$

Thus the relation $\exp(r) = e^r$ holds for all positive rational numbers r. In view of (2.5-3), it actually holds for rational numbers without regard to sign. In real analysis this fact serves to justify the notation $\exp(z) = e^z$ for arbitrary real values of z. Because powers of real numbers with complex exponents have not yet been defined, nothing may keep us from writing e^z in place of $\exp(z)$ also for complex z.

THEOREM 2.5a

The (scalar) *exponential function* e^z *is analytic at every complex point* z_0.

Proof. It is to be shown that for every z_0 there exists $\rho > 0$ such that e^z is represented by a power series in $h := z - z_0$ whenever $|h| < \rho$. By the addition theorem

$$e^z = e^{z_0 + h} = e^{z_0} e^h;$$

thus the desired expansion is

$$e^z = e^{z_0} + \frac{1}{1!} e^{z_0} h + \frac{1}{2!} e^{z_0} h^2 + \cdots,$$

and this actually holds for *all* complex h. ∎

In the following we take for granted the elementary properties of the real sine and cosine functions as well as the representations

$$\cos y = 1 - \frac{y^2}{2!} + \frac{y^4}{4!} - \frac{y^6}{6!} + \cdots,$$

$$\sin y = y - \frac{y^3}{3!} + \frac{y^5}{5!} - \frac{y^7}{7!} + \cdots,$$

which by simple techniques of calculus can be shown to hold for all real values of y. (For a purely arithmetical account of these functions see the appendix of Whittaker and Watson [1927] or pp. 69–73 of Hurwitz and Courant [1929].) Setting $z = iy$ in (2.5-1), where y is real, we get

$$e^{iy} = 1 + i\frac{y}{1!} - \frac{y^2}{2!} - i\frac{y^3}{3!} + \frac{y^4}{4!} + \cdots;$$

ELEMENTARY TRANSCENDENTAL FUNCTIONS

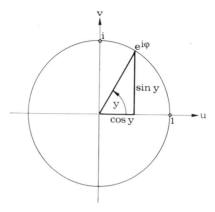

Fig. 2.5a. The number $e^{iy} = u + iv$.

hence

$$e^{iy} = \cos y + i \sin y. \qquad (2.5\text{-}6)$$

By the geometric definition of the sine and cosine functions (see Fig. 2.5a) e^{iy} is a complex number with absolute value 1 and argument y—we have already referred to this fact in §1.1.

We note the special values $e^{\pm i\pi} = -1$, $e^{\pm i\pi/2} = \pm i$, and

$$e^{2\pi i} = 1. \qquad (2.5\text{-}7)$$

If $z = x + iy$, the addition theorem yields

$$e^z = e^{x+iy} = e^x e^{iy} = e^x(\cos y + i \sin y),$$

which shows that $|e^z| = e^{\operatorname{Re} z}$, $\arg e^z = \operatorname{Im} z$. If z is any complex number, then by (2.5-7)

$$e^{z+2\pi i} = e^z e^{2\pi i} = e^z. \qquad (2.5\text{-}8)$$

A function f with the property that there exists a number $p \neq 0$ such that whenever $z \in D(f)$ then also $z + p \in D(f)$ and

$$f(z+p) = f(z)$$

is called **periodic** and p is called a **period** of f. Relation (2.5-8) proves the following theorem:

THEOREM 2.5b

The function e^z is periodic with period $2\pi i$.

The Trigonometric Functions. By adding and subtracting (2.5-6) with the same relation in which y is replaced by $-y$ we get the **formulas of Euler,**

$$\cos y = \frac{1}{2}(e^{iy} + e^{-iy}), \qquad \sin y = \frac{1}{2i}(e^{iy} - e^{-iy}), \qquad (2.5\text{-}9)$$

valid for all real y. We now define the functions cos and sin for all complex z by

$$\cos z := \frac{1}{2}(e^{iz} + e^{-iz}),$$

$$\sin z := \frac{1}{2i}(e^{iz} - e^{-iz}). \qquad (2.5\text{-}10)$$

This definition is in agreement with the conventional definition for z real. By the two theorems just proved the functions cos and sin defined by (2.5-10) are analytic at every point in the complex plane and periodic with the real period 2π. The power series representations at $z=0$ are

$$\cos z = 1 - \frac{z^2}{2!} + \frac{z^4}{4!} - \frac{z^6}{6!} + \cdots,$$

$$\sin z = z - \frac{z^3}{3!} + \frac{z^5}{5!} - \frac{z^7}{7!} + \cdots.$$

Using the addition theorem of the exponential function, we show easily by the definitions (2.5-10) that the usual functional relations for the functions cos and sin also hold for complex values of the arguments. The functions cos and sin have no zeros in the complex plane except the real zeros $z = (n+\frac{1}{2})\pi$ and $z = n\pi$, respectively, where $n = 0, \pm 1, \pm 2,\ldots$. To prove this for the sine function note that $\sin z = 0$ if and only if $e^{2iz} = 1$, i.e., if Re $2iz = 0$ and Im $2iz$ is an integral multiple of 2π. It follows that $z = n\pi$. The proof for the cosine function is similar.

The tangent and cotangent functions are defined for complex z by

$$\tan z := \frac{\sin z}{\cos z}, \qquad \cot z := \frac{\cos z}{\sin z}. \qquad (2.5\text{-}11)$$

Again the usual functional relations hold. Both functions are periodic with the real period π. It follow from Theorem 2.3d that $\tan z$ is analytic

ELEMENTARY TRANSCENDENTAL FUNCTIONS

whenever $\cos z \neq 0$; i.e., for $z \neq (n+\frac{1}{2})\pi$, $n = 0, \pm 1, b+2, \ldots$. Similarly, $\cot z$ is analytic whenever $\sin z \neq 0$; i.e., for $z \neq 0, \pm \pi, \pm 2\pi, \ldots$.

The coefficients of the power series representing $\tan z$ and $\cot z$ near $z = 0$ can be expressed in terms of the **Bernoulli numbers** B_n, formally defined by

$$X(E_1 - I)^{-1} =: \sum_{n=0}^{\infty} \frac{B_n}{n!} x^n \qquad (2.5\text{-}12)$$

(see Problem 4, §1.2). By Theorem 2.3g we have for $|z|$ sufficiently small, $z \neq 0$,

$$\frac{z}{e^z - 1} = \sum_{n=0}^{\infty} \frac{B_n}{n!} z^n. \qquad (2.5\text{-}13)$$

The series $\sum_{n=0}^{\infty} (B_n/n!) z^n + z/2$ represents the function

$$f(z) := \frac{z}{2}\left(\frac{2}{e^z - 1} + 1\right) = \frac{z}{2}\frac{e^z + 1}{e^z - 1},$$

which is easily seen to be even [i.e., to satisfy $f(-z) = f(z)$]. Hence the series contains only even powers of z. There follows

$$B_1 = -\tfrac{1}{2}, \qquad B_{2n+1} = 0, \qquad n = 1, 2, \ldots,$$

and we have

$$\frac{z}{2}\frac{e^z + 1}{e^z - 1} = \sum_{n=0}^{\infty} \frac{B_{2n}}{(2n)!} z^{2n}.$$

In view of

$$\cot z = i \frac{e^{2iz} + 1}{e^{2iz} - 1}$$

we now obtain

$$z \cot z = \sum_{n=0}^{\infty} (-1)^n \frac{B_{2n}}{(2n)!} (2z)^n. \qquad (2.5\text{-}14)$$

The identity

$$\tan z = \cot z - 2i\left(1 + \frac{2}{e^{4iz} - 1}\right)$$

finally yields

$$\tan z = \sum_{n=1}^{\infty} (-1)^{n-1} \frac{B_{2n}}{(2n)!} (2^{2n}-1) 2^{2n} z^{2n-1}$$

$$= z + \tfrac{1}{3} z^3 + \tfrac{2}{15} z^5 + \cdots. \qquad (2.5\text{-}15)$$

The expansions (2.5-13), (2.5-14) and (2.5-15) have been established here only for $|z|$ sufficiently small. We shall see in §3.3 that as a consequence of Theorem 3.3a they hold for $|z| < 2\pi$, $|z| < \pi$ and $|z| < \pi/2$, respectively.

II. The Logarithm

The range of the *real* exponential function, defined for all real x by

$$e^x := 1 + \frac{x}{1!} + \frac{x^2}{2!} + \frac{x^3}{3!} + \cdots,$$

is the set of all positive numbers. The real exponential function is one-to-one; i.e., for every $u > 0$ the equation $e^x = u$ has exactly one solution, called the (real) logarithm of u. We shall henceforth denote this solution as Log u.

The complex exponential function does not assume the value 0, but it does assume every complex value $w \neq 0$, and because it is periodic it assumes every such w an infinite number of times. Let us determine all solutions of the equation

$$e^z = w, \qquad (2.5\text{-}16)$$

if $w \neq 0$. Letting $z = x + iy$ and $w = |w|e^{i\phi}$, where ϕ is any value of $\arg w$, (2.5-16) holds if and only if

$$e^x = |w|, \quad \text{i.e., if } x = \text{Log}|w|,$$

and

$$y = \phi + 2k\pi,$$

where k is any integer. Thus the solution set of (2.5-16) is the set of all numbers

$$z = \text{Log}|w| + i(\arg w + 2\pi k), \qquad k = 0, \pm 1, \pm 2, \ldots$$

We call this set of numbers the **logarithm** of w and denote it by the symbol $\{\log w\}$.

ELEMENTARY TRANSCENDENTAL FUNCTIONS

THEOREM 2.5c

If w_1 and w_2 are any two complex numbers different from zero,

$$\{\log w_1 w_2\} = \{\log w_1\} + \{\log w_2\}. \qquad (2.5\text{-}17)$$

Equation (2.5-17) means that (a) if $z_1 \in \{\log w_1\}$ and $z_2 \in \{\log w_2\}$, then $z_1 + z_2 \in \{\log w_1 w_2\}$; (b) if $z_3 \in \{\log w_1 w_2\}$, then there exist numbers $z_1 \in \{\log w_1\}$ and $z_2 \in \{\log w_2\}$ such that $z_3 = z_1 + z_2$.

Proof. (a) Let $z_1 \in \{\log w_1\}$, $z_2 \in \{\log w_2\}$. Then

$$e^{z_1} = w_1, \qquad e^{z_2} = w_2,$$

hence $e^{z_1} e^{z_2} = w_1 w_2$ or by (2.5-2)

$$e^{z_1 + z_2} = w_1 w_2;$$

i.e., $z_1 + z_2 \in \{\log w_1 w_2\}$. (b) Let $z_3 \in \{\log w_1 w_2\}$. Then

$$e^{z_3} = w_1 w_2.$$

Choose any member z_1 of the set $\{\log w_1\}$. Then by dividing the above by $e^{z_1} = w_1$ the result is

$$e^{z_3 - z_1} = w_2;$$

i.e., $z_2 := z_3 - z_1 \in \{\log w_2\}$, and z_3 is represented as the sum of a member of $\{\log w_1\}$ and a member of $\{\log w_2\}$, as desired. ∎

Since the function e^z is not one-to-one, no inverse function can be defined without restricting the domain of definition. A simple method is as follows: let α be any real number; then in the strip

$$S_\alpha : \alpha - \pi < \operatorname{Im} z < \alpha + \pi$$

e^z assumes every complex number $w \neq 0$ not located on the ray $\arg w = \alpha \pm \pi$ and assumes it exactly once (see Fig. 2.5b). Thus on the set

$$T_\alpha : \alpha - \pi < \arg w < \alpha + \pi$$

the inverse of the restricted exponential function is defined. Its value at a point $w \in T_\alpha$ is given by

$$z := \operatorname{Log}|w| + i\phi, \qquad (2.5\text{-}18)$$

where ϕ is the unique value of $\arg w$ which satisfies $|\phi - \alpha| < \pi$. We call the

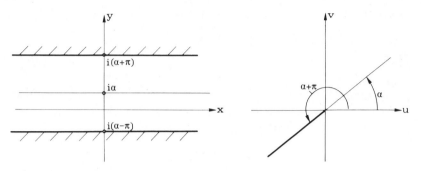

Fig. 2.5b. The exponential function restricted.

function $z = g(w)$ defined by (2.5-18) a **simple branch** of $\{\log w\}$ and denote it by $\log w$ (see Fig. 2.5b). The value of α which is used to determine the simple branch should be understood from the context.

The simple branch determined by $\alpha = 0$ is called the **principal branch** or **principal value** of the logarithm, and is denoted by $\text{Log } w$. Thus, for all $w \neq 0$ that are not real and negative,

$$\text{Log } w = \text{Log}|w| + i\phi, \qquad (2.5\text{-}19)$$

where ϕ is the unique value of $\arg w$ satisfying $-\pi < \phi < \pi$. This use of the symbol Log is consistent with the notation $\text{Log } u$ introduced earlier for positive real u.

THEOREM 2.5d

A simple branch of $\{\log w\}$ is analytic at all points at which it is defined.

Proof. Let $\log w$ denote the simple branch defined by the number α and let w_0 be any point in the domain of definition T of $\log w$. Let $z_0 = x_0 + iy_0$ be the unique point in S that satisfies $e^{z_0} = w_0$. We know that e^z is one-to-one on the set $|z - z_0| < \rho_1 := \min(\alpha + \pi - y_0, y_0 - \alpha + \pi)$. Because the coefficient of h in the expansion of e^z in powers of $h := z - z_0$ is not 0 (see the proof of Theorem 2.5a), there exists, by virtue of Theorem 2.4c, a number $\sigma_1 > 0$ such that e^z assumes every $w \in N(w_0, \sigma_1)$ at some $z \in N(z_0, \rho_1)$; moreover, the inverse function is analytic at w_0. But on $N(w_0, \sigma_1)$ the inverse function agrees with $\log w$. Hence $\log w$ is analytic at w_0. ∎

It remains to determine the power series that represents $\log w$ near w_0. We begin with the special case in which the simple branch is the principal branch and $w_0 = 1$. Then $z_0 = 0$ and since

$$e^z - e^{z_0} = e^z - 1 = E_1(z) - 1$$

ELEMENTARY TRANSCENDENTAL FUNCTIONS

the problem is to invert the series $E_1 - I$. This was solved formally in §1.7 and the reversion was found to be the logarithmic series

$$L = x - \tfrac{1}{2}x^2 + \tfrac{1}{3}x^3 - \cdots.$$

Thus by Theorem 2.4c, if $k := w - 1$ has a sufficiently small modulus, we get

$$\operatorname{Log}(1 + k) = L(k) = k - \tfrac{1}{2}k^2 + \tfrac{1}{3}k^3 - \cdots. \tag{2.5-20}$$

To expand an arbitrary simple branch $\log w$ in powers of $k := w - w_0$, where w_0 is an arbitrary point of T_α, we note that

$$\log w = \log[w_0 + (w - w_0)] = \log\left[w_0\left(1 + \frac{k}{w_0}\right)\right].$$

Hence by Theorem 2.5c

$$\log w = \log w_0 + s,$$

where s is a member of the set $\{\log(1 + k/w_0)\}$. In addition to $w \in T_\alpha$ we now assume that $|w - w_0| < |w_0|$. Then $|\arg w - \arg w_0| < \pi/2$; thus $|\operatorname{Im}(\log w - \log w_0)| < \pi/2$. Thus s must be a value of the logarithm such that $|\operatorname{Im} s| < \pi/2$. The only such value is the principal value; hence

$$\log w = \log w_0 + \operatorname{Log}\left(1 + \frac{k}{w_0}\right).$$

By virtue of (2.5-20) we now find

$$\log(w_0 + k) = \log w_0 + \frac{k}{w_0} - \frac{1}{2}\left(\frac{k}{w_0}\right)^2 + \frac{1}{3}\left(\frac{k}{w_0}\right)^3 - \cdots. \tag{2.5-21}$$

We shall see in §3.3 that the representation (2.5-20) holds for $|k| < 1$. If $|\arg w_0 - \alpha| < \pi/2$, (2.5-21) thus holds for $|k| < |w_0|$.

III. The General Power Function

If a is positive and b rational, $b = m/n$ (m and n integers, $n \neq 0$), we define in algebra $a^b = a^{m/n}$ as the unique positive solution x of the equation $x^n = a^m$. This definition is extended to arbitrary real b by setting

$$a^b := e^{b \operatorname{Log} a}.$$

It is readily seen that this definition is consistent with the definition just given for rational b.

If a and b are complex, $a \neq 0$, we accordingly denote by $\{a^b\}$ the set of all numbers e^{bz}, where z runs through all values of $\{\log a\}$. We consider some special cases.

1. If a is positive and b is real, the set $\{a^b\}$ contains the number $a^b = e^{b \operatorname{Log} a}$ as defined above. In addition, it contains all numbers

$$e^{b(\operatorname{Log} a + 2\pi i k)} = a^b e^{2\pi i k b},$$

where k is any integer. Thus the set $\{a^b\}$ has a finite or infinite number of members, depending on whether b is rational or irrational.

2. If a is complex, $a = |a|e^{i\alpha}$, and b is a positive integer, $b = n$, the set $\{a^b\}$ contains the numbers

$$e^{n(\operatorname{Log}|a| + i\alpha + 2\pi i k)} = \left(e^{\operatorname{Log}|a| + i\alpha}\right)^n e^{2\pi i k n}.$$

Since $e^{2\pi i k n} = 1$, these numbers are all identical with $a^n = a \cdot a \cdots a$ (n factors).

3. If $a = |a|e^{i\alpha}$ is complex and $b = 1/n$, where n is a positive integer, the set $\{a^b\}$ comprises the numbers $\exp[(1/n)(\operatorname{Log}|a| + i\alpha + 2\pi i k)]$ $= |a|^{1/n} e^{i(\alpha/n)} e^{2\pi i k/n}$, where k is any integer. Thus it consists of the n solutions z of the equation $z^n = a$, found in the algebra of complex numbers.

We now let a vary, $a = z$, while keeping b fixed. The problem develops of defining a reasonable function consisting of values of the sets $\{z^b\}$. Among the many possibilities we select the one of putting

$$z^b := e^{b \log z},$$

where $\log z$ is a simple branch of $\{\log z\}$. We call the function z^b so defined a simple branch of $\{z^b\}$. If $\log z$ is the branch for which $|\operatorname{Im} \log z - \alpha| < \pi$, then

$$z^b = e^{b(\operatorname{Log}|z| + i\phi)},$$

where ϕ is the value of $\arg z$ lying in the interval $(\alpha - \pi, \alpha + \pi)$. The simple branch of z^b defined by the principal branch of $\{\log z\}$,

$$z^b := e^{b \operatorname{Log} z},$$

is called the **principal branch** or **principal value** of z^b.

By Theorems 2.5a and 2.5d any simple branch of $\{z^b\}$ is a composition of analytic functions. Thus by Theorem 2.4a we obtain Theorem 2.5e:

ELEMENTARY TRANSCENDENTAL FUNCTIONS

THEOREM 2.5e

Any simple branch of $\{z^b\}$ is analytic at every point of its domain of definition.

To find the power series representation of a simple branch z^b near a point z_0 of its domain of definition $T_\alpha : |\arg z - \alpha| < \pi$ we first consider the special case in which the simple branch is the principal branch and $z_0 = 1$. Letting $h := z - 1$, we have to expand

$$e^{b \operatorname{Log} z} = e^{b \operatorname{Log}(1+h)}.$$

Since $\operatorname{Log}(1+h) = L(h)$, where L is the logarithmic series, we find by Theorem 2.4a that z^b is represented by the series $E_b \circ L(h)$. By Theorem 1.7e this equals $B_b(h)$. Thus, if $|h|$ is sufficiently small,

$$z^b = (1+h)^b \quad \text{(principal branch)}$$

$$= B_b(h) = \sum_{n=0}^{\infty} \frac{(-b)_n}{n!} (-h)^n. \tag{2.5-22}$$

We shall see later (Theorem 3.3a) that this actually holds for $|h| < 1$.

The case in which the simple branch z^b and $z_0 \in T$ are arbitrary is reduced to the above as follows: Letting $z = z_0 + h$, we have, if $\rho > 0$ is so small that $N(z_0, \rho) \subset T_\alpha$,

$$\log z = \log(z_0 + h) = \log\left[z_0\left(1 + \frac{h}{z_0}\right)\right]$$

$$= \log z_0 + \operatorname{Log}\left(1 + \frac{h}{z_0}\right);$$

hence

$$z^b = \exp\left\{b\left[\log z_0 + \operatorname{Log}\left(1 + \frac{h}{z_0}\right)\right]\right\}$$

$$= z_0^b B_b\left(\frac{h}{z_0}\right)$$

$$= z_0^b \sum_{n=0}^{\infty} \frac{(-b)_n}{n!} \left(-\frac{h}{z_0}\right)^n.$$

PROBLEMS

1. Let a,b,c be complex numbers, $a \neq b$, $c \neq 0$, and let

$$w = f(z) := \frac{1}{c}(e^{az} - e^{bz}).$$

Show that the inverse function is near $w=0$ represented by

$$z = f^{[-1]}(w) = \frac{1}{d} \sum_{n=1}^{\infty} \frac{(-1)^{n-1} c^n}{n!} \left(\frac{nb}{d} + 1\right)_{n-1} w^n,$$

where $d := a - b$. Derive as special cases the representations for the inverses of the functions

$$w = \sin z, \quad w = e^z - 1, \quad w = e^{\alpha z} \sin \beta z,$$

and, as the limiting case $a = b + c$, $c \to 0$, $w = ze^{bz}$. [To compute residues use formal integration by parts; see Problem 6, §1.8.]

2. Obtain a closed formula for the sum of the terms in any column of the array

$$1$$
$$1 + z$$
$$1 + 2z + z^2$$
$$1 + 3z + 3z^2 + z^3$$
$$1 + 4z + 6z^2 + 4z^3 + z^4$$

generated by the successive powers of $1 + z$ [Apply Theorem 2.4d.]

3. Express z in powers of w, where

$$w = z \left(\frac{1 + \sqrt{1-z}}{2}\right)^\beta.$$

[To compute residues use Problem 4, §1.9.]

4. Show that the solution of the cubic equation

$$z^3 - 3z - w = 0$$

which tends to zero with w is given by

$$z = \frac{1}{3} w \, {}_2F_1\left(\frac{1}{3}, \frac{2}{3}; \frac{3}{2}; -\frac{w^2}{4}\right).$$

5. Expand in powers of w the solution of the trinomial equation $z^n + nz - w = 0$ which tends to zero with w.

6. *Generating function of the Laguerre polynomials.* If, for complex a and $n = 0, 1, 2, \ldots,$

$$L_n^{(a)}(\xi) := \frac{1}{n!} e^\xi \xi^{-a} \left(\frac{d}{d\xi} \right)^n (e^{-\xi} \xi^{n+a})$$

is the **Laguerre polynomial** of order n, show that

$$\sum_{n=0}^{\infty} L_n^{(a)}(\xi) z^n = (1-z)^{-a-1} \exp\left(-\frac{\xi z}{1-z} \right).$$

[By Taylor's theorem, as proved in the calculus, $e^{-\xi} \xi^a L_n^{(a)}(\xi)$ for $\xi > 0$ equals the coefficient of x^n in the expansion of $e^{-(\xi+x)}(\xi+x)^{n+a}$ in powers of x:

$$e^{-\xi} \xi^a L_n^{(a)}(\xi) = \operatorname{res}\left[x^{n-1} e^{-(\xi+x)} (\xi+x)^{n+a} \right].$$

Now apply the Lagrange-Bürmann formula.]

7. For complex a and b the **Jacobi polynomial** of order n may be defined by

$$P_n^{(a,b)}(\xi) := (1-\xi)^{-a}(1+\xi)^{-b} \frac{(-1)^n}{2^n n!} \left(\frac{d}{d\xi} \right)^n \left[(1-\xi)^{n+a} (1+\xi)^{n+b} \right],$$

$$n = 0, 1, 2, \ldots.$$

Show that

$$\sum_{n=0}^{\infty} P_n^{(a,b)}(\xi) z^n = 2^{a+b} W^{-1} (1 - z + W)^{-a} (1 + z + W)^{-b},$$

where

$$W := (1 - 2z\xi + z^2)^{1/2}.$$

[*Generating function* of the Jacobi polynomials]

8. Let M and ϵ be given real numbers, $|\epsilon| < 1$. The equation for E,

$$M = E - \epsilon \sin E, \qquad (2.5\text{-}23)$$

is called **Kepler's equation**. It arises in the study of the motion of a particle P in a gravitational field of force due to a single celestial body. Under certain conditions the orbit of the particle is an ellipse with center 0, whose one focus F coincides with the center of the gravitational field. In the above equation ϵ is the eccentricity of the ellipse, E is the eccentric anomaly (i.e., the angle formed by the vectors **OP** and **OF**), and M is the mean anomaly, i.e., the time measured in units $(2\pi)^{-1} T$, where T is the period of the motion. The solution

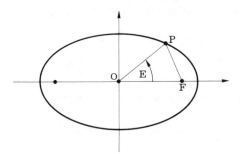

Fig. 2.5c. Notations of Kepler's problem.

to Kepler's equation thus yields the position as a function of time (Fig. 2.5c). Show that for $M \not\equiv 0 \mod \pi$ the unique solution of Kepler's equation is $E := M + z$, where

$$z := \sum_{n=1}^{\infty} b_n \epsilon^n,$$

$$b_n := \frac{1}{n!} \left(\frac{d}{d\xi} \right)^{n-1} (\sin \xi)^n \bigg|_{\xi = M}$$

$$= \frac{1}{2^{n-1} n!} \sum_{k=0}^{[n/2]} \binom{n}{k} (n-2k)^{n-1} \sin[(n-2k)M].$$

[The equation for z reads $z = \epsilon \sin(M+z)$ or $\epsilon = f(z)$, where

$$f(z) := \frac{z}{\sin(M+z)}.$$

This is readily inverted by the Lagrange-Bürmann expansion. For another solution of Kepler's equation see §4.5.]

9. The equation $\tan x = cx$ has for real $c > 0$ a solution x in each of the intervals $[n\pi, (n+\frac{1}{2})\pi]$, $n = 1, 2, \ldots$. This solution tends to $(n+\frac{1}{2})\pi$ for $c \to \infty$. For fixed n expand the solution in powers of $w := 1/c$ and calculate explicitly the first few coefficients of the expansion. [Letting $a := (n+\frac{1}{2})\pi$, $x := a + z$, we find that the equation amounts to $(a+z)\tan z = -w$. Inversion yields

$$z = -\frac{1}{a} w - \frac{1}{a^3} w^2 - \frac{6-a^2}{3a^5} w^3 - \cdots.\bigg]$$

10. Let $f(z) := a_1 z + a_2 z^2 + \cdots$, $a_1 \neq 0$, be a convergent power series. By Theorem 2.4c there exist $\delta > 0$ and $\epsilon > 0$ such that for $|w| < \epsilon$ the equation $f(z) = w$ has

ELEMENTARY TRANSCENDENTAL FUNCTIONS 121

precisely one solution in $|z|<\delta$. If the expansion of this solution in powers of w is truncated after the linear term, show that the approximation obtained is equivalent to applying Newton's method to $f(z)-w=0$ at $z=0$. More generally, by taking more terms in the expansion, show how to obtain "higher order Newton methods" (e.g., see Hildebrand [1956], p. 479).

11. Let $f(z):=a_2z^2+a_3z^3+\cdots$, $a_2\neq 0$, be a convergent power series. Produce the first few terms of the expansion in powers of w of the two solutions of the equation $f(z)=w(1+a_1z)$ close to the origin.

12. Let m and n be integers without a common divisor, $n>m>0$ and let p, q be complex numbers, $q\neq 0$. Show that for $|p|$ sufficiently small the n solutions of the trinomial equation

$$z^n + pz^m - q^n = 0 \qquad (2.5\text{-}24)$$

are given by

$$z_h = r_h q \sum_{k=0}^{\infty} \frac{1}{km+1} \frac{[-(km+1)/n]_k}{k!} \left(\frac{r_h^m p}{q^n}\right)^k, \qquad (2.5\text{-}25)$$

where $r_1, r_2, \ldots, r_n$ are the n nth roots of unity.

13 (continuation). Deduce from (2.5-25) the well-known formula for the solutions of the quadratic equation $z^2+pz-q^2=0$. For which values of p and q does the series converge? What is the meaning of these conditions in terms of the solutions? Does the series converge if the two solutions coincide?

14. Deduce from (2.5-25) Cardano's formulas

$$z_h = r_h \left[Q + \sqrt{Q^2+P^3} \right]^{1/3} + r_h^2 \left[Q - \sqrt{Q^2+P^3} \right]^{1/3}$$

$[h=1,2,3;\ r_h:=\exp(2\pi i h/3)]$ for the solution of the cubic equation $z^3+3Pz-2Q=0$.

15. Let r,s,t be complex. Show that for sufficiently small $|s|$ the four solutions of the biquadratic equation $z^4 - s^2r^2z^2 + str^3z - r^4 = 0$ are given by

$$z_h := r \sum_{m=1}^{3} i^{hm} F_m(s,t), \qquad h=1,2,3,4,$$

where

$$F_m(s,t) := \sum_{n=0}^{\infty} \frac{1}{4n+m} C_{4n-m-1}^{-n-m/4}(t) s^{4n+m-1}.$$

Here C_n^ν denotes the ultraspherical (Gegenbauer) polynomial. See problem 5, §1.6.

§2.6. MATRIX-VALUED FUNCTIONS

In this section we discuss briefly some functions that are analytic at the zero element of the Banach algebra $\mathbb{C}^{n,n}$ of $n \times n$ matrices with complex elements, endowed with the norm.

$$\|\mathbf{A}\| := \|\mathbf{A}\|_2 := \sup_{\mathbf{x} \neq 0} \frac{\|\mathbf{A}\mathbf{x}\|_2}{\|\mathbf{x}\|_2},$$

where $\|\mathbf{x}\|_2$ is the Euclidean norm of the vector $\mathbf{x}$. If f is such a function, then there exists $\rho > 0$ such that, for all $\mathbf{A}$ satisfying $\|\mathbf{A}\| < \rho$, $f(\mathbf{A})$ is defined and represented by a convergent power series in $\mathbf{A}$:

$$f(\mathbf{A}) = a_0 \mathbf{I} + a_1 \mathbf{A} + a_2 \mathbf{A}^2 + \cdots.$$

The best known function is the exponential of a matrix, defined by the exponential series E_t, where t is a complex parameter:

$$e^{t\mathbf{A}} := E_t(\mathbf{A}) = \mathbf{I} + \frac{t}{1!}\mathbf{A} + \frac{t^2}{2!}\mathbf{A}^2 + \cdots. \tag{2.6-1}$$

If t and s are any two complex numbers, then, in view of $E_t E_s = E_{t+s}$, it follows immediately that

$$e^{t\mathbf{A}} e^{s\mathbf{A}} = e^{(t+s)\mathbf{A}}. \tag{2.6-2}$$

Letting $s = -t$, we find

$$e^{t\mathbf{A}} e^{-t\mathbf{A}} = \mathbf{I},$$

where $\mathbf{I} = e^{0\mathbf{A}}$ is the identity matrix. The following statement is now evident:

THEOREM 2.6a

The exponential matrix $e^{t\mathbf{A}}$ is never singular and its inverse is the matrix $e^{-t\mathbf{A}}$.

Concerning products of exponentials of distinct matrices, the reader is warned that the relation

$$e^{t\mathbf{A}} e^{t\mathbf{B}} = e^{t(\mathbf{A}+\mathbf{B})} \tag{2.6-3}$$

holds for all t if and only if the matrices $\mathbf{A}$ and $\mathbf{B}$ commute, i.e., if $\mathbf{AB} = \mathbf{BA}$. Indeed, a proof of (2.6-3) by a power series argument requires, e.g., in the term in t^2, the identity of

$$\mathbf{A}^2 + 2\mathbf{A}\mathbf{B} + \mathbf{B}^2 = (\mathbf{A}+\mathbf{B})^2 = \mathbf{A}^2 + \mathbf{A}\mathbf{B} + \mathbf{B}\mathbf{A} + \mathbf{B}^2.$$

MATRIX VALUED FUNCTIONS

If a matrix $\mathbf{X}=(x_{ij})$ depends on a real parameter τ, $\mathbf{X}=\mathbf{X}(\tau)$, the derivative of $\mathbf{X}$ at the point τ is defined by

$$\mathbf{X}'(\tau) := \lim_{\eta \to 0} \frac{\mathbf{X}(\tau+\eta) - \mathbf{X}(\tau)}{\eta} \qquad (\eta \text{ real}),$$

provided that the limit exists. It is clear that this limit exists if and only if the derivative of each element x_{ij} exists at the point τ and that $\mathbf{X}' = (x'_{ij})$. For $\mathbf{A}$ fixed and τ real the matrix $e^{\tau \mathbf{A}}$ becomes a function of τ.

THEOREM 2.6b

The matrix $\mathbf{X} := e^{\tau \mathbf{A}}$ *satisfies the differential equation*

$$\mathbf{X}' = \mathbf{A}\mathbf{X}. \tag{2.6-4}$$

Proof. By (2.6-2)

$$\mathbf{X}(\tau+\eta) = e^{(\tau+\eta)\mathbf{A}} = e^{\tau\mathbf{A}} e^{\eta\mathbf{A}}.$$

Thus, if $\eta \neq 0$,

$$\frac{1}{\eta}[\mathbf{X}(\tau+\eta) - \mathbf{X}(\tau)] = \frac{1}{\eta}(e^{\eta\mathbf{A}} - \mathbf{I})e^{\tau\mathbf{A}}.$$

The series definition shows that

$$\frac{1}{\eta}(e^{\eta\mathbf{A}} - \mathbf{I}) = \mathbf{A} + \frac{\eta}{2!}\mathbf{A}^2 + \frac{\eta^2}{3!}\mathbf{A}^3 + \cdots,$$

which clearly tends to $\mathbf{A}$ as $\eta \to 0$. ∎

By virtue of Theorem 2.6b the exponential matrix provides the key to the theory of systems of linear differential equations with constant coefficients, a topic developed further in Chapter 9. We mention here only that the unique solution of the system $\mathbf{x}' = \mathbf{A}\mathbf{x}$ which satisfies given initial conditions $\mathbf{x}(0) = \mathbf{b}$ is given by

$$\mathbf{x}(\tau) := \mathbf{X}(\tau)\mathbf{b} = e^{\tau\mathbf{A}}\mathbf{b}.$$

We now return to general matrix-valued analytic functions. We shall show that to some extent the study of these functions can be reduced to the study of the corresponding scalar functions.

We recall that the matrix $\mathbf{B}$ is called **similar** to the matrix $\mathbf{A}$ if there exists a nonsingular matrix $\mathbf{T}$ such that

$$\mathbf{B} = \mathbf{T}\mathbf{A}\mathbf{T}^{-1}. \tag{2.6-5}$$

FUNCTIONS ANALYTIC AT A POINT

The notion of similarity thus defined is reflexive, symmetric—since (2.6-5) implies $A = T^{-1}BT$— and transitive; similarity of matrices thus defines an equivalence relation.

THEOREM 2.6c

Let $a_0, a_1, \ldots$, be complex numbers and let the function f be defined by

$$f(A) := a_0 I + a_1 A + a_2 A^2 + \cdots \qquad (2.6\text{-}6)$$

on the set D of all matrices $A \in \mathbb{C}^{n,n}$ for which the series converges. Then, if $A \in D$, all matrices similar to A belong to D. Furthermore, if $A \in D$ and (2.6-5) holds, then $f(B)$ is similar to $f(A)$, and

$$f(B) = T f(A) T^{-1}. \qquad (2.6\text{-}7)$$

The *proof* is based on two simple lemmas.

LEMMA 2.6d

Let p be a polynomial with complex coefficients. Then, if (2.6-5) holds,

$$p(B) = T p(A) T^{-1}. \qquad (2.6\text{-}8)$$

Proof. Equation (2.6-5) implies

$$B^2 = T A T^{-1} T A T^{-1} = T A^2 T^{-1},$$

and, by induction, $B^k = T A^k T^{-1}$, $k = 0, 1, 2, \ldots$. Furthermore, if a_k is any complex number, then, because scalars commute with matrices,

$$a_k B^k = a_k T A^k T^{-} = T(a_k A^k) T^{-1}.$$

Hence, if

$$p(x) := \sum_{k=0}^{n} a_k x^k,$$

then

$$p(B) = \sum_{k=0}^{n} T(a_k A^k) T^{-1}$$

$$= T \left(\sum_{k=0}^{n} a_k A^k \right) T^{-1} = T p(A) T^{-1}. \qquad \blacksquare$$

MATRIX VALUED FUNCTIONS

LEMMA 2.6e

If $\{\mathbf{P}_n\}$ *is a sequence of matrices converging to the matrix* $\mathbf{P}$ *and* $\mathbf{Q}_n := \mathbf{TP}_n\mathbf{T}^{-1}$, *the sequence* $\{\mathbf{Q}_n\}$ *converges to the matrix* $\mathbf{Q} := \mathbf{TPT}^{-1}$.

Proof: We have

$$\mathbf{Q}_n - \mathbf{Q} = \mathbf{TP}_n\mathbf{T}^{-1} - \mathbf{TPT}^{-1} = \mathbf{T}(\mathbf{P}_n - \mathbf{P})\mathbf{T}^{-1};$$

hence

$$\|\mathbf{Q}_n - \mathbf{Q}\| \leq \|\mathbf{T}\| \, \|\mathbf{T}^{-1}\| \, \|\mathbf{P}_n - \mathbf{P}\|,$$

and $\|\mathbf{P}_n - \mathbf{P}\| \to 0$ implies $\|\mathbf{Q}_n - \mathbf{Q}\| \to 0$. ∎

To prove Theorem 2.6c let

$$f_n(\mathbf{Z}) := a_0\mathbf{I} + a_1\mathbf{Z} + \cdots + a_n\mathbf{Z}^n.$$

By Lemma 2.6d $f_n(\mathbf{B}) = \mathbf{T}f_n(\mathbf{A})\mathbf{T}^{-1}$. Because $f_n(\mathbf{A}) \to f(\mathbf{A})$, the sequence $\{f_n(\mathbf{B})\}$ converges by Lemma 2.6e, implying $\mathbf{B} \in D$. By the same lemma its limit $f(\mathbf{B})$ satisfies $f(\mathbf{B}) = \mathbf{T}f(\mathbf{A})\mathbf{T}^{-1}$. ∎

On the basis of Theorem 2.6c, to compute or study the behavior of $f(\mathbf{A})$ we may replace $\mathbf{A}$ with any matrix $\mathbf{B}$ similar to $\mathbf{A}$. Naturally, we shall endeavor to choose $\mathbf{B}$ in such a manner that $f(\mathbf{B})$ is simple to compute. A particularly happy situation develops if $\mathbf{A}$ is similar to a diagonal matrix

$$\Lambda := \begin{bmatrix} \lambda_1 & & & 0 \\ & \lambda_2 & & \\ & & \ddots & \\ 0 & & & \lambda_N \end{bmatrix} \tag{2.6-9}$$

Because

$$\Lambda^n = \begin{bmatrix} \lambda_1^n & & & 0 \\ & \lambda_2^n & & \\ & & \ddots & \\ 0 & & & \lambda_N^n \end{bmatrix},$$

we have

$$f(\Lambda) = \begin{bmatrix} f(\lambda_1) & & & 0 \\ & f(\lambda_2) & & \\ & & \ddots & \\ 0 & & & f(\lambda_N) \end{bmatrix},$$

and Theorem 2.6f follows:

THEOREM 2.6f

If f and D are defined as in Theorem 2.6c and $\mathbf{A} = \mathbf{T}\Lambda\mathbf{T}^{-1} \in D$, where Λ is the diagonal matrix (2.6-9), then

$$f(\mathbf{A}) = \mathbf{T} \begin{bmatrix} f(\lambda_1) & & & 0 \\ & f(\lambda_2) & & \\ & & \ddots & \\ 0 & & & f(\lambda_N) \end{bmatrix} \mathbf{T}^{-1}.$$

Although certain matrices with special properties (such as all hermitian matrices, all unitary matrices, or more generally all normal matrices) are known to be similar to diagonal matrices, not every matrix can be transformed into diagonal form; for instance,

$$\mathbf{A} = \begin{bmatrix} 0 & 1 \\ 0 & 0 \end{bmatrix}$$

is not similar to any diagonal matrix. It is shown in linear algebra, however, that every matrix $\mathbf{A}$ is similar to a matrix $\mathbf{J}$ which in block form can be written

$$\mathbf{J} = \begin{bmatrix} \mathbf{J}_1 & \mathbf{0} & & \mathbf{0} \\ \mathbf{0} & \mathbf{J}_2 & & \mathbf{0} \\ & & \ddots & \\ \mathbf{0} & \mathbf{0} & & \mathbf{J}_s \end{bmatrix}, \qquad (2.6\text{-}10)$$

where $\mathbf{J}_k$ is a matrix of order n_k $(n_1+n_2+\cdots+n_s=N)$,

$$\mathbf{J}_k = \begin{bmatrix} \lambda_k & 1 & 0 & & 0 & 0 \\ 0 & \lambda_k & 1 & & 0 & 0 \\ 0 & 0 & \lambda_k & & 0 & 0 \\ \cdots & & & \ddots & & \cdots \\ 0 & 0 & 0 & & \lambda_k & 1 \\ 0 & 0 & 0 & & 0 & \lambda_k \end{bmatrix} \qquad (2.6\text{-}11)$$

(the diagonal elements equal λ_k, the elements immediately above the main diagonal are 1, and all other elements are zero). The characteristic polynomial of $\mathbf{J}$ is

$$P(\lambda) = \prod_{k=1}^{s} (\lambda - \lambda_k)^{n_k}.$$

Because similar matrices have identical characteristic polynomials, this is also the characteristic polynomial of $\mathbf{A}$. We thus see that the numbers λ_k are the eigenvalues of $\mathbf{A}$. The λ_k need not be distinct, but if an eigenvalue λ_k has multiplicity one then $\mathbf{J}_k$ is the matrix (λ_k) of order 1. If $\mathbf{A}$ is similar to a diagonal matrix, then all $n_k = 1$ and $s = N$. The matrix $\mathbf{J}$ is called a **Jordan canonical form** of $\mathbf{A}$.

If $\mathbf{J}$ is a Jordan canonical form of $\mathbf{A}$ and $\mathbf{A} = \mathbf{T}\mathbf{J}\mathbf{T}^{-1} \in D$, then by Theorem 2.6c $\mathbf{J} \in D$ and

$$f(\mathbf{A}) = \mathbf{T}f(\mathbf{J})\mathbf{T}^{-1}.$$

Since

$$\mathbf{J}^n = \begin{bmatrix} \mathbf{J}_1^n & & & 0 \\ & \mathbf{J}_2^n & & \\ & & \ddots & \\ 0 & & & \mathbf{J}_s^n \end{bmatrix},$$

we have

$$f(\mathbf{J}) = \begin{bmatrix} f(\mathbf{J}_1) & & & 0 \\ & f(\mathbf{J}_2) & & \\ & & \ddots & \\ 0 & & & f(\mathbf{J}_s) \end{bmatrix},$$

and it remains only to compute $f(\mathbf{J}_k)$, where $\mathbf{J}_k$ is a "Jordan block" of the form (2.6-11).

Let

$$\mathbf{J}_k = \lambda_k \mathbf{I} + \mathbf{E}, \qquad (2.6\text{-}12)$$

where $\mathbf{E}$ denotes the special matrix of order n_k:

$$\mathbf{E} := \begin{bmatrix} 0 & 1 & 0 & & 0 & 0 \\ 0 & 0 & 1 & & 0 & 0 \\ 0 & 0 & 0 & & 0 & 0 \\ & & & \ddots & & \\ 0 & 0 & 0 & & 0 & 1 \\ 0 & 0 & 0 & & 0 & 0 \end{bmatrix}. \qquad (2.6\text{-}13)$$

Since the matrix $\lambda_k \mathbf{I}$ commutes with every matrix, hence with $\mathbf{E}$, we can form powers of $\mathbf{J}_k$ by the binomial theorem to obtain

$$\mathbf{J}_k^n = \sum_{m=0}^{n} \binom{n}{m} \lambda_k^{n-m} \mathbf{E}^m.$$

It is easily verified that if $\mathbf{B}$ is any $n_k \times n_k$ matrix, then the ith row of $\mathbf{EB}$ equals the $(i+1)$st row of $\mathbf{B}$ for $i = 1, 2, \ldots, n_k - 1$, and the n_kth row of $\mathbf{EB}$ is zero. Thus for $m = 1, 2, \ldots$, the matrices $\mathbf{E}^m$ consist of a single diagonal of ones at a distance m from the main diagonal. It follows that $\mathbf{E}^m = \mathbf{0}$ for

MATRIX VALUED FUNCTIONS

$m = n_k$, and letting $r := n_k - 1$ we have

$$\mathbf{J}_k^n = \sum_{m=0}^{r} \binom{n}{m} \lambda_k^{n-m} \mathbf{E}^m$$

$$= \begin{bmatrix} \lambda_k^n & \binom{n}{1}\lambda_k^{n-1} & \cdots & \binom{n}{r}\lambda_k^{n-r} \\ 0 & \lambda_k^n & \cdots & \binom{n}{r-1}\lambda_k^{n-r+1} \\ & & \ddots & \\ 0 & 0 & & \lambda_k^n \end{bmatrix}.$$

Thus we get

$$f(\mathbf{J}_k) = \sum_{n=0}^{\infty} a_n \mathbf{J}_k^n$$

$$= \sum_{m=0}^{r} \left(\sum_{n=m}^{\infty} \binom{n}{m} a_n \lambda_k^{n-m} \right) \mathbf{E}^m.$$

If

$$F := a_0 + a_1 x + a_2 x^2 + \cdots$$

we recognize that

$$\sum_{n=m}^{\infty} \binom{n}{m} a_n x^{n-m} = \frac{1}{m!} F^{(m)},$$

where $F^{(m)}$ is the mth formal derivative of the formal series F. We call the analytic function defined by $F^{(m)}$,

$$f^{(m)}(\mathbf{Z}) := F^{(m)}(\mathbf{Z}), \quad m = 0, 1, 2, \ldots, \quad (2.6\text{-}14)$$

the mth **derivative** of f. Because $\mathbf{J} \in D$, the above shows that the series $f^{(m)}(\lambda_k)$ converges for $m = 0, 1, \ldots, r = n_k - 1$, where n_k is the order of the (largest) Jordan block associated with λ_k, and

$$f(\mathbf{J}_k) = \sum_{m=0}^{r} \frac{1}{m!} f^{(m)}(\lambda_k) \mathbf{E}^m.$$

In summary, we have proved the following:

THEOREM 2.6g

Let f and D be defined as in Theorem 2.6c and let $f^{(m)}$ be defined by (2.6-14). If $\mathbf{A} = \mathbf{TJT}^{-1} \in D$, where $\mathbf{J}$ is a Jordan canonical form of $\mathbf{A}$, defined by (2.6-10) and (2.6-11), the series $f^{(m)}(\lambda_k)$ converges for $m = 0, 1, \ldots, r$, where $n_k = r + 1$ is the order of the Jordan block $\mathbf{J}_k$ ($k = 1, 2, \ldots, s$), and

$$f(\mathbf{A}) = \mathbf{T} \begin{bmatrix} f(\mathbf{J}_1) & & & 0 \\ & f(\mathbf{J}_2) & & \\ & & \ddots & \\ 0 & & & f(\mathbf{J}_s) \end{bmatrix} \mathbf{T}^{-1}, \qquad (2.6\text{-}15)$$

where, with $\mathbf{E}$ defined by (2.6-13),

$$f(\mathbf{J}_k) = f(\lambda_k)\mathbf{I} + \frac{1}{1!}f'(\lambda_k)\mathbf{E} + \cdots + \frac{1}{r!}f^{(r)}(\lambda_k)\mathbf{E}^r.$$

EXAMPLE

If $f(\mathbf{A}) := e^{t\mathbf{A}}$, then $f^{(m)}(\mathbf{A}) = t^m e^{t\mathbf{A}}$, $m = 1, 2, \ldots$, and (2.6-15) holds with

$$f(\mathbf{J}_k) = e^{t\lambda_k}\left(\mathbf{I} + \frac{t}{1!}\mathbf{E} + \cdots + \frac{t^r}{r!}\mathbf{E}^r\right).$$

The Logarithm of a Matrix. In the theory of differential equations there is the problem of determining the "logarithm" of a given matrix $\mathbf{A}$, i.e., of determining a solution $\mathbf{B}$ of the equation

$$e^{\mathbf{B}} = \mathbf{A}. \qquad (2.6\text{-}16)$$

It is clear from Theorem 2.6a that this equation cannot be solved if $\mathbf{A}$ is singular, for $e^{\mathbf{B}}$ is never singular. It turns out that this is the only restriction.

MATRIX VALUED FUNCTIONS

THEOREM 2.6h

Equation 2.6-16 has (infinitely many) *solutions* **B** *for every nonsingular matrix* **A**.

Any solution is called a **logarithm of the matrix A**, and is denoted by $\log \mathbf{A}$.

Proof of Theorem 2.6a. First we show that it suffices to prove the assertion for the Jordan canonical form of **A**, for if $\mathbf{A} = \mathbf{TJT}^{-1}$ and

$$e^{\mathbf{K}} = \mathbf{J}, \qquad (2.6\text{-}17)$$

then by Theorem 2.6c

$$\mathbf{A} = \mathbf{TJT}^{-1} = \mathbf{T}e^{\mathbf{K}}\mathbf{T}^{-1} = e^{\mathbf{TKT}^{-1}},$$

which shows that (2.6-16) is solved by $\mathbf{B} := \mathbf{TKT}^{-1}$.

To solve (2.6-17), where **J** is given by (2.6-10) and (2.6-11), we construct a matrix

$$\mathbf{K} := \begin{bmatrix} \mathbf{K}_1 & & & 0 \\ & \mathbf{K}_2 & & \\ & & \ddots & \\ 0 & & & \mathbf{K}_s \end{bmatrix},$$

where $\mathbf{K}_m$ is a matrix of order n_m such that

$$e^{\mathbf{K}_m} = \mathbf{J}_m, \qquad m = 1, 2, \ldots, s. \qquad (2.6\text{-}18)$$

Because

$$e^{\mathbf{K}} = \begin{bmatrix} e^{\mathbf{K}_1} & & & 0 \\ & e^{\mathbf{K}_2} & & \\ & & \ddots & \\ 0 & & & e^{\mathbf{K}_s} \end{bmatrix},$$

it is then clear that (2.6-17) holds.

To satisfy (2.6-18) we first note that no $\mathbf{J}_m$ is singular (otherwise **A** would be); hence $\lambda_m \neq 0$, $m = 1, 2, \ldots, s$. Let $\log \lambda_m$ denote any (scalar) logarithm of the complex number λ_m. If $n_m = 1$, $\mathbf{J}_m = (\lambda_m)$, then (2.6-18) is clearly satisfied by the 1×1 matrix $\mathbf{K}_m := (\log \lambda_m)$. If $n_m > 1$, we write

$$\mathbf{J}_m = \lambda_m \mathbf{I} + \mathbf{E} = \lambda_m \left(\mathbf{I} + \frac{1}{\lambda_m} \mathbf{E} \right),$$

where $\mathbf{E}$ is defined by (2.6-13). We now put

$$\mathbf{F}_m := \sum_{k=1}^{r} \frac{(-1)^{k-1}}{k} \left(\frac{1}{\lambda_m} \mathbf{E} \right)^m, \qquad (2.6\text{-}19)$$

where $r := n_m - 1$. We assert that

$$e^{\mathbf{F}_m} = \mathbf{I} + \frac{1}{\lambda_m} \mathbf{E}. \qquad (2.6\text{-}20)$$

If this is proved, we may put

$$\mathbf{K}_m := \log \lambda_m \mathbf{I} + \mathbf{F}_m,$$

for then (2.6-20) implies

$$e^{\mathbf{K}_m} = e^{\log \lambda_m \mathbf{I} + \mathbf{F}_m} = e^{\log \lambda_m} \cdot e^{\mathbf{F}_m} = \lambda_m \left(\mathbf{I} + \frac{1}{\lambda_m} \mathbf{E} \right) = \mathbf{J}_m.$$

It remains to prove (2.6-20). Let P denote the polynomial

$$P := x - \tfrac{1}{2} x^2 + \tfrac{1}{3} x^3 - \cdots + (-1)^{r-1} \frac{1}{r} x^r,$$

and let p on the algebra of matrices of order n_m be defined by $p(\mathbf{Z}) := P(\mathbf{Z})$. By Theorem 2.4a, since P is a nonunit and the radius of convergence of the exponential series is infinite, the function $\exp[p(\mathbf{Z})]$ is represented for all $\mathbf{Z}$ by $E_1 \circ P(\mathbf{Z})$, where E_1 is the exponential series. Now P differs from the logarithmic series

$$L = x - \tfrac{1}{2} x^2 + \tfrac{1}{3} x^3 - \cdots$$

only in powers x^n, where $n > r$. Thus by the definition of composition

$$E_1 \circ P = E_1 \circ L + Q,$$

where Q is a series involving powers x^n where $n > r$. By Example **2** of §1.7, $E_1 \circ L = 1 + x$; hence

$$E_1 \circ P = 1 + x + Q.$$

SEQUENCES AT A POINT

Thus

$$e^{P(Z)} = I + Z + Q(Z)$$

holds for all matrices Z. Now evidently

$$F_m = p\left(\frac{1}{\lambda_m} E\right).$$

Thus

$$e^{F_m} = I + \frac{1}{\lambda_m} E + Q\left(\frac{1}{\lambda_m} E\right).$$

However, since $E^n = 0$ for $n > r$, $Q[(1/\lambda_m)E] = 0$ and (2.6-20) is established. ∎

Powers with Matrix Exponents. If z is a complex number $\neq 0$ and A is any matrix, we define the set $\{z^A\}$ as the set of all matrices $e^{A \log z}$, where $\log z$ runs through all values of $\{\log z\}$. A simple branch of the function z^A is defined by selecting for $\log z$ a simple branch of the logarithm. The principal value of z^A is defined for $|\arg z| < \pi$ by

$$z^A := e^{A \operatorname{Log} z}.$$

If $z = \tau > 0$, the function $X(\tau) := e^{A \operatorname{Log} \tau}$ is by virtue of the chain rule seen to satisfy the differential equation

$$X' = \frac{1}{\tau} AX.$$

The facts regarding matrix functions which have been stated in this section are applied in Chapter 9.

§2.7. SEQUENCES OF FUNCTIONS ANALYTIC AT A POINT

Here we consider *sequences* of analytic functions. We show, in brief, that if such a sequence converges uniformly, then the limit function is analytic. A more precise statement is the following:

THEOREM 2.7a (**Weierstrass double series theorem**[1])

In a Banach algebra $\mathcal{B}$ let the functions $f_0, f_1, f_2, \ldots$, be defined for $\|Z\| < \rho$ by convergent power series

$$f_n(Z) := F_n(Z) := a_{n,0} I + a_{n,1} Z + a_{n,2} Z^2 + \cdots,$$

---

[1] The name "double series theorem" derives from the fact that Weierstrass stated the theorem for infinite series (of scalar analytic functions) in place of sequences. The series and the sequence formulation are, of course, equivalent.

hence analytic at **0**. For every $\rho_1 < \rho$ let the sequence $\{f_n(\mathbf{Z})\}$ be uniformly convergent on the set $\|\mathbf{Z}\| \leq \rho_1$. The limits

$$a_m := \lim_{n \to \infty} a_{n,m}, \qquad m = 0, 1, 2, \ldots, \tag{2.7-1}$$

then exist, and the limit function $f := \lim_{n \to \infty} f_n$ for $\|\mathbf{Z}\| < \rho$ is represented by

$$f(\mathbf{Z}) = F(\mathbf{Z}) := a_0 \mathbf{I} + a_1 \mathbf{Z} + a_2 \mathbf{Z}^2 + \cdots, \tag{2.7-2}$$

hence is analytic at **0**.

Proof. Let $\epsilon > 0$ be given and choose $\rho_0 < \rho$ arbitrarily. By uniform convergence there exists $N = N(\epsilon, \rho_0)$ such that for all $k > N$ and all $\mathbf{Z}$ such that $\|\mathbf{Z}\| \leq \rho_0$

$$\|f_k(\mathbf{Z}) - f(\mathbf{Z})\| < \epsilon; \tag{2.7-3}$$

hence, if $n > N$,

$$\|f_n(\mathbf{Z}) - f_k(\mathbf{Z})\| = \|F_n(\mathbf{Z}) - F_k(\mathbf{Z})\| < 2\epsilon.$$

Thus a fortiori, if $|z| = \rho_0$,

$$|F_n(z) - F_k(z)| < 2\epsilon.$$

The Cauchy estimate applied to the series $F_n - F_k$ now yields

$$|a_{n,m} - a_{k,m}| < \frac{2\epsilon}{\rho_0^m} \tag{2.7-4}$$

for each $m = 0, 1, 2, \ldots$. Because $\epsilon > 0$ was arbitrary, this shows that for each fixed m the sequence $\{a_{n,m}\}$ is a Cauchy sequence of complex numbers. Therefore, the limits (2.7-1) all exist.

Now let $\|\mathbf{Z}_1\| = \rho_1 < \rho_0$. We shall show that the series $F(\mathbf{Z})$ converges for $\mathbf{Z} = \mathbf{Z}_1$ and that its sum is $f(\mathbf{Z})$. Since ρ_0, hence ρ_1, may be chosen arbitrarily close to ρ, this will prove the second part of the assertion of Theorem 2.7a. Keeping $k > N$ fixed and putting

$$b_m := a_m - a_{k,m}, \qquad m = 0, 1, 2, \ldots, \tag{2.7-5}$$

letting $n \to \infty$ in (2.7-4) yields

$$|b_m| \leq \frac{2\epsilon}{\rho_0^m}, \qquad m = 0, 1, 2, \ldots. \tag{2.7-6}$$

SEQUENCES AT A POINT

Thus

$$\|a_m \mathbf{Z}_1^m - a_{k,m} \mathbf{Z}_1^m\| \leq |b_m| \|\mathbf{Z}_1\|^m \leq 2\epsilon \left(\frac{\rho_1}{\rho_0}\right)^m.$$

It follows that the series

$$\sum_{m=0}^{\infty} (a_m \mathbf{Z}_1^m - a_{k,m} \mathbf{Z}_1^m)$$

converges and that the norm of its sum is less than

$$\frac{2\epsilon}{1 - \rho_1/\rho_0}.$$

Because the series

$$F_k(\mathbf{Z}_1) = \sum_{m=0}^{\infty} a_{k,m} \mathbf{Z}_1^m$$

is known to be convergent, the convergence of $F(\mathbf{Z}_1)$ is implied and

$$\|F(\mathbf{Z}_1) - F_k(\mathbf{Z}_1)\| \leq \frac{2\epsilon}{1 - \rho_1/\rho_0};$$

but by virtue of (2.7-3), since $|\mathbf{Z}_1| < \rho_0$,

$$\|f(\mathbf{Z}_1) - F_k(\mathbf{Z}_1)\| < \epsilon,$$

and we get

$$\|f(\mathbf{Z}_1) - F(\mathbf{Z}_1)\| < \left(1 + \frac{2}{1 - \rho_1/\rho_0}\right)\epsilon.$$

Because the quantity on the right can be made arbitrarily small after choosing $\|\mathbf{Z}_1\| < \rho$ by making ϵ small enough, it is shown that

$$f(\mathbf{Z}_1) = F(\mathbf{Z}_1)$$

for all $\mathbf{Z}_1$ such that $\|\mathbf{Z}_1\| < \rho$, thus completing the proof of the theorem. ∎

COROLLARY 2.7b

Let F_n' and F' denote the formal derivatives of the series $F_n (n = 0, 1, 2, \ldots)$ and

F, respectively. Then under the hypotheses of Theorem 2.7a the series $F_n'(\mathbf{Z})$ and $F'(\mathbf{Z})$ converge for $\|\mathbf{Z}\| < \rho$, and by letting

$$f_n'(\mathbf{Z}) := F_n'(\mathbf{Z}), \qquad f'(\mathbf{Z}) := F'(\mathbf{Z})$$

for $\|\mathbf{Z}\| < \rho$ the sequence $\{f_n'(\mathbf{Z})\}$ converges to $f'(\mathbf{Z})$ uniformly on every set $\|\mathbf{Z}\| \leq \rho_1$, where $\rho_1 < \rho$.

Proof. The convergence of the series $F_n'(\mathbf{Z})$ and $F(\mathbf{Z})$ follows from the fact that $\rho(G') = \rho(G)$ for any formal series G. To prove the uniform convergence of the sequence $\{f_n'(\mathbf{Z})\}$ for $\|\mathbf{Z}\| \leq \rho_1$ we use the notation of the preceding proof. From

$$f'(\mathbf{Z}) - f_k'(\mathbf{Z}) = \sum_{m=1}^{\infty} m b_m \mathbf{Z}^{m-1}$$

we have, by using (2.7-6),

$$\sigma_k := \sup_{\|\mathbf{Z}\| \leq \rho_1} \|f'(\mathbf{Z}) - f_k'(\mathbf{Z})\| \leq \sum_{m=1}^{\infty} m |b_m| \rho_1^{m-1}$$

$$\leq \frac{2\epsilon}{\rho_0} \sum_{m=1}^{\infty} m \left(\frac{\rho_1}{\rho_0}\right)^{m-1} = \frac{2\epsilon}{\rho_0} \left(1 - \frac{\rho_1}{\rho_0}\right)^{-2}.$$

Because ϵ could be chosen arbitrarily, this shows that $\sigma_k \to 0$ for $k \to \infty$, thus proving uniform convergence on $\|\mathbf{Z}\| \leq \rho_1$. ∎

In view of Corollary 2.7b, the series F_n' satisfies the hypotheses imposed on F_n in Theorem 2.7a, and the following result is obtained by induction.

COROLLARY 2.7c

Under the hypotheses of Theorem 2.7a, the sequence of the functions $f_n^{(k)}$ defined for $\|\mathbf{Z}\| < \rho$ by

$$f_n^{(k)}(\mathbf{Z}) := F_n^{(k)}(\mathbf{Z}), \qquad k = 1, 2, \ldots,$$

converges for $n \to \infty$ to

$$f^{(k)}(\mathbf{Z}) := F^{(k)}(\mathbf{Z})$$

uniformly on every set $|\mathbf{Z}| \leq \rho_1$, where $\rho_1 < \rho$.

The double series theorem and its corollaries can be applied to show that functions obtained from analytic functions by limit processes are themselves analytic. A perfected form of the theorem, valid for scalar analytic functions, is given in §3.4, in which some applications are exhibited.

SEMINAR ASSIGNMENTS

1. Develop an algorithm for the simultaneous determination of all zeros of a polynomial $p(z) := z^n + a_{n-1} z^{n-1} + \cdots + a_0$ by considering $p(z) = 0$ as the special case $w := -1$ of

$$(*) \qquad \frac{z^n}{a_0 + a_1 z + \cdots + a_{n-1} z^{n-1}} = w$$

and inverting * by means of Corollary 2.4g. Apply the method to find the zeros of the polynomials

$$p_{2n}(z) := z^n C_n^\nu \left(\tfrac{1}{2} \left(z + \frac{1}{z} \right) \right),$$

where the C_n^ν are the Gegenbauer polynomials defined in Problem 5, §1.6. Note that

$$\lim_{\nu \to 0} \frac{1}{\nu} p_{2n}(z) = \frac{1}{n} (z^{2n} + 1).$$

2. Let $P := a_0 + a_1 x + a_2 x^2 + \cdots$, $a_0 \neq 0$, be a formal power series with a positive radius of convergence, let k be an integer, and let α be real. It is obvious from a graph that the equation

$$\tan(x + \alpha) = x^k P\left(\frac{1}{x} \right) \qquad (1)$$

has an infinite number of solutions, which for n large lie near the points $x = (n + \tfrac{1}{2})\pi - \alpha$ if $k > 0$, near $n\pi - \alpha$ if $k < 0$, and near $n\pi + \beta - \alpha$, where $\beta := \text{Arctan } a_0$ if $k = 0$. Construct an algorithm to expand these large solutions in terms of powers of $(n\pi)^{-1}$.

[If $k > 0$, consider

$$\frac{1}{\tan(x + \alpha)} = \cot(x + \alpha) = -\tan\left(x + \alpha - \frac{\pi}{2} \right) = -x^{-k} P^{-1}\left(\frac{1}{x} \right).$$

Thus we may assume $k \leq 0$. Then (1) is equivalent to

$$x + \alpha - n\pi = \text{Arctan}\left[x^k P\left(\frac{1}{x} \right) \right] =: A\left(\frac{1}{x} \right),$$

to

$$x + \alpha - A\left(\frac{1}{x} \right) = n\pi,$$

or to

$$\frac{1}{x} \left\{ 1 + \frac{\alpha}{x} - \frac{1}{x} A\left(\frac{1}{x} \right) \right\}^{-1} = \frac{1}{n\pi}.$$

The last equation may be inverted by the Lagrange-Bürmann formula.]
3. Apply the algorithm constructed above (a) to the equation

$$\tan x = \frac{x(\alpha - \beta)}{\alpha\beta + x^2}$$

to determine the natural frequencies of certain vibrating strings (see Collatz [1944], p. 71); (b) to obtain an asymptotic expansion for the large zeros of the Bessel function J_ν (see Chapter 11 or Watson [1944], p. 505) which can be formally obtained from

$$\cot\left(x - \frac{\nu\pi}{2} - \frac{\pi}{4}\right) = \frac{Q(x,\nu)}{P(x,\nu)},$$

where

$$P(x,\nu) := \sum_{m=0}^{\infty} (-1)^m \frac{(\nu, 2m)}{(-2x)^{2m}},$$

$$Q(x,\nu) := \sum_{m=0}^{\infty} (-1)^m \frac{(\nu, 2m+1)}{(-2x)^{2m+1}},$$

and

$$(\nu, k) := (-1)^k \frac{(\tfrac{1}{2} - \nu)_k (\tfrac{1}{2} + \nu)_k}{k!}.$$

Test the accuracy of the results by consulting Bessel function tables.

NOTES

§2.1. For a complete discussion of the abstract concepts dealt with here see Hille & Phillips [1957]. Norms of vectors and matrices are discussed by Householder [1964].

§2.2. For a more general treatment of abstract analytic functions see Hille & Phillips [1957], Chapter V.

§2.4. "New" proofs of the fundamental theorem of algebra continue to be published at a steady rate; e.g., see Boas [1935, 1964], Ankeny [1947], Redheffer [1957, 1964], Feferman [1967], and Wolfenstein [1967]. Some of them are variants of nonconstructive classical proofs. The best known constructive proofs are due to Weierstrass [1903], Brouwer [1924], Weyl [1924], and Rosenbloom [1945]. See also §§6.10, 6.11, and 6.14.

§2.5. For the generating functions of the Laguerre and Jacobi polynomials (Problems 6 and 7) see Szegö [1959]. For the role of Kepler's equation in celestial mechanics see Brouwer & Clemence [1961] or Stiefel & Scheifele [1971].

§2.6. For a more thorough treatment of analytic functions of matrices, including a treatment of Cauchy's theorem, see Gantmacher [1959], Vol. 1. For various ways to compute the exponential of a matrix see Putzer [1966], Kirchner [1967], and Apostol [1969]. The logarithm of a matrix attracted the attention of von Neumann [1929].

§2.7. The proof of Theorem 2.7a is adapted from Hurwitz & Courant [1929], p. 63.

3
ANALYTIC CONTINUATION

§3.1. REARRANGEMENT OF POWER SERIES; DERIVATIVES

From this point on we shall be concerned predominantly with the case in which the Banach algebra $\mathcal{B}$ is the algebra of complex numbers. Thus the functions considered now, called **scalar complex functions**, are complex-valued and are defined on certain sets D of the complex plane. According to the definition in §2.3, a scalar complex function is called *analytic* at some point z_0 of its domain of definition D if there exist a number $\rho > 0$ and a power series F with radius of convergence $\geq \rho$ such that the neighborhood $N(z_0, \rho)$ belongs to D and

$$f(z_0 + h) = F(h) \qquad (3.1\text{-}1)$$

for all $z_0 + h$ in that neighborhood.

Naturally a scalar complex function may be analytic at several points of its domain of definition; for instance, every polynomial

$$p(z) := a_0 + a_1 z + \cdots + a_n z^n$$

is analytic at every point z_0 of the complex plane, as is seen by replacing z by $z_0 + h$ and rearranging in powers of h. It follows by Theorem 2.3d that a rational function is analytic at every point at which its denominator is different from zero.

Can the points of analyticity of a scalar complex function be prescribed arbitrarily? In light of these examples, they can hardly be expected to be. We recall that a set S of points in the complex plane is called **open** if with a point z it also contains a neighborhood of z.

THEOREM 3.1a

The set of points of analyticity of a scalar analytic function is always open.

We shall prove the following quantitative version of the above result:

THEOREM 3.1b

Let the scalar complex function f be analytic at z_0 and let the representation (3.1-1) hold for $|h|<\rho$. Then f is analytic at every point $z_1 \in N(z_0,\rho)$, and the series in powers of $k := z - z_1$ representing f near z_1 converges to $f(z)$ for all k such that $|k|<\rho-|z_1-z_0|$.

Proof. Let

$$F(h) =: a_0 + a_1 h + a_2 h^2 + \cdots.$$

Putting $z_1 - z_0 =: c$, we have $h = z - z_0 = z - z_1 + (z_1 - z_0) = k + c$ (see Fig. 3.1a), and (3.1-1) appears in the form

$$f(z_0 + h) = a_0 + a_1(c+k) + a_2(c+k)^2 + \cdots$$
$$= a_0 + a_1(c+k) + a_2(c^2 + 2ck + k^2) + \cdots. \quad (3.1\text{-}2)$$

We wish to show that this series may be rearranged as a series in powers of k. A basic theorem on series of complex numbers states that such a series may be rearranged in any manner whatever, provided it is absolutely convergent. Thus we have to show that the series

$$a_0 + a_1 c + a_1 k + a_2 c^2 + 2a_2 ck + a_2 k^2 + \cdots \quad (3.1\text{-}3)$$

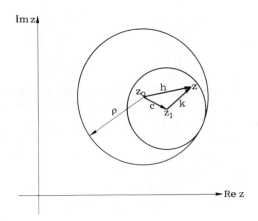

Fig. 3.1a. Power series rearrangement.

is absolutely convergent for $|c|+|k|<\rho$. This follows from the fact that because F has radius of convergence $\geq \rho$ the partial sums of the series

$$|a_0|+|a_1|(|c|+|k|)+|a_2|(|c|+|k|)^2+\cdots$$

are bounded and so are the partial sums of the series obtained by writing each term as a finite sum of positive terms

$$|a_0|+|a_1||c|+|a_1||k|+|a_2||c|^2+2|a_2||c||k|+|a_2||k|^2+\cdots.$$

The latter series is a majorant of the series (3.1-3), which is thus proved absolutely convergent. By writing (3.1-3) as the series of subseries obtained by collecting the terms

$$\left\{ a_n + \binom{n+1}{n} a_{n+1} c + \binom{n+2}{2} a_{n+2} c^2 + \cdots \right\} k^n$$

multiplied by the same power k^n we obtain for $|k|<\rho-|c|$, $c := z_1 - z_0$,

$$f(z_1+k) = b_0 + b_1 k + b_2 k^2 + \cdots, \tag{3.1-4}$$

where

$$b_n := \frac{1}{n!}[(1)_n a_n + (2)_n a_{n+1} c + (3)_n a_{n+2} c^2 + \cdots], \qquad n=0,1,2,\ldots. \tag{3.1-5}$$

This completes the proof of Theorem 3.1b. ∎

Some consequences may be drawn from the proof. By §1.4 the nth formal derivative of the formal power series $F := a_0 + a_1 x + a_2 x^2 + \cdots$ is given by

$$F^{(n)} := (1)_n a_n + (2)_n a_{n+1} x + (3)_n a_{n+2} x^2 + \cdots.$$

Thus we see that b_n equals precisely $1/n!$ times the nth formal derivative of F, evaluated at $x = c$. Because the series defining b_n has already been shown to be convergent, it follows that the radii of convergence of the series $F^{(n)}$, $n = 1, 2, \ldots$, at least equal $\rho(F)$. They cannot be larger, however, for the coefficients are at least equal in modulus to the coefficients of the series

$$x^{-n}(F - a_0 - a_1 x - \cdots - a_{n-1} x^{n-1})$$
$$= a_n + a_{n+1} x + \cdots,$$

which again has radius of convergence $\rho(F)$.

COROLLARY 3.1c

The formal derivatives $F', F'', \ldots$, of a formal power series F have the same radius of convergence as F.

This could also be verified directly by using the Cauchy-Hadamard formula; see Problem 3, §2.2.

Let f and F be related by (3.1-1). Because the series $F', F'', \ldots$, have a radius of convergence $\geq \rho$, they can be used to define functions that are themselves analytic at $z = z_0$. The function defined by $F^{(n)}$ is called the nth **derivative** of f and is usually denoted by $f^{(n)}$. (Derivatives of low order n are customarily denoted by $f', f'', \ldots$.) Thus

$$f^{(n)}(z_0 + h) := F^{(n)}(h), \qquad |h| < \rho. \tag{3.1-6}$$

By Theorem 3.1b the derivatives $f', f'', \ldots$, are themselves analytic at every point of the disk $|z - z_0| < \rho$.

COROLLARY 3.1d

Formula (3.1-4), *which expresses the rearrangement of $F(h)$ in powers of $k := z - z_1$ may be written*

$$f(z_1 + k) = f(z_1) + \frac{1}{1!} f'(z_1) k + \frac{1}{2!} f''(z_1) k^2 + \cdots. \tag{3.1-7}$$

The above series, which converges (at least) for all k such that $|k| < \rho - |c|$, where $c := z_1 - z_0$, is called the **Taylor expansion** of f at the point z_1. The expansion holds for all $z_1 \in N(z_0, \rho)$. Setting $z_1 := z_0$ there follows

$$a_n = \frac{1}{n!} f^{(n)}(z_0), \qquad n = 0, 1, 2 \ldots \tag{3.1-8}$$

where $F := a_0 + a_1 x + a_2 x^2 + \cdots$ is the series representing f at z_0.

We now derive a result that gives meaning to the first derivative f' (and thus by induction to all derivatives) that does not depend on the power series concept. From (3.1-7) we have for $0 < |k| < \rho - |c|$

$$\frac{f(z_1 + k) - f(z_1)}{k} = f'(z_1) + \frac{1}{2!} f''(z_1) k + \cdots.$$

The expression on the right is a power series that converges for $|k| < \rho - |c|$. It represents a function that is continuous at $k = 0$. The expression on the left, which is undefined for $k = 0$, has a limit as $k \to 0$, and the value of the

REARRANGEMENT OF POWER SERIES

limit equals the value of the function on the right for $k=0$. We thus have shown that

$$\lim_{k \to 0} \frac{f(z_1+k)-f(z_1)}{k} = f'(z_1) \qquad (3.1\text{-}9)$$

for every z_1 such that $|z_1 - z_0| < \rho$.

A scalar complex function of a complex variable defined in a neighborhood of $z = z_1$ and having the property that the limit (3.1-9) exists is called **differentiable** at the point z_1.

THEOREM 3.1e

If a scalar complex function is analytic at a point, it is differentiable in a neighborhood of that point.

With this theorem we have reached a pivotal point in our development of the theory of analytical functions and the reader is entitled to some perspective remarks.

The idea of basing the theory of analytic functions on power series was first conceived and fully carried out by Karl Weierstrass (1815–1897). This, with some additional stress on the formal theory of power series and on analytic functions defined on a Banach algebra, is the approach that we have adopted in the early chapters of this book. An alternate possibility consists in basing the theory on the concept of differentiability. This program, which de-emphasizes the computational aspects and leads to a more geometrically oriented theory, was first clearly stated by Bernhard Riemann (1825–1865). In the Riemannian development a function is called **holomorphic** at a point if it is differentiable in a neighborhood of that point. It is then shown (by the theory of complex integration) that a function that is holomorphic at a point is also analytic at that point. This, together with Theorem 3.1e, shows that for scalar functions the two properties of being holomorphic and analytic are equivalent. A sketch of the Riemannian theory is given in §5.5.

The reader may ask why at this point we have abandoned the study of analytic functions defined and with values in a general Banach algebra $\mathcal{B}$. The reason is easy to find if we try to stimulate what we have done in this section. For $\mathbf{Z} \in \mathcal{B}, \|\mathbf{Z}\| < \rho$, let

$$f(\mathbf{Z}) := F(\mathbf{Z}) := a_0 \mathbf{I} + a_1 \mathbf{Z} + a_2 \mathbf{Z}^2 + \cdots.$$

If $\|\mathbf{Z}_1\| < \rho$, $\mathbf{K} := \mathbf{Z} - \mathbf{Z}_1$, $\|\mathbf{K}\| < \rho - \|\mathbf{Z}_1\|$, we have

$$f(\mathbf{Z}_1 + \mathbf{K}) = a_0 \mathbf{I} + a_1(\mathbf{Z}_1 + \mathbf{K}) + a_2(\mathbf{Z}_1 + \mathbf{K})^2 + \cdots.$$

Now, because $\mathfrak{B}$ is not assumed to be commutative, $(\mathbf{Z}_1+\mathbf{K})^2 \neq \mathbf{Z}_1^2 + 2\mathbf{Z}_1\mathbf{K}+\mathbf{K}^2$ in general. All we can say is $(\mathbf{Z}_1+\mathbf{K})^2 = \mathbf{Z}_1^2 + \mathbf{Z}_1\mathbf{K}+\mathbf{K}\mathbf{Z}_1+\mathbf{K}^2$. The series may still be rearranged, for addition is commutative. Thus by collecting terms of equal degree in $\mathbf{K}$ we obtain

$$f(\mathbf{Z}_1+\mathbf{K}) = a_0\mathbf{I} + a_1\mathbf{Z}_1 + a_2\mathbf{Z}_1^2 + \cdots$$

$$+ a_1\mathbf{K} + a_2(\mathbf{Z}_1\mathbf{K}+\mathbf{K}\mathbf{Z}_1) + \cdots$$

$$+ a_2\mathbf{K}^2 + a_3(\mathbf{Z}_1\mathbf{K}^2+\mathbf{K}\mathbf{Z}_1\mathbf{K}+\mathbf{K}^2\mathbf{Z}_1) + \cdots$$

$$+ \cdots .$$

This is no longer a power series in $\mathbf{K}$ with complex coefficients; hence f is no longer analytic (according to §2.3) at point $\mathbf{Z}_1$. Even if $\mathfrak{B}$ is commutative and the rearrangement of factors in products is possible, the coefficients in the resulting power series in $\mathbf{K}$ would be elements of $\mathfrak{B}$ that are not necessarily scalar multiples of $\mathbf{I}$.

PROBLEMS

1. Prove the usual rules for the differentiation of sums, products, and quotients directly from the limit definition (3.1-9) without using power series.
2. Prove the Leibniz rule for the nth derivative of a product

$$(fg)^{(n)} = \binom{n}{0}f^{(n)}g + \binom{n}{1}f^{(n-1)}g' + \cdots + \binom{n}{n}fg^{(n)},$$

using power series.

§3.2. ANALYTIC EXTENSION AND CONTINUATION

A function f is called **analytic on a set** S if it is analytic at every point $z \in S$. The very definition of analyticity at a point requires f to be defined also in a neighborhood of z. Furthermore, if a scalar function is analytic at z, then by Theorem 3.1a it is in fact analytic in some such neighborhood. Thus a scalar function analytic on a set S is automatically analytic in an open set containing S, namely the union of all such neighborhoods $N(z)$ where z runs through S. In the study of scalar functions analytic on sets we may restrict our attention to *open sets*.

In this section we shall study some aspects of the following problem: let S be an open set and let Q be a set (not necessarily open) contained in S. Let the function f be defined on Q. Does there exist a function g, analytic on the larger set S, such that $g(z)=f(z)$ for $z \in Q$? If such a function exists, it is called an **analytic extension** of f from Q to S.

ANALYTIC EXTENSION AND CONTINUATION

Not much of significance can be said if the question is formulated in the above generality. If Q contains only a finite number of points, the Lagrangian interpolating polynomial furnishes an analytic extension of f to the whole plane. This extension is not unique; by adding any analytic function that vanishes at the points of Q without vanishing identically we obtain another analytic extension of f. Interesting results can be expected only if Q has an infinite number of points. Below it is assumed that Q has at least one point of accumulation q and that q belongs to S.

At the other extreme there is the possibility that Q itself is an open set. In this case no analytic extension of f from Q to S exists unless f is analytic on Q, but even if this condition is met the analytic extension is still not necessarily uniquely determined, as the following example shows. Let Q be the set $|z|<1$ and let S denote the union of Q with the set $|z-3|<1$. Then f can be extended from Q to S by defining $f(z)=h(z)$ for $|z-3|<1$, where h is an arbitrary analytic function on $|z-3|<1$. The analytic extension of f is thus to a large degree arbitrary.

More interesting statements are possible if the set S, into which the extension is assumed to be made, is assumed to be *connected*. Intuitively, a set is connected if it consists of a single piece. A strict definition of connectedness will be required only for open sets. The following definition turns out to be most easy to apply: an open set is called **connected** if it cannot be represented as the union of two open sets that are disjoint and nonempty. It can easily be shown (cf. Ahlfors [1966], p. 56) that a nonempty open set is connected if and only if any two of its points can be joined by a polygon that lies in the set.

For brevity, a nonempty, open, connected set is called a **region**. *Examples* of regions are (1) the whole plane, (2) the set of all points satisfying $\operatorname{Re} z > 0$ (the **right half-plane**), (3) the set of all z such that $|\arg z|<\pi$ (referred to as the **plane cut along the negative real axis**), (4) the set of all z such that $0<|z|<1$ (the **punctured unit disk**), and (5) the set of all z such that $z \neq n$ for all nonpositive integers n. A region remains a region if a finite number of its points is removed.

Here it is convenient to introduce one more concept. A **component** of an open set S is a subregion of S (i.e., a region contained in S) that is not contained in any larger subregion of S. The following standard result will be required in §§3.5 and 4.6:

THEOREM 3.2a

Every open set S has a unique decomposition into components.

Proof. First we define a decomposition of S into components and then show that it is unique. Let $c \in S$. We define $C = C(c)$ as the union of all subregions of S that contain c. We wish to show that C is a component of S. Being a union of open sets, C is open. It is implied by the definition of C that C is not contained in any larger subregion of S. Thus it remains to be shown that C itself is connected. Let A, B be two open sets such that $C = A \cup B, A \cap B = \emptyset$. We may assume that $c \in A$ and then have to show that $B = \emptyset$. Indeed, if a point $b \in B$ did exist then, since $b \in C$ by the definition of C, $C_0 \subset C$ containing both b and c would exist. We would then have $C_0 = A_1 \cup B_1$ with $A_1 := A \cap C_0$ and $B_1 := B \cap C_0$ both open and nonempty and $A_1 \cap B_1 = \emptyset$ which contradicts the connectedness of C_0. Thus $B = \emptyset$, and it follows that C is connected; therefore C is a component of S. We have associated with every $c \in S$ a component to which it belongs. We now show that the decomposition of S into components, which has thus been defined, is unique. Let any decomposition of S into components be given, let c be an arbitrary point of S, and let D be the component to which c belongs. Then by definition the set $C(c)$ contains D. On the other hand, D is not contained in any larger subregion of S. It follows that $C \subset D$, hence that $C = D$. The component in which c lies can only be the set $C(c)$, proving the uniqueness of the decomposition into components. ∎

EXAMPLE 6

The set $S := \{z : |z| \neq 1\}$ consists of the two components $\{z : |z| < 1\}$ and $\{z : |z| > 1\}$.

We now state a basic result on analytic continuation.

THEOREM 3.2b

Let R be a region, let Q be a set contained in R with a point of accumulation in R, and let f be an analytic function on Q. There then exists at most one analytic extension of f from Q to R.

Proof. Let g_1 and g_2 be two analytic extensions of f to R. In a neighborhood of each point z_0 of R each of the functions g_1 and g_2 is represented by a power series in $z - z_0$. Let $U \subset R$ denote the set of those $z_0 \in R$ in which these two power series are formally identical and let $V \subset R$ denote the set of those z_0 in which the two power series are formally different. Obviously, $U \cap V = \emptyset$ (the empty set) and $U \cup V = R$. We shall show that both U and V are open. Let z_0 be a point of U. Then by Theorem 3.1a the power series at all points near z_0 can be obtained by rearranging the identical series at z_0 and thus are again identical. This shows that U is open. In order to show that V is open let $z_1 \in V$, and let the nth coefficients of the series in $z - z_1$ representing g_1 and g_2 near z_1 be different. By (3.1-8) this means that

$$g_1^{(n)}(z_1) \neq g_2^{(n)}(z_1).$$

ANALYTIC EXTENSION AND CONTINUATION

Since the functions $g_1^{(n)}$ and $g_2^{(n)}$ are themselves analytic, hence continuous, the above implies $g_1^{(n)}(z) \neq g_2^{(n)}(z)$ also in some neighborhood $N(z_1)$. By (3.1-7) this shows that for $z_2 \in N(z_1)$ the series in powers of $z - z_2$ representing g_1 and g_2 cannot be identical and that V is open. Because R is connected, either U or V must be empty. We shall show that U is not empty. Let $q \in R$ be a point of accumulation of Q. Since $g_1 = g_2$ on Q, the series in powers of $z - q$ representing g_1 and g_2 near q must be identical by Theorem 2.3b. Thus $q \in U$. It follows that V is empty, implying that $U = R$. The series representing g_1 and g_2 are the same throughout R. In particular, their zeroth coefficients are the same, which means that

$$g_1(a_0) = g_2(z_0) \quad \text{for all } z_0 \in R. \quad \blacksquare$$

EXAMPLE 7

Let f for $|z| < 1$ be defined by

$$f(z) := 1 + z + z^2 + \cdots.$$

This function is extended analytically into the region of all $z \neq 1$ by the function

$$g(z) := \frac{1}{1-z}$$

because $g(z) = f(z)$ on the set $|z| < 1$.

As a first application of Theorem 3.2b we prove the following:

THEOREM 3.2c

(i) *Let the function g be analytic on the region $|z - z_0| < \rho$, and* (ii) *let the power series $P(z - z_0)$ representing g near z_0 have radius of convergence $\geq \rho$. Then $g(z) = P(z - z_0)$ throughout the region $|z - z_0| < \rho$.*

The result is not trivially implied by the hypotheses, for by the definition of analyticity these require only that $P(z - z_0) = g(z)$ for $|z - z_0| < \epsilon$, where $\epsilon > 0$ may be arbitrarily small, and not for $|z - z_0| < \rho$. In §3.3 we prove that hypothesis (i) actually implies hypothesis (ii).

Proof of Theorem 3.2c. For $|z - z_0| < \rho$, let $p(z) := P(z - z_0)$. By hypothesis (ii) $p(z) = g(z)$ for $z \in N(z_0, \epsilon)$ for some $\epsilon > 0$. We apply Theorem 3.2b where $Q := N(z_0, \epsilon)$, $R := N(z_0, \rho)$. Both functions p and g are analytic extensions of the restriction of g to Q from Q to R. Since R is connected, $p = g$ throughout R. $\blacksquare$

EXAMPLES

8. By (2.5-20) there holds for sufficiently small values of $|z|$

$$\text{Log}(1+z) = z - \tfrac{1}{2}z^2 + \tfrac{1}{3}z^3 - \cdots. \tag{3.2-1}$$

The function $g(z):=\mathrm{Log}(1+z)$ is analytic for $|z|<1$, and the radius of convergence of the above power series is 1. Thus the representation (3.2-1) actually holds for all z such that $|z|<1$. (By Abel's theorem it also holds for $z=1$; see Example 1 of §2.2.)

9. If b is any complex number, $b\neq 0,1,2,\ldots$, then by (2.5-22) the representation

$$(1+z)^b = \sum_{n=0}^{\infty} \frac{(-b)_n}{n!}(-z)^n \tag{3.2-2}$$

for the principal value of $(1+z)^b$ holds for $|z|$ sufficiently small. Since $g(z):=(1+z)^b$ is analytic for $|z|<1$ and the power series has radius of convergence 1 the representation (3.2-2) actually holds for all z such that $|z|<1$.

If R is a region and Q is a subregion of R, an analytic extension of a function f from Q to R is also called an **analytic continuation** of f. We shall now discuss some simple devices that sometimes permit us to find analytic extensions or continuations.

1. Analytic Continuation by Removing Singularities. It may happen that a function is analytic in a region R with the exception, say, of one point $z_0 \in R$, where f is undefined. Such a point is called an **isolated singularity** of the function f. It may be possible to extend f analytically to R by defining $f(z_0)$ in a suitable manner. If this possibility exists, the point z_0 is called a **removable singularity** of f.

EXAMPLE **10**

Let Q denote the set of all points z such that $z\neq 1$, $z\neq 2$ and define f on Q by

$$f(z) := \frac{z-1}{z^2-3z+2}.$$

This function is analytic in Q, for the denominator does not vanish in Q. Now, since $z^2-3z+2=(z-1)(z-2)$, we also have

$$f(z) = \frac{z-1}{(z-1)(z-2)} = \frac{1}{z-2}, \quad z\in Q.$$

Let R denote the set of all points $z\neq 2$. If on R we define g by $g(z):=(z-2)^{-1}$, then g is analytic on R. Furthermore, $g(z)=f(z)$ for $z\in Q$; hence g represents *the* analytic continuation of f from Q to R. The same example also shows that not every singularity on an analytic function may be removed, for no matter how we attempt to define $g(2)$ the function g will always be unbounded near $z=2$. Were it analytic, it would have to be continuous, hence bounded near $z=2$. In §4.4 we prove the important result, due to Riemann, that an analytic function bounded in a neighborhood of an isolated singularity can always be continued analytically into that singularity.

ANALYTIC EXTENSION AND CONTINUATION

2. Analytic Continuation by Power Series. A simple yet important case of analytic extension or continuation is the following: let Q have a point of accumulation q, and let the function f for $z \in Q$ be represented by a power series in $z - q$. Let this series have radius of convergence ρ, and let the set $R: |z - q| < \rho$ contain points that are not in Q. The power series then represents the analytic extension of f from Q to R, for it defines a function g that is analytic in R and agrees with f on the set Q which has a point of accumulation $q \in R$.

EXAMPLE 11

In real analysis the function e^x can be defined by its Taylor series at $x = 0$. The same series was used in §2.5 to define the exponential function for all complex values of the argument. We now see that the function e^z thus defined is the *only possible* analytic extension of e^x from the real line to the complex plane. Theorem 3.2b also shows that the functions $\cos z$ and $\sin z$, defined in §2.5, are the only possible analytic extensions of $\cos x$ and $\sin x$. (It was not clear a priori that such extensions existed.)

3. Analytic Continuation by Patching. Let R and S be two regions, neither a subregion of the other, whose intersection $R \cap S$ is not empty. Let f be analytic on R and let g be analytic on S. Let it be known that $f(z) = g(z)$ for $z \in R \cap S$. The function h defined by

$$h(z) := f(z), \quad z \in R; \quad h(z) := g(z), \quad z \in S$$

is clearly analytic on the union $R \cup S$ of R and S (which is again a region). Thus h represents both the analytic continuation of f into S, and of g into R. We describe this situation by saying that *f and g are analytic continuations of each other*.

We wish to establish weaker conditions under which we can assert that two functions are analytic continuations of each other. The following example shows that it is not enough to assume that $f(z) = g(z)$ on a set $Q \subset R \cap S$ with a point of accumulation in $R \cap S$.

EXAMPLE 12

Let $R: |\arg z - \pi/2| < \pi$, $S: |\arg z + \pi/2| < \pi$. On R let

$$f(z) := \log z := \text{Log}|z| + i \arg z, \quad \left(-\frac{\pi}{2} < \arg z < \frac{3\pi}{2}\right),$$

and on S let

$$g(z) := \log z := \text{Log}|z| + i \arg z, \quad \left(-\frac{3\pi}{2} < \arg z < \frac{\pi}{2}\right).$$

These functions are analytic on R and S, respectively, and agree on the set $\operatorname{Re} z > 0$, which is contained in $R \cap S$ and certainly has points of accumulation there. Yet these functions are not analytic continuations of each other because they do not even have the same values on the whole set $R \cap S$; for instance, at every $z = x$, where $x < 0$,

$$f(z) = \operatorname{Log}|x| + i\pi \neq g(z) = \operatorname{Log}|x| - i\pi.$$

However, the following theorem holds:

THEOREM 3.2d (Fundamental lemma on analytic continuation)

Let Q be a set with point of accumulation q and let R and S be two regions such that their intersection contains Q and q and is connected. If f is analytic on R, g is analytic on S, and $f(z) = g(z)$ for $z \in Q$, then $f(z) = g(z)$ throughout $R \cap S$ and f and g are analytic continuations of each other.

Proof. Let $A := R \cap S$. As an intersection of open sets, A is open. By hypothesis, A is connected; hence A is a region. The functions f and g both represent analytic continuations of the same function from Q to A. Therefore by Theorem 3.2b $f(z) = g(z)$ for all $z \in A$. Thus there is no ambiguity in defining

$$h(z) := \begin{cases} f(z), & z \in R, \\ g(z), & z \in S, \end{cases}$$

and the conclusion follows. ∎

EXAMPLE 13

Let a, b, and c be complex numbers, $c \neq 0, -1, -2, \ldots$. We recall the formal result (1.6-11):

$$(1-z)^a {}_2F_1(a, b; c; z) = {}_2F_1\left(a, c-b; c; -\frac{z}{1-z}\right),$$

which was proved in §1.6 as an identity between formal power series. It follows from Theorem 2.4a, after verifying the hypotheses, that the identity holds in the sense of equaltiy of complex numbers for every z such that $|z| < \frac{1}{2}$. The function on the left is analytic for $|z| < 1$ and the function on the right is analytic for all z such that $|z/(1-z)| < 1$, i.e., for $\operatorname{Re} z < \frac{1}{2}$. The intersection of the two regions $|z| < 1$ and $\operatorname{Re} z < \frac{1}{2}$ is connected; hence the two functions are analytic continuations of each other.

4. Analytic Continuation by Rearranging Power Series. We return to Theorem 3.1a and assume that a function f is defined for $|z - z_0| < \rho$ by a power series

$$F(h) := a_0 + a_1 h + a_2 h^2 + \cdots,$$

ANALYTIC EXTENSION AND CONTINUATION

where $h := z - z_0$, and that it is not defined anywhere else. Theorem 3.1b states that if $|z_1 - z_0| < \rho$ the rearrangement of the series $F(h)$ in powers of $k := z - z_1$ converges at least for $|k| < \rho - |z_1 - z_0|$ and its sum is $f(z)$. The theorem does not assert, however, that $\rho - |z_1 - z_0|$ is the radius of convergence of the rearranged series. The radius of convergence can be larger, as shown by the following example.

EXAMPLE 14

Let $z_0 = 0$ and for $|z| < 2$ let f be defined by

$$f(z) := \tfrac{1}{2} + \tfrac{1}{4}z + \tfrac{1}{8}z^2 + \cdots . \tag{3.2-3}$$

We have for $|z| < 2$ and $n = 0, 1, 2, \ldots$,

$$f^{(n)}(z) = \frac{1}{2^{n+1}} \sum_{k=0}^{\infty} (k+1)_n \left(\frac{z}{2}\right)^k .$$

With

$$(k+1)_n = \frac{(1)_{n+k}}{k!} = \frac{n!(n+1)_k}{k!},$$

and using the binomial series (3.2-2) this becomes

$$f^{(n)}(z) = \frac{n!}{2^{n+1}} \sum_{k=0}^{\infty} \frac{(n+1)_k}{k!} \left(\frac{z}{2}\right)^k = \frac{n!}{2^{n+1}} \left(1 - \frac{z}{2}\right)^{-n-1}$$

$$= \frac{n!}{(2-z)^{n+1}}, \quad |z| < 2.$$

We now rearrange (3.2-3) in powers of $z - i$. Because

$$\frac{1}{n!} f^{(n)}(i) = \frac{1}{(2-i)^{n+1}},$$

the rearrangement is

$$\frac{1}{2-i} + \frac{1}{(2-i)^2}(z-i) + \frac{1}{(2-i)^3}(z-i)^2 + \cdots . \tag{3.2-4}$$

The rearranged series has radius of convergence $|2 - i| = \sqrt{5}$ and thus defines an analytic function g on the set $|z - i| < \sqrt{5}$. This set is not included in the set $|z| < 2$ (see Fig. 3.2).

We know from Theorem 3.1b that $g(z) = f(z)$ for $|z - i| < 1$. Because the intersection of the sets $|z - i| < \sqrt{5}$ and $|z| < 2$ is connected, it follows that $g(z) = f(z)$ throughout that intersection and that g and f are analytic continuations of each other.

This argument is completely general.

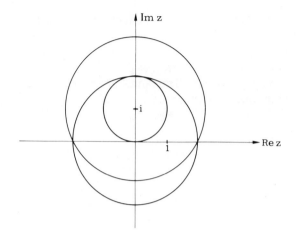

Fig. 3.2a. Analytic continuation.

THEOREM 3.2e

Let the function f for $|z - z_0| < \rho$ *be defined by*

$$f(z) := a_0 + a_1(z - z_0) + a_2(z - z_0)^2 + \cdots. \qquad (3.2\text{-}5)$$

Let $|z_1 - z_0| < \rho$ *and let the rearrangement*

$$g(z) := b_0 + b_1(z - z_1) + b_2(z - z_1)^2 + \cdots \qquad (3.2\text{-}6)$$

of the series (3.2-5) *about the point* z_1 *have radius of convergence* $\rho_1 > \rho - |z_1 - z_0|$. *The functions f and g are then analytic continuations of each other.*

Proof. We apply Theorem 3.2d where $Q := N(z_1, \rho - |z_1 - z_0|)$, $R := N(z_0, \rho)$, $S := N(z_1, \rho_1)$. The set $R \cap S$ is connected and contains the open set Q, hence points of accumulation of Q. According to Theorem 3.2b, $f(z) = g(z)$ for $z \in Q$; but f is analytic on R, g is analytic on S, hence the conclusion of Theorem 3.2d is applicable. ∎

The rearrangement of power series is an important theoretical tool in the theory of analytic continuation because, as we shall see, every analytic continuation can, in principle, be obtained in this manner. It is shown in §3.6 how the method has to be modified to be numerically useful.

5. Analytic Continuation by Exploiting Functional Relationships. Occasionally an analytic continuation of a function f can be obtained by making use of a special functional relationship satisfied by f. Naturally this method is restricted to those functions for which such relationships are known. In this sense, Example 13 exploits a special relation satisfied by the hypergeometric series. Another example is the following:

ANALYTIC EXTENSION AND CONTINUATION

EXAMPLE 15

Let the function g possess the following properties:
(a) g is analytic in the right half-plane $R: \text{Re}\, z > 0$,
(b) For all $z \in R$, $zg(z) = g(z+1)$.
(A function with these properties is constructed in Chapter 8.) We assert that g can be continued analytically into the whole complex plane with the exception of the points $z = 0, -1, -2, \ldots$. We first continue g into $S: \text{Re}\, z > -1$, $z \neq 0$. For $z \in S$, let f be defined by

$$f(z) := \frac{1}{z} g(z+1). \qquad (3.2\text{-}7)$$

For $z \in S$, $\text{Re}(z+1) > 0$. Hence by virtue of (a) f is analytic on S. In view of (b) f agrees with g on the set of R. Since S is a region, f represents the analytic continuation of g from R to S. We note that f satisfies the functional relation $f(z+1) = zf(z)$ on the whole set S.

Denoting the extended function again by g, we may use the same method to continue g analytically into the set $\text{Re}\, z > -2$, $z \neq 0, -1$, and thus step by step into the region $z \neq 0, -1, -2, \ldots$.

Further methods of analytic continuation, such as the *principle of continuous continuation* and the *symmetry principle*, require the complex integral calculus for their justification; see §5.11.

After this discussion of several methods of analytic continuation the question may be raised whether *every* function originally defined by a power series of finite radius of convergence can be continued beyond the disk of convergence of the power series. The following example, due to Weierstrass, shows that this is not the case.

EXAMPLE 16

Let $Q := N(0, 1)$ and let f be defined by

$$f(z) := z + z^2 + z^6 + \cdots + z^{n!} + \cdots, \quad z \in Q.$$

The function f is analytic on Q, the power series having radius of convergence 1. Let T be a region containing Q as a proper subregion. T contains points z_1 such that $|z_1| = 1$, for otherwise T could be covered by the two open sets $|z| < 1$ and $|z| > 1$, contradicting the implied hypothesis that T is connected. If an analytic continuation g of f into T exists, then, in particular, g is analytic, hence continuous, at z_1 and bounded in a neighborhood $N(z_1)$. In the intersection $D := Q \cap N(z_1)$ g must agree with f. The impossibility of analytic continuation is now demonstrated by showing that the values of f are unbounded in D. Let p/q be a rational number such that the ray

$$\arg z = \frac{p}{q} 2\pi =: \theta$$

intersects D. If $z = re^{i\theta}$, then for $n \geq q$, since q divides $n!$,

$$z^{n!} = r^{n!}(e^{2\pi ip/q})^{n!} = r^{n!}$$

and

$$f(re^{i\theta}) = c + r^{q!} + r^{(q+1)!} + \cdots,$$

where $c := z + z^2 + z^6 + \cdots + z^{(q-1)!}$, $|c| \leq q - 1$. Thus

$$\lim_{r \to 1-} |f(re^{i\theta})| = \infty,$$

showing that f cannot be bounded in D.

PROBLEMS

1. Let S be a connected open set and let n be a continuous real function defined on S, assuming integer values only. Show that n is a constant. [Let c be a point of S and define $A := \{z \in S : n(z) = n(c)\}$, $B := \{z \in S : n(z) \neq n(c)\}$. Then $S = A \cup B$, $A \cap B = \emptyset$. Since A and B are both open and $A \neq \emptyset$, B must be empty.]

2. The function f is analytic in the complex plane cut along the negative real axis. It satisfies $[f(z)]^4 = z^3$. Its value for $z = i$ lies in the fourth quadrant. What is its value for $z = -i$?

3. A power series P converges in a disk C. Assume that two rearrangements of P converge in disks C_1 and C_2, both containing points not in C, and let D, the set of points in both C_1 and C_2 but not in C, not be empty. Show that the continuations into D obtained by the two rearranged series are identical.

4. Let f be analytic in $\operatorname{Re} z > 0$:

$$f(z) := \sum_{n=0}^{\infty} a_n(z-1)^n, \quad |z-1| < 1.$$

The value of f at an arbitrary point z such that $\operatorname{Re} z > 0$ can be calculated by

$$f(z) = a_0 + \sum_{p=1}^{\infty} \left(\frac{1-z}{1+z}\right)^p \Delta^{p-1} b_1,$$

where Δ denotes the forward difference operator $[\Delta b_k := b_{k+1} - b_k$, $\Delta^2 b_k := \Delta(\Delta b_k)$, etc.] and $b_k := (-2)^k a_k$, $k = 1, 2, \ldots$.
[The function $g(w) := f(1+w)/(1-w)$ is analytic in $|w| < 1$. Express its Taylor coefficients in terms of those of f at $z = 1$.]

5. Verify the formula in the preceding problem for $f(z) := 1/z$.

6. Use the formula in Problem 4 to compute $\operatorname{Log} 3$ from the series

$$\operatorname{Log} z = z - 1 - \tfrac{1}{2}(z-1)^2 + \tfrac{1}{3}(z-1)^3 - \cdots.$$

THE RADIUS OF CONVERGENCE

7. If f is analytic in the cut plane $S: |\arg z| < \pi$ and

$$f(z) = \sum_{n=0}^{\infty} a_n (z-1)^n, \quad |z-1| < 1,$$

its values at arbitrary points $z \in S$ can be calculated by

$$f(z) = a_0 + \sum_{p=1}^{\infty} c_p \left(\frac{\sqrt{z}-1}{\sqrt{z}+1} \right)^p,$$

where

$$c_p := \sum_{k=0}^{p-1} \frac{(2p-2k)_k}{k!} 4^{p-k} a_{p-k}.$$

8. Let a be any complex number. Show that the functions defined by the series $1 + az + a^2 z^2 + \cdots$ and

$$\frac{1}{1-z} - \frac{(1-a)z}{(1-z)^2} + \frac{(1-a)^2 z^2}{(1-z)^3} - \cdots$$

are analytic continuations of each other.

§3.3. NEW DETERMINATION OF THE RADIUS OF CONVERGENCE

The radius of convergence of a power series was determined in §2.2 by a formula (Cauchy-Hadamard) that depended on the coefficients of the power series. In this section we show that the radius of convergence can also be determined from properties of the analytic function represented by the series. The main result is as follows.

THEOREM 3.3a

Let the function f be analytic at $z = z_0$. The radius of convergence of the power series representing f near z_0 then equals the radius of the largest disk centered at z_0, inside of which f or an analytic continuation of f is analytic.

Proof. Let

$$P(h) = a_0 + a_1 h + a_2 h^2 + \cdots \quad (h := z - z_0)$$

be the power series representing f near z_0 and let ρ_1 be its radius of convergence. Let σ denote the supremum of all numbers $\sigma_0 > 0$ such that f

or an analytic continuation of f (which we again denote by f) is analytic on the set $|z - z_0| < \sigma_0$. It is to be shown that $\rho_1 = \sigma$. The following proof applies if $\sigma < \infty$; if $\sigma = \infty$, it has to be modified only in trivial details.

If we had $\rho_1 > \sigma$, the power series itself would continue f into a disk of radius exceeding σ, contradicting the definition of σ. It follows that $\rho_1 \leq \sigma$. It is shown that assuming $\rho_1 < \sigma$ likewise leads to a contradiction.

Thus let $\rho_1 < \sigma$. Then f is analytic at every point of the circle $\Gamma_1 : |z - z_0| = \rho_1$. For $z_1 \in \Gamma_1$ let $\rho(z_1)$ denote the radius of convergence of the power series representing f near z_1. The function $\rho(z_1)$ is continuous at every point $z_1 \in \Gamma_1$, for if $z_1' \in \Gamma_1$, $|z_1' - z_1| < \rho(z_1)$ the power series at z_1' is a rearrangement of the power series at z_1; hence by Theorem 3.1b $\rho(z_1') \geq \rho(z_1) - |z_1' - z_1|$. Interchanging the roles of z_1 and z_1', we also have $\rho(z_1) \geq \rho(z_1') - |z_1' - z_1|$ and there follows

$$\rho(z_1) - |z_1' - z_1| \leq \rho(z_1') \leq \rho(z_1) + |z_1' - z_1|,$$

establishing continuity. Since $\rho(z_1)$ is positive on the closed bounded set Γ_1, we conclude that

$$\mu := \min_{z_1 \in \Gamma_1} \rho(z_1) > 0.$$

Let $\gamma > 0$ be chosen such that $\rho_1 + \gamma < \sigma$, $\rho_1 + \gamma < \rho_1 + \mu$. Then the set $|z - z_0| \leq \rho_1 + \gamma$ is a closed bounded subset of a set in which f is analytic; hence the maximum of $|f|$ on that subset,

$$\alpha := \sup_{|z - z_0| \leq \rho_1 + \gamma} |f(z)|,$$

is finite.

Let $\Gamma_2 := |z - z_0| = \rho_2$, where $\rho_1 - (\mu - \gamma) < \rho_2 < \rho_1$. Every $z_2 \in \Gamma_2$ is contained in a disk of radius $\rho(z_1)$ about some $z_1 \in \Gamma_1$; hence the power series representing f near every such z_2 is the rearrangement of the series representing f near a suitable $z_1 \in \Gamma_1$ and consequently by Theorem 3.1b its radius of convergence is $> \mu - (\mu - \gamma) = \gamma$. Within its circle of convergence the series represents f; thus by Cauchy's estimate its coefficients satisfy

$$\left| \frac{f^{(n)}(z_2)}{n!} \right| \leq \frac{\alpha}{\gamma^n}, \quad n = 0, 1, 2, \ldots, z_2 \in \Gamma_2. \tag{3.3-1}$$

On the other hand, for every $z_2 \in \Gamma_2$ we also have, again by Theorem 3.1b,

$$\frac{f^{(n)}(z_2)}{n!} = \frac{(1)_n}{n!} a_n + \frac{(2)_n}{n!} a_{n+1} h + \cdots, \tag{3.3-2}$$

THE RADIUS OF CONVERGENCE

where $h := z_2 - z_0$. Applying Cauchy's estimate to (3.3-2) and using (3.3-1), we have

$$\frac{(1+k)_n}{n!} |a_{n+k}| \leq \frac{\alpha}{\gamma^n \rho_2^k}, \quad n, k = 0, 1, 2, \ldots,$$

for every ρ_2 such that $\rho_1 - (\mu - \gamma) < \rho_2 < \rho_1$; hence by letting $\rho_2 \to \rho_1$

$$\frac{(1+k)_n}{n!} |a_{n+k}| \leq \frac{\alpha}{\gamma^n \rho_1^k}, \quad n, k = 0, 1, 2, \ldots.$$

Letting $n + k =: p$, we find that this is the same as

$$\binom{p}{n} \gamma^n \rho_1^{p-n} |a_p| \leq \alpha, \quad n = 0, 1, \ldots, p; \, p = 0, 1, \ldots. \tag{3.3-3}$$

Adding these estimates for $n = 0, 1, \ldots, p$, we get

$$(\rho_1 + \gamma)^p |a_p| \leq (p+1)\alpha$$

by the binomial theorem, followed at once by

$$\limsup_{p \to \infty} |a_p|^{1/p} \leq \frac{1}{\rho_1 + \gamma}.$$

Thus by the Cauchy-Hadamard formula P has a radius of convergence $\geq \rho_1 + \gamma > \rho_1$, contradicting the definition of ρ_1. ∎

Theorem 3.3a, thus established, frequently permits us to indicate the range of validity of a power series representation at a glance without first determining its radius of convergence.

EXAMPLES

1. The identity

$$\text{Log}(1+z) = \sum_{n=1}^{\infty} \frac{(-1)^{n-1}}{n} z^n$$

[compare (2.5-20)] holds for $|z| < 1$ because the function on the left is analytic for $|z| < 1$.

2. From (2.5-22) we deduce that for arbitrary complex b and for $|z| < 1$, if the power has its principal value,

$$(1+z)^b = \sum_{n=0}^{\infty} \frac{(-b)_n}{n!} (-z)^n$$

because $(1+z)^b$ is analytic for $|z|<1$.

3. The representation

$$\tan z = \sum_{n=1}^{\infty} (-1)^{n-1} \frac{B_{2n}}{(2n)!} (2^{2n}-1) 2^{2n} z^{2n-1}, \qquad (2.5\text{-}15)$$

established in §2.5 for $|z|$ sufficiently small, actually holds for $|z|<\pi/2$ because $\tan z$ is analytic for $|z|<\pi/2$.

4. The representation

$$z \cot z = \sum_{n=0}^{\infty} (-1)^n \frac{B_{2n}}{(2n)!} (2z)^{2n} \qquad (2.5\text{-}14)$$

holds for $|z|<\pi$ because, after removing the singularity at $z=0$, the function on the left is analytic for $|z|<\pi$.

5. For a similar reason the representation

$$\frac{z}{e^z-1} = \sum_{n=0}^{\infty} \frac{B_n}{n!} z^n \qquad (2.5\text{-}13)$$

holds for $|z|<2\pi$.

The next examples are somewhat more special.

6. The power series representing the function $f(z):=e^{-z^2}$ near $z=0$ has an infinite radius of convergence because e^{-z^2} is analytic in the whole plane (being the composition of the exponential function with an entire function).

7. The power series representing the function $f(z):=(1+z^2)^{-1}$ (whose graph for real z looks somewhat similar to that of e^{-z^2}!) near $z=0$ has radius of convergence 1 because f has two poles at $z=\pm i$ and is analytic elsewhere. Thus $|z|<1$ is the largest disk about $z=0$ into which f can be continued analytically.

8. For $z\neq 1$, $z\neq 2$ let f be defined by

$$f(z):=\frac{z-1}{z^2-3z+2}$$

and let f be undefined for $z=1,2$. *Problem.* Find the radius of convergence of the power series representing f near $z=0$. Because f is analytic except at $z=1$ and $z=2$, the radius of convergence is at least 1. The question is whether f can be continued analytically into the points $z=1,2$. As shown in Example **10** of §3.2, f can indeed be continued into $z=1$ by defining $f(1)=-1$. Thus the radius of convergence is at least 2. Because the singularity at $z=2$ cannot be removed, the radius of convergence is exactly 2.

The following theorem is an important application of Theorem 3.3a.

SEQUENCES IN A REGION

THEOREM 3.3b (Liouville's theorem)

If a function f is analytic in the whole plane and bounded, it assumes one value only.

Proof. Let $P(z) := a_0 + a_1 z + \cdots$ be the power series representing f near 0. By Theorem 3.3a the radius of convergence of P is infinite and by Theorem 3.2b, $f(z) = P(z)$ for all z. If μ is a bound for $|f(z)|$, Cauchy's estimate yields

$$|a_n| \leq \frac{\mu}{\rho^n}, \qquad n = 0, 1, 2, \ldots,$$

where ρ is arbitrary. For $n > 0$ this is possible only if $a_n = 0$. Thus only the zeroth coefficient of the power series can be different from zero. Because P represents f everywhere, the conclusion follows. ∎

PROBLEMS

1. Determine the radius of convergence of the power series that represents the function $f(z) := (1 + z^2)^{-1}$ near $z = \xi$, ξ real. Verify the result by computing the coefficients of the power series, using partial fractions.

2. Verify the result of Example **8** by direct computation of the power series. Also compute the power series that represents g near $z = 0$, where $g(z) := (z^2 - 3z + 2)^{-1}$.

3. Determine the radii of convergence of the power series which represent the following functions near $z = 0$:

   (a) $\dfrac{e^z - 1}{z}$, (b) $\dfrac{z}{e^z - 1}$, (c) $\dfrac{e^z - 1}{z^2 + 4\pi^2}$,

   (d) $\dfrac{z(z^2 + 4\pi^2)}{e^z - 1}$, (e) e^{e^z}.

4. Show by means of examples that neither of the two hypotheses in Liouville's theorem can be omitted without making the conclusion false.

5. Use Liouville's theorem to prove the fundamental theorem of algebra.

6. Let f be analytic in the whole plane and let there exist a constant $\mu > 0$ and a non-negative integer n such that for all sufficiently large $|z|$

   $$|f(z)| < \mu |z|^n.$$

   Show that f is a polynomial of degree $\leq n$.

§3.4. SEQUENCES OF FUNCTIONS ANALYTIC IN A REGION

Some further notions from point set topology are required. All sets considered are sets of points in the complex plane $\mathbb{C}$. A set S is called

closed if its complement (i.e., the set of points not belonging to S) is open. The set $\mathbb{C}$ by definition is both open and closed; thus the same holds for the empty set. A set S is called **bounded** if there exists a real number ρ such that $|z| < \rho$ for all $z \in S$. A set is called **compact** if it is both closed and bounded.

THEOREM 3.4a (Heine-Borel lemma)

Let T be a compact set and let T be contained in the union of a family of open sets. Then T is already contained in the union of a finite number of these open sets.

The family of open sets whose union contains T is called a **covering** of T, and the theorem states that it is possible to "extract a finite subcovering."

The *proof* of Theorem 3.4a is indirect, hence nonconstructive. Because the set T is bounded, it is contained in a disk, hence in the circumscribing square Q_0 whose sides run parallel to the coordinate axes. Divide Q_0 into four congruent subsquares. If T possesses no finite subcovering, the intersection of T with at least one subsquare will possess no finite subcovering. Call this subsquare Q_1. If we subdivide Q_1 into congruent subsquares, a subsquare Q_2 whose intersection with T possesses no finite subcovering must result. Continuing in this manner, we conclude that there is a sequence of squares $Q_0 \supset Q_1 \supset Q_2 \supset \cdots$, each Q_n having the property that its intersection with T has no finite subcovering. The real coordinates, say, of the left lower corners z_n of Q_n form two bounded increasing sequences; hence $z := \lim_{n \to \infty} z_n$ exists. Since T is closed, $z \in T$. Thus z is contained in an open set of the covering family. By the definition of open sets the same is true for a neighborhood $N(z)$. For n sufficiently large $Q_n \subset N(z)$. Hence Q_n is contained in a single set of the covering family, contrary to the mode of construction of Q_n. This contradiction proves Theorem 3.4a. ∎

Here we wish to consider sequences $\{f_n\}$ of scalar functions, each f_n being defined on the same open set S. Such a sequence is said to **converge locally uniformly** on S if it converges uniformly on every compact subset T of S. If the sequence $\{f_n\}$ converges locally uniformly on S, it converges at every point of S, for every point $z_0 \in S$ is a compact subset of S. Thus the limit function $f = \lim_{n \to \infty} f_n$ is defined on S.

An easy application of Theorem 2.1 will convince the reader of the following fact: if the functions f_n are *continuous* on S and the sequence $\{f_n\}$ converges locally uniformly on S, the limit function f is *continuous* on S. It is one of the fundamental facts of complex analysis that the same statement holds in which the word "continuous" is replaced by "analytic." In fact, the following more general result holds.

SEQUENCES IN A REGION

THEOREM 3.4b

Let the functions $f_0, f_1, f_2, \ldots$ all be analytic on the same open set S and let the sequence $\{f_n\}$ converge locally uniformly on S. Then its limit function f is analytic on S. Moreover, the sequence of derivatives $\{f_n'\}$ converges to f' locally uniformly on S.

By induction it follows immediately that for every positive integer m the sequence $\{f_n^{(m)}\}$ converges to $f^{(m)}$, locally uniformly on S.

Proof of Theorem 3.4b. In a neighborhood of each point $z_0 \in S$ each function f_n is represented by a power series in $z - z_0$, which we call $P_n(z_0; z - z_0)$. By Theorem 3.3a these power series converge to $f_n(z)$ for $|z - z_0| < \rho(z_0)$, where $\rho(z_0)$ denotes the distance of z_0 to the complement of S.

Let z_0 be any point of S. For every $\rho < \rho(z_0)$ the set $Q: |z - z_0| \leq \rho$ is a compact subset of S. By hypothesis the sequence $\{f_n(z)\}$ converges uniformly on every such Q; hence the same holds for the sequence of power series $\{P_n(z_0; z - z_0)\}$. It follows by Theorem 2.7a that the limit function f is analytic at z_0. Since $z_0 \in S$ was arbitrary, we conclude that f is analytic on S.

By using the symbols z_0 and Q as above we learn from Corollary 2.7b moreover that the sequence $\{f_n'(z)\}$ converges to $f'(z)$ uniformly on Q. Now let T be an arbitrary compact subset of S; T evidently is covered by the union of the family of open sets $N(z, \frac{1}{2}\rho(z))$, where z runs through S. According to the Heine-Borel lemma T is already contained in the union of a finite number of these open sets, say of the sets $Q_k := N(z_k, \frac{1}{2}\rho(z_k))$, $k = 1, 2, \ldots, m$. By the above

$$\lim_{n \to \infty} f_n'(z) = f'(z)$$

uniformly on each Q_k. Hence the convergence is uniform also on the union of this finite number of Q_k, hence on T. ∎

Theorem 3.4b is important because it enables us to define analytic functions by tools other than power series. All kinds of limit processes are permissible if only they yield a sequence of analytic functions whose convergence is locally uniform. Among the limit processes of this kind discussed in later chapters we mention integration with respect to a parameter (and, at a second stage, improper integrals with respect to a parameter), infinite products, and continued fraction expansions. Here we point out only the result of applying Theorem 3.4b to the sequence of partial sums of an infinite series.

COROLLARY 3.4c

Let the functions $f_0, f_1, f_2, \ldots$ be analytic on the region S and let the series

$f_0 + f_1 + f_2 + \cdots$ *converge locally uniformly on S. Then its sum*

$$f(z) := \sum_{n=0}^{\infty} f_n(z)$$

is analytic on S. Moreover, if z_0 is any point of S and if

$$f_n(z) =: a_{n,0} + a_{n,1}h + a_{n,2}h^2 + \cdots, \qquad (h := z - z_0)$$

is the Taylor series of f_n at z_0 ($n = 0, 1, 2, \ldots$), the series $a_m := \sum_{n=0}^{\infty} a_{n,m}$, $m = 0, 1, 2, \ldots$ converge and $f(z) = a_0 + a_1 h + a_2 h^2 + \cdots$ is the Taylor series of f at z_0.

EXAMPLES

1. The functions

$$f_n(z) := \frac{z^n}{1 - z^n}, \qquad n = 1, 2, \ldots,$$

are analytic in $|z| < 1$, and if $\rho < 1$ then $|f_n(z)| \leq 2\rho^n$ for $|z| \leq \rho$ and n sufficiently large. The convergence of the series

$$f(z) := \sum_{n=1}^{\infty} \frac{z^n}{1 - z^n}$$

is locally uniform on $|z| < 1$. Its Taylor expansion at $z = 0$ can be found by adding corresponding terms in the expansion of $f_n(z)$:

$$\begin{aligned}
f(z) = z &+ z^2 + z^3 + z^4 + z^5 + z^6 + z^7 + z^8 + \cdots \\
&+ z^2 + z^4 + z^6 + z^8 + \cdots \\
& + z^3 + z^6 + \cdots \\
& + z^4 + z^8 + \cdots \\
& + z^5 + \cdots \\
& + z^6 + \cdots \\
& + z^7 + \cdots \\
& + z^8 + \cdots \\
& + \cdots
\end{aligned}$$

In the column of terms z^n there are as many entries as there are divisors (1 and n included) of the integer n. In number theory the number of divisors of n is denoted by $d(n)$. We thus have

$$f(z) = \sum_{n=1}^{\infty} d(n) z^n.$$

SEQUENCES IN A REGION

2. The series

$$f(z) := \sum_{n=1}^{\infty} \frac{1}{(n-z)^2}$$

converges locally uniformly in the region $S: z \neq 1, 2, 3, \ldots$. (If T is any compact subset of S, let T be contained in $|z| \leqslant \rho$. The terms with $n \leqslant 2\rho$ are then bounded on T and those with $n > 2\rho$ are majorized by the series $\sum_{n \geqslant \rho} n^{-2}$, which converges.) The expansion of the nth term in powers of z is

$$(n-z)^{-2} = n^{-2}\left(1 - \frac{z}{n}\right)^{-2} \sum_{k=0}^{\infty} \frac{(2)_k}{k!}\left(\frac{z}{n}\right)^k$$

$$= \sum_{k=0}^{\infty} (k+1) z^k n^{-k-2}.$$

Thus there follows

$$f(z) = \sum_{k=0}^{\infty} (k+1) c_k z^k,$$

where

$$c_k := \sum_{n=1}^{\infty} \frac{1}{n^{k+2}}.$$

PROBLEMS

1. Let the set S be defined as in Example 2. Show that the series

$$f(z) := \sum_{n=1}^{\infty} \frac{(-1)^n}{n-z}$$

converges locally uniformly on S but does not converge absolutely on any subset of S [A. Pfluger].

2. It can be proved (compare Szegö [1959]) that any function f that is analytic in the interior E of an ellipse with foci ± 1 can be expanded in a series

$$f(z) = \sum_{n=0}^{\infty} a_n P_n(z) \qquad (3.4\text{-}1)$$

converging locally uniformly in E, where

$$P_n(z) := {}_2F_1\left(-n, n+1; 1; \frac{1-z}{2}\right)$$

denotes the nth *Legendre polynomial*. Assuming the expansion (3.4-1), show that for $|z-1|$ sufficiently small

$$f(z) = \sum_{k=0}^{\infty} \frac{2^k (\frac{1}{2})_k}{k!} c_k (z-1)^k,$$

where

$$c_k := \sum_{p=0}^{\infty} \frac{(1+2k)_p}{p!} a_{k+p}, \quad k = 0, 1, 2, \ldots.$$

3. Use the result in Problem 2 to identify the function f when $a_n := t^n$.

4. It can be shown (see Watson [1944], p. 522) that any function f analytic in a disk C centered at the origin can be expanded in a *Neumann series*:

$$f(z) = \sum_{n=0}^{\infty} a_n J_n(z), \qquad (3.4\text{-}2)$$

where J_n denotes the Bessel function

$$J_n(z) := \frac{(z/2)^n}{n!} {}_0F_1\left(n+1; -\frac{z^2}{4}\right).$$

Assuming the expansion (3.4-2), show that

$$f(z) = \sum_{m=0}^{\infty} \frac{b_m}{m!} \left(\frac{z}{2}\right)^m,$$

where

$$b_m := \sum_{k=0}^{[m/2]} \frac{(-m)_k}{k!} a_{m-2k}.$$

5. Use this result to verify the expansions (valid for all z)

$$\cos z = J_0(z) + 2 \sum_{n=1}^{\infty} (-1)^n J_{2n}(z),$$

$$\sin z = 2 \sum_{n=0}^{\infty} (-1)^n J_{2n+1}(z).$$

6. It is shown in the theory of Bessel functions (see Watson [1944], p. 558) that a function f analytic at the origin can for sufficiently small $|z|$ be expanded in a

series of the form

$$f(z) = \sum_{n=0}^{\infty} a_n J_n(nz), \qquad (3.4\text{-}3)$$

called a *Kapteyn series*. Assuming that (3.4-3) converges uniformly for $|z| \leqslant \rho$, where $\rho > 0$, show that

$$f(z) = \sum_{p=0}^{\infty} \frac{b_p}{p!} \left(\frac{z}{2} \right)^p,$$

where $b_0 = a_0$,

$$b_p = \sum_{k=0}^{[(p-1)/2]} \frac{(-p)_k (p-2k)^p}{k!} a_{p-2k}, \qquad p = 1, 2, \ldots.$$

7. Assuming that the Kapteyn series

$$\frac{z}{1-z} = 2 \sum_{n=1}^{\infty} J_n(nz)$$

is valid, prove the identity

$$\sum_{k=0}^{p} \binom{p}{k} (-1)^k (p-2k)^p = 2^p p!, \qquad p = 0, 1, 2, \ldots.$$

§3.5. ANALYTIC CONTINUATION ALONG AN ARC: MONODROMY THEOREM

The concept of a path or arc is of equal importance in the theories of analytic continuation and of complex integration. The word **arc**, synonymously used with **path**, denotes a continuous, complex-valued function of a real variable, defined on a closed bounded interval $[\alpha, \beta]$. Arcs are thus defined in terms of a "parametric representation"

$$z = z(\tau), \qquad \alpha \leqslant \tau \leqslant \beta. \qquad (3.5\text{-}1)$$

If the symbol Γ denotes the arc (3.5-1), the same symbol will also be used to denote its *range*, i.e., the set of points $\{z(\tau): \alpha \leqslant \tau \leqslant \beta\}$ in the complex plane. However, an arc is not determined by its range; in addition, the sense of direction from the **initial point** $z_0 := z(\alpha)$ to the **terminal point** $z_1 := z(\beta)$ is relevant. Any arc with initial point z_0 and terminal point z_1 is called an *arc from z_0 to z_1*.

What matters less is the functional form of the parametric representation. If $\alpha' < \beta'$ and $\phi(\sigma)$ is any continuous, strictly monotonically increasing function on $[\alpha', \beta']$ such that $\phi(\alpha') = \alpha$ and $\phi(\beta') = \beta$, then the arc Γ' defined by $z = z(\phi(\sigma))$, $\alpha' \leq \sigma \leq \beta'$, has the same range and the same initial and terminal points as the arc Γ. Two arcs that can be transformed into each other by such a sense-preserving change of the parameter are called **geometrically equivalent**. Most significant properties of arcs hold for all geometrically equivalent arcs. By a linear change of parameter, $\tau = \gamma \sigma + \delta$ ($\gamma > 0$), every arc is geometrically equivalent to an arc whose parameter ranges over $[0, 1]$.

We introduce some notions that will not be used until Chapter 4. The arc $z = z(\tau) = x(\tau) + iy(\tau)$, $\tau \in [\alpha, \beta]$, is called **differentiable** if the derivatives $x'(\tau)$ and $y'(\tau)$ (according to the real calculus) exist and are continuous. The arc is called **regular** if $z'(\tau) := x'(\tau) + iy'(\tau) \neq 0$ for all $\tau \in [\alpha, \beta]$. The geometric interpretation of these properties is that a regular differentiable arc has a tangent at every point; the tangent at point $z(\tau)$, if oriented in the sense of increasing τ, points in the direction $\arg(z'(\tau))$. An arc is called **piecewise differentiable** or **piecewise regular** if there is a finite subdivision $\alpha = \tau_0 < \tau_1 < \cdots < \tau_n = \beta$ of the parametric interval such that the arc defined by the function $z(\tau)$ on each closed subinterval $[\tau_{k-1}, \tau_k]$ is differentiable or regular. In differentiable arcs only continuously differentiable changes of parameter $\tau = \phi(\sigma)$ are permissible for geometric equivalence; for preserving regularity we require $\phi'(\sigma) \neq 0$.

We return to analytic continuation. Let the function f be analytic at the point z_0 and let Γ be an arc with initial point z_0. We call **analytic continuation of f along Γ** any analytic continuation of f into a region S containing the point set Γ.

THEOREM 3.5a

Let f be analytic at z_0 and let Γ be an arc with initial point z_0. Then all analytic continuations of f along Γ coincide in a neighborhood of every point z^* on Γ.

Proof. Let g_1 and g_2 be two analytic continuations of f along Γ. Let g_i be analytic in the region S_i, $i = 1, 2$. Both regions S_1 and S_2 contain Γ; they also contain a neighborhood Q of z_0, where $g_1 = g_2 = f$. By Theorem 3.2d it would follow that g_1 and g_2 are analytic continuations of each other if it were true that the set $S_1 \cap S_2$ is connected. Unfortunately, this need not be so.

In any case, however, the set $S_1 \cap S_2$ is open and thus by Theorem 3.2a has a unique decomposition into components. Exactly one component contains Γ because Γ connects two points of $S_1 \cap S_2$. Let C be that component. Then C contains Q, and, since $g_1 = g_2 = f$ on Q, it follows from Theorem 3.2b that $g_1 = g_2$ throughout C, and thus in a neighborhood of each $z^* \in \Gamma$. ∎

The main result to be proved in this section, called the monodromy theorem, has to do with conditions under which the analytic continuations of a function along two *different* arcs Γ_1 and Γ_2 from the initial point z_0 to the terminal point z_1 coincide in a neighborhood of z_1. To this end the notion of **homotopy**, a generalization of geometric equivalence, is required. Intuitively, two arcs with the same initial and terminal points are said to be homotopic with respect to a region S containing both of them if they can be continuously deformed into each other (the end point being fixed) without leaving S. It is easy to couch the notion of continuous deformation in analytic terms. If S is a region and z_0 and z_1 are any two points of S, two arcs Γ_0 and Γ_1 from z_0 to z_1 are called **homotopic with respect to** S if there exists a complex-valued function $z = z(\tau, \sigma)$ of two real variables τ and σ, defined on the unit square $K: 0 \leq \tau \leq 1$, $0 \leq \sigma \leq 1$ and continuous on K, such that

(a) $z(\tau, \sigma) \in S$ for all $(\tau, \sigma) \in K$,

(b) $z(0, \sigma) = z_0$, $z(1, \sigma) = z_1$ for all $\sigma \in [0, 1]$,

(c) the arcs $z = z(\tau, 0)$ and $z = z(\tau, 1)$ $(0 \leq \tau \leq 1)$ are geometrically equivalent to the arcs Γ_0 and Γ_1, respectively.

THEOREM 3.5b (*monodromy theorem*)

Let S be a region, let f be analytic at a point $z_0 \in S$, and assume that there exists an analytic continuation of f along every arc emanating from z_0 and lying in S. Then, if z_1 is any point of S and Γ_0 and Γ_1 are any two arcs from z_0 to z_1, the analytic continuations of f along Γ_0 and Γ_1 will agree in a neighborhood of z_1, provided the arcs Γ_0 and Γ_1 are homotopic with respect to S.

Proof. Let the function $z(\tau, \sigma)$ be defined as above and for each $\sigma \in [0, 1]$ let Γ_σ denote the arc

$$z = z(\tau, \sigma), \quad 0 \leq \tau \leq 1,$$

from z_0 to z_1. By hypothesis f can be continued analytically along each arc Γ_σ; i.e., for each $\sigma \in [0, 1]$ there exists a region U_σ containing Γ_σ and a function g_σ analytic in U_σ that agrees with f in a neighborhood of z_0.

Let $\omega \in [0, 1]$. We shall show that there exists $\epsilon > 0$ (which may depend on ω) such that every g_σ such that $\sigma \in [0, 1]$, $|\sigma - \omega| < \epsilon$ agrees with g_ω in a neighborhood of z_1. Indeed, let δ be the distance of Γ_ω from the complement of U_ω. Since the complement is closed and Γ_ω is compact, $\delta > 0$. Because the function $z(\tau,\sigma)$ is continuous on the closed, bounded set K, it is uniformly continuous. Hence there exists $\epsilon > 0$ such that $|z(\tau,\sigma) - z(\tau,\omega)| < \delta$ for $\sigma \in [0,1]$, $|\sigma - \omega| < \epsilon$. It follows that for all such σ, $\Gamma_\sigma \subset U_\omega$, and g_ω may be regarded as an analytic continuation of f along Γ_σ. Hence for all such σ, by virtue of Theorem 3.5a, g_σ coincides with g_ω in a neighborhood of z_1.

The intervals of length $\epsilon = \epsilon_\sigma$ form an open covering of the compact interval $[0,1]$. Hence by the one-dimensional version of the Heine-Borel lemma, there is a finite subcovering. Let $0 = \sigma_0 < \sigma_1 < \cdots < \sigma_n = 1$ be points at which consecutive covering intervals overlap. Then, by the above, $g_{\sigma_{k-1}} = g_{\sigma_k}$ in a neighborhood of z_1, $k = 1, 2, \ldots, n$, and thus $g_0 = g_1$ in the intersection of this finite number of neighborhoods. ∎

Of particular interest is the case in which the analytic continuations along *all* arcs from z_0 to z_1 coincide in a neighborhood of z_1. This will occur if for any two points z_0 and z_1 of S any two arcs going from z_0 to z_1 and lying in S are homotopic with respect to S. A region S with this property is called **simply connected**. Because for these regions the value of f at a point z_1 obtained by analytic continuation no longer depends on the path from z_0 to z_1, we immediately obtain:

THEOREM 3.5c (Monodromy theorem for simply connected regions)

Let the region S be simply connected, let f be analytic at a point $z_0 \in S$, and let an analytic continuation exist along every path emanating from z_0 and lying in S. Then all these analytic continuations are continuations of one another and thus together define the analytic continuation of f into S.

An application of the monodromy theorem occurs in Chapter 9.

We conclude this section by stating some criteria for simple connectedness. A set S is called **convex** if it has the following property: if z_0 and z_1 are any two points of S, then S also contains the straight line segment joining z_0 and z_1, i.e., all points of the form $\sigma z_0 + (1 - \sigma) z_1$ where $0 \leqslant \sigma \leqslant 1$. It is clear that the interior of any circle, ellipse, or rectangle is convex.

THEOREM 3.5d

Every convex region is simply connected.

Proof. Let S be convex, let z_0 and z_1 be any two points of S, and let

$$\Gamma_0 : z = z(\tau), \quad \Gamma_1 : z = w(\tau) \quad (0 \leqslant \tau \leqslant 1)$$

be two arcs from z_0 to z_1 lying in S. It is to be shown that Γ_0 and Γ_1 are

ANALYTIC CONTINUATION ALONG AN ARC 169

homotopic with respect to S. We assert that the function $z(\tau,\sigma)$ required in the definition of homotopy is given by

$$z(\tau,\sigma) := (1-\sigma)z(\tau) + \sigma w(\tau).$$

This evidently satisfies conditions (b) and (c) and, by virtue of the convexity of S, (a). ∎

Many more domains can be shown to be simply connected by the use of maps. A region S in the complex plane is called **homeomorphic** to a region T if there exists a one-to-one mapping g of S onto T [i.e., a one-to-one function g with $D(g) = S$ and $W(g) = T$] such that both g and $g^{[-1]}$ are continuous. (It is not required that g be analytic.) If S is homeomorphic to T, then T is homeomorphic to S and we say that S and T are homeomorphic.

EXAMPLE 1

The complex plane is homeomorphic to the unit disk by virtue of the mapping

$$g : z \to \frac{z}{1+|z|}.$$

THEOREM 3.5e

If T is a simply connected region and S is homeomorphic to T, then S is simply connected.

Proof. Let z_0 and z_1 be two points of S and let

$$\Gamma_0 : z = z(\tau), \qquad \Gamma_1 : z = w(\tau) \qquad (0 \leqslant \tau \leqslant 1)$$

be two arcs from z_0 to z_1. We must construct the function $z(\tau,\sigma)$ that deforms Γ_0 into Γ_1.

Let g be a one-to-one map of S onto T such that both g and $g^{[-1]}$ are continuous. Then

$$\Gamma_0^* : z = g(z(\tau)), \qquad \Gamma_1^* : z = g(w(\tau))$$

are two arcs in T from $z_0^* := g(z_0)$ to $z_1^* := g(z_1)$ and thus by hypothesis can be deformed into each other by a function $z^*(\tau,\sigma)$ satisfying (a), (b), and (c). It is then easily verified that $z(\tau,\sigma) := g^{[-1]}(z^*(\tau,\sigma))$ is a function that satisfies the corresponding conditions for Γ_0 and Γ_1. ∎

It is also true that *every* simply connected region is homeomorphic to the unit disk; see §5.10.

PROBLEMS

1. Show that the conclusion of Theorem 3.5c is false if S is not simply connected.
2. Let $0 \leq \alpha < \beta \leq \infty$. Show that the annulus $A : \alpha < |z| < \beta$ is not simply connected by proving that the analytic continuations of $f(z) := \text{Log} \, z$ (principal value) along the arcs $\Gamma_1 : z = \rho e^{i\tau}, 0 \leq \tau \leq \pi$, and $\Gamma_2 : z = \rho e^{-i\tau}, 0 \leq \tau \leq \pi$, do not coincide.
3. Show that a region S that is homeomorphic to an annulus $A : \alpha < |z| < \beta$ ($0 \leq \alpha < \beta \leq \infty$) is not simply connected.

§3.6. NUMERICAL ANALYTIC CONTINUATION ALONG AN ARC

Let f be analytic at the point z_0 and let the coefficients of its Taylor series at z_0,

$$f(z_0 + h) = \sum_{n=0}^{\infty} a_n^{(0)} h^n, \qquad (3.6\text{-}1)$$

be known. Let Γ be an arc emanating from z_0 whose terminal point z_t lies outside the disk of convergence of the series (3.6-1). Suppose it is known that an analytic continuation of f along Γ exists. Here we consider the problem of computing numerically the value of f, or, more generally, the coefficients of the Taylor expansion of f, at z_t.

The classical method for solving this problem, due to Weierstrass, proceeds as follows: let S be a region containing Γ into which f can be continued. For $z \in S$ we denote by $a_0(z), a_1(z), \ldots$, the Taylor coefficients of f at z. For $|h| < \rho(z)$, where $\rho(z)$ denotes the distance of z from the complement of S, we then have by Theorem 3.3a

$$f(z + h) = \sum_{n=0}^{\infty} a_n(z) h^n. \qquad (3.6\text{-}2)$$

Now let $z_0, z_1, z_2, \ldots, z_t$ be a finite sequence of points on Γ such that for $k = 0, 1, \ldots, t-1$ the subarc of Γ lying between z_k and z_{k+1} is contained in the disk $|z - z_k| < \rho(z_k)$. By hypothesis, the sequence of Taylor coefficients $\{a_n(z_0)\} = \{a_n^{(0)}\}$ is known. Suppose the sequence $\{a_n(z_{k-1})\}$ is known. Then, since f is analytic on the set $|z - z_{k-1}| < \rho(z_{k-1})$, the sequence of Taylor coefficients at z_k can be obtained by rearranging the Taylor series at z_{k-1}. Letting $h_k := z_k - z_{k-1}$, this, by §3.1, yields

$$a_m(z_k) = \sum_{n=m}^{\infty} \binom{n}{m} a_n(z_{k-1}) h_k^{n-m}, \qquad m = 0, 1, 2, \ldots. \qquad (3.6\text{-}3)$$

NUMERICAL CONTINUATION ALONG AN ARC

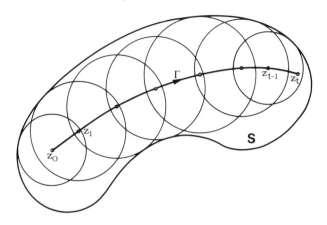

Fig. 3.6a. Weierstrassian analytic continuation.

Repetition of the rearrangement process t times finally yields the desired sequence of Taylor coefficients at z_t (see Fig. 3.6a).

The Weierstrassian method can be described in a compact manner by using vector and matrix notation. For $z \in S$ we define the infinite column vector

$$\mathbf{a}(z) := (a_0(z), a_1(z), a_2(z), \ldots)^T$$

(T denotes the transposed vector or matrix). The Weierstrassian process then successively yields the vectors $\mathbf{a}(z_k)$, $k = 1, 2, \ldots, t$. If for arbitrary complex h we define the infinite triangular matrix

$$\mathbf{M}(h) := \begin{bmatrix} 1 & h & h^2 & h^3 & h^4 & \cdots \\ & 1 & 2h & 3h^2 & 4h^3 & \cdots \\ & & 1 & 3h & 6h^2 & \cdots \\ & & & 1 & 4h & \cdots \\ 0 & & & & 1 & \cdots \\ & \cdots & & & & \end{bmatrix}$$

(whose element in the (m,n) position is $\binom{n}{m} h^{n-m}$ for $n \geqslant m \geqslant 0$ and 0 for

$n < m$), the formulas (3.6-3) are equivalent to

$$\mathbf{a}(z_k) = \mathbf{M}(h_k)\,\mathbf{a}(z_{k-1}), \qquad k = 1, 2, \ldots, t. \tag{3.6-4}$$

Thus, in a sense, the desired coefficient vector $\mathbf{a}(z_t)$ is calculated from the data vector $\mathbf{a}(z_0)$ by

$$\mathbf{a}(z_t) = \mathbf{M}(h_t)\,\mathbf{M}(h_{t-1}) \cdots \mathbf{M}(h_1)\,\mathbf{a}(z_0). \tag{3.6-5}$$

The reader is warned, however, that the product on the right is not associative. Indeed, an easy application of Vandermonde's theorem shows that for any two complex numbers h, k

$$\mathbf{M}(h)\,\mathbf{M}(k) = \mathbf{M}(h + k). \tag{3.6-6}$$

Thus, if the associative law held, the product in (3.6-5) in view of $h_1 + h_2 + \cdots + h_t = z_t - z_0$ would equal

$$\mathbf{a}(z_t) = \mathbf{M}(z_t - z_0)\,\mathbf{a}(z_0).$$

This, however, fails to make sense, since it would amount to substituting $h = z_t - z_0$ into the original series (3.6-1), a value for which it diverges in general.

As described above, the Weierstrassian method is not yet strictly constructive because it fails to yield the desired objects [here the elements of the vector $\mathbf{a}(z_t)$] as limits of sequences of rational functions of the data of the problem [here the coefficients of the vector $\mathbf{a}(z_0)$]. To obtain a constructive process it is necessary to truncate the matrices and vectors involved in a suitable manner. For arbitrary non-negative integers m and n we denote

by $\mathbf{M}_{m,n}(h)$ the finite segment of the matrix $\mathbf{M}(h)$ constituting its first m rows and n columns (last element: $a_{m-1, n-1}$),

by $\mathbf{a}_n(z)$ the finite column vector constituting the first n elements of $\mathbf{a}(z)$,

by $\mathbf{a}_n^{(k)}$ a column vector intended to approximate $\mathbf{a}_n(z_k)$.

At first glance the following would seem to be a reasonable, constructive version of the Weierstrass process. Compute, for $n = 1, 2, \ldots$, the vectors

$$\mathbf{a}_n^{(k)} = \mathbf{M}_{n,n}(h_k)\,\mathbf{a}_n^{(k-1)}, \qquad k = 1, 2, \ldots, t, \tag{3.6-7}$$

starting with $\mathbf{a}_n^{(0)} := \mathbf{a}_n(z_0)$. One hopes that $\mathbf{a}_n^{(t)} \to \mathbf{a}_n(z_t)$ (componentwise) as $n \to \infty$. Unfortunately this is not the case. Because the addition formula

NUMERICAL CONTINUATION ALONG AN ARC

(3.6-6) also holds for the finite segments $\mathbf{M}_{n,n}$ and products of finite matrices are certainly associative, (3.6-7) is equivalent to

$$\mathbf{a}_n^{(t)} = \mathbf{M}_{n,n}(h_t)\mathbf{M}_{n,n}(h_{t-1})\cdots \mathbf{M}_{n,n}(h_1)\mathbf{a}_n(z_0)$$
$$= \mathbf{M}_{n,n}(z_t - z_0)\mathbf{a}_n(z_0),$$

and thus again amounts to nothing more than substituting $h := z_t - z_0$ into the given power series (3.6-1), truncated after the nth term. We have already seen that this, in general, is a divergent process.

It is shown below that a convergent process can be obtained by letting the size of the vectors $\mathbf{a}_n^{(k)}$ decrease with increasing k. To formulate the procedure let γ be a rational fraction, $0 < \gamma < 1$. For those n for which $n\gamma^t$ is an integer, we now compute the vectors

$$\mathbf{a}_n^{(1)} := \mathbf{M}_{n\gamma, n}(h_1)\mathbf{a}_n(z_0),$$

$$\mathbf{a}_{n\gamma}^{(2)} := \mathbf{M}_{n\gamma^2, n\gamma}(h_2)\mathbf{a}_{n\gamma}^{(1)}, \qquad (3.6\text{-}8)$$

$$\cdot \quad \cdot \quad \cdot$$

$$\mathbf{a}_{n\gamma^t}^{(t)} := \mathbf{M}_{n\gamma^t, n\gamma^{t-1}}(h_t)\mathbf{a}_{n\gamma^{t-1}}^{(t-1)},$$

and let $n \to \infty$. Intuitively, this algorithm amounts to shortening the Taylor series at each stage of the computation by a fixed "reduction factor" γ.

In order to study the convergence properties of the above algorithm it is convenient to define the norm of a vector $\mathbf{b} := (b_1, b_2 \ldots b_n)^T$ by

$$\|\mathbf{b}\| := \max_{1 \leq i \leq n} |b_i|,$$

and the norm of a (not necessarily square) matrix $\mathbf{B} := (b_{ij})$ ($1 \leq i \leq m, 1 \leq j \leq n$) by

$$\|\mathbf{B}\| := \max_{1 \leq i \leq m} \sum_{j=1}^{n} |b_{ij}|.$$

It is then easily verified that

$$\|\mathbf{Bb}\| \leq \|\mathbf{B}\| \, \|\mathbf{b}\|,$$

and that for any two compatible matrices $\mathbf{B}, \mathbf{C}$,

$$\|\mathbf{BC}\| \leq \|\mathbf{B}\| \, \|\mathbf{C}\|.$$

THEOREM 3.6

Under the above assumptions there exists a positive number γ_0 such that for all reduction factors $\gamma \in (0, \gamma_0)$ the vectors $\mathbf{a}_{n\gamma'}^{(t)}$ defined by (3.6-8) satisfy

$$\lim_{n \to \infty} \|\mathbf{a}_{n\gamma'}^{(t)} - \mathbf{a}_{n\gamma'}(z_t)\| = 0.$$

Proof. By (3.6-8) the approximate coefficient vectors satisfy

$$\mathbf{a}_{n\gamma}^{(k)} - \mathbf{M}_{n\gamma, n}(h_k) \mathbf{a}_n^{(k-1)} = \mathbf{0} \qquad (3.6\text{-}9)$$

for those n for which they are defined. We shall construct a similar relation for the exact coefficient vectors $\mathbf{a}_{n\gamma}(z_k)$. For $k = 1, 2, \ldots, t$, let

$$|h_k| < \tau_k < \rho(z_{k-1}),$$

and let T denote the union of the closed disks $|z - z_k| \leq \tau_k$, $k = 0, 1, \ldots, t$. If

$$\mu := \sup_{z \in T} |f(z)|,$$

then by Cauchy's coefficient estimate

$$|a_p(z_{k-1})| \leq \mu \tau_k^{-p}, \qquad p = 0, 1, 2, \ldots. \qquad (3.6\text{-}10)$$

Letting

$$\mathbf{a}_{n\gamma}(z_k) - \mathbf{M}_{n\gamma, n}(h_k) \mathbf{a}_n(z_{k-1}) =: \mathbf{f}_{n\gamma}^{(k)}, \qquad (3.6\text{-}11)$$

we find by (3.6-3) that the mth component of $\mathbf{f}_n^{(k)}$ equals

$$a_m(z_k) - \sum_{p=m}^{n-1} \binom{p}{m} a_p(z_{k-1}) h_k^{p-m} = \sum_{p=n}^{\infty} \binom{p}{m} a_p(z_{k-1}) h_k^{p-m}.$$

Using (3.6-10), we majorize the last sum by

$$\sum_{p=n}^{\infty} \binom{p}{m} |h_k|^{p-m} \tau_k^{-p} = \binom{n}{m} \frac{|h_k|^{n-m}}{\tau_k^n} \sum_{q=0}^{\infty} \frac{(n+1)_q}{(n-m+1)_q} \left(\frac{|h_k|}{\tau_k}\right)^q.$$

Because $m < \gamma n$,

$$\frac{(n+1)_q}{(n-m+1)_q} < \frac{1}{(1-\gamma)^q}.$$

NUMERICAL CONTINUATION ALONG AN ARC

Thus, if we assume that γ is small enough to satisfy

$$(1-\gamma)\tau_k > |h_k|, \quad k=1,2,\ldots,t, \qquad (3.6\text{-}12)$$

this sum is in turn majorized by a geometric series with value

$$\chi_k := \frac{(1-\gamma)\tau_k}{(1-\gamma)\tau_k - |h_k|}, \quad k=1,2,\ldots,t.$$

Letting $\chi := \max \chi_k$, we see that the mth component of $\mathbf{f}_n^{(k)}$ is bounded by

$$\mu \chi \binom{n}{m} \frac{|h_k|^{n-m}}{\tau_k^n}, \quad m=0,1,\ldots,n-1. \qquad (3.6\text{-}13)$$

Let

$$\mathbf{r}_n^{(k)} := \mathbf{a}_n(z_k) - \mathbf{a}_n^{(k)};$$

our aim is to estimate $\|\mathbf{r}_n^{(t)}\|$. On subtracting (3.6-9) from (3.6-11) and replacing n with $n\gamma^{k-1}$ we get

$$\mathbf{r}_{n\gamma^k}^{(k)} - \mathbf{M}_{n\gamma^k, n\gamma^{k-1}}(h_k) \mathbf{r}_{n\gamma^{k-1}}^{(k-1)} = \mathbf{f}_{n\gamma^k}^{(k)};$$

hence by induction on k, since $\mathbf{r}_n^{(0)} = \mathbf{0}$,

$$\mathbf{r}_{n\gamma^t}^{(t)} = \sum_{k=1}^{t} \mathbf{P}_{n\gamma^t, n\gamma^k}^{(k)} \mathbf{f}_{n\gamma^k}^{(k)}, \qquad (3.6\text{-}14)$$

where $\mathbf{P}_{n\gamma^t, n\gamma^t}^{(t)} = \mathbf{I}$,

$$\mathbf{P}_{n\gamma^t, n\gamma^k}^{(k)} = \prod_{m=t}^{k+1} \mathbf{M}_{n\gamma^m, n\gamma^{m-1}}(h_m), \quad k=1,2,\ldots,t-1, \qquad (3.6\text{-}15)$$

the index m running from t downward.

To estimate the norms of the summands in (3.6-14), it does not suffice to estimate the norms of each factor in (3.6-15). Instead, it is necessary to factor the matrices $\mathbf{M}_{m,n}(h)$, as follows: let $\mathbf{D}_m(h)$ denote the square diagonal matrix of order m whose diagonal elements are $1, h, h^2, \ldots, h^{m-1}$ and let $\mathbf{M}_{m,n} := \mathbf{M}_{m,n}(1)$. Then, if $h \neq 0$, it is evident that

$$\mathbf{M}_{m,n}(h) = \mathbf{D}_m(h^{-1}) \mathbf{M}_{m,n} \mathbf{D}_n(h). \qquad (3.6\text{-}16)$$

Using this in (3.6-15), we have

$$\mathbf{P}_{n\gamma^t, n\gamma^k}^{(k)} = \left[\prod_{m=t}^{k+1} \mathbf{D}_{n\gamma^m}\left(\frac{h_{m+1}}{h_m}\right) \mathbf{M}_{n\gamma^m, n\gamma^{m-1}} \right] \mathbf{D}_{n\gamma^k}(h_k) \quad (h_{t+1} := 1);$$

hence for the norms of the terms in (3.6-14)

$$\|\mathbf{P}_{n\gamma^l, n\gamma^k}^{(k)} \mathbf{f}_{n\gamma^k}^{(k)}\| \leq \left[\prod_{m=k+1}^{t} \left\| \mathbf{D}_{n\gamma^m}\left(\frac{h_{m+1}}{h_m}\right) \mathbf{M}_{n\gamma^m, n\gamma^{m-1}} \right\| \right] \|\mathbf{D}_{n\gamma^k}(h_{k+1}) \mathbf{f}_{n\gamma^k}^{(k)}\|. \tag{3.6-17}$$

We thus have to estimate, first of all, expressions of the form $\|\mathbf{D}_{n\gamma}(p) \mathbf{M}_{n\gamma,n}\|$, where $p := h_{m+1} h_m^{-1}$ is a complex number. This requires computing the mth row sum of $\mathbf{M}_{n\gamma,n}$ for $m=0,1,\ldots,n\gamma-1$ and finding their maximum when multiplied by $|p|^m$. By Vandermonde's formula the mth row sum equals

$$\binom{m}{m} + \binom{m+1}{m} + \cdots + \binom{n-1}{m} = \binom{n}{m+1}.$$

Thus

$$\|\mathbf{D}_{n\gamma}(p) \mathbf{M}_{n\gamma,n}\| = \max_{0 \leq m \leq n\gamma-1} |p|^m \binom{n}{m+1}.$$

We impose a further restriction on γ which ensures that the maximum occurs for $m = n\gamma - 1$. By forming ratios we readily see that this is the case if

$$\gamma \leq \frac{|p|}{1+|p|} = \frac{|h_{m+1}|}{|h_m| + |h_{m+1}|}. \tag{3.6-18}$$

Under this hypothesis, letting $p := h_{m+1} h_m^{-1}$,

$$\|\mathbf{D}_{n\gamma}(p_m) \mathbf{M}_{n\gamma,n}\| \leq \binom{n}{n\gamma} |p_m|^{n\gamma-1}. \tag{3.6-19}$$

For the last factor in (3.6-17) we obtain in a similar way, using (3.6-13),

$$\|\mathbf{D}_{n\gamma}(h_{k+1}) \mathbf{f}_{n\gamma}^{(k)}\| \leq \mu \chi \binom{n}{n\gamma} |p_k|^{n\gamma} \sigma_k^n, \tag{3.6-20}$$

where $\sigma_k := |h_k| \tau_k^{-1}$, $0 < \sigma_k < 1$. By Stirling's formula (see §8.5), as $n \to \infty$,

$$n! \sim \sqrt{2\pi n} \left(\frac{n}{e}\right)^n;$$

consequently

$$\binom{n}{n\gamma} = \frac{n!}{(\gamma n)![(1-\gamma)n]!} \sim [2\pi\gamma(1-\gamma)n]^{-1/2}[\gamma^{\gamma}(1-\gamma)^{1-\gamma}]^{-n}$$

and thus, if n is sufficiently large,

$$\binom{n}{n\gamma} < \theta^{\gamma n},$$

where

$$\theta := \gamma^{-1}(1-\gamma)^{1-\gamma^{-1}}.$$

Applying this estimate in (3.6-19) and (3.6-20) and using the results in (3.6-17), we obtain

$$\|\mathbf{P}^{(k)}_{n\gamma^t, n\gamma^k} \mathbf{f}^{(k)}_{n\gamma^k}\| \leq \alpha\mu\chi\sigma_k^{n\gamma^{k-1}} \prod_{m=k}^{t} \left[\frac{|p_m|}{\gamma(1-\gamma)^{1/\gamma-1}}\right]^{n\gamma^m},$$

where

$$\alpha := \prod_{m=k+1}^{t} |p_m|^{-1}.$$

We assert that the factors σ_k can be distributed among the factors in the product in such a manner that the term on the right can be replaced by

$$\alpha\mu\chi \prod_{m=k}^{t} \zeta_m^{n\gamma^m},$$

where

$$\zeta_m := |p_m|\sigma_k^{1/\gamma-1}\gamma^{-1}(1-\gamma)^{1-\gamma^{-1}}. \tag{3.6-21}$$

This is so since the number of factors σ_k^n in the latter product equals

$$\left(\frac{1}{\gamma}-1\right)(\gamma^k+\gamma^{k+1}+\cdots+\gamma^t)$$

and thus is less than

$$\left(\frac{1}{\gamma}-1\right)\frac{\gamma^k}{1-\gamma} = \gamma^{k-1},$$

the number of factors in the former expression.

Convergence now hinges on the possibility to choose γ such that

$$\zeta_m < 1, \quad m = 1, 2, \ldots, t. \tag{3.6-22}$$

Writing $\zeta_m = \zeta'_m \zeta'' \zeta'''_k$, where

$$\zeta'_m := |p_m|(1-\gamma)\sigma_k^{-1}, \quad \zeta'' := (1-\gamma)^{-1/\gamma}, \quad \zeta'''_k := \frac{1}{\gamma}\sigma_k^{1/\gamma},$$

we see that

$$\lim_{\gamma \to 0} \zeta'_m = |p_m|\sigma_k^{-1}, \quad \lim_{\gamma \to 0} \zeta'' = e,$$

and, since $|\sigma_k| < 1$,

$$\lim_{\gamma \to 0} \zeta''' = 0.$$

It follows that $\lim_{\gamma \to 0} \zeta_m = 0$. Thus for all sufficiently small γ the expression on the right of (3.6-17) tends to zero. By (3.6-14) it follows that for all such γ

$$\|\mathbf{r}_{n\gamma'}^{(t)}\| \to 0$$

as $n \to \infty$, which is the statement in Theorem 3.6. ∎

For convenience we reformulate the conditions to be satisfied by the reduction factor γ. Denoting the continuation points by $z_0, z_1, z_2, \ldots, z_t$, we have put

$$h_m := z_m - z_{m-1}, \quad p_m := \frac{h_{m+1}}{h_m}, \quad m = 1, 2, \ldots, t.$$

We now let $\sigma_k > |h_k|\rho(z_{k-1})^{-1}$, where $\rho(z)$ denotes the distance to the complement of the domain of analyticity of f, and (3.6-12) is satisfied if

$$\gamma < 1 - \sigma_k, \quad k = 1, 2, \ldots, t. \tag{3.6-23a}$$

In addition, (3.6-18) requires that

$$\gamma < \frac{|p_m|}{1 + |p_m|}, \quad m = 1, 2, \ldots, t, \tag{3.6-23b}$$

and (3.6-22), that

$$\gamma^{-1}\sigma_k^{1/\gamma} \leqslant |p_m|^{-1}\sigma_k(1-\gamma)^{1/\gamma - 1} \tag{3.6-23c}$$

for $1 \leqslant k \leqslant m \leqslant t$.

To understand what these conditions mean in a concrete situation consider the problem of continuing the function $f(z):=\text{Log}\,z$ along the unit circle, starting at $z_0:=1$. Choosing equidistant continuation points

$$z_k := e^{ik\alpha}, \quad k=0,1,2,\ldots,$$

we have

$$|h_m|=2\sin\frac{\alpha}{2}, \quad |p_m|=1, \quad \rho(z_m)=1,$$

hence $\sigma_k > 2\sin(\alpha/2)$. If, for instance, $\alpha = \pi/6$, a numerical check shows that the conditions in (3.6-23) are satisfied for $\gamma = \frac{1}{5}$ but not for $\gamma = \frac{1}{4}$. Because $n\gamma^t$ must be a positive integer, this requires matrices of high order n already when t, the number of continuation steps, is small.

SEMINAR ASSIGNMENTS

1. The series

$$\sum_{n=1}^{\infty} \frac{(-1)^{n-1}}{n^2} z^n$$

has radius of convergence 1; its sum cannot be expressed in terms of elementary functions. Continue the function defined by the series beyond the unit disk and calculate its values for $z = 1.1, 1.2, 1.3$.
2. Make a study (theoretical and experimental) of the numerical version of Weierstrassian analytic continuation, described in §3.6.
3. Implement the method of continuation stated in problems 4 and 7 of §3.2 and apply it to continue the function defined by the hypergeometric series

$$f(z) := {}_2F_1(a,b;c;1-z)$$

for various values of a,b,c.

NOTES

§3.3. Proof of Theorem 3.3a adapted from Hurwitz & Courant [1929], p. 51.

§3.4. For Example 1 and further examples of this kind see Hardy & Wright [1954] Chapter 17.

§3.5. For an elegant treatment of elementary homotopy theory see Redheffer [1969].

§3.6. The method of continuation discussed here was proposed (with an application in fluid dynamics) by Lewis [1960] and formulated in matrix language by Henrici [1966]. The idea can already be found in Painlevé [1899]. For other methods of continuation see van Wijngaarden [1953], Kublanowskaya [1959], Cann & K. Miller [1965], K. Miller [1970] (large bibliography with physical applications).

4

COMPLEX INTEGRATION

The integrals considered in complex analysis are a generalization of the integrals considered in calculus, in two ways: the integrands are complex-valued and the integrals are extended over arcs and closed curves in the complex plane rather than over intervals.

The presentation of complex analysis in preceding chapters has been dominated by algebraical and analytical considerations. For the development of integration theory we must now also draw on some simple notions from topology. In that sense complex integration theory stands on a higher plane than the theory of power series. Not surprisingly, complex integration turns out to be a powerful tool for the further development of the theory. In addition, there are many applications to topics outside complex analysis, such as the computation of improper real integrals, the definition of functions with special properties, and the summation of infinite series.

§4.1. COMPLEX FUNCTIONS OF A REAL VARIABLE

Continuous complex functions of a real variable were introduced and given the name *arc* in §3.5. Here we shall study some of their properties less dependent on the geometric interpretation. If f is such a function, defined, say, on the interval $I := [\alpha, \beta]$, it can always be expressed in terms of two real functions defined on I, giving, respectively, the real and the imaginary part of f:

$$f(\tau) = u(\tau) + iv(\tau), \quad \tau \in I. \tag{4.1-1}$$

Much of the material of the ordinary calculus carries over to these functions. We define the *derivative of f at the point $\tau \in I$* by

$$f'(\tau) := \lim_{\sigma \to \tau} \frac{f(\sigma) - f(\tau)}{\sigma - \tau}, \tag{4.1-2}$$

COMPLEX FUNCTIONS OF A REAL VARIABLE

provided that the limit exists. It is seen immediately that this derivative exists if and only if the derivatives of the real functions u and v exist at the point τ and, if so, that

$$f'(\tau) = u'(\tau) + iv'(\tau). \tag{4.1-3}$$

If the derivative exists at every point of I, f is called *differentiable on I*. The values of the derivative then define a new function called the **derivative** of f and denoted by f'. It is clear that f' is the zero function if and only if f is a constant function.

If f and g are differentiable complex-valued functions on I, we derive from (4.1-2) by the methods used in the calculus [or less directly from (4.1-3) by formal manipulation] the usual differentiation rules

$$(f+g)' = f' + g', \tag{4.1-4}$$

$$(fg)' = f'g + fg', \tag{4.1-5}$$

and, if $f(\tau) \neq 0$ for $\tau \in I$,

$$\left(\frac{1}{f}\right)' = -\frac{f'}{f^2}. \tag{4.1-6}$$

From (4.1-5) there follows by induction for arbitrary integers $n \geq 0$

$$(f^n)' = nf^{n-1}f', \tag{4.1-7}$$

and by (4.1-6) this also holds for integers $n < 0$ if $f(\tau) \neq 0$. Finally, the chain rule holds in the following sense: if f is differentiable on I and ϕ is a real differentiable function defined on some interval J with values in I, the function $f \circ \phi$ defined on J by

$$(f \circ \phi)(\tau) := f(\phi(\tau))$$

is differentiable on J and

$$(f \circ \phi)' = (f' \circ \phi)\phi'. \tag{4.1-8}$$

If f is a continuous complex-valued function on an interval $I := [\alpha, \beta]$, the (Riemann[1]) integral of f over I is defined exactly as in the calculus by forming Riemann sums. Formula (4.1-9) results:

$$\int_\alpha^\beta f(\tau)\,d\tau = \int_\alpha^\beta u(\tau)\,d\tau + i\int_\alpha^\beta v(\tau)\,d\tau, \tag{4.1-9}$$

---

[1] Until Chapter 12 all integrals considered in this work are understood to be Riemann integrals.

which may be taken as an equivalent definition of the integral. It follows that for arbitrary complex constants c,

$$\int_\alpha^\beta cf(\tau)\,d\tau = c\int_\alpha^\beta f(\tau)\,d\tau; \tag{4.1-10}$$

moreover, the triangle inequality applied to the Riemann sums implies

$$\left|\int_\alpha^\beta f(\tau)\,d\tau\right| \leq \int_\alpha^\beta |f(\tau)|\,d\tau. \tag{4.1-11}$$

It follows from (4.1-3) and (4.1-9) that if f is the derivative of another complex-valued function g then

$$\int_\alpha^\beta f(\tau)\,d\tau = g(\beta) - g(\alpha). \tag{4.1-12}$$

If f itself has a continuous derivative, (4.1-7) implies that for $n=0,1,2,\ldots,$

$$\int_\alpha^\beta [f(\tau)]^n f'(\tau)\,d\tau = \frac{1}{n+1}\left\{[f(\beta)]^{n+1} - [f(\alpha)]^{n+1}\right\}. \tag{4.1-13}$$

Another fundamental property that carries over from real integrals is the validity of the rule for changing the variable of integration. Let ϕ be a continuously differentiable, monotonically increasing function mapping $J:=[\gamma,\delta]$ onto I. The chain rule (4.1-8) then implies that

$$\int_\alpha^\beta f(\tau)\,d\tau = \int_\gamma^\delta f(\phi(\sigma))\,\phi'(\sigma)\,d\sigma. \tag{4.1-14}$$

Finally, we require the fact that *a uniformly convergent series of complex continuous functions of a real variable may be integrated term by term.* This again follows from the corresponding result in real calculus.

EXAMPLE 1

Let ω be a real constant. The complex-valued function of the real variable τ,

$$f(\tau) := e^{i\omega\tau} = \cos\omega\tau + i\sin\omega\tau,$$

has the derivative

$$f'(\tau) = -\omega\sin\omega\tau + i\omega\cos\omega\tau = i\omega e^{i\omega\tau}.$$

COMPLEX FUNCTIONS OF A REAL VARIABLE

Thus, if $\omega \neq 0$, $f(\tau) = g'(\tau)$, where $g(\tau) := (1/i\omega)e^{i\omega\tau}$; hence

$$\int_\alpha^\beta e^{i\omega\tau} d\tau = \frac{1}{i\omega}(e^{i\omega\beta} - e^{i\omega\alpha}).$$

As a special case we note that when n is an integer,

$$\int_0^{2\pi} e^{in\tau} d\tau = \begin{cases} 2\pi, & \text{if } n=0, \\ 0, & \text{if } n \neq 0. \end{cases} \qquad (4.1\text{-}15)$$

These results are often used in connection with Euler's formulas (2.5-9) in the evaluation of integrals involving trigonometric functions; for instance, if n is a positive integer,

$$\int_0^{2\pi} (\cos\tau)^{2n} d\tau = \frac{1}{2^{2n}} \int_0^{2\pi} (e^{i\tau} + e^{-i\tau})^{2n} d\tau$$

$$= \frac{1}{2^{2n}} \int_0^{2\pi} \sum_{k=0}^{2n} \binom{2n}{k} e^{i(2n-2k)\tau} d\tau.$$

By virtue of (4.1-15) only the integral of the term where $k = n$ does not vanish, and we obtain

$$\int_0^{2\pi} (\cos\tau)^{2n} d\tau = 2^{-2n} \binom{2n}{n} 2\pi.$$

Analytic Functions Defined by Definite Integrals. Let $[\alpha, \beta]$ be a bounded interval and S, a region in the complex plane, and let $f = f(\tau, z)$ be defined for all $\tau \in [\alpha, \beta]$ and all $z \in S$. If we assume that for every fixed $z \in S$, $f(\tau, z)$ is a continuous function of τ, then for every such z

$$g(z) := \int_\alpha^\beta f(\tau, z) d\tau \qquad (4.1\text{-}16)$$

exists. Moreover, if f is a continuous function of the point (τ, z), then, by a well-known theorem of the calculus, g is a continuous function of z. Of even greater interest, however, is the case in which for each fixed τ, $f(\tau, z)$ is *analytic* as a function of z. This is covered by the following result:

THEOREM 4.1a

If, in addition to the hypotheses made above, $f(\tau, z)$ for each fixed $\tau \in [\alpha, \beta]$ is analytic for $z \in S$, then the function g defined by (4.1-16) is analytic in S.

Moreover, the derivatives of g may be calculated by differentiating under the integral sign:

$$g^{(k)}(z) = \int_\alpha^\beta \frac{\partial^k f}{\partial z^k}(\tau,z)\,d\tau, \qquad k=1,2,\dots$$

Proof. For each positive integer n let

$$\tau_j := \alpha + j\frac{\beta-\alpha}{n}, \qquad j=0,1,\dots,n.$$

For $n=1,2,\dots$ and $z\in S$ we define

$$g_n(z) := \frac{\beta-\alpha}{n}\sum_{j=1}^n f(\tau_j,z).$$

By definition of the Riemann integral

$$\lim_{n\to\infty} g_n(z) = g(z)$$

for each $z \in S$. The assertions of the theorem will follow from Theorem 3.4b and its corollary if we can show that the convergence of the sequence $\{g_n\}$ is locally uniform on S (i.e., uniform on every compact subset of S).

Let T be a compact subset of S. Because the set of all (τ,z) such that $\tau \in [\alpha,\beta]$ and $z \in T$ is compact, f is uniformly continuous on it. Thus, if for $\delta > 0$ we define

$$\omega(\delta) := \sup|f(\tau,z) - f(\tau',z)|,$$

where the supremum is taken with respect to all $z \in T$ and all $\tau, \tau' \in [\alpha,\beta]$ such that $|\tau-\tau'| \leq \delta$, then

$$\lim_{\delta\to 0} \omega(\delta) = 0. \qquad (4.1\text{-}17)$$

Because we may write

$$g(z) = \sum_{j=1}^n \int_{\tau_{j-1}}^{\tau_j} f(\tau,z)\,d\tau,$$

$$g_n(z) = \sum_{j=1}^n \int_{\tau_{j-1}}^{\tau_j} f(\tau_j,z)\,d\tau,$$

COMPLEX FUNCTIONS OF A REAL VARIABLE

we have

$$g(z) - g_n(z) = \sum_{j=1}^{n} \int_{\tau_{j-1}}^{\tau_j} [f(\tau,z) - f(\tau_j,z)] \, d\tau$$

and thus, if $z \in T$,

$$|g(z) - g_n(z)| \leq (\beta - \alpha) \omega\left(\frac{\beta - \alpha}{n}\right),$$

which proves the uniform convergence according to (4.1-17). ∎

EXAMPLE 2

If for all complex z we define

$$g(z) := \frac{1}{\pi} \int_0^{\pi} \cos(z \sin \tau) \, d\tau,$$

then because the integrand for each $\tau \in [0, \pi]$ is an entire function of z the function g is likewise entire. Moreover,

$$g'(z) = -\frac{1}{\pi} \int_0^{\pi} \sin \tau \sin(z \sin \tau) \, d\tau,$$

$$g''(z) = -\frac{1}{\pi} \int_0^{\pi} (\sin \tau)^2 \cos(z \sin \tau) \, d\tau,$$

and so on. In §4.5 we identify g as the Bessel function of order zero, ordinarily denoted by $J_0(z)$.

EXAMPLE 3

If τ^{z-1} is defined by its principal value, the integral

$$h(z) := \int_0^{\infty} e^{-\tau} \tau^{z-1} \, d\tau \qquad (4.1\text{-}18)$$

clearly converges for $\operatorname{Re} z > 0$, since

$$|\tau^{z-1}| = |e^{(z-1)\operatorname{Log}\tau}| = e^{\operatorname{Re}(z-1)\operatorname{Log}\tau} = \tau^{\operatorname{Re} z - 1}.$$

Although for each $\tau > 0$ the integrand is an entire function of z it does not follow from Theorem 4.1a that h is analytic in any region because the interval of integration is infinite. Also, if $\operatorname{Re} z < 1$ the integrand is not continuous at $\tau = 0$. We can show, however, that h is analytic in the right half-plane $\operatorname{Re} z > 0$ by another appeal to Theorem 3.4b. For $n = 1, 2, \ldots$, let

$$h_n(z) := \int_{1/n}^{n} e^{-\tau} \tau^{z-1} d\tau.$$

Each h_n is analytic in $\operatorname{Re} z > 0$ (in fact, entire) by virtue of Theorem 4.1a. If $\operatorname{Re} z > 0$, then clearly $h_n(z) \to h(z)$. The analyticity of h follows if the convergence is locally uniform. Let T be a compact subset of $\operatorname{Re} z > 0$ and let

$$\delta := \inf_{z \in T} \operatorname{Re} z, \qquad \mu := \sup_{z \in T} \operatorname{Re} z.$$

Then $\delta > 0$, $\mu < \infty$; hence for $z \in T$

$$|h(z) - h_n(z)| = \left| \int_0^{1/n} e^{-\tau} \tau^{z-1} d\tau \right| + \left| \int_n^{\infty} e^{-\tau} \tau^{z-1} d\tau \right|$$

$$\leq \int_0^{1/n} e^{-\tau} \tau^{\delta-1} d\tau + \int_n^{\infty} e^{-\tau} \tau^{\mu-1} d\tau.$$

The two integrals on the right are independent of z and tend to zero for $n \to \infty$, proving the uniform convergence of the sequence $\{h_n\}$ on T. It follows that h is analytic for $\operatorname{Re} z > 0$. Moreover,

$$h^{(k)}(z) = \int_0^{\infty} e^{-\tau} (\operatorname{Log} \tau)^k \tau^{z-1} d\tau, \qquad k = 1, 2, \ldots;$$

h is identified in §8.5 as the gamma function, usually denoted by $\Gamma(z)$.

PROBLEMS

1. If $\Gamma : z = z(\tau)$, $\alpha \leq \tau \leq \beta$, is a sufficiently differentiable curve, the *curvature* of Γ (as defined in calculus) at the point $z(\tau)$ is given by

$$\kappa(\tau) := \frac{1}{|z'(\tau)|} \operatorname{Im} \frac{z''(\tau)}{z'(\tau)}.$$

The quantity $\rho(\tau) := 1/\kappa(\tau)$ is known as the *radius of curvature* of Γ at $z(\tau)$. The point located on the normal of Γ through $z(\tau)$ at a distance $\rho(\tau)$ from $z(\tau)$ is called the *center of curvature* of Γ at $z(\tau)$. The locus of all centers of curvature of Γ by definition is the *evolute* of Γ.

Show that the evolute of Γ has the parametric representation

$$z = z(\tau) + \frac{iz'(\tau)}{\operatorname{Im} z''(\tau)/z'(\tau)}, \qquad \alpha \leq \tau \leq \beta.$$

2. Show that the evolute of the cycloid

$$z(\tau) := \mu(i + \tau - ie^{-i\tau}), \qquad -\infty < \tau < \infty,$$

INTEGRAL OF A FUNCTION ALONG AN ARC

is again a cycloid, up to a translation.

3. Let $\alpha > \beta > 0$. Show that the evolute of the ellipse

$$z = \alpha \cos \tau + i\beta \sin \tau, \quad 0 \leqslant \tau \leqslant 2\pi,$$

has the parametric representation

$$z = \frac{\epsilon^2}{\alpha}(\cos \tau)^3 - i\frac{\epsilon^2}{\beta}(\sin \tau)^2,$$

where $\epsilon^2 := \alpha^2 - \beta^2$.

4. Let $\xi > 0$, and let ϕ be real. Evaluate the integral

$$\int_0^\xi \text{Log}(1 + 2\tau \cos \phi + \tau^2)^{1/2} d\tau$$

by considering it as the real part of an appropriate complex integral. As a by-product, compute

$$\int_0^\xi \text{Arctan}\, \tau\, d\tau.$$

§4.2. THE INTEGRAL OF A FUNCTION ALONG AN ARC

We recall the concept of an **arc** which was introduced in §3.5. Let Γ be an arc, given in parametric representation by $z = z(\tau)$, $\alpha \leqslant \tau \leqslant \beta$, and let f be a complex-valued function defined on Γ (i.e., on the set of points z of the form $z(\tau)$ for some $\tau \in [\alpha, \beta]$). A **subdivision** Δ of the parametric interval $[\alpha, \beta]$ is a system of a finite number of points τ_k ($k = 0, 1, \ldots, n$) such that $\alpha = \tau_0 < \tau_1 < \tau_2 < \cdots \tau_n = \beta$. The **norm of the subdivision** Δ is the positive number

$$|\Delta| := \max_{1 \leqslant k \leqslant n} (\tau_k - \tau_{k-1}).$$

A **system of pivotal points** Θ, consistent with the subdivision $\Delta = (\tau_0, \tau_1, \ldots, \tau_n)$, is a set of numbers τ_k' that satisfies $\tau_{k-1} \leqslant \tau_k' \leqslant \tau_k$, $k = 1, 2, \ldots, n$. In accordance with each subdivision Δ and each consistent system of pivotal points Θ we can form the sum

$$S(\Delta, \Theta) := \sum_{k=1}^n f(z(\tau_k'))[z(\tau_k) - z(\tau_{k-1})]. \tag{4.2-1}$$

If there exists a complex number A such that, given any $\epsilon > 0$, a number $\delta = \delta(\epsilon)$ can be found with the property that for all subdivisions Δ satisfying $|\Delta|<\delta$ and all consistent systems of pivotal points Θ,

$$|S(\Delta,\Theta)-A|<\epsilon,$$

then A is called the **integral of f along** Γ, and we write[2]

$$A =: \int_\Gamma f(z)\,dz. \tag{4.2-2}$$

EXAMPLE 1

Let $\Gamma: z=\tau$, $\alpha \leqslant \tau \leqslant \beta$, be a straight line segment on the real axis oriented in the direction of increasing Re z. The integral defined above is then equivalent to the usual definition of the integral

$$\int_\alpha^\beta f(\tau)\,d\tau$$

as a limit of Riemann sums. Thus the integral of a complex function of a real variable is contained in this definition as a special case.

There naturally arises the question of the curves Γ and functions f such that the integral (4.2-2) can be asserted to exist a priori. We mention without proof that the integral always exists if Γ is rectifiable and f is continuous on Γ. Here a curve $\Gamma: z=z(\tau)$, $\alpha \leqslant \tau \leqslant \beta$, is called **rectifiable** if the supremum of the sums

$$\sum_{k=1}^n |z(\tau_k)-z(\tau_{k-1})|,$$

taken with respect to all subdivisions Δ of the parametric interval, is finite. The value of the supremum, if finite, is called the **length** of Γ.

Whenever the integral is known to exist, its value can be computed as the limit of the sums $S(\Delta,\Theta)$ for one particular sequence $\{\Delta_n\}$ of subdivisions such that $|\Delta_n|\to 0$ for $n\to\infty$ and one sequence $\{\Theta_n\}$ of compatible systems of pivotal points.

EXAMPLE 2

To compute

$$\int_\Gamma z\,dz,$$

---

[2] As in calculus, the notation, although suggestive, is redundant. It would suffice to denote the integral by a symbol such as $\int_\Gamma f$.

INTEGRAL OF A FUNCTION ALONG AN ARC

where Γ denotes the positively oriented unit circle, $z = e^{i\tau}$, $0 \leq \tau \leq 2\pi$. For each positive integer n we consider the subdivision defined by the points $\tau_k := 2\pi k/n$, $k = 0, 1, \ldots, n$. Writing $q := \exp(2\pi i/n)$, we have $z(\tau_k) = q^k$. Selection of the pivotal points $\tau'_k := \tau_{k-1}$ yields the Riemann sums

$$S(\Delta_n, \Theta_n) = \sum_{k=0}^{n-1} q^k (q^{k+1} - q^k).$$

Evaluating a geometric sum, we get

$$S(\Delta_n, \Theta_n) = (q-1) \sum_{k=0}^{n-1} q^{2k} = \frac{q^{2n}-1}{q+1} = 0,$$

since $q^n = 1$. Thus all sums are zero and we conclude that

$$\int_\Gamma z\, dz = 0.$$

EXAMPLE 3

To evaluate

$$\int_\Gamma \bar{z}\, dz,$$

where Γ has the same meaning as above, we use the same subdivisions and systems of pivotal points as in Example 2. We now have

$$S(\Delta_n, \Theta_n) = \sum_{k=0}^{n-1} \bar{q}^k (q^{k+1} - q^k);$$

hence, since $q\bar{q} = 1$,

$$S(\Delta_n, \Theta_n) = n(q-1) = n(e^{2\pi i/n} - 1).$$

To compute the limit as $n \to \infty$, we set $h := 2\pi i/n$ and find by using the Taylor series for e^h at $h=0$

$$\lim_{n\to\infty} S(\Delta_n, \Theta_n) = \lim_{h\to 0} \frac{2\pi i}{h} (e^h - 1) = 2\pi i.$$

Thus

$$\int_\Gamma \bar{z}\, dz = 2\pi i.$$

Because $\bar{z} = z^{-1}$ holds for points z on Γ, we also have proved the important result

$$\int_\Gamma \frac{1}{z}\, dz = 2\pi i.$$

To compute complex line integrals by taking limits of Riemann sums is extremely laborious. Fortunately in most applications the arcs one has to deal with are piecewise regular or at least piecewise differentiable (see §3.5 for definitions). In this case there is a much easier method.

THEOREM 4.2a

If $\Gamma: z = z(\tau)$, $\alpha \leqslant \tau \leqslant \beta$, *is a piecewise differentiable arc and f is continuous on* Γ, *there holds*

$$\int_\Gamma f(z)\,dz = \int_\alpha^\beta f(z(\tau))\,z'(\tau)\,d\tau. \qquad (4.2\text{-}3)$$

Proof. Since a piecewise differentiable arc is composed of a finite number of differentiable arcs, it suffices to establish the result for differentiable arcs Γ. Let A^* denote the integral on the right of (4.2-3). By the definition of the integral of a complex function over a real interval A^* is the limit of sums

$$S^*(\Delta, \Theta) := \sum_{k=1}^n f(z(\tau_k'))\,z'(\tau_k')(\tau_k - \tau_{k-1})$$

as $|\Delta| \to 0$, where Δ and Θ have the same meaning as before. Let

$$\mu := \sup_{\alpha \leqslant \tau \leqslant \beta} |f(z(\tau))|;$$

$\mu < \infty$ because f is continuous. The difference between S^* and the corresponding sum $S(\Delta, \Theta)$ is bounded by

$$\mu \sum_{k=1}^n |z(\tau_k) - z(\tau_{k-1}) - z'(\tau_k')(\tau_k - \tau_{k-1})|.$$

Here

$$z(\tau_k) - z(\tau_{k-1}) - z'(\tau_k')(\tau_k - \tau_{k-1}) = \int_{\tau_{k-1}}^{\tau_k} (z'(\tau) - z'(\tau_k'))\,d\tau,$$

which in view of (4.1-11) in absolute value is bounded by $\omega(|\Delta|)(\tau_k - \tau_{k-1})$, where for $\delta > 0$

$$\omega(\delta) := \sup_{|\tau - \tau^*| < \delta} |z'(\tau) - z'(\tau^*)|$$

is the modulus of continuity of z'. Thus

$$|S(\Delta, \Theta) - S^*(\Delta, \Theta)| \leqslant \mu(\beta - \alpha)\,\omega(|\Delta|).$$

INTEGRAL OF A FUNCTION ALONG AN ARC

Since Γ is differentiable, $z'(\tau)$ by definition is continuous on the compact interval $[\alpha,\beta]$; thus $\omega(\delta) \to 0$ as $\delta \to 0$. Hence if $|\Delta|$ is sufficiently small the sums S differ arbitrarily little from the corresponding sums S^*. Because the latter tend to a limit, the former must do likewise and the limits are the same. ∎

EXAMPLE 4

For the integrals considered in Examples 2 and 3 we obtain in view of $z'(\tau) = ie^{i\tau}$

$$\int_\Gamma z\,dz = \int_0^{2\pi} e^{i\tau} \cdot ie^{i\tau} d\tau = i \int_0^{2\pi} e^{2i\tau} d\tau = 0$$

and

$$\int_\Gamma \bar{z}\,dz = \int_0^{2\pi} e^{-i\tau} \cdot ie^{i\tau} d\tau = i \int_0^{2\pi} d\tau = 2\pi i$$

as before.

EXAMPLE 5

More generally, let m be any integer and Γ denote the circle $z = a + \rho e^{i\tau}$, $0 \leq \tau \leq 2\pi$, where $\rho > 0$. Then

$$\int_\Gamma (z-a)^m dz = \int_0^{2\pi} \rho^m e^{im\tau} \rho i e^{i\tau} d\tau$$

$$= i\rho^{m+1} \int_0^{2\pi} e^{i(m+1)\tau} d\tau;$$

$$\int_\Gamma (z-a)^m dz = \begin{cases} 0, & m \neq -1, \\ 2\pi i, & m = -1. \end{cases} \quad (4.2\text{-}4)$$

According to Theorem 4.2a, as long as we are dealing with continuous functions and piecewise regular or differentiable arcs, we may consider the integral of f along Γ to be defined by (4.2-3) which, although carrying less intuitive appeal, is more manageable analytically than the original definition. On this basis we now prove several properties of the integral of a function along an arc.

Let $\Gamma_1 : z = z_1(\tau)$, $\alpha \leq \tau \leq \beta$, and $\Gamma_2 : z = z_2(\sigma)$, $\gamma \leq \sigma \leq \delta$, be two regular, geometrically equivalent arcs. Then by definition (see §3.5) there is an increasing function ϕ from $[\gamma,\delta]$ onto $[\alpha,\beta]$ such that $z_1(\phi(\sigma)) = z_2(\sigma)$, $\gamma \leq \sigma \leq \delta$, and $\phi'(\sigma) = z_2'(\sigma)/z_1'(\phi(\sigma))$ is continuous and different from zero.

Then by (4.1-14), if f is continuous,

$$\int_{\Gamma_1} f(z)\, dz = \int_\alpha^\beta f(z_1(\tau)) z_1'(\tau)\, d\tau$$

$$= \int_\gamma^\delta f(z_1(\phi(\sigma))) z_1'(\phi(\sigma)) \phi'(\sigma)\, d\sigma$$

$$= \int_\gamma^\delta f(z_2(\sigma)) z_2'(\sigma)\, d\sigma = \int_{\Gamma_2} f(z)\, dz.$$

Thus *the integral of f has the same value along all geometrically equivalent regular arcs.*

Let $\Gamma : z = z(\tau)$, $\alpha \leq \tau \leq \beta$, be a differentiable arc. The **opposite arc** is then defined by $z = z(-\tau)$, $-\beta \leq \tau \leq -\alpha$, and denoted by $-\Gamma$. The opposite arc traces the same points as Γ but in the opposite direction. From the definition of the integral by Riemann sums it is clear that

$$\int_{-\Gamma} f(z)\, dz = -\int_\Gamma f(z)\, dz, \tag{4.2-5}$$

which can also be verified in a simple calculation by using (4.2-3).

If Γ is a differentiable arc and c is a complex number, it is clear that

$$\int_\Gamma cf(z)\, dz = c \int_\Gamma f(z)\, dz.$$

Also, if both f and g are continuous functions on Γ, then

$$\int_\Gamma (f(z) + g(z))\, dz = \int_\Gamma f(z)\, dz + \int_\Gamma g(z)\, dz.$$

Thus the integral is additive with respect to the integrated functions. In a sense it is also additive with respect to the path of integration. Let Γ_1 and Γ_2 be two arcs such that the terminal point of Γ_1 equals the initial point of Γ_2. By $\Gamma_1 + \Gamma_2$ we then mean any arc obtained by tracing the arcs Γ_1 and Γ_2 in succession and referred (by a change of parameters, if necessary) to a single parameter interval. If Γ_1 and Γ_2 are piecewise regular or differentiable, so is $\Gamma_1 + \Gamma_2$. If f is continuous on both Γ_1 and Γ_2, (4.2-3) shows that

$$\int_{\Gamma_1 + \Gamma_2} f(z)\, dz = \int_{\Gamma_1} f(z)\, dz + \int_{\Gamma_2} f(z)\, dz. \tag{4.2-6}$$

The integral defined as the limit of the sums $S(\Delta, \Theta)$ (also called the integral with respect to z) is by far the most important line integral that occurs in complex analysis. Two other integrals are of occasional use. The

INTEGRAL OF A FUNCTION ALONG AN ARC

first is the integral with respect to $\bar{z}$, denoted by $\int_\Gamma f(z)\,\overline{dz}$ and defined as the limit of the sums

$$\sum_{k=1}^{n} f(z(\tau'_k))\left(\overline{z(\tau_k)} - \overline{z(\tau_{k-1})}\right)$$

analogous to (4.2-1). It is clear from this definition that for continuous f and rectifiable arcs

$$\int_\Gamma f(z)\,\overline{dz} = \overline{\left(\int_\Gamma \overline{f(z)}\,dz\right)}, \qquad (4.2\text{-}7)$$

and if $\Gamma: z = z(\tau)$, $\alpha \le \tau \le \beta$, is piecewise differentiable,

$$\int_\Gamma f(z)\,\overline{dz} = \int_\alpha^\beta f(z(\tau))\,\overline{z'(\tau)}\,d\tau. \qquad (4.2\text{-}8)$$

Formulas analogous to (4.2-5) and (4.2-6) obviously hold.

In addition there is the **integral with respect to arc length**, denoted by $\int_\Gamma f(z)|dz|$ and defined as limit of the sums

$$\sum_{k=1}^{n} f(z(\tau'_k))|z(\tau_k) - z(\tau_{k-1})|.$$

If Γ is piecewise differentiable, it can be shown that

$$\int_\Gamma f(z)|dz| = \int_\alpha^\beta f(z(\tau))|z'(\tau)|\,d\tau, \qquad (4.2\text{-}9)$$

which may serve as an equivalent definition. Again for a given continuous f this integral has the same value for all geometrically equivalent arcs, but in contrast to (4.2-5) we now have

$$\int_{-\Gamma} f(z)|dz| = \int_\Gamma f(z)|dz|. \qquad (4.2\text{-}10)$$

By applying the triangle inequality to the Riemann sums involved in the definitions, or for differentiable arcs by applying (4.1-11) to (4.2-3), we obtain

$$\left|\int_\Gamma f(z)\,dz\right| \le \int_\Gamma |f(z)||dz|, \qquad (4.2\text{-}11)$$

a fundamental inequality for the absolute value of a complex line integral. It often suffices to use this estimate in the following crude form:

THEOREM 4.2b

Let Γ be a piecewise regular arc and let f be continuous on Γ. If λ denotes the length of Γ and $|f(z)| \leq \mu$ for all $z \in \Gamma$, then

$$\left| \int_\Gamma f(z)\, dz \right| \leq \lambda\mu. \qquad (4.2\text{-}12)$$

PROBLEMS

1. Let $\Gamma: z = z(\tau)$, $\alpha \leq \tau \leq \beta$, be a simple, closed, positively oriented, piecewise regular curve (see §4.5 for definitions). It may be shown that the area of the interior of Γ is given by

$$A = \frac{1}{2i} \int_\Gamma \bar{z}\, dz. \qquad (4.2\text{-}13)$$

Give a calculus-type proof of this formula for the case in which Γ is starlike (see §4.3) with respect to 0. Derive from (4.2-13) the area formulas

$$A = \tfrac{1}{2} \int_\Gamma y\, dx - x\, dy \quad \text{and} \quad A = \tfrac{1}{2} \int_0^{2\pi} \rho^2(\theta)\, d\theta$$

proved in calculus.

2. Let $|a| > |b|$. Show that the curve

$$z = ae^{i\tau} + be^{-i\tau}, \quad 0 \leq \tau \leq 2\pi,$$

is an ellipse and calculate the area of its interior.

3. Calculate the area enclosed by the cardoid

$$z = 2ie^{i\tau} - ie^{2i\tau}, \quad 0 \leq \tau \leq 2\pi.$$

4. A disk of radius $\rho > 0$ rolls on the circumference of a fixed disk of radius $m\rho$ (m a positive integer). Compute the area enclosed by the curve described by a point on the circumference of the small disk (epicycloid).

5. (application of Theorem 4.1a). Let Γ be a piecewise regular curve and let h be a continuous function on Γ. Show that the function

$$f(z) := \int_\Gamma \frac{h(t)}{t - z}\, dt$$

is analytic on the complement of Γ and compute its derivatives.

§4.3. INTEGRALS OF ANALYTIC FUNCTIONS

In the definition of complex line integrals no hypothesis was made on the integrated function except continuity. We now consider integrals of functions f that are *analytic* in a region R. Of special importance are integrals taken along **closed curves**, i.e., arcs whose initial and terminal points coincide (see Fig. 4.3a). (The position of the initial and terminal point on a closed curve is immaterial; two closed curves are considered geometrically equivalent also if they are related by a sense-preserving change of parameter that moves this point along the curve.) All arcs considered henceforth are piecewise differentiable.

We begin by showing that the propositions A, B, and C stated below are equivalent, without asserting that any of them are true.

PROPOSITION A

If Γ is any closed curve in R, then

$$\int_\Gamma f(z)\,dz = 0. \qquad (4.3\text{-}1)$$

PROPOSITION B

For all arcs Γ in R, the value of the integral $\int_\Gamma f(z)\,dz$ depends only on the initial and the terminal point of Γ.

Proof that B implies A. Let B be true and let Γ be a closed curve in R. Then, because Γ and $-\Gamma$ have the same end points,

$$\int_\Gamma = \int_{-\Gamma};$$

but by (4.2-5), because the sense has been reversed,

$$\int_{-\Gamma} = -\int_\Gamma.$$

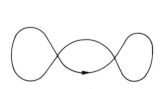

Fig. 4.3a. Piecewise differentiable closed curves.

There follows

$$\int_\Gamma = -\int_{\Gamma'}, \quad \text{i.e.,} \quad \int_\Gamma = 0.$$

Proof that A implies B. Let A be true and let p and q be any two points in R. Then, if Γ_1 and Γ_2 are any two arcs from p to q, $\Gamma_1 - \Gamma_2$ is a closed curve; hence

$$\int_{\Gamma_1 - \Gamma_2} = \int_{\Gamma_1} - \int_{\Gamma_2} = 0$$

or

$$\int_{\Gamma_1} = \int_{\Gamma_2}. \quad\blacksquare$$

We now state a third proposition that is analytically more tractable than A or B.

PROPOSITION C

There exists a function g, analytic in R, such that

$$g'(z) = f(z), \quad z \in R. \tag{4.3-2}$$

Proof that B implies C. Assume that B is true. Select a fixed point $p \in R$ and define

$$g(z) := \int_\Gamma f(t)\,dt,$$

where Γ is any arc from p to z. (By virtue of B the integral for fixed p depends only on z and thus really defines a function of z.) Let z_0 be an arbitrary point of R. We shall show that g is analytic at z_0. Since f is analytic at z_0 we have, for some $\rho > 0$ and for $|h| < \rho$, where $h := z - z_0$ (see Fig. 4.3b),

$$f(z) = a_0 + a_1 h + a_2 h^2 + \cdots.$$

For $|z - z_0| < \rho$ let $g(z)$ be calculated by first integrating from p to z_0 [which integral yields $g(z_0)$] and then from z_0 to z along a straight line segment. We then have

$$g(z) - g(z_0) = \int_{z_0}^{z} f(t)\,dt.$$

INTEGRALS OF ANALYTIC FUNCTIONS

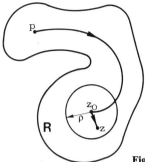

Fig. 4.3b. Choice of path Γ.

We parametrize the straight line segment by setting

$$t = z_0 + \tau h, \quad 0 \leq \tau \leq 1.$$

This yields

$$g(z) - g(z_0) = \int_0^1 \left[a_0 + a_1 \tau h + a_2 (\tau h)^2 + \cdots \right] h \, d\tau.$$

Because of the uniform convergence of the series under the integral, we can integrate term by term and find

$$g(z) - g(z_0) = a_0 h + a_1 \frac{h^2}{2} + a_2 \frac{h^3}{3} + \cdots, \quad |h| < \rho.$$

This expansion shows that (a) g is analytic at z_0 and (b) $g'(z_0) = a_0 = f(z_0)$. Because z_0 was arbitrary, we have shown that B implies C.

To show that C implies B let C be true, let p and q be any two points in R, and let $\Gamma: z = z(\tau)$, $0 \leq \tau \leq 1$, be an arc from p to q. Then

$$\int_\Gamma f(z)\, dz = \int_\Gamma g'(z)\, dz = \int_0^1 g'(z(\tau)) z'(\tau)\, d\tau$$

$$= \int_0^1 \frac{d}{d\tau} (g(z(\tau)))\, d\tau = g(z(1)) - g(z(0))$$

$$= g(q) - g(p),$$

which proves that the integral depends only on the terminal and on the initial points of Γ but not on the shape of Γ. ∎

Altogether we have established:

THEOREM 4.3a

Let the function f be analytic in some region R. Then the propositions A, B, C are equivalent.

As a first application we prove:

THEOREM 4.3b

Let a be a point in the complex plane, let n be an integer, $n \neq -1$, and let Γ be a closed curve which, if $n < -1$, does not pass through a. Then

$$\int_\Gamma (z-a)^n dz = 0. \qquad (4.3\text{-}3)$$

Proof. If $n \geq 0$, let R be the complex plane; if $n < -1$, let R be the complex plane without point a. In either case the function $f(z) := (z-a)^n$ is the derivative of the function

$$g(z) := \frac{1}{n+1}(z-a)^{n+1},$$

which is analytic in R. The assertion now follows from the equivalence of the propositions A and C. ∎

The method of this proof breaks down if $n = -1$, and indeed we know from Example 5 of §4.2 that (4.3-3) does not hold in this case. Far from being a nuisance, this fact serves as the basis of the calculus of residues which is discussed in subsequent sections.

Theorem 4.3b shows that the propositions A, B, C are true for an important class of special functions, namely for all polynomials. Theorem 4.3c below shows that they are true for an important class of special regions.

THEOREM 4.3c (Cauchy's integral theorem for disks)

The propositions A, B, C are true for any analytic f if R is a disk.

Proof. We shall show that proposition C is true. Let $R : |z - z_0| < \rho$, where $0 < \rho \leq \infty$, and let

$$f(z) = a_0 + a_1 h + a_2 h^2 + \cdots, \qquad |h| < \rho,$$

where $h := z - z_0$. The function

$$g(z) := a_0 h + a_1 \frac{h^2}{2} + a_2 \frac{h^3}{3} + \cdots$$

is analytic for $|h| < \rho$ and satisfies $g' = f$. ∎

INTEGRALS OF ANALYTIC FUNCTIONS

Next we wish to establish Cauchy's theorem, i.e., the propositions A, B, C, for all regions R that are *simply connected*. This we do somewhat indirectly. According to the definition given in §3.5, a region R is simply connected if for any two points z_0 and z_1 in R any two arcs from z_0 to z_1 are homotopic with respect to R; i.e., they can be continuously deformed into each other within R. According to this definition, any closed curve Γ in a simply connected region R is homotopic to any point on Γ. Now more generally we call any two *closed* curves homotopic with respect to a region R if they can be continuously deformed into each other in R. We then show that the integral of a function f analytic in a region R (not necessarily simply connected) has the same value along all closed curves that are homotopic with respect to R. Because the integral along a curve reducing to a point has the value zero, this will establish Cauchy's theorem for simply connected regions.

We begin with a formal definition of homotopy for closed curves. If R is a region, two *closed* curves Γ_0 and Γ_1 are called **homotopic with respect to R** if both Γ_0 and Γ_1 lie in R and if there exists a complex-valued function $z = z(\tau, \sigma)$, defined on the unit square $K: 0 \leqslant \tau \leqslant 1, 0 \leqslant \sigma \leqslant 1$ and continuous on K such that

(a) $z(\tau, \sigma) \in R$ for all $(\tau, \sigma) \in K$,
(b) $z(1, \sigma) = z(0, \sigma)$ for all $\sigma \in [0, 1]$,
(c) the closed curves $z = z(\tau, 0)$ and $z = z(\tau, 1)$ ($0 \leqslant \tau \leqslant 1$) are geometrically equivalent to Γ_0 and Γ_1, respectively (see Figs. 4.3c and 4.3d).

By (b), for each $\sigma \in [0, 1]$ the equation $z = z(\tau, \sigma)$, $0 \leqslant \tau \leqslant 1$ defines a closed curve Γ_σ which by (a) lies in R. If Γ_0 and Γ_1 are piecewise differentiable, we shall assume that each of these intermediate curves has the same property and, moreover, that for each $\sigma \in [0, 1]$ the function $z(\tau, \sigma)$ has a continuous derivative with respect to σ.

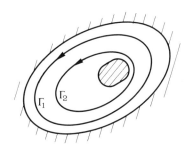
Fig. 4.3c. Homotopic closed curves.

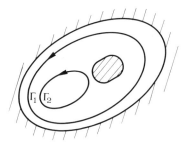
Fig. 4.3d. Nonhomotopic closed curves.

THEOREM 4.3d

Let f be analytic in a region R and let Γ_0 and Γ_1 be two piecewise differentiable closed curves that are homotopic with respect to R. Then

$$\int_{\Gamma_0} f(z)\,dz = \int_{\Gamma_1} f(z)\,dz. \tag{4.3-4}$$

Proof. Let $z(\tau,\sigma)$ denote the function that occurs in the definition of homotopy. The set of points $z(\tau,\sigma)\in R$, where $(\tau,\sigma)\in K$ is the continuous image of a compact set and thus itself compact. Hence it has a distance $\delta>0$ from the boundary of the open set R. Since $z(\tau,\sigma)$ is continuous on the compact set K, it is uniformly continuous, and $\epsilon>0$ exists such that $|z(\tau',\sigma')-z(\tau,\sigma)|<\delta$ whenever $(\tau,\sigma)\in K$, $(\tau',\sigma')\in K$, and $|\tau-\tau'|<\epsilon$, $|\sigma'-\sigma|<\epsilon$. Let n be an integer $>\epsilon^{-1}$. For $k=0,1,\ldots,n-1$ and $j=0,1,\ldots,n-1$, let $\theta_{k,j}$ and $\Lambda_{k,j}$, respectively, denote the arcs

$$z = z\left(\tau, \frac{j}{n}\right), \quad \frac{k}{n}\leqslant \tau \leqslant \frac{k+1}{n},$$

and

$$z = z\left(\frac{k}{n},\sigma\right), \quad \frac{j}{n}\leqslant \sigma \leqslant \frac{j+1}{n}.$$

Then

$$\Gamma_{k,j} := \theta_{k,j} + \Lambda_{k+1,j} - \theta_{k,j+1} - \Lambda_{k,j}$$

is a piecewise differentiable closed curve which, by the definition of ϵ, lies in a disk of radius $<\delta$ centered at $z(k/n,j/n)$. By definition of δ this disk is contained in R; hence by Theorem 4.3c

$$\int_{\Gamma_{k,j}} f(z)\,dz = 0. \tag{4.3-5}$$

We form the sum of all integrals (4.3-5), where k and j run from 0 to $n-1$. By writing the integrals as sums of four integrals according to definition of $\Gamma_{k,j}$

$$\int_{\sum_{k=0}^{n-1}\theta_{k,0}} + \int_{\sum_{j=0}^{n-1}\Lambda_{n,j}} - \int_{\sum_{k=0}^{n-1}\theta_{k,n}} - \int_{\sum_{j=0}^{n-1}\Lambda_{0,j}} = 0$$

remains. But

$$\sum_{k=0}^{n-1}\theta_{k,0} = \Gamma_0, \qquad \sum_{k=0}^{n-1}\theta_{k,n} = \Gamma_1,$$

INTEGRALS OF ANALYTIC FUNCTIONS

and by property (ii) of homotopy $\Lambda_{0,j} = \Lambda_{n,j}$, $j = 0, 1, \ldots, n-1$. Thus the above sum of integrals reduces to (4.3-4). ∎

EXAMPLE 1

Let a be a point in the complex plane and let $\Gamma_0: z = a + e^{i\tau}$, $0 \leq \tau \leq 2\pi$, be the positively oriented circle of radius 1 centered at a. We already know that

$$\int_{\Gamma_0} \frac{1}{z-a} dz = 2\pi i.$$

Now let $\Gamma: z = z_0 + \rho e^{i\tau}$, $0 \leq \tau \leq 2\pi$, be any other positively oriented circle not passing through a ($|z_0 - a| \neq \rho$). Let R denote the domain of analyticity of the function $f(z) := (z-a)^{-1}$, i.e., the complex plane punctured at $z = a$. If the circle Γ does not enclose point a (i.e., if $|z_0 - a| > \rho$), then it can be steadily shrunk to a point in R [e.g., by the function $z(\tau, \sigma) = z_0 + \sigma \rho e^{i\tau}$]. It follows that Γ is homotopic to a point in R; hence

$$\int_{\Gamma} \frac{1}{z-a} dz = 0.$$

If Γ does enclose point a (i.e., if $|z_0 - a| < \rho$), then Γ can be transformed into Γ_0 by the continuous function

$$z(\tau, \sigma) := (1-\sigma)z_0 + \sigma a + [(1-\sigma)\rho + \sigma]e^{i\tau},$$

which yields Γ for $\sigma = 0$, Γ_0 for $\sigma = 1$, and satisfies $z(\tau, \sigma) \neq a$ for all τ and σ under consideration. It follows that Γ is homotopic to Γ_0; therefore

$$\int_{\Gamma} \frac{1}{z-a} dz = 2\pi i.$$

Thus, if Γ is a positively oriented circle not passing through a, we find

$$\int_{\Gamma} \frac{1}{z-a} dz = \begin{cases} 2\pi i, & \text{if } a \text{ is inside } \Gamma, \\ 0, & \text{if } a \text{ is outside } \Gamma. \end{cases} \quad (4.3\text{-}6)$$

Now let R be a simply connected region. Then *every* closed curve Γ in R is homotopic with respect to a point with respect to R, and Theorem 4.3d implies:

THEOREM 4.3e (Cauchy's theorem for simply connected regions)

Let f be analytic in a simply connected region R. Then

$$\int_{\Gamma} f(z) \, dz = 0$$

holds for every closed curve Γ in R.

EXAMPLE 2

Let $R = \mathbb{C}$, $f(z) := e^{-z^2}$ and let $\omega > 0$. We form the integral of f along the boundary of the rectangle with the four corners $\pm \rho$, $\pm \rho + i\omega$, where ρ denotes a parameter that later will be allowed to tend to ∞ (see Fig. 4.3e). By Cauchy's theorem

$$\int_{\Gamma_1} + \int_{\Gamma_2} + \int_{\Gamma_3} + \int_{\Gamma_4} = 0, \qquad (4.3\text{-}7)$$

in which the Γ_i denote the four straight line segments shown in Fig. 4.3e. Parametrizing Γ_1 by $z = \rho$, $-\rho \leq \tau \leq \rho$, we have

$$\int_{\Gamma_1} e^{-z^2} dz = \int_{-\rho}^{\rho} e^{-\tau^2} d\tau.$$

Similarly,

$$\int_{-\Gamma_3} e^{-z^2} dz = \int_{-\rho}^{\rho} e^{-(\tau + i\omega)^2} d\tau = e^{\omega^2} \int_{-\rho}^{\rho} e^{-\tau^2} \cos 2\omega\tau \, d\tau$$

because the integral of the odd function $e^{-\tau^2} \sin 2\omega\tau$ is zero. We now let $\rho \to \infty$. According to (4.2-11) the integrals along Γ_2 and Γ_4 are bounded by $\exp(\omega^2 - \rho^2)$ and thus tend to zero for $\rho \to \infty$. The integral along Γ_1 tends to the value $\sqrt{\pi}$ by a well-known result in calculus. Because (4.3-7) holds for all real ρ, it follows that the integral along $-\Gamma_3$ also tends to $\sqrt{\pi}$. We thus have obtained the nontrivial formula

$$\int_{-\infty}^{\infty} e^{-\tau^2} \cos 2\omega\tau \, d\tau = \sqrt{\pi} \, e^{-\omega^2}. \qquad (4.3\text{-}8)$$

This example hints at the potential of complex analysis as a tool for the evaluation of real integrals. The real power of the method will become apparent only if we are able to integrate functions with singularities made possible by the calculus of residues discussed in §§4.7 to 4.9.

As a theoretical consequence of Cauchy's theorem which will be of occasional use we prove the following theorem:

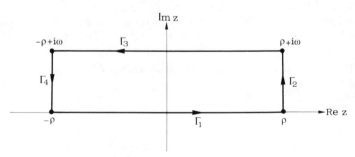

Fig. 4.3e. Path of integration.

INTEGRALS OF ANALYTIC FUNCTIONS

THEOREM 4.3f

Let f be analytic and different from zero in the simply connected region R. Then a single-valued and analytic branch of $\log f$ can be defined in R.

The assertion means that there is a function g, analytic in R, such that

$$e^{g(z)} = f(z) \tag{4.3-9}$$

holds for all $z \in R$. It is clear that for every $z \in R$ a number $g(z)$ can be selected among the infinite number of values of $\{\log f(z)\}$ such that (4.3-9) holds. The point of Theorem 4.3f lies in the fact that all these numbers $g(z)$ can be selected in such a way that the function g thus defined is analytic.

Proof of Theorem 4.3f. Because $f(z) \neq 0$ in R, the function f'/f is analytic in R. Thus by Proposition C (true because proposition A is true) there is a function h, analytic in R, such that

$$h'(z) = \frac{f'(z)}{f(z)}, \quad z \in R.$$

Let z_0 be a point of R and select $\log f(z_0)$ in an arbitrary manner. We assert that the function g, defined by

$$g(z) := h(z) - h(z_0) + \log f(z_0),$$

which is obviously analytic in R, satisfies (4.3-9). Indeed, the function $d(z) := e^{-g(z)} f(z)$ has the value 1 for $z = z_0$, and its derivative

$$d'(z) = -\frac{f'(z)}{f(z)} e^{-g(z)} f(z) + f'(z) e^{-g(z)}$$

vanishes identically. It follows that $d(z) = 1$, which is equivalent to (4.3-9). ∎

In the versions of Cauchy's theorem stated above the closed curve Γ was required to lie in the interior of the region of analyticity. In many applications R is a simply connected region whose boundary[3] Γ can be parametrized as a piecewise differentiable curve, and f is known not only to be analytic in R but also continuous in $R \cup \Gamma$. We then apply Cauchy's theorem with the hope that once again

$$\int_\Gamma f(z)\, dz = 0. \tag{4.3-10}$$

---

[3] A formal definition of the boundary of a set is given in §4.6.

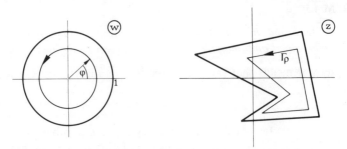

Fig. 4.3f. Piecewise diffeomorphism.

To establish such an extension of Cauchy's theorem, we assume that there exists a continuous one-to-one mapping of the closed unit disk $|w| \leq 1$ onto $R \cup \Gamma$, defined by a function $z = z(\rho, \phi)$ ($0 \leq \rho \leq 1$, $0 \leq \phi \leq 2\pi$), where (ρ, ϕ) are polar coordinates in the w-plane. The boundary curve Γ then has the parametric representation $z = z(1, \phi)$, $0 \leq \phi \leq 2\pi$. We assume that there is a subdivision $0 = \phi_0 < \phi_1 < \cdots < \phi_m = 2\pi$ such that in each sector $0 \leq \rho \leq 1$, $\phi_{k-1} \leq \phi \leq \phi_k$ the partial derivative $z_\phi(\rho, \phi)$ exists and is continuous. It follows that for each fixed ρ, $0 \leq \rho \leq 1$, the curve $\Gamma_\rho : z = z(\rho, \phi)$, $0 \leq \phi \leq 2\pi$, is piecewise differentiable. If the unit disk can be mapped onto the closed set $R \cup \Gamma$ in this manner, the latter is called **piecewise diffeomorphic** to the unit disk (see Fig. 4.3f).

If $R \cup \Gamma$ is piecewise diffeomorphic to the closed unit disk, R is a fortiori homeomorphic to the open disk, and thus simply connected (see §3.5).

THEOREM 4.3g

Let R be a region with boundary Γ such that $R \cup \Gamma$ is piecewise diffeomorphic to the closed unit disk, and let f be analytic in R and continuous in $R \cup \Gamma$. Then (4.3-10) *holds.*

Proof. With the above notations Γ_ρ for $0 \leq \rho < 1$ is a closed curve in a simply connected region; hence by Cauchy's theorem

$$\int_0^{2\pi} f(z(\rho, \phi)) z_\phi(\rho, \phi) \, d\phi = 0, \qquad 0 \leq \rho < 1. \qquad (4.3-11)$$

The function $f(z(\rho, \phi)) z_\phi(\rho, \phi)$ is piecewise continuous on the set $0 \leq \rho \leq 1$, $0 \leq \phi \leq 2\pi$, hence uniformly continuous. Thus we may let $\rho \to 1$ under the integral sign and obtain

$$\int_0^{2\pi} f(z(1, \phi)) z_\phi(1, \phi) \, d\phi = 0,$$

which is the same as (4.3-10). ∎

INTEGRALS OF ANALYTIC FUNCTIONS

EXAMPLE 3

We mention one situation in which the mapping function $z(\rho,\phi)$ is easily constructed. Let the boundary curve Γ have the property that there exists a point z_0 such that every ray emanating from z_0 meets Γ in exactly one point. Such a Γ is called **starlike** with respect to z_0. We denote the point at which Γ meets the ray $\arg(z-z_0)=\phi$ by $z_0+\chi(\phi)e^{i\phi}$. If χ is a piecewise differentiable function, then $z(\rho,\phi):=z_0+\rho\chi(\phi)^{i\phi}$ has the desired mapping properties. Thus the extended form of Cauchy's theorem holds for these starlike regions.

PROBLEMS

1. Compute the integrals

$$\int_0^\infty \cos(\tau^2)\,d\tau, \quad \int_0^\infty \sin(\tau^2)\,d\tau$$

   (**Fresnel's integrals**) by integrating e^{-z^2} along the closed curve shown in Fig. 4.3g and letting $\rho\to\infty$.

2. By integrating the function

$$f(z):=\text{Log}(1-e^{2iz})$$

   along the rectangle with corners $0,\pi,\pi+i\eta,i\eta$ (indented at the corners when necessary) and letting $\eta\to\infty$, show that

$$\int_0^\pi \text{Log}(\sin\tau)\,d\tau = -\pi\text{Log}2.$$

3. Let $\alpha > -1$. The value of the integral

$$B(\alpha):=\int_0^1 \tau^{\alpha-1}(1-\tau)^{\alpha-1}\,d\tau$$

   may be regarded as known (see §8.6). Prove that

$$2^{-\alpha}\int_{-\pi/2}^{\pi/2}\cos\alpha\phi(\cos\phi)^{\alpha-1}\,d\phi = B(\alpha)$$

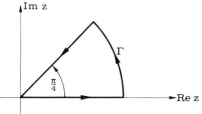

Fig. 4.3g. Path of integration for Problem 1.

by showing that the integral equals

$$\int_\Gamma z^{\alpha-1}(1-z)^{\alpha-1}\,dz,$$

where Γ is the semicircular arc

$$z = \frac{1}{2} - \frac{i}{2}e^{i\phi}, \qquad -\frac{\pi}{2} < \phi < \frac{\pi}{2},$$

and using Cauchy's theorem. Use the result to evaluate

$$\int_0^{\pi/2} \frac{\cos(\phi/2)}{\sqrt{\cos\phi}}\,d\phi = \frac{\pi}{\sqrt{2}}.$$

4. Let $R: z \neq 0$. Give a formal proof that every ellipse

$$z = \alpha\cos\tau + i\beta\sin\tau, \qquad 0 \leq \tau \leq 2\pi,$$

where α and β are real, $\alpha\beta > 0$ is homotopic to the circle $z = e^{i\tau}$, $0 \leq \tau \leq 2\pi$, with respect to R.

5. Let $\alpha > 0$, $\beta > 0$. Evaluate the integral

$$\int_0^{2\pi} \frac{1}{\alpha^2\cos^2\tau + \beta^2\sin^2\tau}\,d\tau$$

by relating it to the integral

$$\int_\Gamma \frac{1}{z}\,dz,$$

taken along the ellipse

$$\Gamma: z = \alpha\cos\tau + i\varphi\sin\tau, \qquad 0 \leq \tau \leq 2\pi.$$

§4.4. THE LAURENT SERIES; ISOLATED SINGULARITIES

If a function f is analytic in a disk D centered at z_0, it is represented in D by a power series in $h := z - z_0$. Here we prove a similar result for functions that are analytic in an annulus, i.e., in a region bounded by two concentric circles centered at z_0. Again f can be represented by a series of powers of $h = z - z_0$, but the series may now involve powers with both negative and positive exponents. This result has a number of striking applications; in particular, it will permit a detailed study of the *isolated singularities* of an analytic function.

Let z_0 be a point in the complex plane, let $0 \leq \alpha < \beta \leq \infty$, and let f be analytic in the annulus $A: \alpha < |z - z_0| < \beta$. For $\alpha < \rho < \beta$ let Γ_ρ denote the

THE LAURENT SERIES

circle $z = z_0 + \rho e^{i\tau}$, $0 \leq \tau \leq 2\pi$. It is clear that these circles are homotopic with respect to A.

Let z be a point of A fixed during the computation that follows. The function g defined in A by

$$g(t) := \begin{cases} \dfrac{f(t) - f(z)}{t - z}, & t \neq z, \\ f'(z), & t = z, \end{cases}$$

is analytic in A. Thus by Theorem 4.3d the integral

$$\int_{\Gamma_\rho} g(t) \, dt$$

is independent of ρ. In particular, if $\alpha < \rho_1 < |z - z_0| < \rho_2 < \beta$, and writing $\Gamma_{\rho_1} =: \Gamma'$, $\Gamma_{\rho_2} =: \Gamma''$,

$$\int_{\Gamma'} g(t) \, dt = \int_{\Gamma''} g(t) \, dt,$$

which is the same as

$$\int_{\Gamma'} \frac{f(t) - f(z)}{t - z} \, dt = \int_{\Gamma''} \frac{f(t) - f(z)}{t - z} \, dt$$

or, because $f(z)$ is a constant,

$$f(z) \left\{ \int_{\Gamma''} \frac{1}{t-z} dt - \int_{\Gamma'} \frac{1}{t-z} dt \right\} = \int_{\Gamma''} \frac{f(t)}{t-z} dt - \int_{\Gamma'} \frac{f(t)}{t-z} dt;$$

but in view of Example 1 of §4.3

$$\int_{\Gamma'} \frac{1}{t-z} dt = 0, \qquad \int_{\Gamma''} \frac{1}{t-z} dt = 2\pi i,$$

and

$$f(z) = \frac{1}{2\pi i} \int_{\Gamma''} \frac{f(t)}{t-z} dt - \frac{1}{2\pi i} \int_{\Gamma'} \frac{f(t)}{t-z} dt \qquad (4.4\text{-}1)$$

follows. This is a representation of the value of f at z in terms of the values on the circles Γ' and Γ''. The representation holds for all z such that $\rho_1 < |z - z_0| < \rho_2$.

We now expand the "Cauchy kernels" $(t-z)^{-1}$ in the integrals (4.4-1) in geometric series. For $t \in \Gamma''$, $|t - z_0| > |z - z_0|$; hence by writing $h := z - z_0$

we have

$$\frac{1}{t-z} = \frac{1}{t-z_0-h} = \frac{1}{t-z_0}\frac{1}{1-h/(t-z_0)} = \sum_{n=0}^{\infty} \frac{h^n}{(t-z_0)^{n+1}}.$$

For $t \in \Gamma'$, $|t-z_0| < |z-z_0|$; hence

$$\frac{1}{t-z} = -\frac{1}{h-(t-z_0)} = -\frac{1}{h}\frac{1}{1-(t-z_0)/h} = -\sum_{n=0}^{\infty}\frac{(t-z_0)^n}{h^{n+1}}.$$

Because these series converge uniformly for $t \in \Gamma''$ and $t \in \Gamma'$, respectively, we may substitute them in the integrals (4.4-1) and integrate term by term. Letting

$$a_n := \frac{1}{2\pi i}\int_{\Gamma''}\frac{f(t)}{(t-z_0)^{n+1}}dt, \quad n = 0, 1, 2, \ldots,$$

$$a_{-n} := \frac{1}{2\pi i}\int_{\Gamma'}f(t)(t-z_0)^n dt, \quad n = 1, 2, \ldots,$$

the result may be written in the simple form

$$f(z) = \sum_{n=-\infty}^{\infty} a_n h^n, \quad h := z - z_0. \tag{4.4-2}$$

The formula for the coefficients may be simplified even further by noting that the functions $f(t)(t-z_0)^k$ are analytic in A for all integers k. Thus by Theorem 4.3d the integrals defining the coefficients may be computed along any circle Γ_ρ and the two formulas can be combined into the single formula

$$a_n = \frac{1}{2\pi i}\int_{\Gamma_\rho}\frac{f(t)}{(t-z_0)^{n+1}}dt, \quad n = 0, \pm 1, \pm 2, \ldots, \tag{4.4-3}$$

where for each n, ρ may be any number such that $\alpha < \rho < \beta$.

If α_1 and β_1 are any two numbers such that $\alpha < \alpha_1 < \beta_1 < \beta$, we assert that the series (4.4-2) converges uniformly for $\alpha_1 \leq |h| \leq \beta_1$. This follows from Theorem 2.2c, for we may write the series as a sum of two power series,

$$\sum_{n=-\infty}^{\infty} a_n h^n = \sum_{n=0}^{\infty} a_n h^n + \sum_{n=1}^{\infty} a_{-n} h^{-n}.$$

THE LAURENT SERIES

The first series on the right is a power series in h with a radius of convergence β; hence it converges uniformly for $|h| \leq \beta_1$; the second series is a power series in h^{-1} with a radius of convergence α^{-1} which converges uniformly for $|h^{-1}| \leq \alpha_1^{-1}$. Altogether we have proved the following theorem:

THEOREM 4.4a (Laurent's theorem)

Let z_0 be a complex number, let $0 \leq \alpha < \beta \leq \infty$, and let f be analytic in the annulus $A: \alpha < |z - z_0| < \beta$. If the coefficients a_n are defined by (4.4-3), then, for all $z \in A$, f is represented by the series (4.4-2). The series converges uniformly on every compact subset of A.

The series (4.4-2) is called the **Laurent expansion** or the **Laurent series** of f in the annulus A. In contrast to the formal Laurent series considered in Chapter 1 the Laurent series occurring here may involve an infinite number of powers with negative exponents.

Next we show that the Laurent expansion of an analytic function f is unique for a given annulus A. Although the coefficients a_n are uniquely determined by (4.4-3), it is conceivable that some other series

$$\sum_{n=-\infty}^{\infty} b_n h^n \qquad (4.4\text{-}4)$$

also represents $f(z)$ in the annulus A. Since the series (4.4-4) is again the sum of two power series, it can be shown as above that it would have to converge uniformly in every compact subannulus $A_1: \alpha_1 \leq |z - z_0| \leq \beta_1$. By subtracting (4.4-2) from (4.4-4) we obtain the uniformly convergent representation

$$0 = \sum_{n=-\infty}^{\infty} (b_n - a_n) h^n \qquad (4.4\text{-}5)$$

of the zero function in A_1. Now assume that $b_m \neq a_m$ for some integer m. Multiplying (4.4-5) by $(z - z_0)^{-m-1} = h^{-m-1}$ and integrating over Γ_ρ, where $\alpha_1 \leq \rho \leq \beta_1$, we have, according to Example 1 of §4.3,

$$0 = (b_m - a_m) 2\pi i,$$

a contradiction. Thus we have proved the following theorem:

THEOREM 4.4b

The Laurent series of an analytic function for a given annulus is unique.

Suppose we have found, by any method whatsoever, a Laurent-like expansion of an analytic function, valid in some annulus A. Theorem 4.4b then permits us to conclude that we have found *the* Laurent expansion and consequently that formula (4.4-3) holds for the coefficients.

EXAMPLE 1

For $0 < |1/z| < \infty$

$$e^{1/z} = 1 + \frac{1}{1!}z^{-1} + \frac{1}{2!}z^{-2} + \cdots.$$

holds. This expansion is like (4.4-2); hence it is *the* Laurent expansion of $e^{1/z}$ in the annulus $0 < |z| < \infty$.

EXAMPLE 2

Let

$$p(z) := a_0 + a_1 z + a_2 z^2 + \cdots + a_n z^n$$

be a polynomial. Its Laurent expansion in any annulus centered at the origin is the polynomial itself.

EXAMPLE 3

For $z \neq 2$, $z \neq 3$ let

$$f(z) := \frac{1}{z-2} + \frac{1}{z-3}.$$

This function is analytic in the annulus $A: 2 < |z| < 3$. We can find its Laurent expansion by using geometric series. For $|z| > 2$

$$\frac{1}{z-2} = \frac{1}{z} \frac{1}{1 - 2/z} = \frac{1}{z} + \frac{2}{z^2} + \frac{2^2}{z^3} + \cdots,$$

and for $|z| < 3$

$$\frac{1}{z-3} = -\frac{1}{3} \frac{1}{1 - z/3} = -\frac{1}{3} - \frac{z}{3^2} - \frac{z^2}{3^3} - \cdots.$$

Thus in the annulus A the Laurent series of f is

$$f(z) = \cdots + \frac{2^2}{z^3} + \frac{2}{z^2} + \frac{1}{z} - \frac{1}{3} - \frac{z}{3^2} - \frac{z^2}{3^3} - \cdots.$$

For the annulus $3 < |z| < \infty$ the same method yields a different Laurent series:

$$f(z) = \frac{1+1}{z} + \frac{2+3}{z^2} + \frac{2^2+3^2}{z^3} + \cdots.$$

This does not contradict Theorem 4.4b, for we are dealing with the Laurent expansions of f in two different annuli.

THE LAURENT SERIES

Laurent's theorem contains the Taylor expansion as a special case, and a remarkable conclusion may be drawn from it. Suppose f is analytic in the disk $|z-z_0|<\beta$ and

$$f(z) = \sum_{n=0}^{\infty} a_n h^n, \quad |h|<\beta. \tag{4.4-6}$$

If the domain of definition of f is restricted to the annulus $A: 0<|z-z_0|<\beta$, then f has a Laurent expansion in A. The expansion (4.4-6) takes the form of a Laurent expansion and by the uniqueness of the expansion is identical with it. It follows that (4.4-3) holds. On the other hand, we know from Corollary 3.1d that

$$a_n = \frac{1}{n!} f^{(n)}(z_0), \quad n=0,1,2,\dots.$$

We conclude

$$f^{(n)}(z_0) = \frac{n!}{2\pi i} \int_{\Gamma_\rho} \frac{f(t)}{(t-z_0)^{n+1}} dt, \quad n=0,1,2,\dots. \tag{4.4-7}$$

This is a version of the **Cauchy integral formula**; see §4.7. By virtue of Cauchy's theorem Γ_ρ can be replaced by any closed curve Γ which is homotopic to Γ_ρ in A. Even the special case $n=0$ of (4.4-7),

$$f(z_0) = \frac{1}{2\pi i} \int_\Gamma \frac{f(t)}{t-z_0} dt, \tag{4.4-8}$$

is remarkable. This formula expresses the value of an analytic function at a point z_0 in terms of the values on an arbitrary curve Γ homotopic to a circle Γ_ρ in A.

The following result generalizes Theorem 2.2f (Cauchy's coefficient estimate for power series) to Laurent series:

THEOREM 4.4c

Let f be analytic in the annulus $A: \alpha<|z-z_0|<\beta$ and for $\alpha<\rho<\beta$ let

$$\mu(\rho) := \max_{|z-z_0|=\rho} |f(z)|. \tag{4.4-9}$$

The coefficients a_n in the Laurent series (4.4-2) then satisfy

$$|a_n| \leq \mu(\rho)\rho^{-n}, \quad n=0,\pm 1,\pm 2,\dots. \tag{4.4-10}$$

Proof. From the integral representation (4.4-3) we have, by using (4.2-11),

$$|a_n| = \frac{1}{2\pi} \left| \int_{\Gamma_\rho} f(t)(t-z_0)^{-n-1} dt \right|$$

$$\leq \frac{1}{2\pi} 2\pi\rho\mu(\rho)\rho^{-n-1},$$

which establishes (4.4-10). ∎

The remainder of this section is devoted to a study of the special case of Laurent's theorem in which $\alpha = 0$. Here, then, we are concerned with functions f that are defined and analytic in a punctured disk $0 < |z - z_0| < \beta$. A point z_0 with this property is called an **isolated singularity** of f. The series (4.4-2) now converges for $0 < |h| < \beta$. The portion of the series that comprises the powers with negative exponents,

$$p(z) := \sum_{n=1}^{\infty} a_{-n} h^{-n}, \qquad (4.4\text{-}11)$$

is called the **principal part** of the function f at the isolated singularity $z = z_0$. Because the power series on the right of (4.4-11) converges for arbitrarily small values of $|h^{-1}|$, it converges for *all* $h \neq 0$.

If z_0 is an isolated singularity, the function $\mu(\rho)$ is defined for $0 < \rho < \beta$. Three kinds of isolated singularities are distinguished, according to the behavior of $\mu(\rho)$ for $\rho \to 0$.

1. *The function $\mu(\rho)$ is bounded as $\rho \to 0$.* The existence of a constant γ such that $\mu(\rho) < \gamma$ for all sufficiently small ρ implies by (4.4-10) that

$$|a_{-n}| \leq \gamma \rho^n$$

for all small ρ, which, if $n > 0$, is possible only if $a_{-n} = 0$, $n = 1, 2, \ldots$. The Laurent series of f near z_0 is thus given by

$$f(z) = \sum_{n=0}^{\infty} a_n h^n \qquad (h := z - z_0). \qquad (4.4\text{-}12)$$

The series is a power series with radius of convergence $\geq \beta > 0$, and from Theorem 2.2d it follows that the limit of $f(z)$ as $z \to z_0$ exists:

$$\lim_{z \to z_0} f(z) = a_0 = \frac{1}{2\pi i} \int_{\Gamma_\rho} \frac{f(t)}{t - z_0} dt.$$

THE LAURENT SERIES

The expansion (4.4-12) shows that if we extend the domain of definition of f by defining

$$f(z_0) := a_0,$$

then f is analytic also at $z = z_0$, hence throughout the disk $|z - z_0| < \beta$. Conversely, if f is analytic in $A: 0 < |z - z_0| < \beta$ and

$$\lim_{z \to z_0} f(z) \text{ exists,}$$

then $\mu(\rho)$ is clearly bounded as $\rho \to 0$. Thus we have proved:

THEOREM 4.4d ("Riemann's theorem")

Let f be analytic in the annulus $0 < |z - z_0| < \beta$ and let $\mu(\rho)$ be defined by (4.4-9). The following four statements are equivalent:

(a) *The function $\mu(\rho)$ is bounded as $\rho \to 0$.*
(b) *The principal part of f at z_0 is zero.*
(c) $a_0 := \lim_{z \to z_0} f(z)$ *exists.*
(d) f *can be continued analytically into z_0.*

By virtue of (d), if any of the above statements obtain, z_0 is called a **removable singularity** of f.

Traditionally, in complex analysis, if a singularity is removable, it is tacitly assumed to have been removed. As an example, consider

$$f(z) := \frac{\sin z}{z}.$$

The analytic expression defines $f(z)$ only for $z \neq 0$. From the Laurent series, however,

$$f(z) = 1 - \frac{z^2}{3!} + \frac{z^4}{5!} - \cdots,$$

we conclude that the singularity at $z = 0$ is removable. Without further comment it is assumed that $f(0) = 1$.

2. *The function $\mu(\rho)$ is unbounded as $\rho \to 0$, but there is a number $\nu > 0$ such that $\rho^\nu \mu(\rho)$ is bounded.* Let ν_0 denote the greatest lower bound of all ν such that $\rho^\nu \mu(\rho)$ is bounded as $\rho \to 0$. By letting $\rho \to 0$ it then follows from Theorem 4.4c that $a_{-n} = 0$ for all integers $n > \nu_0$ and that the principal part of f at z_0 is a polynomial in h^{-1} of degree $m \leq \nu_0$. By the theorem just

proved the function

$$g(z) := (z-z_0)^m f(z) = \sum_{n=-m}^{\infty} a_n h^{n+m} \quad (4.4\text{-}13)$$

has a removable singularity at $z = z_0$; hence $\rho^m \mu(\rho)$ is bounded as $\rho \to 0$ and we conclude that ν_0 necessarily is a positive integer, $\nu_0 = m$. It also follows that $g(z_0) \neq 0$ [otherwise $\nu < m$ would exist such that $\rho^\nu \mu(\rho)$ is bounded] and that the degree of the principal part as a polynomial in h^{-1} is precisely m. Furthermore, we may conclude that

$$\lim_{z \to z_0} f(z) = \infty$$

exists as an improper limit in the sense that, given any $\mu > 0$ (no matter how large), $\delta > 0$ exists such that $0 < |z - z_0| < \delta$ implies $|f(z)| > \mu$. From (4.4-13) and the fact that $g(z_0) \neq 0$ we conclude by Theorem 2.3g that the function

$$f^{-1}(z) = (z-z_0)^m g^{-1}(z)$$

has a removable singularity at z_0 and can be continued analytically into z_0 by being assigned the value 0. Since f does not vanish identically, this implies that $\mu(\rho)$ grows like some negative power of ρ as $\rho \to 0$. The chain of implications is thus closed, and we have proved the following theorem:

THEOREM 4.4e

Let f be analytic in the annulus $0 < |z - z_0| < \beta$ and let $\mu(\rho)$ be defined by (4.4-9). The following four statements are then equivalent:

(a) *The function $\mu(\rho)$ is unbounded as $\rho \to 0$, but there exists $\nu > 0$ such that $\rho^\nu \mu(\rho)$ is bounded.*
(b) *The principal part of f at z_0 is a polynomial in $(z - z_0)^{-1}$ of positive degree.*
(c) $\lim_{z \to z_0} f(z) = \infty$ *exists as an improper limit.*
(d) *By assigning it the value 0 at z_0, the function f^{-1} is continued analytically into the point z_0.*

An isolated singularity of a function f for which any of these statements are true is called a **pole** of f. The degree of the principal part (the integer m in the proof of Theorem 4.4e) is called the **order** of the pole. A function f which at every point of a region R is either analytic or has an isolated singularity that is at most a pole is called **meromorphic** in R.

3. *The function $\mu(\rho)$ as $\rho \to 0$ grows faster than any negative power of ρ.* The logical complement of the two preceding theorems is as follows:

THE LAURENT SERIES

THEOREM 4.4f

Let f be analytic in the annulus $0 < |z - z_0| < \beta$ and let $\mu(\rho)$ be defined by (4.4-9). The following four statements are then equivalent:

(a) *There exists no ν such that $\rho^\nu \mu(\rho)$ is bounded as $\rho \to 0$.*
(b) *The principal part of f at z_0 involves an infinite number of nonzero terms.*
(c) *$\lim_{z \to z_0} f(z)$ exists neither as a proper nor as an improper limit.*
(d) *Neither f nor f^{-1} can be continued analytically into z_0.*

An isolated singularity of a function f which enjoys any of the above properties and thus is neither removable nor a pole is called an **essential singularity** of f. Property (c) suggests that the behavior of a function near an isolated singularity may be somewhat irregular. This impression is amply confirmed by the following result:

THEOREM 4.4g (Casorati-Weierstrass theorem)

Let z_0 be an essential singularity of f and let b be any complex number. Then for every $\delta > 0$ and every $\epsilon > 0$ there exists a point z such that both

$$0 < |z - z_0| < \delta \quad \text{and} \quad |f(z) - b| < \epsilon.$$

Thus, if we pick b at random and draw arbitrarily small circles around z_0 and b, a point exists inside the first circle where f assumes a value inside the second circle.

Proof. The proof of this almost incredible theorem is a simple argument by contradiction. If the assertion were false, then for some b, some $\delta > 0$, and some $\epsilon > 0$ it would be true that $|f(z) - b| \geq \epsilon$ for all z satisfying $0 < |z - z_0| < \delta$. Now consider the function

$$g(z) := \frac{1}{f(z) - b}.$$

Because f is analytic and $\neq b$ for $0 < |z - z_0| < \delta$, g is analytic in the annulus $A : 0 < |z - z_0| < \delta$; moreover, because $|f(z) - b| \geq \epsilon$, it follows that g is bounded (by ϵ^{-1}) in the annulus. By Theorem 4.4d the isolated singularity of g at z_0 is removable. Since g is not identically zero (actually, $g(z) \neq 0$ for all z in A), $g^{-1}(z)$ is meromorphic at z_0. It follows that $f(z) = b + g^{-1}(z)$ is also meromorphic, contrary to the hypothesis. ∎

Isolated Singularities at Infinity. A function that is analytic in an annulus $A : \alpha < |z| < \infty$ is said to have an **isolated singularity at ∞**. If f has an isolated singularity at ∞, the function g defined in the annulus

$A': 0 < |z| < \alpha^{-1}$ by $g(z) := f(z^{-1})$ has an isolated singularity at $z = 0$. The singularity of f at ∞ is called removable, a pole, or an essential singularity according to whether the singularity of g at 0 is removable, a pole, or essential. If the Laurent series of f in A is

$$f(z) = \sum_{n=-\infty}^{\infty} a_n z^n,$$

the Laurent series representing g in A' is

$$g(z) = \sum_{n=-\infty}^{\infty} a_n z^{-n}.$$

The principal part of the latter is $\sum_{n=1}^{\infty} a_n z^{-n}$; consequently we call

$$p(z) := \sum_{n=1}^{\infty} a_n z^n$$

the **principal part** of f at ∞. With these conventions the statements of the last four theorems remain true for isolated singularities at ∞.

An isolated singularity of f at ∞ is removable if and only if the Laurent series in A has the form

$$f(z) = \sum_{n=-\infty}^{0} a_n z^n.$$

We remove the singularity by defining $f(\infty) := a_0$ and with this extension of the definition call f **analytic at** ∞.

EXAMPLE 4

A polynomial p of degree $n > 0$ has a pole of order n at infinity at which its principal part equals the polynomial itself, minus the value at 0.

EXAMPLE 5

An entire function that is not a polynomial has an essential singularity at ∞. As a special example consider $f(z) := e^z$. To illustrate the Casorati-Weierstrass theorem let b be any complex number. If $b = 0$, then e^z comes arbitrarily near to b outside every circle, for $\lim_{\xi \to \infty} e^{-\xi} = 0$. If $b \neq 0$, then e^z actually assumes the value b outside every circle, for $e^z = b$ at the infinite number of values of $\{\log b\}$. The behavior of the exponential function is typical because a celebrated *theorem due to Picard* states that every entire function that is not a constant assumes every value with at most one exception.

THE LAURENT SERIES

EXAMPLE 6

The function $z^{-2}\sin z$ has a pole of order 1 at $z=0$ and an essential singularity at ∞.

EXAMPLE 7

The function $f(z):=(\sin z)^{-2}$ has poles of order 2 at the points $z=n\pi$, $n=0,\pm 1, \pm 2,\ldots$. It has no isolated singularity at ∞, for it is analytic in no annulus $\alpha<|z|<\infty$.

In conclusion we apply some of these results to *rational functions*. A function r is called **rational** if polynomials p and q, $q\neq 0$, exist such that the domain of definition D of r is the set of all z such that $q(z)\neq 0$ and

$$r(z) = \frac{p(z)}{q(z)} \quad \text{for all } z \in D.$$

By the usual division algorithm we can determine a polynomial s_0 (possibly zero) and a polynomial p_0 whose degree is less than the degree of q such that

$$r(z) = s_0(z) + \frac{p_0(z)}{q(z)}. \tag{4.4-14}$$

The function p_0/q is analytic at infinity. It follows that the principal part of r at ∞ is $s_0(z) - s_0(0)$; because s_0 is a polynomial, we conclude that the singularity of r at ∞ is at most a pole. The finite singularities are the zeros of q; because there is only a finite number of such zeros, all singularities are isolated. Let $z_1, z_2, \ldots, z_k$ be the distinct zeros of q. If m_h is the multiplicity of z_h, then $q(z) = (z-z_h)^{m_h} q_h(z)$, where q_h is a polynomial, $q_h(z_h) \neq 0$. We conclude that the finite singularities likewise are at most poles. Let

$$s_h(z) := \sum_{j=1}^{m_h} a_{h,j} (z - z_h)^{-j}$$

be the principal part of r at z_h, $h=1,2,\ldots,k$. Then the function

$$f(z) := r(z) - s_0(z) - \sum_{h=1}^{k} s_h(z)$$

has principal parts zero at all singularities; hence by Theorem 4.4d all singularities are removable and f can be extended to an entire function. However, from

$$\lim_{z \to \infty} (r(z) - s_0(z)) = 0, \quad \lim_{z \to \infty} s_h(z) = 0, \quad h=1,2,\ldots,k,$$

we deduce

$$\lim_{z\to\infty} f(z) = 0,$$

and by Liouville's theorem (Theorem 3.3b) we conclude that f vanishes identically.

$$r(z) = \sum_{h=0}^{k} s_h(z) \qquad (4.4\text{-}15)$$

follows, in words:

THEOREM 4.4h

A rational function differs from the sum of its principal parts at most by a constant.

The representation (4.4-15) of a rational function is called the **partial fraction decomposition** of r. Some practical methods for constructing the partial fraction decomposition are given in §7.1, and some of its applications are discussed in §7.2.

PROBLEMS

1. Show that the totality of functions that are meromorphic in a fixed region R form a field.
2. Prove that if f has only isolated singularities in the extended complex plane and these singularities are poles, then f is rational.
3. Prove that a function that is analytic in the extended complex plane is a constant.
4. Let f have an isolated singularity at z_0 with principal part p and let g be analytic at z_0. Prove that the principal part of $f+g$ is again p.
5. Show that the Casorati-Weierstrass theorem can be improved (trivially) as follows: if f has an essential singularity at z_0, then for every complex number b, every $\delta > 0$, and every $\epsilon > 0$ there is an *infinite number* of points z such that $0 < |z - z_0| < \delta$ and $|f(z) - b| < \epsilon$.
6. Let p be a polynomial of degree n with *distinct* zeros $w_1,\ldots,w_k$ of the respective multiplicities $m_1,\ldots,m_k$. Prove without induction that p can be represented in factored form,

$$p(z) = c \prod_{j=1}^{k} (z - w_j)^{m_j},$$

by showing that the function

$$f(z) := p(z) \prod_{j=1}^{k} (z - w_j)^{-m_j}$$

is entire and bounded.

APPLICATIONS OF LAURENT SERIES 219

7. Why does the following identity not contradict the uniqueness of the Laurent series?

$$0 = \frac{1}{z-1} + \frac{1}{1-z} = \frac{1}{z}\frac{1}{1-1/z} + \frac{1}{1-z}$$

$$= \sum_{n=1}^{\infty} \frac{1}{z^{n+1}} + \sum_{n=0}^{\infty} z^n = \sum_{n=-\infty}^{\infty} z^n.$$

§4.5. APPLICATIONS OF LAURENT SERIES: BESSEL FUNCTIONS, FOURIER SERIES

The *formal* Laurent series discussed in Chapter 1 are distinguished by containing only a finite number of terms that involve powers with negative exponents. Because of this circumstance the formal product of two *formal* Laurent series always exists. Contrary to this, the *nonformal* Laurent series considered in §4.4 may have an infinite number of nonzero terms with negative exponents, and it is therefore not clear that the product of any two such series exists. The following result holds, however.

THEOREM 4.5a

Let the functions f and g be analytic in the annulus $A: \alpha < |z - z_0| < \beta$ and let their Laurent series in A be

$$f(z) = \sum_{n=-\infty}^{\infty} a_n h^n, \quad g(z) = \sum_{n=-\infty}^{\infty} b_n h^n \quad (4.5\text{-}1)$$

($h := z - z_0$). *Then the series*

$$c_n := \sum_{m=-\infty}^{\infty} a_m b_{n-m} = \sum_{m=-\infty}^{\infty} b_m a_{n-m} \quad (4.5\text{-}2)$$

converge for all integers n, and the Laurent series of the function fg in A is given by

$$f(z)g(z) = \sum_{n=-\infty}^{\infty} c_n h^n. \quad (4.5\text{-}3)$$

Proof. The function fg is analytic in A and therefore is represented in A by a Laurent series whose coefficients by (4.4-3) are

$$c_n = \frac{1}{2\pi i} \int_{\Gamma_\rho} \frac{f(t)g(t)}{(t-z_0)^{n+1}} dt,$$

where Γ_ρ denotes a circle $t = z_0 + \rho e^{i\tau}$, $0 \leq \tau \leq 2\pi$, such that $\alpha < \rho < \beta$. The first series (4.5-1) converges uniformly to f on Γ_ρ. Substituting the series and interchanging integration and summation, we get

$$c_n = \frac{1}{2\pi i} \int_{\Gamma_\rho} \left[\sum_{m=-\infty}^{\infty} a_m (t-z_0)^m \right] g(t)(t-z_0)^{-n-1} dt$$

$$= \sum_{m=-\infty}^{\infty} a_m \left[\frac{1}{2\pi i} \int_{\Gamma_\rho} g(t)(t-z_0)^{m-n-1} dt \right].$$

By (4.4-3) the expression in brackets equals b_{n-m}, which establishes (4.5-2). ∎

BESSEL FUNCTIONS

Let w be a complex number. The function

$$f(z) := e^{(w/2)(z - z^{-1})}$$

is analytic in the annulus $0 < |z| < \infty$. Its Laurent series can be found by multiplying the Laurent series

$$e^{(w/2)z} = \sum_{n=-\infty}^{\infty} a_n z^n \quad \text{and} \quad e^{-(w/2)z^{-1}} = \sum_{n=-\infty}^{\infty} b_n z^n,$$

where

$$a_n = \begin{cases} 0, & n < 0 \\ \dfrac{w^n}{2^n n!}, & n \geq 0 \end{cases}, \qquad b_n = \begin{cases} \dfrac{(-w)^{-n}}{2^{-n}(-n)!}, & n \leq 0 \\ 0, & n > 0 \end{cases}.$$

We then have

$$f(z) = \sum_{n=-\infty}^{\infty} c_n z^n,$$

where c_n is given by (4.5-2). If $n \geq 0$, there results

$$c_n = \sum_{m=0}^{\infty} \frac{(-1)^m w^m}{2^m m!} \frac{w^{n+m}}{2^{n+m}(n+m)!}$$

$$= \frac{(w/2)^n}{n!} {}_0F_1\left(n+1; -\frac{w^2}{4}\right);$$

APPLICATIONS OF LAURENT SERIES

for negative indices a similar computation yields

$$c_{-n} = \frac{(-w/2)^n}{n!} {}_0F_1\left(n+1;\ -\frac{w^2}{4}\right) = (-1)^n c_n,$$

$n = 1, 2, \ldots$. For each integer n the coefficient c_n is an entire function of the parameter w. This function is called a **Bessel function of order** n and is universally denoted by the symbol J_n. Thus

$$J_n(z) := \frac{(z/2)^n}{n!} {}_0F_1\left(n+1;\ -\frac{z^2}{4}\right), \quad n = 0, 1, 2, \ldots,$$

$$J_{-n}(z) := (-1)^n J_n(z), \quad n = 1, 2, \ldots. \tag{4.5-4}$$

With a change of notation we have established the expansion

$$e^{(z/2)(t - t^{-1})} = \sum_{n=-\infty}^{\infty} J_n(z) t^n, \tag{4.5-5}$$

which by Theorem 4.4a converges uniformly with respect to t on every compact subset of $0 < |t| < \infty$.

The Bessel functions are required in mathematical physics, especially for the solution of potential and wave propagation problems that involve cylindrical symmetry (see Chapter 15). Here we apply Theorem 4.4a to derive an integral representation that will be useful in a different context. By the formula (4.4-3) for the nth coefficient of a Laurent series

$$J_n(z) = \frac{1}{2\pi i} \int_\Gamma e^{(z/2)(t - t^{-1})} t^{-n-1} dt,$$

where the integral may be extended, for instance, along the circle $\Gamma: t = e^{i\tau}$, $-\pi \leq \tau \leq \pi$. Carrying out the integration, we get according to $t^{-1} dt = i\, d\tau$

$$J_n(z) = \frac{1}{2\pi} \int_{-\pi}^{\pi} e^{iz \sin \tau} e^{-in\tau} d\tau$$

$$= \frac{1}{2\pi} \int_{-\pi}^{\pi} e^{i(z \sin \tau - n\tau)} d\tau$$

or, because the exponent in the integrand is an odd function,

$$J_n(z) = \frac{1}{\pi} \int_0^{\pi} \cos(z \sin \tau - n\tau)\, d\tau. \tag{4.5-6}$$

This integral representation of J_n is due to the German astronomer F. W. Bessel who used it to solve Kepler's equation in celestial mechanics (see below).

We next apply Theorem 4.4c to (4.5-5) to obtain an estimate for $|J_n(z)|$ that depends only on $\rho := |z|$. If, for $\tau > 0$,

$$\mu(\tau) := \max_{|t|=\tau} |e^{(z/2)(t-t^{-1})}|,$$

then evidently

$$\mu(\tau) \leq e^{(\rho/2)(\tau+\tau^{-1})};$$

hence by Theorem 4.4c

$$|J_n(z)| \leq \tau^{-n} e^{(\rho/2)(\tau+\tau^{-1})}, \qquad \tau > 0, \qquad (4.5\text{-}7)$$

for all integers n. By (4.5-4) it suffices to consider $n \geq 0$. Because the inequality (4.5-7) holds for all $\tau > 0$, it holds for the τ minimizing the expression on the right. Differentiation shows that the minimum occurs at $\tau = (1/\rho)(n + \sqrt{n^2 + \rho^2})$, and an evaluation of the minimum yields

$$|J_n(z)| \leq \frac{\rho^n}{\left(n + \sqrt{n^2 + \rho^2}\right)^n} \exp\left[(n^2 + \rho^2)^{1/2}\right], \qquad \rho = |z|. \quad (4.5\text{-}8)$$

An obvious simplification yields

$$|J_n(z)| \leq \left(\frac{\rho e}{2n}\right)^n e^\rho, \qquad \rho = |z|. \qquad (4.5\text{-}9)$$

By Stirling's formula

$$n! \sim \sqrt{2\pi n} \left(\frac{n}{e}\right)^n \qquad (n \to \infty),$$

hence the factor of e^ρ in (4.5-9) is asymptotic to

$$\text{const}\sqrt{n}\, \frac{(\rho/2)^n}{n!}.$$

This shows that the convergence of the series on the right of (4.5-5) is uniform not only with respect to t in every compact subset of $0 < |t| < \infty$ but also uniform with respect to z in every compact subset of the z-plane. It follows by Theorem 3.4b that the series may be differentiated term by

APPLICATIONS OF LAURENT SERIES

term any number of times not only with respect to t but also with respect to z. This result is useful for the purpose of establishing recurrence formulas and differential equations satisfied by the Bessel functions; see Problem 1.

To give another application of Theorem 4.5a, let u and v be any two complex numbers. By multiplying the two Laurent series

$$e^{(u/2)(t-t^{-1})} = \sum_{n=-\infty}^{\infty} J_n(u) t^n, \quad e^{(v/2)(t-t^{-1})} = \sum_{n=-\infty}^{\infty} J_n(v) t^n$$

we obtain

$$e^{(u/2)(t-t^{-1})} e^{(v/2)(t-t^{-1})} = \sum_{n=-\infty}^{\infty} \left[\sum_{m=-\infty}^{\infty} J_m(u) J_{n-m}(v) \right] t^n,$$

which equals

$$\exp\left[\frac{u+v}{2} (t-t^{-1}) \right] = \sum_{n=-\infty}^{\infty} J_n(u+v) t^n.$$

By the uniqueness of the Laurent series we thus obtain the formula, valid for arbitrary complex u, v, and arbitrary integers n,

$$J_n(u+v) = \sum_{m=-\infty}^{\infty} J_m(u) J_{n-m}(v). \tag{4.5-10}$$

This is known as the *addition formula* of the Bessel functions.

FOURIER SERIES

Suppose the complex-valued function p is defined on the real line and has period 2π; i.e.,

$$p(\tau + 2\pi) = p(\tau), \quad -\infty < \tau < \infty.$$

Under certain conditions, e.g., if p is continuous and has a piecewise continuous derivative, p can be expanded in a Fourier series,[4]

$$p(\tau) = \sum_{n=-\infty}^{\infty} a_n e^{in\tau}, \tag{4.5-11}$$

---

[4] For a proof of this expansion we refer to texts in advanced calculus. A short treatment is also given in §10.6, Vol. II.

whose coefficients are given by

$$a_n = \frac{1}{2\pi} \int_0^{2\pi} p(\tau) e^{-in\tau} d\tau, \qquad n = 0, \pm 1, \pm 2, \ldots \quad (4.5\text{-}12)$$

A function with period 2π is perhaps more naturally thought of as being defined on the unit circle than on the real line. We thus define g on $|z|=1$ by

$$g(e^{i\tau}) := p(\tau). \qquad (4.5\text{-}13)$$

Now suppose the function g can be extended to an analytic function in some annulus $A: \alpha < |z| < \beta$, where $\alpha < 1 < \beta$. As shown in §4.4, g then can be expanded in a Laurent series,

$$g(z) = \sum_{n=-\infty}^{\infty} a_n z^n \qquad (4.5\text{-}14)$$

valid for $z \in A$. For $z = e^{i\tau}$ this series reduces to (4.5-11), and indeed the formula for the nth Laurent coefficient,

$$a_n = \frac{1}{2\pi i} \int_\Gamma g(t) t^{-n-1} dt$$

for $\Gamma: t = e^{i\tau}$, $0 \le \tau \le 2\pi$, becomes identical to (4.5-12). We note the result:

THEOREM 4.5b

Let the function p be periodic with period 2π. If the function g defined by $g(e^{i\tau}) := p(\tau)$ can be extended to a function analytic in an annulus A containing $|z|=1$, then the Laurent expansion of g in A is identical to the (complex) Fourier series of p.

In principle, nothing new has been gained by this result. In fact, a function p which gives rise to a g that can be continued analytically into an annulus A automatically has derivatives of all orders, a property not required for the validity of Fourier's expansion. The connection established in Theorem 4.5b, however, can be useful when the direct computation of the Fourier coefficients from (4.5-12) is difficult, whereas the Laurent series can be found by some other methods.

EXAMPLE 1

Let $\gamma > 1$ and let

$$p(\tau) := \frac{1}{\gamma + \cos \tau}.$$

APPLICATIONS OF LAURENT SERIES

The corresponding function g for $|z|=1$ equals

$$g(z) = \frac{1}{\gamma + \frac{1}{2}(z+z^{-1})} = \frac{2z}{z^2 + 2\gamma z + 1}.$$

The last expression, however, is clearly analytic in the annulus $|z_1| < |z| < |z_2|$, where z_1 and z_2 are the two zeros of the denominator:

$$z_1 = -\gamma + \sqrt{\gamma^2 - 1}, \qquad z_2 = -\gamma - \sqrt{\gamma^2 - 1}.$$

The Laurent series is easily obtained from the expansion in partial fractions:

$$g(z) = \frac{z_1}{(z_1 + \gamma)(z - z_2)} + \frac{z_2}{(z_2 + \gamma)(z - z_2)}.$$

Using geometric series, we find

$$g(z) = \frac{z_1}{z_1 + \gamma} \frac{1}{z} \sum_{n=0}^{\infty} \left(\frac{z_1}{z}\right)^n - \frac{1}{z_2 + \gamma} \sum_{n=0}^{\infty} \left(\frac{z}{z_2}\right)^n$$

or, because $z_2 = z_1^{-1}$, $z_1 + \gamma = -(z_2 + \gamma) = \sqrt{\gamma^2 - 1}$,

$$g(z) = \frac{1}{\sqrt{\gamma^2 - 1}} \left\{ 1 + \sum_{n=1}^{\infty} z_1^n (z^n + z^{-n}) \right\}.$$

The desired Fourier series for p is obtained by letting $z = e^{i\tau}$:

$$p(\tau) = \frac{1}{\sqrt{\gamma^2 - 1}} \left\{ 1 + 2 \sum_{n=1}^{\infty} \left(-\gamma + \sqrt{\gamma^2 - 1}\right)^n \cos n\tau \right\}.$$

EXAMPLE 2

Let $|w| < 1$, let ν be real, and let

$$p(\tau) := (1 - 2w \cos \tau + w^2)^{-\nu}$$

(principal value). This function is important in many applications; for $\nu = \frac{1}{2}$, w real, it represents the inverse distance of the points $z = 1$ and $z = we^{i\tau}$ in the complex plane. We have obtained the expansion of p in powers of w in Problem 5, §1.6; let us now calculate its Fourier series. If $p(\tau) = g(e^{i\tau})$, we have for $|z| = 1$

$$g(z) = \left[1 - w\left(z + \frac{1}{z}\right) + w^2 \right]^{-\nu}$$

$$= (1 - wz)^{-\nu} (1 - wz^{-1})^{-\nu},$$

where both powers have their principal values. The first factor is analytic for $|z|<|w^{-1}|$, the second for $|z|>|w|$. It follows that g is analytic in the annulus $|w|<|z|<|w|^{-1}$. The Laurent series can be found by forming the Cauchy product of the two series

$$(1-wz)^{-\nu} = \sum_{n=0}^{\infty} \frac{(\nu)_n}{n!} w^n z^n,$$

$$(1-wz^{-1})^{-\nu} = \sum_{n=0}^{\infty} \frac{(\nu)_n}{n!} w^n z^{-n}.$$

If the nth coefficient in the Laurent expansion of g is denoted by $A_n(w)$, then for $n \geqslant 0$

$$A_n(w) = \sum_{m=0}^{\infty} \frac{(\nu)_m (\nu)_{n+m}}{m!(n+m)!} w^{n+2m}$$

or, using hypergeometric notation,

$$A_n(w) = \frac{(\nu)_n}{n!} w^n {}_2F_1(\nu, \nu+n; n+1; w^2). \qquad (4.5\text{-}15)$$

By reasons of symmetry $A_{-n}(w) = A_n(w)$. The coefficients $A_n(w)$ can also be expressed in terms of Legendre functions; see §9.10. The resulting Fourier series for p may be written in real form as follows:

$$(1 - 2w\cos\tau + w^2)^{-\nu} = A_0(w) + 2\sum_{n=1}^{\infty} A_n(w) \cos n\tau. \qquad (4.5\text{-}16)$$

KEPLER'S EQUATION.

Let ϵ be real, $-1<\epsilon<1$, and let the function ϕ for real η be defined by

$$\phi(\eta) := \eta - \epsilon \sin\eta.$$

Because $\phi'(\eta) = 1 - \epsilon \cos\eta \geqslant 1 - \epsilon > 0$, ϕ defines a one-to-one mapping of the real line onto itself and

$$\phi(\eta) = \theta \qquad (4.5\text{-}17)$$

thus has a unique solution η for every given θ. (4.5-17) is known as **Kepler's equation**. Its significance in celestial mechanics is explained in Problem 8, §2.5, in which the equation was solved by the Lagrange-Bürmann formula. Here we indicate another classical approach to the equation.

For $-\infty < \theta < \infty$ let $\eta(\theta)$ denote the unique solution of (4.5-17). We

APPLICATIONS OF LAURENT SERIES

assert that the function p defined by

$$p(\theta) := \eta(\theta) - \theta = \epsilon \sin \eta(\theta)$$

is periodic with period 2π. Indeed, if $\phi(\eta) = \theta$, then $\phi(\eta + 2\pi) = \eta + 2\pi - \epsilon \sin \eta = \theta + 2\pi$, so that $\eta(\theta + 2\pi) = \eta(\theta) + 2\pi$, and the assertion follows. Since p is odd and has a continuous derivative, it can be expanded in a Fourier sine series,

$$p(\theta) = \sum_{n=1}^{\infty} b_n \sin n\theta,$$

where

$$b_n = \frac{2}{\pi} \int_0^{\pi} p(\theta) \sin n\theta \, d\theta = \frac{2}{\pi} \int_0^{\pi} [\eta(\theta) - \theta] \sin n\theta \, d\theta.$$

Integrating by parts, we find

$$\frac{\pi}{2} b_n = \left[-\frac{1}{n}(\eta(\theta) - \theta) \cos n\theta \right]_0^{\pi} + \frac{1}{n} \int_0^{\pi} [\eta'(\theta) - 1] \cos n\theta \, d\theta$$

$$= \frac{1}{n} \int_0^{\pi} \cos(n\theta) \eta'(\theta) \, d\theta,$$

for $\eta(0) = 0$, $\eta(\pi) = \pi$, and $\int_0^{\pi} \cos n\theta \, d\theta = 0$. Because $\eta'(\theta) > 0$, we may introduce η as a new variable of integration and get

$$b_n = \frac{2}{n\pi} \int_0^{\pi} \cos[n\phi(\eta)] \, d\eta$$

$$= \frac{2}{n\pi} \int_0^{\pi} \cos[n(\eta - \epsilon \sin \eta)] \, d\eta$$

$$= \frac{2}{n} J_n(n\epsilon)$$

by Bessel's integral (4.5-6).

THEOREM 4.5c

The solution η of Kepler's equation (4.5-17) can for every real θ be represented in the form

$$\eta = \theta + \sum_{n=1}^{\infty} \frac{2}{n} J_n(n\epsilon) \sin n\theta. \qquad (4.5\text{-}18)$$

From the known asymptotic behavior of Bessel functions (see §11.8) we may conclude that the series (4.5-18) converges at least like a geometric series with ratio

$$\frac{\epsilon}{1+(1-\epsilon^2)^{1/2}}\exp(1-\epsilon^2)^{1/2}.$$

PROBLEMS

1. Working directly from the generating function (without manipulating the series (4.5-4)), show that for $n = 0, \pm 1, \pm 2, \ldots$,

   (a) $J_n'(z) = \frac{1}{2}(J_{n-1}(z) - J_{n+1}(z))$,

   (b) $J_n(z) = \frac{z}{2n}(J_{n-1}(z) + J_{n+1}(z))$ $\quad (n \neq 0)$,

   (c) $J_n''(z) + \frac{1}{z}J_n'(z) + \left(1 - \frac{n^2}{z^2}\right)J_n(z) = 0$,

   identically in z.

2. Prove that for real values of z

   $$|J_n(z)| \leq 1, \quad n = 0, \pm 1, \pm 2, \ldots.$$

3. Deduce from Bessel's integral that

   $$J_n(z) = \frac{1}{\pi}\int_0^\pi \cos\left(z\cos\tau - \frac{n\pi}{2}\right)\cos n\tau \, d\tau. \qquad (4.5\text{-}19)$$

   Also deduce (4.5-19) from Poisson's integral representation

   $$J_n(z) = \frac{(z/2)^n}{\pi(1/2)_n}\int_{-1}^1 \cos(\mu z)(1-\mu^2)^{n-1/2}\,d\mu$$

   by repeated integration by parts, using Jacobi's transformation (see Problem 8, §1.9).

4. Deduce from the addition formula that

   $$J_0^2(z) + 2\sum_{n=1}^\infty J_n^2(z) = 1,$$

   $$J_0^2(z) + 2\sum_{n=1}^\infty (-1)^n J_n^2(z) = J_0(2z).$$

APPLICATIONS OF LAURENT SERIES

5. Let $|\epsilon|<1$. Deduce from Example 1 that

$$\frac{1}{\pi}\int_0^\pi \frac{1}{1-\epsilon\cos\tau}\,d\tau = \frac{1}{\sqrt{1-\epsilon^2}}$$

and verify the result by expanding in powers of ϵ.

6. Deduce from Example 2 that

$$\frac{1}{\pi}\int_0^\pi \frac{1}{\sqrt{1-\epsilon\cos\tau}}\,d\tau = \left[\frac{1+\omega}{2}\right]^{-1/2} {}_2F_1\left(\frac{1}{2},\frac{1}{2};1;\frac{1-\omega}{1+\omega}\right),$$

where $\omega := (1-\epsilon^2)^{1/2}$. Compute the integral also by expanding in powers of ϵ and conclude that

$$\left[\frac{1+\omega}{2}\right]^{-1/2} {}_2F_1\left(\frac{1}{2},\frac{1}{2};1;\frac{1-\omega}{1+\omega}\right) = {}_2F_1\left(\frac{1}{4},\frac{3}{4};1;\epsilon^2\right).$$

7. Let $\rho>1$. The point $z=\frac{1}{2}(\rho e^{i\tau}+\rho^{-1}e^{-i\tau})$, $0\leq\tau\leq 2\pi$, describes an ellipse. Compute the Fourier series of $|z'(\tau)|$ and use the result to determine the circumference of the ellipse.

8. Find the Fourier expansion of the periodic function

$$p(\tau) := \frac{1}{\cos\tau + (\cos\tau)^{-1}}.$$

9. Prove the expansions

$$\cos(z\sin\tau) = J_0(z) + 2\sum_{n=1}^\infty J_{2n}(z)\cos 2n\tau,$$

$$\sin(z\sin\tau) = 2\sum_{n=0}^\infty J_{2n+1}(z)\sin(2n+1)\tau.$$

10. For $\alpha>0$, ξ real and $-1<\epsilon<1$, let

$$f(\xi) := \int_{-\infty}^\xi e^{\alpha(\eta-\epsilon\sin\eta)}\,d\eta.$$

Show that

$$f(\xi) = e^{\alpha\theta}\sum_{n=0}^\infty \lambda_n J_n(n\epsilon)\frac{\alpha\cos n\theta + n\sin n\theta}{\alpha^2+n^2},$$

where $\theta = \xi - \epsilon\sin\xi$, $\lambda_0 := 1$, $\lambda_n := 2$, $n=1,2,\dots$.

11. By using Euler's formula,

$$\sin \xi = \frac{1}{2i}(e^{i\xi} - e^{-i\xi}),$$

show directly that the solution of Kepler's equation given in Problem 8, §2.5, is identical with the solution of Theorem 4.5c.

§4.6. CONTINUOUS ARGUMENT, WINDING NUMBER, JORDAN CURVE THEOREM

In this section the topological foundation is laid that will be required for a proper statement of the Cauchy integral formula and of the residue theorem to be given in §4.7.

Let $\Gamma: z = z(\tau)$, $\alpha \leq \tau \leq \beta$, be an arc (not necessarily closed or regular, with or without multiple points) and let c be a point not on Γ. For every $\tau \in [\alpha, \beta]$ the complex number $z(\tau) - c$ has an infinite number of arguments. A **continuous argument** of $z - c$ on Γ is a continuous real function ϕ defined on $[\alpha, \beta]$ such that, for every $\tau \in [\alpha, \beta]$, $\phi(\tau)$ equals one of the values of $\arg(z(\tau) - c)$ (see Fig. 4.6a).

THEOREM 4.6a

If Γ is an arc and c is a point not on Γ, a continuous argument of $z - c$ can be defined on Γ. Any two continuous arguments differ by a constant integral multiple of 2π.

Proof. As a point set Γ is compact; hence it has a distance $\eta > 0$ from c. The function $z(\tau)$ is continuous on the compact interval $[\alpha, \beta]$, hence

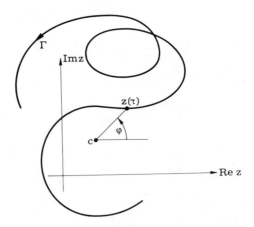

Fig. 4.6a. Continuous argument.

JORDAN CURVE THEOREM

uniformly continuous. Thus there exists $\delta > 0$ such that $|z(\tau)-z(\tau')|<\eta$ whenever $|\tau-\tau'|<\delta$. Let $\alpha=\tau_0<\tau_1<\tau_2<\cdots<\tau_{n-1}<\tau_n=\beta$, where $\tau_{k+1}-\tau_k<\delta$, and let Γ_k denote the subarc corresponding to the parameter subinterval $[\tau_k,\tau_{k+1}]$, $k=0,1,\ldots,n-1$.

For each k, $|z(\tau_k)-c|>\eta$, whereas $|z(\tau)-z(\tau_k)|<\eta$ as long as $\tau \in [\tau_k,\tau_{k+1}]$. Thus for $\tau \in [\tau_k,\tau_{k+1}]$

$$\operatorname{Re}\frac{z(\tau)-c}{z(\tau_k)-c}=\operatorname{Re}\left(1+\frac{z(\tau)-z(\tau_k)}{z(\tau_k)-c}\right)>0,$$

which shows that Γ_k lies in the half-plane

$$H_k:\operatorname{Re}\frac{z-c}{z(\tau_k)-c}>0.$$

In H_k, however, $\arg(z-c)$ can be defined as a continuous function by selecting $\phi_k:=\arg(z(\tau_k)-c)$ arbitrarily and assigning $\arg(z-c)$ the value that differs from ϕ_k by less than $\pi/2$. With this choice of $\arg(z-c)$, the function

$$\phi(\tau):=\arg(z(\tau)-c), \qquad \tau \in [\tau_k,\tau_{k+1}], \qquad (4.6\text{-}1)$$

as a composition of continuous functions, is continuous. Thus it defines a continuous argument of $z-c$ on Γ_k.

To define a continuous argument on Γ we select ϕ_0 arbitrarily and define $\phi(\tau)$ on $[\tau_0,\tau_1]$ by (4.6-1). This determines $\phi_1:=\phi(\tau_1)$, and by applying (4.6-1) for $k=1$ we obtain a continuous argument on Γ_1. By proceeding in the same manner we can define $\phi(\tau)$ as a continuous function on all of $[\alpha,\beta]$.

Now let $\phi_1(\tau)$ and $\phi_2(\tau)$ be any two continuous arguments of $z-c$ on Γ. Both functions ϕ_1 and ϕ_2 are continuous, and for each $\tau \in [\alpha,\beta]$ the difference $\phi_1(\tau)-\phi_2(\tau)$ equals an integral multiple of 2π. Since $[\alpha,\beta]$ is connected, it follows that the difference is a *constant* integral multiple of 2π, and the second part of the theorem is proved. ∎

As a consequence of Theorem 4.6a we note the following fact: if $\Gamma:z=z(\tau)$ is an arc, if c is not on Γ, and if $\phi(\tau)$ is a continuous argument of $z-c$ on Γ, the difference $\phi(\beta)-\phi(\alpha)$ is independent of the choice of ϕ. This is called the **increase of the argument** of $z-c$ along Γ and is denoted by

$$\Delta:=[\arg(z(\tau)-c)]_\Gamma.$$

Clearly, if $\Gamma=\Gamma_1+\Gamma_2+\cdots+\Gamma_n$ and if

$$\Delta_k:=[\arg(z(\tau)-c)]_{\Gamma_k}, \qquad k=1,2,\ldots,n,$$

then

$$\Delta = \Delta_1 + \Delta_2 + \cdots + \Delta_n. \tag{4.6-2}$$

The increase of the argument along piecewise regular arcs can be expressed analytically:

THEOREM 4.6b

If $\Gamma: z = z(\tau)$, $\alpha \leq \tau \leq \beta$, *is a piecewise regular arc and c is not on Γ, then*

$$\int_\Gamma \frac{1}{z-c} dz = \text{Log}\left|\frac{z(\beta)-c}{z(\alpha)-c}\right| + i[\arg(z(\tau)-c)]_\Gamma, \tag{4.6-3}$$

and, in particular,

$$[\arg(z(\tau)-c)]_\Gamma = \text{Im} \int_\Gamma \frac{1}{z-c} dz. \tag{4.6-4}$$

Proof. We subdivide the parametric interval as in the proof of Theorem 4.6a such that for each $k = 0, 1, \ldots, n-1$ the arc Γ_k corresponding to $[\tau_k, \tau_{k+1}]$ lies in the half-plane

$$\text{Re} \frac{z-c}{z(\tau_k)-c} > 0.$$

In this half-plane the function $f(z) := \text{Log}[(z-c)/(z(\tau_k)-c)]$ is analytic and has the derivative

$$f'(z) = \frac{1}{z-c};$$

hence by Cauchy's theorem

$$\int_{\Gamma_k} \frac{1}{z-c} dz = \text{Log} \frac{z(\tau_{k+1})-c}{z(\tau_k)-c}.$$

The expression on the right equals

$$\text{Log}\left|\frac{z(\tau_{k+1})-c}{z(\tau_k)-c}\right| + i[\arg(z(\tau)-c)]_{\Gamma_k},$$

which proves (4.6-3) for a subarc Γ_k. The complete result follows by (4.6-2). ∎

Theorem 4.6b shows that the increase in the argument of $z - c$ along Γ is the same for all curves that are geometrically equivalent to Γ.

As a consequence of the representation (4.6-4) we note the following:

JORDAN CURVE THEOREM

THEOREM 4.6c

Let Γ be a piecewise regular arc and let the function Δ on the complement S of Γ be defined by

$$\Delta(c) := [\arg(z-c)]_\Gamma.$$

Then Δ is continuous on S.

Proof. Let c be any point of S, let $\eta > 0$ be the distance between c and Γ, and let $|c' - c| < \tfrac{1}{2}\eta$. Then by (4.6-4), if λ denotes the length of Γ,

$$|\Delta(c') - \Delta(c)| \leq \left| \int_\Gamma \frac{1}{z-c'} dz - \int_\Gamma \frac{1}{z-c} dz \right|$$

$$= \left| \int_\Gamma \frac{c-c'}{(z-c')(z-c)} dz \right| \leq |c-c'| \frac{2\lambda}{\eta^2}.$$

Thus, if $\epsilon > 0$ is given, it is evidently possible to choose $\delta > 0$ such that $|c' - c| < \delta$ implies $|\Delta(c') - \Delta(c)| < \epsilon$, which establishes continuity at point c. ∎

Of special interest is the case in which the arc $\Gamma: z = z(\tau)$, $\alpha \leq \tau \leq \beta$, is a *closed curve*, i.e., where $z(\beta) = z(\alpha)$. If c is any point not on Γ and $\phi(\tau)$ is a continuous argument of $z-c$ on Γ, then, because both $\phi(\alpha)$ and $\phi(\beta)$ are values of $\arg(z(\alpha) - c)$, the increase in the argument of $z-c$ along Γ is an integral multiple of 2π. Consequently the number

$$n(\Gamma, c) := \frac{1}{2\pi} [\arg(z(\tau) - c)]_\Gamma$$

is an integer. This integer is called the **winding number** of Γ with respect to c.

As a consequence of Theorem 4.6b and because

$$\text{Log} \left| \frac{z(\beta) - c}{z(\alpha) - c} \right| = \text{Log } 1 = 0,$$

we have the following:

THEOREM 4.6d

If Γ is a piecewise regular closed curve and c is not on Γ,

$$n(\Gamma, c) = \frac{1}{2\pi i} \int_\Gamma \frac{1}{z-c} dz. \qquad (4.6\text{-}5)$$

In some applications of the principle of the argument (see §4.10) it is necessary to compute the winding number numerically. Formula (4.6-5) might suggest a simple method of solving this problem by numerical integration. Because the result is known to be an integer, one might be tempted to perform the integration in a crude manner. However, in order to get the right integer, an analysis of the quadrature error would be required anyway, especially if c is close to Γ. Thus the method is not without fallacies. A simple but precise algorithm for determining winding numbers is given in the appendix of this section.

We now state another important property of the winding number. Because the curve Γ, considered as a point set in the complex plane, is closed, its complement S is an open set. As such, it admits, by Theorem 3.2a, a unique decomposition into (open) components, i.e., subregions of S that are not contained in any larger subregion.

THEOREM 4.6e

Let Γ be a piecewise regular closed curve and let S be the complement of Γ. Then the winding number $n(\Gamma,c)$ as a function of c has a constant value on each component of S (see Fig. 4.6b).

Proof. By Theorem 4.6c $n(\Gamma,c)$ is continuous at every point of S. By definition $n(\Gamma,c)$ is integer-valued. Such a function is necessarily constant on each component of its domain of definition (see Problem 1, §3.2). ∎

We are now approaching the celebrated Jordan curve theorem which, although not indispensable in analytic function theory, simplifies the

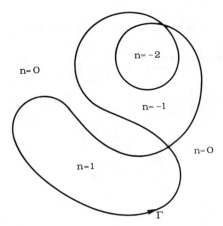

Fig. 4.6b. The winding number.

JORDAN CURVE THEOREM

statement of many results in integration theory. Two more notions are required: the notion of a *Jordan curve* and that of the *boundary of a set*.

A curve $\Gamma: z = z(\tau)$, $\alpha \leq \tau \leq \beta$, is called **simple** if it has no multiple points, i.e., if $z(\tau_1) = z(\tau_2)$ implies $\tau_1 = \tau_2$. If Γ is closed, it is not simple, for by definition it has the multiple point $z(\alpha) = z(\beta)$. A closed curve is called a **Jordan curve** if it has no other multiple points; i.e., if $z(\tau_1) = z(\tau_2)$, $\tau_1 < \tau_2$ implies $\tau_1 = \alpha$, $\tau_2 = \beta$.

If T is any set, a point b is called a **boundary point** of T if every neighborhood of b contains points that are and points that are not in T. The point b itself may or may not belong to T. The set of all boundary points of a set T is called the **boundary** of T. It is easy to see that the boundary of a set is always closed. The boundary of an open set always belongs to the complement of T.

THEOREM 4.6f (Jordan curve theorem)

Let Γ be a piecewise regular Jordan curve. The complement of Γ has exactly two components, each with Γ as its boundary. Exactly one of the two components is unbounded. The winding number $n(\Gamma, c)$ equals zero in the unbounded component and either $+1$ or -1 throughout the bounded component (see Fig. 4.6c).

The bounded component of the complement is called the **interior** of the Jordan curve. A Jordan curve is said to be **positively oriented** if its winding number with respect to the points in its interior is $+1$.

The Jordan curve theorem actually holds without the assumption that Γ is piecewise regular. Although the theorem appears to be intuitively obvious, the proof is not easy, even with this assumption. Several preliminary steps are required.

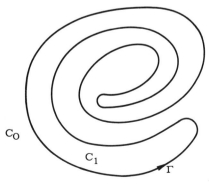

Fig. 4.6c. The Jordan curve theorem.

LEMMA 4.6g

Let $\Gamma: z = z(\tau)$, $\alpha \leq \tau \leq \beta$, be a regular simple arc. Then there exists $\epsilon > 0$ such that for $0 \leq \rho \leq \epsilon$ the equation $|z(\tau) - z(\alpha)| = \rho$ has exactly one solution, $\tau = \tau(\rho)$, such that $\tau \in [\alpha, \beta]$. The portion of Γ lying in $|z - z(\alpha)| \leq \epsilon$ is geometrically equivalent to a curve

$$z = \tilde{z}(\rho) := z(\alpha) + \rho e^{i\phi(\rho)}, \qquad 0 \leq \rho \leq \epsilon, \tag{4.6-6}$$

where $\phi(\rho)$ is a continuous function such that

$$\phi(0) = \arg z'(\alpha). \tag{4.6-7}$$

Proof. By a change of the parameter we may assume $\alpha = 0$, and by a rigid motion of the plane we can achieve $z(0) = 0$, $z'(0) > 0$. Letting $z(\tau) =: x(\tau) + iy(\tau)$, this means that

$$x(0) = y(0) = 0; \qquad x'(0) > 0, \qquad y'(0) = 0. \tag{4.6-8}$$

Now $\rho(\tau) := |z(\tau)| = (x^2 + y^2)^{1/2}$, and computation shows that $\rho'(\tau)$ is continuous for $\tau \geq 0$ and $\rho'(0) = x'(0) > 0$. Thus there exists δ such that $\rho(\tau)$ is strictly increasing for $\tau \in [0, \delta]$, which shows that for $\rho \leq |z(\delta)|$ the equation

$$|z(\tau)| = \rho \tag{4.6-9}$$

has a unique solution $\tau = \tau(\rho)$ such that $\tau \in [0, \delta]$. Let $2\epsilon := \min_{\delta \leq \tau \leq \beta} |z(\tau)|$. Then $\epsilon > 0$, otherwise 0 would be a multiple point of Γ; and it is clear that for $0 \leq \rho \leq \epsilon$ (4.6-9) has no solution other than $\tau(\rho)$ in $[0, \beta]$.

The function $\tau(\rho)$ is the inverse function of a continuous and strictly increasing function; hence it is itself continuous and strictly increasing. Thus ρ may be introduced as a new parameter for the portion of Γ lying in $|z| < \epsilon$. The relations (4.6-8) show that $x(\tau) > 0$ for $\tau > 0$, τ sufficiently small. For ρ less than some ρ_0 we may define

$$\phi(\rho) := \operatorname{Arctan} \frac{y(\tau(\rho))}{x(\tau(\rho))}$$

(principal value), which is continuous for $\rho > 0$ and by L'Hospital's rule has the limit 0 for $\rho \to 0$. For $\rho \geq \rho_0$, $\phi(\rho)$ may be defined as a continuous argument of $\tilde{z}(\rho)$. ∎

The representation (4.6-6) is called the *polar representation* of Γ near its initial point.

JORDAN CURVE THEOREM 237

LEMMA 4.6h

Let $\Gamma: z = z(\tau)$, $\alpha \leq \tau \leq \beta$, be a piecewise regular Jordan curve. If $z(\tau_0)$ is any point on Γ where z' is continuous, the intersection of every neighborhood N of $z(\tau_0)$ with S, the complement of Γ, contains points z_1, z_2 such that

$$|n(\Gamma, z_1) - n(\Gamma, z_2)| = 1. \qquad (4.6\text{-}10)$$

Proof. Again we may assume that $\tau_0 = 0$, $z(\tau_0) = 0$, $z'(\tau_0) > 0$. Because Γ is closed, we may assume that 0 is an interior point of the parametric interval; i.e., $\alpha < 0 < \beta$. Let $\epsilon > 0$ be so small that the conclusions of Lemma 4.6g hold for both the curve $z = z(\tau)$, $0 \leq \tau \leq \beta$ and the curve $z = z(-\sigma)$, $0 \leq \sigma \leq -\alpha$. The two parametric representations may then be combined into one, and Γ_0, the portion of Γ lying in $|z| \leq \epsilon$, may be parametrized by

$$z = \tilde{z}(\rho) := \rho e^{i\phi(\rho)}, \qquad -\epsilon \leq \rho \leq \epsilon,$$

where ϕ is continuous, $\phi(0) = 0$. If ϵ is small enough, $|\phi(\rho)| < \pi/2$ for $|\rho| \leq \epsilon$, and the points $z = \pm i\eta$ do not lie on Γ_0 for $0 < \eta < \epsilon$.

Let Γ_1 be the complementary arc of Γ_0. We compute the winding numbers of $\Gamma = \Gamma_0 + \Gamma_1$ with respect to the points $\pm i\eta$ by adding the increases of the arguments of $z - (\pm i\eta)$ along Γ_0 and Γ_1. For c in the complement of Γ_k let

$$\Delta_k(c) := [\arg(z - c)]_{\Gamma_k}, \qquad k = 0, 1.$$

Then by Theorem 4.6c $\Delta_1(i\eta)$ is continuous at $\eta = 0$, hence

$$\lim_{\eta \to 0+} [\Delta_1(i\eta) - \Delta_1(-i\eta)] = 0. \qquad (4.6\text{-}11)$$

To compute $\Delta_0(i\eta)$ for $\eta > 0$ let H denote the complex plane cut along the imaginary axis from $i\eta$ to $i\infty$. A continuous argument of $z - i\eta$ can be defined in the range $(-3\pi/2, \pi/2)$, and as z travels along Γ_0 the continuous argument of $z - i\eta$ varies from a value close to $\phi(-\epsilon) - \pi$ to a value close to $\phi(\epsilon)$. There follows

$$\lim_{\eta \to 0+} \Delta_0(i\eta) = \phi(\epsilon) - \phi(-\epsilon) + \pi.$$

By similar reasoning we can establish that

$$\lim_{\eta \to 0+} \Delta_0(-i\eta) = \phi(\epsilon) - \phi(-\epsilon) - \pi,$$

and

$$\lim_{\eta \to 0+} [\Delta_0(i\eta) - \Delta_0(-i\eta)] = 2\pi \qquad (4.6\text{-}12)$$

follows. By adding it to (4.6-11) we have, in view of $n(\Gamma)=(1/2\pi)(\Delta_0+\Delta_1)$,

$$\lim_{\eta\to 0+}[n(\Gamma,i\eta)-n(\Gamma,-i\eta)]=1.$$

Because $n(\Gamma,\pm i\eta)$ is an integer, we conclude that for $\eta>0$, η sufficiently small,

$$n(\Gamma,i\eta)-n(\Gamma,-i\eta)=1,$$

which establishes Lemma 4.6h. ∎

LEMMA 4.6i

Let $\Gamma:z=z(\tau)$, $\alpha\leq\tau\leq\beta$, be a piecewise regular Jordan curve and let $z(\tau_0)$ be any point of Γ. Then for every sufficiently small $\eta>0$ the intersection of the set $N:|z-z(\tau_0)|<\eta$ with the complement S of Γ has precisely two components. Each point of $N\cap\Gamma$ is a boundary point of either component. The values of $n(\Gamma,c)$ in the two components differ by 1.

Proof. As above we may suppose that $\tau_0=0$, $z(\tau_0)=0$, $\alpha<0<\beta$. The derivative z' need not exist at $\tau=0$, but since Γ is piecewise regular $\alpha_1<0$ and $\beta_1>0$ exist such that the images Γ_l and Γ_r of $[\alpha_1,0]$ and of $[0,\beta_1]$ under the map $z=z(\tau)$ are regular curves. Let $\epsilon>0$ be such that the conclusion of Lemma 4.6g applies to both $-\Gamma_l$ and Γ_r. The portions of $-\Gamma_l$ and Γ_r lying in $|z|\leq\epsilon$ then have the parametric representations

$$-\Gamma_l:z=\rho e^{i\phi_l(\rho)},\quad 0\leq\rho\leq\epsilon,$$

$$\Gamma_r:z=\rho e^{i\phi_r(\rho)},\quad 0\leq\rho\leq\epsilon,$$

where the functions ϕ_l and ϕ_r are continuous and may be chosen such that $|\phi_l(0)-\phi_r(0)|\leq\pi$. We note that for $0<\rho<\epsilon$ we either always have $\phi_l(\rho)<\phi_r(\rho)<\phi_l(\rho)+2\pi$ or $\phi_r(\rho)<\phi_l(\rho)<\phi_r(\rho)+2\pi$, for otherwise Γ would have a multiple point. Assume for definiteness that $\phi_l(\rho)<\phi_r(\rho)$. The intersection of every set $N:|z|<\eta$, where $0<\eta\leq\epsilon$, with the complement of Γ then consists of the two sets

$$C_1:z=\rho e^{i\phi},\quad 0<\rho<\eta,\quad \phi_l(\rho)<\phi<\phi_r(\rho)$$

and

$$C_2:z=\rho e^{i\phi},\quad 0<\rho<\eta,\quad \phi_r(\rho)<\phi<\phi_l(\rho)+2\pi.$$

Each of these sets is obviously connected, and each point of $N\cap\Gamma$ is a boundary point of both C_1 and C_2. The values of the winding numbers in

JORDAN CURVE THEOREM

C_1 and C_2 differ by 1 because the intersection with S of each sufficiently small neighborhood of a point $z = \rho e^{i\phi r(\rho)}$, where $0 < \rho < \eta$, is contained in $N \cap S$, and by Lemma 4.6h every such neighborhood contains points at which the winding number differs by 1. ∎

We now turn to the *proof of the Jordan curve theorem* proper. Γ as a point set is contained in a disk $|z| \leq \mu$. The set $|z| > \mu$ is connected; hence S contains only one unbounded component, say C_0. If $|c| > \mu$, Cauchy's theorem can be applied to the function $(z-c)^{-1}$ in the disk $|z| < |c|$ with the result that

$$n(\Gamma, c) = \frac{1}{2\pi i} \int \frac{1}{z-c} dz = 0.$$

Thus $n(\Gamma, c) = 0$ for $c \in C_0$.

All components of S are open sets, hence the set of all components is denumerable. (The set of points with rational coordinates is denumerable and each component contains such a point.) Consequently, we number the components $C_0, C_1, C_2, \ldots$, but reserve the index 0 for the unbounded component. On the parametric interval $[\alpha, \beta]$ we now define two real functions ϕ_-, ϕ_+: for $\tau \in [\alpha, \beta]$ let p and q be the numbers of the two components that a sufficiently small neighborhood of the point $z(\tau)$ intersects by virtue of Lemma 4.6i. Then set

$$\phi_-(\tau) := \min(p, q), \qquad \phi_+(\tau) := \max(p, q).$$

The functions ϕ_- and ϕ_+ are obviously integer-valued. By Lemma 4.6i they are also continuous. Since the interval $[\alpha, \beta]$ is connected, it follows that both functions are constant.

Because no component constitutes the whole plane, each component has boundary points. These points can lie only on Γ. Let $z(\tau_0)$ be a boundary point of C_0. Then either $\phi_-(\tau_0)$ or $\phi_+(\tau_0)$ is zero. Consequently, one of the functions ϕ_- and ϕ_+ is identically zero and the other has a constant value $p \neq 0$. It follows that the only components with boundary points on Γ, and therefore the only components of S, are the sets C_0 and C_p. Since $n(\Gamma, c) = 0$ for $c \in C_0$, $n(\Gamma, c)$ for $c \in C_p$ must, by Lemma 4.6i and Theorem 4.6e, have either the constant value $+1$ or the constant value -1. Thus we have completed the proof of the Jordan curve theorem. ∎

APPENDIX: THE NUMERICAL DETERMINATION OF WINDING NUMBERS

Let $\Gamma: z = z(\tau)$, $\alpha \leq \tau \leq \beta$, be any closed curve (not necessarily a Jordan curve or piecewise regular). Here we consider the problem of computing numerically the value of $n(\Gamma,c)$ for a given point c not on Γ. To a certain extent this problem is already solved by the algorithm implicit in the proof of Theorem 4.6a. That algorithm, however, would require the evaluation of inverse trigonometric functions. Moreover, these functions would have to be evaluated with infinite precision. Contrary to this, the algorithm proposed below merely requires a comparison of real numbers, and there is ample allowance for rounding errors.

For $m = 0, 1, \ldots, 7$, let C_m denote the set of all $z \neq 0$ such that

$$m \frac{\pi}{4} \leq \arg z < (m+1) \frac{\pi}{4}$$

for a suitable determination of $\arg z$. Evidently every $z = z + iy \neq 0$ belongs to a unique sector C_m whose index is determined by inequalities on x and y; for instance, $z \in C_0$ if and only if $y \geq 0$ and $x > y$.

If Γ is defined as above, let $\alpha = \tau_0 < \tau_1 < \cdots < \tau_h = \beta$ be a subdivision of the parametric interval such that on each arc $\Gamma_k : z = z(\tau)$, $\tau_{k-1} \leq \tau \leq \tau_k$ ($k = 1, 2, \ldots, h$) $z - c$ is contained in some sector with vertex at 0 and opening $< \pi/2$. We then have

$$|[\arg(z-c)]_{\Gamma_k}| < \frac{\pi}{2}, \qquad k = 1, 2, \ldots, h;$$

therefore, if $z_k := z(\tau_k)$ and the arguments are suitably determined,

$$|\arg(z_k - c) - \arg(z_{k-1} - c)| \leq \frac{\pi}{2}. \tag{4.6-13}$$

Thus, if $z_0 - c \in C_{m_0}$, there exists for $k = 0, 1, \ldots, h$ an integer $m(k)$ such that $m(0) = m_0$,

$$z_k - c \in C_{m_k}, \qquad m_k \equiv m(k) \bmod 8,$$

and $|m(k) - m(k-1)| \leq 2$. The required winding number then equals

$$n(\Gamma, c) = \frac{m(h) - m(0)}{8}. \tag{4.6-14}$$

This construction still requires precise arithmetic which, however, can be avoided, as follows: let the numbers z_k^*, $k = 0, 1, \ldots, h$ be approximations of the z_k in the sense that

$$\left| \frac{z_k - z_k^*}{z_k - c} \right| < \frac{1}{3}, \qquad k = 0, 1, \ldots, h. \tag{4.6-15}$$

RESIDUE THEOREM

It then follows by elementary trigonometry that for a suitable determination of the arguments

$$|\arg(z_k^* - c) - \arg(z_k - c)| < \frac{\pi}{8}, \quad k = 0, 1, \ldots, h, \quad (4.6\text{-}16)$$

hence from (4.6-13),

$$|\arg(z_k^* - c) - \arg(z_{k-1}^* - c)| < \frac{3\pi}{4}, \quad k = 1, 2, \ldots, h.$$

Thus, if the numbers m_k^* are defined by the condition

$$z_k^* - c \in C_{m_k^*}, \quad k = 0, 1, \ldots, h,$$

we can find uniquely determined integers $m^*(k)$ such that $m^*(0) = m_0^*$, $m^*(k) \equiv m_k^* \bmod 8$ and $|m^*(k) - m^*(k-1)| \leq 3$, $k = 1, 2, \ldots, h$. By virtue of (4.6-16)

$$|m^*(k) - m(k)| \leq 1, \quad k = 0, 1, \ldots, h.$$

Hence

$$m^*(h) - m^*(0) = m(h) - m(0) + d,$$

where $|d| \leq 2$, which by (4.6-14) implies the following:

THEOREM 4.6j

Under the above hypotheses $n(\Gamma, c)$ equals the integer nearest to

$$\frac{m^*(h) - m^*(0)}{8}.$$

§4.7. RESIDUE THEOREM: CAUCHY INTEGRAL FORMULA

Let R be a simply connected region and let the function f be analytic in R with the exception of a finite number of isolated singularities at the points $z_1, z_2, \ldots, z_m$. Let Γ be a piecewise regular closed curve in R which does not pass through any of the singular points of f. Our aim is a simple formula for the evaluation of the integral of f along Γ.

For $k = 1, 2, \ldots, m$, let

$$p_k(z) := \sum_{n=-\infty}^{-1} a_n^{(k)} (z - z_k)^{-n} \quad (4.7\text{-}1)$$

be the principal part of f at the point z_k (see §4.4). Each p_k is analytic except at the point z_k. Thus by subtracting the p_k from f we remove the singularities at the points $z_1, \ldots, z_m$ without introducing any new singularities. Hence the function

$$g(z) := f(z) - p_1(z) - \cdots - p_m(z)$$

is analytic in R and by Cauchy's theorem (Theorem 4.3e)

$$\int_\Gamma [f(z) - p_1(z) - \cdots - p_m(z)] \, dz = 0$$

or

$$\int_\Gamma f(z) \, dz = \sum_{k=1}^{m} \int_\Gamma p_k(z) \, dz. \qquad (4.7\text{-}2)$$

The integrals of the p_k are evaluated easily. Since Γ, as a bounded closed set, has a positive distance from points $z_1, \ldots, z_m$, the series (4.7-1) converge uniformly along Γ and we may integrate term by term:

$$\int_\Gamma p_k(z) \, dz = \sum_{n=-\infty}^{-1} a_n \int_\Gamma (z - z_k)^n \, dz.$$

By Theorem 4.3b the integrals are zero for $n = -2, -3, \ldots$, whereas for $n = -1$ the integral can, by virtue of

$$\frac{1}{2\pi i} \int_\Gamma \frac{1}{z - z_k} \, dz = n(\Gamma, z_k),$$

be expressed in terms of the winding number of Γ with respect to z_k. Thus we find

$$\int_\Gamma p_k(z) \, dz = 2\pi i \, n(\Gamma, z_k) a_{-1}^{(k)}. \qquad (4.7\text{-}3)$$

The number $a_{-1}^{(k)}$, i.e., the coefficient of the (-1)st power in the Laurent expansion of f around the singularity z_k, clearly plays a special role. It is called the **residue** of f at z_k and may be denoted by a symbol such as res $f(z_k)$ or, if f is given by a formula, res $f(z)|_{z=z_k}$. By (4.4-3)

$$\mathrm{res} f(z_k) = \frac{1}{2\pi i} \int_{\Gamma_k} f(z) \, dz, \qquad (4.7\text{-}4)$$

where Γ_k denotes a positively oriented circle round z_k not containing any other singularities of f.

RESIDUE THEOREM

With this concept in accordance with (4.7-2) and (4.7-3) we have the following theorem:

THEOREM 4.7a (Residue theorem)
Let f be analytic, up to isolated singularities at the points $z_1,\ldots,z_m$, in the simply connected domain R, and let Γ be a piecewise regular closed curve that does not pass through any of the singular points. Then

$$\int_\Gamma f(z)\,dz = 2\pi i \sum_{k=1}^m n(\Gamma, z_k)\operatorname{res} f(z_k). \qquad (4.7\text{-}5)$$

If Γ is a positively oriented Jordan curve and the points $z_1,\ldots,z_m$ lie in the interior of Γ, then

$$\int_\Gamma f(z)\,dz = 2\pi i \sum_{k=1}^m \operatorname{res} f(z_k). \qquad (4.7\text{-}6)$$

Determination of Residues. To apply the residue theorem in concrete situations a method for evaluating residues is required. Let us now assume that the isolated singularity is at $z = z_0$ and that the Laurent expansion in the disk punctured at z_0 is

$$f(z) =: \sum_{n=-\infty}^\infty a_n (z-z_0)^n. \qquad (4.7\text{-}7)$$

1. z_0 *is a simple pole.* Here (4.7-7) takes the form

$$f(z) = \sum_{n=-1}^\infty a_n (z-z_0)^n.$$

Consequently, the function

$$(z-z_0)f(z) = \sum_{n=0}^\infty a_{n-1}(z-z_0)^n$$

has a removable singularity at z_0 and

$$a_{-1} = \operatorname{res} f(z_0) = \lim_{z\to z_0}(z-z_0)f(z). \qquad (4.7\text{-}8)$$

Let us assume that f is given in the form

$$f(z) = \frac{p(z)}{q(z)},$$

where the functions p and q are analytic, $p(z_0)\neq 0$, and q has a zero of order 1 at z_0; i.e.,

$$q(z_0)=0, \qquad q'(z_0)\neq 0.$$

Then

$$\lim_{z\to z_0}(z-z_0)\frac{p(z)}{q(z)}=p(z_0)\lim_{z\to z_0}\left[\frac{q(z)}{z-z_0}\right]^{-1}$$

$$=p(z_0)\lim_{z\to z_0}\left[\frac{q(z)-q(z_0)}{z-z_0}\right]^{-1}=p(z_0)q'(z_0)^{-1}.$$

Thus

$$\operatorname{res} f(z_0)=\frac{p(z_0)}{q'(z_0)}. \qquad (4.7\text{-}9)$$

The same result can be found, of course, by the Taylor series. We refer to this result as the *standard method* for evaluating residues at a simple pole.

2. z_0 *is a pole of order* $m>1$. Here the series (4.7-7) begins at $n=-m$, and consequently f has the form

$$f(z)=(z-z_0)^{-m}g(z),$$

where

$$g(z):=\sum_{n=0}^{\infty}a_{n-m}(z-z_0)^n$$

is analytic at z_0. The residue a_{-1} thus equals the coefficient of the $(m-1)$st power in the expansion of g in powers of $z-z_0$, i.e.,

$$\operatorname{res} f(z_0)=\frac{1}{(m-1)!}g^{(m-1)}(z_0). \qquad (4.7\text{-}10)$$

If f can be put in the form

$$f(z)=\frac{p(z)}{(z-z_0)^m q_1(z)},$$

where p and q_1 are analytic and do not equal 0 at z_0, it is not generally advisable to carry out the differentiations of $g=p/q_1$ required by (4.7-10). It is usually simpler to expand p and q_1 separately in powers of $h:=z-z_0$

RESIDUE THEOREM

to obtain, say,

$$p(z) =: a_0 + a_1 h + a_2 h^2 + \cdots,$$
$$q_1(z) =: b_0 + b_1 h + b_2 h^2 + \cdots,$$

and then to compute the coefficients c_k in

$$\frac{p(z)}{q_1(z)} =: c_0 + c_1 h + c_2 h^2 + \cdots$$

by the recurrence relation $c_0 = a_0/b_0$,

$$c_k = \frac{1}{b_0}(a_k - b_1 c_{k-1} - b_2 c_{k-2} - \cdots - b_k c_0),$$

$k = 1, 2, \ldots$ (see §1.2). We then have

$$\operatorname{res} f(z_0) = c_{m-1}. \tag{4.7-11}$$

Only the coefficients $a_0, a_1, \ldots, a_{m-1}$ and $b_0, b_1, \ldots, b_{m-1}$ are required for this computation.

There will be ample opportunity in the following two sections to illustrate these formulas with examples.

The following is a simple but important theoretical application of the residue theorem.

THEOREM 4.7b (Cauchy integral formula)

Let R be a simply connected domain, let f be analytic everywhere in R, and let z_0 be a point of R. Then, if Γ is a piecewise regular closed curve in R not passing through z_0,

$$n(\Gamma, z_0) f(z_0) = \frac{1}{2\pi i} \int_\Gamma \frac{f(z)}{z - z_0} dz. \tag{4.7-12}$$

In particular, if Γ is a positively oriented Jordan curve that contains the point z_0 in its interior, then

$$f(z_0) = \frac{1}{2\pi i} \int_\Gamma \frac{f(z)}{z - z_0} dz. \tag{4.7-13}$$

Proof. The function

$$g(z) := \frac{f(z)}{z - z_0}$$

is analytic in R, except for a simple pole at z_0. By the standard method the residue at the pole is found to be $f(z_0)$. The two formulas of the theorem follow directly from the residue theorem. ∎

In place of g we may consider more generally

$$g(z) := \frac{f(z)}{(z-z_0)^{k+1}},$$

where k is any non-negative integer. By (4.7-10) the residue at z_0 equals $(1/k!)f^{(k)}(z_0)$ and the residue theorem yields the following:

COROLLARY 4.7c

Under the hypotheses of Theorem 4.7b, if $k = 0, 1, 2, \ldots$,

$$n(\Gamma, z_0) f^{(k)}(z_0) = \frac{k!}{2\pi i} \int_\Gamma \frac{f(z)}{(z-z_0)^{k+1}} \, dz, \qquad (4.7\text{-}14)$$

and particularly, if Γ is a positively oriented Jordan curve containing z_0 in its interior,

$$\frac{1}{k!} f^{(k)}(z_0) = \frac{1}{2\pi i} \int_\Gamma \frac{f(z)}{(z-z_0)^{k+1}} \, dz. \qquad (4.7\text{-}15)$$

The one tool of integration theory required to obtain the residue theorem was Cauchy's theorem. We know from Theorem 4.3g that this theorem remains valid if Γ is the boundary of a region R such that $R \cup \Gamma$ is piecewise diffeomorphic to the unit disk. Thus we can state the residue theorem and Cauchy's integral formula in the following extended version:

THEOREM 4.7d

Let Γ be a positively oriented Jordan curve with interior R such that the set $R \cup \Gamma$ is piecewise diffeomorphic to the unit disk. Then (4.7-6) remains valid if f is continuous in $R \cup \Gamma$ (excepting the points z_k). Similarly, Cauchy's integral formula (4.7-13) remains valid if f is continuous in $R \cup \Gamma$.

Point z_0 occurring in Cauchy's integral formula for a Jordan curve was an arbitrary point in the interior of Γ. To give graphic emphasis to this fact we may write the formula as follows:

$$\frac{1}{n!} f^{(n)}(z) = \frac{1}{2\pi i} \int_\Gamma \frac{f(t)}{(t-z)^{n+1}} \, dt, \qquad n = 0, 1, 2, \ldots. \qquad (4.7\text{-}16)$$

RESIDUE THEOREM

It is a remarkable feature of the formula that it expresses the value of an analytic function at any point in the interior of a Jordan curve in terms of the values it takes on Γ. Thus the values in the interior are completely determined by the values on Γ, which once again emphasizes the inner coherence of the values of an analytic function.

In developments of complex analysis which do not start from power series Cauchy's integral formula is an important tool for the study of the local properties of an analytic function, in particular for obtaining the Taylor series, mapping properties, and Taylor series of the inverse function. In the present exposition these results have been anticipated in part by the definition of analyticity and in part by the developments in Chapter 2. In the special case in which Γ is a circle, however, we did use Cauchy's formula to establish the Laurent expansion; see §4.4.

PROBLEMS

1. The Legendre polynomial P_n may be defined by

$$P_n(z) := \frac{1}{2^n n!} \frac{d^n}{dz^n} (z^2 - 1)^n.$$

Compare Example 4, §1.9. Use Cauchy's formula to show that if Γ has winding number 1 with respect to z

$$P_n(z) = \frac{1}{2\pi i} \int_\Gamma \frac{(t^2 - 1)^n}{2^n (t - z)^{n+1}} dt.$$

Use this result to compute

$$\sum_{n=0}^{\infty} \rho^n P_n(z)$$

for $|\rho| < 1$, $z = \cos \theta$.

2. *Divided differences.* Let f be analytic in a simply connected region R and let $z_0, z_1, z_2, \ldots$, be a (finite or infinite) sequence of distinct points in R. The **divided differences** $f(z_k, z_{k+1}, \ldots, z_m)$ of order $m - k$ of f are complex numbers defined recursively by

$$f(z_k, z_{k+1}, \ldots, z_m) = \frac{1}{z_k - z_m} \{ f(z_k, \ldots, z_{m-1}) - f(z_{k+1}, \ldots, z_m) \},$$

$$k = 0, 1, \ldots; \ m = k+1, k+2, \ldots$$

If Γ is a positively oriented Jordan curve containing the points $z_k, z_{k+1}, \ldots, z_m$ in its interior, show that

$$f(z_k, z_{k+1}, \ldots, z_m) = \frac{1}{2\pi i} \int_\Gamma \frac{f(t)}{(t - z_k)(t - z_{k+1}) \cdots (t - z_m)} dt.$$

[Use induction with respect to $m-k$. Show that the integrals on the right satisfy the appropriate recurrence relation.]

3. Prove that the divided difference $f(z_k, z_{k+1}, \ldots, z_m)$ is a symmetric function of $z_k, z_{k+1}, \ldots, z_m$.

4. Show that

$$f(z_k, z_{k+1}, \ldots, z_m) = \sum_{l=k}^{m} \frac{f(z_l)}{\prod_{\substack{n=k \\ n \neq l}}^{m} (z_l - z_n)}.$$

[Evaluate the integral of problem 2 by residues.]

5. Show that the unique polynomial p of degree $\leq n$, which satisfies $p(z_k) = f(z_k)$, $k = 0, 1, \ldots, n$, is given by

$$p(z) = \sum_{k=0}^{n} f(z_0, z_1, \ldots, z_k) \prod_{l=0}^{k-1} (z - z_l).$$

6. Let p be a polynomial of degree n. Show that the divided differences of order $>n$ of p (i.e., the differences formed with more than $n+1$ points z_k) vanish identically. [Choose Γ to be a circle of radius ρ and let $\rho \to \infty$.]

7. Let f be analytic and one-to-one in a region R, let $z_0 \in R$, and suppose that $f(R)$ contains the disk $D: |w - w_0| < \rho$, where $w_0 := f(z_0)$. Show that the value of the inverse function $g := f^{[-1]}$ at the point $w \in D$ is given by

$$g(w) = \frac{1}{2\pi i} \int_\Gamma \frac{f'(z)}{f(z) - w} z \, dz. \qquad (4.7\text{-}17)$$

Here Γ is a positively oriented Jordan curve whose interior contains $g(D)$.

8. Let f be analytic in a region R and let $z_0 \in R$. By choosing Γ in Cauchy's integral formula (4.7-16) (where $n=0$) to be a circle round z_0 and by expanding the kernel according to

$$\frac{1}{t-z} = \frac{1}{t-z_0-(z-z_0)} = \frac{1}{t-z_0} \sum_{n=0}^{\infty} \left(\frac{z-z_0}{t-z_0}\right)^n$$

prove that f can be expanded in the Taylor series

$$f(z) = \sum_{n=0}^{\infty} \frac{1}{n!} f^{(n)}(z_0)(z-z_0)^n$$

and show that the series converges in the interior of every disk centered at z_0 and contained in R.

EVALUATION OF DEFINITE INTEGRALS   249

9. If f is meromorphic and has either a zero or a pole at z_0, show that $f'f^{-1}$ has a simple pole at z_0 with residue $+m$ or $-m$, depending on whether z_0 is a zero or a pole of order m.
10. Let Γ be a positively oriented Jordan curve and let p be a polynomial. If it is known that p is different from zero on Γ and has a single zero in the interior of Γ, show that this zero is given by

$$z_0 = \frac{1}{2\pi i} \int_\Gamma \frac{p'(z)}{p(z)} z \, dz.$$

Formulate more general results of this kind.

11. Under the hypotheses of Problem 7 obtain the expansion of g in powers of $w - w_0$ by using

$$\frac{1}{f(z) - w} = \frac{1}{f(z) - w_0 - (w - w_0)} = \sum_{n=0}^{\infty} \frac{(w - w_0)^n}{(f(z) - w_0)^{n+1}},$$

term-by-term integration, and integration by parts. Establish the connection with the Lagrange-Bürmann expansion.

12. Deduce the cases $n = 1, 2, \ldots$ of (4.7-16) from the special case $n = 0$ by means of Theorem 4.1a.

§4.8. APPLICATIONS OF THE RESIDUE THEOREM: EVALUATION OF DEFINITE INTEGRALS

The evaluation of definite integrals by means of the residue calculus is one of the best known applications of complex variable theory. Broadly speaking, this method consists in interpreting a given integral of a function of a real variable as the parametric representation of the integral of an analytic function along a closed curve and in evaluating the latter by means of residues. Although not infallible, the method produces results for certain broad classes of integrals in which the evaluation by other methods would be cumbersome.

I. Rational Functions of Trigonometric Functions

Let $r = r(x, y)$ be a rational function of two variables. We wish to evaluate

$$I := \int_0^{2\pi} r(\cos \phi, \sin \phi) \, d\phi. \tag{4.8-1}$$

By Theorem 4.2a

$$I = \int_\Gamma f(z) \, dz, \tag{4.8-2}$$

where $\Gamma: z = e^{i\phi}$, $0 \leq \phi \leq 2\pi$ is the unit circle and

$$f(z) := -\frac{i}{z} r\left(\frac{1}{2}\left(z + \frac{1}{z}\right), \frac{1}{2i}\left(z - \frac{1}{z}\right)\right).$$

Because f is again rational, the integral (4.8-2) can be evaluated by residues; it equals $2\pi i$ times the sum of the residues of f inside the unit circle.

The transition from (4.8-1) to (4.8-2) is mechanized by expressing both $\sin \phi$ and $\cos \phi$ in terms of $e^{i\phi}$ and by substituting $e^{i\phi} =: z$, $d\phi = -ie^{-i\phi} dz = -(i/z) dz$.

EXAMPLE 1

The textbook example is the integral

$$I := \int_0^\pi \frac{d\phi}{\gamma + \cos \phi},$$

where $\gamma > 1$. Because the integrand is even and periodic with period 2π,

$$I = \frac{1}{2} \int_0^{2\pi} \frac{d\phi}{\gamma + \cos \phi},$$

which has the form considered above; here the function f is

$$f(z) := -\frac{i}{2z} \frac{1}{\gamma + \frac{1}{2}(z + 1/z)} = -\frac{i}{z^2 + 2\gamma z + 1}.$$

The zeros of the denominator are

$$z_1 := -\gamma + \sqrt{\gamma^2 - 1}, \qquad z_2 := -\gamma - \sqrt{\gamma^2 - 1},$$

both simple. Since $z_1 z_2 = 1$ and $|z_2| > 1$, only z_1 is located inside $|z| = 1$. Thus z_1 is the only pole of f located inside $|z| = 1$. The residue is by the standard method found to be

$$-\frac{i}{2z_1 + 2\gamma} = -\frac{i}{2\sqrt{\gamma^2 - 1}}.$$

$$\int_0^\pi \frac{d\phi}{\gamma + \cos \phi} = \frac{\pi}{\sqrt{\gamma^2 - 1}} \quad (\gamma > 1)$$

follows.

EVALUATION OF DEFINITE INTEGRALS

Frequently this method is recommended for the purpose of computing the coefficients of the Fourier expansion

$$r(\cos\phi, \sin\phi) = \sum_{n=-\infty}^{\infty} \alpha_n e^{in\phi}, \qquad (4.8\text{-}3)$$

which are given by

$$\alpha_n = \frac{1}{2\pi} \int_0^{2\pi} r(\cos\phi, \sin\phi) e^{-in\phi} d\phi.$$

If r is rational, then clearly these integrals are of the kind in (4.8-1). It is usually simpler, however, to obtain the Fourier series with the Laurent series method described in §4.5. In fact, an alternate method for evaluating (4.8-1) consists in considering the integral as 2π times the zeroth coefficient of the Fourier series (4.8-3). (Compare Example 1 in §4.5.)

II. Improper Integrals

Let f be a function of a real variable that is Riemann integrable on every finite interval. If the limit

$$\lim_{\rho \to \infty} \int_0^{\rho} f(x) \, dx$$

exists and is finite, its value is called the (improper) integral of f from 0 to ∞ and is denoted by

$$\int_0^{\infty} f(x) \, dx.$$

Similarly, if

$$\lim_{\sigma \to \infty} \int_{-\sigma}^0 f(x) \, dx$$

exists, it is denoted by

$$\int_{-\infty}^0 f(x) \, dx.$$

If both limits exist, then their sum is denoted by

$$I := \int_{-\infty}^{\infty} f(x) \, dx$$

and is called the integral of f from $-\infty$ to ∞. The existence of the "symmetric" limit

$$J := \lim_{\rho \to \infty} \int_{-\rho}^{\rho} f(x)\, dx$$

does not imply the existence of the integral I, but if I exists then J exists and $I = J$.

Another type of improper integral occurs if the integrand is unbounded near a point of the interval of integration. Such integrals are similarly defined as limits of proper, i.e., ordinary, Riemann integrals.

Generally speaking, the evaluation of an improper integral by complex integration involves the following four steps:

1. *Introduce a proper integral* J_ρ, depending on a parameter ρ (or on several such parameters), such that $I = \lim J_\rho$. The path of integration for J is generally a bounded interval γ_ρ on the real axis.

2. *Close the path of integration* by adding to γ_ρ one or several curves $\gamma'_\rho, \gamma''_\rho, \ldots$, such that $\Gamma_\rho := \gamma_\rho + \gamma'_\rho + \gamma''_\rho + \cdots$ is a piecewise regular, positively oriented Jordan curve. Here it is essential that the integrand f be continued to a function that is analytic in the interior of Γ_ρ, with the exception of isolated singularities.

3. *Estimate the contributions of the accessory arcs* $\gamma'_\rho, \gamma''_\rho, \ldots$ to the integral and show that they tend to zero as ρ tends to its limiting value.

4. *Evaluate the integral along* Γ_ρ, usually by residues. Since the contributions to the accessory arcs tend to zero, we have, in an obvious abbreviated notation,

$$I = \lim \int_{\gamma_\rho} f = \lim \int_{\Gamma_\rho} f = 2\pi i \lim \sum \operatorname{res} f,$$

where the sum contains the residues of f at singularities in the interior of Γ_ρ. In many of the more elementary applications the number of singularities is independent of ρ and the evaluation of the last limit is trivial.

We illustrate the method with integrals of several classes:

1. *Rational functions.* Let r be a rational function without real poles. It is known from calculus that the integral

$$I := \int_{-\infty}^{\infty} r(x)\, dx$$

exists if and only if r has a zero of order ≥ 2 at infinity, i.e., if $r = p/q$ where the degree of q exceeds that of p by at least two. In order to evaluate

EVALUATION OF DEFINITE INTEGRALS

the integral when it exists, (a) let

$$J_\rho := \int_{-\rho}^{\rho} r(x)\, dx,$$

so that in the notation used above $\gamma_\rho : z = x$, $-\rho \leq x \leq \rho$; (b) the path of integration is closed by the semicircle $\gamma'_\rho : z = \rho e^{i\phi}$, $0 \leq \phi \leq \pi$ (see Fig. 4.8a); (c) the integral along γ'_ρ is estimated as follows: since r has a zero of order ≥ 2 at infinity, there exist constants $\alpha > 0$ and $\rho_0 > 0$ such that for all $|z| \geq \rho_0$, $|r(z)| \leq \alpha |z|^{-2}$. Hence by Theorem 4.2b, if $\rho \geq \rho_0$,

$$\left| \int_{\gamma'_\rho} r(z)\, dz \right| \leq \alpha \rho^{-2} \pi \rho = \frac{\alpha \pi}{\rho},$$

which tends to zero for $\rho \to \infty$. (d) The integral along the closed curve $\Gamma_\rho := \gamma_\rho + \gamma'_\rho$ equals $2\pi i$ times the sum of the residues of r at the poles inside Γ_ρ, which for ρ large enough involves all poles in the upper half-plane.

THEOREM 4.8a

Let r be a rational function with no poles on the real axis and a zero of order ≥ 2 at infinity. Then

$$\int_{-\infty}^{\infty} r(x)\, dx = 2\pi i \sum_{\operatorname{Im} z > 0} \operatorname{res} r(z), \qquad (4.8\text{-}4)$$

where the sum involves the residues of r at all poles in the upper half-plane.

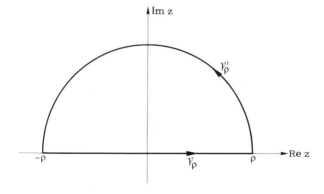

Fig. 4.8a. Closing the path of integration.

EXAMPLE 2

The textbook example is

$$I := \int_{-\infty}^{\infty} \frac{1}{1+x^2}\, dx,$$

which clearly satisfies the hypotheses. The only pole of $r(z) := (1+z^2)^{-1}$ in the upper half-plane occurs at $z = i$ and the residue there, computed by the standard method, equals $(2i)^{-1}$. Hence

$$I = 2\pi i \frac{1}{2i} = \pi,$$

as known from calculus.

EXAMPLE 3

To evaluate

$$I := \int_{-\infty}^{\infty} \frac{1}{(1+x^2)^{n+1}}\, dx$$

for any non-negative integer n we require the residue of the integrand $r(z) := (1+z^2)^{n+1}$ at the sole pole $z = i$ in the upper half-plane. This is easily obtained from the Laurent series at that pole. Letting $z - i =: h$, we have $1 + z^2 = (z-i)(z+i) = h(h+2i)$; hence

$$\frac{1}{(z^2+1)^{n+1}} = (2ih)^{-n-1}\left(1+\frac{h}{2i}\right)^{-n-1}$$

$$= (2i)^{-n-1} h^{-n-1} \sum_{m=0}^{\infty} \frac{(n+1)_m}{m!}\left(-\frac{h}{2i}\right)^m,$$

and the residue, i.e., the coefficient of h^{-1}, turns out to be

$$(2i)^{-n-1}\frac{(n+1)_n}{n!}(-2i)^{-n} = \frac{1}{2i}\frac{(2n)!}{(n!)^2} 2^{-2n} = \frac{1}{2i}\frac{(\frac{1}{2})_n}{n!}.$$

Thus Theorem 4.8a yields at once

$$\int_{-\infty}^{\infty} \frac{1}{(1+x^2)^{n+1}}\, dx = \frac{(\frac{1}{2})_n}{n!}\pi, \qquad n = 0, 1, 2, \ldots.$$

2. *Fourier transforms of rational functions.* The Fourier sine and the Fourier cosine transforms of a function f defined on the real line are for real values

EVALUATION OF DEFINITE INTEGRALS

of ω defined by

$$\int_{-\infty}^{\infty} f(x)\cos\omega x\, dx, \quad \int_{-\infty}^{\infty} f(x)\sin\omega x\, dx,$$

provided that these integrals exist. Here we reconsider the case in which $f = r$, a rational function with no poles on the real axis. If $r = p/q$, where the degrees of p and q are m and n, respectively, it can be shown by integration by parts that these integrals already exist when $n \geq m + 1$. The method of residues is not directly applicable because the functions $\cos\omega z$ and $\sin\omega z$ do not have the appropriate behavior for large $|z|$. If r is real for real x, however, the above integrals can be obtained as the real and imaginary parts of

$$I := \int_{-\infty}^{\infty} r(x) e^{i\omega x}\, dx. \tag{4.8-5}$$

For the purpose of evaluating I we may assume $\omega \geq 0$. (If $\omega < 0$, substitute $x \to -x$.)

As above, $I = \lim_{\rho \to \infty} J_\rho$, where

$$J_\rho := \int_{-\rho}^{\rho} r(x) e^{i\omega x}\, dx.$$

Closing the path of integration as in Fig. 4.8a, we have to estimate the integral over the semicircle γ'_ρ. Because our assumptions now merely imply $|r(z)| \leq \alpha |z|^{-1}$ for sufficiently large $|z|$ a direct application of Theorem 4.2b now no longer suffices for showing that the integral tends to zero. We can, however, apply the following result, which is used repeatedly.

LEMMA 4.8b (Jordan's lemma)

Let P be an unbounded set of positive numbers, let f be continuous on each semicircle $\gamma'_\rho : z = \rho e^{i\phi}$, $0 \leq \phi \leq \pi$, where $\rho \in P$, and let

$$\delta(\rho) := \max_{|z|=\rho} |f(z)|, \quad \rho \in P.$$

If $\delta(\rho) \to 0$ for $\rho \to \infty$, $\rho \in P$, then for each $\omega > 0$

$$J'_\rho := \int_{\gamma'_\rho} f(z) e^{i\omega z}\, dz \to 0 \text{ as } \rho \to \infty, \rho \in P.$$

Proof. By (4.2-11)

$$|J'_\rho| \leq \int_{\gamma'_\rho} |f(z)||e^{i\omega z}||dz| \leq \delta(\rho) \int_0^\pi e^{-\omega\rho\sin\phi} \rho\, d\phi.$$

From a graph of the sine function it is evident that $\sin\phi \geq (2/\pi)\phi$, $0 \leq \phi \leq \pi/2$. Hence

$$\int_0^\pi e^{-\omega\rho\sin\phi}\,d\phi = 2\int_0^{\pi/2} e^{-\omega\rho\sin\phi}\,d\phi \leq 2\int_0^{\pi/2} e^{-2\omega\rho\phi/\pi}\,d\phi$$

$$\leq 2\int_0^\infty e^{-2\omega\rho\phi/\pi}\,d\phi = \frac{\pi}{\omega\rho},$$

and from

$$|J'_\rho| \leq \frac{\pi}{\omega}\delta(\rho),$$

the assertion of the lemma follows in view of $\delta(\rho) \to 0$. ∎

In view of this lemma, step 4 in our general outline is applicable with $f(z) = r(z)e^{i\omega z}$; and we obtain the following theorem:

THEOREM 4.8c

Let r be a rational function with no poles on the real axis and a zero of order ≥ 1 at infinity. Then for $\omega > 0$

$$\int_{-\infty}^\infty r(x)e^{i\omega x}\,dx = 2\pi i \sum_{\operatorname{Im} z > 0} \operatorname{res}\{e^{i\omega z} r(z)\}, \qquad (4.8\text{-}6)$$

where the sum involves the residues at all poles of r in the upper half-plane.

EXAMPLE 4

Since $(1+x^2)^{-1}$ is even,

$$\int_{-\infty}^\infty \frac{\cos\omega x}{1+x^2}\,dx = \operatorname{Re}\int_{-\infty}^\infty \frac{e^{i\omega x}}{1+x^2}\,dx.$$

The residue at the sole pole at $z = i$ is

$$\frac{e^{i\omega i}}{2i} = \frac{e^{-\omega}}{2i};$$

hence for $\omega > 0$ the integral equals $\pi e^{-\omega}$. Since the integral evidently is an even function of ω, there follows

$$\int_{-\infty}^\infty \frac{\cos\omega x}{1+x^2}\,dx = \pi e^{-|\omega|} \quad \text{for all real } \omega.$$

EVALUATION OF DEFINITE INTEGRALS

3. *Integrals with logarithms.* The method of residues is not restricted to rational or meromorphic functions. Using the closed curve of Fig. 4.8a, we may prove the following theorem by considering the function $f(z)$:
$= \text{Log}(z + i\alpha) r(z)$:

THEOREM 4.8d

Let r be a real rational function without real poles and with a zero of order ≥ 2 at infinity and let $\alpha \geq 0$. Then

$$\int_{-\infty}^{\infty} \text{Log}\sqrt{x^2 + \alpha^2}\, r(x)\, dx = \text{Re}\left\{ 2\pi i \sum_{\text{Im}\, z > 0} \text{res}[\text{Log}(z + i\alpha) r(z)] \right\},$$

(4.8-7)

the sum consisting of the residues at all poles in the upper half-plane.

EXAMPLE 5

$$\int_{-\infty}^{\infty} \frac{\text{Log}\sqrt{x^2 + \alpha^2}}{1 + x^2}\, dx = \text{Re}\left\{ 2\pi i \frac{\text{Log}(i + i\alpha)}{2i} \right\} = \pi \text{Log}(1 + \alpha).$$

4. *Fractional powers.* Here we consider

$$I := \int_0^\infty x^\alpha r(x)\, dx,$$

where r again is a rational function and α is real and nonintegral. Assuming that $0 < \alpha < 1$ without loss of generality, I exists if r has no poles on the positive real axis, a pole of order ≤ 1 at 0, and a zero of order ≥ 2 at infinity. If these conditions are met, then

$$I = \lim_{\substack{\delta \to 0 \\ \rho \to \infty}} J_{\delta, \rho},$$

where for $\delta > 0$, $\rho > 0$,

$$J_{\delta, \rho} := \int_\delta^\rho x^\alpha r(x)\, dx.$$

The path of integration here is the segment $z = x$, $\delta \leq x \leq \rho$, of the positive real axis, which in accordance with our general notation we call $\gamma_{\delta, \rho}$. As shown in Fig. 4.8b, we close the path of integration (a) by the large circle $\gamma'_\rho : z = \rho e^{i\phi}$, $0 \leq \phi \leq 2\pi$, (b) by the straight line segment $z = -\tau$, $-\rho \leq \tau \leq -\delta$, (c) by the small circle $\gamma'''_\delta : z = \delta e^{-i\phi}$, $-2\pi \leq \phi \leq 0$. To make the integrand

$z^\alpha r(z)$ single-valued and analytic (with the exception of isolated singularities) in the interior of the closed curve $\Gamma_{\delta,\rho} := \gamma_{\delta,\rho} + \gamma'_\rho + \gamma''_{\delta,\rho} + \gamma'''_\delta$ we cut the complex plane along the positive real axis, thus requiring that $z^\alpha = \rho^\alpha e^{i\alpha\phi}$, where $z = \rho e^{i\phi}$, $\phi = \arg z \in [0, 2\pi]$. This leads to a trivial ambiguity on the positive real axis, which could be avoided by arguing somewhat more pedantically. It is plain that one must choose $\phi = 0$ on $\gamma_{\delta,\rho}$, the "upper edge of the cut," and $\phi = 2\pi$ on $\gamma''_{\delta,\rho}$, the "lower edge of the cut."

Simple applications of Theorem 4.2b show that under the hypotheses made the integrals along γ'_ρ and γ'''_δ tend to zero as $\rho \to \infty$ and $\delta \to 0$. The integral along $\gamma''_{\delta,\rho}$ does not tend to zero but can be related to $J_{\delta,\rho}$ as follows: with the parametrization of $\gamma''_{\delta,\rho}$ given above, $z^\alpha = (-\tau)^\alpha e^{2\pi i \alpha}$ on the lower edge of the cut; hence

$$\int_{\gamma''_{\delta,\rho}} f(z)\,dz = \int_{-\rho}^{-\delta} (-\tau)^\alpha e^{2\pi i\alpha} r(-\tau)\,d\tau$$

$$= -e^{2\pi i\alpha} \int_\delta^\rho x^\alpha r(x)\,dx = -e^{2\pi i\alpha} J_{\delta,\rho}.$$

Letting $\delta \to 0$, $\rho \to \infty$, we thus find that $(1 - e^{2\pi i\alpha})I$ equals $2\pi i$ times the sum

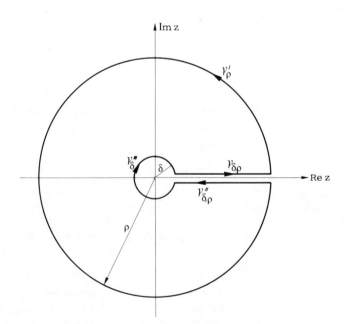

Fig. 4.8b. Closed path of integration for fractional powers.

EVALUATION OF DEFINITE INTEGRALS

of the residues of $z^\alpha r(z)$ in the cut plane. Since α is not an integer, we can solve for I to find the following theorem:

THEOREM 4.8e

Let r be a rational function with no poles on the positive real axis, a pole of order ≤ 1 at 0, and a zero of order ≥ 2 at infinity and let $0 < \alpha < 1$. Then

$$\int_0^\infty x^\alpha r(x)\,dx = \frac{2\pi i}{1 - e^{2\pi i \alpha}} \sum \operatorname{res}\{z^\alpha r(z)\}, \qquad (4.8\text{-}8)$$

where the sum involves the residues at all nonzero poles of r and z^α is computed with $\arg z \in (0, 2\pi)$.

EXAMPLE 6

The conditions of the theorem are met for

$$I := \int_0^\infty \frac{x^\alpha}{x(1+x)}\,dx.$$

The sole nonzero pole occurs at $z = -1 = e^{i\pi}$, and the residue equals $-e^{i\pi\alpha}$. Theorem 4.8e thus yields

$$I = -2\pi i \frac{e^{i\pi\alpha}}{1 - e^{2\pi i \alpha}} = \frac{\pi}{\sin \pi\alpha},$$

making evident the divergence of the integral for $\alpha \to 0$ and $\alpha \to 1$.

III. Principal Value Integrals

Let $[\alpha, \beta]$ be a real interval and let f be a complex-valued function defined on $[\alpha, \beta]$. If f is unbounded near an interior point ξ of $[\alpha, \beta]$, the (Riemann) integral of f over $[\alpha, \beta]$ does not exist. However, the two limits

$$\lim_{\epsilon \to 0+} \int_\alpha^{\xi - \epsilon} f(x)\,dx, \qquad \lim_{\epsilon \to 0+} \int_{\xi + \epsilon}^\beta f(x)\,dx$$

still may exist, and if they do their sum is again called the **improper integral** of f over $[\alpha, \beta]$ and denoted by the ordinary integration symbol $\int_\alpha^\beta f(x)\,dx$.

Even if these two limits do not exist, it may happen that the "symmetric limit,"

$$\lim_{\epsilon \to 0+} \left\{ \int_\alpha^{\xi - \epsilon} f(x)\,dx + \int_{\xi + \epsilon}^\beta f(x)\,dx \right\},$$

exists. If it does, it is called the **principal value integral** of f from α to β and denoted by the symbol

$$\text{PV} \int_\alpha^\beta f(x)\,dx.$$

EXAMPLE 7

$\int_{-1}^1 (1/x)\,dx$ does not exist, but $\text{PV} \int_{-1}^1 (1/x)\,dx = 0$.

Principal value integrals occur in the theory of certain integral transforms and of harmonic functions; in addition, they are of occasional use in the evaluation of ordinary integrals. Here we evaluate principal value integrals of the form

$$\text{PV} \int_{-\infty}^\infty f(x)\,dx,$$

where f denotes a function of one of the types considered in Subsection II.

Let f be analytic, except for isolated singularities, on the closed upper half-plane. One singularity is located at the point $z = \xi$ of the real axis; we assume that it is a pole with principal part

$$p(z) := \sum_{k=1}^n a_k h^{-k}, \qquad h := z - \xi.$$

We wish to calculate

$$I := \lim_{\substack{\rho \to \infty \\ \epsilon \to 0}} J_{\rho,\epsilon}, \qquad (4.8\text{-}9)$$

where

$$J_{\rho,\epsilon} := \int_{-\rho}^{\xi-\epsilon} f(x)\,dx + \int_{\xi+\epsilon}^\rho f(x)\,dx.$$

This may be considered as an integral extended over two straight-line segments on the real axis. We close the path with two semicircles γ_ρ' and γ_ϵ'' of radius ρ and ϵ and centered at 0 and ξ, respectively (see Fig. 4.8c).

We assume that f is such that

$$\lim_{\rho \to \infty} \int_{\gamma_\rho'} f(z)\,dz = 0, \qquad (4.8\text{-}10)$$

which is true for all integrands considered in Subsection II. The integral of the regular part of f along γ_ϵ'' clearly tends to 0 for $\epsilon \to 0$. The integral of the

EVALUATION OF DEFINITE INTEGRALS

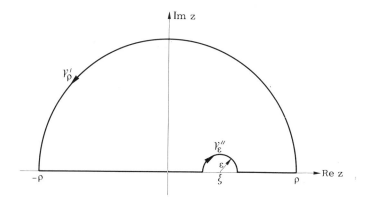

Fig. 4.8c. Closed curve for principal value integral.

singular part may be evaluated by using, say, the parametric representation $z = \xi + \epsilon e^{-i\phi}$, $-\pi \leq \phi \leq 0$. We find

$$\int_{\gamma_\epsilon''} p(z)\, dz = -i\pi a_1 + \sum_{k=2,4,\ldots} \frac{2\epsilon^{1-k}}{1-k} a_k.$$

For $\epsilon \to 0$ this tends to a limit if and only if $a_2 = a_4 = \cdots = 0$; i.e., if the principal part contains odd powers of h only. If it exists, the limit equals $-i\pi \operatorname{res} f(\xi)$. Thus we have found

$$I - i\pi \operatorname{res} f(\xi) = 2\pi i \sum_{\operatorname{Im} z > 0} \operatorname{res} f(z).$$

The same considerations evidently apply if there are several poles on the real axis, and we obtain the following result:

THEOREM 4.8f

Let f be analytic, except for isolated singularities, in the closed upper half-plane and let (4.8-10) hold. If the singularities on the real axis are poles, the principal value integral

$$I := \operatorname{PV} \int_{-\infty}^{\infty} f(x)\, dx$$

exists if and only if the principal parts at these poles involve odd powers only. If I exists,

$$I = 2\pi i \sum_{\operatorname{Im} z > 0} \operatorname{res} f(z) + \pi i \sum_{\operatorname{Im} z = 0} \operatorname{res} f(z). \tag{4.8-11}$$

The singularities straddling the real axis thus enter the formula with weight πi in place of $2\pi i$, i.e., with one leg only.

EXAMPLE 8

The function $(e^{i\omega z}/z)$ $(\omega > 0)$ satisfies the conditions according to Jordan's lemma. Thus

$$\text{PV} \int_{-\infty}^{\infty} \frac{e^{i\omega x}}{x} dx = i\pi.$$

Separating real and imaginary parts, we obtain

$$\text{PV} \int_{-\infty}^{\infty} \frac{\cos \omega x}{x} dx = 0,$$

which is obvious because the integrand is odd, and

$$\text{PV} \int_{-\infty}^{\infty} \frac{\sin \omega x}{x} dx = \pi.$$

However, this integral exists as an ordinary (improper) integral, and thus we have shown that

$$\int_{-\infty}^{\infty} \frac{\sin \omega x}{x} dx = \pi \quad (\omega > 0). \tag{4.8-12}$$

EXAMPLE 9

The integral

$$I := \int_{-\infty}^{\infty} \frac{(\sin \omega x)^2}{x^2} dx$$

may be evaluated by using $(\sin \omega x)^2 = \frac{1}{2}(1 - \cos 2\omega x)$. Thus

$$I = \frac{1}{2} \int_{-\infty}^{\infty} \frac{1 - \cos 2\omega x}{x^2} dz = \frac{1}{2} \text{Re PV} \int_{-\infty}^{\infty} \frac{1 - e^{2i\omega x}}{x^2} dx,$$

provided that the principal value integral exists. The only pole occurs at the origin, and from

$$(1 - e^{2i\omega x}) x^{-2} = -2i\omega x^{-1} + 0(1)$$

we see that the pole is of order 1 with residue $-2i\omega$. There follows

$$\int_{-\infty}^{\infty} \frac{(\sin \omega x)^2}{x^2} dx = \pi \omega \quad (\omega > 0). \tag{4.8-13}$$

EVALUATION OF DEFINITE INTEGRALS

PROBLEMS

1. Evaluate for all integers $m, n \geq 0$

$$I_{m,n} := \int_0^{2\pi} (\cos\phi)^{2m} (\sin\phi)^{2n} d\phi.$$

2. Evaluate by residues

$$\int_0^{2\pi} \frac{1}{(\cos\phi)^2 + (\cos\phi)^{-2}} d\phi.$$

3.
$$\int_0^{2\pi} \frac{1}{(\sin\phi)^4 + (\cos\phi)^4} d\phi = 2\pi\sqrt{2}.$$

4. Let $0 < \epsilon < 1$. If (ρ, ϕ) are polar coordinates in the plane, it is well known that the point $(\rho(\phi), \phi)$, where

$$\rho(\phi) := \frac{1}{1 - \epsilon\cos\phi},$$

describes an ellipse whose one focus lies at the origin. Compute the area enclosed by the ellipse by evaluating the area formula

$$A = \tfrac{1}{2} \int_0^{2\pi} [\rho(\phi)]^2 d\phi$$

by means of the residue calculus. Check the result by means of $A = \pi ab$.

5. Evaluate by residues

$$\int_{-\infty}^{\infty} \frac{1+x}{1+x^3} dx.$$

6. Show that for $m = 2, 3, 4, \ldots$,

$$\int_0^1 \frac{(1-x^2)^m}{1-(-x^2)^m} dx = \frac{2^m \pi}{8m} \sum_{|k| < \frac{m}{2}} (-1)^k \left(\cos\frac{k\pi}{m}\right)^{m-1}.$$

[The integral is invariant under the substitutions $x \to -x$ and $x \to 1/x$ and thus can be expressed as $\tfrac{1}{4}$ of the integral from $-\infty$ to ∞.]

7. Evaluate for $m = 1, 2, \ldots$

$$\int_{-\infty}^{\infty} \frac{1}{1 + x + x^2 + \cdots + x^{2m}} dx$$

and calculate the limit as $m \to \infty$. Explain by means of the graph of the integrand!

8. Show that

$$\int_{-\infty}^{\infty} \frac{1}{(1+x+x^2)^2}\,dx = 4\pi 3^{-3/2}.$$

9. Let p be a polynomial of degree ≥ 2. Prove that the sum of the residues of the rational function $1/p$ is zero.

10. Show that for $\omega \neq 0$, $\alpha \neq 0$,

(a) $$\int_{-\infty}^{\infty} \frac{x \sin \omega x}{x^2 + \alpha^2}\,dx = \pi \operatorname{sgn} \omega e^{-|\omega\alpha|},$$

(b) $$\int_{-\infty}^{\infty} \frac{e^{i\omega x}}{x^4 + 1}\,dx = e^{-|\omega|/\sqrt{2}} \sin\left(\frac{\pi}{4} + \frac{|\omega|}{\sqrt{2}}\right),$$

(c) $$\int_{-\infty}^{\infty} \frac{\cos \omega x}{(x^2+1)^2}\,dx = \frac{\pi}{2}(1+|\omega|)e^{-|\omega|}.$$

11. Evaluate the integrals

(a) $$\int_0^\infty \frac{x^{\alpha-1}}{(x+\beta)(x+\gamma)}\,dx, \qquad 0<\alpha<1, \beta<0, \gamma<0;$$

(b) $$\int_0^\infty \frac{x^{\alpha-1}}{x^2 + 2\beta x \cos\theta + \beta^2}\,dx, \qquad 0<\alpha<1, \beta<0, -\pi<\theta<\pi;$$

(c) $$\int_0^\infty \frac{x^{\alpha-1}}{1+x+x^2+\cdots+x^n}\,dx, \qquad 0<\alpha<1, n=1,2,\ldots.$$

12. Let r be a rational function without poles on the non-negative real axis and a zero of order ≥ 2 at infinity. Show that

$$I := \int_0^\infty r(x)\,dx = \lim_{\alpha \to 0+} \int_0^\infty x^\alpha r(x)\,dx$$

and use this result to develop a method for evaluating I by residues (a) if all poles of r are simple and (b) if there are poles of arbitrary order.

13. Evaluate

$$\int_0^\infty \frac{\operatorname{Log} x}{(1+x^2)^2}\,dx.$$

14. Let $-\pi < \alpha < \pi$. By integrating

$$f(z) := \frac{e^{\alpha z}}{\cosh \pi z}$$

along the rectangle with the corners $\rho, \rho+i, -\rho+i, -\rho$ and letting $\rho \to \infty$, show that

$$\int_0^\infty \frac{\cosh \alpha \tau}{\cosh \pi \tau} d\tau = \frac{1}{2\cos \alpha/2}.$$

15. By considering the principal value integral of

$$f(z) := \frac{e^{i(z^2/2)}}{e^{-\sqrt{\pi} z} - 1}$$

along a parallelogram with the four corners $(1+i)\rho$, $(1+i)\rho + i\sqrt{\pi}$, $-(1+i)\rho + i\sqrt{\pi}$, $-(1+i)\rho$ and letting $\rho \to \infty$, prove without reference to the real calculus that

$$\int_0^\infty e^{-\tau^2} d\tau = \frac{\sqrt{\pi}}{2}.$$

[Srinivasa Rao, *El. Math.* **27** (1972), 88–90.]

§4.9. APPLICATIONS OF THE RESIDUE THEOREM: SUMMATION OF INFINITE SERIES

Not infrequently the residue theorem can be used to find simple expressions for infinite series of the form

$$\sum_{n=-\infty}^{\infty} a_n \qquad (4.9\text{-}1)$$

or of series that can be put in this form. The general idea is as follows: let f be a function that *interpolates* the sequence $\{a_n\}$, i.e., which satisfies

$$f(n) = a_n$$

at all integers n, and let f be analytic in the whole complex plane with the exception of isolated singularities. Let t be a function (a **summatory function**) that has simple poles, with residue $+1$, at all the integers, and is analytic everywhere else. If Γ is a large positively oriented Jordan curve avoiding the integers, the integral

$$\int_\Gamma f(z) t(z) dz$$

COMPLEX INTEGRATION

by the residue theorem equals

$$2\pi i \left\{ \sum_{n \text{ inside } \Gamma} f(n) + \text{residues of } ft \text{ at singularities of } f \text{ inside } \Gamma \right\}.$$

If we are able to show that the integral along Γ tends to zero as Γ is pushed out toward infinity, the sum (4.9-1) is expressed in terms of the residues of ft at the singularities of f. The success of the method depends on the ease with which these residues can be evaluated. By choosing a summatory function s with residue $(-1)^n$ at n, we can similarly evaluate

$$\sum_{n=-\infty}^{\infty} (-1)^n a_n.$$

The method is quite flexible; for the purpose of algorithmic application, however, we need to standardize it somewhat. The closed square curve $\Gamma_m := \gamma_m + \gamma'_m + \gamma''_m + \gamma'''_m$ shown in Fig. 4.9a frequently serves as Γ. The functions s and t defined by

$$s(z) := \frac{\pi}{\sin \pi z}, \quad t(z) := \frac{\pi}{\tan \pi z}$$

are useful summatory functions. Both are meromorphic functions with poles of order 1 at all the integers and nowhere else. The residue of s at

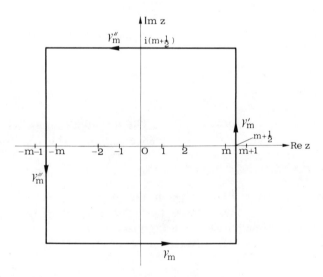

Fig. 4.9a. Jordan curve useful for summing series.

SUMMATION OF INFINITE SERIES

$z = n$ is $(-1)^n$, the residue of t is 1. As to the size of s and t on Γ_m, the representations

$$s(z) = \frac{2\pi i}{e^{i\pi z} - e^{-i\pi z}}, \quad t(z) = i\pi \frac{1 + e^{-2\pi i z}}{1 - e^{-2\pi i z}}$$

easily yield the following crude estimates: if $z = x + iy$, then for $m = 0, 1, 2, \ldots$,

$$\text{on } \gamma_m \text{ and } \gamma_m'' : \begin{cases} |s(z)| \leq 2\pi e^{-\pi m}, \\ |t(z)| \leq 2\pi, \end{cases}$$

$$\text{on } \gamma_m' \text{ and } \gamma_m''' : \begin{cases} |s(z)| \leq \dfrac{\pi}{\cosh \pi y}, \\ |t(z)| \leq \pi \tanh \pi y. \end{cases}$$

For many applications it is sufficient to remember that the absolute values of the functions s and t on Γ_m are bounded by 2π, a constant independent of m.

THEOREM 4.9a

Let r be a rational function with a zero of order ≥ 2 at infinity and with no pole at any integer. Then

$$\sum_{n=-\infty}^{\infty} r(n) = -\sum \text{res}\{r(z)t(z)\}, \tag{4.9-2}$$

$$\sum_{n=-\infty}^{\infty} (-1)^n r(n) = -\sum \text{res}\{r(z)s(z)\}, \tag{4.9-3}$$

where the sums on the right involve the residues at the poles of r.

Proof. Let m be so large that Γ_m encloses all poles of r. The residue theorem then yields

$$\frac{1}{2\pi i} \int_{\Gamma_m} r(z) t(z) \, dz = \sum_{n=-m}^{m} r(n) + \sum \text{res}\, r(z) t(z).$$

As $m \to \infty$ the integral on the left tends to zero by the hypotheses on r and the above estimates on t and the relation (4.9-2) follows. The proof of (4.9-3) is similar. ∎

EXAMPLE **1**

Let $\omega>0$; we wish to compute

$$\sum_{n=1}^{\infty} \frac{1}{n^2+\omega^2}.$$

Let $r(z):=(z^2+\omega^2)^{-1}$. The poles of r are at $z=\pm i\omega$. The residue of rt at $z=i\omega$ is

$$\frac{1}{2i\omega}t(i\omega) = -\frac{\pi}{2\omega}\coth\pi\omega,$$

and because this is real the residue at $z=-i\omega$ is the same. Since $r(n)=r(-n)$,

$$\frac{1}{\omega^2}+2\sum_{n=1}^{\infty}\frac{1}{n^2+\omega^2} = \frac{\pi}{\omega}\coth\pi\omega$$

follows.

EXAMPLE **2**

Let w be a complex number, not an integer. We wish to evaluate

$$\sum_{n=-\infty}^{\infty} \frac{1}{(n-w)^2}.$$

The function $r(z):=(z-w)^{-2}$ satisfies the hypotheses of Theorem 4.9a. It has a pole of order 2 at $z=w$; if $h:=z-w$, the principal part is h^{-2}. To calculate the residue of rt we expand t in powers of h. By Taylor's theorem (Corollary 3.1d) we have

$$t(w+h) = t(w)+t'(w)h+\cdots$$
$$= \pi\cot\pi w - \frac{\pi^2}{(\sin\pi w)^2}h+\cdots,$$

and the residue of rt at $z=w$ is seen to be

$$-\frac{\pi^2}{(\sin\pi w)^2}.$$

Formula (4.9-2) now yields

$$\sum_{n=-\infty}^{\infty}\frac{1}{(n-w)^2} = \left(\frac{\pi}{\sin\pi w}\right)^2. \qquad (4.9\text{-}4)$$

SUMMATION OF INFINITE SERIES

EXAMPLE 3

This example shows that trivial modifications of the situation considered in Theorem 4.9a are sometimes necessary to make the method work. Consider the problem of evaluating

$$\sum_{n=1}^{\infty} \frac{1}{n^{2k}}, \quad k \text{ a positive integer.}$$

The function $r(z) := z^{-2k}$ obviously does not satisfy the hypotheses of the theorem because of the pole at $z=0$. Nevertheless we consider the integral

$$\int_{\Gamma_m} t(z) z^{-2k} dz,$$

where Γ_m again denotes the Jordan curve of Fig. 4.9a. It is obvious that the integral tends to zero as $m \to \infty$. The residues at the nonzero integers yield $2\sum_{n=1}^{m} n^{-2k}$. To determine the residue at $z=0$ we recall (2.5-14) which for $|z|<1$ implies

$$z t(z) = \pi z \cot \pi z = \sum_{k=0}^{\infty} (-1)^k \frac{B_{2k}}{(2k)!} (2\pi z)^{2k}.$$

Here the coefficients B_{2k} are the Bernoulli numbers; a recurrence relation was established in Problem 4, §1.2. Since the principal part of r at $z=0$ is z^{-2k}, the residue of rt is

$$\frac{B_{2k}}{(2k)!} (2\pi)^{2k} (-1)^k.$$

Because the sum of all residues tends to zero as $m \to \infty$,

$$\sum_{n=1}^{\infty} \frac{1}{n^{2k}} = (-1)^{k-1} \frac{2^{2k-1} \pi^{2k}}{(2k)!} B_{2k}. \qquad (4.9\text{-}5)$$

follows. In particular,

$$\sum_{n=1}^{\infty} \frac{1}{n^2} = \frac{\pi^2}{6}, \quad \sum_{n=1}^{\infty} \frac{1}{n^4} = \frac{\pi^4}{90}, \quad \sum_{n=1}^{\infty} \frac{1}{n^6} = \frac{\pi^6}{945}.$$

We next apply the residue theorem to sum Fourier series of the form

$$\sum_{n=-\infty}^{\infty} r(n) e^{in\xi},$$

where r is a rational function that satisfies the hypotheses in Theorem 4.9a. Because of the periodicity of the exponential function, we may suppose that $0 \leq \xi \leq 2\pi$. The obvious idea would be to consider

$$\int_{\Gamma_m} r(z) t(z) e^{iz\xi} dz;$$

this, however, does not work because, for $z \in \Gamma_m$,

$$|e^{iz\xi}| = e^{(m+1/2)\xi},$$

and thus the integral does not necessarily tend to zero as $m \to \infty$. Because

$$\lim_{y \to -\infty} t(x+iy) = i\pi$$

uniformly in x, we use in place of t the new summatory function

$$u(z) := t(z) - i\pi = \frac{2\pi i}{e^{2\pi iz} - 1}.$$

This still has a simple pole with residue 1 at each of the points $z = 0, \pm 1, \pm 2, \ldots$, and the advantage of being small where $e^{iz\xi}$ is large. More precisely,

$$|u(z)| \leq 2\pi e^{-2\pi m} \text{ on } \gamma_m,$$

$$|u(z)| \leq \frac{2\pi}{1+e^{-2\pi y}} \text{ on } \gamma'_m \text{ and } \gamma'''_m,$$

$$|u(z)| \leq 2\pi \text{ on } \gamma''_m.$$

If $0 \leq \xi \leq 2\pi$, these estimates imply that

$$|u(z) e^{iz\xi}| \leq 2\pi e^{\pi}$$

on the whole curve Γ_m, hence that

$$\int_{\Gamma_m} r(z) u(z) e^{iz\xi} dz \to 0$$

as $m \to \infty$. It follows that the sum of all residues of the integrand is zero, and we have proved the following theorem:

THEOREM 4.9b

Let r satisfy the hypotheses of Theorem 4.9a and let $0 \leq \xi \leq 2\pi$. Then

$$\sum_{n=-\infty}^{\infty} r(n) e^{in\xi} = -\sum \text{res}\left\{ r(z) \frac{2\pi i e^{iz\xi}}{e^{2\pi iz} - 1} \right\}, \qquad (4.9\text{-}6)$$

where the sum on the right involves the residues at all poles of r.

SUMMATION OF INFINITE SERIES

EXAMPLE **4**

The sum

$$\sum_{n=1}^{\infty} \frac{\cos n\xi}{n^2+\omega^2},$$

where ξ and ω are positive, obviously equals

$$\frac{1}{2} \sum_{n=-\infty}^{\infty} \frac{e^{in\xi}}{n^2+\omega^2} - \frac{1}{2\omega^2}$$

to which Theorem 4.9b is applicable with $r(z):=(z^2+\omega^2)^{-1}$. The residues of $r(z)u(z)e^{iz}$ at $z = \pm i\omega$ equal

$$\frac{1}{2i\omega} \frac{2\pi i}{e^{-2\pi\omega}-1} e^{-\xi\omega} = -\frac{\pi}{\omega} \frac{e^{-\xi\omega}}{1-e^{-2\pi\omega}}$$

and

$$\frac{1}{-2i\omega} \frac{2\pi i}{e^{2\pi\omega}-1} e^{\xi\omega} = -\frac{\pi}{\omega} \frac{e^{\xi\omega}}{e^{2\pi\omega}-1}.$$

Thus we find

$$\sum_{n=1}^{\infty} \frac{\cos n\xi}{n^2+\omega^2} = \frac{\pi}{2\omega} \frac{\cosh(\pi-\xi)\omega}{\sinh\pi\omega} - \frac{1}{2\omega^2}.$$

We finally consider sums of the form

$$S := \sum_{n=0}^{\infty} f(n) \qquad (4.9\text{-}7)$$

that *cannot* be reduced to (4.9-1) by means of symmetry. Such sums cannot ordinarily be reduced to a finite sum of residues. Under certain conditions, however, they can be expressed by the integral

$$I := \int_0^{\infty} f(z)\, dz \qquad (4.9\text{-}8)$$

(extended along the real axis) plus a correction term which takes the form of an improper integral that converges very rapidly. The result is known as the *Plana summation formula*.

Let f be analytic in the closed right half-plane $\operatorname{Re} z \geq 0$ (further conditions on f will be imposed as we go along). For $x>0$, $y>0$ let $\Gamma(x,y)$

denote the rectangular Jordan curve with the four corners $\pm iy$, $x \pm iy$, oriented positively. Because $\pi \cot \pi z$ has simple poles with residue 1 at the integers,

$$\frac{1}{2\pi i} \int_{\Gamma(m+\frac{1}{2},y)} f(z) \pi \cot \pi z \, dz = \frac{1}{2} f(0) + \sum_{n=1}^{m} f(n) \qquad (4.9\text{-}9)$$

for all integers $m > 0$ and all $y > 0$. (The integral along part of the imaginary axis is a principal-value integral, which causes the factor $\frac{1}{2}$ to appear.) We now let $y \to \infty$, keeping m fixed. In view of

$$\cot \pi z = i \frac{e^{i\pi z} + e^{-i\pi z}}{e^{i\pi z} - e^{-i\pi z}}$$

$$= i \left(1 + \frac{2}{e^{2i\pi z} - 1} \right)$$

$$= -i \left(1 + \frac{2}{e^{-2i\pi z} - 1} \right) \qquad (4.9\text{-}10)$$

we have $\cot \pi (x \pm iy) \to \mp i$ for $y \to \infty$, uniformly in x. Hence the integrals with respect to the horizontal parts of $\Gamma(m+\frac{1}{2},y)$ tend to zero for $y \to \infty$, provided that

$$\lim_{y \to \infty} f(x \pm iy) = 0 \qquad (4.9\text{-}11)$$

uniformly in x on every finite interval $[0,\xi]$. If we assume, in addition, that

$$\int_{-\infty}^{\infty} |f(x+iy)| \, dy$$

exists for all $x \geq 0$, the result is

$$\tfrac{1}{2} f(0) + \sum_{n=1}^{m} f(n) = \tfrac{1}{2} \int_{-\infty}^{\infty} f(m+\tfrac{1}{2}+iy) \cot(m+\tfrac{1}{2}+iy) \pi \, dy$$

$$- \tfrac{1}{2} \text{PV} \int_{-\infty}^{\infty} f(iy) \cot \pi iy \, dy. \qquad (4.9\text{-}12)$$

As already noted,

$$|\cot(m+\tfrac{1}{2}+iy)\pi| = \tanh \pi |y| < 1$$

SUMMATION OF INFINITE SERIES

for all real y. Thus under the final assumption that

$$\lim_{x \to \infty} \int_{-\infty}^{\infty} |f(x+iy)|\,dy = 0, \tag{4.9-13}$$

letting $m \to \infty$ in (4.9-12) yields

$$\sum_{n=0}^{\infty} f(n) = \tfrac{1}{2}f(0) + \frac{i}{2}\,\text{PV} \int_{-\infty}^{\infty} f(iy)\,\frac{e^{\pi y} + e^{-\pi y}}{e^{\pi y} - e^{-\pi y}}\,dy. \tag{4.9-14}$$

This result is not yet satisfactory because it requires $f(x+iy)$ to become small not only for $x \to \infty$ (which is natural because we want the series S to converge) but also for $y \to \pm\infty$, for all x. By giving the computation a different turn the second condition can be replaced by the much weaker condition that

$$\lim_{y \to \infty} |f(x \pm iy)|e^{\mp 2\pi y} = 0, \tag{4.9-15}$$

uniformly in x on every finite interval $[0,\xi]$.

We return to (4.9-9). Denoting by Γ_1 and Γ_2 the portions of $\Gamma(m+\tfrac{1}{2},y)$ lying in the upper and in the lower half-plane, we have, using (4.9-10),

$$\frac{1}{2i}\int_{\Gamma_1} f(z)\cot\pi z\,dz = -\frac{1}{2}\int_{\Gamma_1} f(z)\left(1 + \frac{2}{e^{-2\pi i z}-1}\right)dz$$

$$= \frac{1}{2}\int_0^{m+1/2} f(z)\,dz - \int_{\Gamma_1} f(z)\,\frac{1}{e^{-2\pi i z}-1}\,dz$$

and similarly

$$\frac{1}{2i}\int_{\Gamma_2} f(z)\cot\pi z\,dz = \frac{1}{2}\int_{\Gamma_2} f(z)\left(1 + \frac{2}{e^{2\pi i z}-1}\right)dz$$

$$= \frac{1}{2}\int_0^{m+1/2} f(z)\,dz + \int_{\Gamma_2} f(z)\,\frac{1}{e^{2\pi i z}-1}\,dz.$$

Under the assumption (4.9-15) the integrals along the horizontal parts of Γ_1 and Γ_2 tend to zero for $y \to \infty$. If we assume further that the integrals

$$\int_0^{\infty} |f(x \pm iy)|\,\frac{1}{e^{2\pi y}+1}\,dy$$

exist for all $x \geq 0$ and tend to zero for $x \to \infty$, the integrals along the vertical lines $\operatorname{Re} z = m + \frac{1}{2}$ tend to zero for $m \to \infty$. Supposing that either S or I exists, (4.9-9) therefore implies

$$\sum_{n=0}^{\infty} f(n) = \tfrac{1}{2} f(0) + \int_0^{\infty} f(x)\,dx + i \int_0^{\infty} \frac{f(iy) - f(-iy)}{e^{2\pi y} - 1}\,dy. \quad (4.9\text{-}16)$$

The last integral exists as a proper integral at $y = 0$, since the singularity is removable.

Equation (4.9-16) is known as the **summation formula of Plana**. We summarize the conditions of its validity.

THEOREM 4.9c

Let f be analytic for $\operatorname{Re} z \geq 0$ and let either the series $\sum_{n=0}^{\infty} f(n)$ or the integral $\int_0^{\infty} f(x)\,dx$ be convergent. If, in addition,

(a) $\quad \lim_{y \to \infty} |f(x \pm iy)| e^{-2\pi y} = 0$

uniformly in x on every finite interval and

(b) $\quad \int_0^{\infty} |f(x \pm iy)| e^{-2\pi y}\,dy$

exists for every $x \geq 0$ and tends to zero for $x \to \infty$, the summation formula (4.9-16) holds.

Some of the more spectacular applications of the Plana summation formula occur in connection with the Γ function (§8.5), the Riemann ζ function (§10.8), and the discrete Laplace transform (§10.10). Here we give two simple examples.

EXAMPLE 5

The hypotheses of the theorem are clearly satisfied for $f(z) := (z+1)^{-3}$. The result is

$$\sum_{n=1}^{\infty} \frac{1}{n^3} = 1 + \int_0^{\infty} \frac{6y - 2y^3}{(1+y^2)^3} \frac{1}{e^{2\pi y} - 1}\,dy.$$

SUMMATION OF INFINITE SERIES

EXAMPLE 6

The hypotheses of Theorem 4.9c are not fulfilled for $f(z) := (z+1)^{-1}$ because both the sum S and the integral I diverge. It is well known, however, that the limit

$$\gamma := \lim_{m \to \infty} \left\{ \sum_{n=0}^{m-1} \frac{1}{1+n} - \int_0^{m-1} \frac{1}{1+x} dx \right\}$$

$$= \lim_{m \to \infty} \left(1 + \tfrac{1}{2} + \tfrac{1}{3} + \cdots + \frac{1}{m} - \operatorname{Log} m \right)$$

exists. Its value is known as **Euler's constant** (see §8.4). By grouping terms accordingly, the analysis leading to (4.9-16) is applicable and yields

$$\gamma = \tfrac{1}{2} + 2 \int_0^\infty \frac{y}{1+y^2} \frac{1}{e^{2\pi y} - 1} dy.$$

PROBLEMS

1. Evaluate the sums

   (a) $\displaystyle\sum_{n=1}^{\infty} \frac{1}{n^4+1}$,   (b) $\displaystyle\sum_{n=1}^{\infty} \frac{1}{n^2+n^{-2}}$,

   (c) $1 - \dfrac{1}{2^2} + \dfrac{1}{3^2} - \dfrac{1}{4^2} + \cdots$,   (d) $\dfrac{1}{1 \cdot 3} + \dfrac{1}{5 \cdot 7} + \dfrac{1}{9 \cdot 11} + \cdots$.

2. Evaluate

   $$\sum_{n=-\infty}^{\infty} \frac{1}{(n-w)^3},$$

   where w is complex, not an integer.

3. Using the result of Example 3, show that

   $$(-1)^k \frac{B_{2k}}{(2k)!} \sim \frac{2}{(2\pi)^{2k}}, \quad k \to \infty.$$

4. Verify the formula

   $$1 - \frac{1}{3^3} + \frac{1}{5^3} - \frac{1}{7^3} + \cdots = \frac{\pi^3}{32}.$$

5. Establish the result

   $$\sum_{\substack{n=-\infty \\ n \neq 0}}^{\infty} (-1)^n \left[\frac{1}{w-n} + \frac{1}{n} \right] = \frac{\pi}{\sin \pi w} - \frac{1}{w}.$$

6. Sum the series

$$\sum_{n=-\infty}^{\infty} \frac{e^{in w\xi}}{(n-w)^2},$$

where $0 \leq \xi \leq 2\pi$, $w \neq 0, \pm 1, \pm 2, \ldots$.

7. Let w be complex, not an integer, $-\pi \leq \xi \leq \pi$. Show that

$$\sum_{n=-\infty}^{\infty} (-1)^n \frac{n \sin n\xi}{w^2 - n^2} = \frac{\pi}{2} \frac{\sin \xi w}{\sin \pi w}.$$

8. The Bernoulli polynomials $B_k(\xi)$ are defined by

$$\frac{ze^{\xi z}}{e^z - 1} = \sum_{k=0}^{\infty} \frac{B_k(\xi)}{k!} z^k.$$

Show that for $0 \leq \xi \leq 1$, $k = 1, 2, \ldots$,

$$\sum_{n=1}^{\infty} \frac{\cos 2\pi n\xi}{n^{2k}} = (-1)^{k+1} \frac{(2\pi)^{2k}}{2(2k)!} B_{2k}(\xi),$$

$$\sum_{n=1}^{\infty} \frac{\sin 2\pi n\xi}{n^{2k+1}} = (-1)^{k+1} \frac{(2\pi)^{2k+1}}{2(2k+1)!} B_{2k+1}(\xi).$$

9. Sum the series

(a) $\displaystyle\sum_{n=2}^{\infty} \frac{(-1)^n \cos n\xi}{n(n-1)}$, (b) $\displaystyle\sum_{n=2}^{\infty} \frac{(-1)^n \sin n\xi}{n(n-1)}$,

(c) $\displaystyle\sum_{n=-\infty}^{\infty} \frac{e^{in\xi}}{1 + n + n^2 + \cdots + n^{2k}}$, $k = 1, 2, \ldots$.

§4.10. THE PRINCIPLE OF THE ARGUMENT

If f is analytic in a region R and does not vanish identically, the function f'/f is called the **logarithmic derivative** of f. The isolated singularities of the logarithmic derivative occur at the isolated singularities of f and, in addition, at the zeros of f. The *principle of the argument* results from an application of the residue theorem to the logarithmic derivative.

PRINCIPLE OF THE ARGUMENT 277

Let us calculate the principal part of the logarithmic derivative at an isolated singularity z_0 which arises either from a zero or from a pole of f. Both cases can be dealt with simultaneously by assuming the Laurent series of f in the form

$$f(z) = a_n h^n + a_{n+1} h^{n+1} + \cdots, \quad h := z - z_0, \quad a_n \neq 0,$$

where the integer n is positive or negative. If $n > 0$, then f has a zero of order n at z_0; if $n < 0$, it has a pole of order $-n$. From

$$f'(z) = n a_n h^{n-1} + (n+1) a_{n+1} h^n + \cdots$$

there follows

$$\frac{f'(z)}{f(z)} = \frac{n a_n h^{n-1} + \cdots}{a_n h^n + \cdots} = \frac{n}{h} \frac{1 + \cdots}{1 + \cdots} = \frac{n}{h}(1 + \cdots),$$

and we see that the logarithmic derivative at z_0 has a *simple pole* whose residue is the integer n.

Now let f be analytic in the simply connected domain R, and let f in R have a finite number of zeros $z_1, \ldots, z_k$ with the respective multiplicities $m_1, \ldots, m_k$. Let $\Gamma : z = z(\tau)$, $\alpha \leq \tau \leq \beta$, be a closed curve in R not passing through any z_j. We calculate the integral

$$I := \frac{1}{2\pi i} \int_\Gamma \frac{f'(z)}{f(z)} dz$$

in two different ways. (a) The singularities of the integrand occur precisely at points z_j, and according to the above the residue at z_j is m_j. Thus by the residue theorem (Theorem 4.7a)

$$I = \sum_{j=1}^{k} m_j n(\Gamma, z_j).$$

(b) By making the substitution $w := f(z)$ we have

$$I = \frac{1}{2\pi i} \int_{f(\Gamma)} \frac{1}{w} dw,$$

where $f(\Gamma)$ denotes the curve $w = f(z(\tau))$, $\alpha \leq \tau \leq \beta$, which is again closed and which avoids the origin. By the definition of the winding number the last expression equals $n(f(\Gamma), 0)$, the winding number of $f(\Gamma)$ with respect to 0. By comparing the two results for I we obtain the identity

$$\sum_{j=1}^{k} m_j n(\Gamma, z_j) = n(f(\Gamma), 0). \qquad (4.10\text{-}1)$$

An important special case occurs if Γ is a positively oriented Jordan curve containing all z_j in its interior. Then $n(\Gamma, z_j) = 1$ for all j, and we have the following theorem:

THEOREM 4.10a (principle of the argument)

Let f be analytic in a simply connected region R and let Γ be a positively oriented Jordan curve in R not passing through any zero of f. The number of zeros of f in the interior of Γ (each zero counted according to its multiplicity) equals the winding number of the image curve $f(\Gamma)$ with respect to 0.

By virtue of Theorem 4.7d the principle of the argument remains valid if Γ is the *boundary* of R, provided that f is continuous in $R \cup \Gamma$ and different from zero on Γ.

A generalization of (4.10-1), and accordingly of the principle of the argument, is obtained if f is admitted to have poles $p_1, \ldots, p_l$ in R. If o_i is the order of p_i, then the residue of f'/f at p_i is $-o_i$, and I is augmented by the term

$$-\sum_{i=1}^{l} o_i n(\Gamma, p_i).$$

Hence in place of (4.10-1) we now have

$$\sum_{j=1}^{n} m_j n(\Gamma, z_j) - \sum_{i=1}^{l} o_i n(\Gamma, p_i) = n(f(\Gamma), 0). \qquad (4.10\text{-}2)$$

In particular, if Γ is a positively oriented Jordan curve, the winding number of $f(\Gamma)$ with respect to 0 equals the number of zeros minus the number of poles of f inside Γ, each zero and each pole counted with its multiplicity.

It should be clear that the principle of the argument is a powerful instrument for finding first approximations, however crude, to zeros of analytic functions. (Once such approximations are known, many methods for the accurate determination of the zeros are available; see §6.12.) To determine the number of zeros of an analytic function f inside a Jordan curve Γ one merely has to calculate the winding number $n(f(\Gamma), 0)$. Riemann [1859] used this method to study the distribution of the nontrivial

PRINCIPLE OF THE ARGUMENT

zeros of the ζ function (see §10.8) which is now named after him. In his case it was possible to calculate the winding number approximately by analytic methods. In other cases (see §6.11) a numerical calculation of the winding number, say by an algorithm similar to that given at the end of §4.6, may be feasible.

An important application of the principle of the argument occurs in the theory of automatic control. Here we merely state the mathematical problem; the reader is referred to §10.3 for an explanation of the technical background. Let g be a rational function, analytic at ∞, whose poles are known to lie in the left half-plane. We wish to determine those values of a real parameter κ for which the function

$$f(z) := 1 + \frac{1}{\kappa} g(z)$$

has all its *zeros* in the left half-plane. [f is the denominator of the transfer function of a single-loop feedback system, κ is a gain factor. The condition that f has no zeros in the right half-plane is required for the stability of the system.]

Let $\Gamma := \Gamma_1 + \Gamma_2$ be the Jordan curve shown in Fig. 4.10a; f has no poles inside Γ. Since Γ is negatively oriented, the number of zeros of f inside Γ equals

$$-n(f(\Gamma), 0) = -n\left(1 + \frac{1}{\kappa} g(\Gamma), 0\right)$$

$$= -n\left(\frac{1}{\kappa} g(\Gamma), -1\right) = -n(g(\Gamma), -\kappa).$$

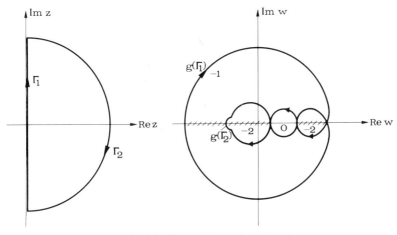

Fig. 4.10a. Nyquist diagram ($////$: forbidden values of $-\kappa$).

Thus the admissible values of $-\kappa$ are those segments of the real axis of the $w := g(z)$ plane that lie in components of the complement of $g(\Gamma)$, where the winding number of $g(\Gamma)$ is zero. In electrical engineering the image curve $g(\Gamma_1)$ can be determined experimentally, for the values $g(i\omega)$ can be measured by exciting the system by a voltage of frequency ω. Since g is regular at infinity, the image $g(\Gamma_2)$ of the semicircle Γ_2 forms a small identation in the image of the imaginary axis which can be ignored. Since $g(z)$ is real for z real, the image $g(\Gamma)$ is symmetric with respect to the real axis, and the values $g(i\omega)$ need be plotted only for $0 \leqslant \omega < \infty$. The resulting plot is known as the **Nyquist diagram**. With but obvious modifications the method of the Nyquist diagram may also be used if g has a known number of poles in the right half-plane or if it has poles and zeros on the imaginary axis.

Returning to theoretical matters, we state a consequence of the principle of the argument that is frequently used.

THEOREM 4.10b (Rouchés theorem)

Let the functions f and g be analytic in the simply connected region R, let Γ be a Jordan curve in R, and let $|f(z)| > |g(z)|$ for all $z \in \Gamma$. Then the functions $f+g$ and f have the same number of zeros in the interior of Γ.

Proof. An equivalent assertion is that the function

$$r(z) := \frac{f(z)+g(z)}{f(z)}$$

has equally many zeros as poles in the interior of Γ. By (4.10-2) the difference between the two numbers equals $\pm n(r(\Gamma), 0)$. For $z \in \Gamma$,

$$|r(z)-1| = \left|\frac{q(z)}{f(z)}\right| < 1;$$

hence $\operatorname{Re} r(z) > 0$, and the winding number of $r(\Gamma)$ with respect to 0 is necessarily zero. ∎

Rouché's theorem asserts that the winding numbers of the curves $f(z)$ and $f(z)+g(z)$ ($z \in \Gamma$) around 0 are the same, which is plausible also because the straight-line segment joining the points $f(z)$ and $f(z)+g(z)$ can never contain the origin.

We present some typical applications of Rouché's theorem.

1. *Another proof of the fundamental theorem of algebra.* Let

$$p(z) := z^n + a_1 z^{n-1} + \cdots + a_n$$

PRINCIPLE OF THE ARGUMENT

be a polynomial. Let

$$\rho := 2 \max_{1 \leq k \leq n} |a_k|^{1/k}.$$

Then

$$|a_k| \leq \left(\frac{\rho}{2}\right)^k, \quad k=1,\ldots,n,$$

and on the circle $|z| = \rho$

$$|a_1 z^{n-1} + a_2 z^{n-2} + \cdots + a_n|$$

$$\leq \rho^n \left(\frac{1}{2} + \frac{1}{2^2} + \cdots + \frac{1}{2^n}\right) = \left(1 - \frac{1}{2^n}\right)\rho^n,$$

whereas $|z|^n = \rho^n$. Thus the hypotheses of Rouché's theorem are satisfied for $f(z) := z^n$, $g(z) := a_1 z^{n-1} + \cdots + a_n$, and it follows that $p = f + g$ has as many zeros (counted with proper multiplicities) inside $|z| = \rho$ as f, namely n.

2. *The continuity of the zeros of a polynomial as a function of the coefficients.* Let $z_1, z_2, \ldots, z_k$ be the distinct zeros of the polynomial

$$p(z) = z^n + a_1 z^{n-1} + \cdots + a_n$$

and let m_i be the multiplicity of z_i. Let $\rho := \min_{i \neq j} |z_i - z_j|$.

THEOREM 4.10c

For each ϵ satisfying $0 < \epsilon < \rho$ there exists $\delta > 0$ such that any polynomial

$$q(z) = z^n + b_1 z^{n-1} + \cdots + b_n$$

whose coefficients satisfy $|b_j - a_j| < \delta$, $j = 1, 2, \ldots, n$, has precisely m_i zeros in each disk

$$|z - z_i| < \epsilon, \quad i = 1, 2, \ldots, k.$$

Proof. We apply Rouché's theorem to the circle $\Gamma_i : |z - z_i| = \epsilon$ with $f := p$, $g := q - p$. On Γ_i the product representation of p implies

$$|f(z)| \leq \epsilon^{m_i} (\rho - \epsilon)^{n - m_i},$$

whereas

$$|g(z)| \leq \delta \mu_i, \quad \mu_i := \sum_{j=0}^{n} (|z_i| + \epsilon)^j.$$

Thus the hypotheses of Rouché's theorem are satisfied if

$$\delta < \mu_i^{-1} \epsilon^{m_i} (\rho - \epsilon)^{n - m_i}, \qquad i = 1, \ldots, k,$$

and we may conclude that the functions $f = p$ and $f + g = p + (q - p) = q$ have an equal number of zeros inside each Γ_i, namely m_i. ∎

Thus we have proved that the zeros of a polynomial are in a well-defined sense continuous functions of its coefficients. The reader is warned, however, not to put too much stock in the numerical aspects of this continuity. In the example $p(z) := z^{10}$, $q(z) := z^{10} - \epsilon$ all zeros will change by 10^{-1} if merely one coefficient is changed by 10^{-10}. This is obviously due to the high multiplicity of the one zero of p. More startling, however, is the following example, due to J. H. Wilkinson, for which there is no such simple explanation. Let

$$p(z) := \prod_{k=1}^{20} (z - k) = \sum_{k=0}^{20} a_k z^k.$$

By careful numerical computations Wilkinson showed that the polynomial

$$q(z) := p(z) - 2^{-23} z^{19}$$

has five pairs of complex conjugate zeros, with imaginary parts ranging from ± 0.64 to ± 2.81!

3. *Zeros of limit functions of sequences of analytic functions.* Let the functions $f_n, n = 0, 1, 2, \ldots,$ be analytic in some fixed region R and let the sequence $\{f_n\}$ converge uniformly on every compact subset of R. We already know that the limit function $f := \lim f_n$ is analytic in R (Theorem 3.4b). If we assume some knowledge of the zeros of f_n, what can be said about the zeros of f?

THEOREM 4.10d

Let Γ be any Jordan curve in R whose interior belongs to R and let $f(z) \neq 0$ for $z \in \Gamma$. Then for all sufficiently large n, the functions f_n have the same number of zeros in the interior of Γ as f.

Proof. Since f is continuous on Γ,

$$\epsilon := \inf_{z \in \Gamma} |f(z)| > 0.$$

Since $\{f_n\}$ converges uniformly on Γ, there exists m such that

$$|f_n(z) - f(z)| < \epsilon \quad \text{for } z \in \Gamma, n > m.$$

PRINCIPLE OF THE ARGUMENT 283

The conclusion follows by an application of Rouché's theorem in which $g := f_n - f$. ∎

COROLLARY 4.10e

If the limit function f does not vanish identically, then each zero of f is a limit of zeros of the functions f_n.

Proof. Let z_0 be a zero of f and let the disk $|z - z_0| \leq \rho_0$ contain no other zeros of f. Let $\rho_k := 2^{-k}\rho_0$. By Theorem 4.10d there exist numbers m_k, $k = 0, 1, 2, \ldots$, such that for $n > m_k$ the functions f_n have zeros $z_{n,k}$ satisfying $|z_{n,k} - z_0| < \rho_k$. We may assume that the sequence $\{m_k\}$ is increasing. Clearly the sequence $\{z_n\}$, where

$$z_n := z_{n,k}, \qquad m_k < n \leq m_{k+1}, \, k = 0, 1, 2, \ldots,$$

converges to z_0. ∎

COROLLARY 4.10f (theorem of Hurwitz)

Let the functions f_n, $n = 0, 1, 2, \ldots$, be analytic and different from zero in a region R and let the sequence $\{f_n\}$ converge uniformly on every compact subset of R. Then the limit function f either vanishes identically or it is different from zero in R.

Proof. If the limit function f had a zero z_0 without vanishing identically, then by Corollary 4.10e z_0 would have to be the limit of zeros of the functions f_n. This is impossible because the functions f_n have no zeros. ∎

The theorem of Hurwitz is required, for instance, in the theory of conformal mapping (Chapter 14).

PROBLEMS

1. Let D be the unit disk and D', its closure. Let f be analytic in D, continuous on D', and let $f(D') \subset D$. Prove that $f(z) = z$ has exactly one solution $z \in D$.

2. Let p be a polynomial of degree n, let $z_0, z_1, \ldots, z_n$ be complex numbers, and let

$$d_k := p(z_0, z_1, \ldots, z_k), \qquad k = 0, 1, \ldots, n,$$

be the divided difference formed with points $z_0, \ldots, z_k$ (see Problem 2, §4.7). Let Γ be a Jordan curve not passing through any z_k. Prove that if for all $z \in \Gamma$

$$\sum_{k=0}^{n-1} |d_k| \prod_{i=0}^{k-1} |z - z_i| < |d_n| \prod_{i=0}^{n-1} |z - z_i|$$

then p has as many zeros inside Γ as there are points z_k inside Γ. [Martin Gutknecht]

3. (continuation). If for a complex number w and a real number τ_0

$$\sum_{k=0}^{n-1} \frac{|d_k|}{\prod_{i=k}^{n-1}||z_i-w|-\tau_0|} \leqslant |d_n|,$$

then p has as many zeros as there are points z_i in the disk $|z-w|\leqslant\tau_0$. [Note that if the points z_k are approximations of the zeros of p, then the d_k are small compared with d_n.]

4. Let w be a complex number and let n be an integer such that $(n+\frac{1}{2})\pi>|w|$. Show that in the strip $|\operatorname{Re} z|\leqslant(n+\frac{1}{2})\pi$ the function $f(z):=z\sin z$ takes the value w precisely $2n+2$ times. [Apply Rouché's theorem, where $g(z):=-w$, to a rectangular region $R:=|\operatorname{Re} z|\leqslant(n+\frac{1}{2})\pi$, $|\operatorname{Im} z|\leqslant k$, where k is sufficiently large.]

5. Let $0<a_0<a_1<\cdots<a_n$. It is known that all zeros of the polynomial

$$p(z):=a_0+a_1z+\cdots+a_nz^n$$

satisfy $|z_i|<1$ (Eneström's theorem; see Problem 5, §6.4). Conclude by means of the principle of the argument that the trigonometric polynomial

$$a_0+a_1\cos\phi+a_2\cos 2\phi+\cdots+a_n\cos n\phi$$

has $2n$ zeros in the interval $[0,2\pi]$.

6. It is well known that

$$e^z=\lim_{n\to\infty}\left(1+\frac{z}{n}\right)^n$$

uniformly on every compact set. Conclude directly from this fact that the exponential function has no zeros.

7. Let $\{f_n\}$ be a sequence of entire functions with real zeros only and let $f=\lim_{n\to\infty}f_n$ exist uniformly on every compact set. Show that f has real zeros only.

8. It is well known that the Legendre polynomials

$$P_n(z):={}_2F_1\left(-n,n+1;\ 1;\ \frac{1-z}{2}\right)$$

have real zeros only. From the fact that

$$J_0(z)=\lim_{n\to\infty}P_n\left(1-\frac{z^2}{2n^2}\right)$$

uniformly on every compact set conclude that the Bessel function J_0 has real zeros only.

SEMINAR ASSIGNMENTS

1. Use the Plana summation formula to compute values of the Riemann zeta function

$$\zeta(z) := \sum_{n=1}^{\infty} n^{-z},$$

near $z=1$. Show that (4.9-16) continues $\zeta(z)$ to a function that is meromorphic in the entire plane, with a single pole, of order 1, at $z=1$. Use the formula to verify numerically that $\zeta(-2n)=0$, $n=1,2,\ldots$. Try to find some of the nontrivial zeros of $\zeta(z)$ on the line $\operatorname{Re} z = \frac{1}{2}$.

NOTES

§4.3. A brief sketch of the history of Cauchy's integral theorem is as follows (see Lindelöf [1905] and Watson [1914]). The theorem appears to have been known to Gauss in 1811. Cauchy [1825] proved the homotopy version of the theorem given here as Theorem 4.3d. Riemann proved our proposition A for simply connected regions by an application of Stokes's theorem (this proof is sketched in §5.5). Both Cauchy and Riemann defined analyticity by differentiability (see §5.5) and moreover assumed the derivative f' to be continuous. Goursat [1900] showed that the local form of Cauchy's theorem can be proved, assuming merely the existence but not the continuity of f'. Variants of Goursat's proof are reproduced in most textbooks; e.g. Ahlfors [1953], p. 88. If, as is done here, analyticity is defined by power series, the local form of the theorem (given here as Theorem 4.2c) becomes almost a triviality. For a stronger form of Theorem 4.3g (Γ a Jordan curve) see Behnke & Sommer [1962], Chapter 2, §2, Satz 5b. Extensions of Theorem 4.3g to arbitrary *starlike* regions are given in a number of recent texts; e.g., Duncan [1968], Levinson & Redheffer [1970].

§4.5. For Kepler's equation see the notes on §2.5. Recurrence relations for special functions such as those established in Problem 1 for the Bessel functions can be used for computing these functions numerically; see Gautschi [1967].

§4.6. For a proof of the Jordan curve theorem in the general case, see Aleksandrov [1956], pp. 39–64. For the essence of the proof given here (after Lemma 4.61) I am indebted to Professor C. Blatter (verbal communication). For another elementary proof for piecewise regular curves see Pederson [1969].

§4.7. The proof of Cauchy's integral *formula* (as opposed to the Cauchy integral *theorem*) raises the question of the index, or of inside and outside. It appears that the early authors dealt with these notions on an intuitive basis. An early attempt to base the integral formula and the residue theorem on a rigorous proof of the Jordan curve theorem was made by Watson [1914]. This became the standard approach in most textbooks until Ahlfors [1953], following a suggestion by Artin, showed that by the notions of homology and winding number a formally satisfactory treatment can be given which avoids the Jordan curve theorem. This approach has been imitated in many texts. For the reasons cited in the preface we have chosen to present the curve theorem approach.

§4.8, §4.9. For nearly all formulas given here see Lindelöf [1905] and Watson [1914]. Lindelöf also contains a history of the Plana summation formula.

§4.10. For an application of the principle of the argument in the theory of prime numbers see Riemann [1859]. Numerous applications to the location of zeros of polynomials are given in Marden [1966]; see also §6.7, 6.8, and 6.11. The role of the Nyquist diagram in control theory is discussed by Effertz & Kolberg [1963]. Wilkinson [1963] comments on the inadequacy of Theorem 4.10e when the zeros of a polynomial are "ill-conditioned."

5

CONFORMAL MAPPING

§5.1. GEOMETRIC INTERPRETATION OF COMPLEX FUNCTIONS

Let D be a region in the complex plane. With each point $z \in D$ let there be associated, by some unambiguous prescription, a complex number w. We then say that a complex-valued **function** is defined on D. Denoting the function by a symbol such as f, we write the number associated with point $z \in D$ as $w = f(z)$.

An important class of such complex-valued functions is the class of *analytic* functions discussed in earlier chapters. Here we are concerned somewhat more with geometric aspects of complex-valued functions, without assuming at first that these functions are analytic.

Let $W := \{f(z) : z \in D\}$ be the set of all points that are values of f, earlier called the **domain of values** of f. The function f may be regarded as a mapping of the set D onto the set W (see Fig. 5.1a). To emphasize this concept the fact that f maps z onto $f(z)$ is frequently described by the notation

$$f : z \to f(z) \qquad (z \in D),$$

which is especially useful when f is given by an explicit formula. In this chapter points in the domain of definition $D = D(f)$ and in the domain of values $W = W(f)$ are frequently denoted by $z = x + iy$ and $w = u + iv$, respectively.

For the geometric aspects of complex variable theory discussed in this chapter it is especially important to have a clear visual picture of the mappings defined by complex functions. A *real* function f of a *real* variable x is ideally represented by its *graph*, i.e., by the set of points $(x, f(x))$, where x ranges in the interval of definition of f. The graph is also a useful tool for the representation of a real function of two real variables

GEOMETRIC INTERPRETATION

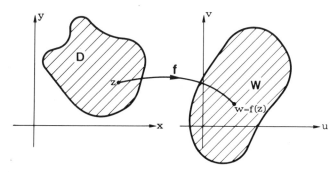

Fig. 5.1a. Mapping defined by function f.

or of one complex variable. It cannot be employed, however, to represent complex functions of a complex variable because here both the domain of definition and the domain of values are two-dimensional.

A commonly used graphical representation of complex functions of a complex variable consists in drawing the domain of definition and the domain of values in separate complex planes. The mapping is visualized by drawing in the image plane (w-plane) the images of suitably chosen geometric configurations in the preimage plane (z-plane). Suitable configurations are the parallels to the coordinate axes or the rays emanating from O together with the concentric circles around O (see Figs. 5.1b and 5.1c).

The present section aims at nothing more ambitious than a discussion of the mapping properties of some simple elementary functions.

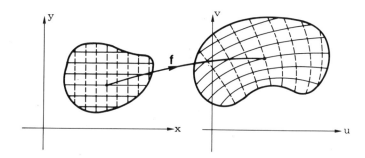

Fig. 5.1b. Mapping the coordinate lines.

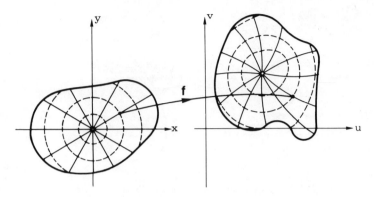

Fig. 5.1c. Mapping concentric circles and orthogonal rays.

EXAMPLE 1

The linear function. Let $a \neq 0$ and b be complex numbers. The function

$$f: z \to az + b$$

is called **linear**. Every linear function is a composition of linear functions of the special types

$$f_1: z \to z + b,$$

$$f_2: z \to az \quad (|a| = 1),$$

$$f_3: z \to cz \quad (c > 0).$$

The mappings f_1 and f_2 represent rigid motions of the complex plane. In particular, f_1 is a translation by the vector b (see Fig. 5.1d) and f_2, if $a = e^{i\phi}$, is a rotation by the angle ϕ about the origin (see Fig. 5.1e).

The mapping defined by f_3 is a *similarity*. It is a stretching by the factor c if $c > 1$ and a contraction by the factor c if $0 < c < 1$. For simplicity's sake, however, we always refer to this mapping as a **stretching** (see Fig. 5.1f).

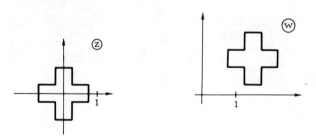

Fig. 5.1d. The translation $z \to z + (\frac{2}{3} + i)$.

GEOMETRIC INTERPRETATION

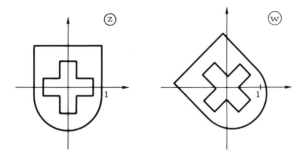

Fig. 5.1e. The rotation $z \to e^{i\pi/4} \cdot z$.

EXAMPLE 2

The quadratic function. Here we discuss the mapping

$$f: z \to az^2 + bz + c,$$

where a, b, and c are complex constants. We may assume that $a \neq 0$, because the linear function has already been discussed. Then we have

$$az^2 + bz + c = a\left(z + \frac{b}{2a}\right)^2 + c - \frac{b^2}{4a}.$$

Thus the mapping f may be regarded as a composition of linear mappings and of the special mapping

$$f_4: z \to z^2.$$

To discuss the mapping defined by f_4 we set $z = x + iy$, $w = z^2 = u + iv$, and trace the images of the straight lines $x = x_0$ and $y = y_0$ in the (u, v)-plane. In view of

$$u = x^2 - y^2, \qquad v = 2xy$$

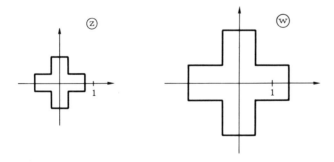

Fig. 5.1f. The stretching $z \to 2z$.

we obtain the following parametric representation of the image of the straight line $y=y_0$, $-\infty<x<\infty$:

$$u=x^2-y_0^2, \quad v=2xy_0.$$

If $y_0=0$, this yields the positive half-line $x=0$, traversed from ∞ to 0 and then from 0 to ∞. If $y_0\neq 0$, elimination of the parameter x yields

$$u=\frac{v^2}{4y_0^2}-y_0^2. \qquad (5.1\text{-}1)$$

This is the equation of a parabola symmetric about the u-axis and open to the right. The focus of the parabola is at the origin. We note that the straight lines $y=y_0$ and $y=-y_0$ are both mapped onto the *same* parabola. This indeed was to be expected, since $(-z)^2=z^2$. In a similar manner we obtain as images of the vertical lines $x=x_0$ the negative half-line (traversed twice) if $x_0=0$ and the curves with the equations

$$u=x_0^2-\frac{y^2}{4x_0^2} \qquad (5.1\text{-}2)$$

if $x_0\neq 0$. The curves (5.1-2) are parabolas with the u-axis as axes of symmetry open to the left (see Fig. 5.1g). The lines $x=x_0$ and $x=-x_0$ are mapped onto the same parabola. All parabolas (5.1-2) have their foci again at O. It can be shown by calculus that each parabola (5.1-1) intersects each parabola (5.1-2) under a right

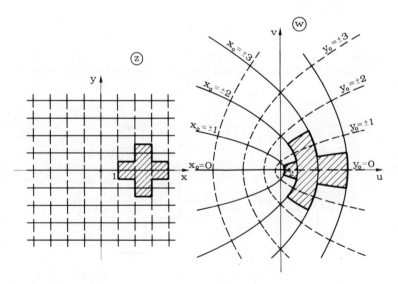

Fig. 5.1g. Mapping $z\to z^2$.

GEOMETRIC INTERPRETATION

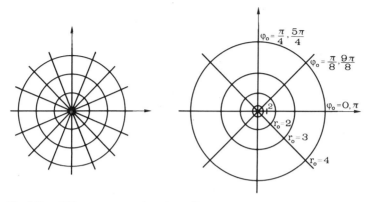

Fig. 5.1h. Different presentation of $z \to z^2$.

angle. Thus the mapping $z \to z^2$ has the property that the right angles between the straight lines $x = x_0$ and $y = y_0$ are preserved everywhere with the exception of the origin.

Another visualization of the mapping $z \to z^2$ is obtained by the use of polar coordinates. If $z = \rho e^{i\phi}$, $w = z^2 = \sigma e^{i\theta}$, it follows from the rules of complex arithmetic that $\sigma = \rho^2$, $\theta = 2\phi$. Thus the circles $\rho = \rho_0$ are mapped onto the circles $\sigma = \rho_0^2$ and the rays $\phi = \phi_0$ are mapped onto the rays $\theta = 2\phi_0$ (see Fig. 5.1h). A sector $-\alpha \leq \phi \leq \alpha$ of the z-plane is mapped onto the sector $-2\alpha \leq \theta \leq 2\alpha$ of the w-plane. If $\alpha \geq \pi/2$, some points in the w-plane are covered more than once by the mapping.

The Inverse Function. Still reviewing some concepts from general function theory, we recall that a function $f: z \to f(z)$ is called **univalent** or **one-to-one** if for every point w of its domain of values $W(f)$ there exists only one point z of its domain of definition $D(f)$ such that $f(z) = w$.

EXAMPLE 3

The linear function $z \to az + b$ ($a \neq 0$) defines a one-to-one mapping of the complex plane onto itself, for if w is any complex number the only point mapped onto w is $z := a^{-1}(w - b)$. The function $z \to z^2$ is not one-to-one in the complex plane because the points z and $-z$ have the same image.

A function f which is not one-to-one can always be made into a one-to-one function by suitably restricting its domain of definition; for instance, $z \to z^2$ is one-to-one in any domain D that does not contain $-z$ if it contains z. Any open half-plane bounded by a straight line through O is such a domain.

Let the function f be one-to-one and let $D(f)$ and $W(f)$ be its domain of definition and domain of values, respectively. Then $f^{[-1]}$, the **inverse function** of f, is the function with domain of definition $W(f)$ whose value at the point $w \in W(f)$ is the unique solution z of the equation $f(z) = w$ (see

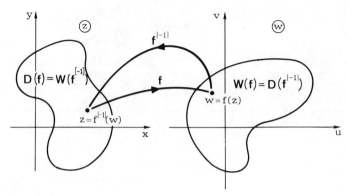

Fig. 5.1i. The inverse function.

Fig. 5.1i). The inverse function by definition satisfies the functional relations

$$f^{[-1]}(f(z)) = z \quad \text{for all } z \in D(f),$$

$$f(f^{[-1]}(w)) = w \quad \text{for all } w \in W(f).$$

EXAMPLE 4

Let α be a real number. By the above the function $f: z \to z^2$ is one-to-one in the domain $D(f): \alpha < \arg z < \alpha + \pi$. Its domain of values is $W(f): 2\alpha < \arg w < 2\alpha + 2\pi$, i.e., the w-plane cut along the ray $\arg w = 2\alpha$. The value of the inverse function at a point $w = \sigma e^{i\theta} \in W(f)$ is

$$z = |w|^{1/2} e^{i\theta/2},$$

where θ is the unique value of $\arg w$ lying in the open interval $(\alpha, \alpha + 2\pi)$. If $\alpha = -\pi/2$ and $w = u + iv$, the value of the inverse function at $w = u + iv$ is also given by

$$z = x + iy = \left(\frac{\sqrt{u^2 + v^2} + u}{2} \right)^{1/2} + i \operatorname{sgn} v \left(\frac{\sqrt{u^2 + v^2} - u}{2} \right)^{1/2}.$$

THE EXPONENTIAL FUNCTION

Let $z = x + iy$. The function defined on $\mathbb{C}$ by

$$z \to e^z := e^x(\cos y + i \sin y)$$

GEOMETRIC INTERPRETATION 293

by definition is called the **exponential function** (see also §2.5). If $w:=e^z$, the images of the straight lines $y=y_0$ have the parametric representation

$$w = e^x e^{iy_0}(-\infty < x < \infty);$$

thus they are the rays $\arg w = y_0$. Any two lines $y = y_0$ whose distance is an integral multiple of 2π are mapped onto the same ray. The exponential function thus certainly is not one-to-one. The images of the straight lines $x = x_0$ are given by

$$w = e^{x_0} e^{iy}(-\infty < y < \infty);$$

these are circles of radius e^{x_0}, run through an infinite number of times (see Fig. 5.1j). Again we see that the right angles between the curves $y = y_0$ and $x = x_0$ are preserved under this mapping.

To discuss inverse functions of the exponential function we consider the equation for z,

$$e^z = w, \qquad (5.1\text{-}3)$$

where w is given. This equation has no solution if $w = 0$. If $w \neq 0$, $w = |w|e^{i\theta}$, §2.5 shows that it has an infinite number of solutions

$$z = \text{Log}|w| + i(\theta + 2\pi k),$$

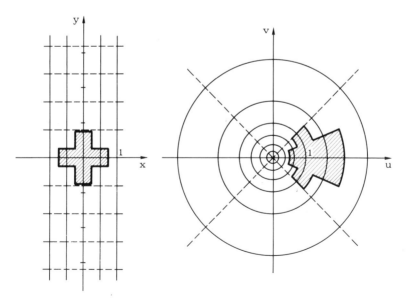

Fig. 5.1j. Map $z \to e^z$.

in which k is an arbitrary integer. It follows that the exponential function is one-to-one only if it is restricted to domains D with the property that if $z \in D$ no point $z + 2\pi i k$ belongs to D for any integer $k \neq 0$. One such domain is the parallel strip $-\pi < \operatorname{Im} z < \pi$. The inverse function of the exponential function thus restricted is called the **principal value of the logarithm** and is denoted by the symbol Log. Its domain of definition W is the w-plane, cut along the negative real axis. For any $w \in W$

$$\operatorname{Log} w = \operatorname{Log}|w| + i \operatorname{Arg} w,$$

where $\operatorname{Arg} w$, the **principal value of the argument**, is chosen in the interval $(-\pi, \pi)$.

THE JOUKOWSKI MAP

Here we shall study the map defined by the **Joukowski function**

$$f: z \to \tfrac{1}{2}\left(z + \frac{1}{z}\right), \qquad z \neq 0. \tag{5.1-4}$$

This function, as well as some of its inverse functions, are of considerable use in several engineering applications.

Setting $z = \rho e^{i\phi}$, we discuss first the images of the curves $\rho = \text{const}$ and $\phi = \text{const}$. If

$$w = f(z) = \tfrac{1}{2}(\rho e^{i\phi} + \rho^{-1} e^{-i\phi}) = u + iv,$$

then

$$u = \tfrac{1}{2}\left(\rho + \frac{1}{\rho}\right)\cos\phi, \qquad v = \tfrac{1}{2}\left(\rho - \frac{1}{\rho}\right)\sin\phi.$$

The image of the circle $\rho = \rho_0$ thus has the parametric representation

$$\phi \to (u, v) = \left[\tfrac{1}{2}\left(\rho_0 + \frac{1}{\rho_0}\right)\cos\phi, \tfrac{1}{2}\left(\rho_0 - \frac{1}{\rho_0}\right)\sin\phi\right], \qquad 0 \leq \phi \leq 2\pi.$$

If $\rho_0 = 1$, this is the straight-line segment joining the points $+1$ and -1, run through twice. If $\rho_0 \neq 1$, elimination of the parameter ϕ yields

$$\frac{u^2}{\left[\tfrac{1}{2}(\rho_0 + \rho_0^{-1})\right]^2} + \frac{v^2}{\left[\tfrac{1}{2}(\rho_0 - \rho_0^{-1})\right]^2} = 1. \tag{5.1-5}$$

GEOMETRIC INTERPRETATION

This is the equation of an ellipse with foci at $(u,v)=(\pm 1,0)$ and semiaxes

$$a := \tfrac{1}{2}(\rho_0 + \rho_0^{-1}), \qquad b := \tfrac{1}{2}|\rho_0 - \rho_0^{-1}|.$$

Evidently the circles $\rho = \rho_0$ and $\rho = \rho_0^{-1}$ are mapped onto the same ellipse. Thus the mapping defined by the Joukowski function is not one-to-one unless the domain of definition is suitably restricted.

The images of the rays $\phi = \phi_0$ are given parametrically by

$$\rho \to (u,v) = \left[\tfrac{1}{2}\left(\rho + \frac{1}{\rho}\right)\cos\phi_0, \tfrac{1}{2}\left(\rho - \frac{1}{\rho}\right)\sin\phi_0 \right], \qquad 0 < \rho < \infty.$$

If $\sin\phi_0 = 0$, $\cos\phi_0 > 0$, this is the half-line $u \geq 1$, $v = 0$, traversed from ∞ to $w = 1$ and back to ∞. If $\sin\phi_0 = 0$, $\cos\phi_0 < 0$, it is the half-line $u \leq -1$, $v = 0$. If $\sin\phi_0 \neq 0$, we obtain an arc of the hyperbola

$$\frac{u^2}{(\cos\phi_0)^2} - \frac{v^2}{(\sin\phi_0)^2} = 1. \tag{5.1-6}$$

The right arc of the hyperbola is obtained if $\cos\phi_0 > 0$, the left, if $\cos\phi_0 < 0$. For $\cos\phi_0 = 0$ the hyperbola degenerates into the straight line $u = 0$, $-\infty < v < \infty$. We note that the rays $\arg w = \pm\phi_0$ are the asymptotes of the image hyperbolas. This is clear because for large values of $|z|$ the Joukowski map is asymptotic to $\tfrac{1}{2}z$ and for small values of $|z|$, to $\tfrac{1}{2}z^{-1}$.

The hyperbolas (5.1-6) again have the common foci $(u,v) = (\pm 1, 0)$. It is easily shown that each hyperbola intersects each ellipse (5.1-5) orthogonally (see Fig. 5.1k).

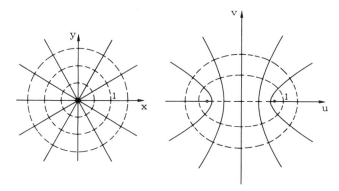

Fig. 5.1k. The Joukowski map $z \to \tfrac{1}{2}(z + 1/z)$.

We shall now study the question of the *inverse map*. If w is a complex number, the point z is mapped onto w by f if and only if it satisfies the quadratic equation

$$z^2 - 2wz + 1 = 0. \tag{5.1-7}$$

The discriminant of this equation is $w^2 - 1$. Thus, if $w = \pm 1$, exactly one point is mapped onto w by f, namely the point $z = \pm 1$. If $w \neq \pm 1$, two distinct points are mapped onto the same point w. According to Vieta the product of these two points is 1. It follows that *the Joukowski function is one-to-one in every domain that together with the point z does not at the same time contain the point z^{-1}*. We discuss two such domains.

(i) $D(f):|z|>1$. Thus restricted, the Joukowski function maps the exterior of the unit disk one-to-one onto the domain W_1 which is obtained by cutting the w-plane between the points $w=1$ and $w=-1$ (see Fig. 5.1*l*). (The values on the cut are not taken by f.)

The value of the inverse function at a point $w \in W_1$ is a solution of the quadratic equation (5.1-7),

$$z = w + (w^2 - 1)^{1/2}; \tag{5.1-8}$$

that value of the square root is to be chosen for which

$$|z| = |w + (w^2 - 1)^{1/2}| > 1. \tag{5.1-9}$$

(ii) $D(f):\operatorname{Im} z > 0$. With this restriction, the Joukowski function maps the upper half-plane one-to-one onto the domain W_2 obtained by cutting

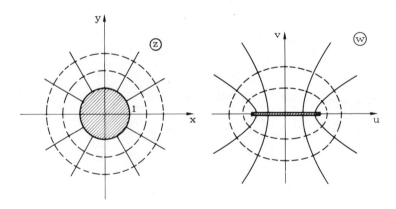

Fig. 5.1l. The Joukowski map $z \to \frac{1}{2}(z + 1/z)$ ($|z| > 1$).

GEOMETRIC INTERPRETATION

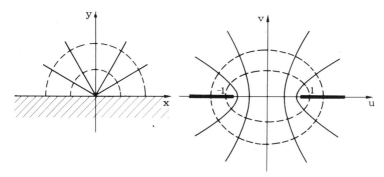

Fig. 5.1m. The Joukowski map $z \to \frac{1}{2}(z + 1/z)$ ($\operatorname{Im} z > 1$).

the w-plane from $w = 1$ to $+\infty$ along the positive real axis and from $w = -1$ to $-\infty$ along the negative real axis (see Fig. 5.1m). (Again the values on the cuts are not taken.) The value of the inverse function at a point $w \in W_2$ is again given by (5.1-8); however, we must now choose that value of the square root for which

$$\operatorname{Im} z = \operatorname{Im}\left[w + (w^2 - 1)^{1/2}\right] > 0. \qquad (5.1\text{-}10)$$

The prescriptions (5.1-9) and (5.1-10) for picking the correct values of z can be rendered somewhat more explicit. The problem is to select the correct value of $(w^2 - 1)^{1/2}$. If we make use of

$$(w^2 - 1)^{1/2} = (w + 1)^{1/2}(w - 1)^{1/2},$$

it suffices to indicate which values of $(w + 1)^{1/2}$ and $(w - 1)^{1/2}$ should be selected. Let, as in Fig. 5.1n,

$$\phi_1 := \arg(w + 1), \qquad \phi_2 := \arg(w - 1).$$

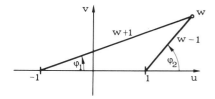

Fig. 5.1n. Arguments used in inverse Joukowski map.

If definite values of ϕ_1 and ϕ_2 are selected, we set

$$(w+1)^{1/2} := |w+1|^{1/2} e^{i\phi_1/2},$$

$$(w-1)^{1/2} := |w-1|^{1/2} e^{i\phi_2/2},$$

which yield

$$(w^2-1)^{1/2} = |w^2-1|^{1/2} e^{i(\phi_1+\phi_2)/2}. \tag{5.1-11}$$

The problem is thus reduced to one of properly selecting the intervals in which the angles ϕ_1 and ϕ_2 should be chosen. The choice has to be made so that the functions $f^{[-1]}$, hence $(w^2-1)^{1/2}$, are continuous in their respective domains of definition.

In (i) we select ϕ_1 and ϕ_2 such that

$$-\pi < \phi_1 \leq \pi, \qquad -\pi < \phi_2 \leq \pi. \tag{5.1-12}$$

This choice yields positive $z > 1$ for positive $w > 1$. Furthermore, it renders $\sqrt{w^2-1}$ continuous, at any rate, in the domain $|\text{Arg}(w-1)| < \pi$. We assert that $(w^2-1)^{1/2}$ is continuous also for real $w < -1$, for if w transverses the negative real axis at a point $w_0 < -1$, say, in the upward direction both ϕ_1 and ϕ_2 increase by 2π. Hence the argument of $(w^2-1)^{1/2}$, defined by (5.1-11), likewise increases by 2π and the root itself is continuous.

In (ii) we first select ϕ_1 in the interval $(-\pi, \pi)$. This yields a value of $(w+1)^{1/2}$ which is continuous in W_2. In order that $(w-1)^{1/2}$ may likewise be continuous in W_2, ϕ_2 must be selected in an interval congruent to $(0, 2\pi)$ modulo 2π. The choice must be such that the resulting values of z have positive imaginary parts. This must be true, in particular, for real w, $-1 < w < 1$. For such values of w, $\text{Im}\, z > 0$ if and only if $\text{Im}(w^2-1)^{1/2} > 0$. Because (with the choice of ϕ_1 already made) $(w+1)^{1/2}$ is positive, we must have

$$\text{Im}(w-1)^{1/2} = |w-1|\sin(\phi_2/2) > 0.$$

This is the case if ϕ_2 is selected in $(0, 2\pi)$ (and not, for instance, in $(-2\pi, 0)$). Thus we obtain the correct values of the inverse function in (ii) if the angles ϕ_1 and ϕ_2 are chosen as follows:

$$-\pi < \phi_1 < \pi, \qquad 0 < \phi_2 < 2\pi. \tag{5.1-13}$$

The formulas (5.1-12) and (5.1-13) now determine the two inverse functions in question in a computationally unambiguous manner.

ALGEBRAIC THEORY

PROBLEMS

1. Let $f: z \to w = z^{-1}$.
   (a) Compute the images of the straight lines $\operatorname{Re} z = \gamma$ (γ real) under the map f and draw them for $\gamma = \frac{1}{2}, 1, 2$.
   (b) Show that every circle $|z - \mu| = (\mu^2 - 1)^{1/2}$ ($\mu > 1$) is mapped onto itself by the map f.

2. Let f denote the map $z \to w = \sqrt{z}$ (principal value). We denote by (ρ, ϕ) and (σ, ψ) polar coordinates in the z-plane and the w-plane, respectively.
   (a) Let $\rho = \rho(\phi)$, $-\pi < \phi < \pi$, be the equation of a curve in the z-plane. Show that its image under f has the equation $\sigma = [\rho(2\psi)]^{1/2}$, $-\pi/2 < \psi < \pi/2$.
   (b) Let $\gamma > 0$. What is the equation of the image under f of the straight line $\rho = \gamma/(\sin\phi)$, $0 < \phi < \pi$, in polar coordinates? In cartesian coordinates?

3. Let $\alpha > 0$. The name *lemniscate* is given to the curve which in the (u, v)-plane satisfies the equation $(u^2 + v^2)^2 = \alpha^2(u^2 - v^2)$. Show that the lemniscate is the image of the straight line $\operatorname{Re} z = \alpha^{-2}$ under the map $z \to \pm (1/z)^{1/2}$.

4. Sketch the image of the strip bounded by the two straight lines $y = x \pm \pi/4$ under the map $z \to w = e^z$.

5. Sketch the image of the semi-infinite strip $-1 < \operatorname{Re} z < 1$, $0 < \operatorname{Im} z < \infty$ under the map $z \to i^z$ (principal value, see §2.5). What is the image of the point $z = i$?

6. We consider the Joukowski map $z \to w = \frac{1}{2}(z + 1/z)$ ($|z| > 1$). It is required to find the equation of the preimages of the straight lines $\operatorname{Im} w = v_0$
   (a) in cartesian coordinates,
   (b) in polar coordinates.
   Find the asymptotes of the preimages and draw the required curves for $v_0 = 0, \pm \frac{1}{4}, \pm \frac{1}{2}, \pm \frac{3}{4}$.

7. Show that the locus of the turning points of the preimages considered in the preceding problem has the equation $(x^2 + y^2)(3x^2 - y^2 - 2) = 1$.

§5.2. MOEBIUS TRANSFORMATIONS: ALGEBRAIC THEORY

In the following three sections we shall study a particularly important family of mappings, namely those defined by *Moebius transformations*. We begin by considering some of the family's algebraic aspects.

Let $\mathfrak{M}$ denote the class of all nonsingular matrices of order two, with complex elements. In $\mathfrak{M}$ we set up an equivalence relation as follows: two matrices M_1 and M_2 in $\mathfrak{M}$ are called **equivalent** if a complex number s (necessarily nonzero) exists such that $M_2 = sM_1$. It is clear that this equivalence relation satisfies the usual postulates of being symmetric, reflexive, and transitive.

With any matrix

$$M = \begin{bmatrix} a & b \\ c & d \end{bmatrix} \quad (ad - cb \neq 0)$$

in $\mathfrak{M}$ we associate the function

$$t_M : z \to \frac{az+b}{cz+d}, \qquad (5.2\text{-}1)$$

which is defined for all $z \neq -dc^{-1}$ if $c \neq 0$ and for all z if $c = 0$. The function t_M is called a *linear transformation, linear fractional transformation* (if $c \neq 0$), *bilinear transformation,* or (more succinctly because there are so many different kinds of linear transformations) a **Moebius transformation.** Moebius transformations play a fundamental role in the more geometrically oriented parts of complex variable theory and also in the theory of continued fractions.

The matrix M uniquely determines the function t_M. However, since we may multiply the four constants a, b, c, d in (5.2-1) by any nonzero constant without changing the function t_M, t_M merely determines the equivalence class of matrices to which M belongs and not the matrix M itself.

For a satisfactory treatment of Moebius transformations it is desirable that, in the case $c \neq 0$, the point $z := -dc^{-1}$, where t_M is undefined, should not play a special role. An elegant way to circumvent the exceptional nature of this point is to treat Moebius transformations as mappings of the complex projective line onto itself. The difficulty can also be resolved in a geometrically convincing fashion by introducing the Riemann number sphere (see §5.3). Here we resolve it in a purely algebraic way.

To this end we enlarge the set of all complex numbers by an element denoted by ∞ and called the **point at infinity.** No arithemetic operators involving ∞ are defined, but nothing keeps us from defining, if $c \neq 0$,

$$t_M\left(-\frac{d}{c}\right) = \infty. \qquad (5.2\text{-}2)$$

Thus we have extended both the domain of definition and the domain of values of the function t_M. The former constitutes the set of all complex numbers; to the latter we have added the element ∞. For symmetry we further extend the domain by defining t_M also for $z = \infty$. If $c \neq 0$, the value assigned to t_M at $z = \infty$ will be chosen to make the function $f : z \to t_M(z^{-1})$ continuous at $z = 0$. If $z \neq 0$,

$$f(z) = t_M(z^{-1}) = \frac{a+bz}{c+dz}.$$

Clearly, f is continuous at $z = 0$ if we put $f(0) := ac^{-1}$; consequently we define

$$t_M(\infty) := \frac{a}{c}. \qquad (5.2\text{-}3)$$

If $c = 0$, we set

$$t_M(\infty) := \infty. \qquad (5.2\text{-}4)$$

ALGEBRAIC THEORY

With these conventions, t_M is defined on a set consisting of all complex numbers and the point at infinity. This set is called the **extended complex plane**. The points of the ordinary complex plane (i.e., all points except ∞) are called the *finite* points of the extended plane.

The basic fact concerning Moebius transformations is as follows:

THEOREM 5.2a

For any $M \in \mathfrak{M}$, the function t_M defines a one-to-one mapping of the extended complex plane onto itself.

Proof. It is to be shown that for any point w in the extended plane the equation

$$t_M(z) = w \tag{5.2-5}$$

has exactly one solution in the extended plane. Assuming that t_M is given by (5.2-1), we distinguish between $c=0$ and $c \neq 0$.

(a) If $c=0$, we have $d \neq 0$ (since M is nonsingular) and

$$t_M(z) = \frac{a}{d}z + \frac{b}{d} \qquad (z \neq \infty).$$

Again, because M is nonsingular, $a \neq 0$. Thus, if w is finite, (5.2-5) has the unique solution

$$z := \frac{dw - b}{a}. \tag{5.2-6}$$

If $w = \infty$, then by (5.2-4) $z = \infty$ is a solution. It is the only solution because $t_M(z)$ is always finite for finite values of z.

(b) If $c \neq 0$ and w is finite, we find that z must satisfy $az + b = (cz + d)w$ or

$$z(a - cw) = -b + dw. \tag{5.2-7}$$

This equation has a unique solution z for all $w \neq a/c$. This gap is filled, however, by the definition (5.2-3) which shows that $z = \infty$ is a solution of (5.2-5) for $w = a/c$. By (5.2-7) no finite z can satisfy (5.2-5) for $w = a/c$; $z = \infty$ is the only solution. Finally, if $w = \infty$, $z = -a/c$ is a solution. Because $t_M(z)$ is finite for all $z \neq -a/c$, $z = -a/c$ is the only solution. ∎

Equation (5.2-7) shows that for $c \neq 0$, $w \neq a/c$ the inverse function of t_M is represented by the formula

$$z = t_M^{[-1]}(w) = \frac{dw - b}{-cw + a}; \tag{5.2-8}$$

i.e., by the Moebius transformation associated with the matrix

$$M^* := \begin{pmatrix} d & -b \\ -c & a \end{pmatrix}. \qquad (5.2\text{-}9)$$

With the definitions (5.2-2) and (5.2-3), (5.2-8) is seen to be correct even when $w = a/c$ and $w = \infty$. It also holds for all w in the extended plane, if $c = 0$. Thus we have

$$(t_M)^{[-1]} = t_{M^*},$$

where M^* is given by (5.2-9). A simple computation shows that

$$MM^* = (ad - bc)I,$$

where I denotes the unit matrix. Thus M^* is equivalent to M^{-1}, and we have proved:

THEOREM 5.2b

The inverse of a Moebius transformation is again a Moebius transformation. Moreover, for any $M \in \mathfrak{M}$,

$$(t_M)^{[-1]} = t_{M^{-1}}. \qquad (5.2\text{-}10)$$

Theorem 5.2b is a mere special case of the following more general fact:

THEOREM 5.2c

The totality of all Moebius transformations t_M associated with the matrices $M \in \mathfrak{M}$, under the operation of composition, form a group. Moreover, if M and N are any two matrices in $\mathfrak{M}$, then

$$t_M \circ t_N = t_{MN}. \qquad (5.2\text{-}11)$$

Proof. We have to verify the four group postulates stated in §1.1:
(a) Let

$$M = \begin{bmatrix} a & b \\ c & d \end{bmatrix}, \quad N = \begin{bmatrix} \alpha & \beta \\ \gamma & \delta \end{bmatrix}$$

be two matrices in $\mathfrak{M}$. We then find that for all finite z, except at most two

ALGEBRAIC THEORY

exceptional values,

$$(t_M \circ t_N)(z) = t_M(t_N(z)) = \frac{a(\alpha z + \beta) + b(\gamma z + \delta)}{c(\alpha z + \beta) + d(\gamma z + \delta)}$$

$$= \frac{(a\alpha + b\gamma)z + (a\beta + b\delta)}{(c\alpha + d\gamma)z + (c\beta + d\delta)} = t_{MN}(z),$$

as desired. Using (5.2-2), (5.2-3), and (5.2-4), we readily see that this also holds for the exceptional values and for $z = \infty$. Thus $t_M \circ t_N$ is again a Moebius transformation and indeed the transformation associated with the matrix MN.

(b) The associative law is generally valid for the composition of functions. Without appealing to this general principle, it readily follows in the present case from the associative law for matrix multiplication, for if M, N, and P are in $\mathfrak{M}$ then by (a) both transformations

$$(t_M \circ t_N) \circ t_P \quad \text{and} \quad t_M \circ (t_N \circ t_P)$$

are equal to t_{MNP}. Note, however, that the group of Moebius transformations is not commutative because the matrices MN and NM need not be equivalent.

(c) The existence of the identity element follows from the fact that the identical mapping $t: z \to z$ can be regarded as a Moebius transformation, namely as t_I. This also keeps ∞ fixed.

(d) The existence of the inverse Moebius transformation was demonstrated in Theorem 5.2b ∎

SPECIAL MOEBIUS TRANSFORMATIONS

Especially simple Moebius transformations are those associated with the matrices

$$\begin{bmatrix} 1 & a \\ 0 & 1 \end{bmatrix}, \quad \begin{bmatrix} b & 0 \\ 0 & 1 \end{bmatrix}, \quad \begin{bmatrix} 0 & 1 \\ 1 & 0 \end{bmatrix},$$

where a and b are complex numbers, $b \neq 0$. The first of these, $z \to z + a$, was called a **translation** in §5.1. The second, $z \to bz$, was called a rotation if $|b| = 1$ and a stretching if $b > 0$; it is now more convenient to call it a **rotation** for all $b \neq 0$. The transformation associated with the third matrix, $z \to z^{-1}$, is called **inversion**. Translations, rotations, and inversions are the building blocks from which any Moebius transformation may be built:

THEOREM 5.2d

Every Moebius transformation can be represented as the composition of at most one translation, one rotation, one inversion, and one more translation.

Proof. Let t be a Moebius transformation and let it be associated with the matrix

$$M = \begin{bmatrix} a & b \\ c & d \end{bmatrix}.$$

By Theorem 5.2c it suffices to show that the matrix M, or a nonzero scalar multiple of it, can be written as a product of matrices of the above standard types in the order indicated by the theorem. If $c=0$, then $d\neq 0$, and the required representation is

$$\frac{1}{d}\begin{bmatrix} a & b \\ 0 & d \end{bmatrix} = \begin{bmatrix} \frac{a}{d} & 0 \\ 0 & 1 \end{bmatrix}\begin{bmatrix} 1 & \frac{b}{a} \\ 0 & 1 \end{bmatrix}.$$

If $c \neq 0$, the assertion follows from the identity

$$\frac{1}{c}\begin{bmatrix} a & b \\ c & d \end{bmatrix} = \begin{bmatrix} 1 & \frac{a}{c} \\ 0 & 1 \end{bmatrix}\begin{bmatrix} \frac{bc-ad}{c^2} & 0 \\ 0 & 1 \end{bmatrix}\begin{bmatrix} 0 & 1 \\ 1 & 0 \end{bmatrix}\begin{bmatrix} 1 & \frac{d}{c} \\ 0 & 1 \end{bmatrix},$$

which is the same as saying that for all but some exceptional values of z

$$\frac{az+b}{cz+d} = \frac{a}{c} + \frac{bc-ad}{c^2}\frac{1}{z+d/c}. \quad \blacksquare$$

These formulas define certain standard forms of Moebius transformations. Other standard forms can be obtained by representing the matrix M in one of the canonical forms of linear algebra (see Problem 5).

PROBLEMS

The first six problems deal with *fixed points* of Moebius transformations.

1. If t is a Moebius transformation, then any point z in the extended complex plane such that $t(z)=z$ is called a **fixed point** of t. Show that the Moebius

ALGEBRAIC THEORY

transformation t_M, where

$$M := \begin{bmatrix} a & b \\ c & d \end{bmatrix}$$

is not equivalent to the unit matrix has either one or two fixed points. It has two fixed points if and only if $(a-d)^2 + 4bc \neq 0$.

2. What are the fixed points (a) of a rotation, (b) of a translation, (c) of the inversion $z \to z^{-1}$?

3. Let the Moebius transformation t have two fixed points z_1, z_2 ($z_1 \neq z_2$). If both fixed points are finite, let

$$s: z \to \frac{z - z_1}{z - z_2}.$$

If one fixed point, say z_2, is at infinity, let

$$s: z \to z - z_1.$$

Show that in either case a rotation

$$r: z \to bz$$

exists such that $t = s \circ r \circ s^{[-1]}$.

4. Let the Moebius transformation t have exactly one fixed point z_1. If z_1 is finite, let

$$s: z \to \frac{1}{z - z_1}.$$

If z_1 is at infinity, let s be the identity. Show that in either case there is a translation $l: z \to z + a$ such that $t = s \circ l \circ s^{[-1]}$.

5. If M is any matrix of order 2 over the complex numbers, then according to linear algebra a matrix $S \in \mathfrak{M}$ exists such that

$$M = SJS^{-1},$$

where the matrix J has one of the forms

$$\begin{bmatrix} \lambda & 0 \\ 0 & \mu \end{bmatrix} (\lambda \neq \mu), \quad \begin{bmatrix} \lambda & 0 \\ 0 & \lambda \end{bmatrix}, \quad \begin{bmatrix} \lambda & 1 \\ 0 & \lambda \end{bmatrix}$$

(Jordan canonical form). If $M \in \mathfrak{M}$, how are the three types of J related to the fixed points of the Moebius transformation t_M?

6. Let the matrix S in Problem 5 be given by

$$S := \begin{bmatrix} s_{11} & s_{12} \\ s_{21} & s_{22} \end{bmatrix}$$

and let

$$z_k := \frac{s_{1k}}{s_{2k}} \quad \text{if } s_{2k} \neq 0,$$

$$z_k := \infty \quad \text{if } s_{2k} = 0 \quad (k = 1, 2).$$

If the Jordan form J of M is diagonal, $\lambda \neq \mu$, show that points z_k are precisely the fixed points of t_M. If J is not diagonal, show that the only fixed point of t_M is z_1.

The following problems are concerned with the n-fold iterate of a Moebius transformation t, $t^{[n]} = t \circ t \circ \cdots \circ t$ (n factors). The best point of attack is the Jordan canonical form of the matrix associated with t.

7. A Moebius transformation t is called an **involution** if $t^{[2]}$ is the identity.
   (a) Show that every involution that is not the identity has precisely two fixed points.
   (b) Show that the set of involutions is identical with the set of Moebius transformations of the form

$$t : z \to \frac{az + b}{cz - a},$$

   where a, b, c are arbitrary complex numbers such that $a^2 + bc \neq 0$.

8. If t is a Moebius transformation with two finite fixed points z_1, z_2, show that there is a complex number $q \neq 0$ such that

$$t^{[n]} : z \to \frac{(z_1 - q^n z_2)z - (1 - q^n)z_1 z_2}{(1 - q^n)z - z_2 + q^n z_1},$$

   where n is an arbitrary integer.

9. Let the Moebius transformation t_M, where

$$M := \begin{bmatrix} a & b \\ c & d \end{bmatrix},$$

   have two finite fixed points. Show that unless $(ad - bc)/(a + d)^2$ is real and $> \frac{1}{4}$ the limit $\lim_{n \to \infty} t^{[n]}(z)$ exists for every z and is equal to a fixed point of t.

10. Let t be a Moebius transformation with precisely one fixed point.
    (a) Find a formula for $t^{[n]}$.
    (b) Show that $\lim_{n\to\infty} t^{[n]}(z)$ exists for every z and is equal to the unique fixed point of t.
11. Show that the following sets of Moebius transformations form subgroups of the group of all Moebius transformations:
    (a) all rotations,
    (b) all translations,
    (c) all transformations t_M in which the elements of M are integers such that $\det M = 1$.
12. What is the smallest group of Moebius transformations containing the two transformations

$$t_1 : z \to z^{-1}, \qquad t_2 : z \to 1-z?$$

§5.3. MOEBIUS TRANSFORMATIONS: THE RIEMANN SPHERE

To dispense with the special role played by point ∞ and in order to elucidate the geometrical properties of Moebius transformations it is convenient at this point to introduce a model of the extended complex plane in which all of its points, including point ∞, have a concrete representative. Such a model is provided by the *Riemann number sphere*.

We identify the complex plane with the (x_1, x_2)-plane of a real, three-dimensional (x_1, x_2, x_3) space. To every point of the sphere we let correspond a point of the sphere $S: x_1^2 + x_2^2 + x_3^2 = 1$. Let $N = (0, 0, 1)$ be the "north pole" of that sphere. The point Z corresponding to $z = x + iy$ is then the second point of intersection of S with the straight line passing through $(x, y, 0)$ and through the north pole N of S. A simple geometric consideration in the plane containing the x_3-axis and the point $(x, y, 0)$ shows that the coordinates of Z are

$$x_1 = \frac{2x}{1+x^2+y^2} = \frac{2\,\mathrm{Re}\, z}{1+|z|^2},$$

$$x_2 = \frac{2y}{1+x^2+y^2} = \frac{2\,\mathrm{Im}\, z}{1+|z|^2}, \qquad (5.3\text{-}1)$$

$$x_3 = \frac{x^2+y^2-1}{1+x^2+y^2} = \frac{|z|^2-1}{|z|^2+1}.$$

It is geometrically clear that this construction sets up a one-to-one correspondence between all finite points of the extended plane and all

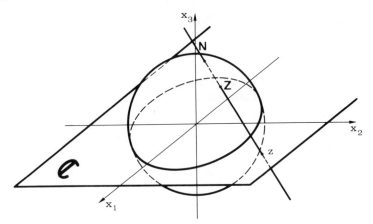

Fig. 5.3a. The Riemann sphere.

points of S except the north pole. Indeed, the point of the plane corresponding to a point $Z = (x_1, x_2, x_3) \neq N$ of S is given by

$$z = \frac{x_1 + ix_2}{1 - x_3}. \tag{5.3-2}$$

We now complete this correspondence by letting the north pole $(0, 0, 1)$ *correspond to the point* ∞. The spherical model of the extended complex plane thus obtained is known as the **Riemann number sphere** (see Fig. 5.3a). The process by which the model has been obtained is the **stereographic projection.**

We wish to determine the images on S of straight lines and circles under stereographic projection. In the extended complex plane it is convenient to include the point at infinity in every straight line. With this convention our construction makes clear that the image of every straight line is a circle passing through the north pole. Indeed, if the straight line is denoted by Λ, then for every $z \in \Lambda$ the projecting line lies in the plane through N and Λ, and this plane intersects S in a circle through N. By similar reasoning the preimage of every (nondegenerate) circle on S through N is a straight line.

We now show, more generally, that the preimage of *every* circle on S (not necessarily passing through N) is a circle or a straight line in the z-plane. To prove this we note that a circle on S can be described as the intersection of S with a plane given by the equation $\alpha_1 x_1 + \alpha_2 x_2 + \alpha_3 x_3 = \alpha_0$, where $\alpha_1^2 + \alpha_2^2 + \alpha_3^2 = 1$ and $0 \leq \alpha_0 < 1$. If a point $Z = (x_1, x_2, x_3)$ on S satisfies this equation, then its preimage by (5.3-1) satisfies

$$\alpha_1(z + \bar{z}) - i\alpha_2(z - \bar{z}) + \alpha_3(z\bar{z} - 1) = \alpha_0(z\bar{z} + 1)$$

THE RIEMANN SPHERE

or, if $z = x + iy$,

$$(\alpha_0 - \alpha_3)(x^2 + y^2) - 2\alpha_1 x - 2\alpha_2 y + \alpha_0 + \alpha_3 = 0. \tag{5.3-3}$$

If $\delta := \alpha_0 - \alpha_3 \neq 0$, this may be written in the form

$$\left(x - \frac{\alpha_1}{\delta}\right)^2 + \left(y - \frac{\alpha_2}{\delta}\right)^2 = \frac{1 - \alpha_0^2}{\delta^2}, \tag{5.3-4}$$

which (since $0 \leq \alpha_0 < 1$) is the equation of a circle. If $\delta = 0$, (5.3-3) reduces to

$$\alpha_1 x + \alpha_2 y = \alpha_3, \tag{5.3-5}$$

the equation of a straight line. Conversely, it is easily verified that given any equation of a circle or of a straight line, there are four uniquely determined constants $\alpha_0, \alpha_1, \alpha_2, \alpha_3$ that satisfy $\alpha_1^2 + \alpha_2^2 + \alpha_3^2 = 1$, $0 \leq \alpha_0 < 1$ such that the equation of the circle appears in the form of (5.3-4) and that of the straight line in the form of (5.3-5), where $\alpha_0 = \alpha_3$. Thus we have obtained the following theorem:

THEOREM 5.3a

The stereographic projection maps the totality of straight lines and circles in the extended complex plane onto the totality of circles on the Riemann sphere. The preimage of a circle is a straight line if and only if the circle passes through N.

GENERALIZED CIRCLES

We have just seen that on the Riemann sphere there is no essential difference between circles and straight lines, both of which are represented by circles. Thus we denote by **generalized circle** in the extended plane any curve that is either an ordinary circle or a straight line (the latter including the point ∞ by definition).

Returning to Moebius transformations, we now ask: What is the image of a generalized circle under a Moebius transformation? The answer is easy if we remember that by Theorem 5.2d any Moebius transformation can be represented as a composition of at most one translation, one rotation, one inversion, and one more translation. It is clear that the image of any generalized circle under a translation or a rotation is again a generalized circle. (Indeed, the image of an ordinary circle is an ordinary circle, and the image of a straight line is a straight line.) What is the image of a generalized circle under the inversion $z \to z' := 1/z$?

For an answer, we turn to the Riemann sphere. Let $Z := (x_1, x_2, x_3)$ correspond to $z = x + iy$ and $Z' := (x'_1, x'_2, x'_3)$ to

$$z' = \frac{1}{z} = \frac{x - iy}{x^2 + y^2}.$$

The formulas in (5.3-1) then show that

$$x'_1 = x_1, \quad x'_2 = -x_2, \quad x'_3 = -x_3.$$

Thus on the Riemann sphere the image of $1/z$ is obtained from the image of z by rotating the sphere $180°$ about the x_1-axis. This geometric interpretation also holds if $z = 0$ or $z = \infty$.

Now let Γ be any generalized circle, let Γ' be its image (in the extended complex plane) under the inversion $z \to 1/z$, and denote by Γ_S and Γ'_S the images of Γ and Γ' on the Riemann sphere. We already know from Theorem 5.3a that Γ_S is a circle, and from the above it follows that Γ'_S, the image of Γ_S under a rigid motion of the sphere, is again a circle. Again using Theorem 5.3a we conclude that Γ' is a generalized circle.

Thus we have shown that the image of a generalized circle under any translation, rotation, or inversion is a generalized circle. The same is true for arbitrary compositions of such transformations. Since any Moebius transformation can be obtained in this manner, we have proved the following theorem:

THEOREM 5.3b

A Moebius transformation carries generalized circles into generalized circles.

The reader is warned, however, that the center of an ordinary circle is in general not mapped onto the center of the image circle; see Problem 5, §5.4.

THE CHORDAL METRIC.

A set Σ is called a **metric space** if for each pair of elements p, q in Σ there is defined a real-valued function $\delta(p, q)$, called the **distance** from p to q, which satisfies the following two postulates:

(a) Positive definiteness: $\delta(p, q) \geq 0$; $\delta(p, q) = 0$ if and only if $p = q$.
(b) Triangle inequality: $\delta(p, q) \leq \delta(p, r) + \delta(r, q)$ for any three points p, q, r.

THE RIEMANN SPHERE 311

For instance, the complex plane becomes a metric space when the distance of two points z, z' is defined by

$$\delta(z,z') := |z - z'|.$$

This is the definition of distance that underlies the concepts of limits and convergence in the complex plane. It cannot be used in the extended complex plane because the distance of any finite point from the point ∞ is undefined.

The extended complex plane can, however, be made into a metric space by means of the Riemann sphere. Let z, z' be two points of the extended plane, and let $Z := (x_1, x_2, x_3)$ and $Z' := (x'_1, x'_2, x'_3)$ be their images on the Riemann sphere. The euclidean distance of these spherical images,

$$\delta(z,z') := \left[(x_1 - x'_1)^2 + (x_2 - x'_2)^2 + (x_3 - x'_3)^2 \right]^{1/2}, \qquad (5.3\text{-}6)$$

is the **chordal distance** of the points z and z'. It is clear that δ satisfies the two postulates of a distance function. The metric defined by it is the **chordal metric**.

It remains to express the chordal distance directly in terms of z and z'. Because both (x_1, x_2, x_3) and (x'_1, x'_2, x'_3) lie on the unit sphere,

$$\delta(z,z')^2 = 2 - 2(x_1 x'_1 + x_2 x'_2 + x_3 x'_3).$$

If both z and z' are finite, we find from (5.3-1)

$$x_1 x'_1 + x_2 x'_2 + x_3 x'_3 = \frac{(z+\bar{z})(z'+\bar{z}') - (z-\bar{z})(z'-\bar{z}') + (z\bar{z}+1)(\bar{z}'z'+1)}{(1+z\bar{z})(1+z'\bar{z}')}$$

$$= 1 - \frac{2|z - z'|^2}{(1+|z|^2)(1+|z'|^2)};$$

hence

$$\delta(z,z') = \frac{2|z - z'|}{\left[(1+|z|^2)(1+|z'|^2)\right]^{1/2}}. \qquad (5.3\text{-}7)$$

If $z' = \infty$, we find analogously

$$\delta(z, \infty) = \frac{2}{(1+|z|^2)^{1/2}}. \qquad (5.3\text{-}8)$$

The chordal distance is used to advantage in those parts of complex variable theory in which functions that take the value of ∞, such as meromorphic functions, are treated.

RIGID MOTIONS OF THE RIEMANN SPHERE

Let

$$M := \begin{bmatrix} a & b \\ c & d \end{bmatrix}$$

be a matrix in the family $\mathfrak{M}$ normalized such that

$$ad - bc = 1 \tag{5.3-9}$$

and let t be the Moebius transformation associated with M. By stereographic projection t induces a one-to-one mapping of the Riemann sphere onto itself. We ask under what conditions can this mapping be regarded as a rigid motion of the Riemann sphere.

[We are already familiar with *some* Moebius transformations which induce rigid motions of the Riemann sphere. The inversion $z \to z^{-1}$ induces a rotation of the sphere by 180° about the x_1-axis. The rotation $z \to e^{i\phi}z$ induces a rotation about the x_3-axis by the angle ϕ.]

It is easy to find a necessary condition for a Moebius transformation t to induce a rigid motion. If two points on the sphere are antipodal (i.e., if they lie on the same straight line through O), then, since distances are preserved, their images must again be antipodal. Now, if z is the preimage of a point (x_1, x_2, x_3) on S, the preimage of its antipodal point $(-x_1, -x_2, -x_3)$ is $-1/\bar{z}$, as is easily verified from (5.3-1). Thus the transformation t will carry antipodal points into antipodal points if and only if

$$t\left(-\frac{1}{\bar{z}}\right) = -\frac{1}{\overline{t(z)}}$$

or

$$\frac{a - b\bar{z}}{c - d\bar{z}} = \frac{\bar{c}z + \bar{d}}{-\bar{a}z - \bar{b}}$$

for all z of the extended plane. This means that the matrices

$$\begin{bmatrix} -b & a \\ -d & c \end{bmatrix} \quad \text{and} \quad \begin{bmatrix} \bar{c} & \bar{d} \\ -\bar{a} & -\bar{b} \end{bmatrix}$$

THE RIEMANN SPHERE

are proportional, which is true if and only if

$$c = -\lambda \bar{b} \quad \text{and} \quad d = \lambda \bar{a},$$

where $|\lambda| = 1$. The normalizing condition (5.3-9) implies that $\lambda = 1$. Thus the matrix M must have the form

$$M = \begin{bmatrix} a & b \\ -\bar{b} & \bar{a} \end{bmatrix},$$

where $|a|^2 + |b|^2 = 1$. Denoting by M^H the transposed conjugate of M, we easily find the $MM^H = I$, which implies that $M^{-1} = M^H$. A matrix with this property is termed **unitary**.

Having shown that the matrix M [normalized by (5.3-9)] is unitary as a necessary condition for the associated Moebius transformation to induce a rigid motion of the Riemann sphere, we now show that this condition is also sufficient. Let z, z' be any two points in the extended complex plane and let $w := t(z)$, $w' := t(z')$, where

$$t: z \to \frac{az + b}{-\bar{b}z + \bar{a}},$$

$a\bar{a} + b\bar{b} = 1$. If none of z, z', w, w' is infinite, computation yields

$$w - w' = \frac{z - z'}{(\bar{a} - \bar{b}z)(\bar{a} - \bar{b}z')},$$

$$1 + w\bar{w} = \frac{1 + z\bar{z}}{(\bar{a} - \bar{b}z)(a - b\bar{z})},$$

$$1 + w'\bar{w}' = \frac{1 + z'\bar{z}'}{(\bar{a} - \bar{b}z')(a - b\bar{z}')},$$

from which it easily follows by (5.3-7) that

$$\delta(w, w') = \delta(z, z').$$

In the exceptional cases the result is similar. Thus t preserves chordal distances and we have proved the following theorem:

THEOREM 5.3c

The Moebius transformation t_M induces a rigid motion of the Riemann sphere if and only if M is a scalar multiple of a unitary matrix.

New proof of Theorem 5.3b. According to a well-known theorem of linear algebra (e.g., Mirsky [1955], p. 307), every matrix over the complex field is unitarily similar to an upper triangular matrix. Thus, if $M \in \mathfrak{M}$, there is a unitary matrix U of order 2 and a matrix

$$T := \begin{bmatrix} \lambda_1 & \mu \\ 0 & \lambda_2 \end{bmatrix},$$

where $\lambda_1 \lambda_2 \neq 0$, such that

$$M = UTU^H.$$

By Theorem 5.2c

$$t_M = t_U \circ t_T \circ t_{U^H}.$$

By Theorem 5.3c, the transformations t_U and t_{U^H} induce rigid motions of the Riemann sphere and thus, by Theorem 5.3a, carry generalized circles into generalized circles. The same is obviously true of

$$t_T : z \to \frac{\lambda_1}{\lambda_2} z + \frac{\mu}{\lambda_2},$$

and therefore of t_M. ∎

PROBLEMS

1. Show that the stereographic projection preserves angles, in the following sense: if $\Gamma_k : z = z_k(\tau)$, $\alpha_k \leq \tau \leq \beta_k$ ($k = 1, 2$) are two regular arcs in the complex plane which at a point z_0 intersect under the angle δ, their images Γ_k^* on the Riemann sphere intersect at the image of z_0 likewise under the angle δ.

2. Show that Moebius transformations preserve angles. [Use problem 1 and either of the two methods used to prove Theorem 5.3b.]

3. The half-disk $|z| < 1$, $\operatorname{Re} z > 0$, is to be mapped onto the full disk $|w| < 1$ such that the boundary points $z = \pm i, 1$ remain fixed (see Fig. 5.3b).
   Solution.
   $$w = \frac{z^2 + 2z - 1}{-z^2 + 2z + 1}.$$

4. Two circular arcs intersect at $z = \pm 1$ under the angle 2δ and are symmetric with respect to the real axis. Map the exterior D of the two arcs onto the

THE RIEMANN SPHERE

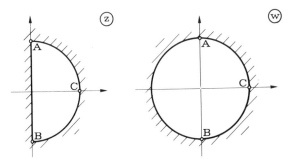

Fig. 5.3b. Mapping the right half of a disk onto a full disk.

w-plane cut between $w = \pm 1$ (generalization of the Joukowski map, see Fig. 5.3c).
Solution. Let t be a linear transformation that sends ± 1 to 0 and ∞, respectively. This maps D onto a wedge-shaped domain. Straighten the angle by means of a suitable fractional power p. The desired map is $t^{[-1]} \circ p \circ t$. With

$$t: z \to \frac{z-1}{z+1}$$

we get

$$w = \frac{1 + [(z-1)/(z+1)]^{\pi/(\pi-\delta)}}{1 - [(z-1)/(z+1)]^{\pi/(\pi-\delta)}}.$$

5. Let $0 \leq \phi < \pi/4$, and let D denote the domain bounded by $|z| = 1$ and by the circular arc intersecting $|z| = 1$ orthogonally at $z = e^{\pm i\phi}$ and lying inside $|z| = 1$ (see Fig. 5.3d). Show that the function mapping D onto the full disk $|w| < 1$ such that the real axis is mapped onto the real axis and $0 \to 0$ is given by

$$f: z \to \frac{z - \cos\phi \cdot z^2}{\cos\phi - z}.$$

Also compute the inverse mapping and expand $g: w \to f^{[-1]}(w)$ in powers of w.

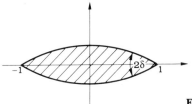

Fig. 5.3c. Mapping the exterior of a lens.

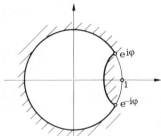

Fig. 5.3d. Mapping region bounded by two orthogonal circles.

§5.4. MOEBIUS TRANSFORMATIONS: SYMMETRY

It is obvious that every *entire* Moebius transformation

$$t: z \to az + b \quad (a \neq 0)$$

preserves symmetry; i.e., if Λ is a straight line in the complex plane and two points z_1 and z_2 lie symmetrically with respect to Λ, then their images $z_i' := t(z_i)$ ($i = 1, 2$) are symmetric with respect to the image Λ' of Λ under the transformation t (see Fig. 5.4a).

Nonentire Moebius transformations do not, in general, carry straight lines into straight lines and thus cannot preserve symmetry in the above sense. However, they always carry straight lines into generalized circles. It will be shown that if symmetry of a pair of points with respect to a circle is defined in an appropriate way all Moebius transformations preserve symmetry. In addition to its intrinsic interest, this is one of the most useful facts for the construction of specific Moebius transformations.

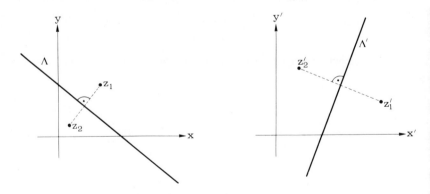

Fig. 5.4a. Symmetry with respect to a straight line.

SYMMETRY

Let c be a complex number, let $\rho > 0$, and denote by Γ the circle with center c and radius ρ whose equation may be written

$$(z-c)(\bar{z}-\bar{c}) = \rho^2. \tag{5.4-1}$$

Two finite points z_1 and z_2 are called **symmetric with respect to** Γ if and only if

$$(z_1-c)(\bar{z}_2-\bar{c}) = \rho^2. \tag{5.4-2}$$

This relation implies that both z_1 and z_2 lie on the same ray emanating from c and that the product of their distances from c equals ρ^2.

To each finite $z_1 \neq c$ there corresponds exactly one point z_2, also finite, which is symmetric to z_1 with respect to Γ. The point z_2 can be found with the elementary construction indicated in Fig. 5.4b. We extend the definition of symmetry to the extended plane by calling points c and ∞ symmetric with respect to Γ.

Every point *on* Γ is symmetric to itself with respect to Γ. Thus on a straight line which in the extended plane by definition contains point ∞ the point symmetric to ∞ with respect to the line is defined as ∞.

With these definitions, if Γ is any *generalized* circle, then to each point z_1 in the extended plane there corresponds exactly one point z_2 in the extended plane that is symmetric to z_1 with respect to Γ.

If Moebius transformations are applied, this definition is not easy to work with in the case of circles because it involves the center of the circle which is not invariant under a Moebius transformation. Also, it has a different analytic form for straight lines and circles. For an effective treatment of symmetry we require a uniform analytic representation of generalized circles.

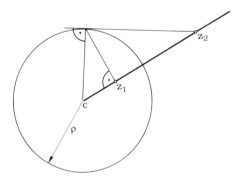

Fig. 5.4b. Symmetry with respect to a circle.

LEMMA 5.4a

If a and b are complex numbers, $a \neq b$, and μ is a positive real number, the equation

$$\left|\frac{z-a}{z-b}\right| = \mu \tag{5.4-3}$$

represents a generalized circle. It represents a straight line if and only if $\mu = 1$.

Proof. Equation (5.4-3) holds if and only if

$$(z-a)(\bar{z}-\bar{a}) = \mu^2 (z-b)(\bar{z}-\bar{b})$$

or if

$$z\bar{z}(1-\mu^2) - (a-\mu^2 b)\bar{z} - (\bar{a}-\mu^2 \bar{b})z + (a\bar{a} - \mu^2 b\bar{b}) = 0. \tag{5.4-4}$$

If $\mu = 1$, this is the equation of a straight line. If $\mu \neq 1$, it is the equation of a circle. Comparison with (5.4-1) shows that the circle has the center

$$c = \frac{a - \mu^2 b}{1 - \mu^2} \tag{5.4-5}$$

and the radius ρ where

$$\rho^2 = \frac{\mu^2 (a-b)(\bar{a}-\bar{b})}{(1-\mu^2)^2}. \quad \blacksquare \tag{5.4-6}$$

THEOREM 5.4b

The points a and b are symmetric with respect to the generalized circle (5.4-3).

Proof. This is obvious if $\mu = 1$. If $\mu \neq 1$, direct computation shows that (5.4-2) is satisfied if c and ρ are given by (5.4-5) and (5.4-6). ∎

Theorem 5.4b has the following converse:

THEOREM 5.4c

Let the two finite points a and b be symmetric with respect to the generalized circle Γ, $a \neq b$. Then there exists a positive number μ such that the equation of Γ is given by (5.4-3).

SYMMETRY

Proof. If Γ is a straight line, its equation is (5.4-3) where $\mu = 1$. If Γ is a circle, let c be its center and ρ, its radius. Since a, b, c are three distinct points lying on a straight line and c does not lie between a and b, there exists $\mu > 0, \mu \neq 1$ such that (5.4-5) holds. With this choice of μ, we find that

$$\rho^2 = (a-c)(\bar{a}-\bar{c}) = \frac{\mu^2}{(1-\mu^2)^2}(a-b)(\bar{a}-\bar{b}).$$

Hence the equation of the circle can be written in the form (5.4-4), which is equivalent to (5.4-3). ∎

We now state the main result concerning symmetry.

THEOREM 5.4d

Let a and b be two points of the extended complex plane which are symmetric with respect to the generalized circle Γ and let t be a Moebius transformation. Then the points $a' := t(a)$ and $b' := t(b)$ are symmetric with respect to Γ', the image of Γ under t.

Proof. By Theorem 5.2d every Moebius transformation is the product (under composition) of translations, rotations, and inversions. Because the assertion is trivial for rotations and translations, it remains to consider the case in which t is the inversion $z \to z^{-1}$.

If a and b coincide, their common value lies on Γ and there is nothing to prove. If a and b are finite and distinct, the equation of Γ may, according to Theorem 5.4c, be written

$$\left|\frac{z-a}{z-b}\right| = \mu,$$

where $\mu > 0$ is a suitable constant. Consequently, the equation of Γ' is

$$\left|\frac{1-aw}{1-bw}\right| = \mu. \tag{5.4-7}$$

If both a and b are different from zero, (5.4-7) may be written

$$\left|\frac{w-a'}{w-b'}\right| = \mu \left|\frac{a'}{b'}\right|,$$

which proves the assertion by Theorem 5.4b. If, say, b is zero, then (5.4-7) is the equation of a circle with center a' which, since $b' = \infty$, again proves the assertion by the definition of symmetry. The case in which $a = 0$ is treated similarly, as are the cases in which exactly one of the points a and b is ∞. ∎

The symmetry principle enunciated in Theorem 5.4d is often useful for the purpose of solving special mapping problems. We give several examples.

EXAMPLE 1 THE MAPPING THEOREM FOR DISKS

Let D be a disk, and let a be a point in D. *Problem.* Find all Moebius transformations t which map D onto the unit disk such that $t(a)=0$.
Solution.   Let Γ be the boundary of D. If a is the center of Γ, the problem can be solved trivially by *entire* Moebius transformations. If a is not the center, let b be the point symmetric to a with respect to Γ. The transformation

$$z \to \frac{z-a}{z-b}$$

maps a into 0, b into ∞, and Γ into a (generalized) circle Γ' with the property that 0 and ∞ lie symmetric with respect to it. Thus Γ' is an ordinary circle about O. We now want a Moebius transformation that maps Γ' onto the unit circle and leaves O fixed. Then ∞ also remains fixed. The only Moebius transformations that leave 0 and ∞ fixed are $w \to cw$. Thus the required transformations are of the form

$$t: z \to c \frac{z-a}{z-b}.$$

The constant $c \neq 0$ is determined by the condition that the image of Γ is the unit circle. It suffices that the image of one point of Γ has modulus 1. Let z_1 be an arbitrary point of Γ; then $|t(z_1)|=1$ is equivalent to

$$|c| = \left| \frac{z_1 - b}{z_1 - a} \right|.$$

Thus D is mapped onto the unit disk, $t(a)=0$, if and only if

$$t: z \to e^{i\alpha} \frac{z_1 - b}{z_1 - a} \frac{z-a}{z-b}, \qquad (5.4\text{-}8)$$

where z_1 is an arbitrary point on the boundary of D and α is an arbitrary real number.

As an application of this result we shall determine the most general Moebius transformation t that maps the unit disk onto itself. Let a be the point mapped onto 0. If $0<|a|<1$, then the point symmetric to a with respect to $|z|=1$ is $(\bar{a})^{-1}$. With $z_1 = 1$ (5.4-8) yields

$$t(z) = e^{i\alpha} \frac{1 - \bar{a}^{-1}}{1-a} \frac{z-a}{z-\bar{a}^{-1}},$$

which may be written

$$t(z) = e^{i\alpha} \frac{1-\bar{a}}{1-a} \frac{z-a}{1-\bar{a}z}.$$

SYMMETRY

Evidently the factor $(1-\bar{a})(1-a)^{-1}$ has modulus one and thus may be absorbed into $e^{i\alpha}$. The resulting formula for t clearly also holds for $a=0$. Thus we have proved:

THEOREM 5.4e

The Moebius transformation t maps the unit disk onto itself if and only if it has the form

$$t: z \to e^{i\alpha} \frac{z-a}{1-\bar{a}z}, \qquad (5.4\text{-}9)$$

where $|a|<1$ and α is real.

We conclude by a somewhat more special application of the symmetry principle.

EXAMPLE 2

Let c and ρ be real, $c > \rho > 0$. *Problem*: Map the domain bounded by the circle $|z-c|=\rho$ and the imaginary axis onto an annular domain bounded by $|w|=1$ and a concentric circle (see Fig. 5.4c).

Solution. We seek a pair of points a,b on the real axis which lies symmetric to $\operatorname{Re} z = 0$ and the circle $|z-c|=\rho$. For symmetry with respect to $\operatorname{Re} z = 0$ it is necessary and sufficient that $b = -a$. We may assume that $a > 0$. The condition for symmetry with respect to the circle then becomes

$$(c+a)(c-a) = \rho^2$$

or

$$a = (c^2 - \rho^2)^{1/2}. \qquad (5.4\text{-}10)$$

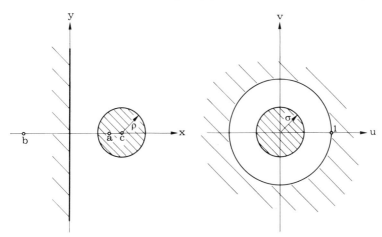

Fig. 5.4c. Mapping considered in Example 2.

With this choice of a the transformation

$$t:z\to \frac{z-a}{z+a}$$

maps the imaginary axis and the circle $|z-c|=\rho$ onto two generalized circles such that 0 and ∞ lie symmetric with respect to both, i.e., onto concentric circles about 0. If we wish $\text{Re}\,z=0$ to go over into $|w|=1$, the image, e.g., of $z=0$ must have absolute value 1. This condition is satisfied by t. The radius of the inner image circle is given by

$$\sigma:=t(c+\rho)=\frac{c+\rho-(c^2-\rho^2)^{1/2}}{c+\rho+(c^2-\rho^2)^{1/2}}.$$

Simplification yields

$$\sigma=\frac{\rho}{c+(c^2-\rho^2)^{1/2}}. \qquad (5.4\text{-}11)$$

As in Example 1 it can be argued that the most general Moebius transformation that solves the above problem is

$$t:z\to e^{i\alpha}\frac{z-a}{z+a}, \qquad (5.4\text{-}12)$$

where a is given by (5.4-10) and α is an arbitrary real number. It is seen that the value of σ is uniquely determined by the mapping problem.

PROBLEMS

1. Let ρ and c be real, $0<c<1-\rho$. Map the domain bounded by the circles $|z|=1$ and $|z-c|=\rho$ onto a domain bounded by two concentric circles such that the outer circle is mapped onto itself. What is the radius of the image of the inner circle?
   [Find a pair of points a,b that is symmetric with respect to both given circles. Their images must be $0,\infty$.]
2. Let H be a half-plane bounded by a straight line Λ and let $a\in H$. Show by means of the symmetry principle that the most general Moebius transformation mapping H onto $|w|<1$ and sending a into the origin is given by

   $$z\to e^{i\alpha}\frac{z-a}{z-b},$$

   where b is symmetric to a with respect to Λ and α is any real number.
3. Prove the following (a) directly by the symmetry principle, (b) by Theorem 5.4e: A Moebius transformation that maps the upper half-plane $\text{Im}\,z>0$ onto itself can be written in the form

   $$t:z\to\frac{az+b}{cz+d},$$

HOLOMORPHIC FUNCTIONS 323

where a, b, c, d are real and $ad - bc > 0$.

4. Find a Moebius transformation that sends the circle $|z+1|=1$ into $|w|=2, 0$ into -2, and i into 0.
   [Which point is mapped onto ∞?]
5. Prove that the Moebius transformation t maps the center of a circle Γ onto the center of $t(\Gamma)$ if and only if ∞ is a fixed point of t.

§5.5. HOLOMORPHIC FUNCTIONS AND CONFORMAL MAPS

It must have become apparent to the attentive reader that all special mapping functions discussed so far enjoy the property of preserving angles; i.e., if $\Gamma_k : z = z_k(\tau)$, $\alpha \leqslant \tau \leqslant \beta$, $k = 1, 2$, are two curves that intersect at a point z_0 under an angle δ and $f : z \to f(z)$ is a mapping whose domain of definition contains the curves Γ_k, then their images $f(\Gamma_k)$ intersect at $f(z_0)$, in many cases again under the angle δ.

It is shown in this section that mapping functions that preserve angles can be characterized analytically by three simple equivalent properties: they are complex-differentiable, they satisfy the Cauchy-Riemann equations, and they possess a complex derivative. Functions satisfying these properties in their domain of definition are called *holomorphic*.

It will emerge as an almost trivial fact that any function that is analytic as defined in Chapter 3 is also holomorphic. It is also true that every holomorphic function is analytic. Thus the classes of holomorphic and analytic functions are identical. For the reasons given in the preface, the proof of this important fact is only sketched.

I. Holomorphic Functions

Let $f : z \to f(z) =: u(z) + iv(z)$ (u, v real) be a complex-valued function in a domain D of the complex plane. To stress the fact that f, u, and v may be regarded as functions of the two real variables x and y, the real and imaginary parts of $z = x + iy$, we also write

$$f(z) =: f(x, y), \qquad u(z) =: u(x, y), \qquad v(z) =: v(x, y).$$

Nothing is assumed about f except that the partial derivatives of the real functions u and v with respect to the real variables x and y, customarily denoted by u_x, u_y, v_x, v_y, exist and are continuous in D.

Under this hypothesis the functions u and v are differentiable (in the real variable sense) at each point $(x_0, y_0) \in D$. By this is meant that if $x = x_0 + dx$, $y = y_0 + dy$ then the linear function of dx and dy defined by

$$du := u_x(x_0, y_0) \, dx + u_y(x_0, y_0) \, dy$$

[called the **differential** of u at (x_0,y_0)] approximates $u(x,y)-u(x_0,y_0)$ in such a fashion that the error ϕ defined by

$$u(x,y) = u(x_0,y_0) + du + \phi(dx,dy) \qquad (5.5\text{-}1)$$

satisfies

$$\phi(dx,dy) = o(dx^2 + dy^2)^{1/2} \qquad (5.5\text{-}2)$$

which is short notation for

$$\lim_{(dx,dy)\to(0,0)} \frac{\phi(dx,dy)}{(dx^2+dy^2)^{1/2}} = 0.$$

A common paraphrase for (5.5-2) is "ϕ tends to zero faster than linearly." Similar statements hold for the differential of v at (x_0,y_0),

$$dv := v_x(x_0,y_0)\,dx + v_y(x_0,y_0)\,dy.$$

On the basis of (5.5-1) and the analogous result for v the function f possesses the differential

$$df := du + i\,dv$$
$$= [u_x(x_0,y_0) + iv_x(x_0,y_0)]\,dx + [u_y(x_0,y_0) + iv_y(x_0,y_0)]\,dy$$

in the sense that

$$f(x,y) = f(x_0,y_0) + df + \psi(dx,dy),$$

where

$$\psi(dx,dy) = o(dx^2 + dy^2)^{1/2}.$$

The combinations of derivatives $u_x + iv_x$, $u_y + iv_y$ appearing in df can be expressed as derivatives of f; for instance,

$$u_x + iv_x = \lim_{h\to 0} \left[\frac{u(x_0+h,y_0) - u(x_0,y_0)}{h} + i\frac{v(x_0+h,y_0) - v(x_0,y_0)}{h} \right]$$

$$= \lim_{h\to 0} \frac{f(x_0+h,y_0) - f(x_0,y_0)}{h}.$$

Thus

$$u_x(x_0,y_0) + iv_x(x_0,y_0) = \frac{\partial f}{\partial x}(x_0,y_0) \qquad (5.5\text{-}3a)$$

HOLOMORPHIC FUNCTIONS

and analogously

$$u_y(x_0,y_0) + iv_y(x_0,y_0) = \frac{\partial f}{\partial y}(x_0,y_0). \qquad (5.5\text{-}3b)$$

The differential of f may thus be written

$$df = \frac{\partial f}{\partial x}dx + \frac{\partial f}{\partial y}dy, \qquad (5.5\text{-}4)$$

as if f were a real function.

If we think of f as a function of the single *complex* variable $z := x + iy$, it seems natural to express the increments dx and dy as increments of z. We therefore set

$$dx + i\,dy =: dz$$

and consequently

$$dx - i\,dy =: \overline{dz}.$$

This implies

$$dx = \tfrac{1}{2}(dz + \overline{dz}), \qquad dy = \frac{1}{2i}(dz - \overline{dz}).$$

The differential of f then becomes

$$df = \tfrac{1}{2}\left(\frac{\partial f}{\partial x} - i\frac{\partial f}{\partial y}\right)dz + \tfrac{1}{2}\left(\frac{\partial f}{\partial x} + i\frac{\partial f}{\partial y}\right)\overline{dz} \qquad (5.5\text{-}5)$$

[all derivatives taken at (x_0, y_0)]. Thus, in general, the differential of a complex function of a complex variable is a linear function of the two variables dz and $\overline{dz}$.

DEFINITION

The function f is called **complex-differentiable** *at a point $z_0 := x_0 + iy_0$ if its differential at this point is independent of $\overline{dz}$, i.e., if it has the form*

$$df = \frac{1}{2}\left(\frac{\partial f}{\partial x} - i\frac{\partial f}{\partial y}\right)dz. \qquad (5.5\text{-}6)$$

A number of important conclusions can be drawn about functions that are complex-differentiable at a point (x_0, y_0).

1. Let f be complex-differentiable at $z_0 = x_0 + iy_0$. The coefficient of $\overline{dz}$ in (5.5-5) then vanishes:

$$\frac{\partial f}{\partial x} + i \frac{\partial f}{\partial y} = 0. \tag{5.5-7}$$

By (5.5-3) this is equivalent to

$$u_x + iv_x + i(u_y + iv_y) = 0$$

or, separating real and imaginary parts,

$$\begin{aligned} u_x &= v_y \\ u_y &= -v_x \end{aligned} \tag{5.5-8}$$

This pair of partial differential equations is known as the **Cauchy-Riemann equations**. Their validity at a point (x_0, y_0) (in addition to the initial assumption that the partial derivatives are continuous) is equivalent to the fact that f is complex-differentiable at $z_0 = x_0 + iy_0$.

2. Let f be complex-differentiable at z_0. Then (5.5-7) implies

$$\frac{\partial f}{\partial x} = -i \frac{\partial f}{\partial y}.$$

Thus the differential (5.5-6) may also be written in one of the following forms:

$$\begin{aligned} df &= \frac{\partial f}{\partial x}(x_0, y_0)\, dz, \\ df &= -i\frac{\partial f}{\partial y}(x_0, y_0)\, dz. \end{aligned} \tag{5.5-9}$$

3. We now discuss the most important consequence of complex differentiability. Let f be complex-differentiable at z_0. By the above this means that $df = c \cdot dz$, where, for instance,

$$c = \frac{\partial f}{\partial x}(x_0, y_0).$$

By the property of the differential we have for all complex $l := dz$, with $|l|$ sufficiently small,

$$f(z_0 + l) = f(z_0) + cl + \psi(l), \tag{5.5-10}$$

HOLOMORPHIC FUNCTIONS

where $\psi(l) = o(l)$, i.e.,

$$\lim_{l \to 0} \frac{\psi(l)}{l} = 0. \qquad (5.5\text{-}11)$$

Here l approaches 0 in an arbitrary manner. From (5.5-10) we have for $l \neq 0$:

$$\frac{f(z_0 + l) - f(z_0)}{l} = c + \frac{\psi(l)}{l}.$$

Hence by (5.5-10) the result is

$$\lim_{l \to 0} \frac{f(z_0 + l) - f(z_0)}{l} = c. \qquad (5.5\text{-}12)$$

The expression whose limit is taken has the form of a difference quotient. Relation (5.5-12) states that for a complex-differentiable function the limit of the difference quotient exists and has the value c. Here the complex increment $l = z - z_0$ may tend to zero in a completely arbitrary manner and, in particular, from any direction.

DEFINITION

The limit (5.5-12) is called the **complex derivative** *(or briefly the* **derivative***) of f at z_0.* Common notations for the complex derivative are

$$f'(z_0) = \frac{df}{dz}(z_0) = \cdots.$$

We have shown that if f is complex-differentiable at z_0 then the complex derivative of f exists at z_0. We now prove the converse: if the complex derivative exists, then, in particular for $l = h$ (h real),

$$f'(z_0) = \lim_{h \to 0} \frac{f(x_0 + h, y_0) - f(x_0, y_0)}{h} = \frac{\partial f}{\partial x}(z_0).$$

Likewise, if $l = ik$ (k real),

$$f'(z_0) = \lim_{k \to 0} \frac{(x_0, y_0 + k) - f(x_0, y_0)}{k} = \frac{1}{i}\frac{\partial f}{\partial y}(z_0).$$

There follows

$$\frac{\partial f}{\partial x} = -i\frac{\partial f}{\partial y},$$

which implies that f is complex-differentiable at z_0. We have proved:

THEOREM 5.5a

Let z_0 be an inner point of the domain of definition of the complex valued function $z \to f(z)$ whose real and imaginary parts have continuous partial derivatives with respect to x and y. The following three statements are equivalent:
(a) f is complex-differentiable at z_0;
(b) the real and imaginary parts of f satisfy the Cauchy-Riemann equations at z_0;
(c) the complex derivative of f exists at z_0.

DEFINITION

A function $f:z \to f(z)$ that satisfies the equivalent conditions (a), (b), (c) of Theorem 5.5a at every point z_0 of its domain of definition D is called **holomorphic** in D.

It is clear that every constant function is holomorphic and that the function $z \to z$ is holomorphic in the entire complex plane. Using property (c), we easily verify that sums, differences, products, and quotients of holomorphic functions are holomorphic where defined. It follows, for instance, that a rational function is holomorphic at all points at which the denominator is not zero. But this is only a special case of the general fact, stated earlier as Theorem 3.1e, that every *analytic* function possesses a complex derivative at every point of its domain of definition. (The proof followed directly from the power series definition of analyticity and was very simple.)

The question naturally arises whether there are holomorphic functions that are not analytic. It is one of the basic facts of complex variable theory that the answer is negative; see the appendix to this section.

II. Conformal Maps

Let $f:z \to f(z)$ be a holomorphic function with domain of definition D and domain of values W, let z_0 be a point of D, and let $\Gamma:\tau \to z(\tau)$, $\alpha \leq \tau \leq \beta$, be a curve lying in D and passing through z_0. We assume that z' exists and is different from zero. If $z_0 = z(\tau_0)$, then the complex number $z'(\tau_0)$, interpreted as a vector, is a tangent vector of Γ at z_0.

The image of Γ under the map f is a curve

$$\Gamma^*:\tau \to w(\tau):=f(z(\tau)),$$

$\alpha \leq \tau \leq \beta$. From the fact that the differential of f at z_0 has the special form

HOLOMORPHIC FUNCTIONS

(5.5-6) it follows that a tangent vector of Γ^* at the image $w_0 := f(z_0)$ of z_0 is given by

$$w'(\tau_0) = f'(z_0) z'(\tau_0). \tag{5.5-13}$$

Assuming that $f'(z_0) \neq 0$, we may conclude

$$\arg w'(\tau_0) = \arg z'(\tau_0) + \arg f'(z_0). \tag{5.5-14}$$

This relation implies that the angle between the tangent vectors to Γ at z_0 and to Γ^* at w_0 is equal to $\arg f'(z_0)$. Hence it is independent of Γ. It follows that any two curves passing through z_0 and forming an angle γ at z_0 are mapped onto two curves forming the angle γ (measured in the same direction) at w_0. A mapping with the property that angles between curves are preserved in size as well as in sense is **conformal**.

Now let f be real-differentiable, but not complex-differentiable, at the point z_0. The differential of f at z_0 then has the form $df = a\,dz + b\,\overline{dz}$, where $b \neq 0$. It follows that the tangent vector of Γ^* at w_0 is given by $w'(\tau_0) = a z'(\tau_0) + b\,\overline{z'(\tau_0)}$. If $z'(\tau_0) = \rho e^{i\phi}$ ($\rho > 0$), the angle between $w'(\tau_0)$ and $z'(\tau_0)$ is

$$\arg \frac{w'(\tau_0)}{z'(\tau_0)} = \arg(a + b e^{-2i\phi}).$$

Because $b \neq 0$, this angle is not independent of ϕ. Thus the mapping defined by f is not conformal. We have obtained:

THEOREM 5.5b

Let the function $f: z \to f(z)$ be defined in a neighborhood of the point z_0 and let it have a nonzero differential at z_0. The mapping defined by f is conformal at z_0 if and only if f is complex-differentiable at z_0.

APPENDIX: HOLOMORPHIC FUNCTIONS ARE ANALYTIC

We know from Theorem 3.1e that all analytic functions are holomorphic, and it has already been mentioned, conversely, that all holomorphic functions are analytic. Here we wish to present a chain of intuitive arguments that leads to this important fact. The stepping stones are Cauchy's integral theorem and the integral formula following from it.

THEOREM 5.5c (Cauchy's integral theorem for holomorphic functions)

Let D be a simply connected region, let $f: z \to f(z)$ be holomorphic in D, and let Γ be a Jordan curve in D. Then

$$\int_\Gamma f(z)\,dz = 0. \tag{5.5-15}$$

Proof. Letting $f = u + iv, z = x + iy$, we have

$$\int_\Gamma f(z)\,dz = \int_\Gamma (u\,dx - v\,dy) + i\int_\Gamma (v\,dx + u\,dy),$$

where the integrals on the right are to be interpreted as real line integrals. Thus the first integral is the integral of the vector $\mathbf{p}: = (u, -v, 0)$ along Γ. By Stokes's theorem this equals the flux of rot $\mathbf{p}$ through D. Calculating rot $\mathbf{p}$, we find by the Cauchy-Riemann equations that rot $\mathbf{p} = \mathbf{0}$. Thus the flux is zero. The second integral equals the flux of rot $\mathbf{q}$ through D, where $\mathbf{q}: = (v, u, 0)$. Again by Cauchy-Riemann, rot $\mathbf{q} = \mathbf{0}$. Thus both integrals on the right vanish, proving (5.5-15). ∎

Remark. This proof simply shows that Cauchy's theorem is a corollary of Stokes's theorem. It naturally uses the continuous differentiability of the functions u and v assumed at the beginning of this section. In many texts on complex variable theory great stress is placed on the fact, first discovered by Goursat, that (5.5-15) can be proved by assuming merely the existence (but not the continuity) of the complex derivative of f. This strengthened version of Cauchy's theorem, although undoubtedly of great mathematico-logical interest, is of little importance in many applications of complex variable theory.

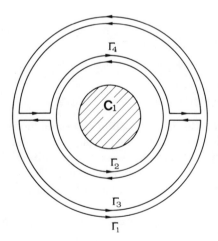

Fig. 5.5. Cauchy's theorem for doubly connected region.

HOLOMORPHIC FUNCTIONS

Let f be holomorphic in a doubly connected domain D and let C_1 denote the bounded component of the complement of D. Let Γ_1 and Γ_2 be two Jordan curves that run around C_1 once in the positive sense, and define Γ_3 and Γ_4 by the cutting process indicated in Fig. 5.5. We then have $\Gamma_3 + \Gamma_4 = \Gamma_1 - \Gamma_2$ and consequently, if $\int f$ is an abbreviation for $\int f(z)\,dz$,

$$\int_{\Gamma_3} f + \int_{\Gamma_4} f = \int_{\Gamma_1} f - \int_{\Gamma_2} f.$$

However, because both Γ_3 and Γ_4 lie in a suitable simply connected domain in which f is holomorphic, the two integrals on the left vanish by Theorem 5.5c and

$$\int_{\Gamma_1} f = \int_{\Gamma_2} f \qquad (5.5\text{-}16)$$

follows. Hence *the integral of f along every curve going around the "hole" C_1 once in the positive sense has the same value.*

We apply this result in the following special context. Let D again be *simply* connected, let Γ be a positively oriented Jordan curve in D, and let a be a point in the interior of Γ. If f is holomorphic in D, the function

$$g: z \to \frac{f(z)}{z-a}$$

is holomorphic in the doubly connected domain $D - \{a\}$. Applying (5.5-16) with $\Gamma_1 := \Gamma$ and $\Gamma_2 : z = a + \rho e^{i\tau}$ ($0 \leq \tau \leq 2\pi$), we have

$$\int_\Gamma \frac{f(z)}{z-a}\,dz = i \int_0^{2\pi} f(a + \rho e^{i\tau})\,d\tau.$$

Letting $\rho \to 0$, the expression on the right tends to $2\pi i f(a)$, and **Cauchy's integral formula** for holomorphic functions follows:

$$f(a) = \frac{1}{2\pi i} \int_\Gamma \frac{f(z)}{z-a}\,dz.$$

Let f be holomorphic in a domain that contains the closed circular disk $D: |z - a| \leq \rho$. If $\Gamma: t = a + \rho e^{i\phi}$, $0 \leq \phi \leq 2\pi$, then by Cauchy's integral formula we have for every interior point z of D

$$f(z) = \frac{1}{2\pi i} \int_\Gamma \frac{f(t)}{t-z}\,dt. \qquad (5.5\text{-}17)$$

Now, as in the proof of Laurent's expansion (§4.4),

$$\frac{1}{t-z} = \frac{1}{t-a-(z-a)} = \frac{1}{t-a} \frac{1}{1-(z-a)/(t-a)} = \sum_{n=0}^{\infty} \frac{(z-a)^n}{(t-a)^{n+1}},$$

and the series converges uniformly for $t \in \Gamma$ and fixed $z \in D$. Hence by substituting in (5.5-17) and integrating term by term,

$$f(z) = \frac{1}{2\pi i} \int_\Gamma \sum_{n=0}^{\infty} \frac{(z-a)^n}{(t-a)^{n+1}} f(t)\, dt = \sum_{n=0}^{\infty} a_n (z-a)^n,$$

where

$$a_n := \frac{1}{2\pi i} \int_\Gamma f(t)(t-a)^{-n-1}\, dt.$$

Thus we have shown that in a neighborhood of the point a f can be expanded in a series of powers of $z-a$. This establishes that f is analytic at a. Since a is an arbitrary point of D, f is analytic in D.

PROBLEMS

1. Let the function $f = u + iv$ be holomorphic in some region D. Show that as a consequence of the Cauchy-Riemann equations the real and imaginary parts of f are *harmonic* in D; i.e., they satisfy

    $$\Delta u = 0, \quad \Delta v = 0,$$

    where $\Delta = \partial^2/\partial x^2 + \partial^2/\partial y^2$.
2. Establish the principle of the maximum (Theorem 2.4h) directly from Cauchy's formula without using power series.
3. Let $f = u + iv$ be holomorphic at z_0. Then the derivative of f at z_0 has the following representations: $f'(z_0) = u_x + iv_x = v_y - iu_y = u_x - iu_y = v_y + iv_x$; all derivatives are taken at z_0.
4. Let f be holomorphic in $|z-a| \leq \rho$ and let

    $$\mu := \max_{|z-a|=\rho} |f(z)|.$$

    If the Taylor coefficients of f at a are

    $$a_n := \frac{1}{n!} f^{(n)}(a), \quad n = 0, 1, 2, \ldots,$$

    derive Cauchy's estimate

    $$|a_n| \leq \mu \rho^{-n}, \quad n = 0, 1, 2, \ldots$$

    (cf. Theorem 2.2f) directly from Cauchy's formula.

CONFORMAL TRANSPLANTS

5. Let f be a (real-) differentiable (but not necessarily holomorphic) function in a region D and define

$$\frac{\partial f}{\partial z} := \frac{1}{2}\left(\frac{\partial f}{\partial x} - i\frac{\partial f}{\partial y}\right), \qquad \frac{\partial f}{\partial \bar{z}} := \frac{1}{2}\left(\frac{\partial f}{\partial x} + i\frac{\partial f}{\partial y}\right).$$

(These "derivatives" cannot be interpreted as limits.) Prove that
(a) the differential of f is

$$df = \frac{\partial f}{\partial z}\,dz + \frac{\partial f}{\partial \bar{z}}\,\overline{dz}\,;$$

(b) if Γ is a positively oriented Jordan curve which together with its interior D_0 lies in D then

$$\frac{1}{2\pi i}\int_\Gamma f(t)\,dt = \frac{1}{\pi}\int\!\!\int_{D_0}\frac{\partial f}{\partial \bar{z}}(z)\,dx\,dy;$$

(c) for every $z \in D_0$

$$f(z) = \frac{1}{2\pi i}\int_\Gamma \frac{f(t)}{t-z}\,dt - \frac{1}{\pi}\int\!\!\int_{D_0}\frac{(\partial f/\partial \bar{t})(t)}{t-z}\,dr\,ds \qquad (t = r + is).$$

[These results are generalizations of Cauchy's integral theorem and integral formula and can be proved by the methods discussed in the appendix.]

§5.6 CONFORMAL TRANSPLANTS

Let D be a region of the z-plane and let $f\colon z \to w = f(z)$ be a holomorphic function defining a one-to-one mapping of D onto a region E of the w-plane (see Fig. 5.6a).

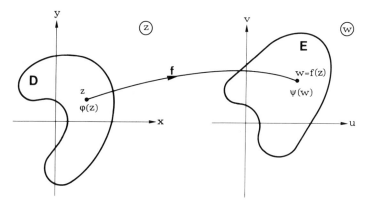

Fig. 5.6a. Conformal transplantation.

Let a real-valued function

$$\phi:(x,y)\to\phi(x,y)=\phi(z)$$

be defined in D. We can then define in E a function ψ as follows: for any $w \in E$ let

$$\psi(w):=\phi(f^{[-1]}(w))=\phi(x(u,v),y(u,v)). \qquad (5.6\text{-}1)$$

Thus the value of ψ at w is equal to the value of ϕ at the preimage of w under the map f. The function ψ is called the **conformal transplant** of ϕ under the mapping f and its process of construction is the **conformal transplantation**.

In physical applications the functions ϕ and ψ frequently signify *potentials* (e.g., electric potentials, flow potentials, or stress potentials), in which case their *gradients* are of interest, i.e., the vectors

$$\left(\frac{\partial\phi}{\partial x},\frac{\partial\phi}{\partial y}\right) \quad \text{and} \quad \left(\frac{\partial\psi}{\partial u},\frac{\partial\psi}{\partial v}\right).$$

How do the gradients behave under conformal transplantation?

By definition

$$\phi(x,y)=\psi(u(x,y),v(x,y)),$$

where $u+iv=w=f(z)=f(x+iy)$ and f is holomorphic. By the chain rule

$$\frac{\partial\phi}{\partial x}=\frac{\partial\psi}{\partial u}\frac{\partial u}{\partial x}+\frac{\partial\psi}{\partial v}\frac{\partial v}{\partial x}, \qquad (5.6\text{-}2a)$$

$$\frac{\partial\phi}{\partial y}=\frac{\partial\psi}{\partial u}\frac{\partial u}{\partial y}+\frac{\partial\psi}{\partial v}\frac{\partial v}{\partial y}. \qquad (5.6\text{-}2b)$$

By the Cauchy-Riemann equations

$$\frac{\partial\phi}{\partial y}=-\frac{\partial\psi}{\partial u}\frac{\partial v}{\partial x}+\frac{\partial\psi}{\partial v}\frac{\partial u}{\partial x}. \qquad (5.6\text{-}2c)$$

The formulas (5.6-2) become more transparent if the following definition is adopted.

DEFINITION

The **complex gradient** *of ϕ is the complex function*

$$\operatorname{grd}\phi:(x,y)\to\frac{\partial\phi}{\partial x}(x,y)+i\frac{\partial\phi}{\partial y}(x,y). \qquad (5.6\text{-}3)$$

CONFORMAL TRANSPLANTS

Analogously,

$$\operatorname{grd}\psi : (u,v) \to \frac{\partial \psi}{\partial u}(u,v) + i\frac{\partial \psi}{\partial v}(u,v).$$

Evidently complex gradients are ordinary gradient vectors interpreted as complex numbers.

By (5.6-2) and (5.6-3)

$$\operatorname{grd}\phi = \left(\frac{\partial \psi}{\partial u}\frac{\partial u}{\partial x} + \frac{\partial \psi}{\partial v}\frac{\partial v}{\partial x}\right) + i\left(-\frac{\partial \psi}{\partial u}\frac{\partial v}{\partial x} + \frac{\partial \psi}{\partial v}\frac{\partial u}{\partial x}\right)$$

$$= \left(\frac{\partial \psi}{\partial u} + i\frac{\partial \psi}{\partial v}\right)\left(\frac{\partial u}{\partial x} - i\frac{\partial v}{\partial x}\right).$$

By §5.5 $f'(z) = \partial u/\partial x + i(\partial v/\partial x)$; hence

$$\frac{\partial u}{\partial x} - i\frac{\partial v}{\partial x} = \overline{f'(z)},$$

and we have obtained:

THEOREM 5.6a

If ψ results from ϕ by conformal transplantation by means of the mapping $w = f(z)$, then

$$\operatorname{grd}\phi(z) = \operatorname{grd}\psi(w) \cdot \overline{f'(z)}. \tag{5.6-4}$$

In many potential problems the quantity

$$\Delta\phi := \frac{\partial^2 \phi}{\partial x^2} + \frac{\partial^2 \phi}{\partial y^2}$$

exists and is known. To emphasize that the derivatives are taken with respect to x and y we also denote it by $\Delta_z\phi$. What connection (if any) is there between $\Delta_z\phi$ and

$$\Delta\psi := \Delta_w\psi := \frac{\partial^2 \psi}{\partial u^2} + \frac{\partial^2 \psi}{\partial v^2}?$$

From (5.6-2a), using the chain rule repeatedly, we get

$$\frac{\partial^2 \phi}{\partial x^2} = \frac{\partial^2 \psi}{\partial u^2}\left(\frac{\partial u}{\partial x}\right)^2 + 2\frac{\partial^2 \psi}{\partial u \partial v}\frac{\partial u}{\partial x}\frac{\partial v}{\partial x} + \frac{\partial^2 \psi}{\partial v^2}\left(\frac{\partial v}{\partial x}\right)^2 + \frac{\partial \psi}{\partial u}\Delta u + \frac{\partial \psi}{\partial v}\Delta v.$$

A similar formula in which the partial derivatives with respect to x are replaced by partial derivatives with respect to y is obtained from (5.6-2b) for $\partial^2\phi/\partial y^2$. There follows

$$\Delta_z\phi = \frac{\partial^2\psi}{\partial u^2}\left[\left(\frac{\partial u}{\partial x}\right)^2 + \left(\frac{\partial u}{\partial y}\right)^2\right] + 2\frac{\partial^2\psi}{\partial u\partial v}\left(\frac{\partial u}{\partial x}\frac{\partial v}{\partial x} + \frac{\partial u}{\partial y}\frac{\partial v}{\partial y}\right)$$

$$+ \frac{\partial^2\psi}{\partial v^2}\left[\left(\frac{\partial v}{\partial x}\right)^2 + \left(\frac{\partial v}{\partial y}\right)^2\right] + \frac{\partial\psi}{\partial u}\Delta_z u + \frac{\partial\psi}{\partial v}\Delta_z v.$$

By the Cauchy-Riemann equations

$$\frac{\partial u}{\partial x}\frac{\partial v}{\partial x} + \frac{\partial u}{\partial y}\frac{\partial v}{\partial y} = 0$$

and

$$\Delta_z u = \Delta_z v = 0.$$

(see Problem 1, §5.5). Furthermore,

$$\left(\frac{\partial u}{\partial x}\right)^2 + \left(\frac{\partial u}{\partial y}\right)^2 = \left(\frac{\partial v}{\partial x}\right)^2 + \left(\frac{\partial v}{\partial y}\right)^2 = \left|\frac{\partial u}{\partial x} + i\frac{\partial v}{\partial x}\right|^2 = |f'(z)|^2.$$

Thus there results:

THEOREM 5.6b

Under the hypotheses of the preceding theorem

$$\Delta_z\phi(z) = \Delta_w\psi(w)|f'(z)|^2. \tag{5.6-5}$$

A particularly important consequence arises if the function ϕ is **harmonic**; i.e., if it satisfies the differential equation

$$\Delta\phi = 0. \tag{5.6-6}$$

The expression on the right of (5.6-5) then vanishes. We now borrow from analytic theory. If the analytic function f defines a one-to-one mapping of D, then $f'(z) \neq 0$ for all $z \in D$ [for, if $f'(z_0) = 0$ for some $z_0 \in D$ then by Theorem 2.4f there would be a neighborhood N of $w_0 := f(z_0)$ such that for

CONFORMAL TRANSPLANTS

every point $w \in N - \{w_0\}$ there would be several $z \in D$ such that $f(z) = w$]. It follows that $\Delta_w \psi = 0$; hence Theorem 5.6b has the following corollary:

COROLLARY 5.6c

The conformal transplant of a harmonic function is harmonic.

A different proof of this result is given in §13.1.

The "transplantation theorems" just discussed form the basis of a method of solving numerous two-dimensional boundary-value problems. A boundary-value problem is given for a domain D of the z-plane, the **physical plane**. By a conformal map the problem is transplanted into a boundary problem for a domain E of the w-plane, the **model plane**, where it can be solved easily. By transplanting back we obtain the solution of the original problem in the physical plane. Several applications of the method are presented in subsequent sections. As a first illustration we consider the **Dirichlet problem** for a simply connected region.

Let D be a simply connected region in the z-plane and let the boundary of D be a piecewise regular Jordan curve Γ. On Γ let there be given a continuous (or piecewise continuous) real function $\phi_0 : z \to \phi_0(z)$ ($z \in \Gamma$). The problem is to find a function ϕ continuous in $D \cup \Gamma$ and twice continuously differentiable in D such that

$$\Delta \phi = 0 \text{ in } D, \qquad (5.6\text{-}7a)$$

$$\phi = \phi_0 \text{ on } \Gamma. \qquad (5.6\text{-}7b)$$

This is known as the **Dirichlet problem for the region D with boundary data ϕ_0**. A physical illustration is the stationary temperature distribution in a cylinder with cross section D. If the cylinder is homogeneous and the function ϕ_0 indicates the values of the temperature (assumed to be independent of the coordinate perpendicular to D) at the surface of the cylinder, the temperature in the interior satisfies (5.6-7) (see Fig. 5.6b).

It is shown in Chapter 13 that under the conditions stated above the function ϕ exists and is uniquely determined by the conditions (5.6-7). Here we show how this function can be constructed. We assume that there is at our disposal a conformal map $f: z \to w = f(z)$ of D onto the unit disk $|w| < 1$ and that f is such that it can be extended to a continuous map of the closure of D onto the closed disk $|w| \leq 1$. By Corollary 5.6c the transplanted solution then satisfies

$$\Delta \psi = 0, \qquad |w| < 1, \qquad (5.6\text{-}8a)$$

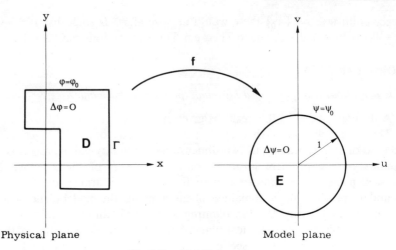

Fig. 5.6b. Dirichlet problem.

and its values of the boundary are

$$\psi(w) = \psi_0(w) = \phi_0\big(f^{[-1]}(w)\big), \qquad |w| = 1. \qquad (5.6\text{-}8b)$$

By expanding ψ_0 in a Fourier series the function ψ can readily be represented by an infinite series, as shown in texts on advanced calculus. Thus ψ may be regarded as known. We now obtain ϕ by transplanting back:

$$\phi(z) = \psi(f(z)). \qquad (5.6\text{-}9)$$

The success of this method evidently depends on the possibility of mapping D conformally onto the unit disk. We shall see in §5.10 that for simply connected D this places no restriction on the scope of the method.

Sometimes in applications the integral of ϕ over D is of interest. This again can be evaluated in the model plane. We use the theory of transformations of variables in real integrals:

$$\int_D \phi(z)\, dx\, dy = \int_E \psi(w)\, \frac{\partial(x,y)}{\partial(u,v)}\, du\, dv,$$

where

$$\frac{\partial(x,y)}{\partial(u,v)} := \begin{vmatrix} \dfrac{\partial x}{\partial u} & \dfrac{\partial x}{\partial v} \\ \dfrac{\partial y}{\partial u} & \dfrac{\partial y}{\partial v} \end{vmatrix}$$

CONFORMAL TRANSPLANTS

is the Jacobian of the mapping $w \to f^{[-1]}(w)$. By the Cauchy-Riemann equations for $f^{[-1]}$, the Jacobian equals

$$\left(\frac{\partial x}{\partial u}\right)^2 + \left(\frac{\partial y}{\partial u}\right)^2 = |f^{[-1]'}(w)|^2,$$

and we have proved:

THEOREM 5.6d

Under the hypotheses of Theorem 5.6a,

$$\int_D \phi(z)\,dx\,dy = \int_E \psi(w)|f^{[-1]'}(w)|^2\,du\,dv. \qquad (5.6\text{-}10)$$

For the sake of illustration let $\phi = 1$. Then the integral on the left becomes

$$A := \int_D dx\,dy,$$

the **area** of D. We now assume that E is the unit disk and that $g := f^{[-1]}$ is given by a power series,

$$g(w) := b_0 + b_1 w + b_2 w^2 + \cdots \qquad (5.6\text{-}11)$$

such that the series for g' converges absolutely for $|w| \leq 1$. Evaluating the integral over $|w| \leq 1$ in polar coordinates we have

$$A = \int_0^1 d\rho \int_0^{2\pi} |g'(\rho e^{i\theta})|^2 \rho\,d\theta.$$

In the product

$$|g'(\rho e^{i\theta})|^2 = g'(\rho e^{i\theta})\,\overline{g'(\rho e^{i\theta})}$$

$$= (b_1 + 2b_2 \rho e^{i\theta} + \cdots)(\overline{b_1} + 2\overline{b_2}\rho e^{-i\theta} + \cdots)$$

the integrals of the cross terms

$$kb_k \rho^{k-1} e^{i(k-1)\theta} m \overline{b_m} \rho^{m-1} e^{-i(m-1)\theta}$$

with respect to θ are zero unless $k = m$. Thus we get

$$A = 2\pi \int_0^1 \sum_{k=1}^\infty k^2 b_k \overline{b_k}\, \rho^{2k-1}\,d\rho,$$

or

$$A = \pi \sum_{k=1}^{\infty} k |b_k|^2. \qquad (5.6\text{-}12)$$

We remark without proof that this result actually holds under the sole hypothesis that the series on the right converges. If the series diverges, then A is infinite.

PROBLEMS

1. If ϕ is a flow potential, the curves which at every point have the direction of the vector $\operatorname{grad}\phi$ are called *streamlines* of the flow defined by ϕ. Show that if f is a conformal map the streamlines are mapped onto streamlines of the flow defined by ψ, the transplant of ϕ. [For the validity of this result it is not required that ϕ or ψ satisfy Laplace's equation.]

2. Let D be a bounded region and let ϕ be continuously differentiable on the closure of D. The functional

$$I(\phi) := \iint_D |\operatorname{grad}\phi|^2 \, dx \, dy$$

is the **Dirichlet integral** of ϕ. Show that the Dirichlet integral is invariant under conformal transplantation, i.e., that with the notations introduced above

$$I(\phi) = I(\psi) := \iint_E |\operatorname{grad}\psi|^2 \, du \, dv.$$

3. Let $\mathbf{q} := (q_1, q_2)$ be a two-dimensional vector field, and let $\Gamma: z = z(\tau)$, $\alpha \leq \tau \leq \beta$, be a differentiable arc lying in the domain of definition of $\mathbf{q}$. Let $\mathbf{n}$ be the unit normal vector of Γ in the direction of the tangent vector rotated by $\pi/2$. The integral

$$\Phi := \int_\Gamma \mathbf{q}\mathbf{n}\, ds = \int_\Gamma q_2 \, dx - q_1 \, dy$$

is the **flux** of $\mathbf{q}$ through Γ in the direction $\mathbf{n}$.

(a) If $\mathbf{q} = \operatorname{grad}\phi$, show that

$$\Phi = -\operatorname{Im}\int_\Gamma \overline{\operatorname{grd}\phi}\, dz.$$

(b) If $\mathbf{q} = \text{grad}\,\phi$, show that the flux is invariant under conformal transplantation. In fact, with the above notation and $\Gamma^* = f(\Gamma)$,

$$\int_\Gamma \overline{\text{grd}\,\phi}\, dz = \int_{\Gamma^*} \overline{\text{grd}\,\psi}\, dw.$$

4. Let D be a simply connected region. In mechanics the quantity

$$J_0 := \iint_D (x^2 + y^2)\, dx\, dy,$$

called the **polar moment of inertia** of D with respect to O, is of interest. If the function g maps $|w| < 1$ conformally onto D and

$$g^2(w) =: \sum_{k=0}^\infty B_k w^k \qquad (|w| \leq 1), \tag{5.6-13}$$

show that

$$J_0 = \frac{\pi}{4} \sum_{k=0}^\infty k |B_k|^2.$$

5. Compute the area and the polar moment of inertia for the image of $|w| \leq 1$ under the map

$$g: w \to Rw + \epsilon w^{n+1},$$

where $R > 0$, n is a positive integer, and $0 \leq (n+1)\epsilon \leq R$.

6. The **center of mass** of a plane domain D, uniformly covered with mass, is given by

$$z_S := \frac{1}{A} \iint_D z\, dx\, dy,$$

where A denotes the area of D. With the assumptions and notations in Problem 4, show that

$$z_S := \frac{\pi}{A} \sum_{k=1}^\infty k B_k \overline{b_k}.$$

7. Let the center of mass of the plane domain D be at the origin and let J_0 be the polar moment of inertia with respect to O. Show that the polar moment of inertia with respect to point $b \neq 0$,

$$J := \iint_D |z - b|^2\, dx\, dy,$$

satisfies

$$J = J_0 + |b|^2 A,$$

where A is the area of D.

8. In the situation sketched in Fig. 5.6b let the transplanted boundary values be represented by the Fourier series

$$\psi_0(e^{i\theta}) = \sum_{n=-\infty}^{\infty} a_n e^{in\theta} \qquad (a_{-n} = \overline{a_n}),$$

where $\sum |a_n| < \infty$. Show that the solution ϕ of the Dirichlet problem in the physical plane is given by

$$\phi(z) = a_0 + 2\operatorname{Re} \sum_{n=1}^{\infty} a_n [f(z)]^n.$$

9. Let ϕ in D satisfy the **Helmholtz equation**

$$\Delta_z \phi + k^2 \phi = 0$$

and let $g: w \to z = g(w)$ map E conformally onto D. Show that the transplanted function ψ satisfies

$$\Delta_w \psi + k^2 |g'(w)|^2 \psi = 0. \tag{5.6-14}$$

Show that this equation can be solved by separating variables in the following three cases:

(a) $g(w) = w^2$, (b) $g(w) = e^w$, (c) $g(w) = \cos w$.

§5.7. PROBLEMS OF PLANE ELECTROSTATICS

A basic problem in *spatial* electrostatics is as follows: a domain D is given whose complement consists of a (usually) finite number of components $C_1, C_2, \ldots, C_r$, called *conductors*. Each conductor C_i is kept at a fixed potential ϕ_i. We are trying to determine the potential ϕ that results in D.

If vacuum prevails in D or the medium filling out D is homogeneous, the mathematical problem is to find a function ϕ, twice continuously differentiable in D and continuous in the closure of D, such that

$$\Delta \phi := \frac{\partial^2 \phi}{\partial x^2} + \frac{\partial^2 \phi}{\partial y^2} + \frac{\partial^2 \phi}{\partial z^2} = 0 \quad \text{in } D, \tag{5.7-1a}$$

$$\phi = \phi_i \quad \text{on the boundary } \Sigma_i \text{ of } C_i. \tag{5.7-1b}$$

Here we are concerned with the two-dimensional problem that results if the conductors C_i are bounded by cylindrical surfaces whose generators are perpendicular to the (x,y)-plane. If Γ_i denotes the indicatrix of these

PROBLEMS OF PLANE ELECTROSTATICS

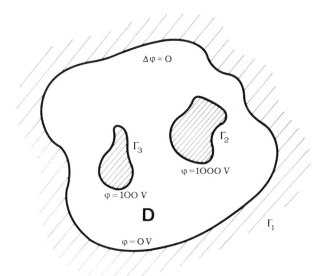

Fig. 5.7a. The basic problem of electrostatics.

cylinders, we now need a function $\phi = \phi(x,y)$ such that (see Fig. 5.7a)

$$\Delta\phi := \frac{\partial^2 \phi}{\partial x^2} + \frac{\partial^2 \phi}{\partial y^2} = 0 \quad \text{in } D, \tag{5.7-2a}$$

$$\phi = \phi_i \quad \text{on} \quad \Gamma_i. \tag{5.7-2b}$$

Once again we use the method of conformal transplantation. We note that (5.7-2) is at once more special and more general than the Dirichlet problem considered in §5.6. It is more special because the prescribed boundary values are constant on each component Γ_i of the boundary of D, and for this reason it is not *desirable* to use as a model problem the case in which D is a circular disk. It is more general because D need not be simply connected. Therefore it is not always *possible* to use the circular disk as a model problem. Rather the model problem must be selected judiciously in each special case.

We illustrate the method with three typical examples.

EXAMPLE 1 FIELD BETWEEN HYPERBOLIC ELECTRODES

Let $a > 0$, $0 < \beta < \pi/2$ and let D be the region bounded by the two branches of the hyperbola

$$\left(\frac{x}{a\sin\beta}\right)^2 - \left(\frac{y}{a\cos\beta}\right)^2 = 1,$$

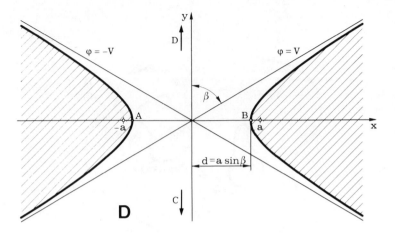

Fig. 5.7b. The physical problem.

whose asymptotes are $x\cos\beta = \pm y\sin\beta$. The left and right branch of the hyperbola are kept at the potentials $-V$ and V, respectively (see Fig. 5.7b). We seek the potential between the electrodes and the resulting field strength

$$\mathbf{E} = -\operatorname{grad}\phi. \tag{5.7-3}$$

The appropriate *model problem* here is the plate condenser, i.e., the field between two infinite plates. The model region in the w-plane is an infinite strip, say the strip $E: -c < \operatorname{Re} w < c$, where c is a suitable positive number (see Fig. 5.7c). The solution of the model problem is trivial:

$$\psi(u,v) = \frac{V}{c}u = \frac{V}{c}\operatorname{Re} w. \tag{5.7-4}$$

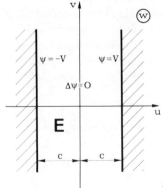

Fig. 5.7c. The model problem.

PROBLEMS OF PLANE ELECTROSTATICS   345

To map D conformally onto E we use some auxiliary elementary mappings. The similarity $z \to z_1 := (1/a)z$ normalizes the scale, bringing the foci of the hyperbola to ± 1. We now apply the Joukowski map

$$z_1 = \tfrac{1}{2}\left(z_2 + \frac{1}{z_2}\right),$$

made one-to-one by stipulating $\operatorname{Re} z_2 > 0$. According to the discussion in §5.1 this maps D onto the wedge-shaped region $\pi/2 - \beta < \arg z_2 < \pi/2 + \beta$. By $z_2 = iz_3$ the wedge is rotated by $-\pi/2$. The map $z_4 = \operatorname{Log} z_3$ maps the wedge onto the strip $-\beta < \operatorname{Im} z_4 < \beta$, which in turn is mapped onto the vertical strip E (where $c = \beta$) by $w = iz_4$ (see Fig. 5.7d). The desired map $z \to w = f(z)$ is given by the composition of all auxiliary maps:

$$w = f(z) = i \operatorname{Log} \frac{z + (z^2 - a^2)^{1/2}}{ia} \tag{5.7-5}$$

It follows from the discussion of the Joukowski map in §5.1 that the correct value of the square root is

$$(z^2 - a^2)^{1/2} = |z^2 - a^2|^{1/2} \exp\left(i\frac{\phi_1 + \phi_2}{2}\right),$$

where $\phi_1 := \arg(z+a)$, $\phi_2 := \arg(z-a)$, and

$$-\pi < \phi_1 < \pi, \quad 0 < \phi_2 < 2\pi.$$

The potential ϕ in the physical plane is now obtained from (5.7-4) by transplanting back:

$$\phi(z) = \psi(f(z)) = \frac{V}{\beta} \operatorname{Re}\left\{ i \operatorname{Log} \frac{z + (z^2 - a^2)^{1/2}}{ia} \right\}. \tag{5.7-6}$$

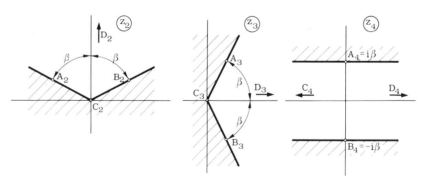

Fig. 5.7d. Auxiliary maps.

This formula makes it possible to calculate $\phi(z)$ for any given $z \in D$. It can also be used to compute the field strength $\mathbf{E} = -\mathrm{grad}\,\phi$.

Alternatively, the field strength can be obtained by use of Theorem 5.6a. If $\mathbf{E} = (E_1, E_2)$, we call the complex number $E := E_1 + iE_2$ the **complex field strength**. In the model plane the complex fields strength is

$$E = -\mathrm{grd}\,\psi = -\frac{V}{\beta}.$$

From (5.7-5) there follows

$$f'(z) = \frac{i}{(z^2 - a^2)^{1/2}},$$

with the above determination of the square root. Equation (5.6-4) thus yields

$$E(z) = -\mathrm{grd}\,\phi(z) = -\frac{V}{\beta}\overline{f'(z)}$$

for the complex field strength in the physical plane. Because $f'(z)$ is real for z real, it follows from the symmetry principle (see §5.11) that $\overline{f'(z)} = f'(\bar{z})$. Thus we get

$$E(z) = -\frac{V}{\beta}\frac{i}{(\bar{z}^2 - a^2)^{1/2}}. \tag{5.7-7}$$

The field strength has its maximum modulus at the vertices of the hyperbola, i.e., for $z = \pm d$, where $d := a\sin\beta$. There we find

$$E(\pm d) = -\frac{V}{\beta}\frac{1}{a\cos\beta} = -\frac{V}{d}\frac{\tan\beta}{\beta}.$$

It is of interest to discuss this quantity as a function of β when $2d$, the distance of the electrodes, is kept fixed. For $\beta \to 0$ the shape of the electrodes approaches that of the plate condenser and indeed $E(\pm d)$ approaches $-V/d$, the field strength of the homogeneous field. For $\beta \to \pi/2$ the tips of the electrodes become sharp edges and $|E(\pm d)| \to \infty$. The factor $\tan\beta/\beta$, which would be called a **form factor** in engineering, reflects the fact that the curvature of the electrodes augments the strength of the field.

EXAMPLE 2 EXCENTRIC CABLE IN CIRCULAR TUBE

Let a, r, R be real, $0 < a < R - r$, and let D be the region bounded by the two circles $|z| = R$ and $|z - a| = r$. The inner circle is at potential V, the outer circle at potential zero (see Fig. 5.7e).

A model problem for the above potential problem is now required. The plate condenser cannot serve as a model problem because it is simply connected and thus cannot be the image of a doubly connected domain D under a conformal map. (It cannot even be the image of a map that is merely required to be continuous.) However, the case in which the cable is located concentrically within the tube presents itself as a simple model problem (see Fig. 5.7f).

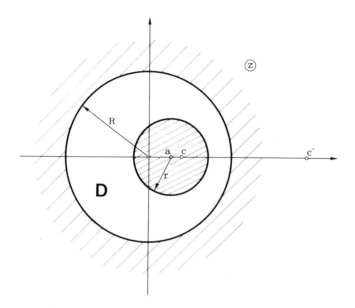

Fig. 5.7e. Eccentric cable.

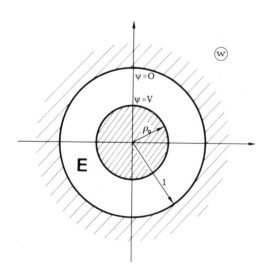

Fig. 5.7f. Concentric cable.

To map D onto the region E bounded by two concentric circles $|w|=1$ and $|w|=\rho_0$ ($0<\rho_0<1$) we apply the symmetry principle discussed in §5.4. To this end we seek a pair of points c,c' ($0<c<R<c'$) which is symmetrical with respect to both circles $|z|=R$ and $|z-a|=r$. Symmetry with respect to $|z|=R$ requires $cc'=R^2$ and symmetry with respect to the inner circle

$$(c-a)(c'-a)=r^2.$$

Eliminating c' and solving for c yields

$$c=\frac{1}{2a}(R^2+a^2-r^2-W), \qquad (5.7\text{-}8)$$

where

$$W:=\left((R^2+a^2-r^2)^2-4a^2R^2\right)^{1/2}$$

$$=\left((R^2-a^2-r^2)^2-4a^2r^2\right)^{1/2}$$

$$=\left((R+a+r)(R+a-r)(R-a+r)(R-a-r)\right)^{1/2}.$$

The Moebius transformation

$$z\to\frac{z-c}{z-c'}=c\frac{z-c}{cz-R^2}$$

maps c onto 0 and c' onto ∞, hence the two circles onto concentric circles about 0. Point $z=R$ is mapped onto $-c/R$. It follows that

$$f:z\to R\frac{z-c}{R^2-cz} \qquad (5.7\text{-}9)$$

is a function mapping $|z|=R$ onto $|w|=1$ and $|z-a|=r$ onto a smaller concentric circle of radius $\rho_0:=f(a+r)$. Computation yields

$$\rho_0=\frac{2Rr}{R^2+r^2-a^2+W}. \qquad (5.7\text{-}10)$$

In the model plane the boundary problem is easily solved because of its rotational symmetry. If $w=\rho e^{i\theta}$ and $\psi(w)=\tilde\psi(\rho,\theta)$, then $\Delta_w\psi=0$ if and only if

$$\frac{\partial^2\tilde\psi}{\partial\rho^2}+\frac{1}{\rho}\frac{\partial\tilde\psi}{\partial\rho}+\frac{1}{\rho^2}\frac{\partial^2\tilde\psi}{\partial\theta^2}=0.$$

The solution that satisfies $\tilde\psi(1,\theta)=0, \tilde\psi(\rho_0,\theta)=V$, being independent of θ, is found to be

$$\tilde\psi(\rho,\theta)=V\frac{\operatorname{Log}\rho}{\operatorname{Log}\rho_0}. \qquad (5.7\text{-}11)$$

PROBLEMS OF PLANE ELECTROSTATICS

Hence

$$\psi(w) = \frac{V}{\operatorname{Log}\rho_0}\operatorname{Log}|w|,$$

and by transplanting back we find the solution in the physical plane:

$$\phi(z) = \frac{V}{\operatorname{Log}\rho_0}\operatorname{Log}\left|R\frac{z-c}{R^2-cz}\right|. \qquad (5.7\text{-}12)$$

The field strength $E = -\operatorname{grd}\phi$ can likewise be found by transplantation. In the model plane $\operatorname{grd}\psi$ is a vector directed radially with radial component $V(\rho\operatorname{Log}\rho_0)^{-1}$. Hence the complex gradient in the model plane is

$$\operatorname{grd}\psi(w) = \frac{V}{\operatorname{Log}\rho_0}\cdot\frac{1}{\bar{w}}.$$

In view of

$$f'(z) = R\frac{R^2-c^2}{(R^2-cz)^2}$$

it follows by virtue of (5.6-4) that

$$E(z) = -\operatorname{grd}\phi(z) = -\frac{V}{\operatorname{Log}\rho_0}\frac{R^2-c^2}{(\bar{z}-c)(R^2-c\bar{z})}.$$

The maximum field strength occurs at $z = a+r$ and is found to be

$$E(a+r) = -\frac{V}{\operatorname{Log}\rho_0}\frac{W}{r(R^2-r^2-a^2)}.$$

CAPACITY

Let D be a bounded domain in 3-space whose complement in addition to the unbounded component C_0 consists of exactly one bounded component C_1. If C_0 and C_1 (or their boundaries) are viewed as conductors, this arrangement in electrostatic theory is called a **condenser**. Let us calculate the amount Q of electric charge that can be held by the condenser if the two conductors are kept at fixed potentials. Let Σ_i denote the surface of the conductor C_i ($i=1,2$), ϵ, the dielectric constant of the medium occupying D ($\epsilon = 1$ in vacuum), **E**, the electric field strength, and **n**, the unit normal vector on Σ_i pointing toward D (see Fig. 5.7g). Then according to electrostatic theory, the electric charge sitting on Σ_i is given by

$$Q_i := \epsilon\int_{\Sigma_i}\mathbf{E}\mathbf{n}\,d\Sigma,$$

where $d\Sigma$ is the area element on Σ_i. Since div $\mathbf{E} = 0$, the divergence theorem shows that $Q_1 = -Q_0$. Now let the conductors C_i be kept at the

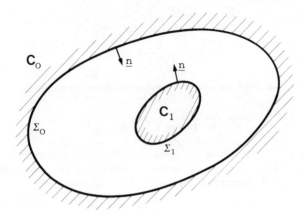

Fig. 5.7g. Condenser.

potentials V_i, $i = 1, 2$. Then $\mathbf{E} = -\mathrm{grad}\,\phi$, where ϕ is the solution of the Dirichlet problem $\Delta\phi = 0$ in D, $\phi = V_i$ on Σ_i. By linearity it is clear that $\phi = V_0 + (V_1 - V_0)\tilde{\phi}$, where $\tilde{\phi}$ is the solution of the Dirichlet problem for the special boundary conditions $\tilde{\phi} = 0$ on Σ_0, $\tilde{\phi} = 1$ on Σ_1. There follows

$$Q_1 = \epsilon(V_1 - V_0)\gamma, \qquad (5.7\text{-}13)$$

where

$$\gamma := -\int_{\Sigma_1} \frac{\partial \tilde{\phi}}{\partial n}\, d\Sigma \qquad (5.7\text{-}14)$$

is a number, called the **capacity** of the condenser, that depends only on the shape of the domain D.

The capacity is not an easy quantity to calculate for many three-dimensional condensers. Ingenious methods, based on variational principles, have been devised for its approximate calculation (see Polya and Szegö [1951]).

In two-dimensional electrostatics we consider condensers bounded by cylindrical surfaces $\sum_i$ with generators perpendicular to the (x, y)-plane, whose intersection with that plane we denote by Γ_i ($i = 0, 1$). If $\tilde{\phi}$ now denotes the solution of the two-dimensional Dirichlet problem for the region D bounded by Γ_0 and Γ_1 and having the values 0 on Γ_0 and 1 on Γ_1, the **capacity per unit length** of the condenser is

$$\gamma = -\int_{\Gamma_1} \frac{\partial \tilde{\phi}}{\partial n}\, ds = \int_{\Gamma_1} \left(\frac{\partial \tilde{\phi}}{\partial x}\, dy - \frac{\partial \tilde{\phi}}{\partial y}\, dx \right).$$

PROBLEMS OF PLANE ELECTROSTATICS

In terms of the complex gradient $\operatorname{grd}\tilde{\phi} = \partial\tilde{\phi}/\partial x + i(\partial\tilde{\phi}/\partial y)$ this is expressed by

$$\gamma = -\operatorname{Im} \int_{\Gamma_1} \overline{\operatorname{grd}\phi(z)} \, dz. \qquad (5.7\text{-}15)$$

Let us now assume that $w = f(z)$ maps D, as in Example 2, on the annulus $E: \rho_0 < |w| < 1$ such that the inner boundary Γ_1 corresponds to $|w| = \rho_0$. If $\tilde{\psi}$ denotes the transplant of $\tilde{\phi}$, then by Theorem 5.6a

$$\overline{\operatorname{grd}\tilde{\phi}(z)} = \overline{\operatorname{grd}\tilde{\psi}(w)} \, f'(z),$$

and since $f'(z)\,dz = dw$ we have

$$\gamma = -\operatorname{Im} \int_{\Gamma_1^*} \overline{\operatorname{grd}\tilde{\psi}(w)} \, dw. \qquad (5.7\text{-}16)$$

For the model problem in which $\tilde{\psi} = 1$ on $|w| = \rho_0$ and $\tilde{\psi} = 0$ on $|w| = 1$ we have already seen in Example 2 that

$$\operatorname{grd}\tilde{\psi}(w) = (\bar{w}\operatorname{Log}\rho_0)^{-1},$$

yielding

$$\gamma = -\frac{1}{\operatorname{Log}\rho_0} \operatorname{Im} \int_{\Gamma_1^*} \frac{1}{w}\,dw.$$

The integral values $2\pi i$ and

$$\gamma = -\frac{2\pi}{\operatorname{Log}\rho_0} \qquad (5.7\text{-}17)$$

follows. The capacity per unit length of a cylindrical condenser thus depends only on ρ_0, the ratio of the inner and the outer radius of the circles bounding the model domain. Because γ is uniquely determined by D, we are forced to conclude that *if a simply connected domain can be mapped onto an annulus the quantity ρ_0 characterizing the annulus is uniquely determined by D.* This conclusion, so far, is based on electrostatics; a mathematical demonstration is given in §5.11.

In three-space consider a homogeneous electric field $E = (-E_0, 0, 0)$ with potential $\phi(x,y,z) = E_0 x$. If a bounded conductor C with surface Σ is placed in the field, the field will be perturbed because the potential must be constant on Σ. The perturbation will be such that at infinity the field is still homogeneous. The problem is to determine the perturbation of the field due to C, called **influence** in electrostatics.

We assume that C is insulated and that the total charge on C is zero, in which case the potential on C alone will assume a certain constant value that cannot be prescribed. (In an airplane flying through a thunderstorm that value may be quite high without disturbing the passengers.) According to our discussion of capacity, if the dielectric constant equals 1, the charge sitting on C is given by

$$Q = -\int_\Sigma \frac{\partial \phi}{\partial n} d\Sigma,$$

n denoting the normal pointing toward D. The mathematical problem now is to find ϕ such that

$$\Delta \phi = 0 \quad \text{on } D, \tag{5.7-18a}$$

$$\phi = \text{const on } \Sigma, \tag{5.7-18b}$$

$$\int_\Sigma \frac{\partial \phi}{\partial n} d\Sigma = 0, \tag{5.7-18c}$$

$$\phi(x,y,z) = E_0 x + O\left(\frac{1}{r}\right), \quad r \to \infty, \tag{5.7-18d}$$

where $r := (x^2 + y^2 + z^2)^{1/2}$.

In the two-dimensional version of the influence problem we assume that Σ is a cylinder with indicatrix Γ. The boundary conditions are the same as above, except that $r = (x^2 + y^2)^{1/2}$. According to our discussion of capacity, Equation (5.7-18c) is now expressed in the form

$$\int_\Gamma \frac{\partial \phi}{\partial n} ds = \text{Im} \int_\Gamma \overline{\text{grd} \phi(z)} \, dz = 0. \tag{5.7-19}$$

To solve the influence problem by conformal transplantation we require an appropriate model problem. The obvious model is a thin vertical sheet whose intersection with the w-plane lies on a straight line $\text{Re} \, w = \text{const}$. This does not disturb the field at all, and the potential in the model plane is given by $\psi(w) = E_0 \text{Re} \, w$. Thus we require a function mapping D, the exterior of Γ, onto the exterior of a vertical slit in the w-plane. Because the field is not to be disturbed at ∞, the mapping function should satisfy

$$f(z) = z + O\left(\frac{1}{z}\right), \quad z \to \infty. \tag{5.7-20}$$

PROBLEMS OF PLANE ELECTROSTATICS

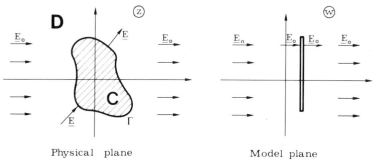

Fig. 5.7h. Influence.

This means that the location of the slit cannot be prescribed. If ψ denotes the solution in the model plane, it follows from the discussion of capacity that

$$\int_\Gamma \overline{\mathrm{grd}\,\phi(z)}\, dz = \int_{\Gamma^*} \overline{\mathrm{grd}\,\psi(w)}\, dw.$$

Because $\mathrm{grd}\,\psi(w) = E_0$, the expression on the right is obviously zero, and (5.7-19) shows that (5.7-18c) is satisfied (see Fig. 5.7h).

EXAMPLE 3

To consider a specific case let Γ be the circle $|z| = a$, where $a > 0$. Here the mapping problem can be solved by the Joukowski function. We require some elementary auxiliary maps to satisfy the condition at infinity (see Fig. 5.7i).
The composition of the mappings

$$z \to z_1 := iz,$$

$$z_1 \to z_2 := \frac{1}{a} z_1,$$

$$z_2 \to z_3 := \tfrac{1}{2}\left(z_2 + \frac{1}{z_2}\right)$$

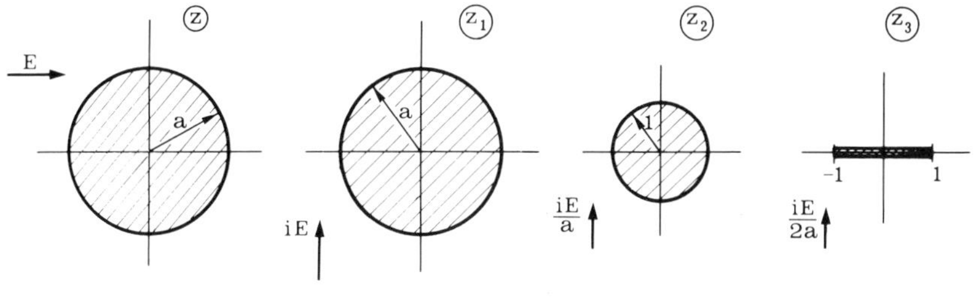

Fig. 5.7i. Auxiliary maps for Example 3.

yields a map

$$z \to \tfrac{1}{2}\left(\frac{iz}{a} + \frac{a}{iz}\right)$$

of the exterior of $|z|=a$ onto the exterior of the slit joining the points $z_3 = \pm 1$. At infinity this map behaves like $iz/2a$. Another map $z_3 \to w := -2iaz_3$ turns the slit around and rectifies the behavior at infinity. Thus the composite mapping $f: z \to z - a^2/z$ satisfies (5.7-20).

From $\psi(w) = E_0 \operatorname{Re} w$ we obtain by transplantation the solution in the physical plane

$$\phi(z) = E_0 \operatorname{Re}\left(z - \frac{a^2}{z}\right). \qquad (5.7\text{-}21)$$

The complex field strength $E = -\operatorname{grd}\phi$ is given (see 5.6-4) by $E = E_0 \overline{f'(z)}$ which yields

$$E(z) = -E_0\left(1 + \frac{a^2}{\bar{z}^2}\right). \qquad (5.7\text{-}22)$$

If C has a more general shape, it is frequently simpler to map the exterior of C onto the exterior of a circle than on the exterior of a slit. In such cases the solution for the exterior of a circle given by (5.7-21) may in turn play the role of a model solution.

PROBLEMS

1. A straight cable with circular cross section is suspended at height h above a plane. The cable is at potential V, the plane at potential 0 (see Fig. 5.7j). Compute the resulting potential distribution and find the maximum field strength. [Map onto a domain bounded by concentric circles, using the symmetry principle.]
2. An elliptical cylinder (semiaxes a,b) is placed into a homogeneous field running in the direction of the semiaxis of length a. Compute the potential

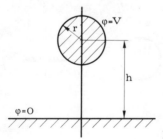

Fig. 5.7j. Geometry of Problem 1.

and show that the maximum field strength is $E_0(1+a/b)$, where E_0 is the field strength at infinity. [Use the circular cylinder as a model problem.]

3. In electrostatic problems having the geometric structure of the plate condenser the potential has the form $\phi(z) = \text{const}\,\text{Re}f(z)$. It follows that the complex field strength is $E = \text{const}\,\overline{f'(z)}$. Conclude that the maximum field strength always occurs on the boundary of the region under consideration.

4. Let D be a doubly connected region and let $f: z \to w = f(z)$ map D onto an annulus $\rho < |w| < 1$.
    (a) If the inner component of the complement of D is at potential V and the outer component at potential 0, show that the solution of the electrostatic problem generally is
    $$\phi(z) = \text{const}\,\text{Log}|f(z)|.$$
    (b) From the fact that the complex field strength is
    $$E(z) = \text{const}\,\overline{\left(\frac{f'(z)}{f(z)}\right)},$$
    conclude by the maximum principle that the maximum field strength always occurs at the boundary of D.

5. The two plates of a plate condenser are at a distance 2d. They are kept at potential zero. A thin plate in the shape of a half-plane which is kept at potential $V \neq 0$ is located in the plane of symmetry (see Fig. 5.7k).
    (a) Determine the potential $\phi(z)$.
    (b) Show that the complex gradient of ϕ is
    $$\text{grd}\,\phi(z) = i\frac{V}{d}(1 - e^{-\pi z/d})^{-1/2},$$
    and study its behavior for $x \to \pm \infty$.

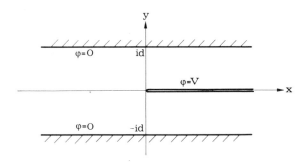

Fig. 5.7k. Geometry of Problem 5.

§5.8. TWO-DIMENSIONAL IDEAL FLOWS

Here we consider two-dimensional flows of an ideal (i.e., inviscid and incompressible) fluid. It is assumed that these flows are irrotational and that there are no sources and sinks. Let

$$\mathbf{q} := \mathbf{q}(x,y) := (q_1(x,y), q_2(x,y))$$

be the velocity vector of the flow at point (x,y). The assumption that there are no sources and sinks then means that

$$\operatorname{div} \mathbf{q} = 0 \quad \text{or} \quad \frac{\partial q_1}{\partial x} + \frac{\partial q_2}{\partial y} = 0. \tag{5.8-1}$$

The assumption that the flow is irrotational means that

$$\operatorname{rot} \mathbf{q} = \mathbf{0} \quad \text{or} \quad \frac{\partial q_2}{\partial x} - \frac{\partial q_1}{\partial y} = 0. \tag{5.8-2}$$

The last relation in turn implies that the differential

$$q_1(x,y)\,dx + q_2(x,y)\,dy$$

is exact. By a basic result of vector analysis this means that in every simply connected domain there is a function $\phi:(x,y)\to\phi(x,y)$ such that

$$\mathbf{q} = \operatorname{grad} \phi$$

or, equivalently,

$$q_1 = \frac{\partial \phi}{\partial x}, \qquad q_2 = \frac{\partial \phi}{\partial y}.$$

(5.8-1) now implies that

$$\frac{\partial^2 \phi}{\partial x^2} + \frac{\partial^2 \phi}{\partial y^2} = 0,$$

i.e., the potential ϕ is a *harmonic* function.

Under the hypotheses stated the determination of a flow in a region D thus can be reduced to the determination of a single scalar function, the **flow potential** ϕ. The function ϕ is subject to the following conditions:

1. In the interior of D it satisfies Laplace's equation

$$\Delta \phi = 0. \tag{5.8-3}$$

2. At the boundary Γ (assuming that there is no absorption of the fluid through the walls) the flow runs in a tangential direction. If **n** is a vector

TWO-DIMENSIONAL IDEAL FLOWS

perpendicular to the wall, we have $\mathbf{n} \cdot \operatorname{grad} \phi = 0$ or

$$\frac{\partial \phi}{\partial n} = 0. \tag{5.8-4}$$

3. In addition, in flow problems, there are usually some conditions at infinity; for instance, if the problem is to determine the flow past an obstacle, the flow must be homogeneous at infinity.

Conditions 1 and 2 are suitable for the method of conformal transplantation. Let f map D onto a model region E and let $\psi : w \to \phi(f^{[-1]}(w))$ be the transplanted function ϕ. Then, by Corollary 5.6c, ψ likewise satisfies Laplace's equation; furthermore, if the mapping is also conformal on the boundary of D, the direction perpendicular to Γ corresponds to the direction perpendicular to the boundary of E and (5.8-4) is equivalent to

$$\frac{\partial \psi}{\partial n} = 0. \tag{5.8-5}$$

Naturally we shall endeavor to choose the model region in such a manner that the solution of the model problem will be simple. We illustrate the method by several examples.

EXAMPLE 1 FLOW THROUGH A CHANNEL OF VARIABLE CROSS SECTION

Here the appropriate model region is clearly a strip parallel, say, to the u-axis (see Fig. 5.8a). An irrotational flow through a channel of constant width is clearly homogeneous. Hence the solution of the problem in the model domain is

$$\psi(u,v) = \operatorname{const} \cdot u.$$

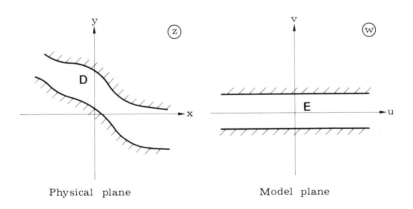

Fig. 5.8a. Flow through a channel.

We delay the discussion of specific examples until our mapping technique has been developed to an extent that will permit the presentation of some problems of technical interest.

EXAMPLE 2 FLOW PAST AN OBSTACLE

Let C, the complement of D, be a simply connected, compact set that represents the cross section of a cylindrical obstacle placed in an originally homogeneous flow that filled the whole space. The generators of the cylinder are assumed to be perpendicular to the direction of the flow (see Fig. 5.8b).

The flow potential ϕ, in addition to (5.8-3) and (5.8-4), must satisfy the following condition at infinity:

$$\lim_{z \to \infty} \operatorname{grd} \phi(z) = v_0 \qquad (5.8\text{-}6)$$

One model situation in which the solution of the problem is trivial occurs when the obstacle consists of an infinitely thin blade parallel to the real axis. Here the solution is $\psi(u,v) = v_0 u$. According to the theory, we must map D onto the domain exterior to a hc. 'zontal slit in the w-plane. The flow pattern is not disturbed at infinity if the mapping function f is such that $f(\infty) = \infty$, $f'(z) \to 1$ as $z \to \infty$. If such an f exists, the flow potential in the physical plane is

$$\phi(z) = v_0 \operatorname{Re} f(z), \qquad (5.8\text{-}7\text{a})$$

and by Theorem 5.6a the complex velocity $q := q_1 + i q_2$ at point z becomes

$$q(z) = v_0 \overline{f'(z)}. \qquad (5.8\text{-}7\text{b})$$

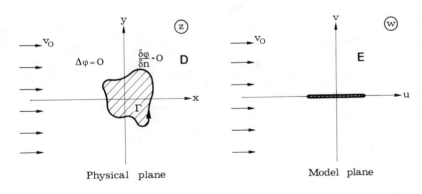

Fig. 5.8b. Flow around an obstacle.

TWO-DIMENSIONAL IDEAL FLOWS 359

Of interest in flow problems are the **streamlines**. These are the curves that at every point z have the direction of the velocity vector q at z. The streamlines in the model plane are the straight lines $v = \text{const}$. Because by Problem 1 of §5.6, streamlines are mapped onto streamlines, the streamlines in the physical plane are the curves

$$\text{Im} f(z) = \text{const}. \qquad (5.8\text{-}8)$$

We emphasize that the three formulas (5.8-7a), (5.8-7b), and (5.8-8) are valid only in the special situation in which the flow in the model domain is a homogeneous flow running in the direction of the u-axis.

To consider a specific case, let the cross section of the obstacle in the physical plane be the circular disk $|z| \leq a$. By considerations similar to those used in Example 3 of §5.7, we find that the function

$$f: z \to z + \frac{a^2}{z}$$

(obviously related to the Joukowski function) maps D, the exterior of the obstacle, onto the w-plane cut between $w = \pm 2a$. It obviously leaves the point at infinity fixed and satisfies $f'(z) \to 1$ as $z \to \infty$. Therefore by denoting the potential in the physical plane by ϕ_0 we have

$$\phi_0(z) = v_0 \text{Re}\left(z + \frac{a^2}{z}\right) = v_0\left(r + \frac{a^2}{r}\right)\cos\theta \qquad (z = re^{i\theta}). \qquad (5.8\text{-}9)$$

The equation of the streamlines is

$$\text{Im}\left(z + \frac{a^2}{z}\right) = \text{const} \quad \text{or} \quad \left(r - \frac{a^2}{r}\right)\sin\theta = \text{const},$$

and by (5.8-7b) for the complex velocity $q = q_1 + iq_2$ we find

$$q(z) = v_0\left(1 - \frac{a^2}{\bar{z}^2}\right) = v_0\left(1 - \frac{a^2}{r^2}e^{2i\theta}\right).$$

At points $z = \pm a$ the velocity is seen to be equal to zero; these are the **stagnation points** of the flow. At points $z = \pm ia$, on the other hand, we have $q(\pm ia) = 2v_0$; these are the points at which the absolute value of the velocity vector reaches its maximum (see Fig. 5.8c).

Somewhat surprisingly it turns out that the flow defined by (5.8-9) is not the only flow past a cylinder that is physically possible, nor does it represent the only solution to the mathematical problem defined by (5.8-1), (5.8-2), (5.8-4), and (5.8-6), the boundary condition at infinity. From Laplace's equation in polar coordinates it follows that $\phi_1(x,y) := K\theta(z$

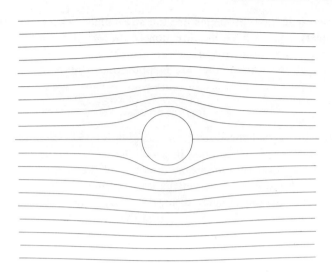

Fig. 5.8c. Flow around a circular cylinder, without circulation.

$=re^{i\theta}$, K real) is a solution in any region in which the argument θ can be defined as a single-valued function. Moreover, because the radial component of $\operatorname{grad}\phi_1$ is zero, ϕ_1 satisfies the required boundary condition on $|z|=a$. Finally, $\operatorname{grad}\phi_1(z)\to 0$ for $z\to\infty$. Thus by forming the gradient of the function

$$\phi(z):=\phi_0(z)+\phi_1(z)=v_0\left(r+\frac{a^2}{r}\right)\cos\theta+K\theta \qquad (5.8\text{-}10)$$

we obtain, for any value of K, a solution to the problem that satisfies all conditions stated. Because the argument θ is not single-valued, the potential ϕ is not single-valued. We should not expect it to be because the domain of definition is not simply connected; $\operatorname{grad}\phi$, however, is defined uniquely. It satisfies (5.8-6) because

$$\operatorname{grd}\phi(z)=v_0\left(1-\frac{a^2}{\bar{z}^2}\right)+\frac{iK}{\bar{z}}. \qquad (5.8\text{-}11)$$

The term $K\theta$ in (5.8-10) has a physical interpretation. Let the cross section C of the cylindrical obstacle be arbitrary, let Γ_0 be a positively oriented Jordan curve enclosing C, and let $\mathbf{q}$ be any two-dimensional vector field such that $\operatorname{rot}\mathbf{q}=\mathbf{0}$. It then follows from Stokes's theorem that the value of the integral

$$c:=\int_{\Gamma_0}\mathbf{q}\cdot d\mathbf{r}$$

TWO-DIMENSIONAL IDEAL FLOWS 361

does not depend on the choice of Γ_0. The constant value of the integral is the **circulation** of the field **q** around the obstacle C.

For the flow defined by (5.8-10), having chosen for Γ_0 a circle of radius $r > a$, we find that the circulation has the value

$$c = 2\pi K. \qquad (5.8\text{-}12)$$

The presence of circulation makes itself felt in the location of the stagnation points of the flow. From (5.8-11) we infer that the points z at which $q(z) = 0$ are the solutions of the equation

$$1 - \frac{a^2}{z^2} - \frac{iK}{v_0 z} = 0.$$

The solutions are

$$z = \frac{iK}{2v_0} \pm \left(a^2 - \frac{K^2}{4v_0}\right)^{1/2}. \qquad (5.8\text{-}13)$$

Fig. 5.8d. Flow around a cylinder with weak circulation ($0 < K < 2av_0$).

If $-2zv_0 < K < 2av_0$, there are two distinct stagnation points at the surface of the cylinder, located at

$$z = ae^{i\alpha} \quad \text{and} \quad z = ae^{i(\pi - \alpha)} \qquad (5.8\text{-}14)$$

(see Fig. 5.8d), where

$$\alpha := \operatorname{Arcsin} \frac{K}{2v_0 a}. \qquad (5.8\text{-}15)$$

For $|K| > 2av_0$ there are no stagnation points at the surface of the cylinder. Instead, there is a point inside the flow at which the velocity is zero (see Figs. 5.8e and 5.8f).

Because it permits a simple treatment of circulation, it is convenient to use the flow around a circular cylinder as a model problem for flows around cylindrical obstacles with arbitrary cross section C. Also, it is

Fig. 5.8e. $K = 2av_0$.

TWO-DIMENSIONAL IDEAL FLOWS

Fig. 5.8f. Flow around a cylinder with strong circulation ($K > 2av_0$).

usually simpler to map D, the exterior of C, onto the exterior of a circle than onto the exterior of a horizontal slit. Thus let C be arbitrary and let the function

$$f: z \to w = f(z)$$

map the exterior D of C onto the region $E: |w| > a$ (see Fig. 5.8g) in such a way that

$$f(z) = z + O(1), \qquad z \to \infty. \tag{5.8-16}$$

According to the general theory of transplants of potential functions and the formulas (5.8-10) and (5.8-11) it follows that the multiple-valued potential

$$\phi(z) := v_0 \operatorname{Re} F(z), \tag{5.8-17}$$

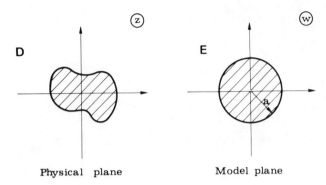

Fig. 5.8g. Flow around an obstacle.

where

$$F(z) := f(z) + \frac{a^2}{f(z)} - i\frac{K}{v_0}\log f(z) \qquad (5.8\text{-}18)$$

defines a flow past C with complex velocity

$$q(z) = \left[v_0\left(1 - \frac{a^2}{f(z)^2}\right) + \frac{iK}{f(z)} \right] \overline{f'(z)}$$

or

$$q(z) = v_0 \overline{F'(z)}. \qquad (5.8\text{-}19)$$

We compute the circulation of this flow. If $\mathbf{q} = \operatorname{grad}\phi$, then

$$\mathbf{q}\cdot d\mathbf{r} = \frac{\partial \phi}{\partial x}dx + \frac{\partial \phi}{\partial y}dy = \operatorname{Re} \overline{\operatorname{grd}\phi}\; dz,$$

but if ψ is the conformal transplant of ϕ under $w = f(z)$ then by Theorem 5.6a

$$\overline{\operatorname{grd}\phi}\; dz = \overline{\operatorname{grd}\psi\, f'(z)}\, dz = \overline{\operatorname{grd}\psi}\; dw,$$

and if Γ_0^* is the image of Γ

$$c = \int_{\Gamma_0} \mathbf{q}\cdot d\mathbf{r} = \operatorname{Re}\int_{\Gamma_0} \overline{\operatorname{grd}\phi}\; dz = \operatorname{Re}\int_{\Gamma_0^*} \overline{\operatorname{grd}\psi}\; dw.$$

TWO-DIMENSIONAL IDEAL FLOWS

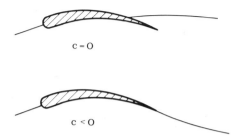

Fig. 5.8h. Flow around an airfoil with and without circulation.

Thus the circulation equals the circulation $c = 2\pi K$ in the model plane and represents another invariant quantity under conformal transplantation.

The stagnation points of the flow are the preimages of the stagnation points in the model plane. If $|K| < 2v_0 a$, these points are

$$w = ae^{i\alpha} \quad \text{and} \quad w = ae^{i(\pi - \alpha)},$$

where α is defined by (5.8-15).

Mathematically, all locations of the stagnation points are equally possible. Which location does nature choose? In one case of practical importance this question is resolved by a famous hypothesis due to Joukowski. Let the obstacle have the shape of an **airfoil**, i.e., let the boundary Γ of C be smooth except for a sharp corner at the trailing edge. Joukowski's hypothesis, which is supported by overwhelming experimental evidence, states that the circulation of the flow around a properly designed airfoil will always adjust itself so that the velocity at the trailing edge is finite. Mathematically this implies that in the model plane one stagnation point coincides with the image point of the trailing edge (see Fig. 5.8h). Thus if $w_0 := ae^{i\beta}$ is this image point, then $\alpha = \beta$, which by (5.8-15) implies $K = 2v_0 a \sin \beta$. The circulation then becomes

$$c = 4\pi v_0 a \sin \beta. \tag{5.8-21}$$

Far from being a mere nuisance, it is precisely the presence of a (negative) circulation that enables an aircraft to fly. Let us compute the total force (per unit length) exerted by the flow on the obstacle. The force exerted on the surface element ds of Γ is $p\mathbf{n}\,ds$, where p is the pressure and $\mathbf{n}$ is the normal unit vector directed toward the exterior of D. We wish to represent the force vector as a complex number. If $dz := dx + i\,dy$ is the line element, the vector $\mathbf{n}\,ds$ is represented by $i\,dz$. Thus the complex force element is $ip\,dz$. By Bernoulli's law, neglecting gravity, the pressure is given as $p = p_0 - \frac{1}{2}\rho|\mathbf{q}|^2$, where ρ is the density and p_0, the constant pressure at velocity zero. Clearly, the constant term p_0 exerts a zero resultant force on the body. Thus we find for the total complex force R

$$R = -i\frac{\rho}{2}\int_\Gamma |\mathbf{q}|^2 dz$$

or, since $q(z) = v_0 \overline{F'(z)}$,

$$R = -\frac{i\rho v_0^2}{2}\int_\Gamma F'(z)\,\overline{F'(z)}\,dz.$$

The line integral has a nonanalytic integrand, but it can nevertheless be evaluated by means of the following observation: on Γ the complex velocity $q(z) = v_0 \overline{F'(z)}$ has the direction of Γ. It follows that

$$\arg F'(z) = -\arg dz + \lambda\pi,$$

where λ is an integer. Comparing arguments, we now find that

$$F'(z)\,\overline{F'(z)}\,dz = \overline{|F'(z)|^2 dz}.$$

Hence

$$R = -\frac{iv_0^2}{2}\overline{\int_\Gamma |F'(z)|^2 dz}. \qquad (5.8\text{-}22)$$

Because F' is analytic in D, we can evaluate the integral along any positively oriented Jordan curve Γ encircling the obstacle. We select as our path of integration the image of a large circle Γ^* under the transformation $w \to z = g(w)$, where $g := f^{[-1]}$. In view of

$$F'(z) = \left\{1 - \frac{a^2}{f(z)^2} - \frac{iK}{v_0 f(z)}\right\} f'(z)$$

we obtain, on introducing w as a variable of integration,

$$\int_\Gamma [F'(z)]^2 dz = \int_{\Gamma^*}\left(1 - \frac{a^2}{w^2} - \frac{iK}{v_0 w}\right)^2 \frac{1}{g'(w)}\,dw.$$

By the residue theorem only the term in w^{-1} contributes to the value of the integral. Since $[g'(w)]^{-1} = 1 + O(w^{-2})$, the integral thus has the value

$$2\pi i\left(-\frac{2iK}{v_0}\right) = \frac{4\pi K}{v_0}.$$

TWO-DIMENSIONAL IDEAL FLOWS 367

Using this result in (5.8-22) and expressing K in terms of the circulation c, we obtain the **formula of Blasius**:

$$R = -i\rho v_0 c. \qquad (5.8\text{-}23)$$

Thus the force acts in the direction perpendicular to the undisturbed velocity and upward if the circulation is negative.

To examine a specific example we consider the Joukowski flow around an airfoil of the shape of a thin blade of length $2d$ tilted by the angle 2δ against the direction of the flow. By means of the auxiliary maps

$$z = e^{-i\delta}dz_1, \qquad z_1 = \tfrac{1}{2}\left(z_2 + \frac{1}{z_2}\right), \qquad z_2 = \frac{2}{d}e^{i\delta}w$$

(see Fig. 5.8i) we find that map f defined by

$$w \to z = f^{[-1]}(w) := w + \frac{d^2 e^{-2i\delta}}{4w} \qquad (5.8\text{-}24)$$

maps $|w| > d/2$ onto the exterior of the blade. It obviously satisfies the required condition at infinity. The image of the trailing edge $z = de^{-i\delta}$ is

$$w = \frac{d}{2}e^{-i\delta}.$$

Thus $\beta = -\delta$ and for the Joukowski circulation (5.8-21) yields $c = -2\pi v_0 d \sin \delta$. By (5.8-23) the lifting force is $R = 2\pi i \rho v_0^2 d \sin \delta$.

EXAMPLE 3 FLOW THROUGH CASCADES OF PROFILES

Further flow problems soluble by conformal maps include flows through straight or circular cascades of blades, as they occur in turbines. Here suitable model problems are the corresponding flows through grids of infinitely thin radial or parallel blades (see Fig. 5.8j).

If we make the assumption that the fluid emanates from a source at the origin, these flows can be generated by a composition of elementary mappings with the Joukowski map.

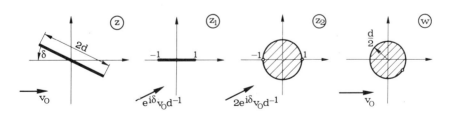

Fig. 5.8i. Auxiliary maps to construct (5.8-24).

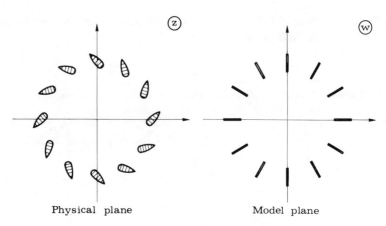

Fig. 5.8j. Flow through a circular arrangement of blades.

EXAMPLE 4 FREE BOUNDARY PROBLEMS

Consider a filled cylindrical container whose cross section is the quarter-plane $\operatorname{Re} z < 0$, $\operatorname{Im} z > 0$. If the container has an opening, represented, say, by the segment $0 < y < a$ of the imaginary axis, the fluid will emanate from the container according to a pattern indicated in Fig. 5.8k.

The problem of determining the flow potential ϕ of the resulting flow is different from any of the problem types previously considered, for here even the domain of definition of ϕ is not known a priori. The location of the free boundary Γ_1 is determined by the condition that the pressure is constant along Γ_1. By Bernoulli's law this means that $|q|$ is constant.

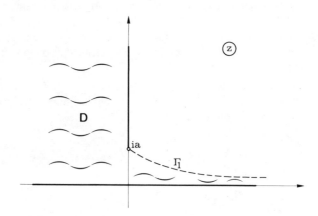

Fig. 5.8k. A free boundary problem.

TWO-DIMENSIONAL IDEAL FLOWS

Problems of this kind can again be solved by complex variable methods, but in place of the model plane we have to work with the **hodograph plane**, i.e., with the image of D under the map $z \to q(z)$, where q is the complex velocity. The theory may become complicated because in some cases this map is not one-to-one.

PROBLEMS

1. Compute the flow (with circulation zero) around an elliptic cylinder (semiaxes a, b) if v_0, the velocity at infinity, has the direction of the semiaxis a. Show that the maximum velocity at the surface of the cylinder is $|v_0|(1 + b/a)$.
2. Let f be the function satisfying (5.8-16) that maps the exterior of the profile C onto $|w| > a$. The mapping function for the profile $e^{-i\delta}C$ (i.e., the profile C turned by δ) is $f_\delta : z \to e^{-i\delta} f(e^{i\delta} z)$.
3. One of the reasons for the success of the mathematical theory of airfoils is that realistic airfoils can be generated by elementary maps; for instance, if $z \to w := (a/2)(z + 1/z)$, then for small $\epsilon > 0$ the image of the circle $|z + \epsilon| = 1 + \epsilon$ generates a reasonable profile (see Fig. 5.8*l*). Verify this fact and compute the flow around the profile if the angle of inclination against the flow is δ. Show that the lifting force is $R = 4\pi i \rho v_0^2 a(1 + \epsilon) \sin \delta$.
4. A complex force $k := k_1 + i k_2$ acting at the point $z = x + iy$ exerts the moment $m = xk_2 - yk_1 = -\text{Im}(z\bar{k})$ with respect to the origin. In the context of the Blasius formula for lift show that the moment exerted by the pressure forces on the airfoil is given by

$$M = -\frac{\rho v_0^2}{2} \text{Re} \int_\Gamma z F'(z)^2 dz. \quad (5.8\text{-}25)$$

5 (continuation). Assuming that the development of the mapping function f at ∞ has the form $f(z) = z + a_0 + a_1/z + \cdots$, prove by the calculus of residues that the moment of the pressure forces about the origin is given by $M = \rho v_0 c \, \text{Re} \, a_0 - 2\pi \rho v_0^2 \, \text{Im} \, a_1$.

6 (continuation). If the profile C is symmetric with respect to the real axis, then, in the Laurent series at infinity of the mapping function, $f(z) = z + a_0 +$

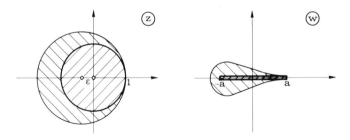

Fig. 5.8l. Joukowski profile.

$a_1/z + a_2/z^2 + \cdots$, all coefficients a_k are real (see §5.11). If C has a sharp trailing edge, show that the moment about O exerted by the Joukowski flow around the profile $e^{-i\delta}C$ is $M = 2\pi\rho v_0^2 (a_1 - aa_0)\sin 2\delta$.

7. By the theory of Laurent series a mapping function f satisfying (5.8-16) can for $|z|$ sufficiently large be expanded in a Laurent series,

$$f(z) = \sum_{n=-1}^{\infty} a_n z^n, \quad a_{-1} = 1.$$

With this series associate the formal Laurent series (according to §1.8) $F := a_{-1}x^{-1} + a_0 + a_1 x + a_2 x^2 + \cdots$. Show that the inverse map $g := f^{[-1]}$ is for $|w| > a$ represented by the Laurent series

$$g(w) = \sum_{n=-1}^{\infty} b_n w^n,$$

where $b_{-1} = 1$, $b_0 = -a_0$, and

$$b_n = -\frac{1}{n}\operatorname{res} x^{-2} F^n, \quad n = 1, 2, \ldots. \quad (5.8\text{-}26)$$

8. *Flow due to finite source around circular cylinder.* Let $0 < a < 1$. The potential of the flow (including circulation) around the cylinder with cross section $|z - 1| \leq a$ due to a source at the origin is

$$\phi(z) := k \operatorname{Log}\left|z\left(1 - \frac{a^2}{1-z}\right)\right| + \frac{c}{2\pi}\arg(z - 1).$$

[A slit on the positive real axis defines a suitable model problem for the flow without circulation. Circulation can be added in the physical plane.]

9. *Flow around arbitrary obstacle due to finite source.* Let C_1 be the cross section of a cylindrical obstacle, $O \notin C_1$, and let f map the exterior of C_1 onto $|w| > a$ such that $f(z) = z + O(1)$, $z \to \infty$. The potential of a flow (with circulation c) around the obstacle due to a source at the origin is

$$\phi(z) := k \operatorname{Log}|F(z)| + \frac{c}{2\pi}\arg f(z),$$

where

$$F(z) = [f(z) - f(0)]\left[1 - \frac{a^2}{f(z)\overline{f(0)}}\right].$$

[Problem 8 serves as a model problem.]

10. (continuation). Let the profile C_1 have a sharp trailing edge at z_0 and let $f(z_0) = ae^{i\beta}$. Show that the Joukowski condition for the flow around C_1 due to

POISSON'S EQUATION

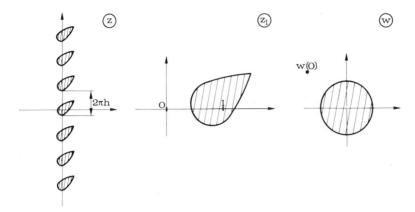

Fig. 5.8m. Flow through a cascade of airfoils.

a source at O is satisfied if and only if the circulation c has the value

$$c = 4\pi a k \frac{\operatorname{Im} e^{-i\beta} f(0)}{|ae^{i\beta} - f(0)|^2}.$$

11. *Flow through circular grid.* Let n profiles of cross section C be arranged symmetrically on a circle as in Fig. 5.8j [if z is in the interior of a profile, so is the point $\exp(2\pi i k/n)z$, $k = 1, 2, \ldots, n$]. Let C_1 be the image of any C under the map $z \to z_1 := z^n$ and let f be the function satisfying (5.8-16) that maps the exterior of C_1 onto $|w| > a$. The potential of a flow through the "turbine wheel" of profiles C due to a source at the origin is

$$\phi(z) = k \operatorname{Log}|F(z^n)| + \frac{c}{2\pi} \arg f(z^n),$$

where F is defined as in Problem 9.

12. *Flow through cascade of airfoils.* Let an infinity of profiles with cross section C be arranged periodically with period $2\pi i h$ (Fig. 5.8m). [If z is in the interior of some profile, then so are the points $z + 2\pi i h k$, $k = \pm 1, \pm 2, \ldots$.] Let C_1 be the image of any C under the map $z \to z_1 := e^{z/h}$ and let f and F be defined as in Problem 11. The potential of a flow homogeneous for $\operatorname{Re} z \to \pm \infty$ through the cascade of profiles C is $\phi(z) := k \operatorname{Log}|F(e^{z/h})| + (c/2\pi) \arg f(e^{z/h})$.

§5.9. POISSON'S EQUATION

Let $\rho : (x, y) \to \rho(x, y)$ be a function defined in a region D. Certain physical problems require the determination of a function ϕ which in D satisfies the **Poisson equation**

$$\Delta \phi = \rho(x, y) \qquad (5.9\text{-}1)$$

and obeys certain conditions on the boundary Γ of D. Often such problems can be simplified by conformal transplantation. Let $f:z \to f(z)$ map D conformally onto some model region E and let $g:w \to f^{[-1]}(w)$ denote the inverse mapping. Let $\sigma(w) := \rho(g(w))$ for $w \in E$; i.e., let σ be the conformal transplant of ρ under the mapping f. If ϕ satisfies (5.9-1), its conformal transplant ψ, according to Theorem 5.6b, satisfies

$$\Delta_w \psi = \sigma(w) |g'(w)|^2. \qquad (5.9\text{-}2)$$

This again is a Poisson equation. It may, however, be easier to solve than (5.9-1), because the boundary conditions may have become simpler.[1]

EXAMPLE: TORSIONAL RIGIDITY

In addition to illustrating Poisson's equation, this example also serves to show the use of power series in conformal mapping problems. Let D be the cross section of a homogeneous cylindrical bar. If no forces are acting in the direction of the bar, the moment required to twist the bar by an angle θ is $M = \gamma \theta \lambda^{-1} \tau$, where γ is a constant that depends only on the material, λ is the length of the bar, and τ is a constant, called **torsional rigidity of the cross section** D and defined as follows: Let ϕ denote the solution of the boundary value problem

$$\Delta \phi = -2 \quad \text{in } D, \qquad (5.9\text{-}3\text{a})$$

$$\phi = 0 \quad \text{on } \Gamma. \qquad (5.9\text{-}3\text{b})$$

Then

$$\tau := \int \int_D \phi(x,y) \, dx \, dy. \qquad (5.9\text{-}4)$$

Our aim is to express τ explicitly in terms of the Taylor coefficients of the inverse mapping function g, where the model region E is the unit disk $|w| < 1$.

We first determine the stress function ϕ. This can evidently be obtained in the form $\phi = \phi_0 + \phi_1$, where ϕ_0 is any particular solution of (5.9-3a) (not necessarily one that satisfies the boundary conditions) and where ϕ_1 is the solution of the Dirichlet problem

$$\Delta \phi_1 = 0 \text{ in } D, \qquad (5.9\text{-}5\text{a})$$

$$\phi_1 = -\phi_0 \text{ on } \Gamma. \qquad (5.9\text{-}5\text{b})$$

To find a particular solution of (5.9-3a) it is not necessary to use conformal transplantation, for the differential equation is already simple enough. Setting $\phi_0(z) = \tilde{\phi}_0(r, \theta)$ ($z = re^{i\theta}$, we find that ϕ_0 is a solution independent of θ if and only if

$$\frac{\partial^2 \tilde{\phi}_0}{\partial r^2} + \frac{1}{r} \frac{\partial \tilde{\phi}_0}{\partial r} = -2.$$

---

[1] Other methods of dealing with Poisson's equation are discussed in §§13.4 and 15.2.

POISSON'S EQUATION

This is satisfied, for instance, for $\tilde{\phi}_0(r,\theta) = -\frac{1}{2}r^2$. Thus for ϕ_0 we may choose

$$\phi_0(z) = -\frac{1}{2}|z|^2.$$

We now determine ϕ_1 by conformal transplantation. Let ψ_i denote the transplant of ϕ_i under the mapping f ($i=0,1$). Then

$$\psi_0(w) = -\frac{1}{2}|g(w)|^2,$$

and by virtue of Corollary 5.6c ψ_1 is the solution of the Dirichlet problem

$$\Delta_w \psi_1 = 0, \quad |w| < 1, \tag{5.9-6a}$$

$$\psi_1(w) = \frac{1}{2}|g(w)|^2, \quad |w| = 1. \tag{5.9-6b}$$

Let the Taylor expansion of g at $w=0$ be

$$g(w) = b_0 + b_1 w + b_2 w^2 + \cdots,$$

and assume that the series converges so strongly for $|w| \leq 1$ (e.g., by terminating) that all subsequent operations are legitimate. (This condition could be enforced by first considering in place of D the image of $|w| \leq 1$ under the map $w \to g(\gamma w)$, where $0 < \gamma < 1$, and then letting $\gamma \to 1$ in the final result.) If $w = \rho e^{i\theta}$, then clearly

$$|g(w)|^2 = g(\rho e^{i\theta}) \overline{g(\rho e^{i\theta})}$$

$$= (b_0 + b_1 \rho e^{i\theta} + \cdots)(\overline{b_0} + \overline{b_1} \rho e^{-i\theta} + \cdots)$$

$$= \sum_{n=-\infty}^{\infty} c_n(\rho) e^{in\theta},$$

where for $n \geq 0$

$$c_n(\rho) := b_n \overline{b_0} \rho^n + b_{n+1} \overline{b_1} \rho^{n+2} + \cdots$$

$$= \sum_{k=0}^{\infty} b_{n+k} \overline{b_k} \rho^{n+2k}$$

and $c_n(\rho) = \overline{c_{-n}(\rho)}$ for $n < 0$. In particular, the boundary values of ψ_0 are

$$\psi_0(e^{i\theta}) = -\frac{1}{2} \sum_{n=-\infty}^{\infty} c_n(1) e^{in\theta}.$$

Using the well-known fact that for the region $|w| < 1$ the solution of the Dirichlet problem with boundary data $e^{in\theta}$ is $\rho^{|n|} e^{in\theta}$ (see Problem 8, §5.6), we find by linear superposition that the solution of the Dirichlet problem (5.9-6) is

$$\psi_1(w) := \frac{1}{2} \sum_{n=-\infty}^{\infty} c_n(1) \rho^{|n|} e^{in\theta}.$$

Thus the transplanted solution $\psi = \psi_0 + \psi_1$ of the original problem (5.9-3) is

$$\psi(w) = \frac{1}{2} \sum_{n=-\infty}^{\infty} [c_n(1)\rho^{|n|} - c_n(\rho)] e^{in\theta}. \qquad (5.9\text{-}7)$$

It remains to evaluate the integral (5.9-4). By Theorem 5.6d,

$$\tau = \iint_{|w| \leq 1} \psi(w) |g'(w)|^2 \rho \, d\rho \, d\theta. \qquad (5.9\text{-}8)$$

We first express $|g'(w)|^2$ as a Fourier series. As above,

$$|g'(w)|^2 = (b_1 + 2b_2 \rho e^{i\theta} + \cdots)(\overline{b_1} + 2\overline{b_2}\rho e^{-i\theta} + \cdots)$$

$$= \sum_{n=-\infty}^{\infty} d_n(\rho) e^{in\theta},$$

where

$$d_n(\rho) = \sum_{k=1}^{\infty} (n+k) k b_{n+k} \overline{b_k} \rho^{n+2k-2}, \qquad n \geq 0,$$

and $d_n(\rho) = \overline{d_{-n}(\rho)}$ for $n < 0$. Integrating in (5.9-8) first with respect to θ, we get by Parseval's relation

$$\tau = \int_0^1 \sum_{n=-\infty}^{\infty} [c_n(1)\rho^{|n|} - c_n(\rho)] \, \overline{d_n(\rho)} \, \rho \, d\rho.$$

The integral of each term in the sum is calculated easily: For $n \geq 0$

$$\int_0^1 [c_n(1)\rho^{|n|} - c_n(\rho)] \, \overline{d_n(\rho)} \, \rho \, d\rho$$

$$= \int_0^1 \sum_{k=0}^{\infty} b_{n+k} \overline{b_k} (\rho^n - \rho^{n+2k}) \sum_{m=0}^{\infty} (n+m) m \, \overline{b_{n+m}} b_m \rho^{n+2m-1} \, d\rho$$

$$= \sum_{k,m=0}^{\infty} (n+m) m b_{n+k} \overline{b_{n+m}} \, \overline{b_k} b_m \int_0^1 (\rho^{2n+2m-1} - \rho^{2n+2m+2k-1}) \, d\rho$$

$$= \frac{1}{2} \sum_{k,m=1}^{\infty} \frac{km}{n+k+m} b_{n+k} \overline{b_{n+m}} \, \overline{b_k} b_m.$$

For $n < 0$ the integral has the complex conjugate value. Thus we find

$$\tau = \pi \sum_{n=0}^{\infty}{}^{*} \sum_{m,k=1}^{\infty} \frac{km}{n+k+m} \operatorname{Re}(b_{n+k} \overline{b_{n+m}} \, \overline{b_k} b_m).$$

POISSON'S EQUATION

The * indicates that the term corresponding to $n=0$ is to be multiplied by $\frac{1}{2}$. Three summations to infinity must be performed in this sum. We can write it as a simply infinite sum if we collect all terms in which the sum of the indices of the coefficients $b_0, b_1, \ldots$, has a fixed value q. Setting $m+k=p$, we first get for the inner sum

$$\sum_{m,k=1}^{\infty} \frac{km}{n+k+m} \operatorname{Re}(\cdots)$$

$$= \sum_{p=2}^{\infty} \frac{1}{n+p} \sum_{m=1}^{p-1} {}' m(p-m) \operatorname{Re}\left(b_{n+p-m} \overline{b_{n+m}} b_m \overline{b_{p-m}}\right).$$

Now, setting $n+p =: q$, we find

$$\tau = \pi \sum_{q=2}^{\infty} \frac{1}{q} \sum_{p=2}^{q} {}^{*} \sum_{m=1}^{p-1} m(p-m) \operatorname{Re}\left(b_{q-m} \overline{b_{q-p+m}} b_m \overline{b_{p-m}}\right), \qquad (5.9\text{-}9)$$

the * indicating that the term with $p=q$ (corresponding to $n=0$) is to be taken with the factor $\frac{1}{2}$. In view of the complicated definition of the torsional rigidity in terms of the solution of a partial differential equation, (5.9-9) is as simple a general formula for τ as can be hoped for.

As a first application of the formula we compute the torsional rigidity of a circular cross section of radius R. Here a suitable mapping is $g(w) := Rw$, i.e., we have $b_0 = 0$, $b_1 = R$, $b_2 = b_3 = \cdots = 0$. Thus only the term with $q=2$, $p=2$, $m=1$ gives a nonzero contribution, and we find $\tau = (\pi/4) R^4$. This result can be verified directly.

As a second application of (5.9-9) we shall prove a famous theorem due to Saint-Venant, which states that among all simply connected cross sections of a given area the circular, and only the circular, cross section has the largest torsional rigidity. Equivalently, denoting the area by A, the theorem states that the ratio τ/A^2 is largest for the circle, where it has the value $1/4\pi$, i.e.,

$$\tau \leq \frac{1}{4\pi} A^2. \qquad (5.9\text{-}10)$$

We shall prove this inequality for all domains D that can be mapped conformally onto the unit disk by a function f whose inverse g possesses a Taylor expansion such that the series (5.6-12) and (5.9-9) converge. From (5.6-12) we have

$$\frac{1}{4\pi} A^2 = \frac{\pi}{4} \sum_{q=2}^{\infty} \sum_{k=1}^{q-1} k(q-k) b_k \overline{b_k} b_{q-k} \overline{b_{q-k}}. \qquad (5.9\text{-}11)$$

We shall prove (5.9-10) by showing that for each $q = 2, 3, \ldots$, the corresponding inequality holds between the qth terms in the series (5.9-9) and (5.9-11). Because

$$(b_{q-m} b_m - b_{q-p+m} b_{p-m})\left(\overline{b_{q-m}} \overline{b_m} - \overline{b_{q-p+m}} \overline{b_{p-m}}\right) \geq 0, \qquad (5.9\text{-}12)$$

we have

$$|2\operatorname{Re}(b_{q-m}\overline{b_{q-p+m}}b_m\overline{b_{q-m}})| \leq B_m + B_{p-m},$$

where

$$B_m := |b_m|^2|b_{q-m}|^2 \quad (m=1,2,\ldots,q-1).$$

Hence the qth term in the series for τ is less than

$$\frac{1}{2q}\sum_{p=2}^{q}\sum_{m=1}^{p-1} m(p-m)(B_m+B_{p-m}) = \frac{1}{q}\sum_{p=2}^{q}\sum_{m=1}^{p-1} m(p-m)B_m. \quad (5.9\text{-}13)$$

Reversing the order of summation, we express (5.9-13) as

$$\frac{1}{q}\sum_{m=1}^{q-1} B_m \sum_{p=m+1}^{q}{}^* m(p-m) = \frac{1}{2q}\sum_{m=1}^{q-1} m(q-m)^2 B_m.$$

Since $B_m = B_{q-m}$, it may also be written

$$\frac{1}{4q}\sum_{m=1}^{q-1}[m(q-m)^2 + (q-m)m^2]B_m = \frac{1}{4}\sum_{m=1}^{q-1} m(q-m)B_m.$$

The last expression, however, is identical with the qth term in (5.9-11).

Our method also permits a discussion of equality. Equality holds in (5.9-10) if and only if it holds in (5.9-12) for all q,p, and m in question, i.e. if

$$b_{q-m}b_m = b_{q-p+m}b_{p-m} \quad (5.9\text{-}14)$$

for $q=2,3,\ldots$; $p=2,\ldots,q$; $m=1,\ldots,q-1$. In particular, this must hold for $m=1$, $q=p+1=2,3,\ldots$:

$$b_p b_1 = b_2 b_{p-1}, \quad p=2,3,\ldots.$$

We know that $b_1 \neq 0$ because the mapping defined by $g(w)=b_0+b_1 w+\cdots$ is conformal at $w=0$. Dividing by b_1 and letting $\epsilon := b_2/b_1$, we find $b_p = \epsilon b_{p-1}$, $p=2,3,\ldots$, hence $b_p = \epsilon^{p-1} b_1$, $p=2,3,\ldots$, as a necessary condition for equality. Thus the mapping function must have the form

$$g(w) = b_0 + b_1(w + \epsilon w^2 + \epsilon^2 w^3 + \cdots).$$

Only for $|\epsilon|<1$ can we obtain a power series that is still convergent for $|w|=1$. The mapping function can then be expressed as

$$g(w) = b_0 + \frac{b_1 w}{1-\epsilon w}.$$

This is a Moebius transformation that maps the unit disk onto a circular disk. Thus in the case of equality in (5.9-10), D is a circular disk.

GENERAL RESULTS ON CONFORMAL MAPS

PROBLEMS

1. Compute the torsional rigidity for the image of $|w| \leq 1$ under the map
$$w \to z := Rw(1+\epsilon w^n),$$
where n is an integer and $0 \leq \epsilon < (1/n+1)$.

2. Conformal mapping is not always the best method of determining torsional rigidity. Compute the torsional rigidity of an elliptical cross section (semiaxes a,b) by suitably determining the constants A,B,C in
$$\phi(x,y) := Ax^2 + By^2 + C.$$

3. Without using conformal maps, compute the torsional rigidity of a circular cross section of radius R which has a crack along a straight line emanating from the center (see Fig. 5.9). [Use $\phi_0(x,y) := -y^2$ and determine ϕ_1 as a Fourier series of the form
$$\phi_1(x,y) = \sum_{n=0}^{\infty} A_n r^{n+1/2} \cos(n+\tfrac{1}{2})\theta\,]$$

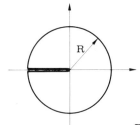

Fig. 5.9. A bar with a crack.

§5.10. GENERAL RESULTS ON CONFORMAL MAPS

The applications of conformal maps, discussed in preceding sections, all require the existence (and availability) of conformal maps of arbitrary "physical" regions onto standardized "model" regions. In this section we state, partly without proof, some general facts concerning the existence and uniqueness of such conformal maps.

To simplify the language, a "map" of a region D onto a region E (unless otherwise specified) is always assumed to be defined by a function f which is analytic in D and which assumes every value w in E at exactly one point z in D. We recall that this f satisfies $f'(z) \neq 0$ at all z in D, implying that the map is conformal. Furthermore, the inverse function $f^{[-1]}$ defines a "map" (in the same strict sense) of E onto D.

To begin with the simplest situation, let the physical region be a region of the (unextended) complex plane and let the intended model region be a disk. The basic facts are as follows.

THEOREM 5.10a (Mapping theorem for simply connected regions)

Let D be a simply connected region and let z_0 be a point of D. (I) If $D = \mathbb{C}$, the complex plane, there exists no function f mapping D onto a bounded disk $|w| < \rho < \infty$. There exists precisely one function f mapping the complex plane onto itself and satisfying the conditions

$$f(z_0) = 0, \qquad f'(z_0) = 1, \qquad (5.10\text{-}1)$$

namely the function $f(z) := z - z_0$. (II) If D does not constitute the whole complex plane, there is precisely one real number $\rho > 0$ and one function f satisfying (5.10-1) and mapping D onto the disk $|w| < \rho$.

Proof of (I). Let f be a function mapping $\mathbb{C}$ onto $|w| < \rho$, where $\rho < \infty$. Then f is entire; it is also bounded, since $|f(z)| < \rho$ for all z. By Liouville's theorem (Theorem 3.3b) f is constant, and, contrary to our assumption, cannot map $\mathbb{C}$ onto a disk. This contradiction shows that $\mathbb{C}$ cannot be mapped onto a bounded disk.

Now let f be a function mapping $\mathbb{C}$ onto itself; f is again entire, hence is represented by a power series in $z - z_0$ with infinite radius of convergence which by (5.10-1) has the form

$$f(z) = (z - z_0) + \sum_{k=2}^{\infty} a_k (z - z_0)^k. \qquad (5.10\text{-}2)$$

Assume that f has an essential singularity at $z = \infty$, i.e., that an infinite number of $a_k \neq 0$. Then by the Casorati-Weierstrass theorem (Theorem 4.4g) there exists, for every integer $n > 0$, a point z_n such that $|z_n| > n$ and $|f(z_n)| < n^{-1}$. Thus, if $w_n := f(z_n)$, the sequence $\{w_n\}$ tends to zero, whereas the sequence $\{f^{[-1]}(z_n)\}$ is unbounded. This contradicts the fact that $f^{[-1]}$ is analytic, hence certainly continuous, at $w = 0$. Therefore f cannot have an essential singularity at $z = \infty$; i.e., f is a polynomial. Because f defines a map, $f'(z) \neq 0$ for all z, and it follows by the fundamental theorem of algebra that the degree of the polynomial f is 1. By virtue of (5.10-2), $f(z) = z - z_0$. ∎

We now turn to the proof of (II) in Theorem 5.10a. The proof of the *existence* of the mapping function is relatively difficult and is postponed until Chapter 14 in which the necessary tools have been prepared. Here, however, we shall prove the *uniqueness* of the mapping function under the conditions stated.

GENERAL RESULTS ON CONFORMAL MAPS

Assume that there are two functions that satisfy (5.10-1), say f_1 and f_2, which map D onto $|w|<\rho_1$ and $|w|<\rho_2$, respectively. Let f be the function mapping $|w|<1$ onto itself, defined by

$$f(w) := \frac{1}{\rho_1} f_1\left(f_2^{[-1]}(\rho_2 w)\right) \quad . \tag{5.10-3}$$

Then $f(0)=0$ and, by the chain rule, $f'(0) = \rho_2/\rho_1$. Without loss of generality we may assume that $\rho_2 \geqslant \rho_1$, since otherwise the roles of f_1 and f_2 could be interchanged. Thus we obtain

$$f'(0) \geqslant 1. \tag{5.10-4}$$

The following theorem, usually called Schwarz's lemma, implies that a function f with the stated properties necessarily equals the function $f(w) := w$.

THEOREM 5.10b (Schwarz's lemma)

Let E denote the unit disk, let f be analytic in E, and let $f(E) \subset E, f(0)=0$. Then (a) $|f(w)| \leqslant |w|$ for all $w \in E$ and consequently $|f'(0)| \leqslant 1$; (b) if $|f'(0)|=1$ or if $|f(w)|=|w|$ for some $w \neq 0$, $w \in E$, then there exists a constant c of modulus 1 such that $f(w) = cw$ for all $w \in E$.

Proof. Let

$$g(w) := \frac{f(w)}{w}, \qquad 0 < |w| < 1.$$

The function g, being a ratio of two analytic functions with a nonvanishing denominator, is analytic. At $w=0$ g has an isolated singularity. Since, however, f is analytic at $w=0$ and $f(0)=0$,

$$\lim_{w \to 0} g(w) = \lim_{w \to 0} \frac{f(w)}{w} = \lim_{w \to 0} \frac{f(w)-f(0)}{w-0} = f'(0)$$

exists. Consequently the singularity is removable and, if we define $g(0) := f'(0)$, g becomes analytic throughout $|w|<1$. We now apply the maximum principle to the set $|w| \leqslant 1-\epsilon$, where ϵ is an arbitrarily small number >0. Since $|f(w)|<1$ for $|w|<1$, we have

$$\max_{|w| \leqslant 1-\epsilon} |g(w)| = \max_{|w|=1-\epsilon} \frac{|f(w)|}{|w|} < \frac{1}{1-\epsilon},$$

CONFORMAL MAPPING

and because this is true for all $\epsilon > 0$

$$|g(w)| \leq 1, \quad w \in E, \qquad (5.10\text{-}5)$$

follows. By the definition of g this implies conclusion (a) of the theorem. If $|f(w)| = |w|$ for some $w \in E$, $w \neq 0$ or if $|f'(0)| = 1$, equality holds in (5.10-5) for some $w \in E$, which by the strengthened form of the maximum principle given as Theorem 2.4h is not possible unless $g(w) = c$ for all $w \in E$, where $|c| = 1$. By the definition of g (b) follows. ∎

To complete the proof of the uniqueness statement in (II) of Theorem 5.10a we apply the Schwarz lemma to the f defined by (5.10-3). From (a) we deduce that equality must hold in (5.10-4), hence that $\rho_1 = \rho_2$. This then means that the hypothesis of (b) is satisfied. There follows $f(w) = cw$, where $|c| = 1$; however, since $f'(0) = 1$, c must equal 1. Hence $f(w) = w$, which in view of $\rho_1 = \rho_2$ implies $f_1 = f_2$. ∎

Not in all applications of conformal mapping is the model domain a disk. The more general situation is easily reduced to that special one.

COROLLARY 5.10c

Let D and E be two simply connected regions that are not the whole complex plane, let $z_0 \in D$, $w_0 \in E$, and let ϕ be a given real number. Precisely one map of D onto E then exists such that the mapping function f satisfies

$$f(z_0) = w_0, \quad \arg f'(z_0) = \phi. \qquad (5.10\text{-}6)$$

Proof. By Theorem 5.10a there exist real numbers ρ_1, ρ_2 and functions f_1, f_2, all uniquely determined, such that f_1 and f_2 map the domains D and E onto the disks $|w| < \rho_1$ and $|w| < \rho_2$, respectively, and that $f_1(z_0) = f_2(w_0) = 0$ and $f_1'(z_0) = f_2'(w_0) = 1$. The function

$$f(z) := f_2^{[-1]}\left(e^{i\phi}\frac{\rho_2}{\rho_1}f_1(z)\right)$$

then has the required properties. Failure of f to be unique would contradict the uniqueness statement of Theorem 5.10a. ∎

The special case of Corollary 5.10c in which E is the unit disk $|w| < 1$ and $\phi = 0$ is usually called the **Riemann mapping theorem**.

Some applications (see §§5.7 and 5.8) require the existence of a function f that maps the complement of a given compact, simply connected set C (the "obstacle" in hydrodynamical applications) onto the exterior of a disk.

GENERAL RESULTS ON CONFORMAL MAPS

The map should be such that at infinity it behaves like the identical mapping; more precisely, the mapping function should have a simple pole at ∞ and its Laurent series should be

$$f(z) = z + c_0 + \frac{c_1}{z} + \cdots. \tag{5.10-7}$$

To construct f let D be the complement of C with respect to the extended plane and let z_0 be a point of C. The auxiliary map

$$g : z \to z^* := g(z) := \frac{1}{z - z_0}$$

maps D onto a simply connected region D^* in the (unextended) plane and $g(\infty) = 0$. By Theorem 5.10a there exists a unique ρ, $0 \leq \rho < \infty$, and a unique function h mapping D^* on a (possibly infinite) disk $E^* : |w^*| < \rho^{-1}$ such that

$$h(0) = 0, \quad h'(0) = 1. \tag{5.10-8}$$

By the inversion

$$k : w^* \to w := k(w^*) = \frac{1}{w^*}$$

E^* is mapped onto the set $E : |w| > \rho$. Thus the function $f := k \circ h \circ g$ maps D onto E and $f(\infty) = \infty$. The Laurent series at ∞ may be computed formally by forming the composition of the Laurent series $g(z) = z^{-1} + z_0 z^{-2} + \cdots$, $h(z) = z + b_2 z^2 + \cdots$, and $k(z) = z^{-1}$. We find

$$f(z) = \left[(z^{-1} + z_0 z^{-2} + \cdots) + O(z^{-2}) \right]^{-1}$$

$$= z + c_0 + O(z^{-1}),$$

as required.

We shall now show that there is only one ρ and one f with the required properties and that the result of the above construction is independent of the choice of z_0 in C. Let f_1 and f_2 be two functions that satisfy (5.10-7) and map D onto the sets $|w| < \rho_1$ and $|w| < \rho_2$, respectively. If z_0 is again an arbitrary point of C and g and k are defined as above, the functions $h_i := k^{[-1]} \circ f_i \circ g^{[-1]}$ ($i = 1, 2$) both map D^* onto disks $|w^*| < \rho_i$ such that (5.10-8) holds. By Theorem 5.10a this can be done in only one way. There follows $h_1 = h_2$, hence $f_1 = f_2$.

We have established the following Corollary:

COROLLARY 5.10d

Let C be a simply connected compact set. A unique number ρ, $0 \leq \rho < \infty$ and a unique function f then exist such that f maps the complement of C onto $|w| > \rho$ and satisfies (5.10-7) *for large $|z|$*.

Proposition (I) of Theorem 5.10a also implies that $\rho = 0$ if and only if C consists of the single point z_0, in which case the mapping function assumes the simple form $f(z) := z - z_0$.

The foregoing theorems all refer to the mapping of *regions* in the strict mathematical sense, i.e., of connected open sets. The functions f which realize the various mappings thus are defined only in the interior of the physical "domains" to be mapped. This does not fully satisfy the requirements of the applications of the conformal maps we have discussed, because it is usually essential that the *boundary* of the region be mapped along with the region itself to allow the mapping to remain continuous (if not necessarily analytic) on the boundary of the domain.

It is suggested by some of the examples that we have considered and thus may seem intuitively obvious that the mapping function can always be extended to the boundary in a continuous manner. The mathematical concept of a simply connected region is a very general one, however, and there are regions for which such an extension of the mapping function is not possible. If, for instance, D is a square with an infinite number of slits, as shown in Fig. 5.10a, then no function mapping D onto $|w| < 1$ can be defined at $z = 0$ in such a way that it will be continuous at $z = 0$.

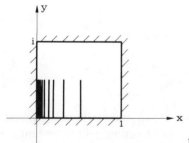

Fig. 5.10a. A simply connected region.

A positive statement is possible in an important special case. We recall the following definitions:

A region is called a **Jordan region** if it is the interior of a Jordan curve.

A map of one set onto another is called **topological** (or a **homeomorphism**) if it is one-to-one and continuous in both directions.

GENERAL RESULTS ON CONFORMAL MAPS

THEOREM 5.10e ("Osgood-Carathéodory theorem")

Let D and D* be two Jordan regions. Any function mapping D conformally and one-to-one onto D* can be extended to a topological map of the closure of D onto the closure of D*.

The *proof* of this theorem is technically difficult (see Chapter 14). At this point we shall be content to prove the following converse:

THEOREM 5.10f (The principle of the boundary map)

Let D be a Jordan region with boundary Γ and let f be analytic in D and continuous on the closure $D \cup \Gamma$ of D. Let $\Gamma^* := f(\Gamma)$ be a positively oriented Jordan curve. Then f maps D conformally and one-to-one onto the interior D^* of Γ^*.

The *proof* is a simple application of the principle of the argument. It is to be shown that f assumes each value w^* in D^* exactly once. By the principle of the argument (Theorem 4.10a) the number of times a value w^* is assumed by f in D equals the winding number

$$n(f(\Gamma), w^*) := \frac{1}{2\pi} [\arg(f(z) - w^*)]_\Gamma;$$

but by hypothesis, since w^* lies in the interior of the positively oriented Jordan curve Γ^*,

$$n(f(\Gamma), w^*) = n(\Gamma^*, w^*) = 1,$$

as was to be shown. ∎

This principle can be useful for the purpose of verifying that a certain analytic function solves a given mapping problem; see §5.13.

If an arbitrary simply connected region D is mapped onto some region E, the mapping function f is determined uniquely by prescribing the image of one point z_0 and the argument of $f'(z_0)$ (Corollary 5.10c). In many physical applications, however, it is more natural to fix the mapping function by prescribing the correspondence of the boundaries. This is possible in Jordan regions.

THEOREM 5.10g

Let D and E be two Jordan regions with boundary curves Γ and Γ^*, both positively oriented. Let z_1, z_2, z_3, and w_1, w_2, w_3 be distinct points on Γ and Γ^*, respectively, ordered in the sense of increasing parameter values. Then there exists exactly one homeomorphism of $D \cup \Gamma$ onto $E \cup \Gamma^*$ which is conformal in the interior and sends the points z_i into w_i ($i = 1, 2, 3$).

Proof. It evidently suffices to consider the case in which E is the unit disk. According to the Riemann mapping theorem there exists a map g of D onto $|w|<1$ which by Theorem 5.10e can be extended to a topological map of the closure of D onto $|w| \leq 1$. The extended map sends the points z_i into points w'_i such that $|w'_i|=1$. The orientation induced by the points w'_i on the boundary $|w|=1$ is the positive one; otherwise the principle of the argument (used in the proof of Theorem 5.10f) would contradict the mapping of D onto $|w|<1$. There now exists a unique Moebius transformation t mapping w'_i onto w_i ($i=1,2,3$). Because both triples lie on $|w|=1$, this transformation maps $|w|=1$ onto $|w|=1$, and since the orientation of the circle is preserved it maps (again by the principle of the argument) $|w| \leq 1$ onto $|w| \leq 1$. The map $f := t \circ g$ has all the required properties. It is unique, for the existence of two such maps would imply the existence of a nontrivial map of $|w| \leq 1$ onto itself, leaving three boundary points fixed. ∎

In spite of its generality the Osgood-Carathéodory theorem does not cover all cases of practical interest. Take, for instance, the region D which consists of the unit disk $|z|<1$, cut along the real axis between $z=\frac{1}{2}$ and $z=1$ (see Fig. 5.10b). The function f mapping D onto $|w|<1$ and satisfying $f(0)=0, f'(0)>0$ is elementary (see Problem 3, §5.11). If

$$z_n := \frac{3}{4} + (-1)^n \frac{1}{n} i, \quad n=1,2,\ldots,$$

then $\lim z_n = \frac{3}{4}$ clearly exists. The limit of the image points $w_n := f(z_n)$ does not exist however. In fact, the sequences of even and odd indexed points approach two different points on the boundary $|w|=1$ of the image region (see Fig. 5.10b). Thus it is not possible to define f at the boundary point $z=\frac{3}{4}$ in a continuous manner.

The example shows that in simply connected regions which are not Jordan regions it is generally not possible to define the mapping function

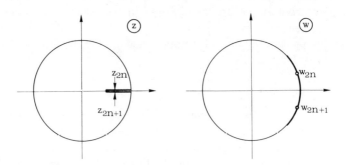

Fig. 5.10b. Mapping a disk with a slit.

GENERAL RESULTS ON CONFORMAL MAPS

continuously on the boundary without taking into account the manner of approach to the boundary. In the above example $\lim f(z_n)$ exists for all sequences $\{z_n\}$ that approach a point on the slit from one side of the slit only.

Carathéodory has created a theory that permits a complete description of the correspondance of the boundaries in conformal maps, based on the notions of *accessible boundary points* and *prime ends*. The generality of Carathéodory's theory of prime ends is seldom required in applications, and no attempt has been made to present the theory in the present context. In many cases of practical interest the behavior of the mapping function near the boundary of non-Jordan regions can be described more elementarily.

Let D be a simply connected domain. We call a **crosscut** of D any simple arc Γ contained in D with the exception of its end points which are boundary points of D (see Fig. 5.10c). Now let D be a simply connected region which is not a Jordan region and suppose that there exists a crosscut Γ of D which divides D into two Jordan regions D_1 and D_2. Let D be mapped onto $E:|w|<1$. Normally the curve $f(\Gamma)$ can be completed at the end points to define a crosscut of E. Then $f(\Gamma)$ divides E into two Jordan regions, E_1 and E_2. Since f now maps each Jordan region D_1 and D_2 onto a Jordan region, the Osgood-Carathéodory theorem is applicable to each region D_i, and it follows that in each subregion f can be extended to a topological map of the region, including its boundary; for instance, in a single slit we find that each edge of the slit is mapped onto a different part of the boundary of the image region. In more complicated regions, if there are several slits, it may be necessary to subdivide D by several (nonintersecting) crosscuts.

So far only conformal maps of simply connected regions have been considered. The theory of conformal maps has also been pushed to a high

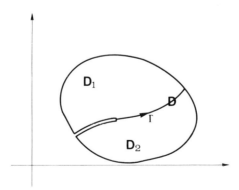

Fig. 5.10c. Cross-cut.

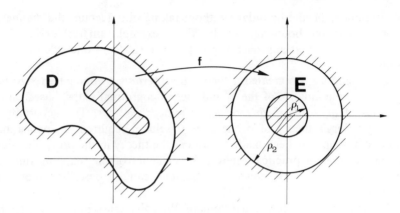

Fig. 5.10d. Mapping a doubly connected region.

degree of perfection for multiply connected regions. Here there is the additional complication that even for regions of a fixed connectivity the number of model domains itself is infinite; it is no longer possible to map all regions on one or two standard regions. We refer to Ahlfors [1966], Chapter 6, for a compact treatment of these matters. It is interesting to observe that the nature of the model regions considered there could be inferred by considerations of electrostatic potential problems.

Constructive methods have been developed for mappings of doubly connected regions; see Chapter 14. The underlying mapping theorem is as follows (see Fig. 5.10d).

THEOREM 5.10h

Let D be a doubly connected region in the complex plane, and let C_1 and C_2 denote the components of D with respect to the extended complex plane. There then exist real numbers ρ_1 and ρ_2, $0 \leq \rho_1 < \rho_2 \leq \infty$, and a function f which maps D onto the annulus $\rho_1 < |w| < \rho_2$. If neither C_1 nor C_2 consists of a single point, then $0 < \rho_1, \rho_2 < \infty$, and the ratio ρ_2/ρ_1 is uniquely determined by D.

The ratio ρ_2/ρ_1 is called the **modulus** of the region D. The proof (save for the exceptional situations mentioned) that the modulus of a doubly connected region is uniquely determined is given in §5.11.

PROBLEMS

1. Prove in the context of Schwarz's lemma (Theorem 5.10b) that $|f(w)| < 1$ for $|w| < 1$ implies $|f'(0)| \leq 1$ regardless of the value of $f(0)$.

2. Prove the following generalization of Schwarz's lemma: if f is analytic in the unit disk E, $f(E) \subset E$, and if $f(0) = f'(0) = \cdots = f^{(n-1)}(0) = 0$ for some integer $n > 0$, then (a) $|f(w)| \leq |w|^n$ and consequently $|f^{(n)}(0)| \leq n!$; (b) if either $|f^{(n)}(0)| = n!$ or $|f(w)| = |w|^n$ for some $w \in E$, $w \neq 0$, then $f(w) = cw^n$, where $|c| = 1$.

3. Prove the following generalization of Schwarz's lemma: let $f : z \to f(z)$ be analytic and $|f(z)| \leq \mu$ for $|z| < \rho$ and let $f(z_0) = w_0$, where $|z_0| < \rho$, $|w_0| < \mu$. Then

$$\left| \frac{\mu [f(z) - w_0]}{\mu^2 - \overline{w}_0 f(z)} \right| \leq \left| \frac{\rho(z - z_0)}{\rho^2 - \overline{z}_0 z} \right|$$

for $|z| < \rho$.

§5.11. SYMMETRY

Although the Riemann mapping theorem asserts the existence of a large class of conformal maps, it does not provide a formula or algorithm for the construction of these maps. Algorithms permitting the construction of the mapping functions with the aid of digital computers are presented in Chapter 14. In the balance of this chapter we discuss some simple but nevertheless important cases in which the mapping problem can be solved by explicit formulas. In this section we state some simple facts on analytic continuation that are required for the derivation of these formulas as well as for other purposes of complex analysis.

I. Analytic Continuation by Continuous Continuation

THEOREM 5.11a

Let D_1 and D_2 be two disjoint regions whose boundaries have a smooth arc Γ in common. Let the functions f_i be analytic in D_i and continuous in $D_i \cup \Gamma$ ($i = 1, 2$), and let $f_1(z) = f_2(z)$ for $z \in \Gamma$. The function f defined by

$$f(z) = \begin{cases} f_1(z), & z \in D_1, \\ f_1(z) = f_2(z), & z \in \Gamma, \\ f_2(z), & z \in D_2, \end{cases} \quad (5.11\text{-}1)$$

is analytic in $D := D_1 \cup \Gamma \cup D_2$.

The point of the theorem lies in the fact that the functions f_1 and f_2, which by hypothesis are merely *continuous* continuations of each other, are in reality *analytic* continuations. Thus where analytic functions are concerned continuous continuation is analytic continuation.

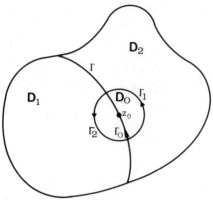

Fig. 5.11a. Continuous continuation.

Proof. It suffices to show that f is analytic at every point of Γ. Let z_0 be a point of Γ, and let ρ be so small that (a) the disk $D_0 \colon |z - z_0| < \rho$ belongs to D, (b) its boundary Γ_ρ meets Γ at exactly two points. We call Γ_i that portion of Γ_ρ which lies in D_i, and Γ_0 that part of Γ which lies in D_0 (Fig. 5.11a).

We now apply Cauchy's integral formula to each of the sets $D_0 \cap D_1$ and $D_0 \cap D_2$, whose boundaries are $\Gamma_1 - \Gamma_0$ and $\Gamma_2 + \Gamma_0$, respectively. If $z \in D_0 \cap D_1$, then

$$\frac{1}{2\pi i}\int_{\Gamma_1 - \Gamma_0} \frac{f_1(t)}{t-z}\,dt = f_1(z), \qquad \frac{1}{2\pi i}\int_{\Gamma_2 + \Gamma_0} \frac{f_2(t)}{t-z}\,dt = 0;$$

if $z \in D_0 \cap D_2$,

$$\frac{1}{2\pi i}\int_{\Gamma_1 - \Gamma_0} \frac{f_1(t)}{t-z}\,dt = 0, \qquad \frac{1}{2\pi i}\int_{\Gamma_2 + \Gamma_0} \frac{f_2(t)}{t-z}\,dt = f_2(z).$$

Because the integrals along Γ_0 cancel each other, adding these integrals according to the definition of f yields

$$\frac{1}{2\pi i}\int_{\Gamma_\rho} \frac{f(t)}{t-z}\,dt = \begin{cases} f_1(z), & z \in D_0 \cap D_1, \\ f_2(z), & z \in D_0 \cap D_2. \end{cases}$$

Since the integrand depends continuously on the parameter z, for $z \in \Gamma_0$ the integral equals the common value of f_1 and f_2. Thus

$$\frac{1}{2\pi i}\int_{\Gamma_\rho} \frac{f(t)}{t-z}\,dt = f(z) \quad \text{for all } z \in D_0.$$

SYMMETRY

In fact, the integrand depends on z not only continuously but analytically. Thus by Theorem 4.1a the integral depends analytically on z. Hence f is analytic in D_0. Of course, the analyticity of f cannot be deduced from the Cauchy integral formula because this formula was derived under the *hypothesis* that f is analytic. ∎

II. Analytic Continuation by Reflection

We recall the concept of symmetry with respect to a generalized circle Γ (see §5.4). It can be used for purposes of analytic continuation:

THEOREM 5.11b (The reflection principle)

Let Γ be a (generalized) circle and let D be a region in the extended complex plane that does not intersect Γ but whose boundary contains a nondegenerate subarc Γ_0 of Γ. Let the function f be analytic in D and continuous in $D \cup \Gamma_0$ and let the values that f takes on Γ_0 all lie on a (generalized) circle Γ^*. Then f can be continued analytically into D', the reflection of D with respect to Γ. If z' is the point symmetric to z with respect to Γ and w^* is the point symmetric to w with respect to Γ^*, the continuation is given by the formula

$$f(z') = f(z)^* \quad (z \in D). \tag{5.11-2}$$

This formula states that the values taken by f at points that are symmetric with respect to Γ are symmetric with respect to Γ^* (Fig. 5.11b).

Proof. Let t and t^* be two Moebius transformations that map Γ and Γ^* onto the real axis. To fix ideas we choose t so that $\Lambda := t(\Gamma_0)$ is an open

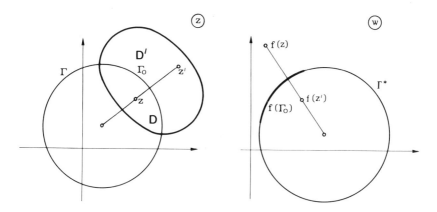

Fig. 5.11b. Symmetric continuation.

interval on the real axis. The function

$$g_1 := t^* \circ f \circ t^{[-1]}$$

is analytic in $E := t(D)$ and continuous in $E \cup \Lambda$ and its values on Λ are real. We now define in $\bar{E} = t(D')$ a function g_2 by reflecting the values of g:

$$g_2(z) := \overline{g(\bar{z})}, \quad z \in \bar{E}. \quad (5.11\text{-}3)$$

Then g_2 is analytic in $\bar{E}$ (this can be determined by the power series or by the Cauchy-Riemann equations) and continuous on $\bar{E} \cup \Lambda$; moreover, for $z \in \Lambda$, $g_1(z) = g_2(z)$. Thus by Theorem 5.11a the function g defined by

$$g(z) = \begin{cases} g_1(z), & z \in E \cup \Lambda, \\ g_2(z), & z \in \bar{E} \cup \Lambda, \end{cases}$$

is analytic in $E \cup \Lambda \cup \bar{E}$. Hence the function

$$f = t^{*[-1]} \circ g \circ t,$$

which in D agrees with the original function f, is analytic in $D \cup \Lambda_0 \cup D'$. Its values at points symmetric with respect to Γ are the images under a Moebius transformation of values that are symmetric with respect to the real axis. Since the real axis is mapped onto Γ^*, they lie symmetric with respect to Γ^*. ∎

Several applications of the **reflection principle** are now presented. Our first application concerns the uniqueness of annular maps.

THEOREM 5.11c

For $k = 1, 2$, let $0 < \rho_k < 1$ and let D_k denote the annulus $\rho_k < |z| < 1$. If the function $f : z \to w := f(z)$ maps D_1 conformally onto D_2 so that the mapping is continuous on the closure of D_1 and $|z| = \rho_1$ is mapped onto $|w| = \rho_2$, then $\rho_1 = \rho_2$, and $f(z) = cz$, where $|c| = 1$.

Proof. Let $D_1 = : D_1^{(0)}$. By the reflection principle f can be continued successively as an analytic function into the annuli

$$D_1^{(k)} : \rho_1^k < |z| < \rho_1^{k+1},$$

$k = 1, 2, \ldots$, always by reflecting at the inner boundary circle. The values taken by f in $D_1^{(k)}$ lie in

$$D_2^{(k)} : \rho_2^k < |w| < \rho_2^{k+1}.$$

SYMMETRY

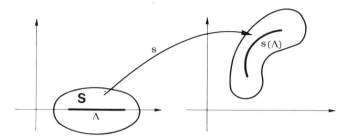

Fig. 5.11c. Analytic arc.

In this way f can be defined as a one-to-one analytic function in the punctured disk $0<|z|<1$. Because f is bounded, the singularity at $z=0$ is removable. Indeed,

$$\lim_{z\to 0} f(z)=0,$$

and thus by defining $f(0)=0$ we obtain a one-to-one mapping of $|z|<1$ onto itself which leaves the origin fixed. By the uniqueness statement of the Riemann mapping theorem the mappings $f: z \to cz$ ($|c|=1$) are the only ones with this property. This implies that $\rho_1 = \rho_2$. ∎

Now let D be any doubly connected region and let D be mapped conformally and one-to-one onto an annulus $\rho_1 < |w| < \rho_2$, where $\rho_1 < \rho_2 < \infty$. The ratio $\mu := \rho_2/\rho_1$, called the **modulus** of D, is uniquely determined by D. If the map is assumed to be continuous on the boundary of D, this is a simple consequence of Theorem 5.11c, for if the functions f and g mapped D onto two annuli with different ratios of the radii, the function $f \circ g^{[-1]}$ would map one annulus onto the other, which is not possible.

The following application of the reflection principle concerns the continuation of the mapping function across analytic arcs. An **analytic arc** Γ by definition is the image of an open segment Λ of the real axis under a conformal one-to-one map s of a domain S containing Λ (see Fig. 5.11c).

THEOREM 5.11d

Let D be a Jordan region whose boundary Γ contains an analytic arc Γ_0. If f maps D onto $|w|<1$ and is continuous on $D \cup \Gamma_0$, then f can be continued analytically across Γ_0.

Proof. Let the symbols s, Λ, and S be used as in the definition of an analytic arc. The function $f \circ s$ is analytic in an open set S^+ lying on one side of the real axis whose boundary contains Λ, and continuous on $S^+ \cup \Lambda$. Its values on Λ lie on $|w|=1$. Thus by the reflection principle $f \circ s$

can be continued analytically across Λ. It follows that $f = f \circ s \circ s^{[-1]}$ can be continued analytically across Γ_0. ∎

As a further application of the symmetry principle, we consider some semielementary mapping problems.

EXAMPLE 1

Let $c > 0$. The curve

$$z = (\tau + ic)^2, \qquad -\infty < \tau < \infty, \qquad (5.11\text{-}4)$$

then describes a parabola with focus at the origin and open to the right. We consider the problem of mapping the *interior* of the parabola onto the right half-plane such that symmetry with respect to the real axis is preserved (see Fig. 5.11d).

The parabola (5.11-4) is the image of the straight line $\operatorname{Im} z_1 = c$ under the map

$$z_1 \to z := z_1^2. \qquad (5.11\text{-}5)$$

In fact, this maps the strip $0 < \operatorname{Im} z_1 < c$ onto the interior of the parabola, cut along the positive real axis (see §5.1). Because a strip is easily mapped onto a half-plane, we might thus be tempted to obtain a solution by inverting the map (5.11-5). This, however, would omit the positive real axis; also, since points sitting opposite each other on the edges of the cut have entirely different preimages, a severe discontinuity would be introduced into the mapping function.

The following strategy is used instead: we map the *upper half* of the parabola onto the first quadrant of the w-plane so that the portion AB of the real axis is mapped onto the positive real axis. The full map is then obtained by the symmetry principle.

A number of intermediate steps are required (see Fig. 5.11e). By $z \to \sqrt{z} =: z_1$ (principal value of the square root) the upper half of the parabola is mapped onto the half-strip $\operatorname{Re} z_1 > 0$, $0 < \operatorname{Im} z_1 < c$. The map $z_1 \to (\pi/c) z_1 =: z_2$ transforms the

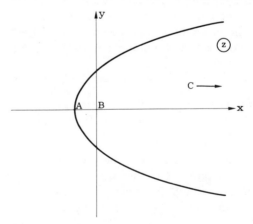

Fig. 5.11d. Mapping the interior of a parabola.

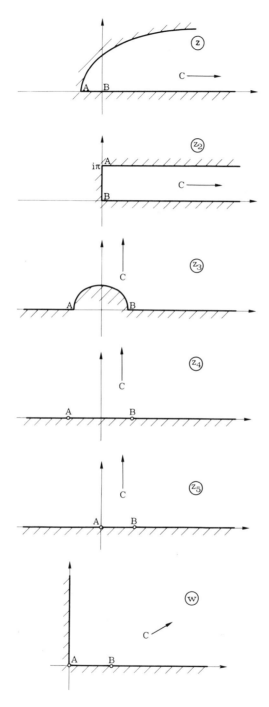

Fig. 5.11e. Auxiliary maps for mapping the interior of a parabola.

strip into one of width π. By $z_2 \to e^{z_2} = :z_3$ the normalized strip is mapped onto the upper half-plane minus the unit disk. The Joukowski map

$$z_3 \to \tfrac{1}{2}\left(z_3 + \frac{1}{z_3}\right) = :z_4$$

maps this strip onto the upper half-plane. Points A and B are now at -1 and $+1$, respectively;

$$z_4 \to \tfrac{1}{2}(z_1 + 1) = :z_5$$

thus sends A into 0, and by $z_5 \to \sqrt{z_5} = :w$ (principal value of the square root) we finally obtain the first quadrant. The line ABC is mapped onto the positive real axis. Composition of these maps yields

$$w = \cosh \frac{\pi \sqrt{z}}{2c}, \qquad (5.11\text{-}6)$$

which by the reflection principle also furnishes the map of the full interior of the parabola onto $\operatorname{Re} w > 0$. An arbitrary positive factor on the right remains undetermined. The principal value of the square root was to be taken originally, but since cosh is an even function the final formula is not sensitive with respect to the choice of the root.

EXAMPLE 2

Let $\rho > 1$. The reader is familiar with the fact that the Joukowski function

$$z = \tfrac{1}{2}\left(z_1 + \frac{1}{z_1}\right)$$

maps the exterior of the circle $|z_1| = \rho$ onto the *exterior* of the ellipse with semiaxes $\tfrac{1}{2}(\rho + \omega^{-1})$ and $\tfrac{1}{2}(\rho - \rho^{-1})$. Here we consider the problem of mapping the *interior* of this ellipse onto the unit disk.

Again this problem cannot be solved simply by inverting the Joukowski map because this would omit the segment $[-1, 1]$ on the real axis. As in the case of the parabola, however, we may try to map the upper half of the ellipse onto the upper half of the unit disk so that the images of the real points are real (see Fig. 5.11f). The function $z \to z + (z^2 - 1)^{1/2} = :z_1$ [choice of root as in (i), p. 298] maps the upper half of the ellipse onto the semiannulus

$$1 < |z_1| < \rho, \qquad \operatorname{Im} z_1 > 0,$$

and by $z_1 \to \operatorname{Log} z_1 = :z_2$ the semiannulus is mapped onto the rectangle

$$0 < \operatorname{Re} z_2 < \operatorname{Log} \rho, \qquad 0 < \operatorname{Im} z_2 < \pi.$$

In the next step this rectangle should be mapped onto a half-plane, which, however, is not possible in terms of elementary functions, as shown in §5.12. Thus the mapping problem for the *interior* of the ellipse cannot be solved elementarily. An explicit solution in terms of infinite series is discussed in Chapter 14.

SYMMETRY

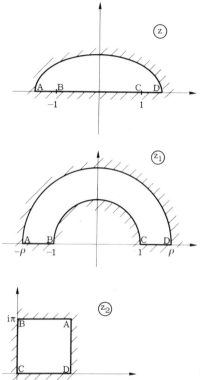

Fig. 5.11f. Auxiliary maps for mapping the interior of an ellipse.

PROBLEMS

1. Let D be the domain exterior to a "star" consisting of n straight line segments emanating from O at the angles $\arg z = 2\pi k/n$, $k = 0, 1, \ldots, n-1$. Show that the domain $|w| > 1$ is mapped onto D by

$$w \to z = \frac{c}{w}(w^n + 1)^{2/n},$$

where $c := 4^{-1/n}a$ (a = length of the straight-line segments). [Generalization of Joukowski map.]

2. The interior of the right branch of the hyperbola

$$\frac{x^2}{a^2} - \frac{y^2}{b^2} = 1$$

($z = x + iy$) is to be mapped onto $\operatorname{Re} w > 0$ such that symmetry with respect to the real axis is preserved. [*Solution.* Let $\theta = \arctan b/a$, $c := (a^2+b^2)^{1/2}$. Then

$$w = C\cosh\left[\frac{\pi}{2\theta}\operatorname{Log}\frac{z+(z^2-c^2)^{1/2}}{c}\right]$$

where $C > 0$.]

3. The mapping of the full disk $|z|<1$ onto the disk $|w|<1$ cut open between $w = 1-\epsilon$ and $w = 1$ (preserving symmetry with respect to the real axis and sending 0 to 0) is given by

$$z \to \frac{1+z-W}{1+z+W},$$

where

$$W := (1 - 2z\cos\phi + z^2)^{1/2},$$

$$\sin\frac{\phi}{2} := \frac{\epsilon}{2-\epsilon}.$$

§5.12. THE SCHWARZ-CHRISTOFFEL MAPPING FUNCTION

In this section the symmetry principle is applied to derive an explicit formula for the function that maps a half-plane onto the interior of a polygon.

To begin with, let D be a simply connected, bounded region whose boundary consists of a finite number, say $n+1$, of straight-line segments (see Fig. 5.12a). Slits are permitted and are to be counted as double edges of the boundary polygon. We denote the corners of the boundary polygon by z_k ($k = 0, 1, \ldots, n$) and the interior angle at the kth corner by $\alpha_k \pi$, where

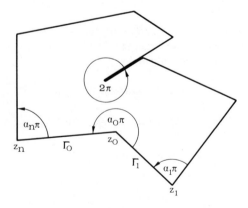

Fig. 5.12a. Notations for a Schwarz-Christoffel map.

SCHWARZ-CHRISTOFFEL MAPPING FUNCTION 397

$0 < \alpha_k \leqslant 2$. The straight line segment joining z_{k-1} and z_k is denoted by Γ_k. The change in direction of the boundary at the kth corner amounts to $(1-\alpha_k)\pi$; since the sum of all changes must amount to 2π, we have $(n+1-\Sigma\alpha_k)\pi = 2\pi$ or

$$\sum_{k=0}^{n} \alpha_k = n - 1. \qquad (5.12\text{-}1)$$

Our goal is the determination of the structure of a function $g: w \to z = g(w)$ mapping the upper half-plane $E: \text{Im } w > 0$ onto D. The formula that will be obtained for g is sufficiently explicit to permit the complete determination of the mapping function in a number of important special cases.

The *existence* of such functions g is assured by the Riemann mapping theorem. By the Osgood-Carathéodory theorem we may assume, in addition, that each such g is continuous and one-to-one for $\text{Im } w \geqslant 0$ (at least, in the case of slits, if the region D is suitably subdivided by crosscuts). We denote the preimage of the corner z_k on $\text{Im } w = 0$ by w_k ($k = 1, 2, \ldots, n$) and assume that $w = \infty$ is mapped onto the corner z_0. (If there is no corner at z_0, we may assume that $\alpha_0 = 1$.) Otherwise, the numbering is chosen such that $w_1 < w_2 < \cdots < w_n$.

Because g maps $\text{Im } w > 0$ conformally, $g'(w) \neq 0$ for $\text{Im } w > 0$, and by Theorem 4.3f it is thus possible to define

$$G(w) := \log g'(w)$$

as an analytic function in $\text{Im } w > 0$.

LEMMA 5.12a

The function $G'(w) = g''(w)/g'(w)$ can be continued to a function that is analytic in the whole w-plane with the exception of isolated singularities at the points $w_1, \ldots, w_n$.

Proof. Let k be an integer, $0 \leqslant k \leqslant n$. Since g is analytic for $\text{Im } w > 0$, continuous for $\text{Im } w \geqslant 0$, and maps the straight line segment $\Lambda_k := (w_{k-1}, w_k)$ onto the straight line segment Γ_k, it can, by the reflection principle (Theorem 5.11b) be continued analytically across Λ_k by assigning to a point $\bar{w}$ such that $\text{Im } \bar{w} < 0$ the value that is symmetrically located to $g(w)$ with respect to Γ_k (see Fig. 5.12b).

The extended function maps a region containing the interior points of Λ_k one-to-one onto a region E_k containing the interior points of the segment Γ_k. Hence $g'(w) \neq 0$ in E_k, and $G = \log g'$ can likewise be defined

CONFORMAL MAPPING

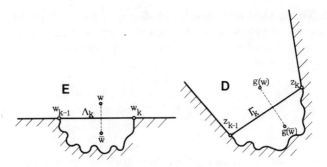

Fig. 5.12b. Application of the reflection principle.

analytically in E_k. On Λ_k, $\arg g'(w) = \text{const}$. Hence

$$\operatorname{Im} G(w) = \text{const}, \quad w \in \Lambda_k,$$

and by differentiation in the direction of the real axis we find

$$\operatorname{Im} G'(w) = 0, \quad w \in \Lambda_k.$$

Thus the function G' is analytic in $E \cup E_k$ and takes on real values for w real, $w \in \Lambda_k$. Again by the reflection principle, the function G' defined in $\operatorname{Im} w < 0$ by

$$G'(w) = \overline{G'(\bar{w})}$$

is analytic in $E \cup \Lambda_k \cup \bar{E}$. Since the above construction can be carried through for each segment Λ_k without affecting the definition of G' for $\operatorname{Im} w < 0$, it follows that G' is analytic in the whole plane except possibly at the points w_k, $k = 1, \ldots, n$. ∎

LEMMA 5.12b

The function G' is rational. It has simple poles at the points w_k with residue $\alpha_k - 1$ and vanishes at infinity. Thus it has the representation

$$G'(w) = \sum_{k=1}^{n} \frac{\alpha_k - 1}{w - w_k}. \qquad (5.12\text{-}2)$$

Proof. Let $1 \leq k \leq n$. For $\operatorname{Im} w \geq 0$, $|w - w_k|$ sufficiently small, we define a continuous argument of $g(w) - z_k$. With this argument we let

$$h(w) := [g(w) - z_k]^{1/\alpha_k}. \qquad (5.12\text{-}3)$$

SCHWARZ-CHRISTOFFEL MAPPING FUNCTION 399

If $|w-w_k|<\epsilon$, say, then h is analytic for $\operatorname{Im} w>0$ and continuous and one-to-one for $\operatorname{Im} w \geq 0$. Furthermore, $h(z_k)=0$. If $\arg(w-w_k)$ increases from 0 to π, $\arg[g(w)-z_k]$ increases by $\alpha_k \pi$. Thus $\arg h(w)$ increases by π, and it follows that the straight line segment $(w_k-\epsilon, w_k+\epsilon)$ is mapped onto a straight line segment through 0. Therefore h can be extended by the reflection principle to a conformal mapping of the whole disk $|w-w_k|<\epsilon$. Hence there exists an expansion of the form

$$h(w) = c_1(w-w_k)\{1+c_2(w-w_k)+\cdots\},$$

where $c_1 \neq 0$. By solving (5.12-3) for g we find that for a suitable choice of the fractional power

$$g(w)-z_k = c_1^*(w-w_k)^{\alpha_k}[1+c_2^*(w-w_k)+\cdots].$$

$c_1^* \neq 0$. From this we get

$$G'(w) = \frac{g''(w)}{g'(w)} = \frac{\alpha_k - 1}{w-w_k} + \text{holomorphic function}.$$

Since the above holds for each finite singular point w_k,

$$G'(w) - \sum_{k=1}^{n} \frac{\alpha_k - 1}{w-w_k}$$

is an entire function. To determine it we study its behavior near $w_0 = \infty$. Let

$$h(w):= \left[g\left(\frac{1}{w}\right)-z_0\right]^{1/\alpha_0}. \tag{5.12-4}$$

For $|w|<\epsilon$, say, h is analytic in $\operatorname{Im} w<0$ and, if $h(0):=0$, continuous in $\operatorname{Im} w \leq 0$. As above, we conclude by the reflection principle that h maps the entire disk $|w|<\epsilon$ conformally, hence has an expansion

$$h(w) = c_1 w\{1+c_2 w+\cdots\},$$

where $c_1 \neq 0$. The same computation now shows that

$$g\left(\frac{1}{w}\right) = z_0 + c_1^* w^{\alpha_0}\{1+c_2^* w+\cdots\}, \tag{5.12-5}$$

and that for $|w|$ sufficiently large
$$g(w) = z_0 + c_1^* w^{-\alpha_0} \{ 1 + c_2^* w^{-1} + \cdots \}.$$
Thus
$$G'(w) = \frac{g''(w)}{g'(w)} = \frac{-\alpha_0 - 1}{w} + O(w^{-2}), \qquad w \to \infty.$$

Thus the entire function mentioned above is zero, by Liouville's theorem, and G' has the form stated in the lemma. ∎

It is now a simple matter of two integrations to determine the mapping function g itself. For $\text{Im } w \geq 0$, $\text{Im } w^* \geq 0$ we define

$$S\begin{bmatrix} w_1, \ldots, w_n & w \\ \alpha_1, \ldots, \alpha_n & w^* \end{bmatrix} := \int_{w^*}^{w} \prod_{k=1}^{n} (w - w_k)^{\alpha_k - 1} dw, \qquad (5.12\text{-}6)$$

where the powers $(w - w_k)^{\alpha_k - 1}$ denote any analytic branches in E. We have obtained:

THEOREM 5.12c

With the notations and under the assumptions stated at the beginning of this section there exists a constant $C \neq 0$ such that for arbitrary w and w^* in the closure of E

$$g(w) - g(w^*) = CS\begin{bmatrix} w_1, \ldots, w_n & w \\ \alpha_1, \ldots, \alpha_n & w^* \end{bmatrix}. \qquad (5.12\text{-}7)$$

The function S making its appearance here is known as the **Schwarz-Christoffel mapping function**. It should be noted that the product in (5.12-6) is extended from $k=1$ to $k=n$, whether or not the polygon D has a corner at z_0. The angle α_0 at that corner does not enter explicitly, although implicitly it is taken into account by virtue of the relation (5.12-1).

The above derivation of the Schwarz-Christoffel formula is based on the Riemann and Osgood-Carathéodory mapping theorems. Although it is possible to dispense with the latter (see Ahlfors [1966], p. 225), the fact remains that we have *assumed* the existence of the mapping function and on this basis have proceeded to show that it has the form (5.12-7). If we did not wish to assume the mapping theorem, we would now have to show that the function (5.12-7) indeed solves the mapping problem posed.

SCHWARZ-CHRISTOFFEL MAPPING FUNCTION

This verification could be based on the *principle of the boundary map* (Theorem 5.10f). From (5.12-6)

$$g'(w) = C \prod_{k=1}^{n} (w - w_k)^{\alpha_k - 1}.$$

If w is real, the argument of the kth factor of the product equals 0 if $w > w_k$ and $\pi(\alpha_k - 1)$ if $w < w_k$. It follows that for w real, $w \neq w_1, \ldots, w_n$,

$$\arg g'(w) = \text{const} + \pi \sum_{w_k > w} (\alpha_k - 1),$$

hence that $\arg g'(w)$ is constant on each segment $\Lambda_k = (w_{k-1}, w_k)$, with a jump of $\pi(1 - \alpha_k)$ as w traverses w_k from the left to the right. We conclude that g maps the real w-axis onto a polygon with interior angles $\alpha_1 \pi, \alpha_2 \pi, \ldots, \alpha_n \pi$ and (by necessity) $\alpha_0 \pi$. It remains to be shown that this polygon is a Jordan curve and indeed that the parameters $w_1, \ldots, w_n$ can be chosen such that any desired polygon with these interior angles can be obtained. We omit this part of the verification, which appears to be tedious. As an intuitive tool the foregoing consideration can nevertheless be useful in the construction of concrete Schwarz-Christoffel maps.

Before turning to examples and applications we shall discuss some generalizations and transformations of the Schwarz-Christoffel formula.

In the first place the validity of the formula is not restricted to bounded polygons. Let D be a polygonal region with one corner, say z_0, at ∞. We number the remaining corners and measure the remaining angles as above. It will be convenient to assign an angular measure also to the corner at ∞. This angle could be measured on the Riemann sphere like an angle at a finite corner. It would turn out, however, that with this definition the basic relation (5.12-1) would no longer hold. It is preferable to define the angle in question in such a way that this relation will remain valid.

Consider Fig. 5.12c. We assume that Γ_0 and Γ_n, the sides which extend to infinity, are not parallel. The straight lines coinciding with these sides then meet at a finite point z^* under an angle that we call $\alpha^* \pi$, $0 < \alpha^* < 2$. Now consider the bounded polygon with the corners $z^*, z_1, \ldots, z_n$. By counting complementary angles we find that

$$(1 - \alpha_1)\pi + \sum_{k=2}^{n-1} (2 - \alpha_k)\pi + (1 - \alpha_n)\pi + \alpha^* \pi$$

$$= \text{sum of angles in a } (n+1)\text{-gon}$$

$$= (n-1)\pi.$$

CONFORMAL MAPPING

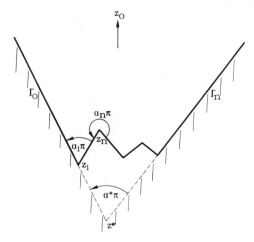

Fig. 5.12c. An unbounded polygon.

It follows that $\alpha^* = \sum_{k=1}^{n} \alpha_k - (n-1)$. If (5.12-1) is to hold, we must put

$$\alpha_0 = -\alpha^*. \tag{5.12-8}$$

If Γ_0 and Γ_n are parallel, we may put $\alpha^* = 0$.

Now let g be a function that maps the standard region $E: \operatorname{Im} w > 0$ onto D. If we assume continuity on the boundary, the proofs of Lemmas 5.12a and 5.12b can be repeated literally, with the exception of the definition of the auxiliary functions h. We distinguish two cases.

1. If $-2 < \alpha_0 < 0$, we use in place of (5.12-4)

$$h(w) := \left[g\left(\frac{1}{w}\right) \right]^{1/\alpha_0}, \tag{5.12-9}$$

$h(0) = 0$. As before it turns out that this maps a half-disk $\operatorname{Im} w \leq 0$, $|w| < \epsilon$ in such a way that the straight line segment $-\epsilon < w < \epsilon$ goes over into a straight line segment through O. Therefore by the reflection principle h can be extended to a map of the whole disk $|w| < \epsilon$, and

$$h(w) = c_1 w [1 + c_2 w + \cdots],$$

$$g\left(\frac{1}{w}\right) = c_1^* w^{\alpha_0} [1 + c_2^* w + \cdots],$$

where $c_1 \neq 0$, $c_1^* \neq 0$. It follows as before that g has a development of the form (5.12-5) and that $G(w) = O(w^{-1})$ as $w \to \infty$.

SCHWARZ-CHRISTOFFEL MAPPING FUNCTION

2. If $\alpha_0 = 0$, let $d > 0$ be the distance between the two parallel sides extending to infinity and let $e^{i\phi}$ ($0 \leq \phi < 2\pi$) be the direction in which they tend to infinity. Let

$$h(w) = \exp\left[-\frac{\pi}{d} e^{-i\phi} g\left(\frac{1}{w}\right)\right], \qquad (5.12\text{-}10)$$

$h(0) = 0$. This has the same mapping properties as the h defined under (1), and according to

$$g\left(\frac{1}{w}\right) = -\frac{d}{\pi} e^{i\phi} \log h(w)$$

we find that g has the expansion

$$g(w) = c_0^* + c_1^* \log \frac{1}{w} + O(w^{-1}),$$

where $c_1^* \neq 0$. There follows

$$G(w) = \frac{g''(w)}{g'(w)} = -\frac{1}{w} + O(w^{-2}),$$

and again $G'(w) \to 0$ for $w \to \infty$.

We conclude that Lemma 5.12b holds in both cases, and that we are able to integrate for g as before.

THEOREM 5.12d

Theorem 5.12c also holds if the corner z_0 of the polygonal region D is at ∞.

Again the angle α_0 does not make its explicit appearance in the mapping function. The reason why the same formula (5.12-7) defines a mapping onto bounded polygonal regions in some cases and onto unbounded regions in others is that for large values of $|w|$ the integrand in (5.12-6) behaves like w^α, where $\alpha = \sum_{k=1}^n \alpha_k - n$. By (5.12-1) $\alpha = \alpha_0 - 1$. If the polygon is bounded, then $\alpha_0 > 0$, and the integral converges for $|w| \to \infty$. If the polygon is unbounded, then $\alpha_0 \leq 0$, which causes the integral to diverge.

In the same manner we can show that Theorem 5.12c also holds if the polygonal region D has several corners at infinity, provided the angles at these corners are measured like the angle α_0.

Next we study the effect of composing the Schwarz-Christoffel function with a Moebius transformation. Let

$$t: s \to \frac{as+b}{cs+d} \qquad (5.12\text{-}11)$$

$(ad-bc \neq 0)$ be a Moebius transformation and let

$$s_k := t^{[-1]}(w_k), \qquad k=0,1,\ldots,n.$$

We furthermore set $E_1 := t^{[-1]}(E)$, where $E: \operatorname{Im} w > 0$. E_1 is either a half-plane or the interior or exterior of a circle. The function $g_1 := g \circ t$, where g is given by (5.12-7), maps E_1 conformally onto D; hence

$$G_1(s) := \log g_1'(s)$$

can be defined analytically in E_1. By the chain rule it is easily verified that

$$G_1'(s) = G'(t(s)) t'(s) + \frac{t''(s)}{t'(s)}.$$

By the explicit form (5.12-2) of G'

$$G_1'(s) = \sum_{k=1}^{n} \frac{t'(s)}{t(s)-t(s_k)} + \frac{t''(s)}{t'(s)}.$$

We then have

$$\frac{t''(s)}{t'(s)} = -\frac{2c}{cs+d},$$

and if s_k is finite an easy computation shows that

$$\frac{t'(s)}{t(s)-t(s_k)} = \frac{cs_k+d}{(cs+d)(s-s_k)}.$$

Thus, if all points $s_0, s_1, \ldots, s_n$ are finite, we find by using $-d/c = s_0$ that

$$G_1'(s) = \frac{1}{s-s_0} \left[\sum_{k=1}^{n} \frac{(\alpha_k-1)(s_k-s)}{s-s_k} - 2 \right].$$

By virtue of the relation following from (5.12-1)

$$\sum_{k=0}^{n} (\alpha_k - 1) = -2,$$

the above simplifies to

$$G'_1(s) = \sum_{k=0}^{n} \frac{\alpha_k - 1}{s - s_k}. \qquad (5.12\text{-}12)$$

This has the same form as the representation (5.12-2) of $G'(w)$, except for the limits of summation, which now include $k=0$. If one of the points $s_k = \infty$, an analogous computation shows that the corresponding term is simply to be omitted from (5.12-12).

By integration we find, as before, that

$$g_1(s) - g_1(s^*) = C_1 S \begin{bmatrix} s_0, s_1, \ldots, s_n & s \\ \alpha_0, \alpha_1, \ldots, \alpha_n & s^* \end{bmatrix}; \qquad (5.12\text{-}13)$$

where $C_1 \neq 0$ is a constant of integration, s^* is an arbitrary point of the closure of E_1, and

$$S \begin{bmatrix} s_0, s_1, \ldots, s_n & s \\ \alpha_0, \alpha_1, \ldots, \alpha_n & s^* \end{bmatrix} := \int_{s^*}^{s} \prod_{k=0}^{n} (s - s_k)^{\alpha_k - 1} ds. \qquad (5.12\text{-}14)$$

The path of integration is arbitrary in E_1, and a factor $(s - s_k)^{\alpha_k - 1}$ where $s_k = \infty$ is to be omitted from the product.

If E_1 is simply connected, it is clear that analytic single-valued branches of the fractional powers $(s - s_k)^{\alpha_k - 1}$ can be defined. If E_1 is the exterior of a circle Γ, this at first sight would not seem to be the case. However, we may define single-valued branches by cutting E_1 along a straight line from a point of Γ to ∞. If s travels along a positively oriented circle enclosing Γ, the winding number with respect to each of the points s_k is 1; hence by (5.12-1) the total increase of the argument of $\prod_{k=0}^{n}(s - s_k)^{\alpha_k - 1}$ equals

$$2\pi \sum_{k=0}^{n} (\alpha_k - 1) = 2\pi(n - 1 - (n+1)) = -4\pi.$$

It follows that the product is, in fact, single-valued in E_1. For $|s|$ large it behaves like s^{-2}, implying that $g(\infty)$ is finite.

For ease of formulation we call **circular** any region whose boundary is a generalized circle. We then can state the following generalization of Theorem 5.12c:

THEOREM 5.12e

Let D be a simply connected region in the complex plane bounded by a polygon with corners $z_0, z_1, \ldots, z_n$ (possibly infinite) and interior angles $\pi \alpha_k$, where $0 < \alpha_k \leq 2$ if z_k is finite and $0 \leq -\alpha_k \leq 2$ if $z_k = \infty$, and let E_1 be any circular region. Then any function g_1 mapping E_1 onto D is given by (5.12-13), where s_k is the preimage of z_k on the boundary of E_1 ($k = 0, 1, 2, \ldots, n$).

The Moebius transformation t in $g_1 = g \circ t$ can be chosen so that three specified corners correspond to three specified points on the boundary of E_1, subject only to the condition that the boundary orientations induced by the corners and by their preimages are compatible.

We mention that a modification of the Schwarz-Christoffel formula also permits the mapping of the *exterior* of a bounded polygon; see Problem 6.

We now give some examples of Schwarz-Christoffel maps.

I. 2-gons

A polygonal region with precisely two corners of necessity has one corner at infinity and thus up to a rigid motion is identical with a point set $0 < \arg z < \alpha\pi$, where $0 < \alpha \leq 2$ (see Fig. 5.12d). We wish to determine a function that maps the upper half-plane $\operatorname{Im} w > 0$ onto D as a special case of the Schwarz-Christoffel formula. As preimages of the corners $z_0 = \infty$ and $z_1 = 0$ we select $w_0 = \infty$ and $w_1 = 0$. Theorem 5.12d then yields the mapping function

$$w \to g(w) = CS \begin{bmatrix} 0 & w \\ \alpha & 0 \end{bmatrix} = C \int_0^w w^{\alpha - 1} \, dw = C_1 w^\alpha.$$

This, of course, agrees with the solution of the same mapping problem obtained earlier by more elementary methods.

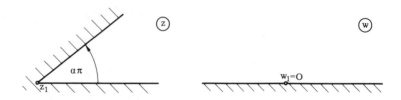

Fig. 5.12d. A two-gon.

II. 3-gons (Triangles)

Let D be a triangle with corners at a, b, c and let the corresponding angles be α, β, γ. We know that $\alpha + \beta + \gamma = 1$. The triangle is bounded if and only if all angles are positive. The corner a, say, is at infinity if and only if $\alpha = 1 - \beta - \gamma \leq 0$. Since there are only three corners, it is possible to select all their preimages, for instance by assigning them the values $\infty, 0, 1$. Then any function g mapping $\operatorname{Im} w > 0$ by Theorem 5.12d satisfies

$$g(w) - g(0) = CS \begin{bmatrix} 0 & 1 & w \\ \beta & \gamma & 0 \end{bmatrix} = C \int_0^w w^{\beta-1}(w-1)^{\gamma-1} dw.$$

The integral can easily be evaluated by series expansions. It can be evaluated in terms of elementary functions, e.g., if the corners are integral multiples of $\pi/2$.

Although triangles are simple geometrical objects, this formula enables us to study a large number of problems of physical interest because we can consider unbounded triangles. The following are some physically motivated examples.

EXAMPLE 1 FLOW AGAINST DAM

Let there be given, in the physical (z-) plane, a "dam" in the shape of a semi-infinite strip $\operatorname{Re} z \geq 0$, $|\operatorname{Im} z| \leq d$. We wish to determine the flow of an ideal fluid past this dam if the velocity at infinity is uniform and parallel to the real axis (see Fig. 5.12e).

The appropriate model problem clearly consists of the uniform flow parallel to the real axis of the w-plane, where the dam has degenerated to the slit $\operatorname{Re} w \geq 0$, $\operatorname{Im} w = 0$. Thus it is necessary to map the w-plane cut along the positive real axis

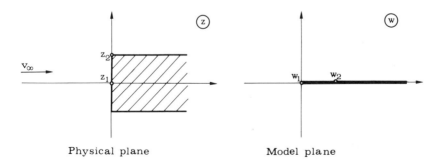

Fig. 5.12e. Flow against a dam.

onto the exterior of the dam in such a way that $w = \infty$ is a fixed point and $g'(\infty)$ is positive. By means of the symmetry principle this can be accomplished by mapping $\operatorname{Im} w > 0$ onto the exterior of the dam in the upper half-plane. Calling this region D, we see that D is a triangle with corners at $a := \infty$, $b := 0$, $c := id$ and corresponding angles $-\pi$, $\pi/2$, and $3\pi/2$. Letting $w_0 = \infty$, $w_1 = 0$, and $w_2 = 1$, the mapping function must satisfy

$$g(w) = C \int_0^w w^{-1/2}(w-1)^{1/2} dw$$

$$= C \left\{ w^{1/2}(w-1)^{1/2} - \operatorname{Log}\left[w^{1/2} + (w-1)^{1/2}\right] \right\} + D,$$

where D is a constant of integration.

The constants C and D are to be determined from the conditions

$$g(0) = 0, \qquad g(1) = id.$$

Now for $w < 1$, $\arg(w-1) = +\pi$, and the logarithm for $w = 0$ assumes the value $\operatorname{Log}(+i) = i(\pi/2)$. Hence the two conditions yield

$$-\frac{i\pi}{2} C + D = 0, \qquad D = id,$$

implying $C = 2d/\pi$. The desired mapping function is

$$g(w) = \frac{2d}{\pi} \left\{ w^{1/2}(w-1)^{1/2} - \operatorname{Log}\left[w^{1/2} + (w-1)^{1/2}\right] + \frac{i\pi}{2} \right\}.$$

The streamlines are the images of the straight lines $\operatorname{Im} w = \operatorname{const}$. Since $g'(\infty) = 2d/\pi$, the uniform velocity in the model plane is $\pi v_\infty / 2d$. The velocity potential in the physical plane is

$$\phi(z) = \frac{2d v_\infty}{\pi} \operatorname{Re} f(z),$$

where $f := g^{[-1]}$. Unfortunately the inverse function has no simple presentation. For the complex gradient, however, we obtain by Theorem 5.6a

$$\operatorname{grd} \phi(z) = \frac{2d v_\infty}{\pi} \overline{f'(z)}.$$

Using the relation $f' = (g' \circ f)^{-1}$ and the explicit form of g', the velocity at any point $z = g(w)$ is

$$\operatorname{grd} \phi(z) = v_\infty \overline{\left[w^{1/2}(w-1)^{-1/2}\right]},$$

where the bar denotes complex conjugation.

SCHWARZ-CHRISTOFFEL MAPPING FUNCTION

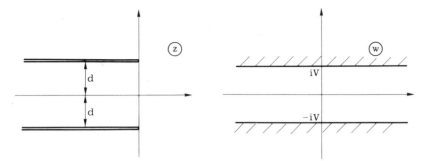

Fig. 5.12f. Semi-infinite plate condenser.

EXAMPLE 2   SEMI-INFINITE PLATE CONDENSER

We consider a condenser formed of two thin semi-infinite parallel plates at distance $2d$. We assume that the plates intersect the z-plane in the rays $\operatorname{Im} z = \pm d$, $\operatorname{Re} z \leq 0$. We wish to determine the electrostatic potential in space if the upper and lower plate are kept at the potentials $\pm V$, respectively. The appropriate model problem here is the infinite plate condenser defined by the two horizontal lines $\operatorname{Im} w = \pm V$; in it the solution is $\psi(w) = \operatorname{Im} w$. The problem will be solved if we can map the strip $|\operatorname{Im} w| < V$ onto the region exterior to the two rays in the z-plane in such a way that $\operatorname{Im} w = V$ will go over into the upper ray (traversed back and forth)

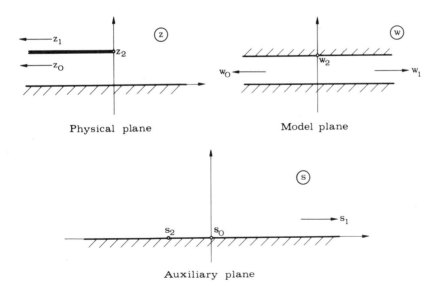

Fig. 5.12g. Auxiliary map.

and $\text{Im}\, w = -V$ into the lower ray (see Fig. 5.12*f*). According to the symmetry principle, it suffices to map the upper half $0 < \text{Im}\, w < V$ of the model domain onto the upper half of the physical region in such a way that the real axis corresponds to the real axis (Fig. 5.12*g*).

Because the model region is not circular, the desired mapping is accomplished via an auxiliary map. We map the model region onto the upper half of an *s*-plane so that $s < 0$ corresponds to $\text{Im}\, w = V$ and $s > 0$ to $\text{Im}\, w = 0$. This is accomplished by

$$w \to s := e^{(\pi/V)w}.$$

We now have to map $\text{Im}\, s > 0$ onto the physical region *D*. This we may view as a triangle with corners $a := \infty$, $b := \infty$, $c := id$ and corresponding angles $0, -\pi, 2\pi$. The boundary of the triangle consists of the line $\text{Im}\, z = 0$ and the two edges of the ray. The physical situation demands that the two edges of the ray correspond to $s < 0$ and the real axis to $s > 0$. This fixes $s_0 = 0$ as the preimage of *a* and $s_1 = \infty$ as the preimage of *b*. We arbitrarily select $s_2 = -1$ as the preimage of *c*. Theorem 5.12e then furnishes the mapping function

$$s \to z := C \int_0^s (s+1)s^{-1}\,ds = C(s + \text{Log}\, s) + D.$$

To determine the constants of integration we note that positive values of $s - 1$ should yield real values of *z*. This is the case if and only if *C* and *D* are both real. Furthermore we have the condition

$$-1 \to id = C(i\pi - 1) + D.$$

Separating real and imaginary parts yields $C = D = d/\pi$.

Composition with the auxiliary map $w \to s$ already found now furnishes the desired map of the model region onto the physical region:

$$z = g(w) = \frac{d}{V}w + \frac{d}{\pi}(e^{(\pi/V)w} + 1).$$

The lines of equal potential are the images of the straight lines $\text{Im}\, w = \text{const}$ and the field lines are the images of the straight line segments $\text{Re}\, w = \text{const}$, $0 < \text{Im}\, w < V$. Since the point $w = iV$ corresponds to $z = id$, the tip of the upper plate, the field line emanating from the tip is given by

$$z = g(iV\tau) = i\tau d + \frac{d}{\pi}(e^{i\pi\tau} + 1),$$

$0 \leq \tau \leq 1$. It meets the real axis at $z = 2d/\pi$. The complex field strength $E(z) = -\text{grd}\,\phi(z)$ is

$$E(z) = -i\,\overline{f'(z)} = -i\,\overline{[g'(w)]^{-1}},$$

where $f := g^{[-1]}$ and $w = f(z)$. Computation yields

$$E(z) = -i\frac{V}{d}(1 + e^{(\pi/V)\bar{w}})^{-1}.$$

SCHWARZ-CHRISTOFFEL MAPPING FUNCTION 411

It is seen that for $\mathrm{Re}\,w \ll 0$, far removed from the edge of the condenser plates, the field strength equals $-iV/d$, the uniform field strength of the doubly infinite condenser. For $\mathrm{Re}\,w \to +\infty$, on the other hand, the field strength tends to zero.

III. The Regular n-gon

Here we consider the problem of mapping the unit disk $|w|<1$ onto a regular n-gon (see Fig. 5.12h for $n=6$) with corners at the points

$$z_k = d e^{2\pi i k/n}, \quad k=0,1,\ldots,n \quad (d>0, z_n = z_0).$$

First we show that a mapping exists with the property that $z=0$ corresponds to $w=0$ and z_k to $w_k := \exp[(2\pi i/n)k]$, $k=1,2,\ldots,n$. Clearly, the triangle with corners $0, z_0, z_1$ can be mapped onto the circular sector $0 < \arg w < 2\pi/n$, $|w|<1$ such that 0 remains fixed and z_0 and z_1 correspond to w_0 and w_1, respectively. By successively reflecting the mapping function on the rays $\arg w = 2\pi k/n$, $k=1,2,\ldots$, the mapping function can be continued to a function mapping the whole disk onto the given regular n-gon such that w_k corresponds to z_k ($k=1,2,\ldots,n$); but then, by Theorem 5.12e, this function must have the form

$$g(w) = C \int_0^w \prod_{k=1}^n (w-w_k)^{-2/n} dw.$$

However,

$$\prod_{k=1}^n (w-w_k) = \prod_{k=1}^n (w - e^{(2\pi i/n)k}) = w^n - 1.$$

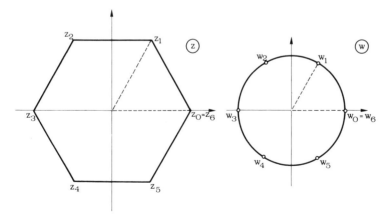

Fig. 5.12h. Mapping a hexagon.

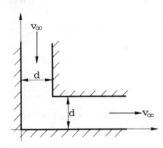

Fig. 5.12i. Channel with a corner.

Thus the mapping function is

$$g(w) = C \int_0^w (1-w^n)^{-2/n} dw,$$

where we have reversed a sign to make C real; C is determined by the condition

$$d = C \int_0^1 (1-w^n)^{-2/n} dw.$$

The integral occurring here can be reduced to a beta integral which can be evaluated in terms of the Γ function (see §8.7). We find

$$C = \frac{\Gamma(1-1/n)}{\Gamma(1+1/n)\Gamma(1-2/n)} d.$$

PROBLEMS

1. Find the potential of a flow through a channel with a 90° corner as shown in Fig. 5.12i, if the flow in remote locations of the channel is homogeneous.
2. *Suddenly narrowing channel.* Determine, for arbitrary positive values of a and $d > a$, the potential of the flow through the channel sketched in Fig. 5.12j if for $\text{Re}\,z \to \pm \infty$ the flow is homogeneous.
3. Solve the potential problem sketched in Fig. 5.12k.

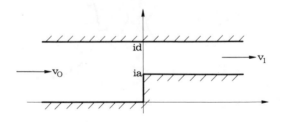

Fig. 5.12j. Suddenly narrowing channel.

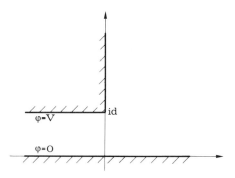

Fig. 5.12k. Potential problem.

4. Determine, for arbitrary positive values of a and d, the potential of the flow through the rectangular turnabout channel shown in Fig. 5.12*l* if the flow is homogeneous in remote parts of the channel.

5. The unit disk is mapped onto the interior of a cross, as shown in Fig. 5.12*m* by the function

$$z = g(w) = C \int_0^w \left[\frac{w^4 + 1}{w^8 + 2w^4 \cos \alpha + 1} \right]^{1/2} dw,$$

where α is appropriately chosen in $(0, \pi)$.

6. *Mapping the exterior of a bounded polygon.* The following problems are concerned with the mapping problem for the exterior D of a bounded polygon. If the polygon has $n+1$ corners z_k with angles $\alpha_k \pi$, measured from the interior of D as shown in Fig. 5.12*n* ($k = 0, 1, \ldots, n$), show that

$$\sum_{k=0}^n \alpha_k = n + 3.$$

7 (continuation). If g maps $E: \mathrm{Im}\, w > 0$ onto D, $g(w_\infty) = \infty$, show that $(w - w_\infty)^2 g'(w)$ is analytic and different from zero in E.

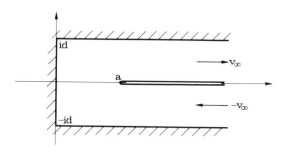

Fig. 5.12l. Turnabout channel.

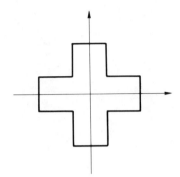

Fig. 5.12m. Swiss cross.

8 (continuation). If $G := \log g'$ and w_k denotes the preimage of z_k, show that

$$G'(w) = \sum_{k=0}^{n} \frac{\alpha_k - 1}{w - w_k} - \frac{2}{w - w_\infty} - \frac{2}{w - \overline{w_\infty}}$$

(a possible term with $w_k = \infty$ omitted) and conclude that the mapping function g satisfies

$$g(w) - g(w^*) = CS\begin{bmatrix} w_0, \ldots, & w_n, & w_\infty, & \overline{w_\infty} & w \\ \alpha_0, \ldots, & \alpha_n, & -2, & -2 & w^* \end{bmatrix}.$$

9 (continuation). Let E_1 be a circular domain with boundary Γ. If g_1 maps E_1 onto D, $g_1(s_k) = z_k$, $k = 0, 1, \ldots, n$, $g(s_\infty) = \infty$, and if s_∞^* is symmetric to s_∞ with respect to Γ then
 (a) if both s_∞ and s_∞^* are finite,

$$g_1(s) - g_1(s^*) = CS\begin{bmatrix} s_0, \ldots, & s_n, & s_\infty, & s_\infty^* & s \\ \alpha_0, \ldots, & \alpha_n, & -2, & -2 & s^* \end{bmatrix};$$

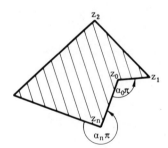

Fig. 5.12n. Mapping the exterior of a bounded polygon.

SCHWARZ-CHRISTOFFEL MAPPING FUNCTION

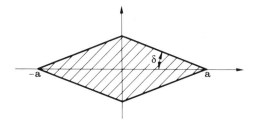

Fig. 5.12o. Mapping the exterior of a prism.

(b) if $s_\infty = \infty$,

$$g_1(s) - g_1(s^*) = CS \begin{bmatrix} s_0, \ldots, & s_n, & s_\infty^* & s \\ \alpha_0, \ldots, & \alpha_n, & -2 & s^* \end{bmatrix}.$$

10. Obtain the Joukowski map as a special case of Problem 9.
11. Map the exterior of the prism shown in Fig. 5.12o onto $|s| > 1$ such that the point at infinity remains fixed. For $\delta \to 0$, obtain the Joukowski map.
12. The mapping of $E_1 : |s| > 1$ onto the exterior of a regular star with n arms (see Fig. 5.12p for $n = 6$) is given by

$$g_1(s) = C \int_1^s (s^n + 1)(s^n - 1)^{(2/n) - 1} \frac{ds}{s^2}.$$

13. The map of $E_1 : |s| > 1$ onto the exterior of the broken line segment shown in Fig. 5.12q is given by

$$z = g_1(s) := \frac{C}{s} (s - s_1)^\alpha (s - s_2)^{2 - \alpha},$$

where C, s_1, and s_2 are appropriately chosen.

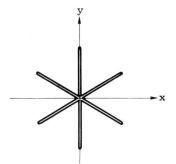

Fig. 5.12p. Mapping the exterior of a star.

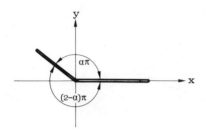

Fig. 5.12q. Mapping the exterior of a broken line segment.

§5.13. MAPPING THE RECTANGLE. AN ELLIPTIC INTEGRAL

As a further application of the Schwarz-Christoffel formula we consider the problem of mapping $\operatorname{Im} w > 0$ onto a rectangle. Let the corners of the rectangle be at the points $z_1 := \frac{1}{2}p_1$, $z_2 := \frac{1}{2}p_1 + ip_2$, $z_3 := -\frac{1}{2}p_1 + ip_2$, $z_4 := -\frac{1}{2}p_1$, where $p_1 > 0$, $p_2 > 0$. The mapping is to be normalized by the condition that $w = 0$ is to be mapped onto $z = 0$ and $w = 1$ onto z_1 and that symmetry with respect to the imaginary axis is to be preserved. It follows that $w_4 = -w_1 = -1$ and that $w_2 = -w_3 = \kappa^{-1}$, where κ is a real number yet to be determined, $0 < \kappa < 1$ (see Fig. 5.13a).

Since all interior angles are $\pi/2$, it follows that the mapping function has the form

$$z = g(w) = C_1 S \left[\begin{array}{cccc} -\dfrac{1}{\kappa}, & -1, & 1, & \dfrac{1}{\kappa} \\ \dfrac{1}{2}, & \dfrac{1}{2}, & \dfrac{1}{2}, & \dfrac{1}{2} \end{array} \middle| \begin{array}{c} w \\ 0 \end{array} \right],$$

which after simplification yields

$$g(w) = C \int_0^w \left[(1-w^2)(1-\kappa^2 w^2) \right]^{-1/2} dw, \qquad (5.13\text{-}1)$$

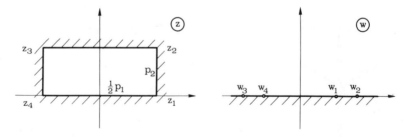

Fig. 5.13a. Mapping a rectangle.

with a different constant C. The square root is to be regarded as $(1-w)^{1/2}(1+w)^{1/2}(1-\kappa w)^{1/2}(1+\kappa w)^{1/2}$; the arguments are determined such that each factor is $+1$ for $w=0$ and continuous for $\text{Im } w \geq 0$. The constant C will then have a positive value.

Equation (5.13-1) does not solve the mapping problem because we have not yet determined the constants C and κ. It is immediately clear that $w_2 = \kappa^{-1}$ can no longer be prescribed arbitrarily, for if it could all rectangles would be similar.

To evaluate κ and C we first note that

$$p_1 = 2CK(\kappa), \qquad (5.13\text{-}2)$$

where

$$K(\kappa) := \int_0^1 [(1-w^2)(1-\kappa^2 w^2)]^{-1/2} dw. \qquad (5.13\text{-}3)$$

The function $K(\kappa)$ is known as a **complete elliptic integral of the first kind**.[1] On setting $w = \sin\phi$, we obtain

$$K(\kappa) = \int_0^{\pi/2} (1 - \kappa^2 \sin^2\phi)^{-1/2} d\phi,$$

and this can be evaluated by expansion in series:

$$K(\kappa) = \int_0^{\pi/2} \sum_{n=0}^{\infty} \frac{(1/2)_n}{n!} \kappa^{2n} (\sin\phi)^{2n} d\phi$$

$$= \frac{\pi}{2} \sum_{n=0}^{\infty} \frac{(1/2)_n (1/2)_n}{n! n!} \kappa^{2n}$$

$$= \frac{\pi}{2} F(\tfrac{1}{2}, \tfrac{1}{2}; 1; \kappa^2), \qquad (5.13\text{-}4)$$

where F denotes the hypergeometric function. By using various transformation formulas for the hypergeometric series (see §§9.9 and 9.10) it is possible to evaluate $K(\kappa)$ rapidly for all κ such that $0 < \kappa < 1$. It is evident from (5.13-4) that $K(\kappa)$ increases monotonically from $\pi/2$ to ∞ as κ increases from 0 to 1.

---

[1] So called for historical reasons. Integrals of this kind were encountered in the seventeenth century in connection with the problem of measuring the circumference of an ellipse.

Next we have

$$ip_2 = C \int_1^{\kappa^{-1}} [(1-w^2)(1-\kappa^2 w^2)]^{-1/2} dw,$$

or, substituting w^{-1} for w,

$$p_2 = C \int_\kappa^1 [(1-w^2)(w^2-\kappa^2)]^{-1/2} dw. \qquad (5.13\text{-}5)$$

It will be shown below that as κ increases from 0 to 1 the value of the integral decreases monotonically from ∞ to $\pi/2$. It follows that as a function of κ the ratio $2p_1/p_2$ of the two sides of the rectangle increases monotonically from 0 to ∞. Thus the value of κ is uniquely determined by that ratio. The constant C merely influences the magnitude of the rectangle; once κ is determined, C can be calculated, e.g., from (5.13-2).

We shall now show that the integral for p_2 can also be transformed into a complete elliptic integral of the first kind. This is a consequence of a more general result. Let $w_1, w_2, w_3, w_4 = w_0$ *be four distinct real numbers, arranged in increasing cyclic order.* (By this we mean that the inequality $w_{i+1} > w_i$ is violated for exactly one of the indices $i = 0, 1, 2, 3$.) *Then the integrals*

$$K_i := \int_{w_i}^{w_{i+1}} \prod_{j=0}^{3} (w - w_j)^{-1/2} dw, \qquad i = 0, 1, 2, 3, \qquad (5.13\text{-}6)$$

can be expressed as complete elliptic integrals of the first kind. (If $w_i > w_{i+1}$, $\int_{w_i}^{w_{i+1}}$ is to be interpreted as $\int_{w_i}^{\infty} + \int_{-\infty}^{w_{i+1}}$.)

To establish this result we subject the variable of integration to a Moebius transformation,

$$t: s \to w := \frac{as+b}{cs+d}, \qquad (5.13\text{-}7)$$

with real a, b, c, d such that

$$\Delta := ad - bc > 0.$$

The points $s_i := t^{[-1]}(w_i)$ are again arranged in increasing cyclic order. We assume that they are all finite. Letting $h(s) := cs + d$, we have

$$\frac{dw}{ds} = \frac{\Delta}{[h(s)]^2},$$

$$w - w_i = \Delta \frac{s - s_i}{h(s) h(s_i)}, \qquad i = 0, 1, \ldots, 4.$$

MAPPING THE RECTANGLE

For a suitable determination of the square root

$$K_i = A \int_{s_i}^{s_{i+1}} \prod_{j=0}^{3} (s - s_j)^{-1/2} ds \qquad (5.13\text{-}8)$$

follows, where

$$A := \frac{(h_0 h_1 h_2 h_3)^{1/2}}{\Delta}, \qquad h_i := h(s_i).$$

It will be convenient to express A in terms of the w_i and s_i. If $w := t(s)$, then

$$\frac{w - w_1}{w - w_3} \frac{w_2 - w_3}{w_2 - w_1} = \frac{s - s_1}{s - s_3} \frac{s_2 - s_3}{s_2 - s_1}. \qquad (5.13\text{-}9)$$

By writing $w_{ij} := w_i - w_j$, $s_{ij} := s_i - s_j$, solving for w yields,

$$w = \frac{(w_1 s_{21} w_{23} - w_3 s_{23} w_{21})s - (s_3 w_1 s_{21} w_{23} - s_1 w_3 s_{23} w_{21})}{(s_{21} w_{32} - s_{23} w_{21})s - (s_3 s_{21} w_{23} - s_1 s_{23} w_{21})}.$$

Some straightforward computation now yields

$$\Delta = s_{21} s_{23} s_{31} w_{12} w_{23} w_{31},$$

$$h_0 = s_{01} s_{23} w_{12} - s_{03} s_{12} w_{23},$$

$$h_1 = s_{31} s_{12} w_{23}, \quad h_2 = s_{12} s_{23} w_{31}, \quad h_3 = s_{23} s_{31} w_{12}.$$

Thus we find

$$A^2 = \frac{s_{01} s_{23} w_{12} - s_{03} s_{12} w_{23}}{w_{12} w_{23} w_{31}}. \qquad (5.13\text{-}10)$$

Without loss of generality we may now assume that the integral K_0 is to be computed, since this can always be accomplished by renumbering the w_i. We pick the Moebius transformation t such that the points w_i correspond to values of s as follows:

w	w_0	w_1	w_2	w_3
s	-1	1	μ	$-\mu$

Here μ is to be a real number >1. It is clear that μ cannot be selected arbitrarily, for a linear transformation is already fully determined by three pairs of corresponding points. The possible value of μ is found from the condition that (5.13-9) must hold when $w=w_0, s=s_0=-1$. This yields

$$\frac{4\mu}{(\mu-1)^2}=-\frac{w_{01}w_{23}}{w_{30}w_{12}}, \qquad (5.13\text{-}11)$$

a quadratic equation with two real solutions whose product equals 1. There is exactly one solution >1. It determines the values of the s_i. With these values, we have

$$K_0 = A\int_{-1}^{1}\left[(1-s^2)(\mu^2-s^2)\right]^{-1}ds,$$

where

$$A=\left[\frac{(1-\mu^2)w_{23}-4\mu w_{12}}{w_{12}w_{23}w_{31}}\right]^{1/2}, \qquad (5.13\text{-}12)$$

or

$$K_0 = 2A\mu^{-1}K(\mu^{-1}). \qquad (5.13\text{-}13)$$

For the special problem of evaluating (5.13-5) the values of the w_i are as follows:

w_0	w_1	w_2	w_3
κ	1	-1	$-\kappa$

Equation (5.13-11) for μ becomes

$$\frac{4\mu}{(\mu-1)^2}=\frac{(1-\kappa)^2}{4\kappa};$$

its solution >1 is

$$\mu=\left(\frac{1+\sqrt{\kappa}}{1-\sqrt{\kappa}}\right)^2.$$

MAPPING THE RECTANGLE

We readily find $A = 2(1 - \sqrt{\kappa})^{-2}$ and thus get

$$\int_\kappa^1 [(1-w^2)(\kappa^2-w^2)]^{-1/2} dw = \frac{4}{(1+\sqrt{\kappa})^2} K\left(\left(\frac{1-\sqrt{\kappa}}{1+\sqrt{\kappa}}\right)^2\right).$$

By using a quadratic transformation formula for the hypergeometric series (see §9.10) we can show that the last expression equals $K(\kappa')$, where $\kappa' := (1-\kappa^2)^{1/2}$. Thus we finally have

$$\frac{p_1}{p_2} = \frac{K(\kappa)}{K(\kappa')}. \tag{5.13-14}$$

The mapping formula (5.13-1) permits a brief glimpse at the theory of elliptic functions, a vast field of classical analysis that had its heyday in the first half of the nineteenth century. The inverse $f := g^{[-1]}$ of the function defined by (5.13-1) maps a rectangle with its four sides $\Gamma_1, \Gamma_2, \Gamma_3, \Gamma_4$ onto the upper half-plane and, in particular, Γ_1 onto a piece of the real axis. By the reflection principle the reflection of the rectangle at Γ_1 is thus mapped onto the lower half-plane and the whole rectangle, together with the reflection, onto the whole w-plane, with the exception of the real intervals $(-\infty, -1)$ and $(1, \infty)$ (see Fig. 5.13b).

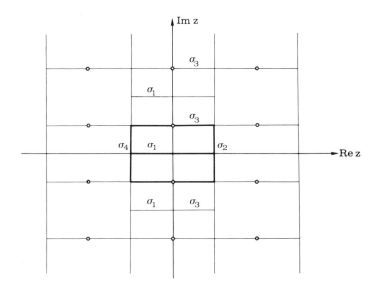

Fig. 5.13b. Doubly periodic function.

The values that f assumes on Γ_3 are also real. Hence we may also reflect f on Γ_3. By similar reflections we can continue f into the entire strip $-\frac{1}{2}p_1 \leq \operatorname{Re} z \leq \frac{1}{2}p_1$. The continued function will be periodic in this strip with period $2ip_2$ and analytic, with the exception of simple poles, at points $z = (2n+1)ip_2$ (n is an integer) corresponding to $w = \infty$. By construction the values taken by f on the straight line segments Γ_2 and Γ_4 and on their straight line continuations are also real. Hence we may reflect also on these vertical lines. By continuing these reflections f can be continued analytically into the whole z-plane, with the exception of isolated singularities at points $mp_1 + (2n+1)ip_2$ (m and n are integers). The resulting function f turns out to be meromorphic in the whole plane and to possess the two independent periods $2p_1$ and $2ip_2$. Thus we have shown geometrically the existence of nonconstant meromorphic, doubly periodic functions. The discovery of these functions is due to Jacobi (1804–1851) who termed them **elliptic functions**.

§5.14. ROUNDING CORNERS IN SCHWARZ-CHRISTOFFEL MAPS

The Schwarz-Christoffel formula, as stated in §5.12, suffers from the drawback that it maps only polygonal domains with sharp corners. Both in electrostatic and in hydrodynamical applications sharp corners are undesirable because they usually give rise to an infinite field strength or an infinite velocity, physical causes, respectively, of electrostatic breakthrough or turbulence. For instance, in Example **1** of §5.12 the velocity is infinite at the point $z = id$; in the semi-infinite plate condenser (Example **2** of §5.12) the field strength is infinite at the rim of the plate. For these reasons it is relevant that by a simple modification of the Schwarz-Christoffel function a rounding of both concave and convex corners can be accomplished.

The derivative of a function mapping $\operatorname{Im} w \geq 0$ onto a polygonal domain D is, up to a constant factor,

$$\frac{dz}{dw} = g'(w) = \prod_{k=1}^{n} (w - w_k)^{\alpha_k - 1}. \qquad (5.14\text{-}1)$$

Let z_m be one of the corners of D and w_m its preimage on the real w-axis. We know from the discussion following Theorem 5.12c that in the above product the factor

$$q(w) := (w - w_m)^{\alpha_m - 1} \qquad (5.14\text{-}2)$$

is responsible for the corner at z_m. As w increases through real values, the argument of the above factor at $w = w_m$ changes discontinuously from

ROUNDING CORNERS IN SCHWARZ-CHRISTOFFEL MAPS

$(\alpha_m - 1)\pi$ to 0. Because all other factors in (5.14-1) are analytic and $\neq 0$ at w_m, this causes the argument of dz/dw to jump discontinuously by the same amount and thus leads to an abrupt change in the direction of the boundary curve at z_m.

All that is necessary to achieve a continuous change of direction therefore is the replacement of $q(w)$ by a factor $q^*(w)$ whose argument changes continuously from $(\alpha_m - 1)\pi$ to 0 as w traverses a real interval containing w_m. In order that integration of the differential quotient dz/dw may still yield a conformal map q^* must, of course, still be analytic in $\text{Im}\,w > 0$ and different from zero in $\text{Im}\,w \geq 0$. For the construction of suitable functions q^* it is necessary to distinguish between $1 < \alpha_m \leq 2$ (concave or re-entrant corner) and $0 < \alpha_m < 1$ (convex corner).

I. Concave (Re-entrant) Corners

A **concave** or **re-entrant corner** is one whose interior angle is $\alpha_m \pi$, where $\alpha_m > 1$. Let $\epsilon > 0$ be sufficiently small so that $w'_m := w_m - \epsilon > w_{m-1}$ and $w''_m := w_m + \epsilon < w_{m+1}$, and let $a > 0$, $b > 0$, $a + b = 1$. Otherwise the notation is as in §5.12.

THEOREM 5.14a

If z_m is a finite concave corner of the polygonal D with interior angle $\alpha_m \pi$, where $1 < \alpha_m < 2$, then replacing the factor (5.14-2) in the Schwarz-Christoffel function with

$$q^*(w) := a(w - w'_m)^{\alpha_m - 1} + b(w - w''_m)^{\alpha_m - 1} \qquad (5.14\text{-}3)$$

rounds the corner at z_m (see Fig. 5.14a).

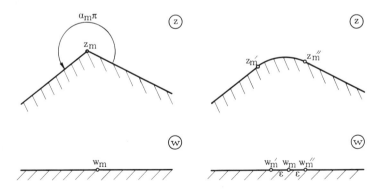

Fig. 5.14a. Rounding a convex corner.

Proof. For $w < w'_m$ both terms on the right of (5.14-3) have the argument $(\alpha_m - 1)\pi$ and for $w > w''_m$ both terms are positive. For $w'_m \leq w \leq w''_m$ $q^*(w)$ may be regarded as the sum of two vectors with the arguments 0 and $(\alpha_m - 1)\pi$. The length of the first vector increases steadily from zero and the length of the second vector decreases to zero. It is then geometrically evident that the argument of the sum changes continuously from $(\alpha_m - 1)\pi$ to zero, as desired. It is clear that q^* is analytic for $\text{Im } w > 0$. Moreover, for $\text{Im } w \geq 0$ both terms in (5.14-3) have arguments in the interval $[0, (\alpha_m - 1)\pi]$ and thus lie in a sector of the complex plane whose opening is less than π. Because the two terms do not vanish simultaneously, their sum cannot be zero. If q is replaced by q^* in (5.14-1), subsequent integration yields a function g that maps the real w-axis onto a curve on which the sharp corner at z_m is replaced by an arc with continuously turning tangent. Since g is analytic in the upper half-plane, Theorem 5.10f (principle of the boundary map) assures us that the image of $\text{Im } w > 0$ is still a polygonal domain with interior angles $\alpha_k \pi$ at certain points z_k^* ($k \neq m$) and a suitably rounded corner near z_m. ∎

Generally, the corners z_k^* are not identical with the corners z_k of the undisturbed polygon, but in view of

$$\lim_{\epsilon \to 0} q^*(w) = q(w) \tag{5.14-4}$$

the z_k^* tend to the z_k for $\epsilon \to 0$ if the same constants of integration are chosen.

The shape of the arc rounding the corner near z_m, as well as the position of its initial and terminal points, can be influenced to some extent by the choice of the parameters a, b, and ϵ. It is clear that only in exceptional cases can the arc be made circular, although if $a = b$ it can become at least infinitesimally circular as $\epsilon \to 0$.

We illustrate Theorem 5.14a by rounding the corners occurring in the first example in §5.12:II.

EXAMPLE 1 FLOW AGAINST ROUNDED DAM

We wish to round the corner at $z_2 := id$ (see Fig. 5.12e) and to let the rounding begin at $z_1 := 0$. In the Schwarz-Christoffel function the factor $(w-1)^{1/2}$ is responsible for the sharp corner at z_2. By trivially modifying the procedure outlined in Theorem 5.14a, we replace the factor with

$$q^*(w) = aw^{1/2} + b(w-1)^{1/2},$$

where $a > 0$, $b > 0$. We drop the condition $a + b = 1$ and thus render the constant C superfluous. The differential quotient of the mapping function thus becomes

$$g'(w) = w^{-1/2}[aw^{1/2} + b(w-1)^{1/2}] = a + bw^{-1/2}(w-1)^{1/2}.$$

ROUNDING CORNERS IN SCHWARZ-CHRISTOFFEL MAPS 425

Integration yields

$$z = g(w) = aw + b\{w^{1/2}(w-1)^{1/2} - \text{Log}[w^{1/2} + (w-1)^{1/2}]\} + c.$$

The constants b and c are determined, as before, by the conditions $g(0) = 0$, $\text{Im}\, g(1) = d$ ($2d$ = thickness of dam), and we find

$$z = aw + \frac{2d}{\pi}\left\{w^{1/2}(w-1)^{1/2} - \text{Log}[w^{1/2} + (w-1)^{1/2}] - \frac{i\pi}{2}\right\}.$$

The end point of the rounding arc is at $g(1) = a + id$, and the parameter a can be used to regulate the "bluntness" of the rounding arc (see Fig. 5.14b).

We now have $g'(\infty) = a + 2d/\pi$. Thus the stream potential is

$$\phi(z) = \left(a + \frac{2d}{\pi}\right) v_\infty \, \text{Re}\, f(z),$$

where $f := g^{[-1]}$, and for its complex gradient we now find

$$\text{grd}\,\phi(z) = \left(a + \frac{2d}{\pi}\right) v_\infty \overline{f'(z)} = v_\infty \left[\pi a + \frac{2d}{\pi a + 2dw^{-1/2}(w-1)^{1/2}}\right],$$

where $w = f(z)$. Considered as a function of the real parameter w, the modulus $v := |\text{grd}\,\phi(z)|$ decreases on the interval $(-\infty, 0]$ from v_∞ to 0, increases on $[0, 1]$ from 0 to its maximum value

$$v_{\max} := v_\infty\left(1 + \frac{2}{\pi}\frac{d}{a}\right),$$

and then decreases again to v_∞. It is a matter of concern in boundary layer theory that

$$\lim_{w \to 1+} \frac{dv}{dw} = \infty.$$

Theorem 5.14a is obviously not useful in a concave corner with interior angle 2π (i.e., a slit), because then $\alpha_m = 1$ and the function $q^*(w)$ defined

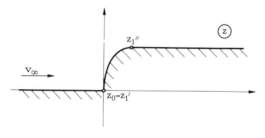

Fig. 5.14b. Flow against a rounded dam.

by (5.14-3) is essentially identical with $q(w)$. By the same reasoning, however, we can prove the following theorem:

THEOREM 5.14b

A concave corner z_m with interior angle 2π is rounded off by replacing the factor $q(w) = w - w_m$ in the Schwarz-Christoffel function by

$$g^*(w) := a(w - w_m) + b(w - w'_m)^{1/2}(w - w''_m)^{1/2}. \quad (5.14\text{-}5)$$

EXAMPLE 2   SEMI-INFINITE PLATE CONDENSER WITH ROUNDED EDGE

This is a modification of Example 2 in §5.12: II. Referring to Fig. 5.12g, we find that the sharp corner at $z_2 := id$, which corresponds to the point $s = -1$ in the auxiliary s-plane, is to be rounded. The factor $q(s) := s + 1$ is responsible for the corner. According to Theorem 5.14b, we replace it with

$$q^*(w) := \tfrac{1}{2}(s + 1 + \omega),$$

where

$$\omega := (s + 1 + \epsilon)^{1/2}(s + 1 - \epsilon)^{1/2} = \left[(s+1)^2 - \epsilon^2\right]^{1/2}.$$

The differential quotient of the mapping function thus becomes

$$\frac{dz}{ds} = Cs^{-1}(s + 1 + \omega),$$

which for purposes of integration we write

$$\frac{dz}{ds} = C\left(1 + \frac{1}{s} + \frac{s+1}{\omega} + \frac{1}{\omega} + \frac{1-\epsilon^2}{s\omega}\right).$$

The integrals of the first four terms are standard; to integrate the last term we substitute $s^{-1} =: t$. This yields

$$z(s) = C\left\{s + \omega + \left[1 + (1 - \epsilon^2)^{1/2}\right]\text{Log}\, s + \text{Log}(s + 1 + \omega)\right.$$
$$\left. - (1 - \epsilon^2)^{1/2}\text{Log}\left[1 + s - \epsilon^2 + (1 - \epsilon^2)^{1/2}\omega\right]\right\} + D.$$

To determine the constants of integration and the parameter ϵ we note, first of all, that $s > 0$ is to be mapped onto real values of z. This implies that C and D are real. The initial point, the point with vertical tangent, and the terminal point of the rounding arc are at

$$z(-1-\epsilon) = C[-1-\epsilon + \text{Log}(1+\epsilon) + (1-\sqrt{1-\epsilon^2})\text{Log}\epsilon] + D$$
$$+ 2i\pi C;$$

$$z(-1) = C\{-1 + [1 - (1-\epsilon^2)^{1/2}\text{Log}\epsilon]\} + D$$
$$+ iC\left[\frac{3\pi}{2} + \epsilon + (1-\epsilon^2)^{1/2}\arccos\epsilon\right];$$

$$z(-1+\epsilon) = C[-1+\epsilon + \text{Log}(1-\epsilon) + (1-\sqrt{1-\epsilon^2})\text{Log}\epsilon] + D$$
$$+ i\pi C(1+\sqrt{1-\epsilon^2}).$$

It follows that the distance between the two plates of the condenser is $2d_1$, where

$$d_1 := \text{Im}\, z(-1+\epsilon) = \frac{\pi C}{2}[1 + (1-\epsilon^2)^{1/2}].$$

The thickness of a plate is

$$d_2 := \text{Im}[z(-1-\epsilon) - z(-1+\epsilon)] = \frac{\pi C}{2}[1 - (1-\epsilon^2)^{1/2}].$$

Thus we find that

$$\frac{d_2}{d_1} = \frac{1 - (1-\epsilon^2)^{1/2}}{1 + (1-\epsilon^2)^{1/2}},$$

which determines ϵ:

$$\epsilon = \frac{2(d_1 d_2)^{1/2}}{d_1 + d_2};$$

C can now be expressed in terms of d_1. We finally let $D = C$, which puts the center of the rounding arc approximately on the imaginary axis. This yields the mapping function (see Fig. 5.14c)

$$z(s) = \frac{d_1}{\pi[1 + (1-\epsilon^2)^{1/2}]}\{s + 1 + \omega + [1 + (1-\epsilon^2)^{1/2}]\text{Log}\, s + \text{Log}(1+s+\omega)$$
$$- (1-\epsilon^2)^{1/2}\text{Log}[1 + s - \epsilon^2 + (1-\epsilon^2 \omega)^{1/2}]\}.$$

For the complete solution of the potential problem the mapping $s \to z(s)$ must be composed with the auxiliary mapping $w \to s := e^{(\pi/V)w}$ to yield the mapping func-

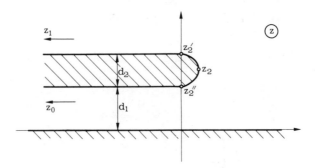

Fig. 5.14c. A semi-infinite plate condenser with a blunt edge.

tion from the model plane onto the physical plane:

$$g: w \to g(w) := z(e^{(\pi/V)w}).$$

Because the potential in the model plane is $\psi(w) = \operatorname{Im} w$, the potential in the physical plane becomes

$$\phi(z) = \operatorname{Im} f(z),$$

where $f := g^{[-1]}$.

The field strength in the model plane is $-i$. Thus the field strength in the physical plane becomes

$$E(z) = -i \overline{f'(z)}.$$

Because $f'(z) = [g'(w)]^{-1}$ and

$$g'(w) = \frac{dz}{ds}(s)\frac{ds}{dw} = \frac{dz}{ds} \cdot \frac{\pi}{V} s = \frac{\pi C}{V}(s+1+\omega),$$

we find a simple formula for the field strength:

$$E(z(s)) = -i\frac{V}{d_1}\overline{\left[\frac{1+(1-\epsilon^2)^{1/2}}{s+1+\omega}\right]}.$$

For $s \to 0$ (i.e., $\operatorname{Re} z \to -\infty$), the field strength is seen to approach the constant value of the ordinary plate condenser. For $s = -1+\eta$, $-\epsilon < \eta < \epsilon$, we obtain the field strength on the rounding arc. It turns out that for these values of s

$$|s+1+\omega| = |\eta + i(\epsilon^2 - \eta^2)^{1/2}| = \epsilon;$$

thus

$$|E(z(s))| = \frac{V}{d_1}\frac{1+(1-\epsilon^2)^{1/2}}{\epsilon} = \frac{V}{(d_1 d_2)^{1/2}}.$$

ROUNDING CORNERS IN SCHWARZ-CHRISTOFFEL MAPS

The rounding arc, although not exactly circular, thus has the interesting property that the field strength has constant magnitude on it.

II. Convex Corners

If the interior angle at point z_m of the polygonal region D is $\alpha_m \pi$, where $0 < \alpha_m < 1$, the corner is **convex**. (The intersection of a circular neighborhood of z_m with D is a convex region.) If the rounding factor $q^*(w)$ defined by (5.14-3) is used at a convex corner, the considerations used to prove Theorem 5.14a do not apply because the components of the vector $q^*(w)$ along the rays $\arg w = 0$ and $\arg w = (\alpha_m - 1)\pi$ are now infinite (instead of zero) for $w = w'_m$ and $w = w''_m$, respectively. This would cause $\arg q^*(w)$ to jump discontinuously from $(\alpha_m - 1)\pi$ to 0 at w'_m, to change continuously back to $(\alpha_m - 1)\pi$, and to jump once again from $(\alpha_m - 1)\pi$ to 0 at w''_m. Thus integration would produce a concave rounding at the convex corner, as shown in Fig. 5.14d.

To achieve a convex rounding we remove the infinite components by letting the weight factors a and b in (5.14-3) depend on w in such a manner that $a = 0$ at $w = w'_m$ and $b = 0$ at $w = w''_m$, whereas $a > 0$, $b > 0$ for $w'_m < w < w''_m$. Choosing the factors as linear functions and retaining the condition $a + b = 1$ then yields

$$a(w) = \frac{1}{2\epsilon}(w - w'_m), \qquad b(w) = \frac{1}{2\epsilon}(w''_m - w).$$

Thus we obtain the rounding factor

$$q^*(w) = \frac{1}{2\epsilon}\left[(w - w'_m)^{\alpha_m} - (w - w''_m)^{\alpha_m}\right]. \tag{5.14-6}$$

It is clear that the $\arg q^*(w) = 0$ for $w > w''_m$. For $w < w'_m$, $\arg(w - w'_m)^{\alpha_m} = \arg(w - w''_m)^{\alpha_m} = \alpha_m \pi$, but, since $|w - w''_m| > |w - w'_m|$, $\arg q^*(w) = (\alpha_m - 1)\pi$, as desired. In between there is a continuous turning as before. Moreover, $q^*(w)$ is clearly analytic for $\mathrm{Im}\, w > 0$ and different from zero for $\mathrm{Im}\, w \geq 0$. For $|w - w_m|$ large $q^*(w)$ behaves like $\mathrm{const} \cdot (w - w_m)^{\alpha_m - 1}$.

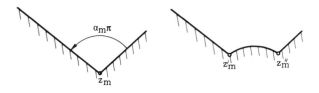

Fig. 5.14d. Concave rounding at a convex corner.

THEOREM 5.14c

A finite convex corner z_m of the polygonal region D may be rounded by replacing the factor (5.14-2) in the Schwarz-Christoffel function with (5.14-6).

EXAMPLE 3 SWEEPING OUT A RIGHT ANGLE

The half-plane $\operatorname{Im} w > 0$ is mapped onto the first quadrant $0 < \arg z < \pi/2$ by the function

$$w \to z = Cw^{1/2}, \quad C > 0,$$

arising from the Schwarz-Christoffel differential quotient

$$\frac{dz}{dw} = \frac{1}{2} C w^{-1/2}. \tag{5.14-7}$$

If the w-plane is endowed with the potential $\psi(w) = \operatorname{Re} w$, which represents a uniform flow parallel to the real axis, the transplanted potential $\phi(z) = \operatorname{Re} f(z) = \operatorname{Re}(z/C)^2$ represents the flow around a convex right angle.

We now wish to round off the corner at $z = 0$. To this end we replace $w^{-1/2}$ in (5.14-7) with

$$q^*(w) = \tfrac{1}{2}\left[(w+1)^{1/2} - (w-1)^{1/2}\right],$$

where we have already selected $\epsilon = 1$. Integration now yields

$$z(w) = C\left[(w+1)^{3/2} - (w-1)^{3/2}\right]$$

with a different value of C. It is clear that $w > 1$ is mapped onto a piece of the real axis and $w < -1$, onto a piece of the imaginary axis. For $-1 < w < 1$, if $z = : x + iy$, we have

$$x = C(1+w)^{3/2}, \quad y = C(1-w)^{3/2},$$

thus the rounding arc is a piece of the **astroid** (see Fig. 5.14e)

$$x^{3/2} + y^{3/2} = 2C^{2/3}.$$

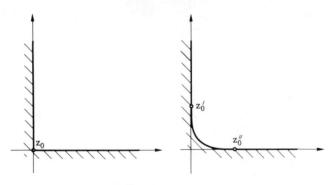

Fig. 5.14e. Convex rounding at a convex corner.

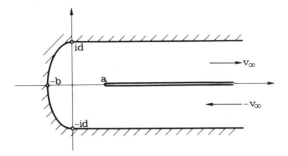

Fig. 5.14f. A turnabout channel (rounded).

The velocity now is

$$\operatorname{grd}\phi(z) = \overline{f'(z)} = \left(\overline{g'(w)}\right)^{-1},$$

where $g'(w) = (3/2)C[(w+1)^{1/2} - (w-1)^{1/2}]$. On the rounding arc,

$$|g'(w)|^2 = \left(\tfrac{3}{2}C\right)^2 (1 + w + 1 - w) = \frac{9C^2}{2}.$$

Thus on the arc the velocity has the constant magnitude $\sqrt{2}/3C$.

PROBLEMS

1. Determine the flow through a turnabout channel, as shown in Fig. 5.14f, if at infinity the flow is homogeneous. Between what limits (in relation to the velocity at infinity) does the flow velocity vary along the curved part of the boundary? [Compare Problem 4, §5.12.]
2. Map $\operatorname{Im} w > 0$ onto the upper half of the z-plane cut between $z = 0$ and $z = ia$ ($a > 0$). Round the corner at $z = ia$ such that the initial and terminal points of the rounding arc lie on the real axis (see Fig. 5.14g). Show that the rounding arc is part of an ellipse. As a special case, obtain the inverse Joukowski map.
3. In Problem 2 of §5.12 round the corner at $z = ia$ with an arc, starting at $z = 0$.
4. In Problem 3 of §5.12 round the corner at $z = id$ and determine the field strength along the curved part of the boundary.
5. In Problem 1 of §5.12 round the re-entrant corner at point $z = (1 + i)d$ and show that this can be done in such a way that the flow will have a constant

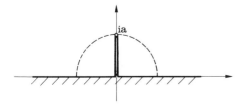

Fig. 5.14g. Rounding a slit in a half-plane.

velocity v_0 along the curved part of the boundary. If r is the approximate radius of curvature of the rounding arc, show that

$$\frac{v_0}{v_\infty} \sim \left(\frac{3\pi}{2}\right)^{1/3} \left(\frac{d}{r}\right)^{1/3} \quad (r \to 0).$$

SEMINAR ASSIGNMENTS

Develop a Schwarz-Christoffel mapping formula for regions bounded by a Jordan curve with a continuously turning tangent (see Paatero [1931]). Use the formula to construct the mapping function for Example 1 in §5.14, in which the rounded corner is a precise circular arc.

NOTES

§5.1. The mapping of the interior of an ellipse onto the interior of the unit circle cannot be accomplished by means of elementary functions; see §5.13 and Chapter 14.

§5.3. Moebius transformations that induce rigid motions of the Riemann sphere were used by Wolfgang Pauli in his theory of spin of the electron; see Bohm [1951], p. 391.

§5.5. The treatment follows Ahlfors [1953]. For the appendix see the notes in §4.3.

§5.7. For a comprehensive treatment of spatial electrostatics see Jackson [1962]. Smythe [1939] in Chapter IV treats plane electrostatics with conformal mapping.

§5.8. Concerning the application of conformal mapping to two-dimensional flows in general, see Milne-Thomson [1952], Betz [1964], Sedow [1965], and Lavrentiev & Schabat [1967]. For the technically important problem of airfoils in cascade see Busemann [1928], Weinig [1935], Traupel [1944, 1966], Scholz [1965], Koch [1968], and Firth [1968].

§5.9. For a different treatment of torsional rigidity by conformal transplantation see Polya & Szegö [1951].

§5.10. For a proof of the Riemann mapping theorem, and a splendidly concise treatment of conformal mapping in general, see Ahlfors [1953, 1966]. Nehari [1952] gives a broad exposition including many special maps. As to boundary behavior, see Golusin [1957]. See also Chapter 14.

§5.11. The very useful Theorem 5.11a is not given in many theoretical texts because Theorem 4.3g is not available. See, however, Hurwitz & Courant [1929], p. 372, and Lavrentiev & Schabat [1967], p. 96.

§5.12. Many special cases of the Schwarz-Christoffel mapping (and of other mappings as well) are discussed in Lavrentiev & Schabat [1967]. Kober [1952] and Churchill [1960] have very useful catalogues (with illustrations) of special mapping functions.

§5.13. For presentations of the theory of elliptic functions, see Chapter 7 of Ahlfors [1966] (very concise), Siegel [1969] (broad, perspective), and Part II (by Hurwitz) of Hurwitz & Courant [1929] (comprehensive, with applications to number theory).

§5.14. Some early references dealing with rounding corners in Schwarz-Christoffel maps are Richmond [1923], Cockroft [1927], Adams [1939]. The treatment given here follows Henrici [1948]. For other methods of rounding corners see Lavrentiev & Schabat [1967], pp. 213–219, and Carrier, Krook & Pierson [1966], pp. 154–157. For curvilinear triangles whose sides are circular arcs the mapping function can be represented in terms of hypergeometric functions, see Hille [1962].

6

POLYNOMIALS

The reader is probably aware of the existence of an extensive algebraic theory of polynomials, which deals with such questions as irreducibility or solvability in terms of roots. In this chapter, however, we deal mainly with the *analytic theory of polynomials*, a theory that has as its central problem the location of the zeros of polynomials over the fields of real or complex numbers. The emphasis is on complex zeros and complex polynomials; real zeros of real polynomials are dealt with briefly in §§6.2 and 6.3. The discussion generally proceeds from rather crude elementary results to sophisticated algorithms that permit the simultaneous determination of all zeros to arbitrary precision.

§6.1. THE HORNER ALGORITHM

For any extended work with polynomials we must be able to compute, as economically as possible, the value of a given polynomial at any given point of the complex plane. Frequently we also require some or all coefficients of the Taylor expansion of the polynomial at a given point.

A given polynomial can be represented analytically in many different ways: it may be defined as an interpolating polynomial (and even as such has many different representations), it may be the characteristic polynomial of a matrix, it may be represented as a sum of Chebyshev polynomials, or it may even be given in factored form. All these representations mathematically define the same function; however, each representation suggests a special algorithm for computing the values of the polynomial and possibly its derivatives.

Here we assume that the polynomial is given by its Taylor series at the origin, which we write in the form

$$p(z) = a_0 z^n + a_1 z^{n-1} + \cdots + a_n. \qquad (6.1\text{-}1)$$

The obvious way to compute $p(z)$ for a given z would seem to consist in building up the powers of z recursively, $z^2:=z\cdot z$, $z^3:=z\cdot z^2,\ldots, z^n=z\cdot z^{n-1}$, and then multiplying by the appropriate coefficients a_k. The number of multiplications required is $2n-1$. (For many purposes it is permissible to ignore additions in this kind of operations count.) The calculation of the first derivative by the same method requires an additional $2n-2$ operations; similarly we find that to calculate the mth derivative we require $2n-2m$ multiplications after the preceding derivatives have been calculated. Thus the calculation of all derivatives would require some $n(n+1)$ multiplications. If the Taylor coefficients are required, we must still divide by suitable factorials, which requires an additional $O(n)$ operations.

We shall now discuss a procedure that permits the calculation of $p(z)$ by n multiplications only and the computation of all Taylor coefficients by $\frac{1}{2}n(n+1)$ multiplications.

By writing (6.1-1) in the form

$$p(z)=(\cdots(((a_0z+a_1)z+a_2)z+\cdots+a_{n-1})z+a_n \qquad (6.1\text{-}1)$$

we see that $p(z)$ can be obtained in n multiplications. Expressing the procedure analytically, we compute numbers $b_0, b_1,\ldots,b_n$ by the recurrence relation

$$b_0=a_0,$$

$$b_k=a_k+zb_{k-1}, \qquad k=1,2,\ldots,n. \qquad (6.1\text{-}2)$$

We then clearly have

$$b_n=p(z); \qquad (6.1\text{-}3)$$

more generally, $b_k=p_k(z)$ $(k=0,1,\ldots,n)$, where

$$p_k(z):=a_0z^k+a_1z^{k-1}+\cdots+a_k. \qquad (6.1\text{-}4)$$

We now subject the numbers $b_k=:b_k^{(0)}$ to the same algorithm as above the a_k by forming $b_k^{(1)}$ according to

$$b_0^{(1)}=b_0^{(0)},$$

$$b_k^{(1)}=b_k^{(0)}+zb_{k-1}^{(1)}, \qquad k=1,2,\ldots,n. \qquad (6.1\text{-}5)$$

THE HORNER ALGORITHM

By the above result, if $k=0,1,\ldots,n,$

$$
\begin{aligned}
b_k^{(1)} &= b_0^{(0)}z^k + b_1^{(0)}z^{k-1} + \cdots + b_k^{(0)} \\
&= a_0 z^k + (a_0 z + a_1)z^{k-1} + (a_0 z^2 + a_1 z + a_2)z^{k-2} + \cdots \\
&\quad + (a_0 z^k + a_1 z^{k-1} + a_2 z^{k-2} + \cdots + a_k) \\
&= (k+1)a_0 z^k + k a_1 z^{k-1} + \cdots + a_k \\
&= p'_{k+1}(z);
\end{aligned}
$$

thus in particular,

$$b_{n-1}^{(1)} = p'(z). \tag{6.1-6}$$

More generally still, we form a two-dimensional array of numbers $b_k^{(m)}$ by repeating this process, which amounts to the following recurrence relations, called the **Horner algorithm**:

$$
\begin{aligned}
b_k^{(-1)} &:= a_k, & k&=0,1,\ldots,n, \\
b_0^{(m)} &:= a_0, & m&=0,1,\ldots,n, \\
b_k^{(m)} &:= b_k^{(m-1)} + a b_{k-1}^{(m)}, & m&=0,1,\ldots,n; \\
& & k&=1,2,\ldots,n-m.
\end{aligned}
\tag{6.1-7}
$$

The following result which generalizes (6.1-3) and (6.1-6) holds:

THEOREM 6.1a

Let p be the polynomial (6.1-1), let z be any complex number, and let

$$c_m := \frac{1}{m!} p^{(m)}(z), \qquad m=0,1,\ldots,n.$$

If the coefficients $b_k^{(m)}$ are constructed by the Horner algorithm (6.1-7), then for $m=0,1,\ldots,n$,

$$b_{n-m}^{(m)} = c_m; \tag{6.1-8}$$

furthermore, if w is any complex number,

$$b_0^{(m)}w^{n-m-1}+b_1^{(m)}w^{n-m-2}+\cdots+b_{n-m-1}^{(m)}$$

$$=\frac{p(w)-c_0-c_1(w-z)-\cdots-c_m(w-z)^m}{(w-z)^{m+1}} \qquad (w\neq z)$$

$$=c_{m+1}+c_{m+2}(w-z)+\cdots+c_n(w-z)^{n-m-1}. \qquad (6.1\text{-}9)$$

The *proof* is by induction with respect to m. We verify the assertion for $m=0$. Here the recurrence relation (6.1-7) may be read

$$a_0=b_0^{(0)}, \qquad a_k=b_k^{(0)}-zb_{k-1}^{(0)}, \qquad k=1,2,\ldots,n.$$

This means that

$$p(w)=a_0w^n+a_1w^{n-1}+\cdots+a_n$$

$$=\left(b_0^{(0)}w^{n-1}+b_1^{(0)}w^{n-2}+\cdots+b_{n-1}^{(0)}\right)(w-z)+b_n^{(0)}$$

holds identically in w. Letting $w=z$ yields (6.1-8) for $m=0$. Assuming that $w\neq z$ and solving for the first parenthesis yields

$$b_0^{(0)}w^{n-1}+b_1^{(0)}w^{n-2}+\cdots+b_{n-1}^{(0)}=\frac{p(w)-c_0}{w-z}.$$

Since the Taylor expansion of p at point z is

$$p(w)=c_0+c_1(w-z)+\cdots+c_n(w-z)^n,$$

the term on the right also equals

$$c_1+c_2(w-z)+\cdots+c_n(w-z)^{n-1},$$

which proves (6.1-9) for $m=0$. Let (6.1-8) and (6.1-9) now be proved for $m-1$ in place of m. The recurrence relation (6.1-7) then shows that

$$b_0^{(m-1)}w^{n-m}+b_1^{(m-1)}w^{n-m-1}+\cdots+b_{n-m}^{(m-1)}$$

$$=\left(b_0^{(m)}w^{n-m-1}+b_1^{(m)}w^{n-m-2}+\cdots+b_{n-m-1}^{(m)}\right)(w-z)+b_{n-m}^{(m)}$$

THE HORNER ALGORITHM

holds identically in w. If (6.1-9) is true for $m-1$, the expression on the left equals

$$c_m + c_{m+1}(w-z) + \cdots + c_n(w-z)^{n-m}.$$

Letting $w = z$ thus yields $b_{n-m}^{(m)} = c_m$, and assuming that $w \neq z$ and solving for the first parenthesis yields (6.1-9) for the index m. ∎

The coefficients $b_k^{(m)}$ constructed by the Horner algorithm are conveniently arranged in the following two-dimensional array, called the **Horner scheme**:

a_0	a_1	a_2	a_3
$b_0^{(0)}$	$b_1^{(0)}$	$b_2^{(0)}$	$\mathbf{b_3^{(0)}}$
$b_0^{(1)}$	$b_1^{(1)}$	$\mathbf{b_2^{(1)}}$	
$b_0^{(2)}$	$\mathbf{b_1^{(2)}}$		
$\mathbf{b_0^{(3)}}$			

The boldface entries are the Taylor coefficients at z.

EXAMPLE 1

Compute the Taylor expansion of the polynomial

$$p(z) := z^4 - 4z^3 + 3z^2 - 2z + 5$$

at $z = 2$. The following Horner scheme results:

1	-4	3	-2	5
1	-2	-1	-4	-3
1	0	-1	-6	
1	2	3		
1	4			
1				

Thus the desired expansion is

$$p(2+h) = h^4 + 4h^3 + 3h^2 - 6h - 3.$$

In particular, $p(2) = -3$, $p'(2) = -6$.

If z is a zero of multiplicity m of p, then $c_0 = c_1 = \cdots = c_{m-1} = 0$. Thus in this situation Theorem 6.1a yields the following:

COROLLARY 6.1b

Under the hypotheses of Theorem 6.1a let z be a zero of multiplicity m of p. Then

$$\frac{p(w)}{(w-z)^m} = b_0^{(m-1)}w^{n-m} + b_1^{(m-1)}w^{n-m-1} + \cdots + b_{n-m}^{(m-1)}.$$

In particular, if $m=1$, then

$$\frac{p(w)}{w-z} = b_0^{(0)}w^{n-1} + b_1^{(0)}w^{n-2} + \cdots + b_{n-1}^{(0)}. \tag{6.1-11}$$

This result is often used in connection with the determination of zeros of polynomials. If all zeros of a polynomial p are to be found and a zero z_0 has been determined, we remove the linear factor $z - z_0$ from p to continue the computation with a polynomial of lower degree and to avoid determining z_0 once more. This process of factor removal is called **deflation**. As shown by the above result, the coefficients of the deflated polynomial are just the coefficients of the row $m=0$ of the Horner scheme, formed with $z = z_0$.

EXAMPLE 2

The polynomial

$$p(z) := z^4 - 2z^3 - 7z^2 + 8z + 12$$

has the zero $z = -2$. Forming the first row of the Horner scheme with $z = -2$ yields

1	-2	-7	8	12
1	-4	1	6	0

We find that $b_4^{(0)} = 0$, thus confirming that $p(-2) = 0$. It follows that the deflated polynomial is

$$\frac{p(z)}{z+2} = z^3 - 4z^2 + z + 6.$$

PROBLEMS

1. (Horner's scheme generalized). Let $\{a_n\}$ and $\{z_k\}$ be two sequences of complex numbers. Let the numbers $b_{k,n} := b_{k,n}(z_0, z_1, \ldots, z_k)$ be defined as follows: $b_{-1,n} = a_n$, $n = 0, 1, \ldots$; $b_{k,-1} = 0$, $k = 0, 1, \ldots$;

$$b_{k,n} = z_k b_{k,n-1} + b_{k-1,n}, \qquad k, n = 0, 1, 2, \ldots.$$

(This reduces to Horner's scheme if all $z_k = z$.) Show that

$$\sum_{n=0}^{\infty} b_{k,n} t^n = [(1 - tz_0) \cdots (1 - tz_k)]^{-1} \sum_{n=0}^{\infty} a_n t^n.$$

2. (continuation). If $z_0 \neq z_k$, show that

$$b_{k,n}(z_0, \ldots, z_k) = \frac{b_{k-1, n+1}(z_0, \ldots, z_{k-1}) - b_{k-1, n+1}(z_1, \ldots, z_k)}{z_0 - z_k},$$

$k = 1, 2, \ldots$; $n = 0, 1, \ldots$. (This characterizes the $b_{k,n}$ as the kth *divided differences* of the functions p_{n+k} defined by (6.1-4); see problem 2, §4.7.)
3. Show that the limit of $b_{k,n}(z_0, z_1, \ldots, z_k)$, as all z_i tend to z, equals $(1/k!) \times p_{n+k}^{(k)}(z)$.

§6.2. SIGN CHANGES. THE RULE OF DESCARTES

In this section we introduce the concept of the number of *sign changes* (or variations of sign) in a given finite sequence of real numbers. This concept is fundamental in subsequent sections. As a first application, we give a proof of the famous Descartes rule that provides an upper bound for the number of positive zeros of a given real polynomial.

Let $\{a_0, a_1, \ldots, a_n\}$ be a finite sequence of real numbers. We shall say that a **sign change** occurs between the elements a_k and a_m, when $0 \leq k < m \leq n$, if

$$a_k a_m < 0$$

and if either $m = k+1$ or $m > k+1$ and

$$a_l = 0, \quad k < l < m.$$

The total number of sign changes that occurs between the elements of a sequence is called the **number of sign changes** of the sequence.

EXAMPLE 1

The sequence $\{1, 0, 2, -3, 0, 0, 2, 1, 2, 2, 0, 0\}$ has two sign changes.

To interpret the number of sign changes graphically we consider the piecewise linear real function f defined on $[0, n]$ whose graph passes through the points (k, a_k) and is linear in between. The number of sign changes equals the number of times the graph crosses the x-axis (see Fig. 6.2).

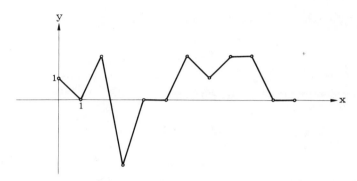

Fig. 6.2. Sign changes.

It is clear that the number of sign changes of a given sequence depends only on the signs of its elements. It is not altered if the sequence is multiplied by a constant $c \neq 0$ or by a sequence of constant sign. We also note without proof the following two propositions:

LEMMA 6.2a

If $a_m \neq 0$, the number of sign changes of the sequence $\{a_0, a_1, \ldots, a_n\}$ equals the sum of the numbers of sign changes of the sequences $\{a_0, a_1, \ldots, a_m\}$ and $\{a_m, a_{m+1}, \ldots, a_n\}$.

LEMMA 6.2b

Let $0 \leq k < m \leq n, a_k a_m \neq 0$. The number of sign changes of the sequence $\{a_k, a_{k+1}, \ldots, a_m\}$ is even (possibly zero) if $a_k a_m > 0$ and odd (thus at least one) if $a_k a_m < 0$.

Along with the given sequence $\{a_k\}$ we consider the sequence $\{b_k\}$ of its differences,

$$b_k := a_k - a_{k-1}, \qquad k = 0, 1, \ldots, n+1,$$

where $a_{-1} := a_{n+1} := 0$. The following lemma is almost clear from the graphical interpretation given above:

LEMMA 6.2c

Let the sequence $\{a_0, a_1, \ldots, a_n\}$ be not identically zero. Then the number of sign changes of the sequence of its differences $\{b_k\}$ exceeds the number of sign changes of the given sequence $\{a_k\}$ by an odd positive number.

Proof. Let there be v sign changes in the given sequence and let them

SIGN CHANGES: RULE OF DESCARTES

occur between the elements

$$a_{k(1)}, a_{m(1)}; \ a_{k(2)}, a_{m(2)}; \ \cdots; \ a_{k(v)}, a_{m(v)}.$$

Furthermore, let $a_{m(0)}$ and $a_{k(v+1)}$ be the first and the last nonvanishing elements in the sequence. Without loss of generality we may assume that $a_{m(0)} > 0$. We then have

$$\operatorname{sign} a_{k(i)} = (-1)^{i-1}, \quad \operatorname{sign} a_{m(i)} = (-1)^{i}, \quad i = 1, 2, \ldots, v,$$

and furthermore

$$(-1)^{i-1} a_{m(i)-1} \geq 0.$$

Hence the signs of the following elements of the sequence $\{b_k\}$ are known:

$$b_{m(0)} = a_{m(0)} > 0,$$

$$b_{m(1)} = a_{m(1)} - a_{m(1)-1} < 0,$$

and generally

$$(-1)^i b_{m(i)} = (-1)^i [a_{m(i)} - a_{m(i)-1}] > 0, \quad i = 1, 2, \ldots, v.$$

Thus

$$\operatorname{sign} b_{m(i)} = (-1)^i, \quad i = 0, 1, \ldots, v,$$

and it follows from Lemma 6.2b that an odd number of sign changes occurs in each of the v subsequences $\{b_{m(i)}, b_{m(i)+1}, \ldots, b_{m(i+1)}\}, i = 0, 1, \ldots, v-1$. Furthermore, $a_{k(v+1)}$ has the same sign as $a_{m(v)}$ and $a_{k(v+1)+1}$ is zero. Hence

$$\operatorname{sign} b_{k(v+i)+1} = (-1)^{v+1},$$

and an odd number of sign changes also occurs in the subsequence $\{b_{m(v)}, \ldots, b_{k(v+1)+1}\}$. It follows from Lemma 6.2a that the total number of sign changes of the sequence $\{b_k\}$ is

$$v \cdot \text{odd} + \text{odd} = v + v \cdot \text{even} + \text{odd} = v + \text{odd},$$

as asserted. ∎

We are now prepared to prove the following classical result.

THEOREM 6.2d (Descartes' rule of signs)

Let $p(x) := a_0 + a_1 x + \cdots + a_n x^n$ be a real polynomial (not the zero polynomial), let v denote the number of sign changes in the sequence $\{a_k\}$ of its coefficients, and let r denote the number of its real, positive zeros (each zero counted with its proper multiplicity). Then $v - r$ is even and non-negative.

Proof (by induction with respect to r). If the polynomial p has no positive zeros, the factor representation shows that the first nonzero coefficient has the same sign as a_n. Hence by Lemma 6.2b the number of sign changes is even and the theorem is true for $r = 0$.

Assume that the theorem is true for some $r \geq 0$. If the real polynomial p has $r+1$ positive zeros, we can write it in the form

$$p(x) = (x - \xi) p_1(x),$$

where $\xi > 0$ and where

$$p_1(x) =: a_0 + a_1 x + a_2 x^2 + \cdots + a_n x^n$$

has r positive zeros. By the induction hypothesis the sequence $\{a_m\}$ has $r + 2k$ sign changes, where k is a suitable non-negative integer. The coefficients of p are

$$b_m := a_{m-1} - \xi a_m, \qquad m = 0, 1, \ldots, n+1,$$

where $a_{-1} = a_{n+1} = 0$. The number v of sign changes of the sequence $\{b_m\}$ is the same as the number of sign changes in

$$\{ -\xi^{m-1} b_m \} = \{ \xi^m a_m - \xi^{m-1} a_{m-1} \}.$$

The latter sequence is the sequence of differences of the sequence $\{\xi^m a_m\}$, which has the same number of sign changes as $\{a_m\}$, namely $r + 2k$. By Lemma 6.2c it follows that

$$v = r + 2k + \text{an odd positive number}$$

$$= r + 1 + \text{an even non-negative number}.$$

The assertion of the theorem thus holds for polynomials with $r+1$ positive zeros, which completes the induction step. ∎

EXAMPLE 2

The polynomial $x^6 - 2x^5 + 3x^4 - 4x^3 + 3x^2 - 2x + 1$ has 6, 4, 2, or 0 positive zeros. (See Example 2 of §6.3.)

CAUCHY INDICES

PROBLEMS

In the following problems p denotes a real polynomial of degree n, and $v(x)$ denotes the number of sign changes in the sequence $\{p(x), p'(x), \ldots, p^{(n)}(x)\}$.

1. If ξ is real and $p(\xi) \neq 0$, show that $v(\xi-) - v(\xi+)$ is an even non-negative integer. [v can change only at points ξ which are zeros of at least one derivative of p. Allowance must be made, however, for the possibility that several derivatives vanish simultaneously and that their zeros may have multiplicities >1. Write out the most general form of Taylor's expansion of p at such a point (which may be taken as 0 without loss of generality). The sign of each derivative is determined by the leading term of its Taylor expansion. Lemma 6.2a may be useful.]

2. If ξ is a real zero of multiplicity $m > 0$ of p, show that

$$v(\xi-) - v(\xi+) = m + \text{an even non-negative integer.}$$

3. Prove that the function v is nonincreasing.
4. Prove the **theorem of Fourier-Budan**. For any real α and β such that $\beta > \alpha$, $p(\alpha) \neq 0$, $p(\beta) \neq 0$ the number of zeros in the interval $[\alpha, \beta]$ (each zero counted with its proper multiplicity) equals $v(\alpha) - v(\beta)$ minus an even non-negative integer.
5. Obtain Descartes' rule of signs as a limiting case of the Fourier-Budan theorem.

§6.3. CAUCHY INDICES: THE NUMBER OF ZEROS OF A REAL POLYNOMIAL IN A REAL INTERVAL

The theorem of Fourier-Budan yields an upper bound for the number of zeros of a real polynomial in a real interval. An exact determination of this number is possible by means of the *Cauchy index*. This is an integer that can be associated with any real rational function r and any interval whose end points are not poles of r. It can be effectively computed by calculating the number of sign changes in certain sequences that are determined by means of the Euclidean algorithm via Sturm sequences. Cauchy indices are used again in §6.7 to determine the number of zeros of a polynomial in a half-plane.

I. Cauchy Indices

Let r be a rational function and let ξ be a real pole of r. Each one-sided limit of r at ξ is then either $+\infty$ or $-\infty$ and all four possible combinations can occur. We shall say that r has the **Cauchy index** $+1$ at ξ if

$$\lim_{x \to \xi-} r(x) = -\infty, \qquad \lim_{x \to \xi+} r(x) = \infty,$$

and the Cauchy index -1 if

$$\lim_{x \to \xi-} r(x) = \infty, \qquad \lim_{x \to \xi+} r(x) = -\infty.$$

In those cases in which the limits from both sides are identical the Cauchy index is defined as zero.

Let the principal part of r (i.e., the terms involving negative powers in the Laurent expansion of r) at ξ be

$$\frac{a_n}{(x-\xi)^n} + \frac{a_{n-1}}{(x-\xi)^{n-1}} + \cdots + \frac{a_1}{x-\xi},$$

where the a_k are real, $a_n \neq 0$. Then the behavior of r at ξ is determined by the leading term $a_n(x-\xi)^{-n}$, and from a discussion of that term it follows that the Cauchy index of r at ξ is

$$\begin{matrix} 0, & \text{if } n \text{ is even,} \\ \operatorname{sign} a_n, & \text{if } n \text{ is odd.} \end{matrix} \qquad (6.3\text{-}1)$$

Now let α and β be two real numbers that are not poles of r, $\alpha < \beta$. The **Cauchy index** of r **for the interval** $[\alpha, \beta]$ is defined as the sum of the Cauchy indices of r for all poles ξ such that $\alpha < \xi < \beta$. It will be denoted by

$$I_\alpha^\beta r \quad \text{or} \quad I_\alpha^\beta r(x).$$

Frequently this number is referred to as "the number of times r jumps from $-\infty$ to $+\infty$ minus the number of times it jumps from $+\infty$ to $-\infty$." For infinite intervals the Cauchy index is defined, for example, by

$$I_\alpha^\infty r := \lim_{\beta \to \infty} I_\alpha^\beta r.$$

Since a rational function has only a finite number of poles, $I_\alpha^\beta r$ is constant for β sufficiently large and the limit always exists.

II. Sturm Sequences

Let $\{f_0, f_1, \ldots, f_m\}$ be a finite sequence of real polynomials. This sequence is called a **Sturm sequence** for the interval $[\alpha, \beta]$ if the following properties hold:

(i) $\quad f_0(\alpha) \neq 0, \quad f_0(\beta) \neq 0,$

(ii) $\quad$ If $f_k(\xi) = 0$, where $0 \leq k \leq m-1$ and $\alpha \leq \xi \leq \beta$, then $f_1(\xi) \neq 0$ if $k = 0$, and

$$f_{k-1}(\xi)f_{k+1}(\xi) < 0, \quad \text{if} \quad 1 \leq k \leq m-1,$$

(iii) $f_m(x) \neq 0$ for $\alpha \leq x \leq \beta$.

If the polynomials $f_0, \ldots, f_m$ form a Sturm sequence for the interval $[\alpha, \beta]$ and g is any real polynomial that does not vanish on $[\alpha, \beta]$, the polynomials $gf_0, \ldots, gf_m$ again form a Sturm sequence for the same interval. Conversely, if the polynomials $f_0, \ldots, f_m$ that form a Sturm sequence for $[\alpha, \beta]$ all contain the same factor h (which does not vanish on $[\alpha, \beta]$, since $f_m \neq 0$), the polynomials $h^{-1}f_0, \ldots, h^{-1}f_m$ still form a Sturm sequence.

For arbitrary real x we denote by $v(x)$ the number of sign changes in the numerical sequence

$$\{f_0(x), f_1(x), \ldots, f_m(x)\}.$$

The function v is related to the Cauchy index by the following result due to Sturm.

THEOREM 6.3a

Let the real polynomials $f_0, f_1, \ldots, f_m$ form a Sturm sequence for the interval $[\alpha, \beta]$, $\alpha < \beta$. Then

$$I_\alpha^\beta \frac{f_1}{f_0} = v(\alpha) - v(\beta). \tag{6.3-2}$$

Proof. Since polynomials are continuous, it is clear that $v(x)$ can change only at points ξ which are zeros of one of the f_k. First we show that it cannot change at zeros that coincide with one of the endpoints α, β. By (i) and (iii), such a zero must be a zero of one of the functions $f_1, \ldots, f_{m-1}$, but if, for instance, $f_k(\alpha) = 0$, where $1 \leq k \leq m-1$, then $f_{k-1}(\alpha)$ and $f_{k+1}(\alpha)$ have opposite signs by (ii); hence the sequence $\{f_i(x)\}$ keeps its one sign change between the indices $k-1$ and $k+1$ for $x - \alpha \geq 0, x - \alpha$ sufficiently small, no matter whether f_k turns to positive or negative values for $x > \alpha$.

Next we show that $v(x)$ does not change at interior zeros of one of the functions $f_k, k > 0$. Indeed, let k and ξ be such that $1 \leq k \leq m-1, \alpha < \xi < \beta, f_k(\xi) = 0$. By continuity and (ii) $f_{k-1}(x)f_{k+1}(x) < 0$ for x sufficiently close to ξ. Hence by Lemma 6.2b the sequence $\{f_i(x)\}$ has an odd positive number of sign changes between the indices $k-1$ and $k+1$. Since this odd number is at most 2, it is equal to 1, hence constant.

It remains to consider the case in which $f_0(\xi) = 0$, $\alpha < \xi < \beta$. By (ii)

$f_1(\xi) \neq 0$, and the function $r := f_1/f_0$ then has a pole at ξ. The order of the pole equals the multiplicity of ξ as a zero of f_0. If the multiplicity is even, then on the one hand the Cauchy index of r at ξ is zero and on the other, since $f(x)$ has the same sign on either side of ξ, no sign change is lost or gained as x passes through ξ. If the multiplicity m is odd, the Cauchy index is either $+1$ or -1. The following four possibilities can occur:

		Left of ξ	Right of ξ	Cauchy Index	Loss of Sign change
(1)	sign f_0	$-$	$+$	1	1
	sign f_1	$+$	$+$		
(2)	sign f_0	$+$	$-$	-1	-1
	sign f_1	$+$	$+$		
(3)	sign f_0	$-$	$+$	-1	-1
	sign f_1	$-$	$-$		
(4)	sign f_0	$+$	$-$	1	1
	sign f_1	$-$	$-$		

Inspection shows that in all four cases the Cauchy index equals the loss of sign changes in the sequence $\{f_k(x)\}$ between the indices 0 and 1 at $x = \xi$. It follows that $v(x)$ decreases or increases by one each time x crosses a pole of Cauchy index $+1$ or -1, thus proving the formula (6.3-2). ∎

III. The Euclidean Algorithm

If p_0 and p_1 are any two polynomials with complex coefficients different from the zero polynomial, we can by the usual long division algorithm determine polynomials q_1 and p_2 such that identically in z

$$p_0(z) = p_1(z) q_1(z) - p_2(z), \qquad (6.3\text{-}3)$$

where p_2 is either the zero polynomial or its degree is less than the degree of p_1. The polynomials q_1 and p_2 are uniquely determined by p_0 and p_1. The polynomial $-p_2$ is called the **remainder** of the division of p_0 by p_1. [For formal reasons it is convenient to attach a minus sign to the remainder in (6.3-3).]

If the degree of p_2 is still positive, we can likewise define $-p_3$ as the remainder when p_1 is divided by p_2, etc. Since the degrees of successive remainders always decrease, we must arrive in a finite number of steps at a point at which the remainder is zero. This algebraic process is known as the **Euclidean algorithm**. It furnishes a finite sequence of polynomials linked by the relations

$$p_0(z) = q_1(z)p_1(z) - p_2(z),$$
$$p_1(z) = q_2(z)p_2(z) - p_3(z),$$
$$\cdots$$
$$p_{k-1}(z) = q_k(z)p_k(z) - p_{k+1}(z), \quad (6.3\text{-}4)$$
$$\cdots$$
$$p_{m-1}(z) = q_m(z)p_m(z).$$

By definition the polynomial p_m is not the zero polynomial. If p_m is not constant, an induction shows that p_m is a factor of all preceding polynomials $p_{m-1}, \ldots, p_0$. [On the other hand, Equations (6.3-4) show that any common factor of p_0 and p_1 is a factor also of all subsequent polynomials and therefore of p_m. It follows that p_m contains every common factor of p_0 and p_1.] Consequently the functions $f_k := p_k/p_m$ are again polynomials.

THEOREM 6.3b

Let p_0 and p_1 be any two real polynomials different from the zero polynomial and let the polynomials $p_2, p_3, \ldots, p_m, 0$ be defined by the Euclidean algorithm. The polynomials $f_k := p_k/p_m$ ($k = 0, 1, \ldots, m$) form a Sturm sequence for any interval $[\alpha, \beta]$ such that $p_0(\alpha)p_0(\beta) \neq 0$.

Proof. It is clear that the sequence $\{f_k\}$ possesses the properties (i) and (iii) of a Sturm sequence. Property (ii) is evident from Equations (6.3-4). ∎

As a corollary of the two preceding theorems we now find that the Cauchy index of the rational function p_1/p_0 for the interval $[\alpha, \beta]$ is given by $v(\alpha) - v(\beta)$, where $v(x)$ denotes the number of sign changes in the numerical sequence $\{f_k(x)\}$. A slight simplification is still possible, however. Since $p_0(\alpha) \neq 0$ by hypothesis, the fact that p_m is a factor of p_0 implies that $p_m(\alpha) \neq 0$. Hence the sequence $\{p_k(\alpha)\} = \{p_m(\alpha) f_k(\alpha)\}$ has exactly the number of sign changes as the sequence $\{f_k(\alpha)\}$. A similar statement holds for $x = \beta$.

THEOREM 6.3c

Let p_0 and p_1 be two real polynomials $\not\equiv 0$, let the sequence $\{p_k\}$ be generated from p_0 and p_1 by the Euclidean algorithm, and let $v(x)$ denote the number of sign changes in the sequence $\{p_k(x)\}$. Then for any real α and β such that $\alpha < \beta$, $p_0(\alpha)p_0(\beta) \neq 0$,

$$I_\alpha^\beta \frac{p_1}{p_0} = v(\alpha) - v(\beta).$$

IV. The Number of Zeros in an Interval

Let p be a polynomial with complex coefficients and *distinct* zeros w_k, $k = 1, 2, \ldots, n$, and let m_k be the multiplicity of w_k. Then by §4.10 the logarithmic derivative of p has the partial fraction decomposition

$$\frac{p'(z)}{p(z)} = \sum_{k=1}^n \frac{m_k}{z - w_k}.$$

Thus, if p is a real polynomial and $\xi_1, \xi_2, \ldots, \xi_j$ are its distinct *real* zeros, then

$$\frac{p'(x)}{p(x)} = \sum_{i=1}^j \frac{m_i}{x - \xi_i} + r(x),$$

where m_i is again the multiplicity of ξ_i and r is a real rational function without real poles. Consequently, if $[\alpha, \beta]$ is any interval such that $p(\alpha)p(\beta) \neq 0$, it follows from (6.3-1) that the number of *distinct* zeros of p in $[\alpha, \beta]$ equals the Cauchy index

$$I_\alpha^\beta \frac{p'}{p}.$$

By Theorem 6.3c this index can be determined by carrying out the Euclidean algorithm (6.3-4) with the starting polynomials $p_0 := p, p_1 := p'$.

The Euclidean algorithm also furnishes information about the multiplicity of the zeros. If z_0 is a zero of multiplicity k of p, it is a zero of multiplicity $k - 1$ of p'. It follows that p_m, the last nonvanishing polynomial produced by the Euclidean algorithm, contains the factor $(z - z_0)^{k-1}$. Conversely, if p_m has a zero z_0 of multiplicity k, then p and p' have the common factor $(z - z_0)^k$ and consequently z_0 is a zero of multiplicity $k + 1$ of p.

THEOREM 6.3d

Let p be a real polynomial, $p \not\equiv 0$, and let $p_0, p_1, \ldots, p_m$ be the sequence of

CAUCHY INDICES   449

polynomials generated by the Euclidean algorithm (6.3-4) *started with* p_0: $=p, p_1:=p'$. *Then for any real interval* $[\alpha,\beta]$ *such that* $p(\alpha)p(\beta)\neq 0$, p *has exactly* $v(\alpha) - v(\beta)$ *distinct zeros in* $[\alpha,\beta]$, *where* $v(x)$ *denotes the number of changes of sign in the sequence* $\{p_i(x)\}$. *A complex number* z_0 *is a zero of multiplicity* k *of* p *if and only if it is a zero of multiplicity* $k-1$ *of* p_m. *Thus all zeros of* p *in* $[\alpha,\beta]$ *are simple if and only if* p_m *has no zeros in* $[\alpha,\beta]$.

EXAMPLE 1

Determine the number of positive zeros of $p(x) := 2x^3 - 7x^2 + 3x - 2$. By Descartes' rule of signs this number is either 3 or 1. The Euclidean algorithm furnishes polynomials which (up to constant factors) are

$$2x^3 - 7x^2 + 3x - 2,$$
$$6x^2 - 14x + 3,$$
$$62x + 15,$$
$$-1.$$

The sequence of signs at $x = 0$ is

$$-\quad +\quad +\quad -$$

and for x large (where the leading terms dominate)

$$+\quad +\quad +\quad -$$

Thus $v(0) = 2, v(\infty) = 1$. There is exactly one positive zero. This zero is simple since p_3 is constant.

EXAMPLE 2

Determine the number of zeros of the polynomial $x^6 - 2x^5 + 3x^4 - 4x^3 + 3x^2 - 2x + 1$ in the interval $[0,2]$. The rule of signs merely tells us that there is an even number of positive zeros (each zero counted with its multiplicity). The Euclidean algorithm furnishes (up to positive constant factors) the sequence of polynomials

$$x^6 - 2x^5 + 3x^4 - 4x^3 + 3x^2 - 2x + 1,$$
$$3x^5 - 5x^4 + 6x^3 - 6x^2 + 3x - 1,$$
$$-x^4 + 3x^3 - 3x^2 + 3x - 2,$$
$$-x^3 + x^2 - x + 1,$$
$$0.$$

The sequence of signs at $x = 0$ is $+\ -\ -\ +$ and $v(0) = 2$; at $x = 2$ it is $+\ +\ -\ -$ and $v(2) = 1$. Thus the number of distinct zeros in $[0,2]$ equals $v(0) - v(2) = 1$. Since $p_3(x) = -x^3 + x^2 - x + 1$ is not constant, that one zero may have a multiplicity

$k > 1$. To determine k we apply the Euclidean algoririthm to p_3 and p_3' to obtain (up to positive constant factors)

$$-x^3 + x^2 - x + 1,$$
$$-3x^2 + 2x - 1,$$
$$x - 2,$$
$$-1.$$

We find $v(0) - v(2) = 1$; thus p_3 has one zero in $[0,2]$ which is simple because the last nonvanishing polynomial in the sequence $\{p_i\}$ is constant. It follows that $k = 2$. In fact the original polynomial is $p(x) = (x^2 + 1)^2(x - 1)^2$.

§6.4. DISKS CONTAINING A SPECIFIED NUMBER OF ZEROS

Here we shall collect some elementary results concerning the following problem: given a polynomial of degree $n \geq 1$ with complex coefficients,

$$p(z) := z^n + a_{n-1}z^{n-1} + \cdots + a_1 z + a_0, \qquad (6.4\text{-}1)$$

and given a point z_0 in the complex plane, we shall construct a disk about z_0 that contains a specified number of zeros. By "specified number" we mean one of the following four cases: (I) no zeros, (II) at least one zero, (III) all zeros, (IV) at least m zeros, where m is a given integer, $1 \leq m \leq n$. Evidently both (II) and (III) are special cases of (IV); furthermore, problems (I) and (III) are in some sense equivalent. It is nevertheless convenient to treat these cases separately.

The zeros of p are denoted consistently by $w_1, w_2, \ldots, w_n$ (multiple zeros occurring repeatedly). We use the notation

$$b_m := \frac{1}{m!} p^{(m)}(z_0), \quad m = 0, 1, \ldots, n \qquad (b_n = 1),$$

consistently, so that

$$p(z_0 + h) = b_0 + b_1 h + b_2 h^2 + \cdots + h^n \qquad (6.4\text{-}2)$$

for all complex h.

I. Disks Containing No Zeros

Here we must necessarily assume that $b_0 = p(z_0) \neq 0$. A number of results are consequences of the following lemma:

DISKS CONTAINING SPECIFIED NUMBER OF ZEROS

LEMMA 6.4a

Let $\lambda_1, \lambda_2, \ldots, \lambda_n$ be positive numbers such that

$$\lambda_1 + \lambda_2 + \cdots + \lambda_n \leq 1, \tag{6.4-3}$$

and let

$$\rho := \min_{\substack{1 \leq m \leq n \\ b_m \neq 0}} \lambda_m^{1/m} \left| \frac{b_0}{b_m} \right|^{1/m}.$$

The open disk $|z - z_0| < \rho$ then contains no zero of p.

Proof. By (6.4-2) $|z - z_0| < \rho$ implies that

$$|p(z)| = \left| \sum_{m=0}^{n} b_m (z - z_0)^m \right| > |b_0| - \sum_{m=1}^{n} |b_m| \rho^m$$

$$= |b_0| \left\{ 1 - \sum_{m=1}^{n} \left| \frac{b_m}{b_0} \right| \rho^m \right\}.$$

If $b_m \neq 0$ in the mth term, we replace ρ^m with

$$\lambda_m \left| \frac{b_0}{b_m} \right|,$$

which is not smaller. This yields

$$|p(z)| > |b_0| \left\{ 1 - \sum_{b_m \neq 0} \lambda_m \right\} \geq |b_0| \left\{ 1 - \sum_{m=1}^{n} \lambda_m \right\} \geq 0,$$

by (6.4-3). Hence $p(z) \neq 0$. ∎

Setting $\lambda_m = 2^{-m}$ ($m = 1, 2, \ldots, n$) and observing that inequality holds in (6.4-3), we obtain the following theorem:

THEOREM 6.4b

Let

$$\alpha(z_0) := \min_{1 \leq m \leq n} \left| \frac{b_0}{b_m} \right|^{1/m}. \tag{6.4-4}$$

The disk $|z-z_0| \leq \tfrac{1}{2}\alpha(z_0)$ contains no zero of the polynomial p.

The following application of Lemma 6.4a will be of interest later.

THEOREM 6.4c

Let

$$\beta(z_0) := \min_{1 \leq m \leq n} \left[\binom{n}{m} \left| \frac{b_0}{b_m} \right| \right]^{1/m}. \tag{6.4-5}$$

No zero of p is contained in the disk $|z-z_0| \leq (1/ne)\beta(z_0)$.

Proof. Let $0 < \gamma < 1$. Lemma 6.4a is applicable [with equality in (6.4-3)] to

$$\lambda_m := \gamma^m (1-\gamma)^{n-m} \binom{n}{m}.$$

It follows that no zero lies in the closed disk with center z_0 and radius

$$\frac{\gamma}{1-\gamma} \min_{1 \leq m \leq n} (1-\gamma)^{n/m} \left[\binom{n}{m} \left| \frac{b_0}{b_m} \right| \right]^{1/m} \geq \gamma(1-\gamma)^{n-1} \beta(z_0).$$

The coefficient $\gamma(1-\gamma)^{n-1}$ becomes largest when $\gamma = 1/n$. By a well-known inequality,

$$\frac{1}{n}\left(1 - \frac{1}{n}\right)^{n-1} > \frac{1}{ne},$$

which implies the result. ∎

To formulate the next result we note that the equation for ρ,

$$|b_0| = |b_1|\rho + |b_2|\rho^2 + \cdots + \rho^n, \tag{6.4-6}$$

has exactly one positive solution. Indeed, for $\rho \geq 0$ the function on the right increases steadily from 0 to ∞.

THEOREM 6.4d

Let ρ_1 denote the positive solution of (6.4-6). No zero of p is contained in the open disk $|z-z_0| < \rho_1$.

DISKS CONTAINING SPECIFIED NUMBER OF ZEROS

Proof. Let

$$\lambda_m := \left|\frac{b_m}{b_0}\right| \rho_1^m, \quad m = 1, 2, \ldots, n.$$

Then $\lambda_1 + \lambda_2 + \cdots + \lambda_n = 1$ and, since

$$\rho_1 = \lambda_m^{1/m} \left|\frac{b_0}{b_m}\right|^{1/m}$$

for all m such that $b_m \neq 0$, the result follows as a corollary of Lemma 6.4a.
∎

II. Disks Containing at Least One Zero

Again we assume that $b_0 = p(z_0) \neq 0$; otherwise z_0 is a zero and every disk with center z_0 trivially contains a zero.

The polynomial (6.4-2), considered as a function of h, has the zeros

$$h_m := w_m - z_0, \quad m = 1, 2, \ldots, n.$$

It follows that the polynomial with the coefficients taken in reverse order,

$$q(h) := 1 + b_{n-1}h + b_{n-2}h^2 + \cdots + b_0 h^n,$$

has the zeros

$$h_m^{-1} = (w_m - z_0)^{-1}, \quad m = 1, 2, \ldots, n.$$

Thus by Vieta's formulas

$$(-1)^{n-m} \frac{b_m}{b_0} = \sum \prod_{l=1}^{m} (w_{k_l} - z_0)^{-1}, \quad m = 1, \ldots, n,$$

the sum being taken with respect to all combinations $(k_1, k_2, \ldots, k_m)$ of m of the integers $(1, 2, \ldots, n)$. There are $\binom{n}{m}$ such combinations.

Now let

$$\rho := \min_{1 \leq m \leq n} |w_m - z_0|, \tag{6.4-7}$$

the distance of z_0 to the nearest zero of p. Then $|w_m - z_0|^{-1} \leq \rho^{-1}$ for all m and

$$\left| \frac{b_m}{b_0} \right| \leq \binom{n}{m} \rho^{-m}, \quad m = 1, 2, \ldots, n, \tag{6.4-8}$$

follows. Consequently

$$\rho^m \leq \binom{n}{m} \left| \frac{b_0}{b_m} \right|, \quad m = 1, 2, \ldots, n. \tag{6.4-9}$$

Thus we have proved:

THEOREM 6.4e

Let $\beta(z_0)$ be defined by (6.4-5). The disk $|z - z_0| \leq \beta(z_0)$ then contains at least one zero of p.

Thus the number $\beta(z_0)$ has the property that, although the circle of radius $(ne)^{-1}\beta(z_0)$ about z_0 does not contain a zero, the circle of radius $\beta(z_0)$ does.

Two weaker forms of Theorem 6.4e are of interest. Considering $m = n$ in (6.4-9), we obtain the following corollary:

COROLLARY 6.4f

The disk $|z - z_0| \leq |p(z_0)|^{1/n}$ contains at least one zero of p.

Similarly by taking $m = 1$ we have Corollary 6.4g.

COROLLARY 6.4g

Let $p'(z_0) \neq 0$. The disk $|z - z_0| \leq n|p(z_0)/p'(z_0)|$ then contains at least one zero of p.

It will be noted that the radius of this disk is just the absolute value of the n-fold Newton correction (see §6.12) at point z_0. This result is improved on in a very elegant manner by Laguerre's theorem 6.5d.

Similar results can be obtained by considering the power sums

$$s_k = s_k(z_0) := -\sum_{m=1}^{n} \frac{1}{(w_m - z_0)^k}, \quad k = 1, 2, \ldots. \tag{6.4-10}$$

DISKS CONTAINING SPECIFIED NUMBER OF ZEROS

To compute these numbers from the Taylor coefficients $b_0, b_1, \ldots$, we note that, if $p(z_0) \neq 0$, $z = z_0 + h$ and $|h|$ sufficiently small,

$$\frac{p'(z)}{p(z)} = \sum_{m=1}^{n} \frac{1}{z - w_m} = -\sum_{m=1}^{n} \frac{1}{w_m - z_0 - h}$$

$$= -\sum_{m=1}^{n} \left(\sum_{k=1}^{\infty} \frac{h^{k-1}}{(w_m - z_0)^k} \right) = \sum_{k=1}^{\infty} s_k h^{k-1}.$$

Thus

$$p'(z_0 + h) = p(z_0 + h) \sum_{k=1}^{\infty} s_k h^{k-1},$$

and by comparing coefficients in

$$b_1 + 2b_2 h + 3b_3 h^2 + \cdots + nb_n h^{n-1}$$

$$= (b_0 + b_1 h + \cdots + b_n h^n)(s_1 + s_2 h + s_3 h^2 + \cdots)$$

we find the recurrence relation

$$s_k = \frac{1}{b_0}(kb_k - b_{k-1}s_1 - b_{k-2}s_2 - \cdots - b_1 s_{k-1}), \quad k = 1, 2, \ldots, \quad (6.4\text{-}11)$$

where $b_k := 0$ for $k > n$.

If ρ is defined by (6.4-7), then obviously

$$|s_k| \leq n\rho^{-k}, \quad k = 1, 2, \ldots,$$

which implies the following:

THEOREM 6.4h

At least one zero of p is contained in each of the disks

$$|z - z_0| \leq \left[\frac{n}{|s_k|} \right]^{1/k}, \quad k = 1, 2, \ldots.$$

Since $s_1 = b_1/b_0$, $k = 1$ again yields Corollary 6.4g. Similarly, we obtain for $k=2$ and $k=3$ the following disks which contain at least one zero:

$$|z-z_0| \leq \sqrt{n} \left|\frac{b_0}{b_1}\right| \left|1 - 2\frac{b_0 b_2}{b_1^0}\right|^{-1/2},$$

$$|z-z_0| \leq \sqrt[3]{n} \left|\frac{b_0}{b_1}\right| \left|1 - 3\frac{b_0 b_1 b_2 - b_0^2 b_3}{b_1^3}\right|^{-1/3}.$$

The interest of these formulas lies in the fact that if p has exactly one zero closest to z_0 then by (6.4-10)

$$\lim_{k \to \infty} |s_k|^{-1/k} = \rho.$$

Since $n^{1/k} \to 1$, we can, by taking k sufficiently large, obtain disks guaranteed to contain a zero whose radius is arbitrarily close to the distance to the nearest zero.

The following result, due to G. D. Birkhoff, complements Theorem 6.4d as Theorem 6.4e complemented Theorem 6.4c.

THEOREM 6.4i

Let ρ_1 denote the positive solution of equation (6.4-6). Then the disk

$$|z - z_0| \leq \frac{1}{2^{1/n} - 1} \rho_1$$

contains at least one zero of p.

Proof. Let ρ be defined by (6.4-7); (6.4-8) then holds, and, by using these inequalities in

$$|b_0| = |b_1|\rho_1 + |b_2|\rho_1^2 + \cdots + \rho_1^n,$$

$$1 \leq \binom{n}{1}\frac{\rho_1}{\rho} + \binom{n}{2}\frac{\rho_1^2}{\rho^2} + \cdots + \binom{n}{n}\frac{\rho_1^n}{\rho^n} = \left(1 + \frac{\rho_1}{\rho}\right)^n - 1$$

follows. This implies that

$$\left(1 + \frac{\rho_1}{\rho}\right)^n \geq 2$$

DISKS CONTAINING SPECIFIED NUMBER OF ZEROS

or

$$\frac{\rho_1}{\rho} \geq 2^{1/n} - 1,$$

which is equivalent to the result to be proved. ∎

III. Disks Containing All Zeros

We now assume that $z_0 = 0$, this being the main case of interest. Our purpose is to determine an **inclusion radius** for the given polynomial p, i.e., a number σ such that the zeros of p satisfy

$$|w_m| \leq \sigma, \quad m = 1, 2, \ldots, n. \tag{6.4-12}$$

(A **proper inclusion radius** is a number σ such that $|w_m| < \sigma$, $m = 1, 2, \ldots, n$.) We assume that $a_0 \neq 0$; otherwise a polynomial of lower degree could be considered. The polynomial with the coefficients taken in reverse order,

$$q(z) := 1 + a_{n-1} z + a_{n-2} z^2 + \cdots + a_0 z^n,$$

then has zeros w_m^{-1}, $m = 1, 2, \ldots, n$. It follows that $\sigma > 0$ is an inclusion radius (a proper inclusion radius) if and only if the disk $|z| < \sigma^{-1}$ (the disk $|z| \leq \sigma^{-1}$) is free of zeros of q. By applying the results of Subsection I we immediately obtain the following theorem:

THEOREM 6.4j

Let $\lambda_1, \lambda_2, \ldots, \lambda_n$ be positive numbers such that

$$\lambda_1 + \lambda_2 + \cdots + \lambda_n \leq 1 \, [\, <1 \,],$$

and let

$$\sigma := \max_{1 \leq m \leq n} \lambda_m^{-1/m} |a_{n-m}|^{1/m}.$$

Then σ is an inclusion radius [a proper inclusion radius] for p.

COROLLARY 6.4k

All zeros of p lie in the open disk with center 0 and radius

$$\sigma := 2 \max_{1 \leq m \leq n} |a_{n-m}|^{1/m}.$$

Transposing Theorem 6.4d yields the following result due to Cauchy:

THEOREM 6.4l

Let σ denote the positive solution of the equation

$$\sigma^n = |a_0| + |a_1|\sigma + \cdots + |a_{n-1}|\sigma^{n-1}.$$

Then σ is an inclusion radius for p.

Comparing the results of Subsections I, II, and III we find that those of Subsections I and III (disks containing no or all zeros) invariably depend on *all* coefficients of the polynomial, whereas those of Subsection II (disks containing at least one zero) may depend only on a few or even on one single coefficient of the polynomial. The following simple consideration purports to show that this difference in complexity is to be expected on quite general grounds. Let

$$E: (a_0, \ldots, a_{n-1}) \to E(a_0, \ldots, a_{n-1}) \subset \mathbb{C}$$

be a function that associates with each n-tuple of complex numbers a proper subset of the complex plane and has the following property: at least m zeros of the polynomial $p(z) := z^n + a_{n-1}z^{n-1} + \cdots + a_0$ are contained in $E(a_0, \ldots, a_{n-1})$. We assert that if $m = n$ the function E necessarily depends on *all* coefficients $a_0, \ldots, a_{n-1}$. Indeed, if E were independent of a_k, say, then a_k could be selected such that p has a zero in the complement of E. Because the radii of the disks in Subsections I and III determine sets containing *all* zeros, they must depend on all coefficients. More generally, a set containing $m > 1$ zeros must depend on at least m coefficients of the polynomial.

IV. Disks Containing at Least m Zeros ($m \geq 1$).

Consider the Taylor expansion (6.4-2) of p at z_0. If $b_0 = 0$, then p has a zero at z_0. The zeros of a polynomial depend continuously on its coefficients (Theorem 4.10c). Thus, if $|b_0|$ is small, p will have a zero "close" to z_0. This vague statement is made precise in Corollary 6.4f: a zero exists in the disk $|z - z_0| \leq |b_0|^{1/n}$. Corollary 6.4g yields a better result if $|b_0|$ is "small compared with $|b_1|$": a zero exists in the disk $|z - z_0| \leq n|b_0 b_1^{-1}|$.

The results to be obtained in this subsection are direct generalizations of the two special results given above. Let m be an integer, $1 \leq m \leq n$. If $b_0 = b_1 = \cdots = b_{m-1} = 0$, then p has a zero of multiplicity m at z_0. Thus, if $|b_0|, \ldots, |b_{m-1}|$ are small, then p can be expected to have m zeros "close" to z_0. This is made precise in the following result due to Montel:

DISKS CONTAINING SPECIFIED NUMBER OF ZEROS

THEOREM 6.4m

Let ρ_m denote the unique non-negative solution of the equation in ρ,

$$\rho^n = \binom{n-1}{m-1}|b_0| + \binom{n-2}{m-2}|b_1|\rho + \cdots + \binom{n-m}{0}|b_{m-1}|\rho^{m-1}.$$

Then at least m zeros of p are contained in the disk $|z - z_0| \leq \rho_m$.

For $m = 1$ this reduces to Corollary 6.4f, for $m = n$ to Cauchy's theorem 6.4l. If $|b_0|, |b_1|, \ldots, |b_{m-1}|$ are small compared with $|b_m|$, it is better to use another result, due to Van Vleck:

THEOREM 6.4n

Let $b_m \neq 0$ and let ρ_m^* denote the unique non-negative solution of

$$|b_m|\rho^m = \binom{n}{m}|b_0| + \binom{n-1}{m-1}|b_1|\rho + \cdots + \binom{n-m+1}{1}|b_{m-1}|\rho^{m-1}.$$

Then at least m zeros of p are contained in the disk

$$|z - z_0| \leq \rho_m^*.$$

For $m = 1$ this reduces to Corollary 6.4g. For $m = n$ we again obtain Cauchy's result.

For the proof of both theorems we may assume that $z_0 = 0$ without loss of generality. Let the zeros of p be numbered such that

$$|w_1| \leq |w_2| \leq \cdots \leq |w_n|.$$

We consider the polynomial

$$q(z) := \frac{p(z)}{(z - w_n)(z - w_{n-1}) \cdots (z - w_{m+1})}$$

$$=: c_0 + c_1 z + \cdots + c_m z^m$$

whose zeros are the m smallest zeros of p. Both theorems will be proved if we can show that the numbers ρ_m and ρ_m^* are inclusion radii for q. This we accomplish by estimating the coefficients c_k and applying a Cauchy type argument as in the proof of Theorem 6.4d.

The polynomial q is obtained by multiplying $p(z)$ by the power series

$$s(z) := \prod_{k=m+1}^{n} (z - w_k)^{-1} =: \sum_{k=0}^{\infty} s_k z^k.$$

Thus we have

$$c_k = \sum_{l=0}^{k} a_l s_{k-l}, \qquad k=0,1,\ldots,m. \qquad (6.4\text{-}13)$$

The coefficients s_k can be obtained explicitly by multiplying the $n-m$ geometric series

$$\frac{1}{z-w_i} = -\frac{1}{w_i}\left(1 + \frac{z}{w_i} + \frac{z^2}{w_i^2} + \cdots\right), \qquad i=m+1,\ldots,n.$$

Thus we find that

$$s_k = \frac{(-1)^{n-m}}{w_{m+1}\cdots w_n} \sum (w_{i_1} w_{i_2} \cdots w_{i_k})^{-1}, \qquad k=0,1,2,\ldots,$$

the sum being taken with respect to all combinations $(i_1,\ldots,i_k)$ (with repetitions) of k of the $n-m$ integers $(m+1,\ldots,n)$. The number of terms in each sum can be calculated by considering the special case in which all $w_i = 1$. We then have

$$s(z) = (z-1)^{-n+m} = (-1)^{n-m} \sum_{k=0}^{\infty} \binom{-n+m}{k}(-z)^k.$$

It follows that the sum s_k contains

$$(-1)^k \binom{-n+m}{k} = \binom{n-m+k-1}{k}$$

terms. Let $|w_m| =: \rho$. Then $|w_i| \geq \rho$, $i > m$, and we obtain

$$|s_k| \leq \binom{n-m+k-1}{k} \rho^{-n+m-k}. \qquad (6.4\text{-}14)$$

Hence from (6.4-13)

$$|c_k| \leq \rho^{-n+m-k} \sum_{l=0}^{k} \binom{n-m+k-l-1}{k-l} |a_l| \rho^l, \qquad k=0,\ldots,m. \qquad (6.4\text{-}15)$$

Because $q(z)=0$ for some z such that $|z|=\rho$,

$$|c_m|\rho^m \leq |c_0| + |c_1|\rho + \cdots + |_{m-1}|\rho^{m-1}.$$

DISKS CONTAINING SPECIFIED NUMBER OF ZEROS

Using (6.4-15) on the right and reversing the order of the summations, we obtain

$$|c_m|\rho^m \leqslant \rho^{-n+m} \sum_{l=0}^{m-1} |a_l|\rho^l \sum_{k=l}^{m-1} \binom{n-m+k-l-1}{k-l}.$$

The sum of the binomial coefficients equals

$$\sum_{k=0}^{m-l-1} \binom{n-m-1+k}{k} = \binom{n-l-1}{m-l-1},$$

by Vandermonde's theorem. Hence

$$|c_m| \leqslant \rho^{-n} \sum_{l=0}^{m-1} \binom{n-l-1}{m-l-1} |a_l|\rho^l. \qquad (6.4\text{-}16)$$

We obtain Theorem 6.4m by noting that $c_m = 1$. The inequality (6.4-16) then shows immediately that $\rho \leqslant \rho_m$.

To prove Theorem 6.4n we substitute the polynomial

$$q^*(z) := w_{m+1} \cdots w_n q(z) =: c_0^* + c_1^* z + \cdots + c_m^* z^m$$

for q. By entirely analogous reasoning it now follows that

$$|c_k^*| \leqslant \rho^{-k} \sum_{l=0}^{k} \binom{n-m+k-l-1}{k-l} |a_l|\rho^l; \qquad (6.4\text{-}17)$$

furthermore we have

$$|c_m^*| \geqslant |a_m| - \sum_{l=0}^{m-1} \binom{n-l-1}{m-l} |a_l|\rho^{l-m}. \qquad (6.4\text{-}18)$$

The fact that $q^*(z) = 0$ for some z such that $|z| = \rho$ again implies

$$|c_m^*|\rho^m \leqslant |c_0^*| + |c_1^*|\rho + \cdots + |c_{m-1}^*|\rho^{m-1},$$

which by use of (6.4-17) and (6.4-18) yields

$$\rho^m |a_m| \leqslant \sum_{l=0}^{m-1} \binom{n-l}{m-l} |a_l|\rho^l.$$

By using monotonicity this inequality shows that $\rho \leqslant \rho_m^*$, and Theorem 6.4n is proved. ∎

PROBLEMS

1. The numbers

$$\beta := \max_{l \leqslant m \leqslant n} (n|a_{n-m}|)^{1/m},$$

$$\gamma := \max_{l \leqslant m \leqslant n} \left[\frac{2^n - 1}{\binom{n}{m}} |a_{n-m}| \right]^{1/m}$$

are inclusion radii for the polynomial (6.4-1).

2. Using Corollary 6.4f, show that

$$\rho_m \leqslant \max_{0 \leqslant k < m} \left[\binom{n}{m-1} |a_k| \right]^{1/(n-k)},$$

$$\rho_m^* \leqslant \max_{0 \leqslant k < m} \left[\binom{n+1}{m} \left| \frac{a_k}{a_m} \right| \right]^{1/(n-k)}.$$

3. Let σ denote the Cauchy inclusion radius for the polynomial $p(z) := a_0 + a_1 z + \cdots + a_n z^n$. Show that at least one zero of p satisfies

$$|w| \geqslant (2^{1/n} - 1)\sigma.$$

(G. D. Birkhoff [1914]. Compare Theorem 6.4c.)

4. Let $a_0 < a_1 < a_2 < \cdots < a_n$. Show that all zeros of the polynomial $p(z) := a_0 + a_1 z + \cdots + a_n z^n$ lie inside the unit circle $|z| = 1$ (**Kakeya-Eneström theorem**).

5. Let

$$\sigma := \min_{l \leqslant k \leqslant n} \left| \frac{n}{s_k} \right|^{1/k}.$$

Theorem 6.4h implies that the disk $|z - z_0| \leqslant \sigma$ contains at least one zero. Show that the disk

$$|z - z_0| \leqslant \frac{\sigma}{2n}$$

contains no zeros.

GEOMETRY OF ZEROS   463

6. Let $p(z) := 1 + a_1 z + a_2 z^2 + \cdots + a_n z^n$, where $|a_1| + |a_2| + \cdots + |a_n| = 1$. Show that the minimum value of $|p(z)|$ on $|z| = 1$ is less than $1 - (2^{1/n} - 1)^n$.

7. Let p_n denote the polynomial of degree n having zeros at the points 2^k, $k = 0, 1, \ldots, n-1$ and satisfying $p_n(0) = 1$. Show that

$$\lim_{n \to \infty} p'_n(0) = -2, \quad \lim_{n \to \infty} p''_n(0) = \frac{8}{3}.$$

[Use the power sums $s_k(0)$.]

§6.5. GEOMETRY OF ZEROS (THEOREMS OF GAUSS-LUCAS, LAGUERRE AND GRACE)

In this section we continue the discussion of results that guarantee that some, or all, zeros of a given polynomial lie in certain circular regions. The main tool is the judicious use of Moebius transformations. Because of its strong geometric appeal, the resulting theory is sometimes called "geometry of zeros."

We begin with a classical result.

THEOREM 6.5a (Gauss-Lucas theorem)

Any closed half-plane that contains all zeros of a polynomial p also contains all zeros of its derivative p'.

Proof. Let the degree of p be n, and let $w_1, w_2, \ldots, w_n$ be the zeros of p. Given any closed half-plane H, there exist a complex number $c \neq 0$ and a real number α such that $z \in H$ if and only if $\operatorname{Re} cz \geq \alpha$. Therefore, if all zeros are in H, then

$$\operatorname{Re} c w_m \geq \alpha, \quad m = 1, 2, \ldots, n.$$

Let z not be in H. We wish to show that $p'(z) \neq 0$. Because $\operatorname{Re} cz < \alpha$, we have

$$\operatorname{Re} c(z - w_m) < 0, \quad m = 1, 2, \ldots, n;$$

hence

$$\operatorname{Re} \frac{\bar{c}}{z - w_m} < 0, \quad m = 1, 2, \ldots, n,$$

and

$$\operatorname{Re} \bar{c} \sum_{m=1}^{n} \frac{1}{z - w_m} < 0.$$

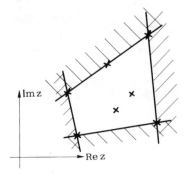

Fig. 6.5. Convex hull of zeros.

There follows

$$\sum_{m=1}^{n} \frac{1}{z - w_m} \neq 0. \qquad (6.5\text{-}1)$$

However, the expression on the left is the logarithmic derivative $p'(z)/p(z)$ of p at z. Consequently $p'(z) \neq 0$. ∎

Theorem 6.5a implies that the zeros of p' are contained in the intersection of all closed half-planes containing the zeros of p. This intersection is also known as the **closed convex hull** of the set of zeros and may be described as the smallest convex polygon containing all the zeros (see Fig. 6.5a).

We shall now obtain inclusion theorems that involve circles rather than straight lines by subjecting the complex plane to linear fractional transformations. To avoid exceptions we find it convenient to make a slight change in the use of the term "polynomial." In this section a function defined by

$$p(z) := a_n z^n + a_{n-1} z^{n-1} + \cdots + a_0$$

is called a **polynomial of degree** n if at least one of the coefficients $a_0, \ldots, a_n$ is different from zero; it is no longer required that $a_n \neq 0$. If we wish to emphasize that $a_n \neq 0$, we say that p has the **exact degree** n. If, for some $k > 0$, $a_n = a_{n-1} = \cdots = a_{n-k+1} = 0$, $a_{n-k} \neq 0$, then p is said to have $z = \infty$ as a zero of multiplicity k. Thus every polynomial of degree n still has n zeros, some of which may be ∞. If all coefficients $a_0, \ldots, a_n$ are zero, p is not a polynomial but is referred to as the **null function**.

To simplify the language we call a **circular region** any set in the extended complex plane that is a closed half-plane, a closed circular disk, or the

GEOMETRY OF ZEROS

closure of the exterior of a circular disk. A basic result of the theory of Moebius transformations (see §5.3) states that the image of a circular region under any such transformation is again a circular region.

It will be helpful to recall the following basic facts. Let f be meromorphic in a region S and let it have the zeros $z_1, z_2, \ldots$, and poles $p_1, p_2, \ldots$, of the respective multiplicities $n_1, n_2, \ldots$, and $o_1, o_2, \ldots$. Let t be a function that maps S conformally onto a region T. The function $f \circ t^{[-1]}$ is then meromorphic in T and has the zeros $t(z_i)$ and poles $t(p_j)$, $i,j = 1, 2, \ldots$, again of the respective multiplicities n_i and o_j. If a polynomial of exact degree n is regarded as a rational function with a pole of order n at infinity, this statement also applies to functions that are meromorphic in the extended complex plane, i.e., rational, and to conformal mappings of the extended complex plane onto itself, i.e., Moebius transformations.

Now let p be a polynomial of degree n and let its zeros $w_1, w_2, \ldots, w_n$ lie in a circular region C. (If the exact degree of p is less than n, this means that $\infty \in C$.) Let c be a boundary point of C, $c \neq \infty$. The linear transformation t defined by

$$s = t(z) = \frac{1}{z-c}$$

then maps C onto a closed half-plane H. Its inverse is

$$z = t^{[-1]}(s) = c + \frac{1}{s}.$$

By the above remark $r := p \circ t^{[-1]}$ is a rational function with zeros $t(w_i)$ all lying in H. Since r has a pole of order $\leq n$ at $s=0$, and no other poles, the function

$$p_1(s) := s^n r(s)$$

is a polynomial of degree n, again with zeros $t(w_1), \ldots, t(w_n)$. (By our convention this is also true if the exact degree of p is $< n$ or if some of the w_i equal c.) By the Gauss-Lucas theorem the zeros of its derivative p_1' also lie in H. Consequently the zeros of the rational function $r_1 := p_1' \circ t$ again lie in C. Because r_1 has a single pole at $z = c$ of order $n-1$ at most, the function

$$p_2(z) := (z-c)^{n-1} r_1(z)$$

is a polynomial of degree $n-1$ with the same zeros as r_1, possibly plus some zeros at c. Thus its zeros are likewise in C.

It is easy to express p_2 in terms of p and thus to obtain a theorem on the location of the zeros of a certain polynomial related to p. We have

$$p_1(s) = s^n p\left(c + \frac{1}{s}\right)$$

and

$$p_1'(s) = s^{n-1}\left[np\left(c + \frac{1}{s}\right) - \frac{1}{s}p'\left(c + \frac{1}{s}\right)\right],$$

$$r_1(z) = (z-c)^{-n+1}[np(z) - (z-c)p'(z)],$$

$$p_2(z) = np(z) - (z-c)p'(z).$$

The function p_2 is called **derivative of the polynomial** p (of degree n) **with respect to the point** c, and is denoted by $D_c p$. (Note that the operator D_c depends on the degree ascribed to the polynomial on which it acts.) If $c = \infty$, then $D_c p$ is defined as the ordinary derivative p'.

It may happen that $D_c p$ is the zero function even if p is a polynomial of degree $n > 0$. If $c = \infty$, this clearly occurs whenever a constant is regarded as a polynomial of degree > 0, which is permissible under our conventions. If c is finite, $D_c p = 0$ means that p_2 and thus r_1 and p_1' are zero. Hence p_1 must be a constant, which means that $r(s) = \text{const} \cdot s^{-n}$ and that $p(z) = \text{const} \cdot (z-c)^n$. Conversely, if $p(z) = \text{const} \cdot (z-c)^n$, calculation shows immediately that $D_c p = 0$. Thus, whether or not c is finite $D_c p = 0$ if and only if all zeros of p coincide with c. If $D_c p$ does not vanish identically, its zeros as a polynomial of degree $n-1$ lie in C.

Almost the same argument can be repeated under the assumption that c is an *exterior* point of the circular region C which contains the zeros of p. If the functions t, r, p_1, r_1, p_2 are defined as above, it follows that the zeros of the polynomial p_1 are contained in the circular region $C_1 := t(C)$. Because the point $\infty = t(c)$ is not contained in C_1, C_1 is a closed circular disk, and because the disk is convex C_1 contains the convex hull of the zeros of p_1, hence by the Gauss-Lucas theorem the zeros of p_1'. It now follows, as above, that the zeros of p_2 are contained in C; $D_c p = 0$ cannot occur because c is exterior to C. Thus we have obtained the following result:

THEOREM 6.5b (theorem of Laguerre)

Let p be a polynomial of degree $n \geq 1$, let C be a circular region that contains all zeros of p, and let c be a point of the closure of the complement of C. Then $D_c p$, the derivative of p with respect to c, either vanishes identically or its

GEOMETRY OF ZEROS

zeros as zeros of a polynomial of degree $n-1$ are located in C. The first possibility occurs if and only if c is a boundary point of C and all zeros of p coincide with c.

EXAMPLE 1

Let $p(z):=z^n+1$. The zeros of p lie in the circular region $C:|z| \geq 1$. The derivative of p with respect to $c:=0$ is $D_0 p(z)=n$. It has the zero $z=\infty$ of multiplicity $n-1$, which is well contained in C.

An immediate consequence of Laguerre's theorem is the following corollary:

COROLLARY 6.5c

If C is a circular region with boundary point c and $D_c p$ has a zero in C without vanishing identically, the polynomial p has at least one zero in C.

Proof. If all zeros of p were contained in the complement of C, then, because their number is finite, they would be contained in a circular region C_1 such that $C \cap C_1 = \emptyset$. Because c is an exterior point of C_1, all zeros of $D_c p$ would also have to be contained in C_1. ∎

Corollary 6.5c permits an interesting application to Newton's method (see §6.12). Let p be a polynomial of degree n and let z_0 be a finite point such that $p(z_0)p'(z_0) \neq 0$. The *Newton correction*

$$h_0 := -\frac{p(z_0)}{p'(z_0)}$$

can then be formed and is different from zero. The next point in the Newton sequence of approximants would be $z_1 := z_0 + h_0$. We now choose c such that $D_c p$ has the zero z_0. This leads to the condition

$$np(z_0) - (z_0 - c)p'(z_0) = 0,$$

i.e.,

$$c = z_0 - n\frac{p(z_0)}{p'(z_0)} = z_0 + nh_0.$$

We apply the corollary to a circular region C with the points z_0 and c on its boundary. This evidently satisfies the hypothesis, and we get:

THEOREM 6.5d

Let p be a polynomial of degree n and let z_0 be a point such that $p'(z_0) \neq 0$.

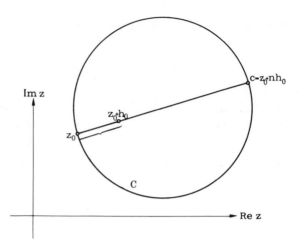

Fig. 6.5. A Newton correction.

Let h_0 denote the Newton correction at z_0. Then any circular region with the points z_0 and $z_0 + nh_0$ on its boundary contains at least one zero of p.

In particular, we may choose for C a circular disk with a straight line segment from z_0 to $z_0 + nh_0$ as its diameter (see Fig. 6.5).

This result should be compared with Corollary 6.4g, which (in our present notation) merely states that the disk

$$|z - z_0| \leqslant n|h_0|$$

contains at least one zero of p.

We shall now determine what can be obtained by repeated applications of Laguerre's theorem. Let $c_1, c_2, \ldots, c_n$ be points (not necessarily distinct) of the extended complex plane. We may then iterate the process of forming the derivative with respect to a point by considering a k-fold derivative as a polynomial of degree $n-k$. Writing $D_i := D_{c_i}$ for brevity, we thus obtain a sequence of functions

$$D^{(k)}p := D_k D_{k-1} \cdots D_1 p, \qquad k = 1, 2, \ldots, n,$$

which either vanish identically or are polynomials of degree $n-k$. Repeated application of Theorem 6.5b shows that *if all zeros of p are contained in a circular region C and none of the points $c_1, c_2, \ldots, c_n$ is contained in C then none of the derivatives $D^{(k)}p$ vanishes identically and all their zeros lie in C.*

From this we obtain Lemma 6.5e by logical contraposition:

LEMMA 6.5e

If one of the derivatives $D^{(k)}p$, $k=1,2,\ldots,n$ vanishes identically, every circular region containing all zeros of p contains at least one of the numbers c_i, $i=1,2,\ldots,n$.

From this lemma we can derive a rather neat result due to Grace by calculating an explicit representation for $D^{(n)}p$ and expressing the fact that $D^{(n)}p = 0$. Direct calculation shows, if both c_1 and c_2 are finite,

$$D_2 D_1 p = n(n-1)p(z) + (n-1)[(c_1-z) + (c_2-z)]p'(z)$$
$$+ (c_1-z)(c_2-z)p''(z).$$

This is symmetric in c_1 and c_2. Thus we have $D_2 D_1 p = D_1 D_2 p$ and conclude that $D^{(k)}p$ does not depend on the order in which the differentiations are performed.

To write a formula for $D^{(k)}p$ we put

$$h_k := c_k - z, \quad k = 1, 2, \ldots, n,$$

and denote for $j = 1, 2, \ldots, k$ by

$$s_j(h_1, h_2, \ldots, h_k) := h_1 h_2 \cdots h_j + \cdots + h_{k-j+1} h_{k-j+2} \cdots h_k$$

the elementary symmetric function of degree j of the k variables $h_1, \ldots, h_k$. For $j = 0$

$$s_0(h_1, h_2, \ldots, h_k) := 1.$$

The following relations are easily verified:

$$s_j(h_1, \ldots, h_{k+1}) = h_{k+1} s_{j-1}(h_1, \ldots, h_k) + s_j(h_1, \ldots, h_k), \qquad (6.5\text{-}2)$$

and if a prime denotes differentiation with respect to z

$$s_j'(h_1, \ldots, h_k) = -(k-j+1)s_{j-1}(h_1, \ldots, h_k), \quad k = 1, 2, \ldots, n;$$
$$j = 1, 2, \ldots, k. \qquad (6.5\text{-}3)$$

Lemma 6.5f

If all c_i are finite, then for $k = 1, 2, \ldots, n$

$$D^{(k)}p(z) = \sum_{j=0}^{k} \frac{(n-j)!}{(n-k)!} s_j(h_1, \ldots, h_k) p^{(j)}(z).$$

Proof. The formula of the lemma is trivial for $k=0$, agrees with the definition for $k=1$ and has already been verified for $k=2$. Assuming its truth for some value of k, where $0<k<n$, we have, by the definition of D_{k+1},

$$D^{(k+1)}p(z) = D_{k+1}(D^{(k)}p)(z)$$

$$= (n-k)D^{(k)}p(z) + h_{k+1}(D^{(k)}p)'(z).$$

By the induction hypothesis this equals

$$\sum_{j=0}^{k} \frac{(n-j)!}{(n-k)!} \Big[(n-k)s_j(h_1,\ldots,h_k)p^{(j)}(z)$$

$$+ h_{k+1}s_j'(h_1,\ldots,h_k)p^{(j)}(z) + h_{k+1}s_j(h_1,\ldots,h_k)p^{(j+1)}(z) \Big].$$

Using (6.5-2) and (6.5-3), we obtain

$$D^{(k+1)}p(z) = \sum_{j=0}^{k} \frac{(n-j)!}{(n-k)!} \Big[(n-k)s_j(h_1,\ldots,h_k)p^{(j)}(z)$$

$$- (k-j+1)h_{k+1}s_{j-1}(h_1,\ldots,h_k)p^{(j)}(z) \Big]$$

$$+ \sum_{j=1}^{k+1} \frac{(n-j+1)!}{(n-k)!} h_{k+1}s_{j-1}(h_1,\ldots,h_k)p^{(j)}(z)$$

$$= \frac{n!}{(n-k-1)!}p(z) + \sum_{j=1}^{k+1} \frac{(n-j)!}{(n-k)!} \Big\{ [s_j(h_1,\ldots,h_{k+1})$$

$$- s_j(h_1,\ldots,h_k)][(n-j+1)-(k-j+1)]$$

$$+ (n-k)s_j(h_1,\ldots,h_k) \Big\} p^{(j)}(z)$$

$$= \sum_{j=0}^{k+1} \frac{(n-j)!}{(n-k-1)!} s_j(h_1,\ldots,h_{k+1}) p^{(j)}(z),$$

thus proving the formula of the lemma with k increased by 1. ∎

We now calculate $D^{(n)}p$. Because this is constant, we may pick an arbitrary value of z. By choosing $z=0$ we have

$$D^{(n)}p(z) = D^{(n)}p(0) = \sum_{j=0}^{n} (n-j)! s_j(c_1,\ldots,c_n) p^{(j)}(0).$$

If

$$p(z) = a_0 + a_1 z + \cdots + a_n z^n,$$

then clearly

$$p^{(k)}(0) = k! a_k.$$

To express $s_j(c_1,\ldots,c_n)$ we define

$$q(z) := \prod_{k=1}^{n} (z - c_k) =: b_0 + b_1 z + \cdots + b_n z^n.$$

We then have, as is well known,

$$s_j(c_1,\ldots,c_n) = (-1)^j b_{n-j}, \qquad j = 0, 1, \ldots, n.$$

Thus there follows

$$D^{(n)}p(0) = \sum_{j=0}^{n} (n-j)! j! (-1)^j a_j b_{n-j}$$

or

$$\frac{1}{n!} D^{(n)}p(0) = \frac{a_0 b_n}{\binom{n}{0}} - \frac{a_1 b_{n-1}}{\binom{n}{1}} + \frac{a_2 b_{n-2}}{\binom{n}{2}} - \cdots . \qquad (6.5\text{-}4)$$

We now consider the case in which some of the c_i, say $c_1,\ldots,c_m$, are infinite. Since $D^{(k)}p$ does not depend on the order in which the differentiations are performed, we may differentiate with respect to $c_1,\ldots,c_m$ first to obtain

$$D^{(m)}p = p^{(m)}.$$

To this polynomial of degree $n-m$ we now apply the remaining differentiations $D_n D_{n-1} \cdots D_{m+1}$. Lemma 6.5f yields

$$D^{(n)}p(z) = \sum_{j=0}^{n-m} (n-m-j)!\, s_j(h_{m+1},\ldots,h_n) p^{(m+j)}(z).$$

If we denote by $b_0, b_1,\ldots,b_n$ the coefficients of a polynomial of degree n with m-fold zero ∞ and finite zeros $c_{m+1},\ldots,c_n$ (so that, according to our convention, $b_n = \cdots = b_{n-m+1} = 0$), the representation (6.5-4) is seen to hold also if some c_i are infinite.

Equation (6.5-4) is symmetric in the two polynomials

$$\begin{aligned} p(z) &:= a_0 + a_1 z + \cdots + a_n z^n, \\ q(z) &:= b_0 + b_1 z + \cdots + b_n z^n; \end{aligned} \qquad (6.5\text{-}5)$$

i.e., the nth derivative of p with respect to the zeros of q equals, up to a factor $(-1)^n$, the nth derivative of q with respect to the zeros of p. The two polynomials p and q are called **apolar** if this nth derivative vanishes, i.e., if

$$\frac{a_0 b_n}{\binom{n}{0}} - \frac{a_1 b_{n-1}}{\binom{n}{1}} + \frac{a_2 b_{n-2}}{\binom{n}{2}} - \cdots + (-1)^n \frac{a_n b_0}{\binom{n}{n}} = 0. \qquad (6.5\text{-}6)$$

According to Lemma 6.5e, this is possible only if every circular region containing all zeros of one polynomial contains at least one zero of the other. Thus we have obtained the result given in Theorem 6.5g.

THEOREM 6.5g (theorem of Grace)

Let p and q be two apolar polynomials, i.e., let them satisfy the relation (6.5-6). Then every circular region that contains all zeros of one polynomial contains at least one zero of the other.

A number of results concerning the location of a zero of a given polynomial p can be deduced from Grace's theorem by appropriately choosing the apolar polynomial q. In addition to satisfying (6.5-6), q must

GEOMETRY OF ZEROS

be such that its zeros are easily calculable.

EXAMPLE 2

We choose q in the form $q(z) := z^n + b_{n-1} z^{n-1}$. This is apolar to p if and only if

$$a_0 - \frac{1}{n} a_1 b_{n-1} = 0,$$

i.e., if $b_{n-1} = n(a_0/a_1)$. The resulting polynomial q has the $(n-1)$-fold zero $z = 0$ and the simple zero $-n(a_0/a_1)$. As a special case of Grace's theorem we thus obtain the corollary of Laguerre's theorem: p has at least one zero in every region bounded by a circle through 0 and $-n(a_0/a_1)$.

EXAMPLE 3

For $q := z^n + b_{n-k} z^{n-k}$ $(k \geq 2)$ the condition of apolarity requires

$$b_{n-k} = (-1)^{k+1} \binom{n}{k} \frac{a_0}{a_5}.$$

The zeros of the resulting q lie on the circular disk of radius

$$\left| \binom{n}{k} \frac{a_0}{a_k} \right|^{1/k}$$

about the origin. Grace's theorem implies that this disk contains at least one zero of p. The result is equivalent to Theorem 6.4e.

EXAMPLE 4

We show that for arbitrary complex c the polynomial $p(z) := 1 - z + cz^n$ has a zero in the disk $|z - 1| \leq 1$. The polynomial $q(z) = b_0 + b_1 z + \cdots + b_n z^n$ is apolar to p if, for instance, $b_0 = 0, b_{n-1} = -n, b_n = 1$. These relations are satisfied by any polynomial that has the zero 0 and whose zeros add up to n. The numbers

$$w_k := 1 - e^{(2\pi i/n)k}, \qquad k = 0, 1, \ldots, n-1$$

form such a set of zeros. They all lie in $|z - 1| \leq 1$.

PROBLEMS

1. Let p be a normalized (highest coefficient 1) polynomial of degree n and let ρ be an inclusion radius for p. Show that for all z

$$|p(z)| \leq (|z| + \rho)^n$$

and for all z such that $|z| \geq \rho$

$$|p(z)| \geq (|z|-\rho)^n.$$

More generally, show that the derivatives of p satisfy for all z

$$|p^{(k)}(z)| \leq \frac{n!}{(n-k)!}(|z|+\rho)^{n-k}, \quad k=1,2,\ldots,n,$$

and for all z such that $|z| \geq \rho$

$$|p^{(k)}(z)| \geq \frac{n!}{(n-k)!}(|z|-\rho)^{n-k}.$$

[Gauss-Lucas theorem, product representation].

2. If all zeros of $p(z) = a_0 + a_1 z + \cdots + a_n z^n$ lie in the sector $S: |\arg z - \alpha| \leq \beta$, where $\beta \leq \pi/2$, then the numbers $b_k := -a_k/a_{k+1}$ ($k=0,1,\ldots,n-1$) also lie in S (Takahashi [1929]).

3. To define $D_c p$, the derivative of the polynomial p with respect to the (finite) point c, the special Moebius transformation t mapping c onto ∞,

$$s = t(z) = \frac{1}{z-c},$$

was used. Show that $D_c p$ does not depend on this special choice, i.e., that any Moebius transformation t such that $t(c) = \infty$ would yield the same result.

4. For what values of c is the polynomial $q(z) := (z-c)^n$ apolar to a given polynomial $p(z) := z^n + a_{n-1} z^{n-1} + \cdots + a_0$? What is the meaning of Grace's theorem in this special case?

5. Let p be a polynomial of degree n and let $p(a) = p(b)$. The derivative p' then has at least one zero in the circular disk with center $(a+b)/2$ and radius $|b-a|\cot \pi/n$.

6. Let the coefficients of the polynomial $p(z) := a_0 + a_1 z + \cdots + a_n z^n$ satisfy

$$\frac{a_0^2}{\binom{n}{0}} - \frac{a_1^2}{\binom{n}{1}} + \frac{a_2^2}{\binom{n}{2}} - \cdots + (-1)^n \frac{a_n^2}{\binom{n}{n}} = 0.$$

Prove that either all zeros of p have modulus 1 or there are zeros inside as well as outside the unit circle.

7. Let p be a polynomial of degree n and let a and b be two points such that $p(a) \neq 0, p(b) \neq 0$. Prove that any circular region containing all solutions of the equation

$$\frac{(z-a)^n}{p(a)} = \frac{(z-b)^n}{p(b)}$$

contains at least one zero of p.

§6.6. CIRCULAR ARITHMETIC

Here we shall continue to study the location of zeros of polynomials by applying Moebius transformations to circular regions, i.e., to closed point sets in the extended complex plane whose stereographic projection on the Riemann sphere is bounded by a circle. It will serve our purposes to consider single points, as well as the full (extended) complex plane, as circular regions. We begin by defining certain arithmetical operations for circular regions. Here and in §6.13 capital letters generally denote circular regions.

If a, b, c, d are complex numbers such that $ad - bc \neq 0$, and Z is any circular region, we define

$$W := \frac{aZ + b}{cZ + d}$$

as the image of the set Z under the mapping

$$z \to \frac{az + b}{cz + d}.$$

By the theory of Moebius transformations W again is a circular region. This, in particular, is true for the sets

$$c + Z = \{c + z : z \in Z\}, \qquad cZ = \{cz : z \in Z\},$$

where c is any complex number, and for

$$\frac{1}{Z} = \left\{\frac{1}{z} : z \in Z\right\}.$$

If Z_1 and Z_2 are circular regions, we let

$$Z_1 + Z_2 := \{z_1 + z_2 : z_1 \in Z_1, z_2 \in Z_2\}.$$

An examination of the various possibilities shows that $Z_1 + Z_2$ is again a circular region. Moreover, the addition of circular regions so defined is associaive and commutative.

The above operations are easily parametrized when the circular regions involved are disks, which is the case of primary computational interest. If Z denotes the disk with center c and radius ρ,

$$Z := \{z : |z - c| \leq \rho\},$$

we write
$$Z = [c; \rho], \quad c =: \operatorname{mid} Z, \quad \rho =: \operatorname{rad} Z$$
for brevity. With this notation we have
$$a + Z = [a + c; \rho],$$
$$aZ = [ac; |a|\rho];$$
and if $0 \notin Z$, as is easily seen,
$$\frac{1}{Z} = \left[\frac{\bar{c}}{c\bar{c} - \rho^2}; \frac{\rho}{c\bar{c} - \rho^2} \right]. \tag{6.6-1}$$

Furthermore, if
$$Z_k := [c_k; \rho_k], \quad k = 1, 2, \ldots, m,$$
then
$$\sum_{k=1}^{m} Z_k = \left[\sum_{k=1}^{m} c_k; \sum_{k=1}^{m} \rho_k \right].$$

In the course of our work we shall also have to form products $Z_1 Z_2$ of disks Z_1 and Z_2. These products will have the property that $z_1 z_2 \in Z_1 Z_2$ whenever $z_1 \in Z_1$ and $z_2 \in Z_2$. Unfortunately, the "true" product $\{z_1 z_2 : z_1 \in Z_1, z_2 \in Z_2\}$ in general is not a disk. To remain within the realm of disks we define the product computationally, as follows: If $Z_k := [c_k; \rho_k]$, $k = 1, 2$, then
$$Z_1 Z_2 := [c_1 c_2; |c_1|\rho_2 + |c_2|\rho_1 + \rho_1 \rho_2]. \tag{6.6-2}$$
The inequality
$$|z_1 z_2 - c_1 c_2| \leq |c_1||z_2 - c_2| + |c_2||z_1 - c_1| + |z_1 - c_1||z_2 - c_2|$$
shows that the set $Z_1 Z_2$ contains all points $z_1 z_2$, where $z_1 \in Z_1$, $z_2 \in Z_2$. Not all points in $Z_1 Z_2$ need to have this form, however.

It is obvious that the product defined by (6.6-2) is commutative, and it can be verified that it is associative. A simple computation shows that the distributive law is replaced by the relation
$$Z_1(Z_2 + Z_3) \subset Z_1 Z_2 + Z_1 Z_3.$$

Thus it is not possible, in general, to "factor out" common factors; however, equality holds if Z_1 is a point.

CIRCULAR ARITHMETIC

It is finally necessary to define the square root of a disk $Z := [c; \rho]$. We distinguish two cases. If $0 \in Z$, we let

$$Z^{1/2} := \left[0; \, (|c| + \rho)^{1/2} \right].$$

If $0 \notin Z$, we define $Z^{1/2}$ as the union of the two disjoint disks

$$\left[\pm c^{1/2}; \, \frac{1}{|c|^{1/2} + (|c| - \rho)^{1/2}} \right],$$

where $c^{1/2}$ is one of the numbers that satisfies $(c^{1/2})^2 = c$. Simple inequalities show that with these definitions any w such that $w^2 \in Z$ satisfies $w \in Z^{1/2}$.

We are now ready to tackle the following question concerning the location of polynomial zeros:

PROBLEM

Let m and n be integers, $0 < m < n$. Let there be given $m+1$ complex numbers $z_0, c_1, c_2, \ldots, c_m$, and $n-m$ circular regions $W_{m+1}, W_{m+2}, \ldots, W_n$. Let the following be known about a polynomial p of degree n with zeros $w_1, w_2, \ldots, w_n$:

(a) $w_l \in W_l$, $l = m+1, m+2, \ldots, n$;

(b) *the values of the functions*

$$q_k(z) := \frac{1}{(k-1)!} \left(\frac{p'(z)}{p(z)} \right)^{(k-1)}, \quad k = 1, 2, \ldots, m$$

at z_0 are

$$q_k(z_0) = c_k, \quad k = 1, 2, \ldots, m. \tag{6.6-3}$$

What can be said about the location of the remaining zeros $w_1, \ldots, w_m$?

The functions q_k are polynomials in $p'p^{-1}, p''p^{-1}, \ldots, p^{(k)}p^{-1}$. In particular, the function

$$q_1(z) = \frac{p'(z)}{p(z)},$$

which in the following we also denote by $q(z)$, is the negative reciprocal of the *Newton correction* $-p/p'$ at z. Thus for $m = 1$ the problem amounts to determining w_1 from an approximate knowledge of the location of $w_2, \ldots, w_n$ and an exact knowledge of the Newton correction at a point. Similarly, for $m > 1$ a solution of the problem may serve to sort out a cluster of zeros by use of the values of m derivatives at a point and the approximate location of the remaining zeros.

For $m = 1$ the solution of the problem is as follows:

THEOREM 6.6a

Let z_0 and c_1 be complex numbers and let $W_2, W_3, \ldots, W_n$ be circular regions. The set of possible values of the zero w_1 of a polynomial p of degree n such that

$$\frac{p'(z_0)}{p(z_0)} = c_1, \qquad (6.6\text{-}4)$$

whose remaining zeros $w_2, \ldots, w_n$ satisfy

$$w_k \in W_k, \qquad k = 2, 3, \ldots, n, \qquad (6.6\text{-}5)$$

is then precisely the circular region

$$W_1 := z_0 - \frac{1}{c_1 - V_1}, \qquad (6.6\text{-}6)$$

where

$$V_1 := \sum_{k=2}^{n} \frac{1}{z_0 - W_k}. \qquad (6.6\text{-}7)$$

Proof. Because $p'(z)/p(z)$ is a logarithmic derivative,

$$c_1 = \frac{p'(z_0)}{p(z_0)} = \sum_{k=1}^{n} \frac{1}{z_0 - w_k};$$

hence

$$\frac{1}{z_0 - w_1} = c_1 - \sum_{k=2}^{n} \frac{1}{z_0 - w_k}$$

and

$$w_1 = z_0 - \left(c_1 - \sum_{k=2}^{n} \frac{1}{z_0 - w_k} \right)^{-1}. \qquad (6.6\text{-}8)$$

CIRCULAR ARITHMETIC

Now, if (6.6-5) holds,

$$\frac{1}{z_0 - w_k} \in \frac{1}{z_0 - W_k}, \quad k = 2, 3, \ldots, n,$$

by the properties of circular arithmetic; hence

$$\sum_{k=2}^{n} \frac{1}{z_0 - w_k} \in V_1$$

and (6.6-8) implies $w_1 \in W_1$. Conversely, if w is any point in W_1, then there exists $v \in V_1$ such that

$$w = z_0 - \frac{1}{c_1 - v}.$$

By definition of V_1 there exist points

$$v_k \in \frac{1}{z_0 - W_k}, \quad k = 2, 3, \ldots, n,$$

such that $v_2 + v_3 + \cdots + v_n = v$, and finally there exist points $w_k \in W_k$ such that

$$v_k = \frac{1}{z_0 - w_k}, \quad k = 2, 3, \ldots, n.$$

Thus (6.6-8), and consequently (6.6-4), holds for a polynomial whose zeros $w_2, \ldots, w_n$ satisfy (6.6-5). We thus have completed the proof by showing that every point of W_1 can be zero of a polynomial of degree n satisfying (6.6-4) whose remaining zeros lie in the circular regions $W_2, \ldots, W_n$. ∎

In view of (6.6-4), (6.6-6) may be written

$$W_1 = z_0 - \frac{p(z_0)}{p'(z_0) - p(z_0) V_1}. \quad (6.6\text{-}9)$$

If z_0 is not contained in any of the sets $W_2, \ldots, W_n$, the circular region V_1 is a disk. If z_0 is close to w_1, then $|p(z_0)|$ will be small, and $p(z_0) V_1$ is a small circular disk close to the origin. In these circumstances (6.6-9) will differ little from Newton's formula

$$z_1 = z_0 - \frac{p(z_0)}{p'(z_0)}$$

to obtain an improved approximation z_1 to a zero from a crude approximation z_0. The fact that $w_1 \in W_1$ may be regarded as a more precise version of Newton's method, obtained on the basis of information on the approximate location of the remaining zeros.

We recall Laguerre's theorem, which in the version given above as Theorem 6.5d stated the following: if p is any polynomial of degree n and z_0 is any complex number, any circular region C whose boundary passes through the points z_0 and $z_0 - np(z_0)/p'(z_0)$ contains at least one zero of p. In slightly strengthened form this may be stated as follows:

COROLLARY 6.6b

If the zeros $w_2, w_3, \ldots, w_n$ are contained in the closure of the complement of a circular region C as described above, then w_1 is contained in C.

Proof (independent of §6.5). Let c_1 be defined by (6.6-4). Then C is a circular region whose boundary passes through z_0 and $z_0 - n/c_1$. Consequently it can be represented in the form

$$C = z_0 - \frac{n}{c_1(1+H)},$$

where H is a suitable half-plane bounded by a straight line through the origin. The closure of the complement of C is obtained by replacing H with $-H$ and thus is

$$W := z_0 - \frac{n}{c_1(1-H)}.$$

We apply Theorem 6.6a where $W_2 = \cdots = W_n = W$. This yields

$$z_0 - W_k = \frac{n}{c_1(1-H)},$$

$$\frac{1}{z_0 - W_k} = \frac{1}{n} c_1(1-H),$$

$$V_1 = \sum_{k=2}^{n} \frac{1}{z_0 - W_k} = \frac{n-1}{n} c_1(1-H),$$

and, since $\alpha H = H$ for every $\alpha > 0$,

$$c_1 - V_1 = c_1 \left[1 - \frac{n-1}{n}(1-H) \right]$$

$$= \frac{c_1}{n}[1 + (n-1)H]$$

$$= \frac{c_1}{n}(1+H);$$

thus finally

$$W_1 = z_0 - \frac{1}{c_1 - V_1} = z_0 - \frac{n}{c_1(1+H)} = C,$$

and the assertion follows from the fact that $w_1 \in W_1$. ∎

We now take up the general problem stated on p. 477 Assuming that the sets W_j are such that none of the sets $(z_0 - W_j)^k$, $k = 1,\ldots,m$; $j = m+1,\ldots,n$ contains the origin, we shall construct disks $B_1, B_2, \ldots, B_m$ with the following property: the zeros $w_1, \ldots, w_m$ of any polynomial p satisfying conditions (a) and (b) of the problem have the form

$$w_k = z_0 + r_k, \qquad k = 1,\ldots,m, \qquad (6.6\text{-}10)$$

where the numbers $r_1, \ldots, r_m$ are the zeros of a polynomial

$$t(r) := 1 + b_1 r + b_2 r^2 + \cdots + b_m r^m \qquad (6.6\text{-}11)$$

such that

$$b_k \in B_k, \qquad k = 1,\ldots,m. \qquad (6.6\text{-}12)$$

The construction of the disks B_k is such that their radii tend to zero if the radii of all disks $W_{m+1}, \ldots, W_n$ tend to zero. With the exception of the special case where $m = 1$, however, which has already been stated as Theorem 6.6a, it is not claimed that every polynomial t of the form (6.6-11) whose coefficients satisfy (6.6-12) has zeros r_k such that the resulting w_k belong to a polynomial p satisfying (a) and (b).

Let $z_0, w_1, \ldots, w_m$ be complex numbers, $z_0 \neq w_l$, $l = 1,\ldots,m$, and let

$$s_k := -\sum_{j=1}^{m} \frac{1}{(w_j - z_0)^k}, \qquad k = 1, 2, \ldots, m. \qquad (6.6\text{-}13)$$

Our construction is based on a property of the numbers s_k, which has already been used in a different way in §6.4: if the numbers $b_1,\ldots,b_m$ are defined recursively by

$$b_k = \frac{1}{k}(s_k + b_1 s_{k-1} + \cdots + b_{k-1} s_1), \qquad k=1,2,\ldots,m, \quad (6.6\text{-}14)$$

then $w_j = z_0 + r_j$, $j=1,\ldots,m$, where $r_1,\ldots,r_m$ are the zeros of the polynomial

$$t(r) := 1 + b_1 r + \cdots + b_m r^m. \qquad (6.6\text{-}15)$$

For the proof we need only observe that (6.6-14) is equivalent to (6.4-11), where $b_0 = 1$.

To apply this fact to our problem let $w_1, w_2, \ldots, w_n$ be the zeros of a polynomial p satisfying (a) and (b). Because

$$q_1(z) = \frac{p'(z)}{p(z)} = -\sum_{j=1}^{n} \frac{1}{w_j - z},$$

differentiation yields

$$q_k(z) = -\sum_{j=1}^{n} \frac{1}{(w_j - z)^k}, \qquad k=1,2,\ldots;$$

hence, in view of (6.6-3), the numbers s_k defined by (6.6-13) satisfy

$$s_k = c_k + \sum_{j=m+1}^{n} \frac{1}{(w_j - z_0)^k}$$

or, in view of (a),

$$s_k \in S_k, \qquad k=1,2,\ldots,m, \qquad (6.6\text{-}16)$$

where the S_k are disks,

$$S_k := c_k + \sum_{j=m+1}^{n} \frac{1}{(W_j - z_0)^k}, \qquad (6.6\text{-}17)$$

the powers $(W_j - z_0)^k$ being defined by the multiplication rule (6.6-2). From the properties of the numbers s_k mentioned above it now follows that the coefficients of the polynomial $t(r) = 1 + b_1 r + \cdots + b_m r^m$ (with zeros $r_j := w_j - z_0$) satisfy (6.6-12), where the disks B_k are defined by

$$B_k = \frac{1}{k}(S_k + S_{k-1} B_1 + \cdots + S_1 B_{k-1}), \qquad k=1,\ldots,m. \quad (6.6\text{-}18)$$

In summary, we have obtained the following theorem:

THEOREM 6.6c

Let $z_0, c_1, \ldots, c_m$ be complex numbers and let $W_{m+1}, \ldots, W_n$ be disks such that none of the disks $(z_0 - W_j)^k$, $k = 1, \ldots, m$; $j = m+1, \ldots, n$ contains the origin. The zeros $w_1, \ldots, w_m$ of a polynomial p satisfying (a) and (b) are then of the form $z_0 + r_k$ ($k = 1, \ldots, m$), where the r_k are the zeros of a polynomial (6.6-11) whose coefficients b_k satisfy (6.6-12), the sets B_k being defined by (6.6-18) and (6.6-17).

As a first application let $m = 1$. Here $S_1 = c_1 - V_1$, where V_1 is defined by (6.6-7), and $B_1 = S_1$. Thus $w_1 = z_0 + r_1$, where $r_1 = -b_1^{-1}$ is the unique solution of an equation $1 + b_1 r = 0$ such that $b_1 \in B_1$; $w_1 \in z_0 - (c_1 - V_1)^{-1}$ follows, in agreement with Theorem 6.6a.

As a second application we consider $m = 2$. In addition to

$$c_1 = \frac{p'(z_0)}{p(z_0)},$$

this requires the value c_2 of

$$q_2(z) = \frac{p(z)p''(z) - [p'(z)]^2}{[p(z)]^2}$$

at z_0. We have

$$S_1 = c_1 + D, \quad S_2 = c_2 + E,$$

where

$$D := \sum_{k=3}^{n} \frac{1}{W_k - z_0}, \quad E := \sum_{k=3}^{n} \frac{1}{(W_k - z_0)^2}.$$

Because $B_1 = S_1$, $B_2 = \frac{1}{2}(S_1^2 + S_2)$, the zeros r_1, r_2 of $t(z) = 1 + b_1 r + b_2 r^2$ are contained in the sets

$$R_{1,2} := \frac{2}{-S_1 \pm (-S_1^2 - 2S_2)^{1/2}}$$

$$= \frac{2}{-c_1 - D \pm [-(c_1 + D)^2 - 2c_2 - 2E]^{1/2}}, \quad (6.6\text{-}19)$$

the square roots being computed by the rules of circular arithmetic. The zeros w_1, w_2 are then contained in $W_{1,2} := z_0 + R_{1,2}$.

This formula can be used to separate or refine two close zeros for which crude approximations have been obtained by some other method.

EXAMPLE

Let
$$p(z) := (z+4.1)(z+3.8)(z+2.05)(z+1.85)(z-1.95)(z-2.15)(z-3.9)(z-4.05).$$
With the data $z_0 = -4$,
$$W_3 = W_4 = [-2; 0.3], \quad W_5 = W_6 = [2; 0.3], \quad W_7 = W_8 = [4; 0.3],$$
the formula yields
$$W_1 = [-4.0999; 0.0013], \quad W_2 = [-3.7985; 0.0052].$$

Continuing with $z_0 = -2$ replaces the double inclusion disks $W_3 = W_4$ with the separate disks
$$W_3 = [-2.0500; 0.00011], \quad W_4 = [-1.8499; 0.00096];$$
$z_0 = 2$ now yields
$$W_5 = [1.9500; 0.0003], \quad W_6 = [2.1502; 0.0027],$$
and with $z_0 = 4$ we finally get
$$W_7 = [3.9000; 0.7 \cdot 10^{-5}], \quad W_8 = [4.0500; 0.2 \cdot 10^{-5}].$$

PROBLEMS

1. Let A, B be two disks not containing the origin. Then the expression $(AB)^{-1}$ can be formed as
$$C_1 = \frac{1}{A} \cdot \frac{1}{B} \text{ (divisions first)}$$
or as
$$C_2 = (AB)^{-1} \text{ (multiplication first)}.$$
Show that $C_2 \neq C_1$ and that $\text{rad } C_2 \leq \text{rad } C_1$.

2. Let the disk $Z_0 := [z_0; \rho]$ contain at most one zero of a polynomial p of degree $n+1$, let $c_1 := p'(z_0)/p(z_0)$, and let $\rho > n|1/c_1|$. Prove that Z_0 contains precisely one zero of p and that zero is actually contained in the disk
$$Z_1 := \left[z_0 - \frac{\rho^2 \overline{c_1}}{\rho^2 |c_1|^2 - n^2}; \frac{n\rho}{\rho^2 |c_1|^2 - n^2} \right].$$

3. Let it be known that the disk $[z_0; \rho]$ contains at most two zeros of a polynomial p of degree n and let $c_k := q_k(z_0)$, $k = 1, 2$. Show that the two zeros of p are contained in the sets

$$z_0 + \frac{2}{[-c_1;\ (n-2)/\rho] \pm [-2c_2 - c_1^2;\ (n-2)(n+2\rho|c_1|)/\rho^2]^{1/2}}.$$

§6.7. THE NUMBER OF ZEROS IN A HALF-PLANE

In §§6.4 to 6.6 we have attacked the following problem with a variety of methods: given a polynomial p, find sets (mostly circular regions or disks) that contain no, some, or all zeros of p. In §§6.7 and 6.8 we shall study a "dual" problem: given a polynomial p and a circular region R, determine the number of zeros of p contained in R.

In this section we deal with the case where R is a half-plane. The tools to be used are the principle of the argument and the theory of Cauchy indices explained in §6.3.

Without loss of generality we may assume that R is the upper half-plane $\operatorname{Im} z \geqslant 0$ because this can always be achieved by a substitution of the form $z \to az + b$. By multiplying the polynomial by a constant factor, we can assume its highest coefficient to be positive. Thus the problem is to find the number m of zeros w such that $\operatorname{Im} w > 0$ of the polynomial

$$p(z) := a_n z^n + a_{n-1} z^{n-1} + \cdots + a_0,$$

where $a_0, a_1, \ldots, a_{n-1}$ are complex numbers, $a_n > 0$.

We first assume that p has no real zeros. Let ρ be a proper inclusion radius for p. The principle of the argument (Theorem 4.10a) may then be applied to the positively oriented Jordan curve $\Gamma = \Gamma_1 + \Gamma_2$ shown in Fig. 6.7. This yields

$$m = n(p(\Gamma), 0), \qquad (6.7\text{-}1)$$

the winding number of $p(\Gamma)$ with respect to 0. To calculate it we keep track of the intervals of length π in which a continuous argument of $p(z)$ lies. For arbitrary integers k let J_k denote the open interval $(k\pi - \pi/2, k\pi + \pi/2)$. For $|z|$ large the behavior of $p(z)$ is dictated by the leading term $a_n z^n$. Thus for $z = \rho$, if ρ is large enough, since $a_n > 0$, the argument of p is approximately 0 (mod 2π). We choose for $\arg p(\rho)$ the value that lies in J_0. If z travels along Γ_1, the continuous argument of $a_n z^n$ increases by $n\pi$. Thus, if ρ is large enough, the value of $\arg p(-\rho)$ obtained by continuous continuation along Γ_1 lies in J_n.

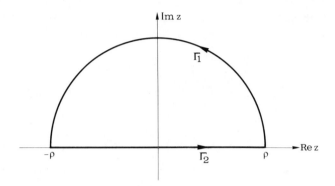

Fig. 6.7. A Jordan curve for counting zeros.

To determine the change in the argument along Γ_2 let

$$a_k =: \alpha_k + i\beta_k, \qquad k = 0, 1, \ldots, n,$$

where α_k, β_k are real, $\beta_n = 0$, and set

$$r(x) := \alpha_0 + \alpha_1 x + \cdots + \alpha_n x^n,$$
$$s(x) := \beta_0 + \beta_1 x + \cdots + \beta_{n-1} x^{n-1}.$$

For x real we then have

$$p(x) = r(x) + is(x). \tag{6.7-2}$$

Since p has no real zeros, the polynomials r and s cannot vanish simultaneously. As x increases from $-\rho$ to ρ, $\arg p(x)$, the continuous argument of $p(x)$, is well defined. The argument changes from an interval J_k to a neighboring interval $J_{k\pm 1}$ only at points that are zeros of odd order of r. If ξ is such a zero, the change is from J_k to J_{k+1} if $r(x)$ changes from positive to negative values and $s(\xi)$ is positive or if $r(x)$ changes from negative to positive values and $s(\xi)$ is negative. In the other cases $\arg p(x)$ changes from J_k to J_{k-1}. The change in index in the interval J_k in which $\arg p(x)$ lies thus equals the negative of the Cauchy index of the real rational function s/r at $x = \xi$. The net change of $\arg p(z)$ as z moves along Γ_2 thus equals $-\pi I_{-\rho}^{\rho}(s/r)$. There follows

$$n(p(\Gamma), 0) = \tfrac{1}{2} n - \tfrac{1}{2} I_{-\rho}^{\rho} \frac{s}{r}. \tag{6.7-3}$$

NUMBER OF ZEROS IN A HALF-PLANE 487

This holds for all sufficiently large ρ, hence for $\rho = \infty$. From (6.7-1) we thus have

$$m = \tfrac{1}{2}\left[n - I_{-\infty}^{\infty} \frac{s}{r} \right]. \qquad (6.7\text{-}4)$$

Now let p be permitted to have zeros on the real axis. Let there be k such zeros, counting multiplicities, and let q be the polynomial formed with these zeros. We write $p = qp^*$, $p^*(x) = r^*(x) + is^*(x)$ (r^*, s^* real). The polynomial p^* is of degree $n - k$ and still has m zeros in the upper half-plane; hence by (6.7-4)

$$m = \tfrac{1}{2}\left[n - k - I_{-\infty}^{\infty} \frac{s^*}{r^*} \right].$$

However,

$$\frac{s^*}{r^*} = \frac{qs^*}{qr^*} = \frac{s}{r}; \quad \text{hence} \quad I_{-\infty}^{\infty} \frac{s^*}{r^*} = I_{-\infty}^{\infty} \frac{s}{r}.$$

By calculating the Cauchy index by Theorem 6.3c we obtain the following theorem:

THEOREM 6.7a

Let $p(z) = a_n z^n + a_{n-1} z^{n-1} + \cdots + a_0$ be a polynomial with complex coefficients, $a_n > 0$, and with k real zeros (counting multiplicities). Let the polynomials $f_0, f_1, \ldots, f_m, 0$ be generated by the Euclidean algorithm (6.3-4), starting with $f_0 = r, f_1 = s$, where $p(x) = r(x) + is(x)$, and denote by $v(x)$ the number of sign changes in the sequence $\{f_j(x)\}$. The number m of zeros w of p satisfying $\operatorname{Im} w > 0$ is then given by

$$m = \tfrac{1}{2}\left[n - k - I_{-\infty}^{\infty} \frac{s}{r} \right], \qquad (6.7\text{-}5)$$

where

$$I_{-\infty}^{\infty} \frac{s}{r} = v(-\infty) - v(\infty).$$

Stable Polynomials

As an application of the preceding theory we consider the following problem of interest in control theory (see Chapter 10): given a polynomial

p of degree n with *real* coefficients

$$p(z) := \alpha_n z^n + \alpha_{n-1} z^{n-1} + \cdots + \alpha_0 \qquad (\alpha_n \neq 0), \qquad (6.7\text{-}6)$$

what are the necessary and sufficient conditions for all zeros of p to have negative real parts? The problem became famous when it was proposed around 1895 by A. Stodola, professor of mechanical engineering in Zurich, to his colleague A. Hurwitz; however, it had been considered earlier by Routh and Liapunov.

A real polynomial with the property that $\operatorname{Re} w < 0$ for all w such that $p(w)=0$ is called **stable**. It is easy to find a necessary condition for a polynomial p to be stable. Let the real zeros of p be denoted by $-\xi_j$ ($j=1,2,\ldots,n-2k$) and the pairs of complex conjugate zeros by $-\xi_j \pm i\eta_j$ ($j=n-2k+1,\ldots,n-k$). If p is stable, we have $\xi_j > 0$ for all j. The representation of p in terms of real linear and quadratic factors,

$$p(z) = \alpha_n \Pi(z+\xi_j) \Pi\left[(z+\xi_j)^2 + \eta_j^2\right],$$

then shows, on carrying out the multiplications, that *all coefficients of a stable polynomial are different from zero and have the same sign* as α_n. This necessary condition for stability can be shown to be sufficient for $n=2$. However, the polynomial

$$p(z) = (z+3)(z^2 - 2z + 10) = z^3 + z^2 + 4z + 30$$

has the zeros -3 and $1 \pm i3$ and thus fails to be stable, although all its coefficients are positive. The above condition thus fails to be sufficient for stability for $n > 2$, and we see that the problem proposed by Stodola is not trivial.

We can easily reduce Stodola's problem to the problem considered in the preceding subsection by letting $z = ix$. For stability it is necessary and sufficient that the number of zeros x such that $\operatorname{Im} x > 0$ equals n. The number of zeros can be calculated by Theorem 6.7a. From

$$i^{-n} p(ix) = \alpha_n x^n - i\alpha_{n-1} x^{n-1} - \alpha_{n-2} x^{n-2} + i\alpha_{n-3} x^{n-3} + \cdots$$

there follows, since the α_k are real,

$$r(x) = \alpha_n x^n - \alpha_{n-2} x^{n-2} + \alpha_{n-4} x^{n-4} - \cdots,$$

$$s(x) = -\alpha_{n-1} x^{n-1} + \alpha_{n-3} x^{n-3} - \alpha_{n-5} x^{n-5} + \cdots.$$

NUMBER OF ZEROS IN A HALF-PLANE

By (6.7-4) we have $k=0$ and $m=n$ if and only if

$$I_{-\infty}^{\infty} \frac{s}{r} = -n.$$

It is convenient in the following to change the sign of s, hence of the index. Thus p is stable if and only if

$$I_{-\infty}^{\infty} \frac{f_1}{f_0} = n, \qquad (6.7\text{-}7)$$

where

$$\begin{aligned} f_0(x) &:= \alpha_n x^n - \alpha_{n-2} x^{n-2} + \cdots, \\ f_1(x) &:= \alpha_{n-1} x^{n-1} - \alpha_{n-3} x^{n-3} + \cdots. \end{aligned} \qquad (6.7\text{-}8)$$

Let $f_0, f_1, f_2, \ldots$, be the sequence of polynomials generated by the Euclidean algorithm (6.3-4) started with the polynomials (6.7-8). Denoting the number of sign changes in the numerical sequence $\{f_k(x)\}$ by $v(x)$, it follows from Theorem 6.7a that (6.7-7) holds, and thus p is stable if and only if

$$v(-\infty) - v(\infty) = n. \qquad (6.7\text{-}9)$$

Because $v(x)$ is a non-negative integer $\leq n$, this relation in turn holds if and only if

$$v(-\infty) = n, \quad v(\infty) = 0. \qquad (6.7\text{-}10)$$

Two conclusions may be drawn. To have n sign changes the sequence $\{f_k(x)\}$ must have its maximum number of $n+1$ elements. Because the degrees of the polynomials f_k are strictly decreasing, this is the case if and only if f_k has the precise degree $n-k$, which, in turn, is equivalent to the statement that the polynomials q_k appearing in (6.3-4) are all of degree 1. For $k=0,1$ the polynomials f_k are by definition even or odd, according to whether $n-k$ is even or odd. By induction it is easily seen that this holds for all k and that the q_k are odd functions. Thus the q_k are of degree 1 if and only if the formulas (6.3-4) take the form

$$f_{k-1}(x) = \gamma_k x f_k(x) - f_{k+1}(x), \quad k=1,2,\ldots,n, \qquad (6.7\text{-}11)$$

where f_n is a constant $\neq 0$, $f_{n+1} = 0$, and the γ_k are certain real numbers, $\gamma_k \neq 0$, $k=1,2,\ldots,n$.

The second conclusion to be drawn from (6.7-9) is as follows: since for $x \to \infty$ the sign of any polynomial is determined by the sign of its leading coefficient, it follows from $v(\infty)=0$ that the leading coefficients of all polynomials f_k must have the same sign. Equation (6.7-11) shows that the leading coefficient of f_{k-1} is obtained by multiplying the leading coefficient of f_k by γ_k, and $v(\infty)=0$ is equivalent to the condition

$$\gamma_k > 0, \quad k = 1, 2, \ldots, n. \tag{6.7-12}$$

Thus we have obtained Theorem 6.7b:

THEOREM 6.7b

The real polynomial $p(z) := \alpha_n z^n + \alpha_{n-1} z^{n-1} + \cdots + \alpha_0$ ($\alpha_n \neq 0$) is stable if and only if the relations connecting the polynomials f_k generated by the Euclidean algorithm (6.3-4) [where f_0 and f_1 are given by (6.7-8)] take the form (6.7-11), where $\gamma_k > 0$, $k = 1, 2, \ldots, n$.

EXAMPLE

For the polynomial $p(z) := z^3 + z^2 + 4z + 30$ we have $f_0(x) = x^3 - 4x$, $f_1(x) = x^2 - 30$. The division algorithm yields $f_2(x) = -26x$, $f_3(x) = 30$. The leading coefficients of the polynomials f_k do not all have the same sign; hence p is unstable.

In Chapter 12 we give a different (and in some ways more elegant) treatment of the stability problem by means of continued fractions. There it is also shown how to express the conditions of stability in terms of certain determinants that depend directly on the coefficients of p.

PROBLEMS

1. The real polynomial $p(z) := \alpha_3 z^3 + \alpha_2 z^2 + \alpha_1 z + \alpha_0$, where $\alpha_3 > 0$, is stable if and only if $\alpha_2 > 0, \alpha_0 > 0, \alpha_1 \alpha_2 > \alpha_0 \alpha_3$.
2. The real polynomial $p(z) := \alpha_4 z^4 + \alpha_3 z^3 + \alpha_2 z^2 + \alpha_1 z + \alpha_0$, where $\alpha_4 > 0$, is stable if and only if $\alpha_3 > 0, \alpha_0 > 0, \alpha_3 \alpha_2 - \alpha_4 \alpha_1 > 0, \alpha_1(\alpha_3 \alpha_2 - \alpha_4 \alpha_1) - \alpha_0 \alpha_3^2 > 0$.
3. If p is a stable polynomial of degree n with zeros $w_1, \ldots, w_n$, the negative number $\sigma := \max_{1 \leq m \leq n} \operatorname{Re} w_m$ is called the **abscissa of stability** of p. Prove that if $|w_m| \leq \rho$ for some m, then $\sigma \geq -\rho$.
4. Let σ be the abscissa of stability of the stable real polynomial $p(z) := \alpha_n z^n + \alpha_{n-1} z^{n-1} + \cdots + \alpha_0$. Show that

$$\sigma \geq -\frac{1}{n-1} \left[\frac{(-1)^n}{\alpha_n} p\left(-\frac{\alpha_{n-1}}{\alpha_n}\right) \right]^{1/n}.$$

(Schrack [1967])

5. Show that any *upper* bound for the abscissa of stability of a stable real polynomial necessarily depends on *all* coefficients of the polynomial.

6. The abscissa of stability σ of the stable real polynomial $p(z) := \alpha_3 z^3 + \alpha_2 z^2 + \alpha_1 z + \alpha_0$ satisfies

$$\sigma < -\min\left[\frac{\alpha_0}{\alpha_1}, \frac{\alpha_2}{3\alpha_3}, \frac{\alpha_1\alpha_2 - \alpha_0\alpha_3}{2(\alpha_1\alpha_3 + \alpha_2^2)}\right].$$

7. Let p be a real polynomial of degree n and let

$$(z-1)^n p\left(\frac{z+1}{z-1}\right) =: b_0 + b_1 z + \cdots + b_n z^n.$$

Prove that $|b_n| > |b_0| + |b_1| + \cdots + |b_{n-1}|$ is sufficient for the stability of p.

8. Let p be a real polynomial of degree $n > 0$, let $\alpha > 0$, and let

$$q(z) := \frac{p(z)p(\alpha) - p(-z)p(-\alpha)}{z + \alpha};$$

q is a real polynomial of degree $n-1$. Prove that (a) if p is stable then q is stable; (b) if q is stable and $|p(1)| > |p(-1)|$ then p is stable.

9. Using the result of the preceding problem, design an algorithm (different from that described in Theorem 6.7b) for deciding whether a given real polynomial is stable.

§6.8. THE NUMBER OF ZEROS IN A DISK

Let p be a polynomial, let z_0 be a complex number, and let $\rho > 0$. Let D denote the closed disk $|z - z_0| \leq \rho$. We wish to consider two obviously related problems:
1. To decide rationally whether p has any zeros in D,
2. To determine the number of zeros in D.

By means of a suitable *fractional* linear transformation these problems can be reduced to one of determining the number of zeros in a half-plane. Here we discuss another solution which does not involve a preliminary fractional transformation. Clearly the number of zeros of p in $|z - z_0| \leq \rho$ equals the number of zeros of the polynomial

$$q(z) := p(z_0 + \rho z)$$

satisfying $|z| \leq 1$. If

$$p(z) = b_0 + b_1(z - z_0) + b_2(z - z_0)^2 + \cdots + b_n(z - z_0)^n,$$

then

$$q(z) = b_0 + b_1 \rho z + \cdots + b_n \rho^n z^n.$$

Granting that the Taylor coefficients of p at z_0 are known, it thus suffices to consider the above-mentioned problem for the *unit disk*.

If D is the unit disk, the two problems can be solved (the second under certain additional hypotheses) by an algorithm due to Schur and Cohn. This algorithm is also the basic tool of Lehmer's search method for determining one zero of a given polynomial to arbitrary accuracy (see §6.10). Furthermore it can be used in a version of Weyl's exclusion algorithm discussed in §6.11.

It is convenient in this section to revive a convention of §6.5 and to call the function

$$p(z) := a_n z^n + a_{n-1} z^{n-1} + \cdots + a_0 \quad (\text{where } n \geqslant 0) \quad (6.8\text{-}1)$$

a *polynomial of degree n* even if some of the leading coefficients are zero. If all coefficients are zero, p is called the zero polynomial. The zero polynomial can have any degree. If $a_n = a_{n-1} = \cdots = a_{n-k+1} = 0$, $a_{n-k} \neq 0$ for some integer k, $0 < k \leqslant n$, the polynomial is said to have the k-fold zero ∞. Thus any polynomial of degree $n > 0$ which is not the zero polynomial always has precisely n zeros.

If p is a polynomial of degree n given by (6.8-1), we define p^*, the **reciprocal polynomial** of p, by the formula

$$p^*(z) := \overline{a_0} z^n + \overline{a_1} z^{n-1} + \cdots + \overline{a_n}. \quad (6.8\text{-}2)$$

The reciprocal polynomial of a polynomial of degree n is always considered to be a polynomial of degree n. For $z \neq 0$ the identity

$$p^*(z) = z^n \overline{p(1/\bar{z})}. \quad (6.8\text{-}3)$$

holds. From this it follows immediately that

$$|z| = 1 \quad \text{implies} \quad |p^*(z)| = |p(z)|. \quad (6.8\text{-}4)$$

Thus any zero of modulus 1 of p is a zero of p^*. More generally we have:

LEMMA 6.8a

If p is a polynomial of degree $n \geqslant 0, p \neq 0$, with zeros $w_1, w_2, \ldots, w_n$ (multiple zeros, including ∞, being listed according to their multiplicity), the reciprocal polynomial p^* is a polynomial of degree n with the zeros $1/\overline{w}_k$, $k = 1, 2, \ldots, n$, where $1/\infty := 0$, $1/0 := \infty$.

NUMBER OF ZEROS IN A DISK

Proof. It is clear from (6.8-3) that $p(w)=0$, $w \neq 0, \infty$, implies that $p^*(1/\bar{w}) = 0$, but this computation does not cover possible zeros 0 or ∞ and it does not prove that multiplicities are preserved. By the product representation the lemma is easily proved for polynomials without the zeros 0 or ∞. If p has the k-fold zero ∞ and the m-fold zero 0, we write $p(z) = z^m q(z)$, where q has the exact degree $n - m - k$. From (6.8-3) we then have $p^*(z) = z^k q^*(z)$, whence the lemma follows. ∎

If p is a polynomial of degree $n \geqslant 1$ given by (6.8-1), the polynomial Tp of degree $n-1$ defined by

$$Tp(z) := \overline{a_0} p(z) - a_n p^*(z)$$

$$= \sum_{k=0}^{n-1} (\overline{a_0} a_k - a_n \overline{a_{n-k}}) z^k \qquad (6.8\text{-}5)$$

is called the **Schur transform** of p. The Schur transform of a polynomial of degree 0 is the zero polynomial. We note that the Schur transform of a polynomial p depends on the degree ascribed to p and not merely on the *function p*. We also note that

$$Tp(0) = \overline{a_0} a_0 - a_n \overline{a_n} = |a_0|^2 - |a_n|^2 \qquad (6.8\text{-}6)$$

is always a real number.

We define the *iterated Schur transforms* $T^2 p, T^3 p, \ldots, T^n p$ by

$$T^k p := T(T^{k-1} p), \qquad k = 2, 3, \ldots, n,$$

where $T^{k-1} p$ is to be regarded as a polynomial of degree $n - k + 1$ even if its leading coefficient is zero. We set

$$\gamma_k := T^k p(0), \qquad k = 1, 2, \ldots, n. \qquad (6.8\text{-}7)$$

THEOREM 6.8b

Let p be a polynomial of degree n, $p \neq 0$. All zeros of p lie outside the closed unit disk $|z| \leqslant 1$ if and only if

$$\gamma_k > 0, \qquad k = 1, 2, \ldots, n. \qquad (6.8\text{-}8)$$

Proof. (a) Suppose p has no zeros in $|z| \leqslant 1$. By Vietà's formulas $|a_0| > |a_n|$ follows; hence $\gamma_1 > 0$. Also by (6.8-4) we have for $|z| = 1$

$$|\overline{a_0} p(z)| > |a_n p^*(z)|.$$

Hence we may apply Rouché's theorem (Theorem 4.10b) to the unit circle with $\overline{a_0}p$ in the role of the "big" function and $a_n p^*$ in the role of the "small" function. It follows that $Tp = \overline{a_0}p - a_n p^*$ has no zeros on the circumference of the unit disk and as many zeros inside as the big function, namely none.

This argument can be repeated. Let

$$T^k p(z) =: a_0^{(k)} + a_1^{(k)} z + \cdots + a_{n-k}^{(k)} z^{n-k} \qquad (6.8\text{-}9)$$

and assume that all zeros of $T^k p$ lie outside the unit disk. Then, according to Vietà, $|a_0^{(k)}| > |a_{n-k}^{(k)}|$; hence $\gamma_{k+1} = |a_0^{(k)}|^2 - |a_{n-k}^{(k)}|^2 > 0$. Applying Rouché's theorem, where $\overline{a_0^{(k)}} T^k p$ and $a_{n-k}^{(k)}(T^k p)^*$ are the big and the small functions, respectively, we find as before that all zeros of $T^{k+1}p = \overline{a_0^{(k)}} T^k p - a_{n-k}^{(k)}(T^k p)^*$ lie outside $|z| \leq 1$. Hence (6.8-8) follows by induction.

(b) Now let $\gamma_k > 0$, $k = 1, 2, \ldots, n$. In the above notation this means that $|a_0^{(k)}| > |a_{n-k}^{(k)}|$, $k = 0, 1, \ldots, n-1$. Hence it follows, as above, that $T^{k+1}p = \overline{a_0^{(k)}} T^k p - a_{n-k}^{(k)}(T^k p)^*$ has an equal number of zeros in $|z| \leq 1$ as $T^k p$. Since $T^n p = \gamma_n$ has no zeros, it follows that $T^0 p = p$ likewise has no zeros in $|z| \leq 1$. ∎

The calculation of the numbers γ_k by (6.8-7) is called the **Schur-Cohn algorithm**. Theorem 6.8b shows that this algorithm can always be used to decide whether a given polynomial is free of zeros in the closed unit disk. By means of an (entire) linear transformation it can be applied to any other disk. Under the additional hypothesis that all $\gamma_k \neq 0$ the algorithm can also be used to determine the exact number of zeros in a disk.

THEOREM 6.8c

Let p be a polynomial of degree n and let the numbers γ_k defined by (6.8-7) satisfy $\gamma_k \neq 0$, $k = 1, 2, \ldots, n$. If those indices k for which $\gamma_k < 0$ are denoted by $k_j, j = 1, 2, \ldots, m$, where $k_1 < k_2 < \cdots < k_m$, then the number $h(p)$ of zeros w of p satisfying $|w| < 1$ (multiple zeros counted with their multiplicity) is given by

$$h(p) = \sum_{j=1}^{m} (-1)^{j-1}(n+1-k_j). \qquad (6.8\text{-}10)$$

Proof. First we show that the hypothesis $\gamma_k \neq 0$ implies that none of the polynomials $T^k p$, $k = 0, 1, \ldots, n$, vanishes on $|z| = 1$. Indeed, let the number $e^{i\theta}$ (θ real) be a zero of some polynomial $T^k p$, where $k \geq 0$. From Lemma

NUMBER OF ZEROS IN A DISK

6.8a it follows that it is also a zero of $(T^k p)^*$, hence $e^{i\theta}$ is a zero of $T^{k+1}p$. By induction it follows that $e^{i\theta}$ is a zero of all $T^m p$, where $m > k$, and thus in particular of $T^n p$. Since $T^n p$ is constant, $T^n p = \gamma_n$, it would follow that $\gamma_n = 0$, contradicting the hypothesis that $\gamma_n \neq 0$.

To prove (6.8-10) let, for any polynomial q, $h(q)$ denote the number of zeros of q inside $|z| = 1$. Rouché's theorem now yields an expression for $h(T^{k+1}p)$ in terms of $h(T^k p)$. We continue to use the notation (6.8-9). Because $\gamma_{k+1} = |a_0^{(k)}|^2 - |a_{n-k}^{(k)}|^2 \neq 0$, Rouché's theorem can be applied with one of the functions $\overline{a_0^{(k)}} T^k p$ and $a_{n-k}^{(k)}(T^k p)^*$ in the role of the "big" function and the other in the role of the "small" function. If $\gamma_{k+1} > 0$, the first function is the big function, and

$$h(T^{k+1}p) = h(T^k p)$$

follows. If $\gamma_{k+1} < 0$, the second function is the big function, and we get

$$h(T^{k+1}p) = h((T^k p)^*).$$

By Lemma 6.8a $h((T^k p)^*)$ equals the number of zeros of $T^k p$ *outside* the unit circle. Since $T^k p$ is of degree $n - k$,

$$h(T^{k+1}p) = n - k - h(T^k p)$$

follows. Reading these relations backward yields

$$h(T^{k-1}p) = \begin{cases} h(T^k p), & \text{if } \gamma_k > 0, \\ n + 1 - k - h(T^k p), & \text{if } \gamma_k < 0. \end{cases}$$

Thus we find

$$h(p) = n + 1 - k_1 - h(T^{k_1}p)$$
$$= n + 1 - k_1 - [n + 1 - k_2 - h(T^{k_2}p)]$$
$$= \cdots = \sum_{j=1}^{m} (-1)^{j-1}(n + 1 - k_j) \pm h(T^n p),$$

and (6.8-10) follows in view of $h(T^n p) = 0$. ∎

There are several other ways to express the quantity $h(p)$ in the above equation. Since $\sum_{j=1}^{m}(-1)^{j-1}(n+1)$ is zero or $n+1$, according to the parity of m, we have

$$h(p) = \begin{cases} \sum_{j=1}^{m} (-1)^j k_j, & \text{if } m \text{ is even,} \\ n + 1 + \sum_{j=1}^{m} (-1)^j k_j, & \text{if } m \text{ is odd.} \end{cases} \quad (6.8\text{-}11)$$

Also, let $\pi_k := \gamma_1 \gamma_2 \cdots \gamma_k$. The number of negative terms in the sequence $\{\pi_k\}$ is given by

$$k_2 - k_1 + k_4 - k_3 + \cdots + \begin{cases} k_m - k_{m-1}, & \text{if } m \text{ is even,} \\ n + 1 - k_m, & \text{if } m \text{ is odd,} \end{cases}$$

which evidently equals (6.8-11). Thus $h(p)$ can be calculated as the number of negative elements of the sequence $\{\pi_k\}$. Finally, let

$$\epsilon_k := \gamma_k \gamma_{k-2} \gamma_{k-4} \cdots, \qquad k = 0, 1, 2, \ldots, n, \qquad (6.8\text{-}12)$$

where $\gamma_j := 1$ for $j \leq 0$. Then $\pi_k = \epsilon_k \epsilon_{k-1}$. Because all ϵ_k are different from zero under the hypotheses of the theorem, the number of negative elements in the sequence $\{\pi_k\}$ clearly equals the number of sign changes in the sequence $1, \epsilon_1, \epsilon_2, \ldots, \epsilon_n$; hence $h(p)$ can also be calculated as a number of sign changes. From a practical point of view it seems pointless to calculate the products π_k or ϵ_k merely to determine (6.8-10), a sum of integers.

Theorem 6.8c cannot be used to count the number of zeros inside the unit circle if some γ_k are zero. This situation not only occurs when p has some zeros *on* the unit circle but also, more generally, whenever one of the polynomials $T^k p$ ($k = 0, 1, \ldots, n$) turns out to be **self-reciprocal**, i.e., when its zeros can be grouped in pairs that lie symmetric to the unit circle. The algorithms that have been devised to deal with this exceptional situation are rather complex and their numerical suitability has not been fully determined.

It is possible to express the numbers γ_k in terms of determinants involving only the coefficients $a_0, \ldots, a_n$ of the polynomial p. From a numerical point of view these determinants are not so easy to use as the formula (6.8-7) which requires merely the recursive computation of the Schur transforms $T^k p$.

PROBLEMS

1. The polynomials

$$p_n(z) = z^n \sum_{k=0}^{n} \frac{(-1)^k}{k+1} (1 - z^{-1})^k$$

are of importance in the numerical integration of ordinary differential equations (Gear [1971]). Show that all zeros w of p_n satisfy $|w| < 1$ for $n = 1, 2, \ldots, 5$, but not for $n = 6$.

2. Let p be a real polynomial whose coefficients (beginning with the constant term) form a decreasing sequence of positive numbers. Show that its Schur transform Tp is a polynomial with the same property and thus obtain a new proof

METHODS FOR DETERMINING ZEROS 497

of the Kakeya-Eneström theorem (Problem 4, §6.4).

3. Under what conditions does the Schur transform of a given polynomial vanish identically?

§6.9. METHODS FOR DETERMINING ZEROS: A SURVEY

In the preceding sections we have learned how to find regions that contain zeros of a given polynomial and how to count the number of its zeros in a given (circular) region. We now turn to the problem of determining the zeros of a polynomial to arbitrary accuracy.

The reader is undoubtedly familiar with Galois' famous theorem to the effect that the only polynomial equations that possess a general solution in terms of "explicit" formulas (i.e., of formulas involving only a finite number of rational operations and of computations of roots) are those of degree not exceeding 4. For the computation of zeros of polynomials of higher degree we must resort to numerical methods. Frequently such methods already are used for polynomials of degree 3 or 4 because the explicit formulas available for the zeros of these polynomials are already fairly complicated.

A numerical method for determining a zero of a polynomial generally takes the form of a prescription to construct one or several sequences $\{z_n\}$ of complex numbers supposed to converge to a zero w of the polynomial. Many prescriptions, or "algorithms," for constructing these sequences are known. Each algorithm usually enjoys its own advantages and disadvantages. The choice of the "best" algorithm for a given problem is not always easy. The selection may depend heavily on extramathematical considerations such as speed and memory of the computing equipment and the required accuracy and trustworthiness of the result.

Any reasonable algorithm must *converge*; i.e., the sequence $\{z_n\}$ generated by it should, under suitable conditions, converge to a zero of the given polynomial. Moreover, an algorithm must be designed to furnish approximations not only to real but also to complex zeros of a polynomial. Even if the given polynomial has real coefficients, it can be proved that in a certain statistical sense, if the degree is high enough, most of its zeros will be complex. In most applications of polynomials (e.g., in the theory of control systems or in differential equations) real and complex zeros are equally relevant.

The following is a partial list of further desirable properties that an algorithm may or may not have and that may be used for a classification of the many available algorithms.

1. Global Convergence. Many well-known algorithms (notably Newton's) can be guaranteed to converge only if the *starting value* z_0 is sufficiently close to a zero of the polynomial. These algorithms are **locally convergent**. Algorithms that do not require a sufficiently close starting value are **globally convergent**.

2. *Unconditional Convergence.* Some algorithms can be expected to converge only if the given polynomial enjoys some special properties; e.g., all its zeros are simple or no two zeros have the same modulus. These algorithms are **conditionally convergent**. If an algorithm is convergent (either locally or globally) for *all* polynomials, it is **unconditionally convergent**.

3. *A posteriori Estimates.* In real life any algorithm must be artificially terminated after a finite number of steps. The approximation at this stage, z_n, say, will in general not be identical to a zero w of the polynomial. Under these circumstances it is desirable that we be able to calculate, from the data provided by the algorithm, a bound β_n for the error $|z_n - w|$ of the last approximation. We then can make the precise statement that there is at least one zero of the polynomial in the disk $|z - z_n| \leq \beta_n$. In some algorithms, such as Weyl's, the calculation of the bound β_n is an essential part of the algorithm; in others, such as Newton's, bounds are easily obtained from the data. There are algorithms, however, for which bounds are hard to come by.

4. *Uniform Convergence.* We have seen in §6.4 that for any given polynomial p it is easy to construct a number ρ such that the disk $|z| < \rho$ contains all zeros of p. This disk may be considered a first crude approximation to the zeros. If an algorithm provides error bounds β_n as described above, we may ask how much work is required to refine this crude first approximation by a given factor $\theta < 1$, i.e., to construct z_n such that $\beta_n / \rho \leq \theta$. If the required amount of work (e.g., the number of times a Horner scheme must be constructed) is bounded by a quantity depending only on θ and the degree of the polynomial, the algorithm is called **uniformly convergent** (in the class of polynomials to which it is applicable).

5. *Speed of Convergence.* As a measure of the ultimate speed of convergence we frequently use the concept of the *order* of an algorithm. The order ν is defined as the supremum of all real numbers α such that

$$\limsup_{n \to \infty} \frac{|z_{n+1} - w|}{|z_n - w|^\alpha} < \infty;$$

for instance, Newton's method has the order of convergence $\nu = 2$, which means asymptotically that the number of correct decimal places is doubled at each step. The higher the order of convergence, the more rapidly $|z_n - w|$ converges ultimately to zero. It must be pointed out, however, that the order of convergence is not always indicative of the initial speed of convergence. Typically, methods with $\nu > 1$ are only conditionally convergent.

6. *Simultaneous Determination of All Zeros.* Most algorithms determine the zeros one at a time. If a zero has been determined with sufficient accuracy, the polynomial is deflated (see Corollary 6.1b) and the algorithm is used again to determine a zero of the deflated polynomial. For practical as well as theoretical reasons (see §6.11) it may be desirable to determine all zeros simultaneously. Some algorithms have this property. If the degree of the given polynomial is n and the zeros are $w_1, w_2, \ldots, w_n$, such an algorithm simultaneously yields n sequences $\{z_k^{(m)}\}$, $m = 1, 2, \ldots, n$, such that $z_k^{(m)} \to w_m$ as $k \to \infty$ for all m.

7. *Cluster Insensitivity.* An obnoxious problem in the numerical determination of zeros is presented by the occasional occurrence of "clusters" of zeros, i.e., sets of several zeros that either coincide or are very close. A simple example of such a polynomial is

$$p(z) := z^n + \epsilon,$$

when ϵ is small. This example shows, incidentally, that an arbitrarily small change in a coefficient may let a zero of multiplicity > 1 explode into a cluster of several zeros. The performance of several otherwise excellent methods (such as Newton's) is severely worsened in the presence of a cluster. What we are looking for, then, are methods that are insensitive to clusters. Ideally, such methods would not even distinguish between a cluster and a multiple zero and would list the cluster as a single zero (with error bound) of the appropriate multiplicity.

8. *Numerical Stability.* In practice all computing is done not in the field of real or complex numbers but in the finite system of discrete and bounded numbers, as provided by the machine on hand. Even the set of numbers provided by floating arithmetic is bounded (exponent overflow). Algorithms originally devised to work in the continuum are adapted to the Procrustean bed of machine numbers by the devices of rounding and scaling. Not all algorithms are equally sensitive to this adaptation. The lack of sensitivity to rounding and scaling operations is generally referred to as **numerical stability**. The subject of numerical stability belongs to computer science rather than mathematical analysis and therefore is not dealt with systematically in this book. The reader is referred to texts in numerical analysis for a more thorough treatment; as to algebraic computations he could hardly do better than to start with Wilkinson [1963].

Keeping the above criteria in mind, we now review briefly the algorithms discussed in subsequent sections. The *methods of search* covered in §6.10 are globally, unconditionally, and uniformly convergent methods of determining one zero at a time. These methods are not sensitive to clusters, but the order of convergence is only 1. The *exclusion algorithms* presented in §6.11 enjoy similar properties but determine all zeros simultaneously. Again convergence is (at best) linear. In §6.12 we turn to methods based on the principle of *iteration*, the best known examples being the methods of Newton and Laguerre. Generally speaking, these methods are only locally convergent, but the order of convergence is higher than 1 ($\nu=2$ for Newton and $\nu=3$ for Laguerre) and can be made arbitrarily high. Commonly, these methods determine one zero at a time, but there are variants, not discussed here, that simultaneously yield two zeros (Bairstow, see Henrici [1964]) or all zeros (Weierstrass [1903], Kerner [1966, 1967]) of a polynomial. In §6.13 we describe a variant of Newton's method, based on *circular arithmetic*, that features built-in error bounds. Its order of convergence is 2, if it is used for determining a single zero, and 3, if all zeros are determined simultaneously. Finally, in §6.14 we consider the *methods of descent*, based on the idea of minimizing $|p(z)|$. In recent years methods of this kind have been developed into powerful algorithms that converge globally, unconditionally, and uniformly. One zero is determined at a time, but the convergence may be quadratic for zeros that are not clustered.

In Chapter 7 we present several methods that are based on the partial fraction decomposition of a rational function with denominator p. Generally speaking, these methods are globally, but only conditionally, convergent. The simplest is *Bernoulli's method* (§7.4), which determines one or two zeros at a time. A modern version is Rutishauser's qd (*quotient-difference*) *algorithm* (§7.6), which under certain conditions produces all zeros simultaneously. Unless special measures are taken, the convergence is only linear and there is a problem of numerical instability. Among the methods not treated in this work we mention Graeffe's root-squaring algorithm (Bareiss [1960]). This is a globally, although not unconditionally, convergent method that in many cases produces all zeros simultaneously with quadratic convergence.

§6.10. METHODS OF SEARCH FOR A SINGLE ZERO

In this section we shall apply the results obtained earlier in the chapter to obtain algorithms that are guaranteed to approximate a zero of a given polynomial with an arbitrarily small error. These algorithms are uniformly convergent, in the following sense: let p be any polynomial of degree n, let σ be any inclusion radius for p, and let $0<\epsilon<1$. The work required to construct a disk of radius $\epsilon\sigma$, which is guaranteed to contain at least one

METHODS OF SEARCH FOR A SINGLE ZERO 501

zero of p, is bounded by a function $\omega(n,\epsilon)$ which depends on n and ϵ but not on the individual polynomial p.

Without loss of generality we may assume that $\sigma = 1$. Let $p(z) = z^n + a_{n-1}z^{n-1} + \cdots + a_0$ be the given polynomial, let $w_1, w_2, \ldots, w_n$ be its zeros, and let σ be an inclusion radius for p. We may assume that $\sigma > 0$, for the case $\sigma = 0$ is trivial. The polynomial

$$q(z) := \sigma^{-n}p(\sigma z) = z^n + \sigma^{-1}a_{n-1}z^{n-1} + \cdots + \sigma^{-n}a_0$$

then has zeros $w_i^* := \sigma^{-1}w_i$, $i = 1, 2, \ldots, n$, and thus the inclusion radius 1. If w^* approximates some zero w_j^* with an error $\leq \epsilon$, the number $w := \sigma w^*$ approximates the corresponding zero w_j of p with an error $\leq \sigma \epsilon$, as required.

In the following we denote, for any positive integer n, by P_n the class of all polynomials $p(z) = z^n + a_{n-1}z^{n-1} + \cdots + a_0$ whose zeros w_i satisfy $|w_i| \leq 1$, $i = 1, 2, \ldots, n$.

I. Proximity Tests

The basic tool of the algorithms to be described is a *proximity test* $T = T(\rho)$, depending on a parameter $\rho > 0$, which can be applied to any polynomial $p \in P_n$ at any point z such that $|z| \leq 1$ and which the polynomial either passes or fails. The test must be such that it is *passed* at all points z close enough to a zero and *failed* at all points far enough away. (There may be an in-between region in which the test is either passed or failed.) The parameter ρ regulates the difficulty of the test. The smaller the ρ, the more difficult it becomes to pass the test.

Speaking formally, a test $T(\rho)$ is called **proximity test** if there are two positive functions ϕ and ψ, defined on some interval $0 < \rho \leq \rho_0$ and having the following properties: if p is any polynomial in P_n and w is any zero of p, then for all $\rho \in (0, \rho_0]$

(a) p passes $T(\rho)$ at all points z such that $|z - w| \leq \phi(\rho)$,
(b) p fails $T(\rho)$ at all points z such that $|z - w| > \psi(\rho)$ (see Fig. 6.10).

The above evidently implies that $\psi(\rho) \geq \phi(\rho)$; we do not require that $\psi = \phi$. The postulate that $T(\rho)$ becomes arbitrarily difficult to pass for $\rho \to 0$ is taken to mean

(c) $\quad \lim_{\rho \to 0} \psi(\rho) = 0$

We further require

(d) The function ψ is continuous and strictly monotonically increasing; its range contains the interval $(0, 1]$.

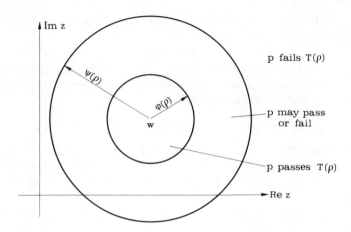

Fig. 6.10. Proximity test.

The functions ϕ and ψ are called, respectively, the **inner** and **outer convergence functions** of the test $T(\rho)$.

The following test, denoted by T_1, may serve as a first example of a proximity test.

$$\text{"}p \text{ passes } T_1(\rho) \text{ at } z\text{"} \Leftrightarrow |p(z)| \leq \rho.$$

To show that this test has the required properties for $0 < \rho \leq 1$, let

$$p(z) = \prod_{i=1}^{n} (z - w_i).$$

If p fails the test at z, then

$$|p(z)| = \prod_{i=1}^{n} |z - w_i| > \rho.$$

Hence for every i

$$|z - w_i| > \rho \prod_{\substack{j=1 \\ j \neq i}}^{n} |z - w_j|^{-1}.$$

Since $|w_j| \leq 1$, $|z| \leq 1$, every factor of the product on the right is at least $\frac{1}{2}$, and we find that

$$|z - w_i| > 2^{-n+1}\rho, \qquad i = 1, 2, \ldots, n.$$

Hence $T_1(\rho)$ cannot be failed for $|z - w_i| \leq 2^{-n+1}\rho$ for some i, and (a) is true for

$$\phi(\rho) := 2^{-n+1}\rho.$$

METHODS OF SEARCH FOR A SINGLE ZERO

If, on the other hand, p passes $T_1(\rho)$ at z, then

$$\prod_{i=1}^{n} |z - w_i| \leq \rho,$$

and it follows that

$$|z - w_i| \leq \rho^{1/n}$$

for at least one index i. Thus the test cannot be passed if $|z - w_i| > \rho^{1/n}$ for all i, and we find that (b) is true for

$$\psi(\rho) = \rho^{1/n}.$$

(By considering a polynomial with a single zero of multiplicity n we see that (b) is not true for any smaller function ψ.) It is clear that ψ has the properties (c) and (d).

Two tests are called **equivalent** if they are defined on the same domain of ρ and if they produce identical results for all polynomials p at all points z and for all values of ρ.

EXAMPLE

The test T_1 is equivalent to a test that is declared passed if and only if $|p(z)|^2 \leq \rho^2$.

Two proximity tests T and T^* are called **similar** if an increasing function ρ^* maps $(0, \rho_0]$ onto an interval $(0, \rho_0^*]$ such that the test $T(\rho)$ is equivalent to $T^*(\rho) := T(\rho^*(\rho))$. Similar tests thus differ only in the choice of the parameter. It is clear that the similarity of tests is also an equivalence relation.

EXAMPLE

The test T_1 is similar to the test $T_1^*(\rho)$ which is passed if and only if $|p(z)| \leq \rho^n$. Convergence functions for T_1^* are $\phi(\rho) := 2^{-n+1}\rho^n$ and $\psi(\rho) := \rho$.

According to (a) every proximity test is similar to a test with outer convergence function $\psi(\rho) := \rho$.

II. The Search Algorithm

We require the notion of an ϵ-**covering**. If ϵ is any positive number and S is any set in the complex plane, an ϵ-covering of S is any system of closed disks of radius $\leq \epsilon$ whose union contains S. The covering is said to be *centered* in S if the centers of the covering disks belong to S. The construction of a *minimal* ϵ-covering of a given bounded set (i.e., a covering that requires the least number of disks) can raise intricate questions of elementary geometry. Of course, we can always use coverings whose centers form a square or a hexagonal grid.

Let $p \in P_n$, let T be a proximity test, and let $\{\gamma_k\}$ be a monotonic sequence of positive numbers converging to zero such that $\gamma_0 = 1$. We shall describe an algorithm for constructing a sequence of points $\{z_k\}$ such that each of the disks

$$D_k := \{z : |z - z_k| \leq \gamma_k\}, \qquad k = 0, 1, 2, \ldots,$$

contains at least one zero of p.

Let $z_0 = 0$. Then D_0 certainly contains a zero, for it contains all zeros. The algorithm now proceeds by induction. Suppose we have found a point z_{k-1} such that D_{k-1} contains a zero. To construct z_k we cover the set $D_{k-1} \cap D_0$ with an ϵ_k-covering centered in it and apply a test $T(\rho_k)$ at the center of each covering disk. The parameters ϵ_k and ρ_k are chosen such that the following two conditions are met:

1. The test is passed at the center of each disk of the covering containing a zero.
2. Any point at which the test is passed is at a distance $\leq \gamma_k$ from a zero.

Condition (1) is satisfied if

$$\epsilon_k \leq \phi(\rho_k)$$

and condition (2) is satisfied if

$$\psi(\rho_k) \leq \gamma_k.$$

Thus both conditions are fulfilled if

$$\begin{aligned} \rho_k &= \psi^{[-1]}(\gamma_k), \\ \epsilon_k &= \phi(\rho_k) = \phi(\psi^{[-1]}(\gamma_k)), \end{aligned} \tag{6.10-1}$$

where $\psi^{[-1]}$ denotes the inverse function of ψ.

At least one of the covering disks contains a zero, since D_{k-1} contains one, and all zeros are contained in D_0. Thus by (1) the test $T(\rho_k)$ is passed at least once. We let z_k be the first center at which the test is passed. There is no assurance that the disk of radius ϵ_k surrounding z_k actually contains a zero but by (2) the disk D_k will.

The whole algorithm may be summarized as follows: *let $z_0 := 0$; having constructed z_{k-1}, cover the set $D_{k-1} \cap D_0$ with an ϵ_k-covering centered in it and apply $T(\rho_k)$ at the center of each covering disk, where ϵ_k and ρ_k are given by* (6.10-1). *Let z_k be the first center that passes the test.*

Provided that identical systems of coverings are used, this algorithm remains unchanged if the test T is replaced by a "similar" test T^*.

III. Convergence

By construction, the centers z_k of successive disks D_k satisfy $|z_{k+1} - z_k| \leq \gamma_k$, where $\gamma_k \to 0$. This in itself does not imply the convergence of the sequence $\{z_k\}$. Nevertheless, Theorem 6.10a holds:

THEOREM 6.10a

The sequence $\{z_k\}$ converges and its limit is a zero of p.

Proof. Let $\delta := \min_{w_i \neq w_j} |w_i - w_j|$ be the minimum distance between *distinct* zeros of p. Let k be an integer such that $3\gamma_k < \delta$. Let $m \geq k$. The disk D_m contains a zero, say w_i. The disk D_{m+1} likewise contains a zero, say w_j. From

$$|z_m - w_i| \leq \gamma_m, \qquad |z_{m+1} - w_j| \leq \gamma_{m+1}$$

it follows by the monotonicity of the sequence $\{\gamma_k\}$ that

$$|w_i - w_j| \leq 2\gamma_m + \gamma_{m+1} \leq 3\gamma_k < \delta,$$

hence that $w_i = w_j$. Thus, for all $m \geq k$, $|z_m - w_i| \leq \gamma_m$, which proves that $\lim_{m \to \infty} z_m = w_i$. ∎

IV. Amount of Work

We measure the amount of work required to approximate a zero with an error $\leq \epsilon$ by estimating the number of applications of the test T required to construct the first disk D_k such that its radius γ_k is less than ϵ. For reasons of simplicity we assume that the centers of the covering disks always form a square grid.

The area of D_{m-1} is $\pi \gamma_{m-1}^2$. In a square ϵ_m-covering the centers of the covering disks must not be more than $\sqrt{2}\, \epsilon_m$ apart. If we neglect boundary effects, approximately $(\pi/2)(\gamma_{m-1}^2/\epsilon_m^2)$ disks of radius ϵ_m are required to cover D_{m-1}. (Working with a hexagonal grid, we could replace the constant $\pi/2$ with $2\pi/3\sqrt{3}$.) Within the same degree of approximation, this also is the maximum number of applications of the test to proceed from z_{m-1} to z_m.

For the given sequence $\{\gamma_k\}$ and for $\epsilon > 0$ let $k(\epsilon)$ denote the smallest k such that $\gamma_k \leq \epsilon$. By the above the total number of applications of the test necessary to approximate a zero within an error $\leq \epsilon$ does not exceed a quantity of the order of

$$\omega(T, \{\gamma_k\}, \epsilon) := \frac{\pi}{2} \sum_{m=1}^{k(\epsilon)} \frac{\gamma_{m-1}^2}{\epsilon_m^2}. \tag{6.10-2}$$

Axiomatically we define the above function ω as the **work function** of the search algorithm based on the proximity test T and the sequence $\{\gamma_k\}$. The work function does not change if T is replaced by a similar test T^*.

From the fact that ω does not depend on p it follows that the search algorithms described above are uniformly convergent in the sense mentioned earlier.

EXAMPLE

Choosing a geometric mode of subdivision ($\gamma_k = \gamma^k, 0 < \gamma < 1, k = 0, 1, 2, \ldots$) for the test T_1, we have in view of $\phi(\rho) = 2^{-n+1}\rho, \psi(\rho) = \rho^{1/n}$,

$$\epsilon_m = \phi(\psi^{[-1]}(\gamma_m)) = 2^{-n+1}\gamma^{mn};$$

hence for $n > 1$

$$\omega(T_1, \{\gamma^k\}, \epsilon) = \frac{\pi}{2} 2^{2n-2} \sum_{m=1}^{k(\epsilon)} \gamma^{2m-2-2mn} \sim \lambda_n \gamma^{(2n-2)k(\epsilon)}, \quad (\epsilon \to 0),$$

where

$$\lambda_n := \frac{2^{2n-3}\pi}{\gamma^2(1-\gamma^{2n-2})}.$$

For the determination of a zero of a polynomial of degree 10 with an error $\leq 10^{-6}$, working with $\gamma = \frac{1}{2}$ (which requires $k = 20$), the function ω yields an upper bound of approximately $2^{397}\pi \doteq 10^{120}$ applications of the test. Since on the average we cannot expect to do much better than to use one-half the maximum number of tests, a search algorithm based on T_1 certainly is not practical.

V. Proximity Tests with Linear Convergence Functions

Suppose the convergence functions of a proximity test T are linear:

$$\phi(\rho) = \alpha\rho, \quad \psi(\rho) = \beta\rho, \quad (0 < \alpha \leq \beta). \tag{6.10-3}$$

Then by (6.10-1)

$$\epsilon_m = \phi(\psi^{[-1]}(\gamma_m)) = \frac{\alpha}{\beta}\gamma_m,$$

and the work function (6.10-2) becomes

$$\omega(T, \{\gamma_k\}, \epsilon) = \frac{\pi}{2}\frac{\beta^2}{\alpha^2} \sum_{m=1}^{k(\epsilon)} \frac{\gamma_{m-1}^2}{\gamma_m^2}. \tag{6.10-4}$$

In particular, if $\gamma_k = \gamma^k$,

$$\omega(T, \{\gamma^k\}, \epsilon) = \frac{\pi \beta^2}{2\alpha^2 \gamma^2} k(\epsilon), \qquad (6.10\text{-}5)$$

and the work necessary to compute a zero to a given accuracy is proportional to the number of decimals required. This convergence behavior is called *linear* in numerical analysis.

We now give several examples of proximity functions with linear convergence functions. As in §6.4, for arbitrary z and h we let

$$p(z+h) = b_0 + b_1 h + b_2 h^2 + \cdots + b_n h^n, \qquad (b_n = 1).$$

It will be convenient to suppress the argument z in the Taylor coefficients b_i.

TEST T_2

Let

$$\beta(z) := \min_{1 \leq k \leq n} \left| \frac{b_0}{b_k} \right|^{1/k}.$$

The polynomial p is said to pass $T_2(\rho)$ at z if and only if $\beta(z) \leq \rho$. To show that T_2 has the required properties (a) to (d) let

$$\rho_0 := \min_{1 \leq i \leq n} |z - w_i|. \qquad (6.10\text{-}6)$$

It follows from Theorem 6.4e that

$$\rho_0 \leq \min_{1 \leq k \leq n} \left| \binom{n}{k} \frac{b_0}{b_k} \right|^{1/k}.$$

Therefore, if $\rho_0 > n\rho$, then $\beta(z) > \rho$ and p fails the test at z. It follows that (b) is true for $\psi(\rho) = n\rho$. On the other hand, by Theorem 6.4b we have $\rho_0 \geq \frac{1}{2}\beta(z)$. Thus if $\rho_0 \leq \frac{1}{2}\rho$, then $\beta(z) \leq \rho$, and p passes the test, showing that (a) is satisfied with $\phi(\rho) = \frac{1}{2}\rho$. The conditions (c) and (d) are always satisfied for linear functions.

For the numerical example considered earlier ($n = 10, \gamma_k = 2^{-k}, \epsilon = 10^{-6}$) (6.10-4) now yields a maximum of some 50,000 applications of the test.

TEST T_3

This test is declared passed if and only if

$$|b_0| \le |b_1|\rho + |b_2|\rho^2 + \cdots + |b_n|\rho^n.$$

Let ρ_0 be defined as above. Then for some h such that $|h| = \rho_0$ we have

$$b_0 = -b_1 h - b_2 h^2 - \cdots - b_n h^n,$$

Consequently, if $\rho \ge \rho_0$

$$|b_0| \le |b_1|\rho + |b_2|\rho^2 + \cdots + |b_n|\rho^n,$$

and p passes $T_3(\rho)$. Thus $\phi(\rho) = \rho$ for this test. On the other hand, Theorem 6.4i implies that the test cannot be passed if $\rho_0 \ge (2^{1/n} - 1)^{-1}\rho$. Thus (b) is satisfied for $\psi(\rho) = (2^{1/n} - 1)^{-1}\rho$.

THE SCHUR-COHN TEST

Let the sequence $\{\gamma_k\}$ be fixed. If the convergence functions ϕ and ψ are linear, the value of the work function ω for a given ϵ is proportional to β^2/α^2 and thus (since $\beta \ge \alpha$) is a minimum for a test such that $\beta = \alpha$. Without loss of generality it may be assumed that $\beta = \alpha = 1$. A test with convergence functions $\phi(\rho) = \psi(\rho) = \rho$ is called a **sharp test**. A sharp test would react positively if and only if the disk of radius ρ about the testing point z contains a zero. Do sharp tests exist?

Clearly, the tests T_2 and T_3 are not sharp. In fact, $\beta/\alpha \to \infty$ as $n \to \infty$ for both. Moreover, it can be shown (see Problem 2) that no test that uses only the absolute values of the Taylor coefficients can be sharp. The Schur-Cohn algorithm discussed in §6.8, however, can evidently be used as a sharp test. The first search method of the type considered here (Lehmer [1961]) is based on the Schur-Cohn algorithm.

In our numerical example $(n = 10, \gamma_k = 2^{-k}, \epsilon = 10^{-6})$ (6.10-5) yields a maximum of 129 tests in an algorithm based on a sharp test. Because of the boundary effects, the true maximum is somewhat higher (see below).

The mere fact that the work function is smallest for the Schur-Cohn test does not in itself imply that this test defines the computationally most efficient algorithm because the work function does not take into account the work required to carry out the test. In the absence of rigorous results concerning the minimum number of arithmetic operations required to administer the various tests, precise results are difficult to obtain. Suffice it to say that all tests described in this section require, among other things, all Taylor coefficients at z. If performed by the Horner algorithm, their computation requires $\frac{1}{2}n^2 + O(n)$ multiplications. The Schur-Cohn algorithm, if programmed in the superior fashion recommended by Stewart [1968], takes another $\frac{1}{2}n^2 + O(n)$ multiplications and divisions. Thus the Schur-Cohn test needs only about twice as much work as T_2 or T_3.

VI. Optimum Choice of $\{\gamma_k\}$

Suppose the search algorithm is based on a test with linear convergence functions (6.10-3). If ϵ is given, for what choice of the sequence $\{\gamma_k\}$ is the work function $\omega(T, \{\gamma_k\}, \epsilon)$ a minimum?

We answer this question first when $k(\epsilon)$ is prescribed. Let $\epsilon > 0$, let k be a given positive integer, and let $\{\gamma_m\}$ be any decreasing sequence such that $\gamma_0 = 1, \gamma_k = \epsilon$. Then, by the inequality of the arithmetic and the geometric mean,

$$\omega(T, \{\gamma_m\}, \epsilon) = \lambda \sum_{m=1}^{k} \frac{\gamma_{m-1}^2}{\gamma_m^2} \qquad \left(\lambda := \frac{\pi \beta^2}{2\alpha^2}\right)$$

$$\geqslant \lambda k \left(\prod_{m=1}^{k} \frac{\gamma_{m-1}^2}{\gamma_m^2}\right)^{1/k}$$

$$= \lambda k \epsilon^{-2/k}$$

$$= \omega(T, \{\epsilon^{m/k}\}, \epsilon),$$

and we have proved Theorem 6.10b.

THEOREM 6.10b

Let $\epsilon > 0$ and $k > 0$ be given. On the space of all monotonic sequences $\{\gamma_m\}_0^k$ such that $\gamma_0 = 1$ and $\gamma_k = \epsilon$ the work function (6.10-4) assumes its smallest value for the geometric sequence $\gamma_m := \epsilon^{m/k}, m = 0, 1, \ldots, k$.

On the basis of this result, we now restrict our attention to geometric sequences, $\gamma_m = \gamma^m$ ($0 < \gamma < 1$) and ask for the optimal value of γ to achieve a given accuracy ϵ. As a function of γ and ϵ, $k(\epsilon)$ is the smallest integer k such that $\gamma^k \leqslant \epsilon$ or

$$k(\epsilon) = \text{smallest integer not exceeded by } \frac{\text{Log}\,\epsilon}{\text{Log}\,\gamma}$$

$$= -\left[-\frac{\text{Log}\,\epsilon}{\text{Log}\,\gamma}\right],$$

where $[x]$ denotes the largest integer $\leq x$. Neglecting a fractional part, we have, approximately,

$$\omega(T,\{\gamma^k\},\epsilon) = \lambda \frac{\mathrm{Log}\,\epsilon}{\gamma^2 \mathrm{Log}\,\gamma}$$

(λ defined as above). By differentiation we easily find that the minimum of this expression is attained for $\gamma = e^{-1/2} = 0.60653\ldots$ and that the value of the minimum is $2e\lambda\,\mathrm{Log}\,1/\epsilon$.

Unfortunately this result does not indicate precisely the maximum number of tests to be applied, for the method of counting the covering disks underlying (6.10-2) becomes increasingly inaccurate (because of the neglect of boundary effects) if the ratio of the covering disks and the disk to be covered approaches 1. To determine the exact maximum let $\nu(\xi)$, for $0 < \xi \leq 1$, denote the minimum number of disks of radius ξ that is required to cover the unit disk. The function ν is nonincreasing, piecewise constant, and continuous from the right; no simple analytical expression for it exists. To proceed from z_m to z_{m+1} in a search algorithm based on a test with linear convergence functions and on a geometric sequence $\{\gamma^m\}$ requires covering a disk of radius γ^m with disks of radius $(\alpha/\beta)\gamma^{m+1}$. Hence, if an optimal covering is used, at most $\nu[(\alpha/\beta)\gamma]$ applications of the test are necessary. The actual number of tests to attain an error $\leq \epsilon$ thus equals

$$\omega(\alpha,\beta,\gamma,\epsilon) := -\nu\left(\frac{\alpha}{\beta}\gamma\right)\left[-\frac{\mathrm{Log}\,\epsilon}{\mathrm{Log}\,\gamma}\right].$$

We shall determine the minimum of ω as a function of γ for the Schur-Cohn test ($\alpha = \beta = 1$).

THEOREM 6.10c

For sufficiently small fixed values of ϵ, the function $\chi(\gamma,\epsilon) := \omega(1,1,\gamma,\epsilon)$ assumes its minimum at $\gamma = \gamma_0 := (1 + 2\cos(2\pi/7))^{-1}$. The value of the minimum is

$$\chi(\gamma_0,\epsilon) = -8\left[-\frac{\mathrm{Log}\,\epsilon}{\mathrm{Log}\,\gamma_0}\right] \doteq -8\left[\frac{\mathrm{Log}\,\epsilon}{0.8096}\right].$$

Proof. First we determine the minimum of the function

$$\chi_1(\gamma) := \nu(\gamma)\frac{\mathrm{Log}\,\epsilon}{\mathrm{Log}\,\gamma}.$$

Let the points of discontinuity of ν be, in decreasing order, $1 = \xi_0 > \xi_1 > \xi_2 > \cdots$, and let the constant value of ν in the interval $\xi_m \leq \xi < \xi_{m-1}$ be denoted by ν_m ($m = 1, 2, \ldots$). Then $\chi_1(\gamma)$ is increasing in each of the intervals $\xi_m \leq \xi < \xi_{m-1}$ and has a downward jump at the point ξ_m ($m = 1, 2, \ldots$). It is smallest where

$$\chi_1(\xi_m) = \nu_m \frac{\operatorname{Log} \epsilon}{\operatorname{Log} \xi_m}$$

is smallest. It can be shown that $\xi_1 = \sin(\pi/3), \nu_1 = 3$,

$$\xi_m = \left(2 \cos \frac{\pi}{m+2}\right)^{-1}, \quad \nu_m = m+2 \quad \text{for} \quad m = 2, 3;$$

$$\xi_m = \left(1 + 2 \cos \frac{2\pi}{m+2}\right)^{-1}, \quad \nu_m = m+3 \quad \text{for} \quad m = 4, 5, 6.$$

From these values and from the trivial estimate $\nu(\xi) \geq \xi^{-2}$ it follows by computation that the minimum is assumed only at $\gamma_0 = \xi_5 = (1 + 2\cos(2\pi/7))^{-1} \doteq 0.44504$ and that it has the value

$$\chi_1(\gamma_0) = 8 \frac{\operatorname{Log} \epsilon}{\operatorname{Log} \gamma_0} \doteq 9.882 \log \epsilon^{-1}.$$

The function χ has the form

$$\chi(\gamma) = \nu(\gamma) \mu(\gamma),$$

where

$$\mu(\gamma) := -\left[-\frac{\operatorname{Log} \epsilon}{\operatorname{Log} \gamma}\right].$$

The function μ is piecewise constant, nondecreasing, and continuous from the left. We denote its points of discontinuity by $0 < \mu_0 < \mu_1 < \mu_2 < \cdots$. Evidently, $\chi(\gamma) \geq \chi_1(\gamma)$, with equality holding if and only if $\gamma = \mu_n$ for some n. Let n^* be the smallest index n such that $\mu_n \geq \gamma_0$. For sufficiently small values of ϵ the points μ_n are arbitrarily dense, hence $\mu_{n^*} < \xi_4$, and furthermore

$$\chi(\mu_{n^*}) < \chi_1(\xi_m), \quad m \neq 5.$$

It follows that $\chi(\mu_{n^*})$ is the smallest value of χ. If $\mu_{n^*} = \gamma_0$, the theorem is established. If $\mu_{n^*} > \gamma_0$, the theorem follows from the fact that $\chi(\gamma)$ is constant for $\gamma_0 \leq \gamma \leq \mu_{n^*}$. ∎

The optimal covering of the unit disk by eight disks of radius γ_0 consists of a disk centered at the origin, surrounded by seven disks centered at the points
$$z_k := \rho e^{2\pi i k/7}, \qquad k=0,1,\ldots,6,$$
where
$$\rho = \frac{2\cos(\pi/7)}{1+2\cos(2\pi/7)} \doteq 0.80194.$$

So far we have discussed coverings with disks of equal radius. It can be shown that from a probabilistic point of view it may be more efficient to use disks whose radii are not all equal. Lehmer [1969], for instance, covers the unit disk with a disk of radius $\frac{1}{2}$ centered at O, surrounded by eight disks of radius $\frac{2}{3}$ centered on a circle of radius $\frac{3}{4}\cos\pi/8$. It can be shown that the Lehmer covering is probabilistically not optimal. Optimal coverings, from both a deterministic and a probabilistic point of view, are discussed by Friedli [1972].

PROBLEMS

1. Let s_k be defined by (6.4-10) and let
$$\sigma := \min_{1 \leq k \leq n} \left| \frac{n}{s_k} \right|^{1/k}.$$

   Define the test $T(\rho)$ to be passed if $\sigma \leq \rho$. Show that this is a proximity test with
$$\phi(\rho) = \frac{\rho}{2n}, \qquad \psi(\rho) = \rho.$$

2. Prove that if T is a proximity test depending only on the absolute values of the Taylor coefficients and the convergence functions of T are of the form (6.10-2) then
$$\frac{\beta}{\alpha} \geq (2^{1/n} - 1)^{-1}.$$

   [This is a consequence of the fact that the constant in Birkhoff's theorem (6.4i) is best possible.]

3. Using probability arguments, show that the *average* number of applications of the test in a search algorithm, where $\phi(\rho) = \psi(\rho) = \rho$, $\gamma_k = 2^{-k}$, necessary to improve the solution by the decimal place lies between 11 and 12, depending on the order in which the disks are searched.

SIMULTANEOUS DETERMINATION OF ALL ZEROS 513

4. Determine the sequence $\{\gamma_k\}$ that minimizes $\omega(T_1, \{\gamma_k\}, \epsilon)$ for given ϵ. Show that the minimizing sequence is not geometric and that for $n = 10$, $\epsilon = 10^{-6}$ the minimum of ω is about 10^{12} times smaller than the value obtained in the text for $\gamma_k := 2^{-k}$.

§6.11. METHODS OF SEARCH AND EXCLUSION FOR THE SIMULTANEOUS DETERMINATION OF ALL ZEROS

In the preceding section (see also §§6.12 and 6.14) we discussed methods for the determination of one zero of a given polynomial. If all zeros are wanted, these algorithms are usually employed as follows: after a zero has been determined to sufficient accuracy, the corresponding linear factor is removed from the polynomial by the Horner algorithm (see §6.1) and the process is started afresh on the "deflated" polynomial whose degree is now lowered by 1.

The method of successive removal of linear factors has the following disadvantage. In many practical applications, it is not necessary to determine the zeros to any great accuracy. If the method of successive deflations is used, evidently it is not possible to take advantage of this special situation, for if the early linear factors are very inaccurate the polynomials obtained by successive deflations may be falsified to an extent that makes the remaining approximate zeros meaningless.

In practice the above difficulty is overcome by determining all zeros, especially the early ones, to full working accuracy. Thus the advantage inherent in the requirement of low accuracy is lost. To avoid build-up of rounding error, it is sometimes recommended that a zero be "purified" by performing the last stages of an iteration algorithm on the complete, undeflated polynomial. This purification may fail in the presence of clusters of zeros because the algorithm might converge to a zero that has already been determined.

In this section we consider a class of globally, unconditionally, and uniformly convergent methods that avoid the difficulties discussed by the simple device of determining all zeros simultaneously. (Other methods for the simultaneous determination of all zeros are the qd algorithm discussed in §7.6, which is only conditionally convergent, and a method due to Kerner based on Weierstrass' proof of the fundamental theorem of algebra, which converges only locally.) The safety of the methods presented here must be paid for by rather slow convergence; however, once crude approximations to all zeros are known, the zeros can be determined rapidly and with great precision by the circular arithmetic algorithm described in §6.13.

I. Statement of the Problem

Let the class P_n be defined as in §6.10. Without loss of generality it can be assumed that a polynomial of degree n whose zeros are sought belongs to P_n. The problem of determining all zeros $w_1, w_2, \ldots, w_n$ of a polynomial $p \in P_n$ can be split in two parts.

PROBLEM 1 (Location of zeros)

Given $\epsilon > 0$, construct a set $S(\epsilon)$ such that

$$w_i \in S(\epsilon), \qquad i = 1, 2, \ldots, n, \qquad (6.11\text{-}1)$$

and

$$z \in S(\epsilon) \Rightarrow |z - w_i| \leq \epsilon \text{ for some } i. \qquad (6.11\text{-}2)$$

The first condition means that the set $S(\epsilon)$ contains all zeros, and the second, that every point of $S(\epsilon)$ is at a distance $\leq \epsilon$ from a suitable zero.

A set $S(\epsilon)$ with these properties is called an **ϵ-inclusion set** for the polynomial p.

The solution to Problem 1 is not yet equivalent to the complete (approximate) factorization of the polynomial because in the case of multiple zeros their multiplicities are not determined. More generally, if several zeros are clustered together and are contained in the same component of the set $S(\epsilon)$, the number of zeros in the cluster remains unknown. The solution to Problem 1 must therefore be followed by treating Problem 2:

PROBLEM 2 (Multiplicity of zeros)

Determine the number of zeros in each complonent of the set $S(\epsilon)$.

II. Simultaneous Search for All Zeros

Here we shall solve Problem 1 by a method related to the search algortihm discussed in §6.10. Let T be a proximity test with convergence functions ϕ and ψ and let $\{\gamma_k\}$ be a monotonic sequence of positive numbers converging to zero such that $\gamma_0 = 1$. We revive the use of the symbol $[z; \rho]$ introduced in §6.6 to denote the closed disk of radius ρ centered at z.

Given any polynomial $p \in P_n$, we construct a sequence of sets $S_0, S_1, \ldots$, as follows:

Let $S_0 := [0; 1]$. Having constructed S_{k-1}, let ρ_k and ϵ_k be defined by the relations

$$\rho_k = \psi^{[-1]}(\gamma_k),$$
$$\epsilon_k = \phi(\rho_k) = \phi(\psi^{[-1]}(\gamma_k)), \qquad (6.11\text{-}3)$$

identical with (6.10-1). We cover the set S_{k-1} with disks of radius ϵ_k centered in it and apply the test $T(\rho_k)$ at the center of each covering disk. Let $z_{k1}, z_{k2}, \ldots, z_{kn_k}$ be the centers at which the proximity test is passed. We then let

$$S_k := \bigcup_{i=1}^{n_k} [z_{ki}; \gamma_k] \cap S_0.$$

THEOREM 6.11a

For $k = 0, 1, 2, \ldots$, the sets S_k are $2\gamma_k$-inclusion sets for p.

Proof. The unit disk $[0; 1]$ contains all zeros; hence (6.11-1) and (6.11-2) are true for the set $S_0 = [0; 1]$, where $\epsilon = 2\gamma_0 = 2$. Now assume that these relations are true for the set S_{k-1} with $\epsilon = 2\gamma_{k-1}$. Then all zeros are contained in S_{k-1}; hence each zero is contained in at least one of the covering disks of radius ϵ_k. The test $T(\rho_k)$ is passed at all centers at a distance $\leq \phi(\rho_k)$ from a zero, hence at all centers of covering disks containing a zero. Because $\gamma_k \geq \epsilon_k$, S_k contains all zeros and (6.11-1) holds for S_k. To show that (6.11-2) holds, let z be any point of S_k. Then z belongs to a disk $[z_{kj}; \gamma_k]$ such that $T(\rho_k)$ was passed at z_{kj}. This is possible only if z is at a distance $\leq \psi(\rho_k) = \gamma_k$ from z_{kj}. Since $[z_{kj}; \gamma_k]$ contains a zero, z is at a distance $\leq 2\gamma_k$ from a zero. ∎

If the sequence $\{\gamma_k\}$ converges to zero, it follows from Theorem 6.11a that ϵ-inclusion sets can be constructed for arbitrarily small ϵ in a large variety of ways. Because the algorithm given above offers no completely satisfactory method for solving problem 2, we forego a discussion of the amount of work and of optimal proximity tests and sequences $\{\gamma_k\}$. This discussion could be carried out much along the lines of §6.10.

III. Principle of the Argument

A completely different approach to the solution of Problems 1 and 2 can be based on the principle of the argument. Let Γ be a positively oriented Jordan curve and let S be the interior of Γ. If f is analytic on the closure of S and $\neq 0$ on Γ, the principle of the argument (see §4.10) states that the number of zeros of f inside Γ (each zero counted with proper multiplicity) is given by the winding number

$$n(f(\Gamma), 0) = \frac{1}{2\pi} [\arg f(z)]_\Gamma, \qquad (6.11\text{-}4)$$

where the expression in brackets denotes the increase on a continuous argument of $f(z)$ as z travels once along Γ. An algorithm for determining the winding number numerically was given in §4.6.

For non-negative integers h consider the grid of points $(l+im)2^{-h}$, where l and m are integers, and denote by Q_h any square (of side length 2^{-h}) spanned by four adjacent points of the grid. If $p \in P_n$, then all zeros of p are contained in the union of the four squares of type Q_0 which have a common corner at the origin. Assuming that p does not vanish on the boundary of any of these squares, we can apply the principle of the argument to each Q_0 and identify those squares that contain zeros. A square containing zeros is called **pregnant** for brevity.

For $k = 0, 1, 2, \ldots$, let S_k be the union of all pregnant squares of type Q_k. Clearly, S_k satisfies the conditions (6.11-1), (6.11-2) for $\epsilon = 2^{1/2-k}$. If $k > 0$, any pregnant square Q_k is contained in a pregnant square Q_{k-1}. The following formula holds:

$$S_k = \bigcup_{\substack{Q_k \in S_{k-1} \\ Q_k \text{ pregnant}}} Q_k. \tag{6.11-5}$$

If no zero of p lies on the boundary of any Q_k ($k = 0, 1, \ldots$), the pregnancy of any Q_k can be tested by applying the principle of the argument to its boundary. The sets S_k then can be constructed recursively by means of (6.11-5): Each square Q_{k-1} contained in S_{k-1} is divided into its four subsquares of type Q_k, and the principle of the argument is applied to each such Q_k. Since S_{k-1} contains at most n squares Q_{k-1}, the principle of the argument has to be applied at most $4n$ times to step from S_{k-1} to S_k. The above process for constructing the sequence of inclusion sets S_k thus converges linearly. At the same time, the principle of the argument furnishes the exact number of zeros in each pregnant Q_k and thus solves Problem 2.

In place of a square pattern of subdivision, one obviously could also use some other pattern, such as a pattern of triangles. Also, instead of subdividing the larger unit into four smaller units, one could also subdivide at each step into a larger number of subunits. Again we do not wish to entertain the considerations of optimality that arise here.

An obvious handicap of the algorithm sketched above is the hypothesis that no zero lies on a potential line of subdivision. Theoretically this could be circumvented by testing the sides of all squares with a Sturm sequence algorithm (see §6.3). It is not certain whether this could always be done in a numerically stable fashion, and in any case the resulting algorithm would be cumbersome.

A further difficulty arises if p has a zero w of multiplicity $m \geq 1$ at a distance ϵ from Γ, where ϵ is small. Unless compensated by corresponding zeros on the other side, $\arg p(z)$ then changes by $m(\pi/2)$ over a segment of length $O(\epsilon)$. Thus to guarantee the applicability of the algorithm of §4.6

arbitrarily short subarcs must be selected. The amount of work required could become arbitrarily large. The algorithm described above thus cannot be uniformly convergent on the space P_n.

As pointed out in §4.6, computing the winding number by the formula

$$n(p(\Gamma),0) = \frac{1}{2\pi i} \int_\Gamma \frac{p'(z)}{p(z)} dz \qquad (6.11\text{-}6)$$

and evaluating the integral by numerical quadrature is sometimes recommended. This procedure is redundant because the real part of the integral is known a priori to have the value zero. Moreover, the difficulty just mentioned is still present: if p has zeros close to Γ, an arbitrarily small integration step may be required to guarantee that the error of the numerical integration is less than π.

To prevent large errors in the evaluation of (6.11-4) or (6.11-6) we could try to detect zeros close to Γ by applying one of the proximity tests of §6.10 at selected points of Γ. Weyl's method, which we discuss next, systematically implements this point of view.

IV. Weyl's Exclusion Algorithm

Suppose that a test X, which, given any polynomial $p \in P_n$, can be applied to any closed square Q in the complex plane, is at our disposal. The square either passes or fails the test. This test must have several properties, among which we mention the following:

(A) If Q passes the test, it contains no zeros of p.

A square that does not pass the test is called **suspect**. It is not required that a suspect square actually contain a zero. Any test with property (A) is called an **exclusion test**.

Given a polynomial $p \in P_n$ and an exclusion test X, we construct a nested sequence of inclusion sets S_k, $k = -1, 0, 1, \ldots$. Using the symbol Q_k, as above, we define S_{-1} as the union of the four squares Q_0 having O as a common corner. We now apply X to each of these Q_0 and let S_0 be the union of those Q_0 that are found to be suspect. More generally, having determined S_{k-1} as a union of suspect squares Q_{k-1}, we define S_k as the union of all those squares Q_k in S_{k-1} that are found to be suspect on application of the test X:

$$S_k = \bigcup_{\substack{Q_k \subset S_{k-1} \\ Q_k \text{ suspect}}} Q_k, \qquad k = 0, 1, 2, \ldots. \qquad (6.11\text{-}7)$$

The complement of each S_k consists of the complement of S_{-1} (which contains no zero of p), and of squares Q_h ($h \leq k$), which at one stage or another have passed the test. By property (A) the complement of S_k is thus free of zeros of p and all zeros are contained in S_k.

Let these zeros be $w_1, w_2, \ldots, w_n$, and let

$$\rho_k^{X,p} := \max_{z \in S_k} \min_{1 \leq i \leq n} |z - w_i|.$$

Because the disks of radius $\rho_k^{X,p}$ about the w_i cover S_k, S_k is an ϵ-inclusion set, where $\epsilon = \rho_k^{X,p}$. The algorithm defined by (6.11-7) produces ϵ-inclusion sets for every $\epsilon > 0$ and for every $p \in P_n$ if the **convergence function** of X defined by

$$\rho_k^X := \sup_{p \in P_n} \rho_k^{X,p}$$

has the property that

(B) $$\lim_{k \to \infty} \rho_k^X = 0.$$

An exclusion test satisfying the above (B) is called **convergent**. The following is an example of a convergent exclusion test. We denote by z_0 the center of the square to be tested and by δ, its semidiagonal. Q is said to pass the test X_1 if and only if

$$|p(z_0)| > \kappa_n \delta,$$

where

$$\kappa_n := n(1 + \sqrt{2})^{n-1}. \qquad (6.11\text{-}8)$$

The significance of the constant κ_n lies in the fact that

$$\kappa_n = \sup_{p \in P_n} \max_{z \in S_{-1}} |p'(z)|,$$

as follows readily from the Gauss-Lucas theorem. If Q passes X_1, then for each $z \in Q$

$$|p(z)| \geq |p(z_0)| - |p(z) - p(z_0)|$$

$$> \kappa_n \delta - \left| \int_{z_0}^{z} p'(t) \, dt \right|$$

$$\geq \kappa_n \delta - \kappa_n \delta = 0;$$

hence $p(z) \neq 0$, which shows that X_1 has the exclusion property. To show

SIMULTANEOUS DETERMINATION OF ALL ZEROS

that X_1 is convergent assume that Q_k is suspect. From

$$|p(z_0)| \leq \kappa_n 2^{-1/2-k}$$

it follows, by Corollary 6.4f, that

$$|z_0 - w_i| \leq (2^{-1/2}\kappa_n)^{1/n} 2^{-k/n}$$

for at least one zero w_i. Hence

$$\rho_k^X \leq (2^{-1/2}\kappa_n)^{1/n} 2^{-k/n} + 2^{-1/2-k},$$

and (B) is clearly satisfied.

Any test that passes Q, provided that the disk $[z_0; \delta]$ contains no zeros, clearly is an exclusion test. Hence proximity tests can be used indirectly as exclusion tests. Let T be a proximity test with convergence functions ϕ and ψ. We assume that T enjoys the properties (a) and (b) stated in §6.10 at all points of the complex plane and not merely in $[0; 1]$, as required in §6.10. (This additional requirement is satisfied by the tests T_2 and T_3 in §6.10 and also by the Schur-Cohn test.) It then follows from (a) that the disk $[z_0; \delta]$ certainly does not contain any zeros of p if p fails the test $T(\rho)$ at z_0, where $\phi(\rho) \geq \delta$. If, on the other hand, p passes $T(\rho)$, then by (b) $|z - w_i| \leq \psi(\rho)$ for at least one zero w_i of p. Hence we obtain:

THEOREM 6.11b

Let T be a proximity test such that the properties (a) *and* (b) *of §6.10 hold at all points of the complex plane and let $\phi^{[-1]}$ exist. Let the test X be defined as follows: the square Q with center z_0 and semidiagonal δ passes X if and only if p fails $T(\phi^{[-1]}(\delta))$ at z_0. Then X is a convergent exclusion test, and*

$$\rho_k^X = \psi\big(\phi^{[-1]}(2^{-1/2-k})\big) + 2^{-1/2-k}, \qquad k = 0, 1, \ldots. \quad (6.11\text{-}9)$$

EXAMPLES

1. Test $T_2(\rho)$ described in §6.10 is a proximity test with convergence functions $\phi(\rho) = \frac{1}{2}\rho$ and $\psi(\rho) = n\rho$. It gives rise to the convergent exclusion test X_2: Q passes X_2 if and only if $\beta(z_0) > 2\delta$, where $\beta(z)$ is defined by (6.10-6). For this test

$$\rho_k^{X_2} = (2n+1) 2^{-1/2-k}.$$

2. Similarly, $T_3(\rho)$ gives rise to the convergent exclusion test X_3: Q passes X_3 if and only if

$$|b_0| > |b_1|\delta + |b_2|\delta^2 + \cdots + |b_n|\delta^n,$$

where $b_k := (1/k!) p^{(k)}(z_0)$. Here (6.11-9) yields

$$\rho_k^{X_3} = (1 - 2^{-1/n})^{-1} 2^{-1/2-k}.$$

3. The proximity test based on the Schur-Cohn algorithm can be reformulated as an exclusion test X_4: Q passes X_4 if and only if the disk $[z_0; \delta]$ is free of zeros of p. For this test

$$\rho_k^{X_4} = 2^{1/2-k}.$$

Given any convergent exclusion test X, the algorithm (6.11-7) furnishes ϵ-inclusion sets S for arbitrarily small ϵ, hence solves Problem 1. The solution to Problem 2, the determination of the actual number of zeros in each component of an inclusion set S_k, remains.

Let us assume here that the zeros of p are contained in the *interior* of S_{-1}. Then, if k is large enough, each component of S_k is surrounded by squares $Q_h (h \leq k)$ that at one stage of the process have passed the test X. Let C be a typical component of S_k. We may assume that C is simply connected by filling up any possible "holes." The boundary Γ of C is a Jordan curve consisting of straight line segments which we assume are positively oriented. The number of zeros of p in C is again given by

$$n(p(\Gamma), 0) = \frac{1}{2\pi} [\arg p(z)]_\Gamma. \qquad (6.11\text{-}10)$$

The winding number could again be determined by the algorithm in §4.6; however, in the present case there is a simpler method.

Let us assume that the exclusion test X has the following additional property:

(C) If the square Q passes X, the set $p(Q)$ lies in an open half-plane bounded by a straight line through O.

A test X with the above property is called **one-sided**. In the tests X_1 and X_3 mentioned above the image of a nonsuspect square is contained in a circle of radius $< |p(z_0)|$ about the point $p(z_0)$. These tests are obviously one-sided. The test X_4, on the other hand, is not.

Now let Γ, the boundary of the component C of S_k, be represented in the form

$$\Gamma = \sum_{j=1}^{m} \sigma_j,$$

where each σ_j is a complete side of some nonsuspect square (Fig. 6.11). We denote the initial and terminal points of σ_j by z_{j-1} and z_j, respectively; $z_0 = z_m$.

SIMULTANEOUS DETERMINATION OF ALL ZEROS

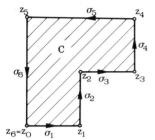

Fig. 6.11. Boundary of component C in inclusion set S_k.

By (4.6-2) we have

$$[\arg p(z)]_\Gamma = \sum_{j=1}^{m} [\arg p(z)]_{\sigma_j}.$$

By (C), since each σ_j belongs to a nonsuspect square, each arc $p(\sigma_j)$ is contained in a half-plane bounded by a straight line through O. Hence the variation of the argument of $p(z)$ along σ_j is less than π, and there follows

$$[\arg p(z)]_{\sigma_j} = \operatorname{Arg}\frac{p(z_j)}{p(z_{j-1})}, \qquad j=1,2,\ldots,m,$$

where Arg denotes the principal value of the argument. It follows that the number of zeros of p in C is given by

$$n(p(\Gamma),0) = \frac{1}{2\pi}\sum_{j=1}^{m} \operatorname{Arg}\frac{p(z_j)}{p(z_{j-1})}. \tag{6.11-11}$$

We summarize the above results on the exclusion algorithm.

THEOREM 6.11c

Let X be a one-sided, convergent exclusion test. Then each set S_k determined by (6.11-7) is a ρ_k^X-inclusion set; $\rho_k^X \to 0$ as $k \to \infty$. The number of zeros in each (simply connected) component C of S_k is given by (6.11-11), where the underlying subdivision of the boundary Γ of C consists of complete sides of the nonsuspect squares bordering on C.

AMOUNT OF WORK

As in §6.10, we compute an upper bound for the number of applications of the test X required to construct an ϵ-inclusion set for a given $\epsilon > 0$. This amounts to constructing the set S_k, where $k = k(\epsilon, X)$ is the smallest integer such that $\rho_k^X \leq \epsilon$. Each set S_{h-1} can be covered by at most n disks of radius ρ_{h-1}^X. Its area cannot exceed $n\pi(\rho_{h-1}^X)^2$ and thus it contains at most $2^{2h}n\pi(\rho_{h-1}^X)^2$ squares Q_h of side length 2^{-h} to which X must be applied in order to construct S_h. For the total number of applications of the test we obtain the rigorous upper bound

$$\omega(X, n, \epsilon) := n\pi \sum_{h=1}^{k(\epsilon, X)} 2^{2h}(\rho_{h-1}^X)^2. \tag{6.11-12}$$

Let $T(\rho)$ be a proximity test with linear convergence functions (6.10-3), and let X be the corresponding exclusion test defined in Theorem 6.11b. Then

$$\rho_h^X = \left(1 + \frac{\beta}{\alpha}\right) 2^{-1/2 - h}$$

and

$$\omega(X, n, \epsilon) = \frac{n\pi}{2}\left(1 + \frac{\beta}{\alpha}\right)^2 k(\epsilon, X). \tag{6.11-13}$$

Since

$$k(\epsilon, X) = \frac{\text{Log}\,\epsilon^{-1}}{\text{Log}\,2} + O(1), \quad \epsilon \to 0,$$

an exclusion algorithm based on such a proximity test is seen to be linearly convergent.

Even for an arbitrary exclusion test X the expression (6.11-12) is independent of the polynomial p. Hence the exclusion algorithm (6.11-7) is uniformly convergent on the space P_n in the following sense: the work required to construct an ϵ-inclusion set is bounded by a quantity that depends on ϵ and n but not on the polynomial $p \in P_n$.

PROBLEMS

1. Instead of dividing a suspect square into $4 = 2^2$ equal subsquares in Weyl's algorithm, we could also divide it into m^2 subsquares, where m is any integer ≥ 2. Show, however (perhaps heuristically), that for a linear convergent test it is most efficient to use the value $m = 2$.

2. For $z = x + iy$, let $\|z\| := \sqrt{2}\,\max(|x|, |y|)$. (This is a multiplicative norm in the Banach algebra of complex numbers in which the "unit sphere" is a square.) Show that the test $X:Q$ is suspect if and only if (in the usual notation)

$$\|b_0\| = \|b_1\|\delta + \|b_2\|\delta^2 + \cdots + \|b_n\|\delta^n$$

is a convergent, one-sided exclusion test.

§6.12. FIXED POINTS OF ANALYTIC FUNCTIONS: ITERATION

Let S be a region in the complex plane and let the function f be analytic in S. Any complex number z satisfying the equation

$$z = f(z) \tag{6.12-1}$$

is called a **fixed point** of f. In this section we are concerned with conditions under which f can be expected to have at least one, or precisely one, fixed point. Also, if f has precisely one fixed point w, we wish to know conditions under which the **iteration sequence** $\{z_n\}$, defined by

$$z_{n+1} = f(z_n), \quad n = 0, 1, 2, \ldots, \tag{6.12-2}$$

where z_0 is an arbitrary point of S, converges to w.

The problem of fixed points is ordinarily studied in real analysis. If f is a real continuous function defined on a closed finite interval I, then it is shown in calculus that the following is a sufficient condition for f to have fixed points: all values of f lie in I. For the existence of a unique fixed point ξ, and for the convergence of the iteration sequence to it, it is sufficient that there be a constant λ, $0 \leq \lambda < 1$, such that for all x_1 and x_2 of I

$$|f(x_1) - f(x_2)| \leq \lambda |x_1 - x_2|. \tag{6.12-3}$$

If f is a continuous mapping of a simply connected, compact set T of the Euclidean plane onto a set T_1 of the same plane, a fixed point always exists if $T_1 \subset T$. The fixed point is unique and the iteration sequence converges if the mapping is a contraction mapping, i.e., if there is a constant $\lambda < 1$ such that for any two points p_1 and p_2 of T

$$\|f(p_1) - f(p_2)\| \leq \lambda \|p_1 - p_2\|, \tag{6.12-4}$$

where $\|\cdot\|$ denotes Euclidean distance.

We now return to our initial problem in which f is an analytic function defined on a domain S of the complex plane. By interpreting f as a mapping of a set of the real Euclidean plane it follows that f certainly has a fixed point if S is simply connected and bounded, if f is defined continuously also on the closure T and S, and if $f(T) \subset T$. The condition (6.12-4), which guarantees uniqueness of the fixed point and convergence of the iteration sequence, written in complex notation, requires that for any two points z_1 and z_2 of T.

$$|f(z_1) - f(z_2)| \leq \lambda |z_1 - z_2|. \tag{6.12-5}$$

These are the results obtained from real variable theory. They hold for any continuous function f; its analyticity has not been used at all. If f is analytic, it is clear that condition (6.12-5) is fulfilled if T is convex and if for the same constant $\lambda < 1$

$$|f'(z)| \leq \lambda, \quad z \in S. \tag{6.12-6}$$

Remarkably enough, however, it turns out that by making a more efficient use of properties of holomorphic functions the uniqueness of the fixed point and the convergence of the iteration sequence can be proved without making any hypotheses such as (6.12-5) or (6.12-6). Indeed, the following result holds:

THEOREM 6.12a

Let f be analytic in a simply connected region S and continuous on the closure S' of S, and let $f(S')$ be a bounded set contained in S. Then f has exactly one fixed point and the sequence defined by (6.12-2) converges to the fixed point for arbitrary $z_0 \in S'$.

Clearly, functions exist that satisfy these hypotheses for which f' is unbounded; e.g., $f(z) = \frac{1}{2}(z+1)^{1/2}$ (principal value) in $|z| < 1$.

Proof of Theorem 6.12a. Instead of requiring that f be continuous on S', we shall prove the assertion under the weaker hypothesis that the closure of $f(S)$ is a bounded set contained in S. We first consider the case in which S is the unit disk, $S = \{z : |z| < 1\}$. Here the hypothesis implies

$$\mu := \sup_{|z|<1} |f(z)| < 1. \tag{6.12-7}$$

The point w is a fixed point of f if and only if it is a zero of the function $z - f(z)$. To prove the existence of a zero we apply Rouché's theorem (Theorem 4.10b) to a circle $|z| = \rho$ where $\mu < \rho < 1$ with z in the role of the "big" function and $f(z)$ in the role of the "small" function. On $|z| = \rho$, $|f(z)| \leq \mu < \rho = |z|$; hence the hypotheses of Rouché's theorem are satisfied. It follows that $z - f(z)$ has exactly as many zeros inside $|z| = \rho$ as z, namely one. Obviously there are no zeros in the annulus $\rho \leq |z| < 1$. Thus f has exactly one fixed point w and $|w| \leq \mu$.

To prove the convergence of the iteration sequence $\{z_n\}$ defined by (6.12-2), in which z_0 is arbitrary, we use the device of conformal transplantation. Let t be a Moebius transformation which maps $|z| \leq 1$ onto itself and sends w into 0; for instance

$$t : z \to \frac{z-w}{1-\bar{w}z}. \tag{6.12-8}$$

FIXED POINTS OF ANALYTIC FUNCTIONS 525

The function $g := t \circ f \circ t^{[-1]}$ has the fixed point 0. It is analytic in $|z| < 1$ and, since t maps closed subsets of $|z| < 1$ onto closed subsets,

$$\kappa := \sup_{|z|<1} |g(z)| < 1. \qquad (6.12\text{-}9)$$

In fact, since for every ρ, $0 \leqslant \rho \leqslant 1$,

$$\max_{|z|=\rho} |t(z)| = \frac{\rho + |w|}{1 + |w|\rho},$$

there actually holds, if μ is defined by (6.12-7),

$$\kappa \leqslant \frac{2\mu}{1+\mu^2}. \qquad (6.12\text{-}10)$$

We may assume that $\kappa > 0$, for otherwise g, and consequently f, is constant, and convergence takes place in one step. The function $\kappa^{-1}g$ vanishes at 0 and is bounded by 1; hence by Schwarz's lemma (Theorem 5.10b) $|\kappa^{-1}g(s)| \leqslant |s|$ and consequently

$$|g(s)| \leqslant \kappa |s| \qquad (6.12\text{-}11)$$

for all s such that $|s| < 1$. Let $s_n := t(z_n)$. Since

$$s_{n+1} = t(z_{n+1}) = t(f(z_n)) = t(f(t^{[-1]}(s_n))) = g(s_n),$$

(6.12-11) shows that $|s_{n+1}| \leqslant \kappa |s_n|$, hence

$$|s_n| \leqslant \kappa^n |s_0|, \quad n = 0, 1, \ldots, \qquad (6.12\text{-}12)$$

implying that $s_n \to 0$ ($n \to \infty$). By the continuity of $t^{[-1]}$

$$z_n = t^{[-1]}(s_n) \to t^{[-1]}(0) = w.$$

Now let S be an arbitrary simply connected region. If S is the complex plane, then f is an entire bounded function. By Liouville's theorem (Theorem 3.3b), f must be constant and the assertions of the theorem are obvious. If S is not the complex plane, let g be a function that maps S conformally and one-to-one onto $D : |z| < 1$. The existence of such a g is assured by the Riemann mapping theorem (Theorem 5.10a). The assertion that w is a fixed point of f is equivalent to the assertion that $u := g(w)$ satisfies $g^{[-1]}(u) = f(g^{[-1]}(u))$, that $u = g(f(g^{[-1]}(u)))$, or that u is a fixed point of $h := g \circ f \circ g^{[-1]}$. The function h is analytic in D and maps D onto a set whose closure is contained in D. Thus h satisfies the hypotheses of the theorem in the special case in which S is the unit disk and has exactly one fixed point. Thus the same is true for the function f.

Let $s_n := g(z_n)$, $n = 0, 1, \ldots$. It is easily seen that $s_{n+1} = h(s_n)$, $n = 0, 1, \ldots$. Since by the above $\{s_n\}$ converges to the unique fixed point of h, the sequence $\{z_n\}$ converges, by the continuity of $g^{[-1]}$, to the unique fixed point of f. ∎

We add some remarks concerning the **speed of convergence**. In the general case, in which S is arbitrary, let m be the smallest positive integer such that $f^{(m)}(w) \neq 0$. (Ordinarily, $m = 1$.) We know that $z_n \to w$. Let $e_n := z_n - w$. By Taylor's expansion we have, using $f(w) = w$,

$$e_{n+1} = f(z_n) - w = f(w + e_n) - f(w) = \frac{1}{m!} f^{(m)}(w) e_n^m + O(e_n^{m+1}),$$

$n \to \infty$. Hence, if the iteration does not terminate in a finite number of steps,

$$\lim_{n \to \infty} \frac{e_{n+1}}{e_n^m} \text{ exists.} \qquad (6.12\text{-}13)$$

Thus in the terminology introduced in §6.9 the algorithm defined by (6.12-2) is of the order m. One also says that f is an **iteration function** of order m.

A more explicit, nonasymptotic statement about the error is possible in the special case in which S is the disk $|z| < 1$ (or any disk). Let t be defined by (6.12-8). The function $g := t \circ f \circ t^{[-1]}$ then satisfies

$$g(0) = g'(0) = \cdots = g^{(m-1)}(0) = 0, \qquad g^{(m)}(0) \neq 0.$$

By a simple generalization of Schwarz's lemma (see Problem 2, §5.10) we have

$$|g(s)| \leq \kappa |s|^m \qquad (6.12\text{-}14)$$

in place of (6.12-11), and consequently, as is easily verified by induction,

$$|s_n| \leq \kappa^{(m^n - 1)/(m - 1)} |s_0|^{m^n},$$

where the first exponent is to be interpreted as n for $m = 1$. Now

$$z_n - w = t^{[-1]}(s_n) - t^{[-1]}(0) = s_n \frac{1 - w\bar{w}}{1 + \bar{w} s_n}.$$

FIXED POINTS OF ANALYTIC FUNCTIONS

Snce $|s_n| < 1$, we have

$$\left|\frac{1-w\bar{w}}{1+\bar{w}s_n}\right| \leq \frac{1-|w|^2}{1-|w|} = 1 + |w| \leq 1 + \mu.$$

Using the estimate (6.12-10) for κ, it follows that for an arbitrary choice of z_0

$$|z_n - w| \leq (1+\mu)\left(\frac{2\mu}{1+\mu^2}\right)^{(m^n-1)/(m-1)}, \qquad n = 0, 1, 2, \ldots.$$

For an arbitrary disk this result reads as follows:

THEOREM 6.12b

Let f be analytic in $S: |z-a| < \sigma$, continuous in $S': |z-a| \leq \sigma$, and let $f(S')$ be contained in $|z-a| \leq \rho$ where $\rho < \sigma$. Let w denote the unique fixed point of f and let $f'(w) = \cdots = f^{(m-1)}(w) = 0$, $f^{(m)}(w) \neq 0$. Then for arbitrary $z_0 \in S'$ the iteration sequence defined by (6.12-2) approaches w in such a manner that

$$|z_n - w| \leq \begin{cases} (\rho + \sigma)\gamma^n, & \text{if } m = 1, \\ (\rho + \sigma)\gamma^{(m^n-1)/(m-1)}, & \text{if } m > 1, \end{cases} \qquad (6.12\text{-}15)$$

where

$$\gamma := \frac{2\rho\sigma}{\rho^2 + \sigma^2} < 1.$$

Both the asymptotic relation (6.12-13) and the above error estimate show that the convergence of the iteration sequence to the fixed point w will be especially rapid if at least the first derivative of the iteration function vanishes at the fixed point.

In theoretical investigations we frequently *assume* the existence of a fixed point w. It is clear by continuity that if $|f'(w)| < 1$ then for sufficiently small values of ρ $|z-w| \leq \rho$ implies $|f(z) - w| \leq \mu(\rho)$, where $\mu(\rho) < \rho$. Thus the iteration sequence $\{z_n\}$ converges to w if $|z_0 - w| < \rho$.

To determine the admissible values of ρ more explicitly let f be analytic in $|z - w| < \rho_0$ and let the integer m be defined as above. For $0 \leq \rho < \rho_0$ let

$$\mu_m(\rho) := \sup_{|z-w| \leq \rho} |f^{(m)}(z)|;$$

μ_m is a continuous, increasing function. By a form of Taylor's theorem

$$f(z)-w = \frac{1}{(m-1)!} \int_w^z (z-t)^{m-1} f^{(m)}(t)\, dt.$$

There follows for $|z-w| \leq \rho < \rho_0$

$$|f(z)-w| \leq \frac{1}{m!} \rho^m \mu_m(\rho),$$

and the hypotheses of Theorem 6.12b are certainly satisfied for every $\rho > 0$ such that

$$\mu_m(\rho) < \frac{m!}{\rho^{m-1}}. \tag{6.12-16}$$

We now apply the above to the problem of determining a *zero* of an analytic function g under the assumption that a "sufficiently close" approximation to the zero w is already known. The methods developed here are applicable to arbitrary equations of the form $g(z)=0$, where g is analytic.

NEWTON'S METHOD

The number w is a zero of g if and only if it is a fixed point of $f(z) := z - kg(z)$, where $k \neq 0$, or more generally of the function

$$f(z) := z - h(z)g(z),$$

where h is an arbitrary nonzero analytic function. We shall try to determine h such that $f'(w)=0$, thus ensuring the quadratic convergence of the iteration sequence if the initial approximation is close enough. We have

$$f'(z) = 1 - h'(z)g(z) - h(z)g'(z)$$

and, since $g(w)=0$,

$$f'(w) = 1 - h(w)g'(w).$$

It follows that $f'(w)=0$ if and only if $h(w)g'(w)=1$. A simple way to satisfy this condition is by putting

$$h(z) := \frac{1}{g'(z)}.$$

FIXED POINTS OF ANALYTIC FUNCTIONS

Thus we obtain the **Newton iteration function**

$$f(z) = z - \frac{g(z)}{g'(z)}. \qquad (6.12\text{-}17)$$

If $g'(w) \neq 0$, Theorem 6.12b shows that the corresponding iteration sequence

$$z_{n+1} := z_n - \frac{g(z_n)}{g'(z_n)}, \qquad n = 0, 1, 2, \ldots,$$

converges to w whenever z_0 is close enough to w and that the order of convergence is (at least) 2.

SCHRÖDER'S ITERATION FUNCTIONS

We now construct iteration functions of arbitrary order for the solution of $g(z) = 0$. Some results from the theory of power series will be required. Let g be analytic in some region S. Then, in the neighborhood of every point $z \in S$, g can be expanded in a Taylor series,

$$g(z+h) = g(z) + \sum_{n=1}^{\infty} b_n(z) h^n,$$

where

$$b_n(z) := \frac{1}{n!} g^{(n)}(z), \qquad n = 1, 2, \ldots. \qquad (6.12\text{-}18)$$

The series

$$P := \sum_{n=1}^{\infty} b_n(z) h^n,$$

considered as a formal power series in h with coefficients depending on the parameter z, at every point z where $b_1(z) \neq 0$ possesses a formal reversion

$$P^{[-1]} = \sum_{n=1}^{\infty} c_n(z) h^n$$

whose coefficients may be computed by the Lagrange-Bürmann formula (see §1.9)

$$c_n = \frac{1}{n} \operatorname{res}(P^{-n}), \qquad n = 1, 2, \ldots. \qquad (6.12\text{-}19)$$

Since the coefficients of P^{-n} are rational expressions in the functions $b_1, b_2, \ldots$, whose denominators are powers of $b_1 = g'$, the same is true of the c_n. They are analytic in every subregion of S where $g'(z) \neq 0$.

For $m = 2, 3, \ldots$, we now define

$$f_m(z) := z + \sum_{k=1}^{m-1} c_k(z)[-g(z)]^k. \tag{6.12-20}$$

THEOREM 6.12c

Let g be analytic in a region S and let $g'(z) \neq 0$ for $z \in S$. Then the functions $f_2, f_3, \ldots$, are analytic in S; moreover, for every $w \in S$ such that $g(w) = 0$,

$$\begin{aligned} f_m(w) &= w, \\ f'_m(w) &= f''_m(w) = \cdots = f_m^{(m-1)}(w) = 0. \end{aligned} \tag{6.12-21}$$

Proof. Since the functions c_k are analytic in S, the f_m are finite sums of analytic functions, hence themselves analytic. To prove the relations (6.12-21) let $g(w) = 0$. Since $g'(z) \neq 0$ in S, the inverse function $g^{[-1]}$ by Theorem 2.4c is defined and analytic in a certain neighborhood of $0 = g(w)$, and

$$\sum_{n=1}^{\infty} c_n(z) u^n = g^{[-1]}(g(z) + u) - g^{[-1]}(g(z))$$

$$= g^{[-1]}(g(z) + u) - z. \tag{6.12-22}$$

The Cauchy coefficient estimate (Theorem 2.2f) now shows that the convergence of this series is uniform with respect to both z and u if $|z - w|$ and $|u|$ are sufficiently small. Substituting $u := -g(z)$ thus yields for $|z - w|$ sufficiently small a uniformly convergent series of analytic functions,

$$s(z) := \sum_{k=1}^{\infty} c_k(z)[-g(z)]^k,$$

the sum of which by (6.12-22) equals $g^{[-1]}(0) - z = w - z$. Since

$$s(z) - \{f_m(z) - z\} = [-g(z)]^m \sum_{k=m}^{\infty} c_k(z)[-g(z)]^{k-m}$$

is an analytic function with a zero of order $\geq m$ at $z = w$, the derivatives at $z = w$ of $f_m(z) - z$ agree with those of $s(z)$ up to and including the order $m - 1$, proving the relations (6.12-21). ∎

FIXED POINTS OF ANALYTIC FUNCTIONS

The iteration functions f_m were introduced by Schröder [1870] and are called the **Schröder iteration functions**. The function f_m constitutes the first m terms of the series

$$f(z) = z - \frac{1}{b_1(z)} g(z) - \frac{b_2(z)}{[b_1(z)]^3} [g(z)]^2$$

$$- \frac{2[b_2(z)]^2 - b_1(z)b_3(z)}{[b_1(z)]^5} [g(z)]^3 - \cdots . \qquad (6.12\text{-}23)$$

We see that f_2 is identical with the Newton iteration function. The evaluation of f_m requires the first $m-1$ derivatives of g at the point z.

LAGUERRE ITERATION

Let g be analytic in some region R, let $w \in R$ be a zero of g, and let $g'(w) \neq 0$. Let ν be a real number, $\nu \neq 0, 1$. There then exists a neighborhood D of w such that

$$\left| \frac{\nu}{\nu-1} \frac{g(z)g''(z)}{[g'(z)]^2} \right| < 1, \qquad z \in D.$$

Consequently the square root

$$r(z) := \left\{ 1 - \frac{\nu}{\nu-1} \frac{g(z)g''(z)}{[g'(z)]^2} \right\}^{1/2}$$

is analytic in D and can be defined by its principal value; furthermore

$$r(z) = 1 - \frac{\nu}{2(\nu-1)} \frac{g(z)g''(z)}{[g'(z)]^2} + O((z-w)^2). \qquad (6.12\text{-}24)$$

For $z \in D$ we define

$$f(z) := z - \frac{g(z)}{g'(z)} \frac{\nu}{1 + (\nu-1)r(z)} \qquad (6.12\text{-}25)$$

and assert the following:

THEOREM 6.12d

For every $\nu \neq 0, 1$ the function f defined by (6.12-25) is an iteration function of order 3 for solving $g(z) = 0$.

Proof. In view of (6.12-24)

$$\frac{\nu}{1+(\nu-1)r(z)} = 1 + \frac{1}{2} \frac{g(z)g''(z)}{[g'(z)]^2} + O((z-w)^2);$$

hence $f(z) = f_3(z) + O((z-w)^3)$, where f_3 is the Schröder iteration function of order 3. It follows that $f(z) - z$ has a zero of order $\geqslant 3$ at w; we omit the verification that 3 (in general) is the exact order of the zero. ∎

The function f is known as the **Laguerre iteration function**. By algebraic manipulation it can be put in the form

$$f(z) = z - \frac{\nu g(z)}{g'(z) + \{(\nu-1)^2 [g'(z)]^2 - \nu(\nu-1)g(z)g''(z)\}^{1/2}},$$

where the argument of the root is to be chosen to differ by less than $\pi/2$ from the argument of $(\nu-1)g'(z)$.

Judging merely from the above analysis, the Laguerre iteration function offers no advantages over Schröder's f_3, which is also somewhat easier to compute. From a practical point of view, however, the presence of the square root is said to have the desirable effect that if g is a real polynomial the iteration automatically branches out into the complex plane if no real roots are found. Moreover, if g is a real polynomial of degree $n \geqslant 2$, the choice $\nu = n$ furnishes remarkable inclusion theorems for the real zeros.

PROBLEMS

1. Show by examples that one or both conclusions of Theorem 6.12a become false if any of the following hypotheses are violated: (a) $f(S)$ bounded; (b) $f(S')$ contained in S.
2. Let c be a complex number, $c \neq 0$. Discuss Newton's method for computing a square root of c,
3. Suppose that f is analytic on a set containing the disk $D := [z_0; \rho]$ and assume that for some λ, $0 < \lambda < 1$,

    (a) $|f(z) - f(z')| \leqslant \lambda |z - z'|$ for all $z, z' \in D$;

    (b) $|z_0 - f(z_0)| < (1 - \lambda)\rho$.

    Show that f has a unique fixed point w in D and that the iteration sequence $\{z_n\}$ converges to w if started at z_0. [Apply Theorem 6.12a.]
4. Let w be a zero of order $k > 1$ of g. Then the iteration functions of Newton and Schröder have an isolated singularity at $z = w$. Show that the singularity is

FIXED POINTS OF ANALYTIC FUNCTIONS 533

removable and that, if continued into the singularity, the functions (6.12-20) satisfy

$$f_m'(w) = 1 - \frac{1}{k},$$

thus still producing convergence, although only at a linear rate.

5. Let w be a zero of arbitrary multiplicity of g. Show that the iteration function

$$f(z) = z - \frac{g(z)g'(z)}{[g'(z)]^2 - g(z)g''(z)}$$

always produces quadratic convergence to w.

6. Let p be a polynomial of degree n with zeros $w_1, \ldots, w_n$. Iteration functions for solving $p(z) = 0$ can be obtained via the power sums

$$s_k(z) := -\sum_{i=1}^{n} (w_i - z)^{-k},$$

calculable from the Taylor coefficients at z via (6.4-11).

(a) Assuming $|w_1 - z| \ll |w_i - z|$, $i = 2, \ldots, n$, show that

$$w_1 - z \sim -\frac{1}{s_1(z)}$$

and obtain the iteration function

$$f(z) = z - \frac{1}{s_1(z)}.$$

Show that this is identical to Newton's method.

(b) Assume that p has one zero w_1 closest to z of unknown multiplicity. Show that

$$w_1 - z \sim \frac{s_1(z)}{s_2(z)}$$

and obtain the iteration function

$$f(z) = z + \frac{s_1(z)}{s_2(z)}.$$

Compare with the iteration function obtained in Problem 5.

(c) Assume that $w_2 = w_3 = \cdots = w_n$, $|w_1 - z| < |w_2 - z|$. Show that

$$w_1 - z \sim \frac{n}{-s_1(z) \pm \{(n-1)[-ns_2(z) - \{s_1(z)\}^2]\}^{1/2}},$$

the sign being chosen to make the absolute value smaller, and obtain the iteration function

$$f(z) = z - \frac{n}{s_1(z) \pm \{(n-1)[-ns_2(z) - \{s_1(z)\}^2]\}^{1/2}}.$$

Show that it is identical to the Laguerre iteration function, where $\nu = n$.

7. Let w be a zero of arbitrary multiplicity of the polynomial p, let m be a positive integer, and let the functions s_k be defined as above. Show that the iteration function $f(z) := z + h$, where h denotes the solution of smaller absolute value of

$$(s_1 s_3 - s_2^2)h^2 + (ms_3 + s_1 s_2)h - s_1^2 - ms_2 = 0,$$

always has order 3 (Maehly [1954]).

8. In the notation of the preceding problem, let $f(z) := z + h$, where h denotes the zero of smaller absolute value of

$$(s_2 s_4 - s_3^2)h^2 - (s_1 s_4 - s_2 s_3)h + s_1 s_3 - s_2^2 = 0.$$

Show that this iteration function always possesses the order 4. (Maehly [1954].)

9. Let p be a polynomial of degree n and let f be the Laguerre iteration function formed with $g = p$ and $\nu = n$. Show that for each complex number z there is a zero w of p such that $|w - z| \leqslant \sqrt{n} \, |f(z) - z|$. (Kahan [1967].)

10. Show that the sequence $\{z_n\}$ generated by

$$z_{n+1} = z_n + \frac{f(z_n)}{f'(z_n)}$$

(Newton correction with wrong sign) for suitable z_0 converges to the *poles* of a meromorphic function f (Rutishauser [1969]).

§6.13. NEWTON'S METHOD FOR POLYNOMIALS

Here again we use the conventions and notations of *circular arithmetic* introduced in §6.6. We recall Theorem 6.6a: if p is a polynomial of degree n with zeros $w_1, w_2, \ldots, w_n$, if $W_2, \ldots, W_n$ are circular regions such that $w_k \in W_k$, $k = 2, \ldots, n$, and if $p'(z_0)/p(z_0) = c_1$, then

$$w_1 \in W_1 := z_0 - \frac{1}{c_1 - V_1}, \tag{6.13-1}$$

where

$$V_1 := \sum_{k=2}^{n} \frac{1}{z_0 - W_k}.$$

There are several ways to develop this result into an algorithm for constructing inclusion disks of arbitrarily small diameter for one, several, or all zeros of a polynomial. We mention two possibilities.

I. Determining a Single Zero

Let it be known that the disk $W^{(0)} := [z_0; \epsilon_0]$ contains precisely one zero, say w_1, of p and that no other zero is contained in the open disk $|z - z_0| < \rho_0$, where $\rho_0 > \epsilon_0$. (This information may have been obtained, for instance, as a by-product of a method of search that used the Schur-Cohn algorithm, or by an application of Rouché's theorem.) We denote by $U := \{z : |z - z_0| \geq \rho_0\}$ the circular region containing the remaining zeros. The result quoted above suggests the following Newton-like algorithm for determining w_1:

ALGORITHM 6.13a

For $m = 0, 1, 2, \ldots$, let

$$z_m := \mathrm{mid}\, W^{(m)},$$

$$V^{(m)} := \frac{n-1}{z_m - U},$$

$$W^{(m+1)} := z_m - \frac{1}{q(z_m) - V^{(m)}},$$

where $q(z) := p'(z)/p(z)$.

By Theorem 6.6a the zero w_1 is contained in all circular regions $W^{(m)}$ so constructed. Concerning convergence of the algorithm, we prove Theorem 6.13b:

THEOREM 6.13b

Let $\rho_0 := (n-1)\eta_0$. If

$$6\epsilon_0 \leq \eta_0, \qquad (6.13\text{-}2)$$

then all circular regions $W^{(m)}$ are disks and their radii $\epsilon_m := \mathrm{rad}\, W^{(m)}$ tend to zero such that

$$\epsilon_{m+1} \leq \frac{3}{2\eta_0} \epsilon_m^2, \qquad m = 0, 1, 2, \ldots. \qquad (6.13\text{-}3)$$

Proof. From

$$q(z_0) = \sum_{k=1}^{n} \frac{1}{z_0 - w_k}$$

there follows

$$|q(z_0)| \geq \epsilon_0^{-1} - \eta_0^{-1}.$$

Since $z_0 - U = \{z : |z| \geq \rho_0\}$,

$$V^{(0)} = \frac{n-1}{z_0 - U} = [0; \eta_0^{-1}].$$

Hence $q(z_0) - V^{(0)}$ is a disk of diameter η_0^{-1} contained in $\{z : |z| \geq \epsilon_0^{-1} - 2\eta_0^{-1}\}$. It follows that $W^{(1)}$ is a disk, and by (6.6-1) its diameter satisfies

$$\epsilon_1 = \mathrm{rad}(q(z_0) - V^{(0)})^{-1} = \frac{\eta_0^{-1}}{(\lambda - \eta_0^{-1})^2 - \eta_0^{-2}},$$

where $\lambda \geq \epsilon_0^{-1}$; hence

$$\epsilon_1 \leq \frac{\epsilon_0^2}{\eta_0 - 2\epsilon_0}. \tag{6.13-4}$$

By virtue of (6.13-2) this proves (6.13-3) for $m = 0$. A repetition of this argument yields

$$\epsilon_2 \leq \frac{\epsilon_1^2}{\eta_1 - 2\epsilon_1}, \tag{6.13-5}$$

where η_1 is such that $\rho_1 := (n-1)\eta_1$ satisfies $|w_k - z_1| \geq \rho_1$ for $k = 2, \ldots, n$. This is also true for $\rho_1 = \rho_0 - \delta_0$, where

$$\delta_0 := |z_1 - z_0|,$$

and with this choice of ρ_1, if $n > 1$,

$$\eta_1 \geq \eta_0 - \delta_0. \tag{6.13-6}$$

To estimate δ_0 we use

$$z_0 - z_1 = \mathrm{mid}(q(z_0) - V^{(0)})^{-1}.$$

Although $\mathrm{mid}(Z^{-1}) = (\mathrm{mid}\, Z)^{-1}$ holds for no disk, $\mathrm{mid}(Z^{-1})$ is always the

NEWTON'S METHOD FOR POLYNOMIALS

inverse of some point of Z; hence

$$\delta_0 \leq \frac{\epsilon_0 \eta_0}{\eta_0 - 2\epsilon_0}. \tag{6.13-7}$$

In order to establish (6.13-3) for $m=1$, it remains, in view of (6.13-5), to be shown that

$$\eta_1 - 2\epsilon_1 \geq \eta_0 - 2\epsilon_0. \tag{6.13-8}$$

According to (6.13-4), (6.13-6), and (6.13-7) we have

$$\eta_1 - 2\epsilon_1 \geq \eta_0 - \epsilon_0 \left(\frac{\eta_0}{\eta_0 - 2\epsilon_0} + \frac{2\epsilon_0}{\eta_0 - 2\epsilon_0} \right),$$

and by (6.13-2) the expression in parentheses does not exceed 2. Together with (6.13-4) this also implies $\eta_1 \geq 6\epsilon_1$. Thus the argument may now be repeated indefinitely to prove (6.13-3) for all m. ∎

II. Simultaneous Determination of all Zeros

Algorithm 6.13a determines one zero at a time and does so with quadratic convergence. We may hope to do better by iterating on all zeros simultaneously. Assume that we have found an array of n nonoverlapping disks $(W_1^{(0)}, W_2^{(0)}, \ldots, W_n^{(0)})$ such that $w_k \in W_k^{(0)}$, $k=1,2,\ldots,n$. (This array may have been obtained by one of the algorithms described in §6.11.) Theorem 6.6a then suggests the following procedure for determining all zeros simultaneously:

ALGORITHM 6.13c

For $m = 0, 1, 2, \ldots$, let for $k = 1, 2, \ldots, n$

$$z_k^{(m)} := \text{mid } W_k^{(m)},$$

$$V_k^{(m)} := \sum_{\substack{j=1 \\ j \neq k}}^{n} \frac{1}{z_k^{(m)} - W_j^{(m)}},$$

$$W_k^{(m+1)} := z_k^{(m)} - \frac{1}{q(z_k^{(m)}) - V_k^{(m)}}.$$

By Theorem 6.6a each zero w_k is contained in all circular regions $W_k^{(m)}$.

Let

$$\epsilon_0 := \max_{1 \leq k \leq n} \operatorname{rad} W_k^{(0)},$$

the maximum radius of the initial array of disks, and

$$\rho_0 := \min_{i \neq j} \{|z - z_i^{(0)}| : z \in W_j^{(0)}\},$$

the minimum distance of the center of any disk from any of the remaining disks. The convergence properties of Algorithm 6.13c are then described by Theorem 6.13d:

THEOREM 6.13d

Let $\rho_0 = (n-1)\eta_0$. If

$$6\epsilon_0 \leq \eta_0, \tag{6.13-2}$$

then all circular regions $W_k^{(m)}$ are disks and

$$\epsilon_m := \max_{1 \leq k \leq n} \operatorname{rad} W_k^{(m)},$$

the maximum radius of the mth array, tends to zero such that

$$\epsilon_{m+1} \leq \frac{3}{\rho_0 \eta_0} \epsilon_m^3, \quad m = 0, 1, 2, \ldots. \tag{6.13-9}$$

Proof. As in the proof of Theorem 6.13b,

$$|q(z_k^{(0)})| \geq \epsilon_0^{-1} - \eta_0^{-1}, \quad k = 1, 2, \ldots, n.$$

Since $z_k^{(0)} - W_j^{(0)}$ for $j \neq k$ is a disk or radius $\leq \epsilon_0$ exterior to the circle $|z| = \rho_0$, $(z_k^{(0)} - W_j^{(0)})^{-1}$ is contained in $[0; \rho_0^{-1}]$; hence

$$V_k^{(0)} \subset [0; \eta_0^{-1}], \quad k = 1, 2, \ldots, n;$$

furthermore

$$\operatorname{rad} V_k^{(0)} \leq (n-1) \frac{\epsilon_0}{(\rho_0 + \epsilon_0)^2 - \epsilon_0^2} \leq \frac{\epsilon_0}{\rho_0 \eta_0}.$$

Thus $q(z_k^{(0)}) - V_k^{(0)}$ is a disk of radius $\leq \epsilon_0 (\rho_0 \eta_0)^{-1}$ exterior to $|z| = \epsilon_0^{-1} -$

NEWTON'S METHOD FOR POLYNOMIALS

$2\eta_0^{-1}$; hence $W_k^{(1)}$ is a disk and

$$\epsilon_1 \leq \frac{\epsilon_0^3}{\rho_0(\eta_0 - 4\epsilon_0)}. \tag{6.13-10}$$

In view of (6.13-2) this establishes (6.13-9) for $m=0$. Defining η_1 by $(n-1)\eta_1 = \rho_1$, where

$$\rho_1 := \min_{i \neq j} \{|z - z_i^{(1)}| : z \in W_j^{(1)}\},$$

we obtain by a repetition of the above argument,

$$\epsilon_2 \leq \frac{\epsilon_1^3}{\rho_1(\eta_1 - 4\epsilon_1)};$$

(6.13-9) will be proved for $m=1$ if it is shown that

$$\rho_1(\eta_1 - 4\epsilon_1) \geq \rho_0(\eta_0 - 4\epsilon_0). \tag{6.13-11}$$

To estimate ρ_1 note that since the disks $W_k^{(0)}$ and $W_k^{(1)}$ overlap

$$\rho_1 \geq \rho_0 - \delta_0 - 2\epsilon_1,$$

where

$$\delta_0 := \max_{1 \leq k \leq n} |z_k^{(1)} - z_k^{(0)}|.$$

As above,

$$\delta_0 = \max_{1 \leq k \leq n} |\text{mid } q(z_k^{(0)} - v_k^{(0)})^{-1}|;$$

hence

$$\delta_0 \leq \frac{\epsilon_0 \eta_0}{\eta_0 - 2\epsilon_0}.$$

By using (6.13-10) and giving away a factor $(n-1)^{-1}$ we obtain

$$\eta_1 \geq \eta_0 - \epsilon_0 \left[\frac{\eta_0}{\eta_0 - 2\epsilon_0} + \frac{2\epsilon_0^2}{\rho_0(\eta_0 - 4\epsilon_0)} \right];$$

hence

$$\eta_1(\eta_1 - 4\epsilon_1) \geq \eta_0^2 - \eta_0 \epsilon_0 \left[\frac{2\eta_0}{\eta_0 - 2\epsilon_0} + \frac{8\epsilon_0^2}{\eta_0(\eta_0 - 4\epsilon_0)} \right],$$

and (6.13-11) is established if the expression in brackets is shown not to exceed 4. This is an easy consequence of (6.13-2), which also implies $6\epsilon_1 \leq \eta_1$. The argument may now be repeated to prove (6.13-9) for all m. ∎

In Algorithm 6.13c the disks of each new array are computed in parallel and the convergence, as we have shown, is cubic. Even more rapid convergence can be achieved by computing the new disks serially and using the new disks $W_k^{(m+1)}$ to compute the auxiliary disks $V_k^{(m)}$ as soon as they are available. This amounts to replacing $V_k^{(m)}$ with

$$V_k^{(m)'} := \sum_{j=1}^{k-1} \left[z_k^{(m)} - W_j^{(m+1)} \right]^{-1} + \sum_{j=k+1}^{n} \left[z_k^{(m)} - W_j^{(m)} \right]^{-1} \quad (6.13\text{-}12)$$

EXAMPLE

For

$$p(z) := z^3 - 2z^2 - z + 2 = (z-2)(z-1)(z+1),$$

using the initial disks

$$W_1^{(0)} := [2.2; 0.3], \qquad W_2^{(0)} := [0.9; 0.2], \qquad W_3^{(0)} := [-0.9; 0.3],$$

we find that one cycle of Algorithm 6.13c [using (6.13-12)] produces the inclusion disks

$$W_1^{(1)} = [2.00110; 0.00604],$$

$$W_2^{(1)} = [0.99955; 0.00099],$$

$$W_3^{(1)} = [-1.000000; 0.00001].$$

With the same starting values, the ordinary Newton method merely yields the approximations (without error bounds) $w_1 \sim 2.03729, w_2 \sim 0.99631, w_3 \sim -1.00954$.

PROBLEMS

1. If the values $q(z_m)$ are not calculated exactly, but (e.g., due to rounding errors) are merely known to lie in disks Q_m, the formula for $W^{(m+1)}$ in Algorithm 6.13a must be replaced by

$$W^{(m+1)} := z_m - \frac{1}{Q_m - V_m}.$$

What is the maximum order of magnitude of $\text{rad}\, Q_m$ in terms of ϵ_m if the quadratic character of the convergence is to be preserved?

2. Give a short proof of the qualitative aspect of Theorem 6.13d by noting that as a direct consequence of the hypotheses

$$\frac{p(z_0)}{p'(z_0)} = O(\epsilon_0), \qquad \text{rad}\, V_k^{(0)} = O(\epsilon_0),$$

which implies $\epsilon_1 = O(\epsilon_0^3)$.

3. Show by a qualitative argument that Algorithm 6.13c still converges quadratically if the iteration is restricted to $l < n$ zeros of p.

§6.14. METHODS OF DESCENT

Let p be a polynomial of degree $n \geq 1$ and let the real function ϕ be defined by

$$\phi(z) := |p(z)|.$$

Let ϕ assume a local minimum at the point z_0. We assert that the minimum has the value 0. Indeed, if this were not so, the nonconstant function

$$\frac{1}{\phi(z)} = \left|\frac{1}{p(z)}\right|$$

would have a local maximum at z_0, an interior point of its domain of definition, which contradicts the principle of the maximum (Theorem 2.4h).

The above simple observation shows that the search for the zeros of p is equivalent to a search for the local minima of ϕ. To conduct this search the idea of descent offers itself. Starting from a point z_0, where $\phi(z_0) > 0$, we look for a point z_1 such that $\phi(z_1) < \phi(z_0)$. Continuing in the same fashion, we construct a sequence $\{z_n\}$ such that $\phi(z_{n+1}) < \phi(z_n)$. We hope that $w := \lim z_n$ exists and that $\phi(w) = 0$. The process can be viewed geometrically by considering the graph of the function ϕ. From a point $(z_n, \phi(z_n))$ we would always descend to a point $(z_{n+1}, \phi(z_{n+1}))$ of lower altitude.

For a given polynomial p the descent process is formalized if, given any point z at which $p(z) \neq 0$, there is a prescription for finding a point z' such that $|p(z')| < |p(z)|$. If $p(z) = 0$, we set $z' = z$. Writing $z' =: f(z)$, we thus require a function f, defined on a nonempty point set T of the complex plane and having the following properties:

$$f(T) \subset T; \tag{6.14-1}$$

$$p(z) \neq 0 \Rightarrow |p(f(z))| < |p(z)|; \tag{6.14-2}$$

$$p(z) = 0 \Rightarrow f(z) = z. \tag{6.14-3}$$

A function f with these properties is called a **descent function** for the polynomial p. A descent function is a special kind of iteration function. Given any $z_0 \in T$, the prescription

$$z_{n+1} := f(z_n), \qquad n = 0, 1, 2, \ldots, \tag{6.14-4}$$

determines a sequence of points $\{z_n\}$ such that the sequence of real numbers $\{|p(z_n)|\}$ is strictly decreasing unless $p(z_n) = 0$ for some $n = n_0$, in which case $z_n = z_{n_0}$ for all $n > n_0$.

Without assuming the descent function to be continuous, we can assert the following:

THEOREM 6.14a

Let f be a descent function for the polynomial p, defined on a closed set T. Then T contains a zero of p.

Proof. Let c be a point of T and let S denote the set $\{z : |p(z)| \leq |p(c)|\}$. Then S is closed by the continuity of p and bounded, for $|p(z)| \to \infty$ for $z \to \infty$ (no descent function can be defined for a constant polynomial). Hence $T \cap S$ is closed and bounded, and by the Weierstrass theorem $|p(z)|$ takes its minimum value at some point $z_0 \in T \cap S$. By the definition of S, $|p(z_0)|$ is also the minimum value taken by $|p(z)|$ on T; hence

$$|p(z)| \geq |p(z_0)| \quad \text{for all } z \in T. \tag{6.14-5}$$

Assume that $|p(z_0)| > 0$. Let $z_1 = f(z_0)$. Then $z_1 \in T$ by (6.14-1), and $|p(z_1)| < |p(z_0)|$ by (6.14-2). This contradicts (6.14-5), and thus we must have $p(z_0) = 0$; T contains a zero of p. ∎

We now examine what can be achieved by continuous descent functions. It turns out that continuity imposes a severe restriction on the domain of definition of f.

THEOREM 6.14b

Let T be a closed set that contains a path joining two distinct zeros of the polynomial p. Then no continuous descent function exists on T for p.

METHODS OF DESCENT

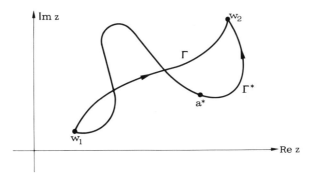

Fig. 6.14. Proof of Theorem 6.14a.

Proof (see Fig. 6.14). Let w_1 and w_2 be two distinct zeros of p. We connect w_1 to w_2 by an arc $\Gamma \in T$ chosen so that

$$\mu := \max_{z \in \Gamma} |p(z)|$$

is as small as possible. (The proof that such a Γ exists is omitted. It would make use of the closedness of T and of the fact that $|p(z)| \to \infty$ for $z \to \infty$. In the mountainous region represented by the graph of $|p(z)|$, Γ traverses the lowest of the available passes.) Let f be a continuous descent function for p on T. Then by (6.14-1) and (6.14-3) the curve $\Gamma^* := f(\Gamma)$ is likewise a path running in T from w_1 to w_2, but in view of the choice of Γ

$$\mu^* := \max_{z \in \Gamma^*} |p(z)| \geq \mu. \tag{6.14-6}$$

Let a^* be a point on Γ^*, where $|p(z)|$ attains its maximum and $|p(a^*)| = \mu^*$. If $a^* = f(a)$, then by (6.14-6) $|p(a)| \leq \mu \leq \mu^* = |p(a^*)|$. Hence f does not possess the descent property at the point a, contradicting (6.14-2). ∎

We conclude from the above that the domain of definition T of a continuous descent function, if closed, contains precisely one zero or is disconnected in the sense that no two zeros can be joined by arc lying in T. In either case we have the following theorem.

THEOREM 6.14c

Let p be a nonconstant polynomial and let f be a continuous descent function for p, defined on a closed set T. Then for any $z_0 \in T$, the sequence $\{z_n\}$ defined by (6.14-4) converges to a zero of p.

Proof. In view of (6.14-3), $|p(z_n)| \leq |p(z_0)|$ for all n; hence all z_n lie in the closed bounded set of all z such that $|p(z)| \leq |p(z_0)|$. Thus the sequence

$\{z_n\}$ has at least one limit point; i.e., there exists w such that for every $\epsilon > 0$, $|z_n - w| < \epsilon$ for an infinite number of n. Let $\mu := \inf |p(z_n)|$. By the continuity of p, $\mu = |p(w)|$. Suppose that $\mu > 0$ and let $w^* := f(w)$. By the descent property $|p(w^*)| = \mu - \eta$, where $\eta > 0$. Since p is continuous, there exists $\epsilon > 0$ such that $|z - w| < \epsilon$ implies $|p(z) - p(w^*)| < \eta$, hence $|p(z)| < \mu$. Because f is also continuous, there exists $\delta > 0$ such that $|z - w| < \delta$ implies $|f(z) - f(w)| < \epsilon$ which is the same as $|f(z) - w| < \epsilon$. Since w is a limit point of $\{z_n\}$, there exists n such that $|z_n - w| < \delta$. It follows that $|f(z_n) - w^*| = |z_{n+1} - w^*| < \epsilon$, hence $|p(z_{n+1})| < \mu$, contradicting the definition of μ. The assumption that $\mu > 0$ thus is ill-founded. It follows that any limit point w of the sequence $\{z_n\}$ is a zero of p.

Let the distinct zeros of p be denoted by $w_1, w_2, \ldots, w_k$. If $k = 1$, the sequence $\{z_n\}$ has w_1 as its only limit point and, because it is bounded, converges to w_1. If $k > 1$, then for any $\epsilon > 0$ there exists $m = m(\epsilon)$ such that for all $n > m$, $|z_n - w_{i(n)}| < \epsilon$ for a suitable zero $w_{i(n)}$. Because f is continuous at the zeros and $f(w_i) = w_i$, it follows that all $i_n = i$ for some fixed i if n is sufficiently large, hence that $z_n \to w_i$. ∎

A continuous descent function can be constructed by using the principle of *steepest descent*. If $p(z) = u(x,y) + iv(x,y)$, the gradient of $\varphi(x,y) := |p(z)|$ is

$$\text{grad}\, \varphi = \left(\frac{uu_x + vv_x}{(u^2 + v^2)^{1/2}}, \frac{uu_y + vv_y}{(u^2 + v^2)^{1/2}} \right).$$

By the Cauchy-Riemann equations

$$\text{grad}\, \varphi = \frac{1}{\varphi} (uu_x + vv_x, -uv_x + vu_x).$$

Since $u_x + iv_x = p'(z)$, the *complex* gradient $\text{grd}\, \varphi := \varphi_x + i\varphi_y$ at the point z is given by

$$\text{grd}\, \varphi(z) = \frac{1}{\varphi} p(z) \overline{p'(z)} = \frac{|p'(z)|^2}{\varphi} \frac{p(z)}{p'(z)}.$$

As is well known, the gradient points in the direction of strongest increase. If $p(z)p'(z) \neq 0$, the direction of strongest *decrease* (i.e., of *steepest descent* on the graph of φ) is given by $-p(z)/p'(z)$. This is the Newton correction at point z, and we recognize that this correction always points in the direction of steepest descent. It cannot be asserted, however, that using the full Newton correction always produces a descent [see the example given in (2) below]. Ideally, if z_n is a point in the iteration sequence, the best

METHODS OF DESCENT

subsequent point in the direction of steepest descent would be one on the straight line

$$z = z_n - \tau \frac{p(z_n)}{p'(z_n)}, \quad -\infty < \tau < \infty,$$

for which $|p(z)|$ is smallest; but to determine the corresponding value of τ itself requires the solution of an algebraic equation. We must therefore be content to construct a real function $\lambda = \lambda(z)$ such that the function

$$f(z) := z - \lambda(z) \frac{p(z)}{p'(z)} \tag{6.14-7}$$

always produces a descent, if not an optimal descent.

THEOREM 6.14d

Let p be a nonconstant polynomial, let $\mu > 0$, and let T be the set of all z such that $|p(z)| \leq \mu$ and $p'(z) \neq 0$. If

$$\delta_2 > \sup_{z \in T} |p''(z)|, \quad \lambda(z) := \min\left(1, \frac{|p'(z)|^2}{|p(z)|\delta_2}\right), \tag{6.14-8}$$

then the function (6.14-7) is a continuous descent function for p on T; in fact,

$$|p(f(z))| \leq \left(1 - \frac{1}{2}\lambda(z)\right)|p(z)|, \quad z \in T. \tag{6.14-9}$$

Proof. It is clear that $f(w) = w$ at all zeros $w \in T$ of p. The function λ is continuous at all z, hence f is likewise continuous. To verify the descent property let $z \in T, p(z) \neq 0$ and let $q := p(z)/p'(z)$. By the gradient property mentioned above

$$|p(z - \tau q)| \leq |p(z)| \tag{6.14-10}$$

for all sufficiently small $\tau > 0$. Let τ^* be the infimum of all $\tau > 0$ such that (6.14-10) is violated. Let $0 < \tau_0 \leq \tau^*$. By a version of Taylor's formula

$$p(z - \tau_0 q) = p(z) - \tau_0 q p'(z) + \int_z^{z - \tau_0 q} (z - \tau_0 q - t) p''(t) \, dt$$

$$= (1 - \tau_0) p(z) + (\tau_0 q)^2 \int_0^1 (1 - \sigma) p''(z - \sigma \tau_0 q) \, d\sigma.$$

By the definition of τ^*, $|p(t)| \leq \mu$ for all t on the path of integration, hence

$|p''(t)|<\delta_2$, and by estimating the integral there follows
$$p(z-\tau_0 q)=h(\tau_0)p(z), \qquad (6.14\text{-}11)$$
where
$$h(\tau_0):=1-\tau_0+c\tau_0^2\frac{q^2\delta_2}{2p(z)}$$

for some complex number c such that $|c|<1$. We wish to show that $\lambda(z)<\tau^*$. Because $|h(\tau^*)|=1$, we conclude from (6.14-11) that either $\tau^*>1$ or
$$\left|c\tau^{*2}\frac{q^2\delta_2}{2p(z)}\right|\geq \tau^*,$$
which, since $|c|<1$, is possible only if
$$\tau^*>\frac{2|p(z)|}{\delta_2|q|^2}=\frac{2|p'(z)|^2}{\delta_2|p(z)|}>\lambda(z).$$

Thus for $\tau_0=\lambda(z)$ a descent is achieved, and (6.14-9) follows because $h(\lambda(z))$ is contained in a disk of radius $\leq \frac{1}{2}\lambda(z)$ with center $1-\lambda(z)$. It remains to verify (6.14-1), i.e., that $p'(f(z))\neq 0$ for all $z\in T$. If τ^* is defined as above, then for all τ_0 such that $0<\tau_0\leq \tau^*$
$$p'(z-\tau_0 q)=p'(z)+\int_z^{z-\tau_0 q}p''(t)\,dt=p'(z)+c\tau_0 q\delta_2$$

for some complex number c such that $|c|<1$. Thus, if $p'(z-\tau_0 q)=0$, then
$$\tau_0>\frac{|p'(z)|}{|q|\delta_2}=\frac{|p'(z)|^2}{|p(z)|\delta_2}\geq \lambda(z),$$

proving that $p'(z-\lambda(z)q)\neq 0$. ∎

If $p'(z)\neq 0$ for all z such that $|p(z)|\leq \mu$, the set T is closed and Theorem 6.14c is applicable. If $p'(z_0)=0$ for some z_0 such that $0<|p(z_0)|\leq \mu$, T is not closed. It may then happen that the iteration sequence generated by f converges to a zero of p' in place of a zero of p. Although this difficulty can be overcome, descent functions based on the idea of steepest descent have gained little favor in practice because their convergence can be slow.

We now consider descent functions that are defined in the whole complex plane, and thus by Theorem 6.14b by necessity have points of discontinuity. Among others, the following proposals for such descent functions are found in the literature.

METHODS OF DESCENT

1. A simple proposal (Ward [1957]) puts
$$f(z) := z + i^k \lambda \qquad (i^2 = -1),$$
where $\lambda > 0$ is a sufficiently small parameter and where the integer k is chosen to achieve maximum descent. However, e.g., for $p(z) := 1 + z^4$ this function does not possess the descent property at $z = 0$, no matter how λ or k are chosen.

2. A much more sophisticated proposal is due to Nickel [1966]; see also Dejon and Nickel [1969]. Let
$$p(z+h) = b_0(z) + b_1(z)h + \cdots + b_n(z)h^n \qquad (6.14\text{-}12)$$
be the Taylor expansion of p at z. Let k be an index such that
$$\left|\frac{b_0(z)}{b_k(z)}\right|^{1/k} = \min_{1 \leq m \leq n} \left|\frac{b_0(z)}{b_m(z)}\right|^{1/m} \qquad (6.14\text{-}13)$$

Nickel defines
$$f(z) := z + c_k(z),$$
where
$$c_k(z) := \left[-\frac{b_0(z)}{b_k(z)} \right]^{1/k},$$
the value of the kth root being chosen to produce maximum descent. If $k = 1$, the correction thus defined is identical with the correction defined by Newton's method. In any case f produces an exact zero of p if the Taylor expansion has the special form
$$p(z+h) = b_0(z) + b_k(z)h^k.$$
Nevertheless, the above descent function does not always possess the descent property, not even if k is allowed to vary instead of being fixed by (6.14-13); for instance, if $z = 0$ and
$$p(h) := 4 - 2h + h^2,$$
then $k = 1$ yields $c_1(0) = 2, p(c_1(0)) = 4$, and $k = 2$ yields $c_2(0) = \pm 2i, p(c_2(0)) = \pm 4i$, and we have $|p(c_k(0))| = |p(0)|$ in both cases. To deal with such situations an "emergency step" is recommended by Nickel, which consists in successively halving the $c_k(z)$ obtained with various values of k. It can be proved that this procedure will ultimately produce a descent such that
$$|p(f(z))| \leq \gamma |p(z)|, \qquad (6.14\text{-}14)$$

where γ is a constant, $0<\gamma<1$, depending only on the degree of p.

3. Another discontinuous descent function, which combines the ideas of descent and of search, is due to Kellenberger [1971]. It is based on the following result suggested by Rutishauser:

THEOREM 6.14e

Let

$$q(h) := 1 + \sum_{j=1}^{n} c_j h^j, \qquad (6.14\text{-}15)$$

where

$$\alpha := \sum_{j=1}^{n} |c_j| < 1, \qquad (6.14\text{-}16)$$

let k and N be positive integers, $N > kn$, and let $\phi := \dfrac{2\pi}{N}$. Then

$$\frac{1}{N} \sum_{m=0}^{N-1} \frac{1}{|q(e^{im\phi})|^2} \geq 1 + \frac{1}{n} \frac{\alpha^2}{(1+\alpha)^2} - \frac{2\alpha^{k+1}}{(1-\alpha^k)(1-\alpha)^2}. \qquad (6.14\text{-}17)$$

We omit the proof of Theorem 6.14e which involves some rather intricate estimates of the coefficients d_j in the expansion

$$\frac{1}{q(h)} = 1 + \sum_{j=1}^{\infty} d_j h^j. \qquad (6.14\text{-}18)$$

To show how the theorem can be used to construct a descent function let the Taylor expansion of p at a point z such that $p(z) \neq 0$ again be given by (6.14-12) and let $\lambda > 0$. Then the polynomial

$$q(h) := \frac{p(z + \lambda h)}{p(z)}$$

has the form (6.14-15), where

$$c_j = \frac{b_j(z)}{b_0(z)} \lambda^j, \qquad j = 1, 2, \ldots, n.$$

If

$$\lambda := \frac{1}{2} \min_{1 \leq j \leq n} \left| \frac{b_0(z)}{b_j(z)} \right|^{1/j},$$

then $|c_j| \leq 2^{-j}$, and (6.14-16) is clearly satisfied. The point of Theorem 6.14e lies in the fact that given any $\alpha \in (0,1)$, k can be chosen so that the term on the right of (6.14-17) is >1. If k is such that

$$\frac{1}{10n} \frac{\alpha^2}{(1+\alpha)^2} > \frac{2\alpha^{k+1}}{(1-\alpha^k)(1-\alpha)^2}, \qquad (6.14\text{-}19)$$

then (6.14-17) implies

$$\max_{0 \leq m < N} \left[|q(e^{im\phi})|^2 \right]^{-1} > 1 + \frac{9}{10n} \frac{\alpha^2}{(1+\alpha)^2}.$$

Hence, if m is an integer for which the maximum is attained, then letting

$$f(z) := z + \lambda e^{im\phi} \qquad (6.14\text{-}20)$$

produces a decrease in the absolute value of p of the form (6.14-14), where

$$\gamma = \gamma(\alpha, n, N) := \left[1 + \frac{9}{10n} \frac{\alpha^2}{(1+\alpha)^2} \right]^{-1/2}.$$

To get an idea of the numbers involved let $n = 100$, $\alpha = 0.75$. Then $k = 41$ is necessary to satisfy (6.14-19), requiring $N = 4101$. The result is $\gamma = 0.999$. The evaluation of q at the many points $e^{im\phi}$ is facilitated by the *fast Fourier transform* of Cooley and Tukey [1965].

In extensive numerical tests run by their respective authors, both Nickel's and Kellenberger's descent methods have been found to be fast and successful, with Kellenberger's method enjoying an edge over Nickel's in the numerical performance for certain pathological polynomials. In regard to convergence, it is clear that any descent method that satisfies (6.14-14) for a fixed value of γ, $0 < \gamma < 1$, produces a sequence $\{z_m\}$ of approximations such that $p(z_m) \to 0$. This in itself does not imply that the sequence $\{z_m\}$ converges, but it is not hard to show that if the descent function is continuous merely at the zeros of p the sequence $\{z_m\}$ produced by it converges. Since Kellenberger's descent function always produces a new point that is closer to the old point than the nearest zero, it is obviously continuous at the zeros, hence defines a convergent method in the sense of §6.9.

PROBLEMS

1. *Proof of Theorem 6.14e.* Let the coefficients d_j be defined by (6.14-18) and let

$$f_k := \sum_{r=0}^{\infty} d_{k+r} \overline{d_r}, \qquad k = 0, 1, 2, \ldots; \qquad f_{-k} := \overline{f_k}.$$

Then
$$[|q(e^{im\phi})|^2]^{-1} = \sum_{k=-\infty}^{\infty} f_k e^{ikm\phi}$$
and consequently
$$\frac{1}{N} \sum_{m=0}^{N-1} [|q(e^{im\phi})|^2]^{-1} = \sum_{k=-\infty}^{\infty} f_{kN}.$$
Show that
$$\sum_{j=1}^{\infty} |d_j| \leq \frac{\alpha}{1-\alpha},$$
$$\sum_{j=1}^{\infty} |d_j|^2 \leq \frac{1}{n} \frac{\alpha^2}{(1+\alpha)^2},$$
$$\sum_{j=N}^{\infty} |d_j| \leq \frac{\alpha^{k+1}}{1-\alpha} \qquad (N > kn);$$
hence
$$f_0 = 1 + \sum_{j=1}^{\infty} |d_j|^2 \geq 1 + \frac{1}{n} \frac{\alpha^2}{(1+\alpha)^2},$$
$$f_N \leq \frac{\alpha^{k+1}}{(1-\alpha)^2} (N > kn),$$
and deduce (6.14-17).

2. Let $q(z) := 1 + \sum_{j=1}^{n} c_j z^j$, $\sum_{j=1}^{n} |c_j| = 1$. Show that
$$\min_{|z|=1} |q(z)| \leq \left[1 + \frac{1}{4n}\right]^{-1/2}.$$

[In Theorem 6.14e let $k \to \infty$ and then $\alpha \to 1$.]

SEMINAR ASSIGNMENTS

1. Study practical regions of convergence for various iteration functions. Is the Laguerre function superior to the third order Schröder function?

2. Write a subroutine package for rigorous circular arithmetic, taking into account rounding errors.

3. Compare the efficiency of optimized polynomial routines based on (a) Weyl's exclusion algorithm and (b) successive deflation, using any of the available methods for determining a single zero. The former is likely to be more efficient if all zeros of high-degree polynomials are desired to low precision, the latter if all zeros of low-degree polynomials are desired to high precision. Using a large statistical sample of polynomials, determine the crossover point experimentally.

4. We consider polynomials $p(z,\epsilon)$ of degree n that depend analytically, or at least linearly, on a parameter ϵ: $p(z,\epsilon)=p_0(z)+p_1(z)\epsilon(+\cdots)$. Develop a routine that *plots* all n zeros in the complex plane as a function of ϵ.

5. Construct a comprehensive program for the complete factorization of a polynomial, using first a Weyl exclusion algorithm for the crude approximation of the zeros and then refining them by the simultaneous Newton algorithm.

NOTES

For comprehensive treatments of the analytic theory of polynomials see Dieudonné [1938], Specht [1958], and Marden [1966]. Householder [1970b] contains a complete bibliography on the solution of nonlinear equations.

§6.1. Knuth [1969], pp. 422–444, deals extensively with the numerical evaluation of polynomials. Belaga [1958], Pan [1959], and Eve [1964] deal with schemes requiring only about $\frac{1}{2}n$ multiplications for the repeated evaluation of the same nth degree polynomial. See also Knuth [1962]. For the stability of Horner's algorithm see Stewart [1971].

§6.4. For Theorem 6.4i see Birkhoff [1914]. This was shown to be a special case of a general principle by Batschelet [1944, 1945, 1946]. Buckholtz [1967] complements Theorem 6.4h as follows: if

$$\sigma := \max_{1 \leq k \leq n} \left| \frac{s_k}{n} \right|^{1/k},$$

then the annulus

$$\frac{\sqrt{2}-1}{2\sigma^{-1}} \leq |z-z_0| \leq \sigma^{-1}$$

always contains a zero of p. (The remarkable fact is that the ratio of the two radii does not depend on n, contrary to the annuli provided by the pairs of Theorems 6.4c and e and 6.4d and i.) Results permitting inference of the proximity of a zero from the smallness of the value of the polynomial were proved by Börsch-Supan [1963, 1970] and improved by Schmidt & Dressel [1967]. See also Nickel [1970]. Tables of the functions s_k are given by David, Kendall, & Barton [1966]. For inclusion radii see Sluis [1970].

§6.6. Circular arithmetic was introduced by Gargantini & Henrici [1972].

§6.7. For a different treatment of the stability problem see Householder [1968, 1970a]. Talbot [1960] deals fully with the situation in which some elements of the Routh scheme are zero: the zeros may be ignored.

§6.8. Stewart [1968] deals with the computational implementation of the Schur-Cohn algorithm.

§6.9. Wilkinson [1963] is an important reference for the practical solution of polynomial equations. In addition to some of the methods dealt with here, he recommends Graeffe's method, which is treated in most textbooks on numerical analysis: for a computational implementation see Bareiss [1960]. For an extension of Graeffe's method to certain transcendental equations see Polya [1968] and Dirschmied [1969]. Ostrowski [1940] in the main does *not* deal with Graeffe's method. Kerner's method [1966, 1967] is based on a constructive proof of the fundamental theorem of algebra due to Weierstrass [1903]. See also the notes on Chapter 7.

§6.10. Search procedures are implicit in the proofs of the fundamental theorem of algebra by Brouwer [1924] (see also Brouwer & de Loer [1924]) and Rosenbloom [1945]. They were proposed for the actual determination of zeros by Lehmer [1961, 1969] and improved by Rasmussen [1964]. Friedli [1973] discusses questions of optimality.

§6.11. The use of the principle of the argument for the determination of zeros is discussed by Spira [1967] and Delves & Lyness [1967]. The method of exclusion is implicit in Weyl's proof of the fundamental theorem of algebra [1924]. Computational implementation by Henrici & Gargantini [1969]: see also Henrici [1968].

§6.12. The origin of Theorem 6.12a is shrouded in mystery; in spite of its simplicity the theorem has been largely ignored by writers on the subject (Ostrowski [1966], Isaacson & Keller [1966], Householder [1970a]). For S multiply connected the existence of a fixed point was proved by Ritt [1920], for mappings defined by functions of several complex variables by Hervé [1963] and Reiffen [1965], and for infinitely many variables by Earle & Hamilton [1970]. See also M. Heins [1940]. Special aspects of iteration are dealt with by Julia [1918] and Siegel [1942]. For Schröder's iteration functions see Schröder [1870]. Derr [1959] deals with Newton's method in the presence of multiple zeros. For Laguerre's iteration function see Maehly [1954], Parlett [1964] and Kahan [1967]. See also Pomentale [1971].

§6.13. See Gargantini & Henrici [1972].

§6.14. Simple methods of descent were proposed by Ward [1957] and Ellenberger [1960]. Ellenberger's method was commented on and improved by Alexander [1961]. Kohfeld [1962], Cohen [1962], and Svejgaard [1967].   method of steepest descent dates back to Cauchy [1847]. For a modern treatment see Goldstein [1967]. Computational implementations for the case of polynomials were given by Bachmann [1963] and Ostrowski [1969]. Nasitta [1964] provided an idea that was used by Nickel [1966] and improved by Dejon & Nickel [1969], Soukup [1969], Bauhuber [1970]. For Kellenberger's algorithm see Kellenberger [1970, 1971]. The fast Fourier transform was made popular by Cooley & Tukey [1965].

7
PARTIAL FRACTIONS

§7.1. CONSTRUCTION OF THE PARTIAL FRACTION REPRESENTATION OF A RATIONAL FUNCTION

Let p and q be polynomials, the degree of p being less than the degree of q, and let $z_1, z_2, \ldots, z_k$ be the distinct zeros of q, z_i having multiplicity m_i. Then at each point z_i the rational function

$$r(z) := \frac{p(z)}{q(z)} \tag{7.1-1}$$

has a pole of order $\leq m_i$. If the principal part of r at the pole z_i is denoted by s_i,

$$s_i(z) = \sum_{j=1}^{m_i} a_{i,j}(z - z_i)^{-j}, \tag{7.1-2}$$

then, since $r(\infty) = 0$, it follows from Theorem 4.4h that

$$r(z) = \sum_{i=1}^{k} s_i(z). \tag{7.1-3}$$

This is called the **partial fraction representation** of r. It holds for all complex z, the poles of r naturally excepted.

Before mentioning some applications of the partial fraction decomposition, we shall discuss several algorithms for its construction. It is assumed that the poles z_i as well as their multiplicities m_i are known.

METHOD A. UNDETERMINED COEFFICIENTS

This is the method recommended in many calculus texts. Normalize p and

q to make the leading coefficient of q equal to 1. Multiply the identity

$$\frac{p(z)}{q(z)} = \sum_{i=1}^{k} \sum_{j=1}^{m_i} a_{i,j}(z-z_i)^{-j}$$

by

$$q(z) = \prod_{i=1}^{k} (z-z_i)^{m_i}.$$

The result is

$$p(z) = \sum_{i=1}^{k} \prod_{\substack{l=1 \\ l \neq i}}^{k} (z-z_l)^{m_l} \sum_{j=1}^{m_i} a_{i,j}(z-z_i)^{m_i-j}, \qquad (7.1\text{-}4)$$

an identity between two polynomials. It can be used in several ways:

1. If the products on the right are multiplied out and the resulting expression arranged in powers of z, the coefficient of any fixed power z^j is a linear expression in the constants $a_{i,j}$. It must equal the coefficient of z^j in p. The result is a system of $m := m_1 + m_2 + \cdots + m_k$ linear equations for the m constants $a_{i,j}$. Since the $a_{i,j}$ are known to be uniquely determined, the system is nonsingular, hence solvable. The solution by Gaussian elimination requires some $\frac{1}{3}m^3$ multiplications and divisions.

2. Two polynomials of degree $<m$ are identical if they agree at m points. Thus we may substitute in (7.1-4) m distinct but otherwise completely arbitrary values of z. Again the result is a system of m linear equations for the m unknowns $a_{i,j}$. The method is particularly simple if all $m_i = 1$. Then $k = m$ and the identity (7.1-4) takes the form

$$p(z) = \sum_{i=1}^{m} a_{i,1} \prod_{\substack{l=1 \\ l \neq i}}^{m} (z-z_l).$$

For $z = z_j$ the terms where $i \neq j$ vanish and we find

$$a_{j,1} = \frac{p(z_j)}{\prod_{\substack{l=1 \\ l \neq j}}^{m} (z_j - z_l)} \qquad (7.1\text{-}5)$$

Evidently this may also be written

$$a_{j,1} = \frac{p(z_j)}{q'(z_j)}. \qquad (7.1\text{-}6)$$

CONSTRUCTION OF THE PARTIAL FRACTION

Mixtures of (1) and (2) are also possible and are recommended if some $m_i > 1$. Additional equations can also be obtained by differentiating the identity (7.1-4) and then giving z a specific value.

METHOD B. DIRECT DETERMINATION OF PRINCIPAL PARTS

The principal part of an analytic function at a finite isolated singularity constitutes the terms that involve negative powers of the Laurent series representation near the singularity. For rational functions the Laurent series at a pole z_i is easily determined. In the framework of the above notation let

$$q(z) =: (z-z_i)^{m_i} q_i(z).$$

Then

$$r(z) = \frac{1}{(z-z_i)^{m_i}} \frac{p(z)}{q_i(z)}.$$

Since $q_i(z_i) \neq 0$, the ratio p/q_i is analytic, hence can be expanded in a series of non-negative powers of $h := z - z_i$,

$$\frac{p(z_i+h)}{q_i(z_i+h)} = \sum_{n=0}^{\infty} c_{i,n} h^n. \qquad (7.1\text{-}7)$$

There follows

$$r(z_i+h) = \sum_{n=0}^{\infty} c_{i,n} h^{n-m_i}$$

and consequently, by the uniqueness of the Laurent series,

$$a_{i,j} = c_{i, m_i - j}, \qquad j = 1, 2, \ldots, m_i. \qquad (7.1\text{-}8)$$

The calculation of the $c_{i,n}$ by differentiation,

$$c_{i,n} = \frac{1}{n!} \frac{d^n}{dz^n} \left[\frac{p(z)}{q_i(z)} \right]_{z=z_i},$$

is frequently recommended. It appears much simpler, however, to calculate them by forming the quotient of the power series for p and q_i—truncated after the $(m_i - 1)$st power and easily calculated by the Horner algorithm (see §6.1)—according to the method given in §1.3. The complete procedure for one principal part requires somewhat less than $m_i(2m + m_i)$ multiplications or divisions and consequently for all principal parts less than $2m^2 + \sum m_i^2$ such operations.

EXAMPLE 1

Calculate the principal part of
$$r(z) := \frac{4z^5 - 2z^4 + 2z^3 - z^2 - 8z - 9}{(3z^3 - 2z^2 + 5z + 1)(z-1)^3}$$
at $z=1$. Since the order of the pole is 3, three rows of the Horner schemes at $z=1$ for $p(z) := 4z^5 - 2z^4 + 2z^3 - z^2 - 8z - 9$ and $q_1(z) := 3z^3 - 2z^2 + 5z + 1$ are required. The required Horner schemes are

4	−2	2	−1	−8	−9
4	2	4	3	−5	−14
4	6	10	13	8	
4	10	20	33		

3	−2	5	1
3	1	6	7
3	4	10	
3	7		

From the identity
$$(7 + 10h + 7h^2 + \cdots)(c_0 + c_1 h + c_2 h^2 + \cdots)$$
$$= -14 + 8h + 33h^2 + \cdots$$
we now find
$$c_0 = -2, \quad c_1 = 4, \quad c_2 = 1.$$

Thus the required singular part is
$$s_1(z) = -\frac{2}{(z-1)^3} + \frac{4}{(z-1)^2} + \frac{1}{z-1}.$$

METHOD C: INCOMPLETE DECOMPOSITION

In some applications r is a real rational function. It may still have complex poles, of course, but they appear in complex conjugate pairs and the respective principal parts have complex conjugate coefficients. Their sum is a rational function whose denominator is some power of the quadratic factor whose zeros are the complex conjugate poles. For some applications the **real partial fraction decomposition** obtained in this manner is of interest (see, however, subsection I in §7.2).

To obtain the real partial fraction decomposition by Method B requires complex arithmetic. Here we shall discuss an algorithm that permits, among other things, the calculation of the real partial fraction decomposition in real arithmetic.

CONSTRUCTION OF THE PARTIAL FRACTION

The general problem solved by the algorithm is as follows: let $r := p/q$ be a rational function (not necessarily real but satisfying $r(\infty) = 0$) and let $q = fg$, where the polynomials f and g have no common zeros. It follows from the partial fraction decomposition that there exist polynomials h and k whose respective degrees are lower than those of f and g such that

$$\frac{p}{fg} = \frac{h}{f} + \frac{k}{g}. \tag{7.1-9}$$

The polynomials h and k are unique, because, if they were not, we could obtain by subtraction

$$0 = \frac{h_1}{k} + \frac{k_1}{g},$$

where h_1 and k_1 are polynomials, not both identically zero, whose respective degrees are less than those of f and g. There follows $0 = h_1 g + k_1 f$. Since f and g have no common zeros, every zero of f must be a zero of the same multiplicity as h_1 and every zero of g must be a zero of the same multiplicity as k_1. This contradicts the statement about the respective degrees of h_1 and k_1. Hence both h_1 and k_1 vanish identically, which proves the uniqueness of h and k.

The decomposition (7.1-9) is called the **incomplete partial fraction decomposition** of $r = p/q$ for the factoring $q = fg$ of the denominator. Our problem now is to construct the polynomials

$$h(z) =: h_0 z^{m-1} + h_1 z^{m-2} + \cdots + h_{m-1},$$

$$k(z) =: k_0 z^{n-1} + k_1 z^{n-2} + \cdots + k_{n-1},$$

if f, g, and p are given in the form

$$f(z) =: z^m + f_1 z^{m-1} + \cdots + f_m,$$

$$g(z) =: g_0 z^n + g_1 z^{n-1} + \cdots + g_n,$$

$$p(z) =: p_0 z^{m+n-1} + p_1 z^{m+n-2} + \cdots + p_{m+n-1}.$$

Clearly the above problem could again be solved by the method of undetermined coefficients, which requires (among other things) the solution of a system of $n + m$ linear equations with $n + m$ unknowns and thus at least $\frac{1}{3}(n+m)^3$ multiplications or divisions. More easily, the problem may be solved as follows:

THEOREM 7.1

For $i=0,1,\ldots,m+n-1$, let a_i and b_i be defined by the recurrence relations

$$\begin{cases} a_i := p_i - f_1 a_{i-1} - \cdots - f_m a_{i-m}, \\ b_i := g_i - f_1 b_{i-1} - \cdots - f_m b_{i-m}, \end{cases} \quad (7.1\text{-}10)$$

where $a_i = b_i = 0$ for $i<0$ and $g_i = 0$ for $i>n$. Then the coefficients $h_0,\ldots,h_{m-1}$ of the polynomial h in the decomposition (7.1-9) form the solution of the nonsingular system of linear equations

$$\begin{cases} b_n h_0 & +b_{n-1}h_1 & +\cdots & +b_{n-m+1}h_{m-1} & =a_n \\ b_{n+1}h_0 & +b_n h_1 & +\cdots & +b_{n-m+2}h_{m-1} & =a_{n+1} \\ \cdots\cdots\cdots\cdots\cdots\cdots\cdots\cdots\cdots\cdots\cdots\cdots\cdots\cdots \\ b_{n+m-1}h_0 & +b_{n+m-2}h_1 & +\cdots & +b_n h_{m-1} & =a_{n+m-1} \end{cases}$$
$$(7.1\text{-}11)$$

The coefficients $k_0, k_1, \ldots, k_{n-1}$ of $k(z)$ satisfy

$$k_i = a_i - b_i h_0 - b_{i-1} h_1 - \cdots - b_{i-m+1} h_{m-1}. \quad (7.1\text{-}12)$$

Proof. With the a_i and b_i defined by (7.1-10), let

$$a(z) := a_0 z^{n-1} + a_1 z^{n-2} + \cdots + a_{n-1} + a_n z^{-1} + \cdots + a_{n+m-1} z^{-m},$$

$$b(z) := b_0 z^{n-m} + b_1 z^{n-m-1} + \cdots + b_{n+m-1} z^{-2m+1}.$$

By virtue of the relations (7.1-10)

$$\begin{aligned} p(z) &= a(z)f(z) + O(z^{-1}), \\ g(z) &= b(z)f(z) + O(z^{-m}), \end{aligned} \quad (7.1\text{-}13)$$

where here and below $O(z^{-k})$ denotes a rational function with a zero of order k at $z=\infty$. From (7.1-9),

$$k(z)f(z) = p(z) - h(z)g(z).$$

By (7.1-13)

$$\begin{aligned} k(z)f(z) &= a(z)f(z) + O(z^{-1}) - [b(z)f(z) + O(z^{-m})]h(z) \\ &= [a(z) - b(z)h(z)]f(z) + O(z^{-1}). \end{aligned}$$

Dividing by $f(z)$ produces in $O(z^{-1})$ a zero of order $m+1$ at infinity:

$$k(z) = a(z) - b(z)h(z) + O(z^{-m-1}).$$

CONSTRUCTION OF THE PARTIAL FRACTION

Considering the Laurent series at $z = \infty$ and comparing the coefficients of $z^{-1}, z^{-2}, \ldots, z^{-m}$ yields the equations (7.1-11). Comparing coefficients of non-negative powers yields (7.1-12).

It remains to prove the nonsingularity of the system (7.1-11). We have just seen that the system possesses at least one solution. Reversing the steps, we see that *any* solution of the system produces a polynomial h such that (7.1-9) holds if k is defined by (7.1-12). Since we know that h is unique, it follows that (7.1-11) has precisely one solution, proving nonsingularity. ∎

AMOUNT OF WORK

The computation of the sequences $\{a_i\}$ and $\{b_i\}$ each requires $(n+m)m$ multiplications and the coefficients k_i are computed in nm multiplications. In addition, there is the work needed for the solution of the system (7.1-11) of m linear equations with m unknowns. If n is large compared with m, which should be considered the normal situation, the algorithm requires only about $3nm$ multiplications.

SPECIAL CASES

(a) *Linear Factors.* If $m=1$, $f(z) := z - z_1$, the recurrence relations (7.1-10) reduce to $a_i = p_i + z_1 a_{i-1}$, $b_i = g_i + z_1 b_{i-1}$ and thus are identical with the Horner algorithm for computing $a_n = p(z_1)$ and $b_n = g(z_1)$. The linear system (7.1-11) reduces to $g(z_1)h_0 = p(z_1)$ in agreement with the usual formula for calculating the residue at a simple pole.

(b) *Quadratic Factors.* If $m=2$, $f(z) := z^2 + f_1 z + f_2$, the recurrence relations (7.1-10) are

$$a_i = p_i - f_1 a_{i-1} - f_2 a_{i-2},$$
$$b_i = g_i - f_1 b_{i-1} - f_2 b_{i-2}, \quad (7.1\text{-}14)$$

$i = 0, 1, \ldots, n+1$, and the system (7.1-11) reads

$$b_n h_0 + b_{n-1} h_1 = a_n,$$
$$b_{n+1} h_0 + b_n h_1 = a_{n+1}. \quad (7.1\text{-}15)$$

These formulas are useful for extracting the partial fraction corresponding to a pair of complex poles of a real rational function, using real arithmetic only. If the quadratic factor has been determined by Bairstow's method (see Henrici [1964], p. 108), the numbers g_i and b_i are already available.

EXAMPLE 2

To determine the polynomials h and k in the decomposition

$$\frac{7z^4-8z^3+12z^2-z+7}{(3z^3-2z^2+5z+1)(z^2+2z+2)} = \frac{h(z)}{z^2+2z+2} + \frac{k(z)}{3z^3-2z^2+5z+1}.$$

The recurrence relations (7.1-10) yield

p_i	a_i	g_i	b_2
7	7	3	3
-8	-22	-2	-8
12	42	5	15
-1	-41	1	-13
7	5	0	-4

The resulting system (7.1-11),

$$-13h_0 + 15h_1 = -41$$
$$-4h_0 - 13h_1 = 5$$

has the solution $h_0 = 2, h_1 = -1$. The formulas (7.1-12) now yield $k_0 = 1$, $k_1 = -3$, $k_2 = 4$. Hence the required polynomials are

$$h(z) = 2z - 1, \qquad k(z) = z^2 - 3z + 4.$$

Method C is equally useful for determining the principal parts belonging to multiple poles. In contrast to method B, it carries the option of determining the coefficients k_i and continuing the decomposition with a rational function of lower degree. By way of illustration we consider the problem of Example 1:

EXAMPLE 3

Determine h and k in

$$\frac{4z^5-2z^4+2z^3-z^2-8z-9}{(3z^3-2z^2+5z+1)(z-1)^3} = \frac{h(z)}{(z-1)^3} + \frac{k(z)}{3z^3-2z^2+5z+1}.$$

From (7.1-10), with $f_1 = -3, f_2 = 3, f_3 = -1$, we get

p_i	a_i	g_i	b_i
4	4	3	3
-2	10	-2	7
2	20	5	17
-1	33	1	34
-8	41	0	58
-9	35	0	89

CONSTRUCTION OF THE PARTIAL FRACTION

which yields the system
$$34h_0 + 17h_1 + 7h_2 = 33$$
$$58h_0 + 34h_1 + 17h_2 = 41$$
$$89h_0 + 58h_1 + 34h_2 = 35$$

with the solution $h_0 = 1$, $h_1 = 2$, $h_2 = -5$. From (7.1-12) we get $k_0 = 1$, $k_1 = -3$, $k_2 = 4$. Thus the given rational function equals
$$\frac{z^2 + 2z - 5}{(z-1)^3} + \frac{z^2 - 3z + 4}{3z^3 - 2z^2 + 5z + 1}.$$

To decompose further the principal part at $z = 1$, we expand $h(z)$ in powers of $z - 1$ by the complete Horner algorithm, to obtain $z^2 + 2z - 5 = -2 + 4(z-1) + (z-1)^2$. Hence
$$\frac{z^2 + 2z - 5}{(z-1)^3} = -\frac{2}{(z-1)^3} + \frac{4}{(z-1)^2} + \frac{1}{z-1}.$$

For this computation the coefficients k_i are not needed.

Multiple *quadratic* factors can be dealt with in the same way.

If a rational function with denominator degree n is to be decomposed into j partial fractions with denominator degrees $m_1, \ldots, m_j$ ($\sum m_i = n$), Method C can be used in two ways:

1. Start j times from the full rational function, always splitting off a different factor. The polynomials $k(z)$ are then not needed. Since n in the above formula is to be replaced by $n - m_i$, this process requires $2\sum nm_i = 2n^2$ multiplications, plus the work required for solving j linear systems of the respective orders $m_1, \ldots, m_j$.

2. Reduce successively the denominator degree of the rational function being dealt with by continuing with the function k/g after splitting off the partial fraction h/f. This requires the computation of k. At the ith stage n in the formula is to be replaced by $n - m_1 - m_2 - \cdots - m_i = m_{i+1} + \cdots + m_j$. The number of multiplications required by the process is now

$$2\sum_{i=1}^{j-1}(m_i + m_{i+1} + \cdots + m_j)m_i + \sum_{i=1}^{j-1}(m_{i+1} + \cdots + m_j)m_i$$

$$= 3\sum_{k>i} m_k m_i + 2\sum_{i=1}^{j} m_i^2$$

$$= \frac{3}{2}\left(\sum_{i=1}^{j} m_i\right)^2 + \frac{1}{2}\sum_{i=1}^{j} m_i^2$$

$$= \frac{3}{2}n^2 + \frac{1}{2}\sum_{i=1}^{j} m_i^2,$$

plus the work for solving the linear systems as before. This is somewhat less than what we found under (1), but there may be more propagation of rounding error. In any case, the work is much less than that required for the method of undetermined coefficients (Method A). Method B also requires about $2n^2$ multiplications but is applicable only if the denominator factors are powers of linear factors.

PROBLEMS

1. Verify by means of method C

$$\frac{10}{z^{10}-1} = \frac{1}{z-1} - \frac{1}{z+1} - \frac{z^3+2z^2+3z+4}{z^4+z^3+z^2+z+1} + \frac{z^3-2z^2+3z-4}{z^4-z^3+z^2-z+1}.$$

2. Another algorithm for the incomplete decomposition in partial fractions is as follows: using the notation in (7.1-9), we want the numerator of

$$\frac{p-hg}{fg} = \frac{p}{fg} - \frac{h}{f}$$

to be divisible by f. Hence, if

$$p = q_1 f + r_1, \qquad g = q_2 f + r_2,$$

where the degrees of r_1 and r_2 are less than that of f, we want $r_1 - hr_2 = q_3 f$. This is an identity between two polynomials of degree $\leqslant 2m-2$, where h and q_3 are unknown. Show that the determination of the coefficients of h requires solving a system of order $2m-1$. Do Examples 2 and 3 by this method.

§7.2. PARTIAL FRACTIONS: MISCELLANEOUS APPLICATIONS

Here the following rather straightforward applications of partial fractions are considered: I. Integration of rational functions. II. Summation of certain infinite series. III. Taylor expansions of rational functions. IV. Interpolation by sums of exponentials. We mention at this point that the partial fraction decomposition is also an important tool in electric circuit theory, particularly in the computation of inverse Laplace transforms (see §10.4).

I. Integration of Rational Functions

In calculus partial fractions are introduced for the purpose of finding indefinite integrals of rational functions. We do not plan to dwell unduly on this well-worn subject, but it is perhaps not superfluous to point out that the point of view of complex analysis offers some advantages even in the integration of *real* rational functions.

PARTIAL FRACTIONS

The partial fraction decomposition of a real rational function consists of the principal parts at the real poles which have the form

$$\sum_{k=1}^{m} \frac{\alpha_k}{(z-\xi)^k}$$

(α_k, ξ real) and of the principal parts at the nonreal poles which occur in conjugate pairs, each pair having the form

$$\sum_{k=1}^{m} \left[\frac{c_k}{(z-z_0)^k} + \frac{\bar{c}_k}{(z-\bar{z}_0)^k} \right]$$

(c_k, z_0 complex, $\operatorname{Im} z_0 \neq 0$).

As in calculus, we restrict z to real values by writing $z=x$. There is nothing to be said about the integration of the real partial fractions (except, perhaps, that the integral of $(x-\xi)^{-1}$ is $\operatorname{Log}|x-\xi|$ and not $\operatorname{Log}(x-\xi)$, as given in some older texts). Regarding the complex fractions, we first consider the case in which $k=1$. Evidently

$$\int \left(\frac{c_1}{x-z_0} + \frac{\bar{c}_1}{x-\bar{z}_0} \right) dx = 2 \operatorname{Re} \int \frac{c_1}{x-z_0} dx$$

$$= 2 \operatorname{Re} c_1 \operatorname{Log}(x-z_0),$$

omitting a constant of integration. By definition

$$\operatorname{Log}(x-z_0) = \operatorname{Log}|x-z_0| + i \operatorname{Arg}(x-z_0),$$

where Arg denotes the principal value of the argument. Let $z_0 = \xi + i\eta$, where $\eta > 0$. Then

$$\operatorname{Log}|x-z_0| = \operatorname{Log}\left[(x-\xi)^2 + \eta^2 \right]^{1/2},$$

and

$$\operatorname{Arg}(x-z_0) = \operatorname{Arg}[i(x-z_0)] - \frac{\pi}{2} = \operatorname{Arctan} \frac{x-\xi}{\eta} - \frac{\pi}{2}.$$

Hence, if $c_1 = \alpha + i\beta$, omitting a constant that may be absorbed into the constant of integration,

$$\int \left(\frac{c_1}{x-z_0} + \frac{\bar{c}_1}{x-\bar{z}_0} \right) dx$$

$$= 2\alpha \operatorname{Log}\left[(x-\xi)^2 + \eta^2 \right]^{1/2} - 2\beta \operatorname{Arctan} \frac{x-\xi}{\eta}. \quad (7.2\text{-}1)$$

The compactness of this formula compares favorably with the cumbersome treatment of simple complex conjugate poles by real methods.

The integrals of the terms where $k>1$ are simply

$$\int \left[\frac{c_k}{(x-z_0)^k} + \frac{\bar{c}_k}{(x-\bar{z}_0)^k} \right] dx = \frac{2}{1-k} \operatorname{Re} \frac{c_k}{(x-z_0)^{k-1}}. \quad (7.2\text{-}2)$$

Again this should be compared with the real variable treatment where the student is taught to assume the partial fraction decomposition in the form

$$\sum_{k=1}^{m} \frac{\alpha_k x + \beta_k}{(x^2 + \gamma x + \delta)^k}$$

and complicated recurrence relations are derived for the integrals

$$I_k := \int \frac{x}{(x^2+\gamma x+\delta)^k} dx, \quad J_k := \int \frac{1}{(x^2+\gamma x+\delta)^k} dx.$$

II. Summation of Certain Infinite Series

Let r be a rational function. Once again we consider (compare §4.9) the problem of summing the infinite series

$$s := \sum_{n=1}^{\infty} r(n), \quad (7.2\text{-}3)$$

$$t := \sum_{n=1}^{\infty} (-1)^{n-1} r(n) \quad (7.2\text{-}4)$$

in finite terms. It is not assumed that $r(-z) = r(z)$; hence the method of residues employed in §4.9 is not applicable. We shall see, however, that in many cases the series s and t can be summed explicitly via the partial fraction decomposition of r.

For the series s and t to be defined the function r may not have any poles at the positive integers. For the series s to be convergent r must have a zero of order $\geqslant 2$ at infinity; for the convergence of the series t a zero of order 1 is sufficient. Although the method could be extended, we consider only the case in which all poles of r are simple. If the poles are denoted by z_k, $k = 0, 1, \ldots, m$, the partial fraction expansion of r has the form

$$r(z) = \sum_{k=0}^{m} \frac{c_k}{z - z_k}; \quad (7.2\text{-}5)$$

PARTIAL FRACTIONS

r has a zero of order ≥ 2 at infinity if and only if

$$c_0 + c_1 + \cdots + c_m = 0. \tag{7.2-6}$$

The method consists in finding an integral representation for the power series

$$s(\xi) := \sum_{n=1}^{\infty} r(n)\xi^n, \qquad t(\xi) := \sum_{n=1}^{\infty} (-1)^{n-1} r(n)\xi^n.$$

Since the series converge for $\xi = 1$, we have according to Abel's theorem (Theorem 2.2e)

$$s = \lim_{\xi \to 1-} s(\xi), \qquad t = \lim_{\xi \to 1-} t(\xi).$$

We shall carry out the computations for the series $s(\xi)$; for the series $t(\xi)$ they are analogous. By (7.2-5), if $|\xi| < 1$,

$$s(\xi) = \sum_{n=1}^{\infty} \left(\sum_{k=0}^{m} \frac{c_k}{n - z_k} \right) \xi^n = \sum_{k=0}^{m} c_k \sum_{n=1}^{\infty} \frac{\xi^n}{n - z_k}.$$

We now make the additional hypothesis that

$$\operatorname{Re} z_k < 1, \qquad k = 0, 1, \ldots, m. \tag{7.2-7}$$

This can always be realized by renormalizing the variable and omitting a finite number of terms from the series s and t. We then have for $0 < \xi < 1$

$$\frac{\xi^n}{n - z_k} = \xi^{z_k} \int_0^{\xi} \tau^{n - z_k - 1} d\tau,$$

where $\xi^{z_k} := e^{z_k \operatorname{Log} \xi}$; hence

$$\sum_{n=1}^{\infty} \frac{\xi^n}{n - z_k} = \xi^{z_k} \int_0^{\xi} \frac{\tau^{-z_k}}{1 - \tau} d\tau$$

and

$$s(\xi) = \int_0^{\xi} \frac{1}{1 - \tau} \sum_{k=0}^{m} c_k \xi^{z_k} \tau^{-z_k} d\tau. \tag{7.2-8}$$

In an analogous manner we find

$$t(\xi) = \int_0^{\xi} \frac{1}{1 + \tau} \sum_{k=0}^{m} c_k \xi^{z_k} \tau^{-z_k} d\tau. \tag{7.2-9}$$

In the above integrals we now let $\xi \to 1-$. (In view of (7.2-6), this is permissible in (7.2-8) in spite of the apparent singularity at $\xi=1$.) This yields:

THEOREM 7.2a

Let r be a rational function whose partial fraction expansion is (7.2-5) where the poles z_k satisfy (7.2-7). Then

$$\sum_{n=1}^{\infty} (-1)^{n-1} r(n) = \int_0^1 \frac{1}{1+\tau} \sum_{k=0}^{m} c_k \tau^{-z_k} d\tau, \qquad (7.2\text{-}10)$$

and if (7.2-6) holds

$$\sum_{n=1}^{\infty} r(n) = \int_0^1 \frac{1}{1-\tau} \sum_{k=0}^{m} c_k \tau^{-z_k} d\tau. \qquad (7.2\text{-}11)$$

EXAMPLE **1**

Let

$$r(z) = \frac{1}{z(z+1)} = \frac{1}{z} - \frac{1}{z+1}.$$

Theorem 7.2a yields

$$\frac{1}{1 \cdot 2} - \frac{1}{2 \cdot 3} + \frac{1}{3 \cdot 4} - \cdots = \int_0^1 \frac{1-\tau}{1+\tau} d\tau = 2\log 2 - 1,$$

$$\frac{1}{1 \cdot 2} + \frac{1}{2 \cdot 3} + \frac{1}{3 \cdot 4} + \cdots = \int_0^1 \frac{1-\tau}{1-\tau} d\tau = 1.$$

If the numbers z_k are rational,

$$z_k = \frac{p_k}{q}, \qquad k = 0, 1, \ldots, m,$$

where the p_k and q are integers, the substitution $\tau = \sigma^q$ transforms the integrals (7.2-10) and (7.2-11) into the integrals

$$q \int_0^1 (1 \pm \sigma^q)^{-1} \sum_{k=0}^{m} c_k \sigma^{q-p_k-1} d\sigma,$$

which can be evaluated by expanding in partial fractions. Thus in this case the series s and t can always be evaluated in finite terms.

PARTIAL FRACTIONS

The evaluation is elementary if the poles of r are given by $z_k = z_0 - k$, $k = 0, 1, \ldots, m$. By the Horner algorithm (Theorem 6.1a) we have

$$(\tau - 1)^{-1} \sum_{k=0}^{m} c_k \tau^k = c_m \tau^{m-1} + (c_m + c_{m-1}) \tau^{m-2} + \cdots$$
$$+ (c_m + c_{m-1} + \cdots + c_1),$$

which in view of (7.2-6) may also be written

$$(\tau - 1)^{-1} \sum_{k=0}^{m} c_k \tau^k = -[c_0 + (c_0 + c_1)\tau + \cdots$$
$$+ (c_0 + c_1 + \cdots + c_{m-1}) \tau^{m-1}].$$

Substituting in (7.2-11) and carrying out the integration yields

$$\sum_{n=1}^{\infty} r(n) = \frac{c_0}{1 - z_0} + \frac{c_0 + c_1}{2 - z_0} + \cdots + \frac{c_0 + c_1 + \cdots + c_{m-1}}{m - z_0}. \quad (7.2\text{-}12)$$

Similarly from

$$(\tau + 1)^{-1} \sum_{k=0}^{m} c_k \tau^k = c_m \tau^{m-1} + (c_{m-1} - c_m) \tau^{m-2} + \cdots$$
$$+ c_1 - c_2 + \cdots + (-1)^{m-1} c_m + \frac{c_0 + c_1 + \cdots + (-1)^m c_m}{\tau + 1}$$

we get

$$\sum_{n=1}^{\infty} (-1)^{n-1} r(n) = (c_0 - c_1 + \cdots + (-1)^m c_m) \int_0^1 \frac{\tau^{-z_0}}{1 + \tau} d\tau$$
$$+ \frac{c_1 - c_2 + \cdots + (-1)^{m-1} c_m}{1 - z_0} + \frac{c_2 - c_3 + \cdots + (-1)^{m-2} c_m}{2 - z_0} + \cdots + \frac{c_m}{m - z_0}.$$

$$(7.2\text{-}13)$$

EXAMPLE 2

The series

$$s := \frac{8}{1 \cdot 3 \cdot 5} + \frac{16}{3 \cdot 5 \cdot 7} + \frac{24}{5 \cdot 7 \cdot 9} + \cdots$$

is of the form (7.2-3), where

$$r(z) := \frac{z}{(z-\frac{1}{2})(z+\frac{1}{2})(z+\frac{3}{2})}.$$

Here $z_k = z_0 - k$, where $z_0 = \frac{1}{2}$. In view of

$$r(z) = \frac{\frac{1}{4}}{z-\frac{1}{2}} + \frac{\frac{1}{2}}{z+\frac{1}{2}} - \frac{\frac{3}{4}}{z+\frac{3}{2}}$$

(7.2-12) yields

$$s = \frac{\frac{1}{4}}{1-\frac{1}{2}} + \frac{\frac{3}{4}}{2-\frac{1}{2}} = 1.$$

It is interesting to note that Equation (7.2-12) may also be derived in the following completely elementary manner: the pth partial sum of s,

$$s_p := \sum_{n=1}^{p} r(n) = \sum_{n=1}^{p} \left(\sum_{k=0}^{m} \frac{c_k}{n-z_0-k} \right),$$

by rearranging and introducing $n+k$ as a new index of summation may be written

$$s_p = \sum_{k=0}^{m} c_k \left(\sum_{j=1+k}^{1+m} \frac{1}{j-z_0} + \sum_{j=2+m}^{p-1} \frac{1}{j-z_0} \right) + v_p,$$

where

$$v_p := \sum_{k=0}^{m} c_k \sum_{j=p}^{p+k} \frac{1}{j-z_0}.$$

In view of (7.2-6) the above simplifies to

$$s_p = \sum_{k=0}^{m} c_k \sum_{j=1+k}^{1+m} \frac{1}{j-z_0} + v_p,$$

and, since $v_p \to 0$ as $p \to \infty$, we find

$$s = \sum_{k=0}^{m} c_k \sum_{j=k}^{m} \frac{1}{j+1-z_0}.$$

Rearranging the order of summation yields (7.2-12).

EXAMPLE 3

Once again we consider

$$s := \frac{1}{1 \cdot 2} + \frac{1}{2 \cdot 3} + \frac{1}{3 \cdot 4} + \cdots .$$

Here

$$s_p = (1 - \tfrac{1}{2}) + (\tfrac{1}{2} - \tfrac{1}{3}) + \cdots + \left(\frac{1}{p} - \frac{1}{p+1}\right)$$

$$= 1 - (\tfrac{1}{2} - \tfrac{1}{2}) - (\tfrac{1}{3} - \tfrac{1}{3}) - \cdots - \frac{1}{p+1}$$

$$= 1 - \frac{1}{p+1},$$

making evident that $s_p \to 1$ as $p \to \infty$.

III. Taylor Series Representing Rational Functions

For the remainder of this chapter the most important application of partial fractions is the derivation of a formula for the coefficients of the Taylor series that represents a rational function near a point that is not a pole. Let r be a rational function. The principal part at infinity, being a polynomial, can easily be expanded in powers of $z - z_0$ for any z_0 by the Horner algorithm. Thus we may assume that r is zero at infinity. If the distinct finite poles of r are $z_1, \ldots, z_n$ and the order of z_i is m_i, r can then be represented in the form

$$r(z) = s_1(z) + \cdots + s_n(z),$$

where s_i, the principal part of r at z_i, is given by

$$s_i(z) := \sum_{h=1}^{m_i} \frac{a_{i,h}}{(z - z_i)^h}, \qquad i = 1, \ldots, n. \qquad (7.2\text{-}14)$$

Now let $z_0 \neq z_i$, $i, \ldots, n$. We wish to calculate the Taylor series of r at z_0. By the binomial series, if $|z - z_0| < |z_i - z_0|$,

$$\frac{1}{(z - z_i)^h} = (-1)^h [z_i - z_0 - (z - z_0)]^{-h}$$

$$= (-1)^h (z_i - z_0)^{-h} \left(1 - \frac{z - z_0}{z_i - z_0}\right)^{-h}$$

$$= (-1)^h \sum_{k=0}^{\infty} \frac{(h)_k}{k!} \frac{(z - z_0)^k}{(z_i - z_0)^{h+k}};$$

hence

$$s_i(z) = \sum_{k=0}^{\infty} p_i(k)\left(\frac{z-z_0}{z_i-z_0}\right)^k,$$

where

$$p_i(k) = \sum_{h=1}^{m_i} (-1)^h a_{i,h}(z_i-z_0)^{-h}\frac{(h)_k}{k!}.$$

In view of

$$\frac{(h)_k}{k!} = \frac{(k+1)_{h-1}}{(h-1)!}$$

we also have the representation

$$p_i(k) = \sum_{h=1}^{m_i} (-1)^h a_{i,h}\frac{(z_i-z_0)^{-h}}{(h-1)!}(k+1)_{h-1}, \qquad (7.2\text{-}15)$$

making evident that p_i is a polynomial of degree m_i-1 in k.

Summing the contributions of the different s_i, we thus find that the Taylor coefficients a_k of r at z_0 are

$$a_k = \sum_{i=1}^{n} p_i(k)(z_i-z_0)^{-k}, \qquad k=0,1,2,\ldots, \qquad (7.2\text{-}16)$$

where the polynomials p_i are given by (7.2-15). Thus we have proved the first half of the following theorem:

THEOREM 7.2b

Let r be a rational function with distinct poles $z_1, z_2, \ldots, z_n$ and principal parts (7.2-14). The Taylor series of r at a point $z_0 \neq z_k$, $k=1,\ldots,n$, has the form

$$r(z) = \sum_{k=0}^{\infty} a_k(z-z_0)^k,$$

where the a_k are given by (7.2-16), with p_i defined by (7.2-15). Conversely, let $z_0, z_1, \ldots, z_n$ be distinct complex numbers, let $p_1, \ldots, p_n$ be nonzero polynomials of the degrees $m_1-1, \ldots, m_n-1$, and let the coefficients a_k be defined by (7.2-16). Then the power series

$$s(z) = \sum_{k=0}^{\infty} a_k(z-z_0)^k$$

PARTIAL FRACTIONS

has radius of convergence $\rho := \min_{1 \leq i \leq n} |z_i - z_0| > 0$ and represents a rational function with poles of order m_i at the points z_i, $i = 1, 2, \ldots, n$.

Proof of the Second Half of the Theorem. The statement concerning the radius of convergence is a direct consequence of the Cauchy-Hadamard formula (Theorem 2.2a). An induction argument shows that any polynomial p_i of degree $m_i - 1$ can be represented in the form

$$p_i(x) = \sum_{h=1}^{m_i} b_{i,h}(x+1)_{h-1},$$

where $b_{i,m_i} \neq 0$. If we now define $a_{i,h}$ such that

$$(-1)^h a_{i,h} = (h-1)!(z_i - z_0)^h b_{i,h},$$

the above computation shows that $s(z)$ coincides with the Taylor series at z_0 of a rational function with poles at $z_1, \ldots, z_n$ whose respective principal parts are given by (7.2-14). ∎

The connection between the coefficients of the power series representing a certain analytic function and the location of the singularities of that function is discussed intensively in the following sections.

IV. Interpolation by Sums of Exponentials

Let f be a complex-valued function, defined, say, on an interval I, and let $\tau_1, \ldots, \tau_n$ be n distinct points of I. It is a well-known fact of numerical analysis that there exists a unique polynomial p of degree $\leq n-1$, the *interpolating polynomial*, such that $p(\tau_k) = f(\tau_k)$, $k = 1, 2, \ldots, n$. A number of algorithms are known for the explicit construction of this polynomial.

Interpolating polynomials are much exploited for computational purposes, such as numerical integration or differentiation or the numerical integration of differential equations. This use is founded on the belief that the polynomial not only interpolates but also approximates. Although this is usually true in the small, there obviously are functions whose properties in the large are not reproduced well by polynomials. It suffices to call to mind a function f that tends to a limit as $\tau \to \infty$ or has an approximately periodic behavior. In such cases it might be better to approximate f by a sum of exponentials, i.e., by a function of the form

$$g(\tau) = \sum_{m=1}^{n} c_m e^{s_m \tau}. \tag{7.2-17}$$

Here we shall examine the possibility of choosing the $2n$ parameters c_m and s_m $(m=1,2,\ldots,n)$ such that g interpolates f at $2n$ distinct points, i.e., that

$$g(\tau_k) = f(\tau_k) \tag{7.2-18}$$

for $2n$ distinct values τ_k.

Unlike the corresponding problem for polynomial interpolation, this problem is nonlinear and difficulties must be expected both with regard to the existence and the construction of the interpolating function g. Indeed, even if $n=1$ there exists no g of the form (7.2-17) such that

$$g(\tau_0) = 0, \qquad g(\tau_1) = 1, \tag{7.2-19}$$

because the first condition forces $c_1 = 0$ which automatically makes $g(\tau_1) = 0$.

We restrict ourselves to the case in which the τ_k are equidistant, $\tau_k = kh$, $k = 0, 1, \ldots, 2n-1$, $h > 0$, and shall derive a set of sufficient conditions for the existence of an interpolating function g of the type (7.2-17).

To begin with, assume that g exists. Writing $f(\tau_k) =: f_k$ and

$$e^{s_m h} =: u_m, \qquad m = 1, 2, \ldots, n, \tag{7.2-20}$$

the relations (7.2-18) take the form

$$\sum_{m=1}^{n} c_m u_m^k = f_k, \qquad k = 0, 1, \ldots, 2n-1. \tag{7.2-21}$$

Without loss of generality we may assume that the numbers u_m are all different, for otherwise some c_m may be put equal to zero. Since all $u_m \neq 0$, (7.2-21) by virtue of Theorem 7.2b means that the numbers $f_0, f_1, \ldots, f_{2n-1}$ are the first $2n$ Taylor coefficients at $z = 0$ of the rational function

$$r(z) := \sum_{m=1}^{n} \frac{c_m}{1 - u_m z}. \tag{7.2-22}$$

Since r vanishes at infinity and has at most n finite poles, none zero, r must have the form

$$r(z) = \frac{b_0 + b_1 z + \cdots + b_{n-1} z^{n-1}}{1 + a_1 z + a_2 z^2 + \cdots + a_n z^n}, \tag{7.2-23}$$

and the fact that $r(z) = \sum f_k z^k$ means that

$$(1 + a_1 z + \cdots + a_n z^n)(f_0 + f_1 z + f_2 z^2 + \cdots)$$
$$= b_0 + b_1 z + \cdots + b_{n-1} z^{n-1}$$

PARTIAL FRACTIONS

or that

$$f_k + f_{k-1}a_1 + f_{k-2}a_2 + \cdots + f_{k-n}a_n = 0, \quad k = n, n+1, \ldots, 2n-1 \quad (7.2\text{-}24)$$

and

$$f_k + f_{k-1}a_1 + \cdots + f_0 a_k = b_k, \quad k = 0, 1, \ldots, n-1. \quad (7.2\text{-}25)$$

Equations (7.2-24) represent a system of n nonhomogeneous equations for the n unknowns $a_1, a_2, \ldots, a_n$, and we are now able to recognize some conditions that guarantee the existence of an interpolating function g of the form (7.2-17). If the system (7.2-24) has a solution $a_1, \ldots, a_n$ and the b_k are determined from (7.2-25), the rational function r defined by (7.2-23) has $f_0, f_1, \ldots, f_{2n-1}$ as its first $2n$ Taylor coefficients at $z = 0$. If r vanishes at ∞ and the poles of r are simple, then r can be written as in (7.2-22) and its Taylor coefficients will be given by (7.2-21). Thus, if all these conditions are met, the numbers u_m are the reciprocals of the poles of the rational function (7.2-23) defined by (7.2-24) and (7.2-25) and the c_m can be found from the residues $-c_m u_m^{-1}$ at these poles.

To guarantee the existence of a solution of the system (7.2-24) we may assume that its determinant is $\neq 0$, i.e., that

$$H_n^{(0)} := \begin{vmatrix} f_0 & f_1 & \cdots & f_{n-1} \\ f_1 & f_2 & & f_n \\ \vdots & & & \vdots \\ f_{n-1} & f_n & & f_{2n-2} \end{vmatrix} \neq 0. \quad (7.2\text{-}26a)$$

To guarantee that r vanishes at ∞ we may postulate that $a_n \neq 0$, which by Cramer's rule is the case if

$$H_n^{(1)} := \begin{vmatrix} f_1 & f_2 & \cdots & f_n \\ f_2 & f_3 & \cdots & f_{n+1} \\ \vdots & & & \vdots \\ f_n & f_{n+1} & \cdots & f_{2n-1} \end{vmatrix} \neq 0. \quad (7.2\text{-}26b)$$

Finally, it is not necessary to postulate that the poles of r are simple. If r has the *distinct poles* u_j^{-1} ($j = 1, 2, \ldots, m, m \leq n$) of the respective orders $m_j, m_1 + \cdots + m_m = n$, then by Theorem 7.2b the Taylor coefficients at $z = 0$ will have the form

$$\sum_{j=1}^{m} c_j(k) u_j^k,$$

where c_j is a polynomial of degree $m_j - 1$. Thus it will still be possible to interpolate the data f if we permit the sum in (7.2-21) to involve only $m < n$ distinct exponentials which are multiplied by polynomials in τ whose degrees add up to $n - m$. We note that these polynomials are uniquely determined by the principal parts of r, hence by r itself.

Altogether we have obtained:

THEOREM 7.2c

Let $h > 0$, let $n > 0$ be an integer, and let $f_0, f_1, \ldots, f_{2n-1}$ be complex numbers such that both determinants (7.2-26) are different from zero. Then there exist an integer m, $1 \leqslant m \leqslant n$, distinct complex numbers $u_j = e^{s_j h}$ ($j = 1, 2, \ldots, m$), and polynomials $p_j(\tau)$ of the respective degrees $m_j - 1$ ($j = 1, 2, \ldots, m$), where $m_1 + \cdots + m_m = n$, all uniquely determined, such that the function

$$g(\tau) := \sum_{j=1}^{m} p_j(\tau) e^{s_j \tau} \qquad (7.2\text{-}27)$$

satisfies

$$g(hk) = f_k, \qquad k = 0, 1, \ldots, 2n - 1.$$

It is seen that the data implied by (7.2-19) do not satisfy the conditions (7.2-26). On the other hand, the conditions imposed by the theorem are not necessary for the existence of an interpolating function (7.2-27), for the data $f_0 = f_1 = \cdots = f_{2n-1} = 0$ are obviously interpolated by the function $g(\tau) = 0$.

The determinants (7.2-26) are special cases of Hankel determinants, which will be studied in great detail in §7.5 et seq.

The actual construction of the interpolating function g involves the following steps:

1. A test of the conditions (7.2-26).
2. The solution of the system (7.2-24) for the a_k and the calculation of the b_k from (7.2-25).
3. The determination of the poles and principal parts of the function r defined by (7.2-23).
4. The determination of the Taylor coefficients at the origin of each principal part, using (7.2-15).

If the poles of r are simple, which can normally be expected, the determination of the principal parts just amounts to a calculation of residues, and Step (4) is trivial. We shall see in §7.7 how in the case of simple poles Steps (1) through (3) can be carried out elegantly by means of the quotient-difference algorithm.

PARTIAL FRACTIONS

EXAMPLE 4

Interpolate the data $\{f_k\} = \{2, 2, 1, 1\}$ by a sum of two exponentials. The determinants (7.2-26) are

$$H_2^{(0)} = \begin{vmatrix} 2 & 2 \\ 2 & 1 \end{vmatrix}, \quad H_2^{(1)} = \begin{vmatrix} 2 & 1 \\ 1 & 1 \end{vmatrix}$$

and are seen to be $\neq 0$. Equations (7.2-24) read

$$1 + 2a_1 + 2a_2 = 0,$$
$$1 + 1a_1 + 2a_2 = 0;$$

the solution is $a_1 = 0$, $a_2 = -\frac{1}{2}$. Equations (7.2-25) now yield

$$b_0 = f_0 = 2, \quad b_1 = f_1 + f_0 a_1 = 2.$$

The rational function (7.2-23) thus is

$$r(z) = \frac{2 + 2z}{1 - \frac{1}{2}z^2}.$$

Its poles are $z_{1,2} = \pm \sqrt{2}$. The partial fraction decomposition by the standard method is found to be

$$r(z) = -\frac{\sqrt{2} + 2}{z - \sqrt{2}} - \frac{2 - \sqrt{2}}{z + \sqrt{2}}.$$

Thus the quantities $c_m = -r_m u_m$ are (r_m = residue at z_m)

$$c_1 = 1 + \sqrt{2}, \quad c_2 = 1 - \sqrt{2}.$$

This yields the interpolating function

$$f_k = (\sqrt{2} + 1)\left(\frac{1}{\sqrt{2}}\right)^k - (\sqrt{2} - 1)\left(-\frac{1}{\sqrt{2}}\right)^k.$$

It is seen that the sequence $\{f_k\}$, continuing the trend set by the data, tends to 0 for $k \to \infty$.

PROBLEMS

1. It is well known that

$$1 - \frac{1}{3} + \frac{1}{5} - \frac{1}{7} + \cdots = \frac{\pi}{4}.$$

Show that

$$\frac{1}{1\cdot 3\cdot 5} - \frac{1}{7\cdot 9\cdot 11} + \frac{1}{13\cdot 15\cdot 17} - \cdots = \frac{\pi}{48}.$$

2. Show that

$$1 + \frac{1}{3} - \frac{1}{5} - \frac{1}{7} + \frac{1}{9} + \frac{1}{11} - \cdots = \frac{\pi}{4}\sqrt{2}.$$

3. For $m = 1, 2, \ldots$, let

$$s_m := 2^{-m} \sum_{k=0}^{\infty} \frac{(-1)^{km}}{(\frac{1}{2} + km)_m}.$$

Show that

$$s_m = \frac{\pi}{4m!} \sum_{|k| < \frac{m}{2}} (-1)^k \left(\cos \frac{k\pi}{m} \right)^{m-1}.$$

4. Express in rational form the sum of the series

$$\sum_{n=1}^{\infty} n^2 z^n, \quad |z| < 1.$$

5. Determine the sum of the series

$$\sum_{k=1}^{\infty} k \cos(k\phi + \epsilon) z^k,$$

where $|z| < 1, 0 < \phi < \pi$ and ϵ is real.

6. Let $f(z) := \operatorname{Arctan} z$. Prove that $f^{(8)}(1) = 0$. Which other derivatives of f vanish at $z = 1$?

7. Let p be a polynomial of degree $m \geq 0$. Prove that the formal power series $\sum_{k=0}^{\infty} p(k) z^k$ is the Taylor series at $z = 0$ of a rational function with a single pole of order $m + 1$ at $z = 1$.

8. A rational function r is analytic in $|z| < 1$ and has at least one pole on the circle $|z| = 1$. Prove that the coefficients of the Taylor series of r at $z = 0$ form a bounded sequence if and only if all poles of r on $|z| = 1$ have order 1.

9. Prove the following special case of a theorem due to Hadamard: let r and s be two rational functions with poles $z_1, \ldots, z_k$ and $u_1, \ldots, u_m$, respectively, and for $|z|$ sufficiently small let

$$r(z) = \sum_{n=0}^{\infty} a_n z^n, \quad s(z) = \sum_{n=0}^{\infty} b_n z^n.$$

PARTIAL FRACTIONS

Then the formal power series

$$\sum_{n=0}^{\infty} a_n b_n z^n$$

is the Taylor series of a rational function whose poles are to be found among the numbers $z_i u_j$, $1 \leq i \leq k$, $1 \leq j \leq m$.

10. Find the sum of three exponentials interpolating the data $\{f_k\} = \{11, 12, 8, 6, 6, 7\}$. [Solution: $f_k = 8 + 4[(1+i/2)^k + 4[(1-i)/2]^k]$.]
11. The data $\{f_k\} = \{3, 3, 2, 2, 1, 1\}$ are interpolated by the generalized exponential sum

$$f_k = \frac{13}{4} - \frac{1}{2}n - \frac{1}{4}(-1)^n.$$

What is the explanation for the linear term in n?

12. The method of interpolation by sums of exponentials is especially suited to periodic sequences. Let $\{f_k\}$ be an infinite sequence of period $2n$ ($f_{k+2n} = f_k$ for all $k \geq 0$), and let $p(z) := f_0 + f_1 z + \cdots + f_{2n-1} z^{2n-1}$. Then the generating function of the sequence $\{f_k\}$ is

$$f(z) := \sum_{k=0}^{\infty} f_k z^k = (1 + z^{2n} + z^{4n} + \cdots)p(z) = \frac{p(z)}{1-z^{2n}}.$$

The poles are $z_k = w^k$, $k = 0, 1, \ldots, 2n-1$, where $w := \exp(i\pi/n)$, and the residues are easily calculated by the standard method. Thus show that f_m for all $m \geq 0$ is represented by a finite Fourier series,

$$f_m = \frac{1}{2n} \sum_{k=1}^{2n} p(w^{-k}) w^{km}.$$

13. Show that the periodic sequence $\{f_k\} = \{2, 2, 1, 1, 2, 2, 1, 1, \ldots\}$ is interpolated by the exponential sum

$$f_m = \frac{3}{2} + \frac{1}{2}\left(\cos\frac{m\pi}{2} + \sin\frac{m\pi}{2}\right).$$

14. For n an integer, $n \geq 2$, let the sequence $\{f_k\}$ be defined by

$$f_0 = f_n = 0, \quad f_1 = \cdots = f_{n-1} = 1, \quad f_{n+1} = \cdots = f_{2n-1} = -1,$$

periodically repeated. Show that for all integers m

$$f_m = \frac{1}{n} \sum_{\substack{k=1,3,\ldots}}^{2n-1} \cot\frac{k\pi}{2n} \cdot \sin\frac{mk\pi}{n}.$$

Letting $n \to \infty$, $m \to \infty$ such that $m/n =: \phi$ is fixed, obtain the well-kown Fourier series for the function of period 2, $f(\phi) = (-1)^{[\phi]}$.

§7.3. SOME APPLICATIONS TO COMBINATORIAL ANALYSIS

The branch of mathematics that goes under the name of combinatorial analysis is concerned, among other things, with counting the number of ways some given objects can be arranged according to certain rules. Questions of this kind are important in such diverse fields as the theory of probability and operations research. Complex analysis—and in particular the theory of power series—often proves to be an efficient tool for solving combinatorial problems. In this section, without attempting to give a systematic introduction to the subject, we discuss some typical applications.

We begin with the simplest possible problem. In how many ways can p indistinguishable objects be put into n drawers, if each drawer is to contain at most one object? Let the number of possible ways be called c_p. It is clear that $c_p > 0$ only if $0 \leq p \leq n$ ($c_0 := 1$) and that $c_p = 0$ for $p > n$.

For each possible arrangement, let numbers ϵ_m ($m = 1, 2, \ldots, n$) be defined as follows:

$$\epsilon_m := \begin{cases} 1, & \text{if the } m\text{th drawer contains an object,} \\ 0, & \text{if the } m\text{th drawer does not contain an object.} \end{cases}$$

To each possible arrangement there corresponds a vector $(\epsilon_1, \epsilon_2, \ldots, \epsilon_n)$ whose components are either 0 or 1 and add up to p. The total number of arrangements, c_p, clearly equals the total number of these vectors.

The decisive step toward solving problems of this kind now consists in introducing the formal power series

$$F := c_0 + c_1 x + c_2 x^2 + \cdots.$$

If this series happens to have a positive radius of convergence and thus defines an analytic function f, f is called the **generating function** of the sequence $\{c_p\}$. In the present case F terminates and the generating function is a polynomial.

By the above c_p is equal to the number of factorings

$$x^p = \prod_{i=1}^{n} x^{\epsilon_i},$$

where the exponents ϵ_i are 0 or 1 and $\sum \epsilon_i = p$. These are precisely the factorings that occur when the product

$$\underbrace{(1+x)(1+x) \cdots (1+x)}_{n \text{ factors}} = (1+x)^n$$

APPLICATIONS TO COMBINATORIAL ANALYSIS

is formed by multiplying out the factors on the left. The generating function thus equals

$$f(x) = (1+x)^n = \binom{n}{0} + \binom{n}{1}x + \cdots + \binom{n}{n}x^n,$$

and it follows that

$$c_p = \binom{n}{p}.$$

Most problems that follow can be solved by applying this principle. For instance, let us change the conditions of the drawer problem. Each drawer may not contain more than two objects. The number c_p of arrangements of p objects under this new condition evidently is equal to the number of vectors $(\epsilon_1, \epsilon_2, \ldots, \epsilon_n)$ in which each ϵ_i has one of the values $0, 1, 2$, and $\sum \epsilon_i = p$. It follows that the coefficients of the generating function for the new problem,

$$f(x) := \sum_{p=0}^{\infty} c_p x^p,$$

equal the number of factorings

$$x^p = \prod_{i=1}^{n} x^{\epsilon_i},$$

where $\epsilon_i = 0, 1,$ or 2 and $\sum \epsilon_i = p$. These are the factorings that occur when forming

$$\underbrace{(1+x+x^2)(1+x+x^2)\cdots(1+x+x^2)}_{n \text{ factors}} = (1+x+x^2)^n.$$

Thus

$$f(x) = (1+x+x^2)^n.$$

The required coefficients c_p can easily be found by the J. C. P. Miller formula (Theorem 1.6c).

An explicit solution is possible if we admit that any drawer may contain an arbitrary number of objects. The generating function then is

$$f(x) = (1+x+x^2+\cdots)^n = \left(\frac{1}{1-x}\right)^n,$$

and we find

$$c_p = \frac{(n)_p}{p!}.$$

An apparently different class of problems is represented by the following. A cigarette machine accepts nickels (n), dimes (d), and quarters (q). In how many ways can we pay for a pack of cigarettes worth 45 cents? (If coins of the same denomination are inserted in a different order, this will be considered a different form of payment.) If the number of coins is fixed, this will be a drawer problem of the type already considered; for instance, if we require payment with three coins, the generating function for c_p, the number of ways of paying $5p$ cents, will be

$$f(x) = (x + x^2 + x^5)^3,$$

and we would find that $c_9 = 3$ (the three possibilities are qdd, dqd, ddq). If the number of coins is arbitrary, however, we obtain the correct generating function by adding up the number of possibilities if $0, 1, 2, 3, \ldots$ coins are used. This yields

$$f(x) = 1 + (x + x^2 + x^5) + (x + x^2 + x^5)^2 + \cdots,$$

the composition of the geometric series with $x + x^2 + x^5$. Thus

$$f(x) = \frac{1}{1 - x - x^2 - x^5}.$$

The identity $(1 - x - x^2 - x^5)\sum c_p x^p = 1$ yields the recurrence relation $c_0 = 1$,

$$c_p = c_{p-1} + c_{p-2} + c_{p-5}, \qquad p = 1, 2, \ldots$$

($c_p = 0$ for $p < 0$), which easily permits the construction of the following table:

p	0	1	2	3	4	5	6	7	8	9
c_p	1	1	2	3	5	9	15	26	44	75

Thus there are $c_9 = 75$ ways to pay 45 cents.

Up to now we have considered drawer problems in which the same conditions have been imposed on all drawers. Let us now impose different conditions on different drawers. If in the mth drawer we permit α_{m1} or α_{m2} or $\alpha_{m3} \ldots$ objects, the generating function for the number c_p of different arrangements of p objects is

$$f(x) = (x^{\alpha_{11}} + x^{\alpha_{12}} + x^{\alpha_{13}} + \cdots)(x^{\alpha_{21}} + x^{\alpha_{22}} + \cdots)$$
$$\cdots (x^{\alpha_{n1}} + x^{\alpha_{n2}} + \cdots). \qquad (7.3\text{-}1)$$

APPLICATIONS TO COMBINATORIAL ANALYSIS

Whether or not this expression simplifies depends on our conditions. An interesting case is obtained for

$$\alpha_{m1}=0, \quad \alpha_{m2}=2^{m-1}, \quad \alpha_{mk}=0, \quad k>2.$$

Here the mth drawer must contain either no or precisely 2^{m-1} objects ($m=1,2,\ldots,n$). The generating function then becomes

$$f(x)=(1+x)(1+x^2)(1+x^4)\cdots(1+x^{2^{n-1}}).$$

By induction we easily verify that

$$f(x)=1+x+x^2+x^3+\cdots+x^{2^n-1}.$$

This means that under the given rules p objects can be distributed in n drawers in exactly one way if $p<2^n$ and in no way if $p\geqslant 2^n$. This result is equivalent to the well-known fact that each non negative integer $<2^n$ has a unique representation as a binary number with at most n digits.

Choosing

$$\alpha_{m1}=0, \quad \alpha_{m2}=m, \quad \alpha_{mk}=0, \quad k>2,$$

we obtain the function

$$f(x)=(1+x)(1+x^2)(1+x^3)\cdots(1+x^n)$$

generating the number of ways a positive integer can be represented as a sum of distinct integers $\leqslant n$. If the size of the summands is not restricted, the appropriate generating function is

$$f(x)=(1+x)(1+x^2)(1+x^3)\cdots.$$

This function cannot be simplified in terms of other, more familiar functions. An interesting result is obtained, however, by observing the identity

$$(1+x)(1+x^2)(1+x^3)\cdots=[(1-x)(1-x^3)(1-x^5)\cdots]^{-1}$$

which is established in Chapter 8 (Theorem 8.2a). The function

$$g(x):=[(1-x)(1-x^3)(1-x^5)\cdots]^{-1}$$
$$=(1+x+x^2+\cdots)(1+x^3+x^6+\cdots)(1+x^5+x^{10}+\cdots)$$

is the generating function for the number of representations of a given integer as a sum of equal or distinct odd integers. Thus we have obtained a result due to Euler: *the number of representations of a positive integer n as a sum of distinct positive integers equals the number of representations of n as a sum of equal or distinct odd positive integers.*

EXAMPLE 1

The number $n:=6$ can be represented as a sum of distinct integers as follows:

$$6, \quad 1+5, \quad 2+4, \quad 1+2+3.$$

Thus there are four representations. This is also the number of representations of a sum of odd integers (which may be repeated):

$$1+5, \quad 3+3, \quad 1+1+1+3, \quad 1+1+1+1+1+1.$$

Diophantine Equations. In the special case of (7.3-1) in which, for given positive integers ω_m,

$$\alpha_{mk} = (k-1)\omega_m, \qquad k=1,2,\ldots,$$

each factor is a geometric series and we find

$$f(x) = (1 + x^{\omega_1} + x^{2\omega_1} + \cdots)(1 + x^{\omega_2} + x^{2\omega_2} + \cdots)$$
$$\cdots (1 + x^{\omega_n} + x^{2\omega_n} + \cdots)$$
$$= \left[(1-x^{\omega_1})(1-x^{\omega_2})\cdots(1-x^{\omega_n})\right]^{-1}.$$

This is a rational function whose Taylor expansion at $x=0$ can be found by Theorem 7.2b. From the combinatorial point of view the coefficient c_p in this expansion is equal to the number of solutions of the equation

$$k_1\omega_1 + k_2\omega_2 + \cdots + k_n\omega_n = p$$

in non-negative integers $k_1, k_2, \ldots, k_n$. Such equations are called **diophantine equations** and are given considerable attention in number theory.

EXAMPLE 2

Let us calculate c_p, the number of solutions of the diophantine equation

$$k_1 + 2k_2 = p.$$

According to the above, the generating function is

$$f(x) := \sum_{p=0}^{\infty} c_p x^p = \frac{1}{(1-x)(1-x^2)}.$$

APPLICATIONS TO COMBINATORIAL ANALYSIS

There is a pole of order 2 at $x=1$ and one of order 1 at $x=-1$. From the expansion ($h:=x-1$)

$$f(1+h) = \frac{1}{h^2(2+h)} = \frac{1}{2}\frac{1}{h^2}\frac{1}{1+h/2} = \frac{1}{2}h^{-2}(1-\frac{1}{2}h+O(h^2))$$

the principal part at $x=1$ is found to be

$$s_1(x) = \frac{1}{2(x-1)^2} - \frac{1}{4(x-1)}.$$

The residue at $x=-1$ is $\frac{1}{4}$; hence the principal part

$$s_2(x) = \frac{1}{4(x+1)}.$$

Thus according to Theorem 4.4h

$$f(x) = \frac{1}{2(x-1)^2} - \frac{1}{4(x-1)} + \frac{1}{4(x+1)}$$

and expanding each term separately we get

$$c_p = \frac{1}{2}\frac{(2)_p}{p!} + \frac{1}{4} + \frac{1}{4}(-1)^p = \frac{1}{2}(p+1) + \frac{1}{4}(1+(-1)^p);$$

for example, $c_8 = \frac{9}{2} + \frac{1}{2} = 5$, corresponding to the five solutions

$$1\cdot 0 + 2\cdot 4, \quad 1\cdot 2 + 2\cdot 3, \quad 1\cdot 4 + 2\cdot 2, \quad 1\cdot 6 + 2\cdot 1, \quad 1\cdot 8 + 2\cdot 0.$$

PROBLEMS

1. A group of n people selects a committee of p people from among their midst. In how many ways can this be done?
2. A total of n silver dollars shall be distributed among n people. In how many ways can this be done?
3. One dollar is to be distributed among 10 boys under the side condition that each boy will get an amount payable in precisely one silver or nickel coin. In how many ways can this be done? [The generating function $\sum c_p x^p$ for the number of ways of distributing $5p$ cents is $f(x) = (x + x^2 + x^5 + x^{10} + x^{20})^{10}$. We are looking for c_{20}.]
4. In how many ways can 60 identical objects be put away in four drawers if each drawer is to contain a number of objects that is an odd prime?
5. Set up the generating function for the slot machine problem if it is required that the number of coins
   (a) be odd,
   (b) be less than 5.
   Compute c_9 in both cases.

5. A letter sent abroad from Switzerland requires 50 cents postage. In how many ways can the required stamps be placed (horizontally) on the letter if stamps for 5, 10, 15, 20 cents are available? (If stamps of different denominations are placed on the letter in a different order, this will be considered a different way.)

6. Show that every positive integer n can be represented in 2^{n-1} ways as a sum of positive integers if representations in which the order of terms is different are considered different.
   EXAMPLE. $n=4$ has eight representations:

   $$4, \quad 3+1, \quad 2+2, \quad 1+1+2, \quad 1+1+1+1.$$
   $$1+3 \qquad\qquad 1+2+1$$
   $$\qquad\qquad\qquad 2+1+1$$

7. In how many ways can a dollar be changed into nickels, dimes, quarters, and half dollars? [We have to find the coefficient of x^{20} in the expansion of $f(x)=[(1-x)(1-x^2)(1-x^5)(1-x^{10})]^{-1}.$]

8. The number of non-negative solutions (k_1,k_2,k_3) of the diophantine equation $k_1+2k_2+3k_3=p$ is equal to the integer closest to

   $$(p+3)^2/12.$$

9. Let $\omega_1,\omega_2,\ldots,\omega_n$ be mutually prime positive integers. If n is fixed, the number c_p of non-negative solutions $(k_1,k_2,\ldots,k_n)$ of the diophantine equation $k_1\omega_1+k_2\omega_2+\ldots+k_n\omega_n=p$ satisfies $\lim_{p\to\infty}p^{1-n}c_p=[(n-1)!\omega_1\omega_2\cdots\omega_n]^{-1}.$

10. In Chapter 12 we establish the formula

    $$\sum_{m=1}^{\infty}\frac{x^m}{(1-x^m)(1-x^{m+1})}=\frac{x}{(1-x)^2}.$$

    Find a combinatorial interpretation.

§7.4. DIFFERENCE EQUATIONS

Let $a_0,a_1,\ldots,a_k$ be given complex numbers, $a_0 a_k \neq 0$, and let $\{s_n\}_k^\infty$ be a given sequence of complex numbers. The following problem arises, e.g., in mathematical economics and in numerical analysis: determine all sequences $\{x_n\}_0^\infty$ such that

$$a_0 x_n + a_1 x_{n-1} + \cdots + a_k x_{n-k} = s_n \tag{7.4-1}$$

for $n=k,k+1,\ldots$. Equation (7.4-1) is called a **linear difference equation** (with constant coefficients) of order k. The equation is called *homogeneous* if $\{s_n\}$ is the null sequence; otherwise it is called *inhomogeneous*. A

DIFFERENCE EQUATIONS

sequence $\{x_n\}$ satisfying (7.4-1) for $n \geq k$ is called a **solution** of the difference equation.

Clearly, for every given set of k numbers $x_0, x_1, \ldots, x_{k-1}$ there exists precisely one solution $\{x_n\}$ of the difference equation (7.4-1) having these numbers as its initial elements. This solution can be constructed numerically by considering (7.4-1) as a recurrence relation. Our interest, however, lies in representing the solutions in closed analytic form, which will make it possible, for instance, to determine the structure of the set of all solutions and to study their behavior as $n \to \infty$. Although an elementary treatment of linear difference equations is possible, the above questions are answered most rapidly by the method of generating functions. The method also carries over to certain difference equations of infinite order; see Problems 5, 6, 7.

First we consider the homogeneous equation

$$a_0 x_n + a_1 x_{n-1} + \cdots + a_k x_{n-k} = 0, \qquad (7.4\text{-}2)$$

$n = k, k+1, \ldots$. Let $\{x_n\}$ denote the solution of (7.4-2) with starting values $x_0, x_1, \ldots, x_{k-1}$. We introduce the (terminating) formal power series

$$A := a_0 + a_1 z + \cdots + a_k z^k$$

and

$$S := s_0 + s_1 z + \cdots + s_{k-1} z^{k-1},$$

where

$$s_m := a_0 x_m + a_1 x_{m-1} + \cdots + a_m x_0, \qquad m = 0, 1, \ldots, k-1. \qquad (7.4\text{-}3)$$

If we now let

$$X := x_0 + x_1 z + x_2 z^2 + \cdots,$$

(7.4-2) and (7.4-3) express the fact that

$$AX = S \qquad (7.4\text{-}4)$$

as an identity between formal power series. Since $a_0 \neq 0$, A is a unit in the integral domain of formal power series, and we may write

$$X = A^{-1} S. \qquad (7.4\text{-}5)$$

The series A and S both terminate and thus have an infinite radius of convergence. They represent analytic functions, which happen to be polynomials. If we denote these polynomials by a and s, respectively, it follows from Theorem 2.3d that the formal series X is the power series that represents the rational function

$$x(z) := \frac{s(z)}{a(z)} := \frac{s_0 + s_1 + \cdots + s_{k-1} z^{k-1}}{a_0 + a_1 z + \cdots + a_k z^k}$$

near $z=0$. Conversely, every power series X that represents a rational function x of the above form satisfies (7.4-5), hence (7.4-4), and is a generating function of a solution of (7.4-2).

If $x(z)$ has the distinct poles $z_1, z_2, \ldots, z_h$ of the respective orders $m_1, m_2, \ldots, m_h$ ($\sum m_i = k$), then by Theorem 7.2b the coefficients x_n of the series X, and thus the general solution of the homogeneous difference equation (7.4-2), have the form

$$x_n = \sum_{i=1}^{h} p_i(n) z_i^{-n}, \qquad n = 0, 1, 2, \ldots, \tag{7.4-6}$$

where p_i is a polynomial of degree $m_i - 1$. Particular solutions of the difference equation are the k sequences

$$x_n^{(i,j)} := n^j z_i^{-n}, \qquad i = 1, 2, \ldots, h; \ j = 0, 1, \ldots, m_i - 1. \tag{7.4-7}$$

It is well known and easy to see that the totality of solutions of (7.4-2) forms a linear space of dimension k. We can now readily prove the result, somewhat tedious to establish by elementary methods, that *the k sequences (7.4-7) are linearly independent and form a basis of the solution space*. This is the same as saying that the identity

$$\sum_{i=1}^{h} \left(\sum_{j=1}^{m_i - 1} c_{ij} n^j \right) z_i^{-n} = 0 \tag{7.4-8}$$

can hold for all n only if all $c_{ij} = 0$. By Theorem 7.2b the expression on the left of (7.4-8) is the nth Taylor coefficient at $z=0$ of a rational function r which at points $z_1, \ldots, z_h$ has poles with the respective principal parts

$$\sum_{j=1}^{m_i} b_{i,j} (z - z_i)^{-j},$$

where

$$\sum_{j=1}^{m_i} (-1)^j b_{i,j} \frac{z_i^{-j}}{(j-1)!} (n+1)_{j-1} = \sum_{j=0}^{m_i - 1} c_{ij} n^j$$

for $n = 0, 1, 2, \ldots$. Since (7.4-8) states that all Taylor coefficients are zero, r is the zero function. Hence all $b_{i,j}$ are zero, which by a simple induction argument proves that all c_{ij} are zero.

EXAMPLE 1

The **Fibonacci sequence** $\{x_n\} = \{0, 1, 1, 2, 3, 5, 8, 13, \ldots\}$ is defined as the solution of the difference equation $x_n = x_{n-1} + x_{n-2}$ having the starting values $x_0 := 0, x_1 := 1$. Here $a(z) = 1 - z - z^2$, $s(z) = z$; hence the generating function of the sequence is $x(z) := z/(1 - z - z^2)$. The zeros of the denominator are

$$z_1 := \tfrac{1}{2}(\sqrt{5} - 1), \qquad z_2 := \tfrac{1}{2}(-\sqrt{5} - 1),$$

and thus the expansion in partial fractions is

$$x(z) = \frac{z_1}{-1 - 2z_1} \frac{1}{z - z_1} + \frac{z_2}{-1 - 2z_2} \frac{1}{z - z_2}$$

$$= \frac{1}{1 + 2z_1} \frac{1}{1 - z/z_1} + \frac{1}{1 + 2z_2} \frac{1}{1 - z/z_2}.$$

By expanding in geometric series we easily find

$$x_n = \frac{1}{1 + 2z_1} z_1^{-n} + \frac{1}{1 + 2z_2} z_2^{-n}$$

$$= \frac{1}{\sqrt{5}} \left(\frac{\sqrt{5} + 1}{2} \right)^n - \frac{1}{\sqrt{5}} \left(\frac{-\sqrt{5} + 1}{2} \right)^n.$$

The method of generating functions can also be used to solve the nonhomogeneous equation (7.4-1). Defining $s_0, s_1, \ldots, s_{k-1}$ by (7.4-3), we now let

$$S := s_0 + s_1 z + s_2 z^2 + \cdots,$$

where s_n for $n \geq k$ is the nonhomogeneous term in (7.4-1). If A and X still have the same meaning, then (7.4-1) and (7.4-3) again express the fact that

$$X = A^{-1} S$$

as an identity between formal power series. If the coefficients s_n grow too rapidly, the series S may have radius of convergence zero and not much is gained by our method. If, however, S has a positive radius of convergence and the analytic function defined by S can somehow be identified, then analytical techniques can be used to determine the x_n, as above.

Bernoulli's Method. The results on homogeneous difference equations can easily be transformed into a numerical method for finding the zero of smallest absolute value of a polynomial if there is a single such zero. Denote by $z_1, z_2, \ldots, z_h$ the distinct zeros of the polynomial

$$a(z) := a_0 + a_1 z + \cdots + a_k z^k$$

and suppose that

$$0 < |z_1| < |z_2| \leqslant \cdots \leqslant |z_h|.$$

If the starting values $x_0, x_1, \ldots, x_{k-1}$ of a solution $\{x_n\}$ of the homogeneous difference equation

$$a_0 x_n + a_1 x_{n-1} + \cdots + a_k x_{n-k} = 0 \qquad (7.4\text{-}9)$$

are chosen such that z_1 is not a zero of the polynomial $s(z)$ defined above, then (7.4-6) shows that

$$x_n = p_1(n) z_1^{-n} + t_n, \qquad (7.4\text{-}10)$$

where p_1 is a polynomial of degree $m_1 - 1$ (m_1 being the multiplicity of z_1), and the numbers t_n are such that for every ρ satisfying $|z_1| < \rho < |z_2|$ there exists a constant μ with the property that $|t_n| \leqslant \mu \rho^{-n}$ for all n. From (7.4-10) there follows

$$\lim_{n \to \infty} \frac{x_{n+1}}{x_n} = z_1^{-1};$$

more precisely, as $n \to \infty$,

$$\frac{x_{n+1}}{x_n} = \frac{1}{z_1}\left(1 + \frac{m_1 - 1}{n}\right) + O(\gamma^n),$$

where $\gamma := |z_1|/\rho$.

The zero z_1 thus can be determined numerically by forming the ratios of successive elements of a suitably started solution of the difference equation formed with the coefficients of the given polynomial. This algorithm is known as **Bernoulli's method**. The sequence of quotients converges geometrically if z_1 has multiplicity 1 and with an error $O(n^{-1})$ if the multiplicity is greater than 1.

Bernoulli's method furnishes only one zero at a time, the zero of smallest modulus. In subsequent sections we discuss the quotient-difference algorithm, a modern extension of Bernoulli's method, which under certain conditions furnishes all zeros simultaneously.

DIFFERENCE EQUATIONS

PROBLEMS

1. Solve the difference equation
$$x_n - 5x_{n-1} + 6x_{n-2} = 0, \qquad x_0 = 3, \; x_1 = 8.$$

2. The sequence $\{x_n\}$ satisfies the difference equation
$$x_n - 2x_{n-1} + 2x_{n-2} = 0, \qquad x_0 = 1, \; x_1 = 2.$$
Show that $x_{4n} = (-4)^n$ and find expressions for $x_{4n+1}, x_{4n+2}, x_{4n+3}$.

3. Let the polynomial $a(z)$ satisfy $a_0 \neq 0$. Prove that if the starting values for Bernoulli's method are chosen to be
$$x_0 = x_1 = \cdots = x_{k-2} = 0, \qquad x_{k-1} = 1,$$
the polynomial $s(z)$ cannot have a zero at $z = z_1$.

4. For the convergence of Bernoulli's method it is undesirable that the rational function s/a have poles of order >1. Show that all poles have order 1 if the starting values are chosen according to the algorithm
$$x_0 = \frac{a_1}{a_0},$$
$$x_1 = \frac{1}{a_0}(2a_2 - a_1 x_0),$$
$$x_2 = \frac{1}{a_0}(3a_3 - a_2 x_0 - a_1 x_1),$$
$$\cdots$$
$$x_{k-1} = \frac{1}{a_0}(ka_k - a_{k-1}x_0 - \cdots - a_1 x_{k-2}).$$

5. Solve the difference equations of infinite order,

   (a) $x_n = x_{n-1} + x_{n-2} + \cdots + x_0,$

   (b) $x_n = x_{n-1} + 2x_{n-2} + \cdots + n x_0,$

   under the initial condition $x_0 = 1$.

6. If the sequence $\{x_n\}$ satisfies the difference equation of infinite order,
$$x_0 = 1,$$
$$x_n = 1 \cdot x_{n-1} + \frac{1}{2} x_{n-2} + \cdots + \frac{1}{n} x_0,$$

show that
$$\lim_{n\to\infty} \frac{x_{n+1}}{x_n} = \frac{e}{e-1}.$$

7. Let the sequence $\{x_n\}$ satisfy the difference equation of infinite order.
$$x_0 = 1,$$
$$x_n = x_{n-1} - \frac{1}{1!}x_{n-2} + \frac{1}{2!}x_{n-3} + \cdots + (-1)^{n-1}\frac{1}{(n-1)!}x_0.$$
Show that
$$\lim_{n\to\infty} x_n = 0.$$

8. Denoting by f_n the nth Fibonacci number, show that
$$\sum_{n=0}^{\infty} f_n^2 z^n = \frac{z - z^2}{1 - 2z - 2z^2 + z^3}.$$

9. In certain questions of numerical integration the coefficients
$$c_n := \int_{-1}^{0} \frac{(s)_n}{n!} ds, \quad n = 0, 1, 2, \ldots,$$
are of interest. Consider the generating function
$$c(z) := \sum_{n=0}^{\infty} c_n z^n$$
and prove the recurrence relation $c_0 = 1$:
$$c_n = -\frac{1}{2}c_{n-1} - \frac{1}{3}c_{n-2} - \cdots - \frac{1}{n+1}c_0.$$

10. Show that the generating function of the numbers
$$a_n := \int_{-1}^{1} \frac{s}{s+n}\binom{s+n}{2n} ds$$
(also of interest in numerical integration) is
$$a(z) := \sum_{n=0}^{\infty} a_n z^n = \frac{z(1+z^2/4)^{1/2}}{\text{Log}[z/2 + (1+z^2/4)^{1/2}]}$$
and derive a recurrence relation for the a_n.

§7.5. HANKEL DETERMINANTS

Here we begin the study of certain determinants, called **Hankel determinants**, which can be formed with the coefficients of any formal power series and which enjoy some remarkable formal and analytic properties. They can be used to localize the poles of analytic functions, especially in connection with the quotient-difference algorithm. For the benefit of readers not sufficiently acquainted with determinants in general we begin by summarizing their definition and some of their less well known properties.

I. Determinants

It is necessary first to say a few words about *permutations*. Let $n > 0$ be an integer and let $N_n := \{1, 2, \ldots, n\}$, the set of the first n positive integers. A **permutation** π of n elements is a one-to-one function from N_n to N_n (see §2.1 for a general explanation of this concept). It is specified by its set of values $(\pi(1), \pi(2), \ldots, \pi(n))$. There are precisely $n!$ such functions. The set of all permutations forms a (multiplicative) group under the operation of composition.

A permutation π is said to contain the **inversion** (i, j) if $i < j$ and $\pi(i) > \pi(j)$. By examining each pair (i, j), where $i < j, i, j \in N_n$, we can easily count the total number of inversions contained in a permutation.

EXAMPLES

Determine the number k of inversions in the permutations (a) $(1, 3, 2)$, (b) $(3, 1, 2)$.
(a) Since $1 < 3$, $1 < 2$, $3 > 2$, $k = 1$. (b) Since $3 > 1$, $3 > 2$, $1 < 2$, $k = 2$.

A **transposition** is a permutation π such that $\pi(k) = k$ for all $k \in N_n$ save for precisely two integers, say i and j, where (by necessity) $\pi(i) = j$, $\pi(j) = i$. Every permutation can be written as a product of transpositions. (By a first transposition bring 1 to the correct place, by a second, 2, and so on.) Although this representation is not unique, it can be shown that the number of factors in any representation of a permutation π as a product of transpositions always has the same parity as the number of inversions contained in π. It thus is always even or always odd. A permutation is called **even** if the number of factors is even and **odd** if the number of factors is odd.

EXAMPLES

(a) The permutation $(1, 3, 2)$ is odd because it *is* a transposition. It also contains precisely one inversion. (b) The permutation $(3, 1, 2)$ is even because it is the product of the two transpositions $(2, 1, 3)$ and $(3, 2, 1)$. As we have seen, it contains two inversions.

Now let $\mathbf{A} = (a_{ij})$ be a square matrix of order n whose elements a_{ij} are taken from some integral domain. The **determinant** of $\mathbf{A}$, denoted by $\det \mathbf{A}$ or by $|\mathbf{A}|$, is defined as the number

$$\sum_\pi \epsilon_\pi a_{1,\pi(1)} a_{2,\pi(2)} \cdots a_{n,\pi(n)},$$

where the sum is taken with respect to all permutations of N_n and where ϵ_π is $+1$ or -1, according to whether the permutation π is even or odd. We also write

$$\det \mathbf{A} = \begin{vmatrix} a_{11} & a_{12} & \cdots & a_{1n} \\ a_{21} & a_{22} & \cdots & a_{2n} \\ \cdots & \cdots & \cdots & \cdots \\ a_{n1} & a_{n2} & \cdots & a_{nn} \end{vmatrix}.$$

Assuming that the reader is familiar with the basic properties of determinants (such that $\det \mathbf{A}^\mathrm{T} = \det \mathbf{A}$ or $\det \mathbf{A} = 0$ if $\mathbf{A}$ contains two equal rows or columns), we shall discuss some less elementary results.

THE LAPLACIAN EXPANSION OF A DETERMINANT

A **minor** m of the determinant of a matrix $\mathbf{A}$ is obtained by deleting from $\mathbf{A}$ any number of rows and an equal number of columns and forming the determinant of the elements contained in both the remaining rows and columns.

EXAMPLE

$$\begin{vmatrix} a_{11} & a_{13} \\ a_{41} & a_{43} \end{vmatrix} \text{ is a minor of } \begin{vmatrix} a_{11} & a_{12} & a_{13} & a_{14} \\ a_{21} & a_{22} & a_{23} & a_{24} \\ a_{31} & a_{32} & a_{33} & a_{34} \\ a_{41} & a_{42} & a_{43} & a_{44} \end{vmatrix}$$

The **complement** $c(m)$ of a minor m of $\mathbf{A}$ is the minor obtained by deleting precisely the rows and columns that have been retained in forming m.

EXAMPLE

The complement of the above minor is

$$\begin{vmatrix} a_{22} & a_{24} \\ a_{32} & a_{34} \end{vmatrix}.$$

HANKEL DETERMINANTS

The **algebraic complement** of a minor m is defined as $+c(m)$ if the sum of the indices of the rows and columns deleted from $\mathbf{A}$ in forming m is even and $-c(m)$ if that sum is odd.

EXAMPLE

In the above example the indices of the deleted rows are 2 and 3 and of the deleted columns, 2 and 4. Thus the algebraic complement of the exhibited minor is

$$(-1)^{2+3+2+4}\begin{vmatrix} a_{22} & a_{24} \\ a_{32} & a_{34} \end{vmatrix} = -\begin{vmatrix} a_{22} & a_{24} \\ a_{32} & a_{34} \end{vmatrix}.$$

With these definitions the following rule for calculating the determinant of a matrix $\mathbf{A}$ (**Laplacian expansion** of $\det \mathbf{A}$) holds:

1. Pick $k < n$ rows of $\mathbf{A}$.
2. Form all possible minors of order k from these k rows and multiply them by their algebraic complements.
3. The sum of all these products equals $\det \mathbf{A}$.

EXAMPLES

(a) $\begin{vmatrix} a_{11} & a_{12} \\ a_{21} & a_{22} \end{vmatrix} = a_{11}a_{22} - a_{12}a_{21}.$

(b) $\begin{vmatrix} a_{11} & a_{12} & a_{13} \\ a_{21} & a_{22} & a_{23} \\ a_{31} & a_{32} & a_{33} \end{vmatrix} = a_{11}\begin{vmatrix} a_{22} & a_{23} \\ a_{32} & a_{33} \end{vmatrix} - a_{12}\begin{vmatrix} a_{21} & a_{23} \\ a_{31} & a_{33} \end{vmatrix} + a_{13}\begin{vmatrix} a_{21} & a_{22} \\ a_{31} & a_{32} \end{vmatrix}.$

THE DETERMINANT OF A SUM OF MATRICES

The problem here is to write the determinant of a sum of matrices $\mathbf{A}_1 + \mathbf{A}_2 + \cdots + \mathbf{A}_r$ as a sum of determinants. Let $\mathbf{A}_i$ be expressed in terms of its columns:

$$\mathbf{A}_i = (\mathbf{a}_{i1}, \mathbf{a}_{i2}, \ldots, \mathbf{a}_{in}), \quad i = 1, 2, \ldots, r.$$

It is then shown in determinant theory that

$$\det(\mathbf{A}_1 + \mathbf{A}_2 + \cdots + \mathbf{A}_r) = \sum_{\kappa} \det \mathbf{A}^{(\kappa)},$$

where

$$\mathbf{A}^{(\kappa)} := (\mathbf{a}_{\kappa(1),1}, \mathbf{a}_{\kappa(2),2}, \ldots, \mathbf{a}_{\kappa(n),n}),$$

and the sum is taken with respect to all *combinations* κ of r integers from the set N_n, i.e., with respect to all r^n functions κ from N_n to N_r.

EXAMPLE

$$\begin{vmatrix} a_{11}+b_{11} & a_{12}+b_{12} \\ a_{21}+b_{21} & a_{22}+b_{22} \end{vmatrix} = \begin{vmatrix} a_{11} & a_{12} \\ a_{21} & a_{22} \end{vmatrix} + \begin{vmatrix} a_{11} & b_{12} \\ a_{21} & b_{22} \end{vmatrix} + \begin{vmatrix} b_{11} & a_{12} \\ b_{21} & a_{22} \end{vmatrix} + \begin{vmatrix} b_{11} & b_{12} \\ b_{21} & b_{22} \end{vmatrix}.$$

THE DETERMINANT OF A PRODUCT OF MATRICES

The determinant of a product of matrices equals the product of the determinants, i.e.,

$$\det \prod_{i=1}^{r} \mathbf{A}_i = \prod_{i=1}^{r} \det \mathbf{A}_i$$

for any $n \times n$ matrices $\mathbf{A}_1, \mathbf{A}_2, \ldots, \mathbf{A}_r$.

II. Hankel Determinants

Let $F := a_0 + a_1 x + a_2 x^2 + \cdots$ be a formal power series with coefficients taken from some integral domain. We set $a_m := 0$ for $m < 0$. For arbitrary integers n and for integers $k \geq 0$ we define the determinants $H_k^{(n)}$ by $H_0^{(n)} := 1$,

$$H_k^{(n)} := \begin{vmatrix} a_n & a_{n+1} & \cdots & a_{n+k-1} \\ a_{n+1} & a_{n+2} & \cdots & a_{n+k} \\ \vdots & & & \vdots \\ a_{n+k-1} & a_{n+k} & \cdots & a_{n+2k-2} \end{vmatrix}, \quad k = 1, 2, \ldots.$$

These determinants are called the **Hankel determinants** associated with the formal power series F. The remainder of this section is devoted to an exposition of various properties of these determinants. We begin with a formal result.

HANKEL DETERMINANTS

THEOREM 7.5a (Jacobi's identity)

For all integers n and for $k \geq 1$

$$\left(H_k^{(n)}\right)^2 - H_k^{(n-1)}H_k^{(n+1)} + H_{k+1}^{(n-1)}H_{k-1}^{(n+1)} = 0. \tag{7.5-1}$$

Proof. We define the column vectors of order $k+1$

$$\mathbf{h}_n := \begin{bmatrix} a_n \\ a_{n+1} \\ \vdots \\ a_{n+k} \end{bmatrix}, \quad \mathbf{e}_1 := \begin{bmatrix} 1 \\ 0 \\ \vdots \\ 0 \end{bmatrix}, \quad \mathbf{e}_{k+1} := \begin{bmatrix} 0 \\ 0 \\ \vdots \\ 1 \end{bmatrix}$$

and evaluate the determinant of the matrix of order $2k+2$,

$$\mathbf{A} := \begin{bmatrix} \mathbf{e}_1 & \mathbf{h}_n & \mathbf{h}_{n+1} \cdots \mathbf{h}_{n+k-2} & \mathbf{h}_{n+k-1} & \cdot & \mathbf{h}_{n-1} & \mathbf{0} \cdots \mathbf{0} & \mathbf{e}_{k+1} \\ \mathbf{e}_1 & \mathbf{0} & \mathbf{0} & \cdots & \mathbf{0} & \mathbf{h}_{n+k-1} & \cdot & \mathbf{h}_{n-1} & \mathbf{h}_n \cdots \mathbf{h}_{n+k-2} & \mathbf{e}_{k+1} \end{bmatrix}.$$

in two different ways. Subtracting the last $k+1$ rows from the corresponding first $k+1$ rows, we obtain the determinant

$$\begin{vmatrix} \mathbf{0} & \mathbf{h}_n \cdots \mathbf{h}_{n+k-2} & \mathbf{0} & \cdot & \mathbf{0} & -\mathbf{h}_n \cdots -\mathbf{h}_{n+k-2} & \mathbf{0} \\ \mathbf{e}_1 & \mathbf{0} \cdots \mathbf{0} & \mathbf{h}_{n+k-1} & \cdot & \mathbf{h}_{n-1} & \mathbf{h}_n \cdots \mathbf{h}_{n+k-2} & \mathbf{e}_{k+1} \end{vmatrix}.$$

This determinant is zero, for if we expand in terms of minors formed with the first $k+1$ rows then at least one column in every minor or its complement is zero. On the other hand, when $\det \mathbf{A}$ is calculated directly by expanding in terms of minors formed with the first $k+1$ rows, we obtain six nonzero products that are equal in pairs. When collected and interpreted properly, their sum equals twice the expression on the left of (7.5-1). ∎

If the determinants $H_k^{(n)}$, where $n \geq 0$ are arranged in a triangular array,

$$\begin{array}{ccccccc}
1 & & & & & & \\
1 & H_1^{(0)} & & & & & \\
1 & H_1^{(1)} & H_2^{(0)} & & & & \\
& & | & & & & \\
1 & H_1^{(2)} & \text{---} H_2^{(1)} \text{---} & H_3^{(0)} & & & \\
& & | & & & & \\
1 & H_1^{(3)} & H_2^{(2)} & H_3^{(1)} & H_4^{(0)} & & \\
\cdot & \cdot & \cdot & \cdot & \cdot & \cdot & \cdot
\end{array}$$

the terms linked together in Equation (7.5-1) appear in a starlike configuration. Because the first two columns in the array are trivial, (7.5-1) may be used to calculate the Hankel determinants recursively, proceeding from left to right.

We now discuss an analytical property of the Hankel determinants associated with the power series that represents a meromorphic function.

THEOREM 7.5b

Let the function f be analytic at $z=0$ and meromorphic in the disk $D: |z|<\sigma$ and let its poles $z_i = u_i^{-1}$ in D, which may be finite or infinite in number, be numbered such that

$$0 < |z_1| \leqslant |z_2| \leqslant \cdots < \sigma,$$

each pole occuring as many times in the sequence $\{z_k\}$ as indicated by its order. Let $H_k^{(n)}$ denote the Hankel determinants associated with the power series

$$f(z) = \sum_{n=0}^{\infty} a_n z^n$$

which represents f near 0. Then for each m such that

$$|z_m| < |z_{m+1}| \qquad (7.5\text{-}2)$$

there exists a constant $c_m \neq 0$ independent of n such that, for any ρ satisfying $|u_m| > \rho > |u_{m+1}|$,

$$H_m^{(n)} = c_m (u_1 u_2 \cdots u_m)^n \left\{ 1 + O\left(\left(\frac{\rho}{|u_m|} \right)^n \right) \right\} \qquad (7.5\text{-}3)$$

as $n \to \infty$.

Proof. We first prove the theorem under the additional hypothesis that the poles of f are simple. Let m be an index that satisfies (7.5-2). By extracting the principal parts, we can write

$$f(z) = \frac{r_1}{z_1 - z} + \frac{r_2}{z_2 - z} + \cdots + \frac{r_m}{z_m - z} + g(z), \qquad (7.5\text{-}4)$$

where g is analytic in $|z| < |z_{m+1}|$. If

$$g(z) = \sum_{n=0}^{\infty} b_n z^n,$$

HANKEL DETERMINANTS

then by the Cauchy coefficient estimate there is a number $\mu > 0$ such that

$$|b_n| \leq \mu \rho^n, \quad n = 0, 1, 2, \ldots,$$

for every $\rho \in (|u_{m+1}|, |u_m|)$. By expanding in geometric series

$$a_n = r_1 z_1^{-n-1} + r_2 z_2^{-n-1} + \cdots + r_m z_m^{-n-1} + b_n$$
$$= r_1 u_1^{n+1} + r_2 u_2^{n+1} + \cdots + r_m u_m^{n+1} + b_n \qquad (7.5\text{-}5)$$

follows from (7.5-4). By the addition formula for determinants we can now write

$$H_m^{(n)} = \sum_\kappa D_m^{(n)} + \sum_\kappa \hat{D}_m^{(n)}, \qquad (7.5\text{-}6)$$

where in the first sum

$$D_m^{(n)} := \begin{vmatrix} r_{\kappa(1)} u_{\kappa(1)}^{n+1} & r_{\kappa(2)} u_{\kappa(2)}^{n+2} & \cdots & r_{\kappa(m)} u_{\kappa(m)}^{n+m} \\ r_{\kappa(1)} u_{\kappa(1)}^{n+2} & r_{\kappa(2)} u_{\kappa(2)}^{n+3} & & r_{\kappa(m)} u_{\kappa(m)}^{n+m+1} \\ \cdots & \cdots & \cdots & \cdots \\ r_{\kappa(1)} u_{\kappa(1)}^{n+m} & r_{\kappa(2)} u_{\kappa(2)}^{n+m+1} & \cdots & r_{\kappa(m)} u_{\kappa(m)}^{n+2m-1} \end{vmatrix},$$

and the summation is taken with respect to all functions κ from N_m to N_m. In the second sum $\hat{D}_m^{(n)}$ denotes a determinant of the same form as $D_m^{(n)}$ except that at least one of the columns

$$\begin{bmatrix} r_{\kappa(i)} u_{\kappa(i)}^{n+i} \\ r_{\kappa(i)} u_{\kappa(i)}^{n+i+1} \\ \vdots \\ r_{\kappa(i)} u_{\kappa(i)}^{n+i+m-1} \end{bmatrix}$$

is replaced by a corresponding column

$$\begin{bmatrix} b_{n+i-1} \\ b_{n+i} \\ \vdots \\ b_{n+i+m-2} \end{bmatrix}.$$

The summation here is taken with respect to all functions κ from N_m to N_{m+1} that assume the value $m+1$ at least once. If $\kappa(i)=m+1$, the ith r-column is to be replaced by a b-column.

To calculate the first sum we note that if κ assumes a value of the set N_m twice two columns of the determinant $D_m^{(n)}$ are proportional and $D_m^{(n)}=0$. It thus suffices to take the sum with respect to those functions κ that assume each value of N_m exactly once, i.e., with respect to the permutations. Each of the remaining determinants then contains the factor

$$\prod_{i=1}^{m} r_i u_i^{n+1},$$

and we have

$$\sum_{\kappa} D_m^{(n)} = \prod_{i=1}^{m} r_i u_i^{n+1} \cdot p_m, \qquad (7.5\text{-}7)$$

where

$$p_m := \sum_{\pi} \begin{vmatrix} u_{\pi(1)}^0 & u_{\pi(2)}^1 & \cdots & u_{\pi(m)}^{m-1} \\ u_{\pi(1)}^1 & u_{\pi(2)}^2 & \cdots & u_{\pi(m)}^m \\ \cdots & \cdots & \cdots & \cdots \\ u_{\pi(1)}^{m-1} & u_{\pi(2)}^m & \cdots & u_{\pi(m)}^{2m-2} \end{vmatrix}, \qquad (7.5\text{-}8)$$

the summation being taken with respect to all permutations. We shall show that $p_m \neq 0$. We note that $p_m = p_m(u_1, u_2, \ldots, u_m)$ is a homogeneous polynomial of degree $m(m-1)$ in the variables $u_1, u_2, \ldots, u_m$. It vanishes when any two u_i are equal and does not change its value when two u_i's are interchanged. It must therefore contain all factors $(u_i - u_j)^2$, where $1 \leq i < j \leq m$. Because the product of all these factors has degree $m(m-1)$, we must have

$$p_m = d_m \prod_{1 \leq i < j \leq m} (u_i - u_j)^2, \qquad (7.5\text{-}9)$$

where d_m is a numerical constant. We shall show that $d_m = 1$ by induction with respect to m. Evidently $d_1 = 1$. Assume that $d_{m-1} = 1$. We calculate $p_m(0, u_2, \ldots, u_m)$ from (7.5-8). Only the determinants where $\pi(1) = 1$ do not vanish and their sum adds up to

$$p_m(0, u_2, \ldots, u_m) = (u_2 \cdots u_m)^2 p_{m-1}(u_2, \ldots, u_m).$$

By the induction hypothesis this equals

$$(u_2 \cdots u_m)^2 \prod_{2 \leq i < j \leq m} (u_i - u_j)^2.$$

Comparison with (7.5-9) now yields $d_m = 1$. This proves that in (7.5-7) $p_m \neq 0$, hence that

$$\sum_\kappa D_m^{(n)} = c_m (u_1 u_2 \cdots u_m)^n, \qquad (7.5\text{-}10)$$

where

$$c_m := p_m \prod_{i=1}^m r_i u_i \neq 0;$$

c_m is independent of n.

To estimate the second sum in (7.5-6) we note that only those determinants $\hat{D}_m^{(n)}$ are $\neq 0$ where the values $\kappa(i) \neq m+1$ are distinct. Because at least one column of r's is replaced by a column of b's, it follows that for $n \to \infty$

$$\hat{D}_m^{(n)} = O\big((u_1 u_2 \cdots u_{m-1} \rho)^n\big). \qquad (7.5\text{-}11)$$

Because the sum of $\hat{D}$'s has only a finite number of terms, it is of the same order. Together with (7.5-10) this proves Theorem 7.5b in the special case in which the poles of f are all simple.

To deal with the general case in which f may have poles of order > 1, we use a *method of confluence* which we shall now explain. To begin with, let f have a single pole of order $k > 1$, at z_1 say, and let

$$s(z) := \sum_{h=1}^k \frac{s_h}{(z - z_1)^h} \qquad (s_k \neq 0) \qquad (7.5\text{-}12)$$

be the principal part of f at z_1. We shall represent $s(z)$ as the limit of a sum of the singular parts at k simple, distinct poles.

LEMMA 7.5c

Let $s(z)$ be defined by (7.5-12), let $w := \exp(2\pi i / k)$, and let, for $\epsilon > 0$,

$$s(z, \epsilon) := \sum_{h=1}^k \frac{r_h(\epsilon)}{z - (z_1 + \epsilon w^h)} \qquad (7.5\text{-}13)$$

where

$$r_h(\epsilon) := \frac{1}{k} \sum_{j=1}^{k} (\epsilon w^h)^{1-j} s_j, \qquad h=1,\ldots,k. \qquad (7.5\text{-}14)$$

Then

$$\lim_{\epsilon \to 0} s(z,\epsilon) = s(z) \qquad (7.5\text{-}15)$$

uniformly with respect to z on every compact set not containing z_1.

Assuming the truth of Lemma 7.5c, we complete the proof of Theorem 7.5b as follows. We denote by $f(z,\epsilon)$ the function $f(z)$, where the singular part $s(z)$ (involving a pole z_1 of order k) is replaced by the function $s(z,\epsilon)$ (involving k poles of order 1, located at a distance ϵ from z_1), and by $a_n(\epsilon)$ the Taylor coefficients at $z=0$ of $f(z,\epsilon)$. Expressing the Taylor coefficients by Cauchy's integral (Corollary 4.7c) extended along a small circle around $z=0$, we get in view of the uniform convergence of $f(z,\epsilon)$ to $f(z)$

$$\lim_{\epsilon \to 0} a_n(\epsilon) = a_n, \qquad n=0,1,2,\ldots.$$

Hence if $H_m^{(n)}(\epsilon)$ denotes the Hankel determinant formed with the coefficients $a_n(\epsilon)$,

$$\lim_{\epsilon \to 0} H_m^{(n)}(\epsilon) = H_m^{(n)}.$$

On the other hand, since the poles of $f(z,\epsilon)$ are now simple, $H_m^{(n)}(\epsilon)$ may be calculated by (7.5-6), where the numbers $z_1, z_2, \ldots, z_k$ are to be replaced by $z_h(\epsilon) := z_1 + \epsilon w^h$ ($h=1,\ldots,k$), the numbers $u_1,\ldots,u_k$ by

$$u_h(\epsilon) := z_h(\epsilon)^{-1} = u_1(1 - \epsilon u_1 w^h + O(\epsilon^2)), \qquad (7.5\text{-}16)$$

and the numbers $r_1,\ldots,r_k$ by the $r_h(\epsilon)$ defined by (7.5-14). In computing the limit of the sum (7.5-10) as $\epsilon \to 0$, it is necessary to write

$$c_m(\epsilon) = c'_m(\epsilon) c''_m(\epsilon),$$

where

$$c'_m(\epsilon) := \prod_{1 \leq i < j \leq k} (u_i(\epsilon) - u_j(\epsilon))^2 \prod_{h=1}^{k} r_h(\epsilon), \qquad (7.5\text{-}17)$$

and $c''_m(\epsilon)$ contains all the remaining factors. Clearly, $\lim_{\epsilon \to 0} c''_m$ exists and is different from zero. By (7.5-14) and (7.5-16) we have, since the first product in (7.5-17) contains $k(k-1)$ factors ϵ,

$$c'_m(\epsilon) = u_1^{2k(k-1)} \prod_{1 \leq i < j \leq k} (w^i - w^j)^2 \prod_{h=1}^{k} \left(\frac{1}{k} \sum_{j=1}^{k} s_j w^{h-hj} \epsilon^{k-j} \right).$$

In the limit $\epsilon \to 0$ the only terms in the sums that are $\neq 0$ are those in which $j = k$. It thus follows that $\lim_{\epsilon \to 0} c'_m$ exists and is $\neq 0$. Thus (7.5-10) holds with a constant $c_m \neq 0$.

Each of the determinants $\hat{D}_m^{(n)}$ is obtained from a determinant $D_m^{(n)}$ by replacing at least one column with a column of smaller order of magnitude as $n \to \infty$. This does not suffice to establish (7.5-11) by letting $\epsilon \to 0$, for the coefficients $r_h(\epsilon)$ are unbounded as $\epsilon \to 0$. However, by grouping together those terms in $\sum \hat{D}_n^{(n)}$ that involve the quantities $u_1(\epsilon), \ldots, u_k(\epsilon)$ in the same columns and applying Laplace's expansion, we can see that the negative powers are cancelled; hence

$$\sum_K \hat{D}_m^{(n)} = O\left(\left(u_1^k u_{k+1} \cdots u_{m-1} \rho \right)^n \right)$$

holds in the limit $\epsilon \to 0$.

In this argument the assumption that z_1 is a pole of order $k > 1$ was just a matter of notational convenience. Nowhere did we use the fact that z_1 is a pole of smallest modulus. Also, for several higher order poles we evidently can argue similarly by generating them one after the other by the coalescence of simple poles. Thus Theorem 7.5b is completely proved, subject to a proof of Lemma 7.5c.

Proof of Lemma 7.5c. This is a simple application of the Lagrangian interpolation formula. Letting

$$p(z, \epsilon) := \prod_{i=1}^{k} (z - z_1 - \epsilon w^i) = (z - z_1)^k - \epsilon^k,$$

we have

$$s(z, \epsilon) = \frac{q(z, \epsilon)}{p(z, \epsilon)},$$

where

$$q(z, \epsilon) := \sum_{h=1}^{k} r_h(\epsilon) \prod_{\substack{i=1 \\ i \neq h}}^{k} (z - z_1 - \epsilon w^i)$$

is a polynomial of degree $\leq k-1$, assuming at the k points $z = z_1 + \epsilon w^h$ ($h = 1, 2, \ldots, k$) the values

$$r_h(\epsilon) p'(z_1 + \epsilon w^h, \epsilon) = r_h(\epsilon) \cdot k (\epsilon w^h)^{k-1}$$

$$= \sum_{j=1}^{k} (\epsilon w^h)^{k-j} s_j.$$

Thus it assumes the same values as the polynomial

$$q_1(z) := \sum_{j=1}^{k} s_j (z - z_1)^{k-j}.$$

Since a polynomial of degree $\leq k-1$ is uniquely determined by its values at k distinct points, it follows that $q = q_1$, hence that

$$s(z, \epsilon) = \sum_{j=1}^{k} \frac{s_j (z - z_1)^{k-j}}{(z - z_1)^k - \epsilon^k}.$$

The conclusion of the lemma is now evident. ∎

The proof of Theorem 7.5b being complete, we note two immediate consequences of the theorem.

COROLLARY 7.5d

Under the hypotheses of Theorem 7.5b, for each m such that (7.5-2) holds, if n is sufficiently large,

$$H_m^{(n)} \neq 0; \tag{7.5-18}$$

furthermore

$$\lim_{n \to \infty} \frac{H_m^{(n+1)}}{H_m^{(n)}} = u_1 u_2 \cdots u_m. \tag{7.5-19}$$

The corollary solves, in principle, the problem of determining the poles of a meromorphic function from the coefficients of its Taylor expansion at $z = 0$, provided that the moduli of the poles are distinct. We say more about this subject in §7.6.

We now consider the special case of Theorem 7.5b in which f is rational.

HANKEL DETERMINANTS

THEOREM 7.5e

Let the f of Theorem 7.5b be a rational function with a pole of order h at infinity and let the sum of the orders of all its finite poles be k. Then for $n > h$ and $m = k$ (7.5-3) is exact without the O-term, i.e.,

$$H_k^{(n)} = c_k (u_1 u_2 \cdots u_k)^n, \qquad n > h, \qquad (7.5\text{-}20)$$

where $c_k \neq 0$; furthermore,

$$H_m^{(n)} = 0 \quad \text{for all } n > h \text{ and all } m > k. \qquad (7.5\text{-}21)$$

Proof. We have $f = p + f_1$, where p is a polynomial of degree h and f_1 is a rational function vanishing at infinity with the same finite singularities as f. If all finite poles of f have order 1, then for $m = k$ and $n > h$ (7.5-5) is an exact representation of a_n with $b_n = 0$. Thus the determinants $\hat{D}_k^{(n)}$ are zero for $n > h$ and $H_k^{(n)}$ is given by (7.5-10). Equation (7.5-21) follows from the fact that for $m > k$ the determinants $D_m^{(n)}$ by necessity have two identical columns, hence are all zero. If f has finite poles of order > 1, the method of confluence of singularities can be carried through as in the proof of Theorem 7.5b. ∎

Theorem 7.5e has the following converse:

THEOREM 7.5f

Let $F := a_0 + a_1 x + a_2 x^2 + \cdots$ be a formal power series and let there exist integers h and $k \geq 0$ such that the Hankel determinants $H_m^{(n)}$ associated with F satisfy

$$H_k^{(n)} \neq 0, \quad H_{k+1}^{(n)} = 0 \quad \text{for all } n > h. \qquad (7.5\text{-}22)$$

Then F is the Taylor series at $z = 0$ of a rational function with a pole of order $\leq h$ at infinity and finite poles of total order k.

If $h < 0$, a "pole of order $\leq h$" means a zero of order $\geq -h$.

Proof. We introduce the row vectors

$$\mathbf{r}_n^T := (a_n, a_{n+1}, \ldots, a_{n+k}), \qquad n = 0, 1, \ldots$$

Since $H_{k+1}^{(n)} = 0$ for $n > h$, any $k+1$ consecutive vectors $\mathbf{r}_n^T, \mathbf{r}_{n+1}^T, \ldots, \mathbf{r}_{n+k}^T$ are linearly dependent for $n > h$. Since $H_k^{(n)} \neq 0$, any k consecutive vectors $\mathbf{r}_n^T, \ldots, \mathbf{r}_{n+k-1}^T$ are, however, linearly independent. It follows that for any $n > h$ the vector $\mathbf{r}_{n+k}^T$ is a linear combination of $\mathbf{r}_n^T, \ldots, \mathbf{r}_{n+k-1}^T$. We now consider the homogeneous system of k linear equations with $k+1$ unknowns:

$$\mathbf{r}_{h+1}^T \mathbf{c} = 0,$$

$$\mathbf{r}_{h+2}^T \mathbf{c} = 0, \qquad (7.5\text{-}23)$$

$$\cdots\cdots$$

$$\mathbf{r}_{h+k}^T \mathbf{c} = 0,$$

where

$$\mathbf{c} := \begin{bmatrix} c_k \\ c_{k-1} \\ \vdots \\ c_0 \end{bmatrix}$$

is the vector of unknowns. By a theorem of linear algebra the system (7.5-23) has a nontrivial solution $\mathbf{c}$. In this solution $c_k \neq 0$ and $c_0 \neq 0$, for otherwise we would have $H_k^{(h+1)} = 0$ or $H_k^{(h+2)} = 0$, thus contradicting (7.5-22). Since $\mathbf{r}_{h+k+1}$ is a linear combination of $\mathbf{r}_{h+1}, \ldots, \mathbf{r}_{h+k}$, (7.5-23) implies

$$\mathbf{r}_{h+k+1}^T \mathbf{c} = 0,$$

and by induction $\mathbf{r}_n^T \mathbf{c} = 0$ for all $n > h$ follows. This result means that the product of the formal power series

$$(a_0 + a_1 x + a_2 x^2 + \cdots)(c_0 + c_1 x + c_2 x^2 + \cdots + c_k x^k)$$

is a polynomial of degree $< h+k$ and thus that the series F arises by dividing a polynomial of degree $< h+k$ by $c_0 + c_1 x + \cdots + c_k x^k$, a polynomial of precise degree k. ∎

Theorem 7.5f exhibits a further connection between the coefficients of a Taylor series of an analytic function and the nature and location of the singularities of that function.

Theorem 7.5e mentions an example in which a certain column of the array of Hankel determinants associated with a given power series contains only nonzero entries. In the applications that deal with the quotient-difference algorithm it is important to know whether this statement can be made about all columns or at least about all columns up to a certain one.

HANKEL DETERMINANTS

A formal power series is called **normal** if its associated Hankel determinants satisfy $H_m^{(n)} \neq 0$ for all $m \geq 0$ and all $n \geq 0$; it is called k-**normal** if $H_m^{(n)} \neq 0$ for $m = 0, 1, \ldots, k$ and for all $n \geq 0$. Also, a series is termed **ultimately normal** (ultimately k-normal) if for every $m \geq 0$ (every m such that $0 \leq m \leq k$) there exists $n(m)$ such that $H_m^{(n)} \neq 0$ for $n > n(m)$.

Corollary 7.5d implies that the Taylor series at $z = 0$ of a meromorphic function with an infinite number of poles, all simple, whose moduli are all distinct, is ultimately normal. By the same corollary and by Theorem 7.5e the Taylor series which represents a rational function with k simple poles, all having distinct moduli, is ultimately m-normal for $m \leq k$ and not for any larger m.

We mention without proof a sufficient condition for normality.

THEOREM 7.5g

Let ψ be a real, nondecreasing, bounded function defined on $[0, \infty)$ and let the Stieltjes integrals

$$a_n := \int_0^\infty \tau^n d\psi(\tau), \qquad n = 0, 1, 2, \ldots, \tag{7.5-24}$$

all exist. Then the series $F = a_0 + a_1 x + a_2 x^2 + \cdots$ is k-normal if ψ has at least k points of increase; F is normal if ψ has an infinite number of points of increase.

For a discussion of Stieltjes integrals and for a proof of this theorem see Chapter 12 (Vol. 2). A special case of the representation (7.5-24) (where ψ has precisely k points of increase) is

$$a_n := \sum_{j=1}^k \gamma_j \beta_j^n, \qquad n = 0, 1, 2, \ldots, \tag{7.2-25}$$

where $\gamma_j > 0$ and where the β_j are distinct positive numbers ($j = 1, 2, \ldots, k$). By Theorem 7.2b this means that F is the Taylor series at 0 of the rational function

$$r(z) := \sum_{j=1}^k \frac{\gamma_j}{1 - \beta_j z}.$$

If p is a polynomial with k distinct zeros, all positive, its logarithmic derivative $p'p^{-1}$ is such a function.

The definition of the Hankel determinants $H_m^{(n)}$ given earlier is valid also for negative n. If, however, $n \leq -m$, then $H_m^{(n)}$ contains a row of zeros;

hence

$$H_m^{(n)} = 0, \quad n \leq -m. \tag{7.5-26}$$

The first determinant that is possibly different from zero is

$$H_m^{(-m+1)} = \begin{vmatrix} 0 & \cdots & 0 & a_0 \\ 0 & \cdots & a_0 & a_1 \\ \cdots & \cdots & \cdots & \cdots \\ a_0 & \cdots & a_{m-2} & a_{m-1} \end{vmatrix} = (-1)^{\epsilon_m} a_0^m, \tag{7.5-27}$$

where

$$\epsilon_m = \frac{m(m-1)}{2} \tag{7.5-28}$$

is the number of column interchanges that bring the a_0 to the main diagonal. Incidentally,

$$H_m^{(-m+2)} = (-1)^{\epsilon_m} \begin{vmatrix} a_1 & a_2 & \cdots & a_{m-1} & a_m \\ a_0 & a_1 & \cdots & a_{m-2} & a_{m-1} \\ 0 & a_0 & \cdots & a_{m-3} & a_{m-2} \\ & & \cdots & & \\ 0 & 0 & \cdots & a_0 & a_1 \end{vmatrix} \tag{7.5-29}$$

is a determinant connected with the coefficients of the reciprocal series F^{-1} by means of the Wronski formula (Theorem 1.3).

We call a formal power series F **hypernormal** if the associated Hankel determinants satisfy

$$H_m^{(n)} \neq 0 \quad \text{for} \quad m = 1, 2, \ldots, \text{ and all } n > -m; \tag{7.5-30}$$

we call it (k, h)-**hypernormal** if

$$H_m^{(n)} \neq 0 \text{ whenever } 0 \leq m \leq k \text{ and/or } -m+1 \leq n \leq -m+h. \tag{7.5-31}$$

It is clear that hypernormality implies normality, and (k, h) normality implies k-normality for every $h \geq 0$. We state without proof a theorem due to Schönberg and Karlin (for a proof see Gragg [1972]) which contains sufficient conditions for hypernormality.

HANKEL DETERMINANTS

THEOREM 7.5h

Let $\gamma \geq 0$ and let $\{\alpha_i\}_1^\infty$ and $\{\beta_i\}_1^\infty$ be two infinite sequences of non-negative numbers such that $\sum_{i=1}^\infty (\alpha_i + \beta_i) < \infty$ and $\alpha_i \neq \beta_j$ for all i and j such that $\alpha_i \beta_j \neq 0$. Let F be the Taylor series at O of the meromorphic function

$$f(z) := e^{\gamma z} \frac{\prod_{i=1}^\infty (1 + \alpha_i z)}{\prod_{j=1}^\infty (1 - \beta_j z)}. \tag{7.5-32}$$

Then F is hypernormal if $\gamma > 0$, or if at least one of the sequences $\{\alpha_i\}$ and $\{\beta_j\}$ has an infinite number of nonzero elements. If $\gamma = 0$ and both sequences have only a finite number of nonzero elements, then F is (k,h)-hypernormal, where k and h are the numbers of nonzero elements in the sequences $\{\beta_j\}$ and $\{\alpha_i\}$, respectively.

It can be shown that many important functions of mathematical physics, such as the resolvents of many self-adjoint eigenvalue problems, can be brough to the form (7.5-32) by elementary transformations. A typical example is

$$f(z) := z^{\nu/2} [J_\nu(2\sqrt{z})]^{-1},$$

where J_ν is the Bessel function and $\nu > -1$.

It should be emphasized that the sufficient conditions for normality given by Theorems 7.5g and 7.5h are by no means necessary. There is overwhelming numerical and some analytical evidence that many power series are normal (or hypernormal) without satisfying any of the above sets of conditions, for instance real power series representing meromorphic functions with complex conjugate poles.

PROBLEMS

1. Obtain the traditional expansion of a determinant of order n in terms of determinants of order $n-1$ as a special case of Laplace's expansion.
2. Write the complete expansion of a fourth-order determinant in terms of minors formed with the first two rows.
3. Show that in the definition of the sign of the algebraic complement of a minor it makes no difference whether the sums of the indices of the *deleted* rows and columns or the sums of the indices of the *retained* rows and columns are counted. Show also that it does not matter whether the rows and columns are numbered 1 to n or 0 to $n-1$.
4. What statements can be made about the Hankel determinants of a power series known to represent an entire function?

5. Using Theorem 7.2a and the method of confluence of singularities, show that

$$\sum_{n=1}^{\infty} \frac{1}{n^k} = \frac{1}{(k-1)!} \int_0^1 \frac{(\log \tau)^{k-1}}{1-\tau} \, d\tau, \quad k = 2, 3, \ldots.$$

§7.6. THE QUOTIENT-DIFFERENCE ALGORITHM

Let $\{a_n\}$ be a sequence of elements of an additive (abelian) group; in applications this is usually the group of real or complex numbers. It is generally understood what is meant by the (forward) **difference scheme** of the sequence $\{a_n\}$. We may think of it as a two-dimensional array of numbers $d_m^{(n)}$, arranged in the form

$$\begin{array}{cccc}
d_1^{(0)} & & & \\
& d_2^{(0)} & & \\
d_1^{(1)} & & d_3^{(0)} & \\
& d_2^{(1)} & & \ddots \\
d_1^{(2)} & & d_3^{(1)} & \\
& d_2^{(2)} & & \vdots \\
d_1^{(3)} & & \vdots & \\
\vdots & & &
\end{array}$$

and defined by the recurrence relations

$$d_1^{(n)} := a_{n+1} - a_n, \quad n = 0, 1, 2, \ldots; \tag{7.6-1}$$

$$d_{m+1}^{(n)} := d_m^{(n+1)} - d_m^{(n)}, \quad m = 1, 2, \ldots; \quad n = 0, 1, 2, \ldots. \tag{7.6-2}$$

We have reversed the usual role of subscripts and superscripts to stress the analogy with the notation commonly used for the **quotient-difference scheme** (qd scheme). This scheme is built up similarly from a given sequence $\{a_n\}$, with the exception that the forming of differences now alternates with the forming of quotients; moreover, both quotients and differences are modified in a certain simple manner. Since quotients must be formed, it must now be assumed that the a_n are members of a field; again, in applications this is usually the field of real or complex numbers. Modified quotients and differences are denoted by $q_m^{(n)}$ and $e_m^{(n)}$, respectively, and are conventionally arranged as in Table 7.6a.

QUOTIENT-DIFFERENCE ALGORITHM

The rules defining the qd scheme are

(a) the initial conditions [analogous to (7.6-1)]:

$$e_0^{(n)} := 0, \quad n = 1, 2, \ldots,$$

$$q_1^{(n)} := \frac{a_{n+1}}{a_n}, \quad n = 0, 1, 2, \ldots;$$

(7.6-3)

(b) the continuation relations (analogous to (7.6-2):

$$e_m^{(n)} := [q_m^{(n+1)} - q_m^{(n)}] + e_{m-1}^{(n+1)}, \quad m = 1, 2, \ldots; \quad n = 0, 1, 2, \ldots, \quad (7.6\text{-}4a)$$

$$q_{m+1}^{(n)} := \frac{e_m^{(n+1)}}{e_m^{(n)}} q_m^{(n+1)}, \quad m = 1, 2, \ldots; \quad n = 0, 1, 2, \ldots. \quad (7.6\text{-}4b)$$

Table 7.6a The Quotient-Difference Scheme

		$q_1^{(0)}$				
0			$e_1^{(0)}$			
		$q_1^{(1)}$		$q_2^{(0)}$		
0			$e_1^{(1)}$		$e_2^{(0)}$	
		$q_1^{(2)}$		$q_2^{(1)}$		
0			$e_1^{(2)}$		$e_2^{(1)}$	$\cdot$ $\cdot$
		$q_1^{(3)}$		$q_2^{(2)}$		
0			$e_1^{(3)}$		$e_2^{(2)}$	
		$\vdots$		$q_2^{(3)}$		$\vdots$
			$\vdots$			

Each of the rules (7.6-4) connects four adjacent elements of the qd scheme. The rule (7.6-4a) states that in any rhombuslike configuration of four elements centered in a q-column the *sums* of the two NE and the two SW elements are equal. Rule (7.6-4b) states that in any rhombuslike configuration centered in an e-column the *products* of the NE and the SW elements are equal. For this reason Equations (7.6-4) are frequently referred to as the **rhombus rules**.

In applications the starting elements of the qd scheme are usually the coefficients of some formal power series (which may even converge). The

scheme we have defined is called the *qd* scheme **associated** with the formal power series
$$F := a_0 + a_1 x + a_2 x^2 + \cdots.$$

EXAMPLE 1

For the power series
$$0! + 1!x + 2!x^2 + 3!x^3 + \cdots$$
the following scheme results:

```
        1
   0         1
        2         2
   0         1         2
        3         3         3
   0         1         2         3
        4         4         4
   0         1         2         3
        5         5         5
```

In this special case it would be possible to give a formula for the elements of the scheme. In general, this is not so, even if the a_n are given analytically, which is not always the case.

It is plain that the difference scheme of a sequence $\{a_n\}$ always exists. With the quotient-difference scheme the question of existence is not so trivial, since one of the denominators in (7.6-3) or (7.6-4) may happen to be zero. Indeed, if one of the coefficients a_n is zero, the very first q-column fails to exist. The question of existence is clarified by the following theorem:

THEOREM 7.6a

Let F be a formal power series and let $H_k^{(n)}$ be the Hankel determinants associated with F. If there exists a positive integer k such that F is k-normal (see §7.5 for definition) the columns $q_m^{(n)}$ of the qd scheme associated with F exist for $m = 1, 2, \ldots, k$, and

$$q_m^{(n)} = \frac{H_m^{(n+1)} H_{m-1}^{(n)}}{H_m^{(n)} H_{m-1}^{(n+1)}}, \qquad (7.6\text{-}5)$$

$$e_m^{(n)} = \frac{H_{m+1}^{(n)} H_{m-1}^{(n+1)}}{H_m^{(n)} H_m^{(n+1)}} \qquad (7.6\text{-}6)$$

QUOTIENT-DIFFERENCE ALGORITHM

for $m = 1, 2, \ldots, k$ and all $n \geq 0$.

Proof. We denote the quantities defined by the expressions on the right of (7.6-5) and (7.6-6) by $\hat{q}_m^{(n)}$ and $\hat{e}_m^{(n)}$, respectively, and set $\hat{e}_0^{(n)} := 0$. Since $H_0^{(n)} = 1$, $H_1^{(n)} = a_n$,

$$\hat{q}_1^{(n)} = \frac{a_{n+1}}{a_n},$$

hence

$$\hat{q}_1^{(n)} = q_1^{(n)}$$

follows for all $n \geq 0$. By definition $\hat{e}_0^{(n)} = e_0^{(n)}$. Thus the quantities $\hat{e}_m^{(n)}$ and $\hat{q}_m^{(n)}$ satisfy the correct initial conditions (7.6-3), and the theorem is proved if it can be shown that they also satisfy the continuation rules (7.6-4). The rule (7.6-4b) follows easily, since

$$\hat{q}_{m+1}^{(n)} \hat{e}_m^{(n)} = \hat{q}_m^{(n+1)} \hat{e}_m^{(n+1)} = \frac{H_{m-1}^{(n+1)} H_{m+1}^{(n+1)}}{\left[H_m^{(n+1)}\right]^2}.$$

The rule (7.6-4a) follows from the definitions by two applications of Jacobi's identity (Theorem 7.5a). Because the continuation rules determine the $q_m^{(n)}$ and $e_m^{(n)}$ uniquely, we have $q_m^{(n)} = \hat{q}_m^{(n)}$ and $e_m^{(n)} = \hat{e}_m^{(n)}$ for $m = 1, 2, \ldots, k$ and all $n \geq 0$, establishing the formulas (7.6-5) and (7.6-6). ∎

The following examples also serve as illustrations for Theorems 7.5g and 7.5h.

EXAMPLE 2

The series $F := 0! + 1!x + 2!x^2 + \cdots$ is normal by the criterion of Theorem 7.5g because

$$n! = \int_0^\infty \tau^n e^{-\tau} d\tau, \quad n = 0, 1, 2, \ldots.$$

We have already seen in Example 1 that the entire qd scheme exists.

EXAMPLE 3

The exponential series

$$e^x = 1 + \frac{1}{1!}x + \frac{1}{2!}x^2 + \cdots$$

is normal by the criterion of Theorem 7.5h. Hence the entire qd scheme exists. It is easily verified that

$$\begin{cases} q_k^{(n)} = \dfrac{n+k-1}{(n+2k-2)(n+2k-1)}, \\ e_k^{(n)} = -\dfrac{k}{(n+2k-1)(n+2k)}. \end{cases} \quad (7.6\text{-}7)$$

We now state an important analytical property of the qd scheme.

THEOREM 7.6b

Let F be the Taylor series at $z=0$ of a function f meromorphic in the disk $D:|z|<\sigma$ and let the poles $z_i = u_i^{-1}$ of f in D be numbered such that

$$0 < |z_1| \leq |z_2| \leq \cdots < \sigma,$$

each pole occurring as many times in the sequence $\{z_k\}$ as indicated by its order. If F is ultimately k-normal for some integer $k>0$, then the qd scheme associated with F has the following properties:

(a) For each m such that $0 < m \leq k$ and

$$|z_{m-1}| < |z_m| < |z_{m+1}| \quad (7.6\text{-}8)$$

(where $z_0 := 0$ and, if f has only k poles, $z_{k+1} := \infty$),

$$\lim_{n \to \infty} q_m^{(n)} = u_m; \quad (7.6\text{-}9)$$

(b) For each m such that $0 < m \leq k$ and

$$|z_m| < |z_{m+1}|, \quad (7.6\text{-}10)$$

$$\lim_{n \to \infty} e_m^{(n)} = 0. \quad (7.6\text{-}11)$$

Proof. Since the hypotheses of the theorem include those of Theorem 7.5b, we may apply (7.5-3), and (7.6-9) follows immediately by virtue of (7.6-5). By (7.6-6) the relation (7.5-3) enables us to prove (7.6-11) if in addition to (7.6-10) we assume $|z_{m-1}| < |z_m| < |z_{m+1}| < |z_{m+2}|$. To prove (7.6-11) merely under the hypothesis (7.6-10) we require an asymptotic relation for the Hankel determinants which is analogous to but more

QUOTIENT-DIFFERENCE ALGORITHM

complicated than that of Theorem 7.5b. It is convenient to postpone statement and proof of this relation to §7.9, where it is also shown how to deal with the case in which several poles have the same modulus. ∎

For Taylor series representing rational functions, the statement (7.6-11) may be strengthened as follows:

THEOREM 7.6c

Let the f of Theorem 7.6b be a rational function having a pole of order h at infinity, and let the sum of the orders of all its finite poles be k. Then if the series F is k-normal,

$$e_k^{(n)} = 0 \quad \text{for all } n > h. \tag{7.6-12}$$

Proof. This is an immediate consequence of Theorem 7.5e and of (7.6-6). ∎

Judging from the above theorems the **qd algorithm**, as defined by (7.6-4), would appear to be an ingenious tool for determining, under certain conditions, the poles of a meromorphic function f directly from its Taylor series at the origin. The algorithm would proceed as follows: presupposing normality (or k-normality) of the power series, the qd scheme would be built up from the left to the right. Any q-column corresponding to a simple pole of isolated modulus would tend to the reciprocal value of that pole. It would be flanked by e-columns that tend to zero, which behavior is easily recognized numerically. If f is rational, the last e-column would be zero, which could serve as a test on the accuracy of the computation.

Unfortunately the test just mentioned would almost certainly be failed because the qd algorithm as just described is numerically unstable. Generally speaking, an algorithm is called **numerically unstable** if the result depends in an overly sensitive way on rounding errors; for a more thorough discussion of the concept we must refer the reader to a numerical analysis text. In the present case the reason for the numerical instability is not hard to find. We suppose, for simplicity, that the condition (7.6-10) is satisfied for all $m \leq k$. The relations (7.6-9) and (7.6-11) then hold for all $m \leq k$; moreover, it is an easy result of Theorem 7.5b that

$$e_m^{(n)} \sim \left(\frac{u_{m+1}}{u_m}\right)^n \quad \text{as } n \to \infty. \tag{7.6-13}$$

If the qd scheme is constructed on a machine working with a fixed number of binary digits, the elements in the first q-column are determined (since they tend to a nonzero limit) with an error that can be as large as 2^{-s}, say, for arbitrarily large values of n. The column $e_1^{(n)}$, consisting of differences of consecutive q's, has absolute errors that can be as large as 2^{1-s}. When, however, the column $q_2^{(n)}$ is formed by correcting $q_1^{(n+1)}$ by the factor

$e_1^{(n+1)}/e_1^{(n)}$, the relative errors of the $e_1^{(n)}$ matter. By (7.6-13) these relative errors can be as large as

$$\left(\frac{u_1}{u_2}\right)^n 2^{1-s},$$

which tends to infinity if $|u_1|>|u_2|$, no matter how large the s, i.e., how accurate the machine. Thus already the second q-column cannot be generated accurately, let alone the further columns of the scheme. These pessimistic conclusions are borne out in full by numerical experiments.

The error study carried out above points a way to a more stable manner of constructing the qd scheme. The rhombus rules (7.6-4) defining the scheme may be rearranged as follows:

$$q_m^{(n+1)} := [e_m^{(n)} - e_{m-1}^{(n+1)}] + q_m^{(n)}, \tag{7.6-14a}$$

$$e_m^{(n+1)} := \frac{q_{m+1}^{(n)}}{q_m^{(n+1)}} e_m^{(n)}. \tag{7.6-14b}$$

Written in this way, the formulas permit the construction of the qd scheme row by row, proceeding from top to bottom. In this manner the *absolute* errors of the small quantities $e_m^{(n)}$ come into play, and quotients are formed only with the quantities $q_m^{(n)}$ which do not tend to zero and thus have bounded relative errors. Again these qualitative considerations are confirmed by numerical experiments.

The row-wise generation of the qd scheme is called the **progressive form of the qd algorithm**. It obviously poses the problem of getting started. The algorithm could be started, for instance, if a first diagonal of the scheme were known (Fig. 7.6a). Together with the column $e_0^{(n)}$, which is known to be zero, this knowledge suffices to generate further diagonals successively by means of (7.6-14). It is shown in Chapter 12 how to obtain the first diagonal with a continued fraction expansion of f.

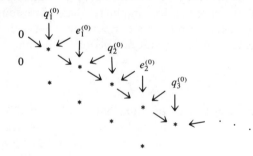

Fig. 7.6a. Starting a qd scheme from a diagonal. The arrows indicate entries that are needed to construct further entries.

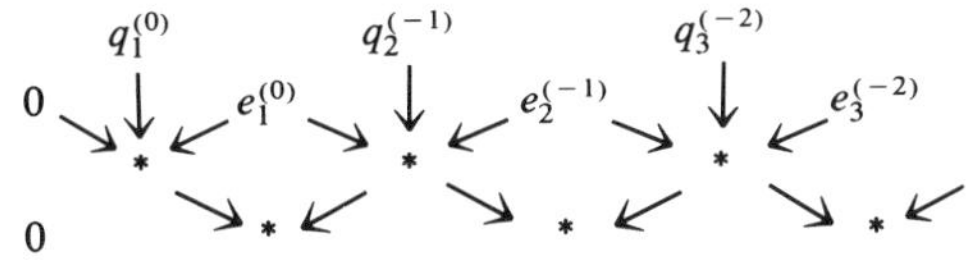

QUOTIENT-DIFFERENCE ALGORITHM

Fig. 7.6b. Starting a qd scheme from the first two horizontal rows. Arrows are used as in Fig. 7.6a.

Another, simpler approach consists in starting the scheme from the first two horizontal rows of q's and e's (see Fig. 7.6b). This also requires the definition of $q_m^{(n)}$ and $e_m^{(n)}$ for certain negative values of n. The recurrence relations (7.6-4) cannot be used, for no starting values are known. As long as no denominator is zero, however, $q_m^{(n)}$ and $e_m^{(n)}$ may be defined by (7.6-5) and (7.6-6) for negative n. The scheme thus obtained is called the **extended qd scheme**. The rhombus rules (7.6-4) also hold in the extended scheme, for they are based solely on formal identities between Hankel determinants.

If the series $F := a_0 + a_1 x + a_2 x^2 + \cdots$ is *hypernormal* (see §7.5), the denominators in $q_m^{(n)}$ and $e_m^{(n)}$ are $\neq 0$ for $n > -m$, $m = 1, 2, \ldots$, and in particular for the first two rows of the scheme shown in Fig. 7.6b. By virtue of (7.5-26) it turns out that

$$q_m^{(-m+1)} = 0, \quad m = 1, 2, \ldots. \quad (7.6\text{-}15)$$

The quantities $e_m^{(-m+1)}$ are related to the qd scheme of the reciprocal series F^{-1}, as shown by the following result.

THEOREM 7.6d

Let the series $F := a_0 + a_1 x + a_2 x^2 + \cdots$ *be hypernormal and let*

$$\hat{F} := b_0 + b_1 x + b_2 x^2 + \cdots := (a_0 + a_1 x + a_2 x^2 + \cdots)^{-1} = F^{-1},$$

the reciprocal of F. *Then* $\hat{F}$ *is hypernormal also, and the following relations hold between the elements* $(q_m^{(n)}, e_m^{(n)})$ *and* $(\hat{q}_m^{(n)}, \hat{e}_m^{(n)})$ *of the qd schemes associated with series* F *and* $\hat{F}$:

$$\hat{q}_1^{(0)} = -q_1^{(0)}; \quad (7.6\text{-}16\text{a})$$

$$\hat{q}_m^{(n)} = e_{n+m-1}^{(1-n)}, \quad \hat{e}_m^{(n)} = q_{n+m}^{(1-n)}, \quad n > -m; \quad (7.6\text{-}16\text{b})$$

$$q_m^{(n)} = \hat{e}_{n+m-1}^{(1-n)}, \quad e_m^{(n)} = \hat{q}_{n+m}^{(1-n)}, \quad n > -m. \quad (7.6\text{-}16\text{c})$$

In particular,

$$e_m^{(1-m)} = \hat{q}_1^{(m)} = \frac{b_{m+1}}{b_m}, \quad m = 1, 2, \ldots.$$

The relations (7.6-16) mean that the *qd* scheme of series $\hat{F}$ can be obtained from the extended scheme of series F by reflection at the diagonal $n = \frac{1}{2}$. The only exception occurs in the element $\hat{q}_1^{(0)}$.

Proof of Theorem 7.6d. Let the quantities $\hat{q}_m^{(n)}$ and $\hat{e}_m^{(n)}$ be *defined* by (7.6-16). We shall show that they satisfy (a) the correct initial conditions and (b) the correct continuation relations for the *qd* scheme of $\hat{F}$. The assertion then follows from the fact that the *qd* scheme of a formal power series is uniquely characterized by these properties.

As to (a), we note that by (7.6-15)

$$\hat{e}_0^{(n)} = q_n^{(1-n)} = 0, \quad n = 1, 2, \ldots,$$

which agrees with (7.6-3). Furthermore, using (7.6-6),

$$\hat{q}_1^{(n)} = e_n^{(1-n)} = \frac{H_{n+1}^{(1-n)} H_{n-1}^{(2-n)}}{H_n^{(1-n)} H_n^{(2-n)}}, \quad n = 1, 2, \ldots.$$

The Hankel determinants appearing here can be expressed differently. By (7.5-27)

$$H_n^{(1-n)} = (-1)^{\epsilon_n} a_0^n, \quad n = 1, 2, \ldots,$$

where $\epsilon_n := \binom{n}{2}$, and by (7.5-29), combined with the Wronski formula (Theorem 1.3),

$$H_n^{(2-n)} = (-1)^{\epsilon_n + n} a_0^{n+1} b_n, \quad n = 1, 2, \ldots.$$

A short computation now yields

$$\hat{q}_1^{(n)} = \frac{b_{n+1}}{b_n}, \quad n = 1, 2, \ldots,$$

again in accordance with (7.6-3).

As to (b) we have, using (7.6-4),

$$\hat{q}_m^{(n)} + \hat{e}_m^{(n)} = e_{n+m-1}^{(1-n)} + q_{n+m}^{(1-n)}$$

$$= e_{n+m}^{(-n)} + q_{n+m}^{(-n)}$$

$$= \hat{q}_m^{(n+1)} + \hat{e}_{m-1}^{(n+1)},$$

which establishes the first rhombus rule for $\hat{q}$ and $\hat{e}$. [The exception for $n=0$ and $m=1$ is taken care of by the reversal of sign in (7.6-16a).] The second rhombus rule is similarly established. ∎

EXAMPLE 4

The reciprocal of the exponential series considered in Example 3 is, of course,

$$e^{-x} = 1 - \frac{1}{1!}x + \frac{1}{2!}x^2 - \cdots.$$

Changing the sign of x reverses all signs in the qd table. Hence we expect

$$\hat{q}_m^{(n)} = -\frac{n+m-1}{(n+2m-2)(n+2m-1)},$$

$$\hat{e}_m^{(n)} = +\frac{m}{(n+2m-1)(n+2m)}.$$

On the other hand, by Theorem 7.6d these quantities should equal $e_{n+m-1}^{(1-n)}$ and $q_{n+m}^{(1-n)}$, respectively, as defined by (7.6-7). It is readily verified that this is the case.

The method of proof of Theorem 7.6d also yields the following:

COROLLARY 7.6e

If the series F is (k,h)-hypernormal (see §7.5), then $\hat{F} := F^{-1}$ is (h,k)-hypernormal. The relations (7.6-16) still hold where the qd schemes can be constructed.

By virtue of Theorem 7.6d and its corollary it is now a simple matter to start the progressive form of the qd algorithm. We discuss two typical problems.

ZEROS OF POLYNOMIALS

Let $p(z) := b_0 + b_1 z + \cdots + b_k z^k$ ($b_0 b_k \neq 0$) be a polynomial of degree k. Its zeros can be found as the poles of the rational function $r := p^{-1}$. We

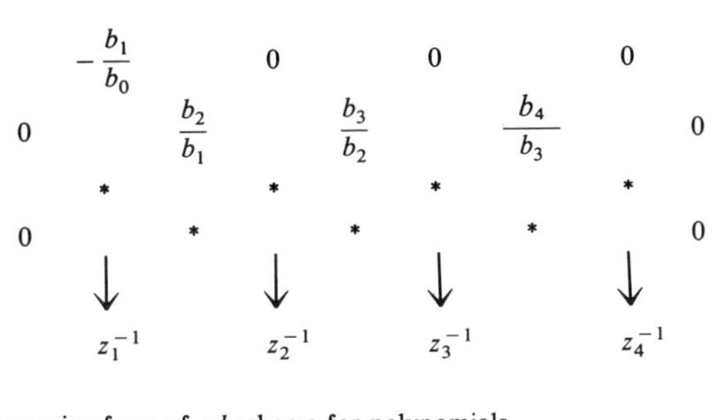

$$\begin{array}{ccccccccc}
& & -\dfrac{b_1}{b_0} & & 0 & & 0 & & 0 \\
0 & & & \dfrac{b_2}{b_1} & & \dfrac{b_3}{b_2} & & \dfrac{b_4}{b_3} & & 0 \\
& * & & * & & * & & * & \\
0 & & * & & * & & * & & 0 \\
& \downarrow & & \downarrow & & \downarrow & & \downarrow & \\
& z_1^{-1} & & z_2^{-1} & & z_3^{-1} & & z_4^{-1} &
\end{array}$$

Fig. 7.6c Progressive form of qd scheme for polynomials.

assume that the Taylor series representing r at $z=0$ is $(k,0)$-hypernormal, which in particular implies that all $b_i \neq 0$. (By Theorem 7.5h this is true, for instance, if the zeros are positive and simple.) The first two rows of the extended qd scheme of r can be constructed from the coefficients of $r^{-1} = p$. Moreover, since r has a zero of order k at infinity, $e_k^{(n)} = 0$ for $n > -k$. Thus the qd scheme is flanked on both sides by a column of zeros (see Fig. 7.6c for $k=4$).

If the zeros of p (i.e., the poles of r) are denoted by z_m, $m=1,2,\ldots,k$, and the moduli of these zeros are all different, then by Theorem 7.6b the mth q-column of the scheme tends to z_m^{-1}. By reversing the order of the coefficients of p [i.e., by considering $p^*(z) := z^k p(z^{-1})$ in place of p] we may also construct a scheme whose q-columns tend to the reciprocals of the zeros of p^*, i.e., to $(z_m^{-1})^{-1} = z_m$, the zeros of p.

EXAMPLE 5

The following scheme results for the polynomial

$$p(z) := \frac{1}{120}(120 - 600z + 600z^2 - 200z^3 + 25z^4 - z^5),$$

the Laguerre polynomial of degree 5:

25.000000	0.000000	0.000000	0.000000	0.000000	
0.000000	−8.000000	−3.000000	−1.000000	−0.200000	0.000000
17.000000	5.000000	2.000000	0.800000	0.200000	
0.000000	−2.352941	−1.200000	−0.400000	−0.050000	0.000000
14.647059	6.152941	2.800000	1.150000	0.250000	
0.000000	−0.988424	−0.546080	−0.164286	−0.010870	0.000000
13.658635	6.595285	3.181795	1.303416	0.260870	
0.000000	−0.477276	−0.263448	−0.067299	−0.002175	0.000000
13.181358	6.809113	3.377943	1.368540	0.263045	
0.000000	−0.246547	−0.130694	−0.027266	−0.000418	0.000000
12.934811	6.924966	3.481372	1.395388	0.263463	
↓	↓	↓	↓	↓	
12.640801	7.085810	3.596426	1.413403	0.263560	

After 28 q-lines the errors in all zeros are less than 10^{-6}; after 56 q-lines, less than 10^{-12}. In the last q-column the convergence is especially rapid because of the favorable ratio u_5/u_4.

ZEROS OF ENTIRE FUNCTIONS

Let

$$g(z) := b_0 + b_1 z + b_2 z^2 + \cdots$$

be an entire function and let the series be hypernormal. Then the zeros z_i of g (if there are any) can be found as the poles of the meromorphic function $f := g^{-1}$. The first two rows of the qd scheme can be constructed, as indicated above, from the coefficients of $f^{-1} = g$, but since all rows are now infinite the scheme can no longer be generated row by row. The stable progressive form of the algorithm can still be used, however, to construct the scheme by diagonals slanted upward, as shown in Fig. 7.6d. Each diagonal is built from the top right to the bottom left.

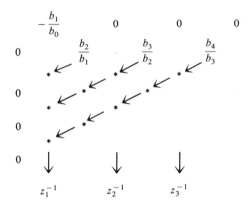

Fig. 7.6d. The progressive form of qd scheme for entire functions. The arrows indicate the order of computation.

EXAMPLE 6

The zeros of the Bessel function of order $\nu > -1$,

$$J_\nu(x) := \frac{(x/2)^\nu}{\Gamma(\nu+1)} \sum_{n=0}^{\infty} \frac{(-x^2/4)^n}{n!(\nu+1)_n}$$

(see §8.3 for the definition of the function Γ) may be computed by considering the power series

$$f(z) := \Gamma(\nu+1) z^{-\nu/2} J_\nu(2\sqrt{z}) = \sum_{n=0}^{\infty} \frac{(-z)^n}{n!(\nu+1)_n},$$

which is hypernormal by Theorem 7.5h. Here the required ratios of coefficients are simply

$$\frac{b_{n+1}}{b_n} = -\frac{1}{(n+1)(n+\nu+1)}.$$

The following scheme results, e.g., for $\nu = \frac{1}{3}$:

0.750000		0.000000		0.000000		0.000000
0.000000	−0.214286		−0.100000		−0.057692	
	0.535714		0.114286		0.042308	
0.000000	−0.045714		−0.037019			
	0.490000		0.122981			
0.000000	−0.011473					
	0.478527					
0.000000	↓		↓		↓	
	0.474777		0.109908		0.047563	

If the limits of the q-columns are denoted by u_m ($m = 1, 2, \ldots$), the zeros of the Bessel function are $x_m = 2/\sqrt{u_m}$. From the above limits the result is

$$x_1 = 2.90268 \qquad x_2 = 6.0327, \qquad x_3 = 9.1705.$$

ZEROS AND POLES OF MEROMORPHIC FUNCTIONS

Let $F := a_0 + a_1 x + a_2 x^2 + \cdots$ be the hypernormal Taylor series at $z = 0$ of a meromorphic function f with simple poles z_i and simple zeros w_i that satisfy

$$0 < |z_1| < |z_2| < |z_3| < \cdots,$$

$$0 < |w_1| < |w_2| < |w_3| < \cdots.$$

Since the first two columns of the qd scheme are known, the scheme may be constructed again by diagonals slanted upward, proceeding from the bottom left to the top right. In addition to the fact, already established in Theorem 7.6b, that

$$\lim_{n \to \infty} q_m^{(n)} = z_m^{-1}, \qquad m = 1, 2, \ldots, \qquad (7.6\text{-}17)$$

it follows from Theorem 7.6d in conjunction with Theorem 7.6b that

$$\lim_{n \to \infty} e_{n+m-1}^{(1-n)} = w_m^{-1}, \qquad m = 1, 2, \ldots; \qquad (7.6\text{-}18)$$

i.e., the horizontal e-rows of the scheme tend to the reciprocals of the zeros

QUOTIENT-DIFFERENCE ALGORITHM

of f. Thus in theory the qd scheme furnishes the zeros and poles of a meromorphic function simultaneously. However, since the scheme is generated from left to right and the q-columns tend to limits, this form of the algorithm is again unstable.

PROBLEMS

1. Show that the qd scheme associated with
$$F := 1 + \tfrac{1}{2}x + \tfrac{1}{3}x^2 + \tfrac{1}{4}x^3 + \cdots$$
is given by
$$q_k^{(n)} = \frac{(n+k)^2}{(n+2k-1)(n+2k)}, \qquad e_k^{(n)} = \frac{k^2}{(n+2k)(n+2k+1)}.$$

2. Let s be a complex number, $s \neq 0, -1, -2, \ldots$. Construct the qd scheme associated with the formal power series
$$F := \sum_{n=0}^{\infty} (s)_n x^n.$$

3. If, in the qd scheme associated with a k-hypernormal series F, the column $e_k^{(n)}$ is identically zero, show that the sum
$$q_1^{(n)} + q_2^{(n-1)} + \cdots + q_k^{(n-k+1)}$$
is independent of n and determine its value.

4. What statements can be made about the qd scheme of an entire function whose Taylor expansion at 0 is normal? Verify for $f(z) = e^z$.

5. If the function f in Theorem 7.6b has precisely two poles of smallest modulus, $|z_1| = |z_2| < |z_3|$, show by using the representation (7.5-5) for the Taylor coefficients that
$$\lim_{n \to \infty} [q_1^{(n+1)} + q_2^{(n)}] = u_1 + u_2, \qquad (7.6\text{-}19)$$
$$\lim_{n \to \infty} q_1^{(n)} q_2^{(n)} = u_1 u_2. \qquad (7.6\text{-}20)$$

6. The coefficients c_n in
$$-\frac{x}{\text{Log}(1-x)} = c_0 + c_1 x + c_2 x^2 + \cdots$$
are required in the theory of numerical integration. Determine them numerically from the qd scheme associated with
$$F(x) := -\frac{\text{Log}(1-x)}{x} = 1 + \tfrac{1}{2}x + \tfrac{1}{3}x^2 + \cdots,$$
by using the symmetry property expressed in Theorem 7.6d.
[*Caution*: The formulas given in Problem 1 hold only for $n \geq 0$.]

§7.7. HADAMARD POLYNOMIALS

The properties of the qd scheme established in §7.6 still leave several questions unanswered; for instance, we have not yet shown how to deal with the case in which several poles of the function f have the same modulus. We also might ask for an estimate of the error $|q_m^{(n)} - u_m|$ of the approximate values of the poles furnished by the qd algorithm. Then there is the problem of determining the residues of a rational or meromorphic function with simple poles. Finally, we must consider the problem, occurring in the theory of interpolation by exponentials [see (7.2-27)], of recovering the denominator of a rational function from its Taylor series at 0. All these questions can be attacked in the study of certain *polynomials* that can be constructed from the qd scheme.

Let $F := a_0 + a_1 x + a_2 x^2 + \cdots$ be a formal power series. In addition to the Hankel determinants $H_k^{(n)}$, we define the **Hankel polynomials** $H_0^{(n)}(u) := 1$,

$$H_k^{(n)}(u) := \begin{vmatrix} a_n & a_{n+1} & \cdots & a_{n+k-1} & 1 \\ a_{n+1} & a_{n+2} & \cdots & a_{n+k} & u \\ \cdots & \cdots & \cdots & \cdots & \cdots \\ a_{n+k} & a_{n+k+1} & & a_{n+2k-1} & u^k \end{vmatrix}, \quad (7.7\text{-}1)$$

$$n = 0, \pm 1, \pm 2, \ldots; \quad k = 1, 2, \ldots.$$

(It is convenient here to abandon strict functional notation and to denote by $H_k^{(n)}(u)$ the function itself and not only the value of the function.) These polynomials enjoy properties similar to those of the Hankel determinants studied in §7.5.

THEOREM 7.7a

For all integers n, all $k \geq 1$, and all u

$$H_{k-1}^{(n+1)} H_k^{(n)}(u) - u H_k^{(n)} H_{k-1}^{(n+1)}(u) + H_k^{(n+1)} H_{k-1}^{(n)}(u) = 0. \quad (7.7\text{-}2)$$

Proof. The proof is similar to that of Jacobi's identity (Theorem 7.5a). In addition to the $(k+1)$-dimensional vectors $\mathbf{h}_n, \mathbf{e}_1$, and $\mathbf{e}_{k+1}$ introduced there, we require the vector

$$\mathbf{u} := \begin{bmatrix} 1 \\ u \\ \vdots \\ u^k \end{bmatrix}$$

and compute the determinant of the matrix

$$\begin{bmatrix} \mathbf{e}_1 & \mathbf{h}_n & \cdots & \mathbf{h}_{n+k-2} & \mathbf{h}_{n+k-1} & \mathbf{u} & 0 & \cdots & 0 & \mathbf{e}_{k+1} \\ \mathbf{e}_1 & 0 & \cdots & 0 & \mathbf{h}_{n+k-1} & \mathbf{u} & \mathbf{h}_n & \cdots & \mathbf{h}_{n+k-2} & \mathbf{e}_{k+1} \end{bmatrix}$$

in the two ways indicated in the proof of Theorem 7.5a. ∎

THEOREM 7.7b

Let f satisfy the hypotheses of Theorem 7.5b and let $H_k^{(n)}(u)$ denote the Hankel polynomials associated with the Taylor series of f at 0. Then for every m such that $|z_m| < |z_{m+1}|$, uniformly for u in any bounded set,

$$H_m^{(n)}(u) = c_m(u_1 u_2 \cdots u_m)^n \left\{ \prod_{i=1}^m (u - u_i) + O\left(\left(\frac{\rho}{|u_m|}\right)^n\right) \right\}, \quad (7.7\text{-}3)$$

where c_m and ρ are the same as in Theorem 7.5b.

Proof. The proof is analogous to that of Theorem 7.5b. Assuming that the poles are simple, we write

$$H_m^{(n)}(u) = \sum_\kappa D_m^{(n)}(u) + \sum_\kappa \hat{D}_m^{(n)}(u), \quad (7.7\text{-}4)$$

where in the first sum

$$D_m^{(n)}(u) := \begin{vmatrix} r_{\kappa(1)} u_{\kappa(1)}^{n+1} & r_{\kappa(2)} u_{\kappa(2)}^{n+2} & \cdots & r_{\kappa(m)} u_{\kappa(m)}^{n+m} & 1 \\ r_{\kappa(1)} u_{\kappa(1)}^{n+2} & r_{\kappa(2)} u_{\kappa(2)}^{n+3} & \cdots & r_{\kappa(m)} u_{\kappa(m)}^{n+m+1} & u \\ \cdots & \cdots & \cdots & \cdots & \cdots \\ r_{\kappa(1)} u_{\kappa(1)}^{n+m+1} & r_{\kappa(2)} u_{\kappa(2)}^{n+m+2} & \cdots & r_{\kappa(m)} u_{\kappa(m)}^{n+2m} & u^m \end{vmatrix},$$

and the summation is taken with respect to all functions κ from N_m to N_m. In the second sum $\hat{D}_m^{(n)}(u)$ denotes a determinant of the same structure as $D_m^{(n)}(u)$, except that at least one of the columns

$$\begin{bmatrix} r_{\kappa(i)} u_{\kappa(i)}^{n+i} \\ r_{\kappa(i)} u_{\kappa(i)}^{n+i+1} \\ \vdots \\ r_{\kappa(i)} u_{\kappa(i)}^{n+i+m} \end{bmatrix}$$

is replaced by the column

$$\begin{bmatrix} b_{n+i-1} \\ b_{n+i} \\ \vdots \\ b_{n+i+m-1} \end{bmatrix};$$

the coefficients b_m are defined as in §7.5. The rules given in that section concerning the mode of summation also apply here.

To evaluate the first sum we observe that all determinants for which κ twice assumes the same value are zero; hence

$$\sum_\kappa D_m^{(n)}(u) = \prod_{i=1}^m r_i u_i^{n+1} \cdot p_m(u),$$

where

$$p_m(u) := \sum_\pi \begin{vmatrix} u_{\pi(1)}^0 & u_{\pi(2)}^1 & \cdots & u_{\pi(m)}^{m-1} & 1 \\ u_{\pi(1)}^1 & u_{\pi(2)}^2 & \cdots & u_{\pi(m)}^m & u \\ \cdots & \cdots & \cdots & \cdots & \cdots \\ u_{\pi(1)}^m & u_{\pi(2)}^{m+1} & \cdots & u_{\pi(m)}^{2m-1} & u^m \end{vmatrix},$$

the summation being taken with respect to all permutations of N_m. By a combinatorial consideration similar to that of §7.5 we can show that

$$p_m(u) = \prod_{i=1}^m (u - u_i) \prod_{1 \le i < j \le m} (u_i - u_j)^2.$$

If c_m is defined as in (7.5-10),

$$\sum_\kappa D_m^{(u)} = c_m (u_1 u_2 \cdots u_m)^n \prod_{i=1}^m (u - u_i) \qquad (7.5\text{-}5)$$

follows.

To estimate the second sum in (7.7-4) we can show that each term, hence the sum, is $O((u_1 u_2 \cdots u_{m-1}\rho)^n)$ as $n \to \infty$. Clearly the estimate is uniform in u for u in any bounded set. With (7.7-5) this establishes (7.7-3) where all poles of f are simple. The formula can then be extended to poles of order >1 by the method of confluence of singularities. ∎

The analog of Theorem 7.5e is as follows:

HADAMARD POLYNOMIALS

THEOREM 7.7c

Let the f of Theorem 7.7b be a rational function with a pole of order h at infinity and let the sum of the orders of all its finite poles be k. Then for $n > h$, $m = k$ (7.7-3) is exact without the O-term, i.e.,

$$H_k^{(n)}(u) = c_k(u_1 u_2 \cdots u_k)^n \prod_{i=1}^{k} (u - u_i). \qquad (7.7\text{-}6)$$

Proof. As in §7.5, the proof follows from the fact that for $m = k$ and $n > h$ (7.5-5) is exact with $b_n = 0$. The determinants $\hat{D}_k^{(n)}(u)$ are all zero and $H_k^{(n)}(u)$ is identical to (7.7-5). ∎

Let now F, the formal power series from which the Hankel polynomials $H_m^{(n)}(u)$ are constructed, be k-normal. We can then define the **Hadamard polynomials** $p_m^{(n)}(u)$ by

$$p_0^{(n)}(u) := 1, \qquad p_m^{(n)}(u) := \frac{H_m^{(n)}(u)}{H_m^{(n)}}, \qquad m = 1, 2, \ldots. \qquad (7.7\text{-}7)$$

This is a polynomial of degree m with leading coefficient 1. As a result of the normalization introduced by the Hankel determinants, the properties of Hankel polynomials discussed above easily lead to the following properties of Hadamard polynomials.

THEOREM 7.7d

For all values of the indices for which the polynomials are defined,

$$p_{m+1}^{(u)}(u) = u p_m^{(n+1)}(u) - q_{m+1}^{(u)} p_m^{(u)}(u). \qquad (7.7\text{-}8)$$

Proof. This is an immediate result of Theorem 7.7a and of (7.6-5). ∎

The identity (7.7-8) permits the construction of the Hadmard polynomials directly from the qd scheme, according to the rule in Fig. 7.7a.

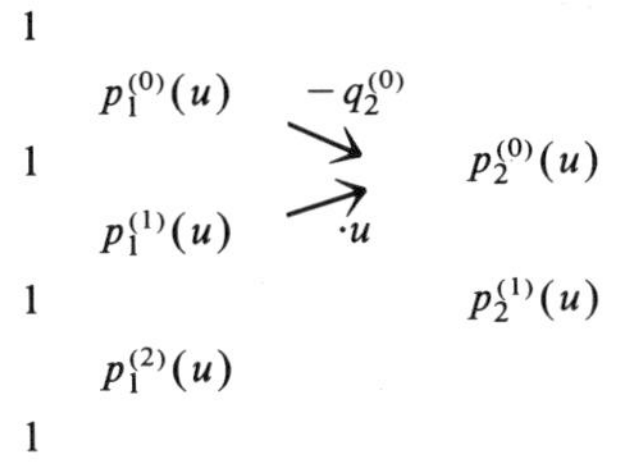

```
        1
             p_1^{(0)}(u)    -q_2^{(0)}
        1                ↘
                               p_2^{(0)}(u)
             p_1^{(1)}(u)    ·u ↗
        1
                               p_2^{(1)}(u)
             p_1^{(2)}(u)
        1
```

Fig. 7.7a Construction of the Hadamard polynomials.

The following is an immediate result of Theorems 7.5b and 7.7b:

THEOREM 7.7e

Let f satisfy the hypotheses of Theorem 7.5b, and let the Taylor series of f at 0 be normal. Then for every m such that $|z_m|<|z_{m+1}|$ the Hadamard polynomials associated with the Taylor series satisfy

$$\lim_{n\to\infty} p_m^{(n)}(u) = (u-u_1)(u-u_2)\cdots(u-u_m), \tag{7.7-9}$$

uniformly in u in every bounded set.

Similarly, we have from the Theorems 7.5e and 7.7c:

THEOREM 7.7f

Let f be a rational function as in Theorem 7.7c and let its Taylor series at 0 be k-normal. The associated Hadamard polynomials $p_k^{(n)}$ are then identical for all n:

$$p_k^{(n)}(u) = (u-u_1)(u-u_2)\cdots(u-u_k), \quad n=0,1,2,\ldots. \tag{7.7-10}$$

This result resolves in a rather elegant manner the problem of recovering the denominator of a rational function r from the Taylor expansion of r at 0 (this was required, e.g., for the interpolation of a given function by a sum of exponentials; see §7.2). Let

$$r(z) := \frac{b_0 + b_1 z + \cdots + b_{k-1} z^{k-1}}{1 + a_1 z + \cdots + a_k z^k} = \sum_{h=0}^{\infty} f_h z^h$$

be a rational function that vanishes at infinity with poles $z_1, z_2, \ldots, z_k$ (multiple poles counted repeatedly). Then

$$1 + a_1 z + \cdots + a_k z^k = \left(1 - \frac{z}{z_1}\right)\left(1 - \frac{z}{z_2}\right)\cdots\left(1 - \frac{z}{z_k}\right)$$

$$= (1 - u_1 z)(1 - u_2 z)\cdots(1 - u_k z).$$

Hence, if the series $\sum f_k z^k$ is k-normal and

$$p_k^{(0)}(u) = u^k + c_1 u^{k-1} + \cdots + c_k$$

denotes the associated Hadamard polynomial, then evidently

$$1 + a_1 z + \cdots + a_k z^k = z^k p_k^{(0)}(z^{-1}) \tag{7.7-11}$$

HADAMARD POLYNOMIALS

or $a_m = c_m, m = 1, \ldots, k$. Thus to recover the denominator it suffices to construct the polynomial $p_k^{(0)}$ and to take the coefficients in reverse order.

The construction of the qd table for the calculation of $p_k^{(0)}$ requires $8k^2 + O(k)$ multiplications or divisions. The construction of $p_k^{(0)}$ by the recurrence relation (7.7-8), however, requires $\frac{1}{6}k^3$ multiplications, which is of the same order of magnitude as the solution of (7.2-24) by Gaussian elimination. In §7.8 it is shown how $p_k^{(0)}$ can be calculated in only $O(k^2)$ operations.

We continue to discuss Hadamard polynomials associated with general power series.

LEMMA 7.7g

Let F be a k-normal formal power series and let $p_k^{(n)}$ denote any associated Hadamard polynomial. The coefficients c_m in the formal Laurent series

$$F(z) p_k^{(n)}\left(\frac{1}{z}\right) =: \sum_{m=-k}^{\infty} c_m z^m \qquad (7.7\text{-}12)$$

are zero for $m = n, n+1, \ldots, n+k-1$; furthermore,

$$c_{n+k} = \frac{H_{k+1}^{(n)}}{H_k^{(n)}}. \qquad (7.7\text{-}13)$$

Proof. The coefficient of z^m in the formal Laurent series

$$F(z) H_k^{(n)}\left(\frac{1}{z}\right)$$

is easily seen to be

$$\begin{vmatrix} a_n & a_{n+1} & \cdots & a_{n+k-1} & a_m \\ a_{n+1} & a_{n+2} & \cdots & a_{n+k} & a_{m+1} \\ \cdots & \cdots & \cdots & \cdots & \cdots \\ a_{n+k} & a_{n+k+1} & \cdots & a_{n+2k-1} & a_{m+k} \end{vmatrix}.$$

If $n \leq m \leq n+k-1$, the determinant is zero because of repeated columns. For $m = n+k$ the determinant equals $H_{k+1}^{(n)}$, which implies (7.7-13). ∎

The quantity

$$s_k^{(n)} := \frac{H_{k+1}^{(n)}}{H_k^{(n)}}$$

can by virtue of Theorem 7.6a be computed from the qd scheme by means of the formula

$$s_k^{(n)} = a_n q_1^{(n)} e_1^{(n)} \cdots q_k^{(n)} e_k^{(n)}. \tag{7.7-14}$$

Lemma 7.7g implies that the coefficients of $z^{-1}, 1, z, \ldots, z^{k-2}$ of the formal Laurent series

$$z^{-n-1} F(z) p_k^{(n)}\left(\frac{1}{z}\right)$$

are zero and that the coefficient of z^{k-2} equals $s_k^{(n)}$. Consequently the residue of

$$z^{-n-1} F(z) p_k^{(n)}\left(\frac{1}{z}\right) p_m^{(n)}\left(\frac{1}{z}\right)$$

is equal to zero for $0 \leqslant m < k$ and equal to $s_k^{(n)}$ if $m = k$. Thus, if F is the Taylor series at 0 of a function f analytic at 0 and Γ denotes a circle of sufficiently small radius surrounding the origin, the residue theorem implies

$$\frac{1}{2\pi i} \int_\Gamma z^{-n-1} f(z) p_k^{(n)}\left(\frac{1}{z}\right) p_m^{(n)}\left(\frac{1}{z}\right) dz = \begin{cases} s_k^{(n)}, & m = k, \\ 0, & m < k. \end{cases} \tag{7.7-15}$$

If the series F is normal and not merely k-normal, this integral is defined for all non-negative integers m and k and is always zero for $m \neq k$. Making the transformation $z^{-1} =: u$ in (7.7-15) and setting $u^{-1} f(1/u) =: g(u)$, we obtain a result that expresses some kind of orthogonality of the polynomials $p_m^{(n)}$.

THEOREM 7.7h

Let the function g be analytic at infinity, let it be represented by

$$g(u) = \frac{a_0}{u} + \frac{a_1}{u^2} + \frac{a_2}{u^3} + \cdots \tag{7.7-16}$$

for $|u| > \rho$, and let the series $F := a_0 + a_1 x + a_2 x^2 + \cdots$ be normal (k-normal). Then, if Γ is a positively oriented Jordan curve containing the set $|u| \leqslant \rho$ in its interior, the Hadamard polynomials $p_m^{(n)}$ associated with F satisfy

$$\frac{1}{2\pi i} \int_\Gamma u^h g(u) p_h^{(n)}(u) p_m^{(n)}(u) du = \begin{cases} 0, & h \neq m, \\ s_m^{(n)}, & h = m, \end{cases} \tag{7.7-17}$$

where $h, m = 0, 1, 2, \ldots$ ($h, m = 0, 1, \ldots, k$); $n = 0, 1, 2, \ldots$.

HADAMARD POLYNOMIALS

Readers familiar with the theory of orthogonal functions will recognize in (7.7-17) a sort of **orthogonality relation** for the polynomials $p_k^{(n)}$ (n fixed, $k = 0, 1, 2, \ldots$). Orthogonality, however, does not obtain with respect to a real weight function but with respect to an analytic function, and the integration is not carried out over a real interval but over a closed curve. It is easy to see how the classical orthogonality (over a finite interval) is contained in Theorem 7.7h as a special case. Let ω be a non-negative continuous real function defined on a bounded real interval $[\alpha, \beta]$, $\omega \neq 0$, let S denote the exterior of the straight line segment $[\alpha, \beta]$, and let

$$g(u) := \int_\alpha^\beta \frac{\omega(\tau)}{u - \tau} d\tau \qquad (7.7\text{-}18)$$

for $u \in S$. Then, according to Theorem 4.1a, g is analytic in S, and by using the expansion

$$\frac{1}{u - \tau} = \frac{1}{u} + \frac{\tau}{u^2} + \frac{\tau^2}{u^3} + \cdots,$$

valid for $|u| > \max(|\alpha|, |\beta|)$, and by integrating term by term we see that g is analytic at ∞ and is represented by a series of the form (7.7-16), where

$$a_n := \int_\alpha^\beta \tau^n \omega(\tau) d\tau, \qquad n = 0, 1, 2, \ldots. \qquad (7.7\text{-}19)$$

If $0 \leq \alpha < \beta$, this series is normal according to the criterion of Theorem 7.5g, where $d\psi(\tau) = \omega(\tau) d\tau$; hence the qd scheme and the polynomials $p_m^{(n)}$ exist for all $n \geq 0$ and all $m \geq 0$. Using the representation (7.7-18) in (7.7-17) and interchanging the order of the integrations, we have

$$\frac{1}{2\pi i} \int_\Gamma u^n g(u) p_h^{(n)}(u) p_m^{(n)}(u) du$$

$$= \int_\alpha^\beta \left\{ \frac{1}{2\pi i} \int_\Gamma \frac{u^n p_h^{(n)}(u) p_m^{(n)}(u)}{u - \tau} du \right\} \omega(\tau) d\tau.$$

By Cauchy's integral formula the inner integral equals $\tau^n p_h^{(n)}(\tau) p_m^{(n)}(\tau)$, and we find

$$\int_\alpha^\beta \tau^n \omega(\tau) p_h^{(n)}(\tau) p_m^{(n)}(\tau) d\tau = \begin{cases} 0, & h \neq m, \\ s_m^{(n)}, & h = m. \end{cases} \qquad (7.7\text{-}20)$$

We thus have established:

THEOREM 7.7i

Let g be defined by (7.7-18) where $0 \leq \alpha < \beta < \infty$ and ω is continuous and non-negative. Then the Laurent series representing g near ∞ is normal, and for $n = 0, 1, 2, \ldots$, the Hadamard polynomials $p_m^{(n)}$ associated with it are the orthogonal polynomials, normalized to have highest coefficient 1, for the weight function $\tau^n \omega(\tau)$.

Another application of Theorem 7.7h is as follows: let g be a rational function vanishing at ∞, with k finite poles $u_1, \ldots, u_k$ of order 1 and respective residues $r_1, \ldots, r_k$; hence

$$g(u) = \frac{r_1}{u - u_1} + \frac{r_2}{u - u_2} + \cdots + \frac{r_k}{u - u_k}.$$

The integral (7.7-17) can then be evaluated by residues to yield

$$\sum_{i=1}^{k} r_i u_i^n p_h^{(n)}(u_i) p_m^{(n)}(u_i) = \begin{cases} 0, & h \neq m, \\ s_m^{(n)}, & h = m. \end{cases} \qquad (7.7\text{-}21)$$

These relations, which hold for each fixed n and for $h, m = 0, 1, \ldots, k-1$, can be regarded as a system of equations for the residues r_i. Introducing the diagonal matrices

$$\mathbf{R} := \begin{bmatrix} r_1 u_1^n & & & 0 \\ & r_2 u_2^n & & \\ & & \ddots & \\ 0 & & & r_k u_k^n \end{bmatrix},$$

$$\mathbf{S} := \begin{bmatrix} s_0^{(n)} & & & 0 \\ & s_1^{(n)} & & \\ & & \ddots & \\ 0 & & & s_{k-1}^{(n)} \end{bmatrix}$$

and the full matrix

$$\mathbf{P} := \begin{bmatrix} p_0^{(n)}(u_1) & p_0^{(n)}(u_2) & \cdots & p_0^{(n)}(u_k) \\ p_1^{(n)}(u_1) & p_1^{(n)}(u_2) & \cdots & p_1^{(n)}(u_k) \\ \cdots & \cdots & \cdots & \cdots \\ p_{k-1}^{(n)}(u_1) & p_{k-1}^{(n)}(u_2) & \cdots & p_{k-1}^{(n)}(u_k) \end{bmatrix},$$

we can write Equations (7.7-21) in the form

$$\mathbf{PRP}^T = \mathbf{S}.$$

This equation may be solved for $\mathbf{R}^{-1}$ without inverting $\mathbf{P}$:

$$\mathbf{R}^{-1} = \mathbf{P}^T \mathbf{S}^{-1} \mathbf{P}.$$

Looking at the diagonal elements, we find

$$u_i^n r_i = \left\{ \sum_{m=0}^{k-1} \frac{1}{s_m^{(n)}} [p_m^{(n)}(u_i)]^2 \right\}^{-1}, \qquad i=1,2,\ldots,k. \qquad (7.7\text{-}22)$$

Thus the residues of the rational function g can be evaluated with the help of the qd table without ever computing the numerator polynomial.

We conclude this section by applying some of its results to obtain an algorithmic solution to the problem of interpolating a given sequence $\{f_n\}$ of $2k$ complex numbers by a sum of k exponentials. It will be recalled from the discussion in §7.2 that under certain conditions there exists a rational function f, analytic at 0, vanishing at ∞, and of denominator degree $\leq k$, whose first $2k$ Taylor coefficients at 0 are just the given f_n. If the poles $z_1,\ldots,z_k$ of f are distinct and the partial fraction decomposition of f is

$$f(z) = \sum_{m=1}^{k} \frac{r_m}{z - z_m},$$

the required representation is

$$f_n = \sum_{m=1}^{k} c_m u_m^n, \qquad (7.7\text{-}23)$$

where $u_m := z_m^{-1}$, $c_m := -r_m u_m$, $m = 1, 2, \ldots, k$.

Now let us assume that the *qd* scheme of the given sequence $\{f_n\}$ exists as far as it can be constructed. (This takes the form of a triangular array having $2k - m$ elements in its mth column.) It is then possible by (7.7-8) to construct the polynomials $p_m^{(n)}(u)$ for $m = 0, 1, \ldots, k$ and $n = 0, 1, \ldots, 2k - 2m - 1$ and thus, in particular, the polynomial $p_m^{(0)}(u)$, whose zeros are the u_m required in (7.7-23). (These zeros could be found by continuing the *qd* scheme downward, as discussed in §7.6.) Furthermore, the function $g(u)$ has the partial fraction expansion

$$g(u) = \frac{1}{u} \sum_{m=1}^{k} \frac{r_m}{u^{-1} - u_m^{-1}} = \sum_{m=1}^{k} \frac{-r_m u_m}{u - u_m} = \sum_{m=1}^{k} \frac{c_m}{u - u_m}.$$

Thus the quantities c_m needed in (7.7-23) are just the residues of g and can be determined from (7.7-22).

EXAMPLE 1

Once again we consider Example 4 of §7.2 with the data $\{f_k\} = \{2, 2, 1, 1\}$. The *qd* table turns out as follows:

f_n	$q_1^{(n)}$	$e_1^{(n)}$	$q_2^{(n)}$
2			
	1		
2		$-\frac{1}{2}$	
	$\frac{1}{2}$		$-\frac{1}{2}$
1		$\frac{1}{2}$	
	1		
1			

This, in turn, yields the *p*-table

$p_0^{(n)}$	$p_1^{(n)}$	$p_2^{(n)}$
1		
	$u - 1$	
1		$u^2 - \frac{1}{2}$
	$u - \frac{1}{2}$	
1		

$[u^2 - \frac{1}{2} = u(u - \frac{1}{2}) + \frac{1}{2}(u - 1).]$ The zeros of $p_2^{(0)}$ are $u_1 = 2^{-1/2}$, $u_2 = -2^{-1/2}$. From the first diagonal of the *qd* table we obtain the *s*-values

$$s_0^{(0)} = f_0 = 2, \qquad s_1^{(0)} = f_0 q_1^{(0)} e_1^{(0)} = -1.$$

HADAMARD POLYNOMIALS

Now (7.7-22) furnishes

$$c_1 = 2^{1/2}+1, \qquad c_2 = -2^{1/2}+1.$$

The resulting interpolating function,

$$f_n = (2^{1/2}+1)(2^{-1/2})^n + (-2^{1/2}+1)(-2^{-1/2})^n,$$

of course agrees with that found in §7.2.

PROBLEMS

1. By expanding the determinant of the matrix of order $2k+2$

$$\begin{bmatrix} h_n & h_{n+1} & \cdots & h_{n+k-1} & e_{k+1} & . & 0 & \cdots & 0 & h_{n+k} & u \\ h_n & 0 & \cdots & 0 & e_{k+1} & . & h_{n+1} & \cdots & h_{n+k-1} & h_{n+k} & u \end{bmatrix},$$

prove the identity

$$H_k^{(n)} H_k^{(n+1)}(u) - H_k^{(n+1)} H_k^{(n)}(u) + H_{k+1}^{(n)} H_{k-1}^{(n+1)}(u) = 0.$$

Show that whenever the polynomials involved are defined

$$p_k^{(n+1)}(u) - p_k^{(n)}(u) = -e_k^{(n)} p_{k-1}^{(n+1)}(u). \qquad (7.7\text{-}24)$$

2. Give an alternate proof of the result of Problem 5, §7.6, by applying the statement of Theorem 7.7e to the polynomials $p_2^{(n)}(u)$.
3. Solve Problem 10 in §7.2 by the qd method.
4. Let $F := 0! + 1!x + 2!x^2 + \cdots$.
   (a) Show that the Hadamard polynomials associated with F are

   $$p_m^{(n)}(u) := \sum_{k=0}^{m} \binom{m}{k} (-n-m)_k u^{m-k}.$$

   (b) Show that the $p_m^{(n)}$ are orthogonal on $[0, \infty)$ with respect to the weight function

   $$\omega(\tau) := \tau^n e^{-\tau}.$$

   This almost follows from Theorem 7.7i; why only almost?
   (c) Show that for m fixed, $n \to \infty$ the m zeros $u_1, \ldots, u_m$ of $p_m^{(n)}$ satisfy

   $$u_i = n + O(n^{1-1/m}), \qquad i = 1, \ldots, m.$$

   (Verify in the case $m=2$.) Why does this not contradict Theorem 7.7e? [Consider the polynomial $q_m^{(n)}(v) := n^{-m} p_m^{(n)}(nv)$ and apply Theorem 4.10c.]

§7.8. MATRIX INTERPRETATION

We continue our study of the Hadamard polynomials associated with a given formal power series F, assumed to be k-normal for some positive integer k. It will be seen that these polynomials may be regarded as the characteristic polynomials of certain matrices. We employ well-known results on the location of eigenvalues of matrices that will yield inclusion theorems for the zeros of polynomials that use data already contained in the qd scheme.

We begin by establishing a new recurrence relation for Hadamard polynomials with the same superscript.

THEOREM 7.8a

The Hadamard polynomials associated with a k-normal series F satisfy the identity

$$p_{m+1}^{(n)}(u) - (u - q_{m+1}^{(n)} - e_m^{(n)})p_m^{(n)} + q_m^{(n)}e_m^{(n)}p_{m-1}^{(n)}(u) = 0 \qquad (7.8\text{-}1)$$

for $n = 0, 1, 2, \ldots$ and $m = 1, 2, \ldots, k-1$.

Proof. For typographical simplicity we omit the argument u and the superscript (n) and replace $(n+1)$ with $+$. By Theorem 7.7d

$$p_m - up_{m-1}^+ + q_m p_{m-1}^+ = 0 \qquad (7.8\text{-}2)$$

holds for $m = 1, 2, \ldots, k$. According to the identity (7.7-24),

$$p_m = p_m^+ + e_m p_{m-1}$$

likewise holds for $m = 1, 2, \ldots, k$. The expression on the left of (7.8-1) thus equals

$$p_{m+1} - u(p_m^+ + e_m p_{m-1}^+) + (q_{m+1} + e_m)p_m + q_m e_m p_{m-1}$$

$$= [p_{m+1} - up_m^+ + q_{m+1} p_m] + e_m[p_m - up_{m-1}^+ + q_m p_{m-1}].$$

The two expressions in brackets are zero by (7.8-2), which proves the desired result. ∎

If $p_{-1}^{(n)} := 0$, (7.8-1) by virtue of (7.8-2) obviously holds for $m = 0$. If the nth diagonal of the qd scheme is known (including the products $q_m e_m$, which already occur in the rhombus rules), (7.8-1) makes it possible to compute either a *value* of $p_m^{(n)}$ by using $2m - 3$ multiplications or the *Taylor expansion* of $p_m^{(n)}$ by using $(m-1)^2$ multiplications.

MATRIX INTERPRETATION

It is convenient to introduce the abbreviations

$$g_{m+1}^{(n)} := q_{m+1}^{(n)} + e_m^{(n)}, \qquad h_m^{(n)} := q_m^{(n)} e_m^{(n)} \qquad (7.8\text{-}3)$$

for the constants appearing in the recurrence relations (7.8-1) so that it will take the form

$$p_{m+1}^{(n)}(u) = (u - g_{m+1}^{(n)}) p_m^{(n)}(u) - h_m^{(n)} p_{m-1}^{(n)}(u), \qquad m = 0, 1, \ldots, k-1. \qquad (7.8\text{-}4)$$

We now introduce the tridiagonal matrices

$$\mathbf{M}_m^{(n)} := \begin{bmatrix} g_1^{(n)} & h_1^{(n)} & & & 0 \\ 1 & g_2^{(n)} & h_2^{(n)} & & \\ & 1 & g_3^{(n)} & h_3^{(n)} & \\ & & \ddots & \ddots & \ddots \\ 0 & & & 1 & g_m^{(n)} \end{bmatrix} \qquad (7.8\text{-}5)$$

and assert the following:

THEOREM 7.8b

For $n = 0, 1, 2, \ldots$, and $m = 1, \ldots, k$, $p_m^{(n)}$ equals the characteristic polynomial of $\mathbf{M}_m^{(n)}$,

$$p_m^{(n)}(u) = \det(u\mathbf{I} - \mathbf{M}_m^{(n)}). \qquad (7.8\text{-}6)$$

Proof. The proof is by induction with respect to m. Obviously the statement is true for $m = 1$, and (7.8-4), where $m = 1$, shows that it is true for $m = 2$. Assuming the truth of (7.8-6) for two integers m and $m-1$, where $m > 2$, we find by expanding $\det(u\mathbf{I} - \mathbf{M}_{m+1}^{(n)})$ in terms of the elements of the last row, and again omitting the superscript (n),

$$\det(u\mathbf{I} - \mathbf{M}_{m+1}) = (u - g_{m+1}) \begin{vmatrix} u - g_1 & -h_1 & & & 0 \\ -1 & u - g_2 & -h_2 & & \\ & \ddots & \ddots & \ddots & \\ & & -1 & u - g_{m-1} & -h_{m-1} \\ 0 & & & -1 & u - g_m \end{vmatrix}$$

$$+\begin{vmatrix} u-g_1 & -h_1 & & & 0 \\ -1 & u-g_2 & -h_2 & & \\ \cdots & \cdots & \cdots & & \\ & & -1 & u-g_{m-1} & 0 \\ 0 & & & -1 & -h_m \end{vmatrix}$$

$$= (u - g_{m+1})p_m(u) - h_m p_{m-1}(u).$$

By using (7.8-4) we find that this equals $p_{m+1}(u)$, which completes the induction step. ∎

According to this theorem the zeros of the polynomial $p_m^{(n)}$ coincide with the eigenvalues of the matrix $\mathbf{M}_m^{(n)}$. Any result on the location of the eigenvalues of $\mathbf{M}_m^{(n)}$ thus implies a result on the zeros of $p_m^{(n)}$. This is of particular interest if the underlying series F is the Taylor series at 0 of a rational function r vanishing at infinity and having k finite poles, since by Theorem 7.7f all polynomials $p_k^{(n)}$ are identical and their zeros are the reciprocals of the poles of r.

We state some simple results on eigenvalues of matrices that the reader may have missed in his linear algebra course.

THEOREM 7.8c

There exists a function $\delta = \delta_n(\epsilon, \mu)$ *defined for* $\epsilon > 0$, $\mu > 0$, *and* $n = 1, 2, \ldots$, *with the following property: if* $\mathbf{A} := (a_{ij})$ *is any matrix of order n that satisfies* $|a_{ij}| \leq \mu$ $(i, j = 1, \ldots, n)$ *with eigenvalues* $u_1, \ldots, u_n$, *then the eigenvalues* $v_1, \ldots, v_n$ *of any matrix* $\mathbf{B} := (b_{ij})$ *satisfying* $|b_{ij} - a_{ij}| < \delta_n(\epsilon, \mu)$ $(i, j = 1, \ldots, n)$ *can be numbered such that* $|v_i - u_i| < \epsilon$, $i = 1, 2, \ldots, n$.

In short, the eigenvalues of a matrix are uniformly continuous functions of the elements of the matrix. For a fixed $\mathbf{A}$ this follows very simply from the fact that the coefficients of the characteristic polynomial are continuous functions of the elements of the matrix and that the zeros of a polynomial with leading coefficient 1 are continuous functions of the coefficients (Theorem 4.10c). The existence of a universal function $\delta_n(\epsilon, \mu)$, which permits the same conclusion for all matrices $\mathbf{A}$ with $|a_{ij}| \leq \mu$, follows from a theorem of Ostrowski [1966], in which an explicit expression for δ is also given.

THEOREM 7.8d (the Gershgorin disk theorem)

Let $\mathbf{B} := (b_{ij})$ *be a complex matrix of order n and denote by D_i the set*

$$|u - b_{ii}| \leq \rho_i,$$

where
$$\rho_i := \sum_{\substack{j=1 \\ j \neq i}}^{n} |b_{ij}|.$$

Then each eigenvalue of **B** *is contained in at least one of the disks* D_i.

Proof. Let u be an eigenvalue of **B** and let

$$\mathbf{x} := \begin{bmatrix} x_1 \\ x_2 \\ \vdots \\ x_n \end{bmatrix}$$

be a corresponding eigenvector, i.e., a nontrivial solution of $(u\mathbf{I} - \mathbf{B})\mathbf{x} = \mathbf{0}$. Let i be chosen such that
$$|x_i| = \max_{1 \leq j \leq n} |x_j|.$$
Then $|x_i| > 0$, and from
$$(u - b_{ii})x_i = \sum_{\substack{j=1 \\ j \neq i}}^{n} b_{ij} x_j$$
there follows
$$|u - b_{ii}||x_i| \leq \sum_{\substack{j=1 \\ j \neq i}}^{n} |b_{ij}||x_j| \leq |x_i| \sum_{\substack{j=1 \\ j \neq i}}^{n} |b_{ij}| = |x_i|\rho_i,$$

or $|u - b_{ii}| \leq \rho_i$, i.e., $u \in D_i$. ∎

Theorem 7.8d does not assert that each disk contains an eigenvalue. However, there holds:

THEOREM 7.8e

Under the hypotheses of the preceding theorem, each component of the set $\cup D_i$ *contains as many eigenvalues of* **B** *as points* b_{ii}.

Proof. We consider the family of matrices

$$\mathbf{B}_\tau := \begin{bmatrix} b_{11} & \tau b_{12} & \cdots & \tau b_{1n} \\ \tau b_{21} & b_{22} & \cdots & \tau b_{2n} \\ \vdots & & & \vdots \\ \tau b_{n1} & \tau b_{n2} & \cdots & b_{nn} \end{bmatrix},$$

where $0 \leq \tau \leq 1$. Clearly $\mathbf{B}_1 = \mathbf{B}$, whereas $\mathbf{B}_0$ is the diagonal matrix with diagonal elements b_{ii}. For each $\tau \in [0,1]$, if u is an eigenvalue of $\mathbf{B}_\tau$, u belongs to one of the disks $D_{i\tau}$ defined by $|u - b_{ii}| \leq \tau \rho_i$, and thus, a fortiori, to one of the disks D_i. Now let $\mu := \max |b_{ij}|$ and let $\epsilon > 0$ be less than the distance between any two components of the set $\cup D_i$. (This distance is positive, since the disks D_i are closed sets.) Let $\delta = \delta_n(\epsilon, \mu)$ be the value of the function so denoted in the statement of Theorem 7.8c. Let $0 \leq \tau_0 < \tau_1 < \cdots < \tau_m = 1$ be a subdivision of the interval $[0,1]$ such that $\tau_{k+1} - \tau_k < \delta/\mu$ for $k = 0, 1, \ldots, m-1$. We assert that for each matrix $\mathbf{B}_{\tau(k)}$ ($k = 0, 1, \ldots, m$) each component of $\cup D_i$ contains as many eigenvalues as disks. This statement is clearly true for $k = 0$. We assume its truth for some k such that $0 \leq k < m$. Let C be a component of $\cup D_i$ and let it consist of h disks; it then contains exactly h eigenvalues of $\mathbf{B}_{\tau k}$. By Theorem 7.8c the eigenvalues of $\mathbf{B}_{\tau(k+1)}$ can differ by at most ϵ from the eigenvalues of $\mathbf{B}_{\tau(k)}$. Thus an eigenvalue of $\mathbf{B}_{\tau(k+1)}$ can be found within an ϵ-neighborhood of each eigenvalue of $\mathbf{B}_{\tau(k)}$ in C. By Theorem 7.8d and the definition of ϵ all these eigenvalues belong to C. Thus C contains at least as many eigenvalues of $\mathbf{B}_{\tau(k+1)}$ as of $\mathbf{B}_{\tau(k)}$. Because this is true for every component, it contains exactly as many. It now follows by induction that each component of $\cup D_i$ contains as many eigenvalues of $\mathbf{B} = \mathbf{B}_1$ as of $\mathbf{B}_0$, i.e., as points b_{ii}. ∎

Theorem 7.8e is of special interest if the disks D_i are disjoint. Each component of $\cup D_i$ then consists of exactly one disk, and we can assert that each disk contains exactly one eigenvalue.

To apply these results to the matrices $\mathbf{M}_m^{(n)}$ derived from a qd scheme let us suppose that the scheme is associated with a normal series F, which is the Taylor series at $z = 0$ of a rational function satisfying the conditions of Theorem 7.7c, and let its k simple poles satisfy the strict inequalities

$$0 < |z_1| < |z_2| < \cdots < |z_k|. \tag{7.8-7}$$

It then follows from Theorem 7.6b that

$$\lim_{n \to \infty} e_m^{(n)} = 0, \quad m = 1, 2, \ldots, k-1,$$

and that the matrices $\mathbf{M}_m^{(n)}$ thus resemble increasingly the matrices

$$\begin{bmatrix} q_1^{(n)} & 0 & 0 \cdots 0 & 0 \\ 1 & q_2^{(n)} & 0 \cdots 0 & 0 \\ 0 & 1 & q_3^{(n)} \cdots 0 & 0 \\ & & \cdots & \\ 0 & 0 & 0 \cdots 1 & q_k^{(n)} \end{bmatrix}$$

MATRIX INTERPRETATION

which have eigenvalues $q_m^{(n)}$, $m = 1, 2, \ldots, k$. Since the matrices $\mathbf{M}_m^{(n)}$ are known to have the eigenvalues $u_1, \ldots, u_k$, once again it follows that the $q_m^{(n)}$ tend to the u_m as $n \to \infty$.

For applying Gershgorin's theorem the matrices $\mathbf{M}_k^{(n)}$ are not suitable, since the elements below the main diagonal do not approach zero. However, for any choice of h_m', h_m'' such that

$$h_m' h_m'' = h_m, \quad m = 1, 2, \ldots, k-1$$

the matrix

$$\hat{\mathbf{M}}_k^{(n)} := \begin{bmatrix} g_1 & h_1' & & & & 0 \\ h_1'' & g_2 & h_2' & & & \\ & h_2'' & g_3 & h_3' & & \\ & & \ddots & \ddots & \ddots & \\ & & & h_{k-2}'' & g_{k-1} & h_{k-1}' \\ 0 & & & & h_{k-1}'' & g_k \end{bmatrix}$$

(we again omit the superscript n) is similar to $\mathbf{M}_k$, hence has the same set of eigenvalues. To make both the elements above and below the main diagonal small, we choose h_m' and h_m'' such that

$$|h_m'| = |h_m''| = |h_m|^{1/2}.$$

This yields the following estimate for the eigenvalues u_i:

THEOREM 7.8f

Let F be the k-normal Taylor series at 0 of a rational function r vanishing at infinity and having k finite poles z_i, $i = 1, \ldots, k$. In terms of the associated qd scheme let

$$D_m^{(n)} : |u - g_m^{(n)}| \leq |h_{m-1}^{(n)}|^{1/2} + |h_m^{(n)}|^{1/2}, \quad m = 1, \ldots, k,$$

where $g_m^{(n)}$ and $h_m^{(n)}$ are defined by (7.8-3). Then each number $u_i := z_i^{-1}$ is contained in at least one of the disks $D_m^{(n)}$. If a disk $D_m^{(n)}$ is isolated from all other disks, it contains precisely one number u_i.

If the poles of r satisfy (7.8-7), then

$$g_m^{(n)} \to u_m, \quad h_m^{(n)} \to 0 \quad \text{as} \quad n \to \infty,$$

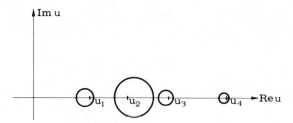

Fig. 7.8. Gershgorin disks.

and the diameters of the Gershgorin disks $D_m^{(n)}$ all tend to zero. Since the u_m are distinct, the disks will ultimately be disjoint (see Fig. 7.8), and the assertion of the Theorem can be strengthened to

$$u_m \in D_m^{(n)}, \qquad m=1,2,\ldots,k. \qquad (7.8\text{-}8)$$

Thus in the situation described the qd algorithm furnishes inclusion regions for the poles of a rational function that become arbitrarily small as $n \to \infty$.

PROBLEMS

1. Prove that the eigenvalues of the matrix

$$\begin{bmatrix} 2 & -1 & & & \\ -1 & 2 & -1 & & \\ & -1 & 2 & -1 & \\ & & -1 & 2 & -1 \\ & & & -1 & 2 \end{bmatrix}$$

   have non-negative real parts.

2. Without computing the characteristic polynomial, show that all eigenvalues of the nonsymmetric real matrix

$$\begin{bmatrix} 2 & -1 & -1 \\ 0 & 5 & 1 \\ 2 & 13 & 21 \end{bmatrix}$$

   are real.

3. Let f be a rational function vanishing at infinity and of denominator degree m and let the power series F representing f at 0 be m-normal. Show that the matrix $\mathbf{M}^{(n)} := \mathbf{M}_m^{(n)}$ defined by (7.8-5) can be factored,

$$\mathbf{M}^{(n)} = \mathbf{L}^{(n)} \mathbf{R}^{(n)}, \qquad (7.8\text{-}9)$$

where

$$\mathbf{L}^{(n)} = \begin{bmatrix} q_1^{(n)} & & & 0 \\ 1 & q_2^{(n)} & & \\ & \ddots & \ddots & \\ 0 & & 1 & q_m^{(n)} \end{bmatrix}, \quad \mathbf{R}^{(n)} = \begin{bmatrix} 1 & e_1^{(n)} & & 0 \\ & 1 & e_2^{(n)} & \\ & & \ddots & \ddots \\ 0 & & & 1 \end{bmatrix}.$$

4. Using the notation of Problem 3, show that

$$\mathbf{M}^{(n+1)} = \mathbf{R}^{(n)} \mathbf{L}^{(n)}.$$

[Thus by factoring $\mathbf{M}^{(n)}$ as indicated in (7.8-9) and then multiplying the factors in reverse order, we obtain (if the moduli of the poles of f are distinct) a sequence of matrices whose diagonal elements tend to the u_i. This is the genesis of the L-R algorithm, also due to Rutishauser, which under certain conditions is applicable to arbitrary matrices.]

§7.9. POLES WITH EQUAL MODULI

The properties of the qd algorithm which have been developed so far do not permit the numerical determination of the poles of a meromorphic function f if several of these poles have the same modulus. This situation occurs, for instance, if a function f which is real on the real axis has nonreal poles. By the symmetry principle the complex poles occur in complex conjugate pairs. Obviously both poles of a pair have the same modulus. Our aim is to describe a modification of the qd algorithm that can be used to determine distinct poles of equal modulus.

As before, we assume that F is the Taylor series at $z = 0$ of a meromorphic function f with a finite or infinite number of poles at points z_k, numbered such that

$$0 < |z_1| \leq |z_2| \leq |z_3| \leq \cdots.$$

A pole will occur as many times in the sequence $\{z_k\}$ as indicated by its order. Any index m such that the strict inequality

$$|z_m| < |z_{m+1}| \qquad (7.9\text{-}1)$$

holds is called a **critical index** of f. (The critical indices of a function do not depend on the order in which the poles of equal modulus are numbered.) As already mentioned but not completely proved in §7.6, if m is a critical index and F is m-normal, then

$$\lim_{n\to\infty} e_m^{(n)} = 0. \tag{7.9-2}$$

Thus the qd table is divided into subtables by the e-columns tending to zero. If a subtable contains h q-columns, the presence of h poles of equal modulus is indicated. If $h=1$, the q-column converges to the reciprocal of the corresponding pole, as shown in §7.6. The question now is how to determine the poles if $h > 1$. This is answered by Theorem 7.9a.

THEOREM 7.9a. (Rutishauser's rule)

Let m and $m+h$ be two consecutive critical indices of f and let F be $(m+h)$-normal. If the polynomials $\tilde{p}_k^{(n)}$ are defined by

$$\tilde{p}_0^{(n)}(u) := 1,$$

$$\tilde{p}_{k+1}^{(n)}(u) := u\tilde{p}_k^{(n+1)}(u) - q_{m+k+1}^{(n)}\tilde{p}_k^{(n)}(u),$$

$$n = 0, 1, 2, \ldots; \ k = 0, 1, \ldots, h-1, \tag{7.9-3}$$

then under hypothesis (H) noted below there exists an infinite set $\mathfrak{N}$ of positive integers such that

$$\lim_{\substack{n\to\infty \\ n\in\mathfrak{N}}} \tilde{p}_h^{(n)}(u) = \tilde{p}_h(u), \tag{7.9-4}$$

where

$$\tilde{p}_h(u) := (u - u_{m+1})(u - u_{m+2}) \cdots (u - u_{m+h}). \tag{7.9-5}$$

If $m = 0$, the polynomials $\tilde{p}_k^{(n)}$ are identical with the Hadamard polynomials $p_k^{(n)}$ and the above reduces to Theorem 7.7e. If $m > 0$, the algorithm (7.9-3) for constructing the $\tilde{p}_k^{(n)}$ is the same as the algorithm of Theorem 7.7d for constructing the $p_k^{(n)}$, except that it is now applied only to the qd subtable bounded by the columns $e_m^{(n)}$ and $e_{m+h}^{(n)}$.

POLES WITH EQUAL MODULI

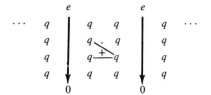

Fig. 7.9. Sums and products used in (7.9-6).

As an illustration, consider $h=2$. Here (7.9-3) yields

$$p_1^{(n)}(u) = u - q_{m+1}^{(n)},$$

$$p_2^{(n)}(u) = u^2 - [q_{m+1}^{(n+1)} + q_{m+2}^{(n)}]u + q_{m+1}^{(n)} q_{m+2}^{(n)}.$$

Theorem 7.9a thus states that the two limits

$$a := \lim_{\substack{n \to \infty \\ n \in \mathfrak{N}}} [q_{m+1}^{(n+1)} + q_{m+2}^{(n)}], \qquad (7.9\text{-}6a)$$

$$b := \lim_{\substack{n \to \infty \\ n \in \mathfrak{N}}} q_{m+1}^{(n)} q_{m+2}^{(n)} \qquad (7.9\text{-}6b)$$

exist and that the numbers u_{m+1}, u_{m+2} are the zeros of the polynomial

$$p_2(u) := u^2 - au + b.$$

In applications it is frequently possible to ignore the restriction $n \in \mathfrak{N}$ in the two formulas (7.9-6). Figure 7.9 indicates which sums and products have to be formed in (7.9-6):

EXAMPLE 1

Determine the zeros of $p(u) := x^4 - 4x^3 + 8x^2 + 24x + 36$. The first two rows of the qd scheme (arranged as in Fig. 7.6c to produce the zeros directly and not the reciprocals of the zeros) are

	4		0		0		0	
0		−2		3		$\tfrac{3}{2}$		0

In decimal notation the resulting qd scheme begins as follows (omitting the columns $e_0^{(n)}$ and $e_4^{(n)}$, which are zero):

4.00000		0.00000		0.00000		0.00000	
	−2.00000		3.00000		1.50000		
2.00000		5.00000		−1.50000		−1.50000	
	−5.00000		−0.90000		1.50000		
−3.00000		9.10000		0.90000		−3.00000	
	15.16667		−0.08901		−5.00000		
12.16667		−6.15568		−4.01099		2.00000	
	−7.67352		−0.05800		2.49315		
4.49315		1.45984		−1.45984		−0.49315	
	−2.49315		0.05800		0.84222		
2.00000		4.01099		−0.67562		−1.33537	
	−5.00000		−0.00977		1.66463		
−3.00000		9.00122		0.99878		−3.00000	
	15.00203		−0.00108		−5.00000		
12.00203		−6.00190		−4.00014		2.00000	
	−7.50212		−0.00072		2.49992		
4.49992		1.49950		−1.49950		−0.49992	

...

Evidently the column $e_2^{(n)}$ tends to zero, whereas the columns $e_1^{(n)}$ and $e_3^{(n)}$ do not. This indicates the presence of two pairs of complex conjugate zeros. Accordingly, we form the limits (7.9-6) for $m=0$ and $m=2$:

$q_1^{(n+1)}+q_2^{(n)}$	$q_1^{(n)}q_2^{(n)}$	$q_3^{(n+1)}+q_4^{(n)}$	$q_3^{(n)}q_4^{(n)}$
10.00000	20.00000	−3.00000	0.00000
6.10000	18.20000	−2.10000	4.50000
6.01099	18.46703	−2.01099	1.80000
5.95299	17.76138	−1.95299	1.97802
6.01099	18.02198	−2.01099	1.94942
6.00122	18.00244	−2.00122	2.02687
6.00014	18.00569	−2.00014	1.99756
5.99941	17.99702	−1.99941	1.99973
↓	↓	↓	↓
6.00000	18.00000	−2.00000	2.00000

POLES WITH EQUAL MODULI 645

Indeed, the given polynomial can be factored as follows:
$$p(x) = (x^2 - 6x + 18)(x^2 + 2x + 2).$$

Proof of Theorem 7.9a. In order to understand Rutishauser's rule we require, first of all, information analogous to Theorem 7.5b concerning the asymptotic behavior of the Hankel determinants $H_k^{(n)}$ for noncritical indices k. It is instructive to begin the discussion with the situation in which the critical index next lower to k is 0 and in which f, in fact, is a rational function with h simple poles $z_1, \ldots, z_h$, all of the same modulus. In that case the Taylor coefficients of f at 0 have the form

$$a_n = r_1 u_1^{n+1} + \cdots + r_h u_h^{n+1},$$

and if $1 \leq k \leq h$ we find, as in the proof of Theorem 7.5b,

$$H_k^{(n)} = \sum_\kappa \begin{vmatrix} r_{\kappa(1)} u_{\kappa(1)}^{n+1} & \cdots & r_{\kappa(k)} u_{\kappa(k)}^{n+k} \\ \cdots & \cdots & \cdots \\ r_{\kappa(1)} u_{\kappa(1)}^{n+k} & \cdots & r_{\kappa(k)} u_{\kappa(k)}^{n+2k-1} \end{vmatrix}, \qquad (7.9\text{-}7)$$

the summation being taken with respect to all functions κ from N_h to N_k. Since determinants where κ takes the same value twice are zero, the summation may be restricted to κ whose values are all different. We split the sum in (7.9-7) according to the domains of values of the functions κ. Let $S := \{i_1, i_2, \ldots, i_k\}$ be an arbitrary subset of k elements of N_h. The sum of those terms in (7.9-7), where the domain of values of the function κ is S, clearly equals

$$r_{i_1} \cdots r_{i_k} (u_{i_1} \cdots u_{i_k})^{n+1} \sum_\pi \begin{vmatrix} 1 & u_{\pi(i_2)} & \cdots & u_{\pi(i_k)}^{k-1} \\ u_{\pi(i_1)} & u_{\pi(i_2)}^2 & \cdots & u_{\pi(i_k)}^k \\ \cdots & \cdots & \cdots & \cdots \\ u_{\pi(i_1)}^{k-1} & u_{\pi(i_2)}^k & \cdots & u_{\pi(i_k)}^{2k-1} \end{vmatrix},$$

where the sum is taken with respect to all permutations π of the set S. It was shown in §7.5 by a symmetry argument that this sum equals

$$\prod_{\substack{i,j \in S \\ i < j}} (u_i - u_j)^2.$$

Summing with respect to all sets S, we find that

$$H_k^{(n)} = F_k^{(n)}(u_1, u_2, \ldots, u_h), \tag{7.9-8}$$

where $F_k^{(n)}$ denotes the homogeneous function of degree k of the h variables $u_1^n, \ldots, u_h^n$ defined by

$$F_k^{(n)}(u_1, \ldots, u_h) := \sum_S c_S \left[\prod_{i \in S} u_i \right]^n \tag{7.9-9}$$

and

$$c_S := \prod_{i \in S} r_i u_i \prod_{\substack{i,j \in S \\ i<j}} (u_i - u_j)^2, \tag{7.9-10}$$

where $c_S \neq 0$ for all sets S. We note that all terms in (7.9-9) have the same order of magnitude; hence in general the asymptotic behavior of $H_k^{(n)}$ for $n \to \infty$ is not so simple as that described in Theorem 7.5b. If $k = h$, however, there is only one subset S of N_h, namely N_h itself, and (7.9-8) reduces to Theorem 7.5e.

We turn next to the case in which f has poles in addition to $z_1, \ldots, z_h$ and

$$a_n = r_1 u_1^{n+1} + \cdots + r_h u_h^{n+1} + b_n,$$

where $b_n = O((\gamma \omega)^n)$ with some $\gamma < 1$, ω denoting the common value of $|u_k|$, $k = 1, \ldots, h$. As in the proof of Theorem 7.5b, we find that

$$H_k^{(n)} = F_k^{(n)}(u_1, \ldots, u_h) + \epsilon_n \omega^{kn}, \tag{7.9-11}$$

where

$$\epsilon_n = O(\gamma^n). \tag{7.9-12}$$

We finally consider the general case in which m and $m+h$ are any two consecutive critical indices, $m \geq 0$, and we wish to study the asymptotic behavior of $H_{m+k}^{(n)}$ for $1 \leq k \leq h$ as $n \to \infty$. Using the representation

$$a_n = r_1 u_1^{n+1} + \cdots + r_{m+h} u_{m+h}^{n+1} + b_n, \tag{7.9-13}$$

where $b_n = O((\gamma\omega)^n)$, $\gamma < 1$, $\omega := |u_{m+1}| = \cdots = |u_{m+h}|$, we find by the addition formula for determinants that the dominant contribution to $H_{m+k}^{(n)}$ is

$$\sum_{\kappa} \begin{vmatrix} r_{\kappa(1)} u_{\kappa(1)}^{n+1} & \cdots & r_{\kappa(m+k)} u_{\kappa(m+k)}^{n+m+k} \\ \cdots & \cdots & \cdots \\ r_{\kappa(1)} u_{\kappa(1)}^{n+m+k} & \cdots & r_{\kappa(m+k)} u_{\kappa(m+k)}^{n+m+2k-1} \end{vmatrix},$$

the sum being taken with respect to all one-to-one functions κ from N_{m+k} to N_{m+h} that assume every value in N_m. These functions assume exactly k values of the set $N_{m+h} - N_m := \{m+1, \ldots, m+h\}$. We split the sum again according to the subsets S which κ assumes in $N_{m+h} - N_m$. For each fixed S the resulting sum can be evaluated as in the proof of Theorem 7.5b. The result may be written

$$c_m c_S (u_1 u_2 \cdots u_m)^n \left(\prod_{j \in S} u_j \right)^n, \tag{7.9-14}$$

where c_m is the same constant as in (7.5-3),

$$c_S := \prod_{i \in S} \tilde{r}_i u_i \prod_{\substack{i,j \in S \\ i<j}} (u_i - u_j)^2, \tag{7.9-15}$$

and

$$\tilde{r}_i := r_i \prod_{j=1}^{m} (u_i - u_j)^2. \tag{7.9-16}$$

The neglected determinants either involve columns of b's or they arise through functions κ which assume more than k values in $N_{m+h} - N_m$. In either case they are of smaller order than (7.9-14). Summing with respect to all sets S, we have Lemma 7.9b:

LEMMA 7.9b

Let m and $m+h$ denote two consecutive critical indices of the function f and let ω denote the common value of $|u_{m+1}|, \ldots, |u_{m+h}|$. Then for $1 \leq k \leq h$, if $n \to \infty$,

$$H_{m+k}^{(n)} = c_m (u_1 \cdots u_m)^n \{ F_k^{(n)}(u_{m+1}, \ldots, u_{m+h}) + \epsilon_n \omega^{kn} \}, \tag{7.9-17}$$

where c_m is the same as in (7.5-3), $\epsilon_n = O(\gamma^n)$ for some $\gamma < 1$, and

$$F_k^{(n)}(u_{m+1},\ldots,u_{m+h}) = \sum_S c_S \left[\prod_{i \in S} u_i \right]^n. \qquad (7.9\text{-}18)$$

The summation here is taken with respect to all subsets S of k elements of $\{m+1,\ldots,m+h\}$, and c_S is defined by (7.9-15) and (7.9-16).

Equation (7.9-17) is interesting only if it can be shown that the first term in curly brackets has a larger order of magnitude than the second. Each term in the sum (7.9-18) clearly has modulus const·ω^{kn} but there is the theoretical possibility that different terms cancel. Our next concern is to show that for an infinite number of n this cancellation cannot occur. For this we require Lemma 7.9c.

LEMMA 7.9c

Let $w_1, w_2, \ldots, w_l$ be distinct numbers of modulus 1 and let

$$s_n := c_1 w_1^n + c_2 w_2^n + \cdots + c_l w_l^n, \qquad (7.9\text{-}19)$$

$n = 0, 1, 2, \ldots$, where the c_i are complex constants. If

$$\lim_{n \to \infty} s_n = 0, \qquad (7.9\text{-}20)$$

then all $c_i = 0$.

Proof. Writing (7.9-19) for l consecutive indices $n, n+1, \ldots, n+l-1$, we obtain a system of l linear equations for the c_i. The determinant of the system is $(w_1 \cdots w_l)^n$ times the Vandermondian of the w_i and thus has constant absolute value. The elements of the inverse matrix of the matrix of the system are bounded. According to (7.9-20) the conclusion follows. ∎

We wish to apply the Lemma 7.9c to the sums

$$s_n := \omega^{-kn} F_k^{(n)}(u_{m+1},\ldots,u_{m+h}),$$

which are of the form (7.9-19), with each term $c_j w_j^n$ corresponding to a subset S and

$$w_j = \prod_{i \in S} (\omega^{-1} u_i), \qquad c_j = c_S.$$

Since all $c_S \neq 0$, Lemma 7.9c implies that s_n does not tend to zero, provided that the numbers w_j are all distinct. This is clearly the case if $k=1$ (because the u_i are distinct) and if $k=h$ (because there is then only one set S). If

POLES WITH EQUAL MODULI

$1 < k < h$, it could happen, theoretically, that although the u_i are all distinct,

$$\prod_{i \in S_1} u_i = \prod_{i \in S_2} u_i$$

for two distinct sets S_1 and S_2 and thus that the corresponding terms in (7.9-18) cancel. It is easy to show that this cannot occur if $h = 3$ or 4. In all these cases the s_n cannot tend to zero, which means that constants $\delta_k > 0$ exist such that

$$|\omega^{-kn} F_k^{(n)}(u_{m+1}, \ldots, u_{m+h})| > \delta_k \qquad (7.9\text{-}21)$$

for $n \in \mathfrak{N}_k$, where $\mathfrak{N}_k$ is an unbounded set of positive integers.

Hypothesis (H) of Theorem 7.9a is now defined to mean the following:

(i) The statement (7.9-21) holds for $k = 1, 2, \ldots, h$,
(ii) Denoting by $\mathfrak{N}_k^{(l)}$ the set $\mathfrak{N}_k$ with all elements increased by l, the intersection

$$\mathfrak{N} := \bigcap_{\substack{k=1,2,\ldots,h \\ 0 \leq l \leq h-k+1}} \mathfrak{N}_k^{(l)} \qquad (7.9\text{-}22)$$

is still a set with infinitely many integers.

For any integer n the quotients of the qd scheme required for forming the polynomials $\tilde{p}_k^{(n)}$, $k = 1, 2, \ldots, h$, in view of the recurrence relations (7.9-3), are

$$q_{m+k}^{(n+l)}, \qquad k = 1, 2, \ldots, h; \qquad l = 0, 1, \ldots, h-k.$$

By means of the fundamental relations (7.6-5) these q's can be expressed in terms of the determinants

$$H_{m+k}^{(n+l)}, \qquad k = 0, 1, \ldots, h; \qquad l = 0, 1, \ldots, h-k+1.$$

By virtue of (7.9-17) these determinants can in turn be expressed by the functions $F_k^{(n+l)}$ plus correction terms. If $n \in \mathfrak{N}$, then by definition of $\mathfrak{N}$ the relations (7.9-21) are applicable to all $F_k^{(n+l)}$ involved. Thus, if n is large enough, the correction terms may be neglected, and we find

$$q_{m+k}^{(n+l)} \sim \frac{F_k^{(n+l+1)} F_{k-1}^{(n+l)}}{F_k^{(n+l)} F_{k-1}^{(n+l+1)}}, \qquad n \to \infty, \quad n \in \mathfrak{N}, \quad k = 1, 2, \ldots, h;$$

$$l = 0, 1, \ldots, h-k, \qquad (7.9\text{-}23)$$

where the arguments in each function F are $u_{m+1}, \ldots, u_{m+h}$, $F_0^{(n+l)} := 1$.
Now let us imagine the qd scheme of the rational function

$$\tilde{f}(z) := \frac{\tilde{r}_{m+1}}{z_{m+1} - z} + \cdots + \frac{\tilde{r}_{m+h}}{z_{m+h} - z},$$

where the $\tilde{r}_i$ are defined by (7.9-16). The Hankel determinants associated with the Taylor series at 0 by (7.9-8) are

$$\tilde{H}_k^{(n)} = F_k^{(n)}(u_{m+1}, \ldots, u_{m+h}),$$

where the $F_k^{(n)}$ are defined by (7.9-18). By the fundamental relations (7.6-5) the quotients $\tilde{q}_k^{(n+l)}$ of the corresponding qd scheme precisely equal the expression on the right of (7.9-23). By Theorem 7.7f the polynomials $\tilde{p}_h^{(n)}(u)$ formed according to (7.9-3), but with the $q_k^{(n)}$ in place of the $q_{m+k}^{(n)}$, precisely equal the polynomial $\tilde{p}_h(u)$ defined by (7.9-5). Since the $q_{m+k}^{(n)}$ are asymptotically equal to the $\tilde{q}_k^{(n)}$ for $n \to \infty$, $n \in \mathfrak{N}$, the assertion in Theorem 7.9a follows. ∎

As a further application of Lemma 7.9b we prove the statement, already contained in Theorem 7.6b but not yet proved, that

$$\lim_{n \to \infty} e_m^{(n)} = 0 \qquad (7.9\text{-}24)$$

for each critical index m. Using the fundamental formula (7.6-6) and expressing the Hankel determinants by (7.5-3) or (7.9-17) we have, if $m-1$ is likewise a critical index,

$$e_m^{(n)} = \text{const} \cdot u_m^{-n} \{ F_1^{(n)}(u_{m+1}, \ldots, u_{m+h}) + \epsilon_n \omega^n \}$$

and thus

$$e_m^{(n)} = 0\left(\left(\frac{\omega}{u_m} \right)^n \right) \to 0. \qquad (7.9\text{-}25)$$

If $m-1$ is not a critical index, let $j := m - i$ be the critical index preceding m. By (7.9-17)

$$H_{m-1}^{(n)} = (u_1 \cdots u_j)^n \{ F_{i-1}^{(n)}(u_{j+1}, \ldots, u_m) + \epsilon_n |u_m|^{(i-1)n} \}, \qquad \epsilon_n = 0(\gamma^n),$$

which may be turned around to read

$$H_{m-1}^{(n)} = (u_1 \cdots u_m)^n \{ F_1^{(n)}(u_{j+1}^{-1}, \ldots, u_m^{-n}) + \epsilon_n |u_m|^{-n} \}.$$

Using this in (7.6-6), (7.9-24) follows as above. ∎

POLES WITH EQUAL MODULI

In the numerical determination of poles of functions defined by power series a situation that is almost as unpleasant as the case of several poles of *equal* modulus is the occurence of poles whose moduli are *nearly* equal. It is implicit in the above analysis that sets of poles of nearly equal modulus can be treated by Rutishauser's rule as if all poles had the same modulus.

Whether the *qd* method is used to determine single poles (Theorem 7.6b) or sets of several poles of equal modulus (Theorem 7.9a), it is clear that the algorithm, unlike Newton's method, for instance, is not self-checking. Even if the scheme is generated by one of the stable versions of the *qd* algorithm it is subject to the building up of rounding errors. Thus, in general, the algorithm should be expected to furnish close approximations to the poles rather than precise values and the final values generated by the algorithm should be checked by an independent calculation.

PROBLEMS

1. *Complex zeros of the Bessel function J_ν.* For real ν, $\nu \neq -1, -2, \ldots$, the function

$$f(z) := {}_0F_1(\nu+1; -z) = \sum_{k=0}^{\infty} \frac{(-z)^k}{(\nu+1)_k k!} \qquad (7.9\text{-}26)$$

is related to J_ν, the Bessel function of order ν (see §§7.6 and 9.7) by the formula

$$f(z) = \text{const} \cdot z^{-\nu/2} J_\nu(2\sqrt{z}), \qquad (7.9\text{-}27)$$

where the value of the constant is irrelevant. Although the function J_ν has real zeros only for $\nu > -1$, it is known to have nonreal zeros for $\nu < -1$, whose precise number as a function of ν has been determined by Hurwitz (see Watson [1944], p. 483); for instance, if $-3 < \nu < -2$, there are precisely four nonreal zeros, corresponding to two complex conjugate zeros of f. Determine these zeros for $\nu := -\frac{7}{3}$.

Solution. The *qd* scheme of the power series representing $1/f$ can be constructed as in Example 6 of §7.6. The resulting triangular array starts out as follows:

```
-0.750000              0.000000              0.000000           0.000000
0.000000    1.500000             -0.500000            -0.150000
         0.750000             -2.000000              0.350000
0.000000   -4.000000              0.087500
        -3.250000              2.087500
0.000000    2.569231
        -0.680769
0.000000
```

...

Further computation confirms what is already suggested by this brief excerpt, namely that the column $e_1^{(n)}$ does not converge to zero, whereas the column $e_2^{(n)}$ does. This indicates the presence of a pair of complex conjugate zeros whose moduli are smaller than those of all other zeros. According to Rutishauser's rule, we form the sequences of the sums $q_1^{(n+1)} + q_2^{(n)}$ and products $q_1^{(n)} q_2^{(n)}$:

$q_1^{(n+1)} + q_2^{(n)}$	$q_1^{(n)} q_2^{(n)}$
-0.750000	
-1.250000	1.500000
-1.162500	1.565625
-1.152844	1.534244
-1.157137	1.537166
-1.156677	1.537607
$\ldots$	$\ldots$

The limits, to six digit accuracy, are -1.156696 and 1.537467. Thus the reciprocals u_1 and u_2 of the zeros of f in question are the zeros of the polynomial $\tilde{p}_2(u) := u^2 + 1.156696 u + 1.537467$. Solving the quadratic yields $u_{1,2} = -0.587348 \pm i \cdot 1.096805$. By (7.9-27) the corresponding zeros of $J_{-7/3}$ are

$$j := \frac{2}{\pm (u_{1,2})^{1/2}}.$$

Extracting the complex square roots yields the four complex zeros $j = \pm 1.537957 \pm i \cdot 0.927704$.

2. Let the function f be real on the real line, so that the complex poles of f occur in conjugate pairs. Let m, $m+2$ be two consecutive critical indices of f, and let $u_{m+1} = \rho e^{i\varphi}$, $u_{m+2} = \rho e^{-i\varphi}$, $0 < \varphi < \pi$. Demonstrate that $F_1^{(n)}(u_{m+1}, u_{m+2}) = \rho^n \cos(n\varphi + \alpha)$, where α is some fixed angle and consequently

$$q_{m+1}^{(n)} \sim \rho \frac{\cos[(n+1)\varphi + \alpha]}{\cos(n\varphi + \alpha)}, \quad q_{m+2}^{(n)} \sim \rho \frac{\cos(n\varphi + \alpha)}{\cos[(n+1)\varphi + \alpha]}. \quad (7.9\text{-}28)$$

Use these formulas to verify Rutishauser's rule.

3. Under the hypotheses of Problem 2, let $V_i^{(n)}$ denote the number of negative elements in the sequence $q_{m+i}^{(1)}, q_{m+i}^{(2)}, \ldots, q_{m+i}^{(n)}$, $i = 1, 2$; $n = 1, 2, \ldots$. Prove that

$$\lim_{n \to \infty} \frac{V_i^{(n)}}{n} = \frac{\varphi}{\pi}, \quad i = 1, 2.$$

[Thus from a glance at the qd table it is possible to estimate $\arg u_{m+1}$. Compare Example 1.]

4. Prove under the hypotheses of Problem 2 that if either sequence $\{q^{(n)}_{m+i}\}$, $i=1,2$, is bounded, then φ/π is rational.

5. The classical Bernoulli method for determining the dominant zero of a polynomial (see §7.4) can be viewed as a special case of the qd algorithm. Show that the usual way of dealing with a pair of dominant zeros in Bernoulli's method (see, e.g., Chapter 7 of Henrici's *Elements of Numerical Analysis* [1964]) can be viewed as a special case of Rutishauser's rule.

6. If m is a critical index of f, show that

$$\lim_{n\to\infty} q^{(n)}_m e^{(n)}_m = 0. \qquad (7.9\text{-}29)$$

[Apply Lemma 7.9b.]

The remaining problems are concerned with the qd scheme of a rational function r vanishing at infinity and having precisely m finite poles, all of order 1. We denote by $0 = m_0 < m_1 < \cdots < m_s = m (s \leq m)$ the sequence of critical indices of r.

7. Show that the matrix $\mathbf{M}^{(n)}_m$ defined by (7.8-5) is similar to a matrix $\mathbf{A}$ which in partitioned form can be written as follows:

$$\mathbf{A} = \begin{bmatrix} \mathbf{A}_1 & \mathbf{V}_1 & & & & \mathbf{0} \\ \mathbf{L}_2 & \mathbf{A}_2 & \mathbf{V}_2 & & & \\ & \cdots & \cdots & \cdots & & \\ & & & \mathbf{L}_{s-1} & \mathbf{A}_{s-1} & \mathbf{V}_{s-1} \\ \mathbf{0} & & & & \mathbf{L}_s & \mathbf{A}_s \end{bmatrix}.$$

For $k=1,2,\ldots,s$, the diagonal block $\mathbf{A}_k$ is a tridiagonal matrix of order $m_k - m_{k-1}$ whose eigenvalues for $n\to\infty$ approximate the kth set of eigenvalues of equal modulus of $\mathbf{M}^{(n)}_m$. The matrices $\mathbf{L}_k$ and $\mathbf{V}_k$ have the form

$$\mathbf{L}_k = \begin{bmatrix} 0 & \cdots & l_k \\ & \cdots & \\ 0 & \cdots & 0 \end{bmatrix}, \qquad \mathbf{V}_k = \begin{bmatrix} 0 & \cdots & 0 \\ & \cdots & \\ v_k & \cdots & 0 \end{bmatrix}.$$

Each of these matrices has only the one nonzero element shown. These elements satisfy

$$l_{k+1} v_k = q^{(n)}_{m(k)} e^{(n)}_{m(k)};$$

they can be made arbitrarily small for $n\to\infty$. [Use Problem 6.]

For Problems 8, 9, 10 we assume that the only sets of equimodular poles of r are pairs of complex conjugate poles, the order of each submatrix $\mathbf{A}_k$ thus being at most 2.

8. If
$$\mathbf{A}_k := \begin{bmatrix} \alpha_k & \gamma_k \\ -\gamma_k & \beta_k \end{bmatrix},$$

where $p := m_{k-1}$ and

$$\alpha_k := q^{(n)}_{p+1} + e^{(n)}_p, \qquad \beta_k := q^{(n)}_{p+2} + e^{(n)}_{p+1}, \qquad \gamma^2_k := -q^{(n)}_{p+1} e^{(n)}_{p+1},$$

show that the eigenvalues of $\mathbf{A}_k$ are $\mu_k \pm i\omega_k$, where

$$\mu_k := \tfrac{1}{2}(\alpha_k + \beta_k), \qquad \omega_k := (\gamma_k^2 - \delta_k^2)^{1/2}, \qquad \delta_k := \tfrac{1}{2}(\alpha_k - \beta_k).$$

9. Show that $\mathbf{A}_k$ is diagonalized by

$$\mathbf{T}_k := \sigma_k \begin{bmatrix} z_k & \bar{z}_k \\ -\bar{z}_k & -z_k \end{bmatrix},$$

where

$$z_k := e^{i\varphi_k/2}, \qquad \cos\varphi_k := \frac{\delta_k}{\gamma_k}, \qquad 0 < \varphi_k < \pi;$$

$$\sigma_k := (2\sin\varphi_k)^{-1/2} = \left(\frac{\gamma_k}{2\omega_k}\right)^{1/2}.$$

10. Let o_k denote the order of $\mathbf{A}_k$. Let σ_k be defined as above if $o_k = 2$ and $\sigma_k := 1$ if $o_k = 1$; $\zeta_k := \sigma_k \sigma_{k+1}$, $\pi_k := |l_{k+1} v_k|^{1/2}$ $(k = 1, 2, \ldots, s)$. By applying Gerschgorin's theorem to the matrix $\mathbf{T}\mathbf{A}\mathbf{T}^{-1}$, where

$$\mathbf{T} := \begin{bmatrix} \mathbf{T}_1 & & & 0 \\ & \mathbf{T}_2 & & \\ & & \ddots & \\ 0 & & & \mathbf{T}_s \end{bmatrix}$$

($\mathbf{T}_k$ is the identity matrix of order 1 if $o_k = 1$), show that for every choice of the numbers $\lambda_k > 0$ ($k = 0, 1, \ldots, s-1$) every reciprocal pole u_i of r is contained in at least one of the disks

$$|u - \mu_k \pm i\omega_k| \leq \lambda_{k-1}^{-1} \pi_{k-1} \zeta_{k-1} o_{k-1} + \lambda_k \pi_k \zeta_k o_{k+1}$$

PARTIAL FRACTION EXPANSIONS

($k = 1, 2, \ldots, s$). If any disk is isolated from all others, it contains precisely one u_i. [By applying this result to the qd scheme of $r := 1/p$, where p is a polynomial, we obtain arbitrarily accurate inclusion regions for the zeros of p.]

§7.10. PARTIAL FRACTION EXPANSIONS OF MEROMORPHIC FUNCTIONS

There has been much ado in preceding sections about the basic fact that the difference of a rational function and the sum of its principal parts is a polynomial. Regarding meromorphic functions (i.e., functions with isolated singularities that are at most poles) we have proved and used only the following: if f is meromorphic in a region D and f has a finite number of poles z_m in D with corresponding singular parts s_m, then $g := f - \sum s_m$ has removable singularities at the z_m and can be extended to a function that is analytic in D. If f is meromorphic in the whole plane and has only a finite number of poles, it follows immediately that the difference between f and the sum of its principal parts is an entire function. Is this statement also true for meromorphic functions with an infinite number of poles, such as $\tan z$?

When dealing with this question, we must determine whether the phrase "the sum of its principal parts" is meaningful for a meromorphic function with an infinite number of poles. A first obvious requirement is that the infinite number of poles be denumerable. This easily follows from the much stronger assertion that *the poles of a function meromorphic in the whole plane cannot have a finite accumulation point*. Indeed, if z_0 were a point of accumulation of poles, then z_0 would be neither a point of analyticity nor an isolated singularity of f; f would not be meromorphic at z_0. It follows from this assertion that any closed disk $|z| \leq \rho$ can contain only a finite number of poles. Thus the poles can be numbered $z_1, z_2, \ldots$, and, whatever numbering is chosen, $z_m \to \infty$ holds for $m \to \infty$.

If s_m is the singular part of f at the pole z_m, the series $\sum s_m$ can always be set up. The difficulty, however, now arises that the sum need not converge. [A concrete example of such a function is the gamma function (see §8.4).] Fortunately it turns out that convergence can be obtained by adding to each s_m a suitable "convergence generating" polynomial. This is the essential content of the following theorem.

THEOREM 7.10a (Mittag-Leffler theorem)

Let f be meromorphic in the complex plane, with poles z_m and corresponding singular parts s_m ($m = 1, 2, \ldots$). Polynomials p_m ($m = 1, 2, \ldots$) then exist such that the series

$$s(z) := \sum_{m=1}^{\infty} [s_m(z) - p_m(z)]$$

converges uniformly on every bounded set not containing any of the points z_m and $f - s$ is an entire function.

The function $f - s$ is entire if s is *any* meromorphic function whose poles and principal parts agree with those of f. Theorem 7.10a thus is amply implied by the following:

THEOREM 7.10b

Let $\{z_m\}$ be any sequence of distinct complex numbers without accumulation point and let $\{s_m\}$ be a sequence of analytic functions such that s_m is analytic for all $z \neq z_m$ and has an isolated singularity (which may be essential) at z_m. Polynomials p_m $(m = 0, 1, \ldots)$ then exist such that the series

$$s(z) := \sum_{m=0}^{\infty} [s_m(z) - p_m(z)] \qquad (7.10\text{-}1)$$

converges uniformly on every bounded set not containing any z_m. The function s is analytic everywhere except at the points z_m; its principal part at z_m is identical with that of s_m.

Proof. The proof given below contains an algorithm for constructing the "convergence-generating" polynomials p_m. Without loss of generality, we may assume that $z_0 = 0$, for if 0 is not among the z_m we may set $s_m = 0$. Furthermore we may assume that the z_m are numbered such that

$$0 = |z_0| < |z_1| \leq |z_2| \leq \cdots. \qquad (7.10\text{-}2)$$

For $m = 1, 2, \ldots$, let

$$\mu_m := \max_{|z| = \frac{1}{2}|z_m|} |s_m(z)|,$$

and let the integers n_m be chosen such that the series

$$t(z) := \sum_{m=1}^{\infty} \mu_m \left(\frac{2z}{|z_m|} \right)^{n_m + 1} \qquad (7.10\text{-}3)$$

has an infinite radius of convergence. (By the Cauchy–Hadamard formula

PARTIAL FRACTION EXPANSIONS

(Theorem 2.2a) this is the case if

$$\lim_{m \to \infty} \mu_m^{1/n_m} |z_m|^{-1} = 0$$

or, since $z_m \to \infty$, if the numbers μ_m^{1/n_m} form a bounded sequence, i.e., if $n_m \geq \kappa \log \mu_m$ for some $\kappa > 0$.) For $m = 1, 2, \ldots$, we now define p_m as the partial sum of degree n_m of the Taylor expansion of s_m at $z = 0$. By virtue of Cauchy's estimate (Theorem 2.2f) the kth Taylor coefficient of s_m is bounded by

$$\frac{\mu_m}{|z_m/2|^k}, \quad k = 0, 1, 2, \ldots,$$

and the difference $s_m(z) - p_m(z)$ for $|z| \leq \frac{1}{4}|z_m|$ can be estimated as follows:

$$|s_m(z) - p_m(z)| \leq \sum_{k=n_m+1}^{\infty} \frac{\mu_m}{|z_m/2|^k} |z|^k$$

$$= \mu_m \left|\frac{2z}{z_m}\right|^{n_m+1} \frac{1}{1 - |2z/z_m|} \leq 2\mu_m \left|\frac{2z}{z_m}\right|^{n_m+1}. \quad (7.10\text{-}4)$$

To show that the series (7.10-1) converges let $\rho > 0$ be arbitrary. Let the integer j be determined by the condition that

$$|z_m| < 4\rho \quad \text{for } m \leq j, \qquad |z_m| \geq 4\rho \quad \text{for } m > j.$$

We set

$$s(z) = \sum_{m=0}^{j} [s_m(z) - p_m(z)] + \sum_{m=j+1}^{\infty} [\cdots],$$

where $p_0 := 0$. The first sum represents an analytic function with a finite number of isolated singularities z_m ($m = 0, 1, \ldots, j$); the principal part at z_m evidently agrees with that of s_m. The second series is a series of functions that is analytic in $|a| < \rho$; by construction it is majorized by the corresponding terms of the series (7.10-3). Thus the second series represents an analytic function in $|z| < \rho$. We conclude that, for each $\rho > 0$, s is an analytic function in $|z| < \rho$, up to isolated singularities at the points z_m that satisfy $|z_m| < \rho$, where its principal parts agree with those of s_m. Since this is true for every $\rho > 0$, the assertion of Theorem 7.10b is established. ∎

EXAMPLE

We carry out the above algorithm to construct the partial fraction expansion of

$$f(z) := \frac{2\pi}{e^{2\pi z} - 1}.$$

The poles are at

$$z_0 := 0, \quad z_{2n-1} := in, \quad z_{2n} := -in, \quad n = 1, 2, \ldots.$$

The principal part at 0 is z^{-1}, and, since f is periodic with period i, the principal part at z_m is

$$s_m(z) := \frac{1}{z - z_m}.$$

We have

$$\mu_m = \max_{|z| = \frac{1}{2}|z_m|} |s_m(z)| = \frac{2}{|z_m|} = \frac{4}{m};$$

hence $\log \mu_m = \log 4/m \leq 0$ for $m \geq 4$, and $n_m := 0$ $(m = 1, 2, \ldots)$ will do. Indeed, since $|z_m| \geq m/2$ the series

$$\sum_{m=1}^{\infty} \frac{4}{m} \left| \frac{2z}{z_m} \right| \leq 16|z| \sum_{m=1}^{\infty} \frac{1}{m^2}$$

converges for all z. Hence it suffices to take as p_m the zeroth (constant) term of the Taylor series of s_m at $z = 0$, $p_m(z) := -z_m^{-1}$. The construction thus yields

$$s(z) = \sum_{m=0}^{\infty} [s_m(z) - p_m(z)] = \frac{1}{z} + \sum_{\substack{n = -\infty \\ n \neq 0}}^{\infty} \left(\frac{1}{z - in} + \frac{1}{in} \right),$$

or, by grouping together the terms belonging to $-n$ and n,

$$s(z) = \frac{1}{z} + \sum_{n=1}^{\infty} \frac{2z}{z^2 + n^2}. \tag{7.10-5}$$

By Theorem 7.10b s is a meromorphic function whose poles and principal parts agree with those of f. Hence

$$\frac{2\pi}{e^{2\pi z} - 1} = \frac{1}{z} + \sum_{n=1}^{\infty} \frac{2z}{z^2 + n^2} + g(z),$$

where g is entire.

PARTIAL FRACTION EXPANSIONS

It remains to determine g. As already noted, f is periodic with period i. We assert that the same holds for s. Indeed, if for $n = 1, 2, \ldots$,

$$t_n(z) := \frac{1}{z} + \sum_{m=1}^{n} \frac{2z}{z^2 + m^2} = \sum_{m=-n}^{n} \frac{1}{z - im},$$

then

$$t_n(z+i) - t_n(z) = \frac{1}{z + i(n+1)} - \frac{1}{z - in},$$

hence as $n \to \infty$

$$s(z+i) - s(z) = 0$$

for every z such that $s(z)$ is defined. We conclude that s, and consequently $g = f - s$, is periodic with period i.

We next wish to show that g is bounded. This, by Liouville's theorem (Theorem 3.3b), will imply that g is constant. Since g is periodic, it suffices to establish the boundedness in a period strip, say in the strip $0 \leqslant \mathrm{Im}\, z \leqslant 1$. As an analytic function g is certainly bounded on the compact set $-1 \leqslant \mathrm{Re}\, z \leqslant 1, 0 \leqslant \mathrm{Im}\, z \leqslant 1$. For $|\mathrm{Re}\, z| > 1$ we use the representation $g = f - s$ and show that each term on the right is bounded. It follows straight from the definition that

$$\lim_{x \to \infty} f(x + iy) = 0, \qquad (7.10\text{-}6)$$

the limit holding uniformly in y for $0 \leqslant y \leqslant 1$. To show the boundedness of $s(z)$, we first let $z = x > 0$. We then have

$$s(x) = \frac{1}{x} + \sum_{n=1}^{\infty} \frac{2x}{x^2 + n^2} = \frac{1}{x} + \frac{1}{x} \sum_{n=1}^{\infty} \frac{2}{1 + n^2 x^{-2}}. \qquad (7.10\text{-}7)$$

The last sum may be regarded as a Riemann sum for the integral

$$\int_0^{\infty} \frac{2}{1 + \tau^2} \, d\tau,$$

calculated with the step $\Delta\tau := 1/x$, the integrand being evaluated at $\tau := \tau_n := n\Delta\tau$, the right end point of each subinterval. Since the integrand is monotonically decreasing and the integral is known to have the value π, there follows

$$s(x) < \frac{1}{x} + \pi$$

and, incidentally,

$$\lim_{x \to \infty} s(x) = \pi. \qquad (7.10\text{-}8)$$

For complex z such that $0 \leqslant \mathrm{Im}\, z \leqslant 1$ and $\mathrm{Re}\, z \geqslant 1$ the trivial estimates $|z| \leqslant 2x$,

$|z^2+n^2| \geq \frac{1}{2}(x^2+n^2)$ ($n=1,2,\ldots$) suffice to show that

$$|s(z)| \leq 4s(x),$$

implying the boundedness of $s(z)$ in the period strip for $x \to \infty$. For $x \to -\infty$ the boundedness of $s(z)$ follows from the fact that s is odd, $s(-z) = -s(z)$. We thus may conclude that g is bounded, hence constant.

The constant value of g can be found by letting $x \to \infty$ in the relation $g(x) = f(x) - s(x)$ and observing (7.10-6) and (7.10-8). There follows

$$g(z) = -\pi \tag{7.10-9}$$

Thus we have finally established the expansion

$$\frac{2\pi}{e^{2\pi z} - 1} = \pi + \frac{1}{z} + \sum_{n=1}^{\infty} \frac{2z}{z^2 + n^2}. \tag{7.10-10}$$

We make several applications of this result.

As already noted, s is odd. Thus the function $f-g$ is odd, which fact becomes apparent if we note that $f-g$ equals

$$\pi \frac{e^{2\pi z} + 1}{e^{2\pi z} - 1} = \pi \coth \pi z.$$

Thus the above result may be written

$$\pi \coth \pi z = \frac{1}{z} + \sum_{n=1}^{\infty} \frac{2z}{z^2 + n^2} = \lim_{n \to \infty} \sum_{m=-n}^{n} \frac{1}{z - im}.$$

On replacing z with iz and multiplying by i, this becomes

$$\pi \cot \pi z = \frac{1}{z} + \sum_{n=1}^{\infty} \frac{2z}{z^2 - n^2} = \lim_{n \to \infty} \sum_{m=-n}^{n} \frac{1}{z - m}, \tag{7.10-11}$$

another interesting result. Because the series in (7.10-11) converges uniformly on every compact set avoiding the integers, we may differentiate term by term to obtain

$$\frac{\pi^2}{(\sin \pi z)^2} = \sum_{n=-\infty}^{\infty} \frac{1}{(z-n)^2}. \tag{7.10-12}$$

As another application of (7.10-10) let us recalculate the Taylor expansion of the

PARTIAL FRACTION EXPANSIONS

function

$$\frac{2\pi}{e^{2\pi z}-1}+\pi-\frac{1}{z}=\pi\coth\pi z-\frac{1}{z}$$

at $z=0$. Because the series in (7.10-10) converges uniformly, say, for $|z|\leq\frac{1}{2}$, we can apply the double series theorem (Theorem 2.7a). Term-by-term expansion and interchanging summations according to

$$\frac{2z}{z^2+n^2}=2\sum_{k=1}^{\infty}(-1)^{k+1}\frac{z^{2k-1}}{n^{2k}}$$

yields

$$\pi\coth\pi z-\frac{1}{z}=2\sum_{k=1}^{\infty}(-1)^{k+1}z^{2k-1}\sum_{n=1}^{\infty}\frac{1}{n^{2k}}.$$

On the other hand, it follows from the definition (2.5-12) of the Bernoulli numbers that

$$\pi\coth\pi z-\frac{1}{z}=\sum_{k=1}^{\infty}\frac{B_{2k}}{(2k)!}(2\pi)^{2k}z^{2k-1}.$$

Comparing coefficients, we again obtain the formula

$$\sum_{n=1}^{\infty}\frac{1}{n^{2k}}=(-1)^{k+1}\frac{(2\pi)^{2k}}{2\cdot(2k)!}B_{2k}, \quad k=1,2,\ldots, \qquad (7.10\text{-}13)$$

already obtained in §4.9 by using summatory functions and the calculus of residues.

PROBLEMS

1. Use Mittag-Leffler's method to find the partial fraction expansion of $\pi^2(\sin\pi z)^{-2}$ directly.
2. Use the identity

$$\frac{1}{\sin\pi z}=\frac{1}{2}\cot\frac{\pi z}{2}-\frac{1}{2}\cot\frac{\pi(z-1)}{2}$$

to show that

$$\frac{\pi}{\sin\pi z}=\lim_{n\to\infty}\sum_{m=-n}^{n}\frac{(-1)^m}{z-m}.$$

SEMINAR ASSIGNMENTS

1. Study (experimentally and/or theoretically) the numerical stability of the various versions of algorithm C in §7.1. Also compare with the algorithm of Problem 2 in §7.1, which has been claimed to be numerically superior.

2. Many methods for the numerical integration of an ordinary differential equation $Y' = f(x,y)$ are based on integrating a *polynomial* interpolating the function $y'(x) = f(x, y(x))$. Construct an algorithm based on interpolation by *exponential sums* in place of polynomials. (Such an algorithm would integrate linear differential equations with constant coefficients exactly.)

3. Construct an algorithm optimizing (by means of similarity transformations) the inclusion disks for polynomial zeros given by the Theorems 7.8f and 7.9d (see Henrici [1963a]).

4. The formal results on Hankel determinants are valid for formal power series with coefficients taken from any field, thus in particular from the field of formal Laurent series (say in a parameter ϵ, see §1.8). If $F := c_0 + c_1 z + c_2 z^2 + \cdots$ is the Taylor expansion at 0 of a rational function satisfying the hypotheses of Theorem 7.5e, let $F(a) = c_0(\epsilon) + c_1(\epsilon) z + \cdots := c_0 + c_1(z + \epsilon) + c_2(z + \epsilon)^2 + \cdots$. Prove or disprove that the series $F(\epsilon)$ is always k-normal.

5. Qualitative information is available about the nonreal zeros of $J_\nu(z)$ for $\nu < 0$ (see Watson [1944], p. 483). Make a numerical study of these zeros and graph them as a function of ν, using the extended form of the qd algorithm as in §7.9.

6. Study the real and the nonreal zeros of the series ${}_1F_1(\alpha; \beta; z)$ as a function of the real parameters α and β. Compare with the qualitative results of Kienast [1921].

NOTES

§7.1. For some recent treatments of partial fraction decomposition from an engineering point of view see Kuo [1962], Pottle [1964], Brugia [1965]. Ahlfors ([1953], pp. 44–45; [1966], pp. 31–32) determines singular parts in an elegant manner by using Moebius transformations. Computationally, however, his method is less efficient than Method B. Method C is due to Henrici [1971]. The variant of Method C stated in Problem 2 was communicated to the author independently by M. Gutknecht and M. A. Budin.

§7.2. Interpolation by sums of exponentials is dealt with briefly by Rutishauser [1957]. See also Meinardus [1964], p. 169; Meinardus & Schwedt [1964], Rice [1962].

§7.3. See Netto [1927] or Riordan [1958] for comprehensive treatments of combinatorics. Whitworth [1942] has many specific problems. Apart from its intrinsic interest, combinatorics provides an excellent opportunity to illustrate the basic concepts of function theory (see §§2.1 and 2.4) by means of functions defined on finite sets.

§7.4. For difference equations see Nörlund [1924], Goldberg [1958], or Hildebrand [1968]. Bernoulli's method is treated elementarily by Henrici [1964b]. Sebastiao e Silva [1941], Bauer [1956], Samelson [1959], and Traub [1966] have generalized Bernoulli's method in a direction different from the qd algorithm. The connection between these various methods is made clear by Householder [1970a], who gives more complete references.

§7.5. The theory of determinants tends to be de-emphasized in modern treatments of linear algebra. The material required here is treated adequately by Nering [1963] and Mirsky [1955]. For Theorem 7.5b see Aitken [1926]. The theory of Hankel determinants and of the qd algorithm is closely linked with the Padé table; see Gragg [1972], Baker & Gammel [1970].

§7.6. The qd algorithm was introduced by Rutishauser [1954, 1956, 1957, 1963] and further discussed by Henrici [1958, 1963a, 1967] and Householder [1970a]. Stewart [1971] gives a succinct exposition that to some extent avoids determinants. The special case $m=1$ in Theorem 7.6b dates back to König [1884]. For an acceleration of the convergence of the qd algorithm by continued fractions see Rutishauser [1962].

§7.7. Hadamard polynomials were introduced by Hadamard [1892].

§7.8. The continuity of solutions of algebraic equations is discussed in Appendix A of Ostrowski [1966]. See also §4.10.

§7.9. For the bounds given here see Henrici [1963b], in which the Gershgorin bounds are improved by diagonal similarity transformations; see also Varga [1964], Todd [1965], Medley & Varga [1968].

BIBLIOGRAPHY

Sections to which a reference is relevant are numbered at the end of each reference. $S=$ seminar assignment.

Adams, E. P. [1939]. Note on a problem in electrostatics. *Quart. J. Math., Oxford Ser.* **10**, 241–246. §5.14.

Ahlfors, L. V. [1953]. *Complex analysis*. McGraw-Hill, New York. §§4.3, 4.7, 4.8, 4.10, 5.10, and 7.1.

⸺ [1966]. *Complex analysis*, 2nd ed. McGraw-Hill, New York. §§5.12, 5.13.

Aitken, A. C. [1926]. On Bernoulli's numerical solution of algebraic equations. *Proc. Roy. Soc. Edinburgh Ser.* **A46**, 289–305. §7.5.

Aleksandrov, P. S. [1956]. *Combinatorial topology*, Vol. I. Graylock, Rochester. §4.6.

Alexander, W. J. [1961]. Certification of Algorithm 30. *Comm. ACM* **3**, 238. §6.14.

Ankeny, N. C. [1947]. One more proof of the fundamental theorem of algebra. *Amer. Math. Monthly* **54**, 464. §2.4.

Apostol, T. M. [1969]. Some explicit formulas for the exponential matrix e^{tA}. *Amer. Math. Monthly* **76**, 289–292. §2.6.

Bachmann, K.-H. [1960]. Lösung algebraischer Gleichungen nach der Methode des stärksten Abstiegs. *Z. angew. Math. Mech.* **40**, 132–135. §6.14.

Bailey, W. N. [1935]. *Generalized hypergeometric series*. Cambridge Tracts in Mathematics and Mathematical Physics No. 32. Cambridge University Press, Cambridge. §1.5.

Baker, G. A., and J. L. Gammel [1970]. The Padé approximants in theoretical physics. Academic, New York. §§7.5, 7.6.

Bauer, F. L. [1956]. Das Verfahren der abgekürzten Iteration für algebraische Eigenwertprobleme, insbesondere zur Nullstellenbestimmung eines Polynoms. *Z. angew. Math. Phys.* **7**, 17–32. §7.4.

Bauhuber, F. [1970]. Direkte Verfahren zur Berechnung zur Nullstellen von Polynomen. *Computing* **5**, 97–118. §6.14.

Bareiss, E. [1960]. Resultant procedure and the mechanization of the Graeffe process. *J. Assoc. Comp. Mach.* **7**, 346–386. §6.9.

Batschelet, E. [1944]. Untersuchungen über die absoluten Beträge der Wurzeln algebraischer, insbesondere kubischer Gleichungen. *Verh. Naturf. Ges. Basel* **55**, 158–179. §6.4

———— [1945]. Ueber die absoluten Beträge der Wurzeln algebraischer Gleichungen. *Acta Math.* **76**, 253–260. §6.4.

———— [1946]. Ueber die Abschätzung der Wurzeln algebraischer Gleichungen. *El. Math.* **1**, 73–81. §§6.4, 6.5.

Behnke, H., and F. Sommer [1962]. *Theorie der analytischen Funktionen einer komplexen Veränderlichen*, 2. Aufl. Springer, Berlin. §4.3.

Belaga, E. G. [1958]. Some problems involved in the calculation of polynomials. *Dokl. Akad. Nauk SSSR* **123**, 775–777. §6.1.

Betz, A. [1964]. *Konforme Abbildung*, 2. Aufl. Springer, Berlin. §5.8.

Birkhoff, G., and S. MacLane [1941]. *A survey of modern algebra*. Macmillan, New York. §1.1.

Birkhoff, G. D. [1914]. An elementary double inequality for the roots of an algebraic equation having greatest absolute value. *Bull. Amer. Math. Soc.* **21**, 494–495. §6.4.

Boas, R. P. [1935]. A proof of the fundamental theorem of algebra. *Amer. Math. Monthly* **42**, 501–502. §2.4.

———— [1964]. Yet another proof of the fundamental theorem of algebra. *Amer. Math. Monthly* **71**, 180. §2.4.

Bohm, D. [1951]. *Quantum theory*. Prentice-Hall, Englewood Cliffs, New Jersey. §5.3.

Börsch-Supan, W. [1963]. A posteriori error bounds for polynomials. *Numer. Math.* **5**, 380–398. §6.4.

———— [1970]. Residuenabschätzung für Polynom-Nullstellen mittels Lagrange-Interpolation. *Numer. Math.* **14**, 287–296. §6.4.

Bourbaki, N. [1950]. *Algèbre*, Chapter IV, Polynomes et fractions rationelles. Hermann, Paris. §§1.2, 1.4, 1.6–1.9.

Broucke, R.A. [1971]. Construction of rational and negative powers of a formal series. *Comm. ACM* **14**, 32–35.§1.7.

Brouwer, D., and G. M. Clemence [1961]. Methods of celestial mechanics. Academic, New York. §§2.5, 4.5.

Brouwer, L. E. J. [1924]. Intuitionistische Ergänzung des Fundamentalsatzes der Algebra. *Amsterdam Konigl. Akad. van Wetenschapen, Proc.* **27**, 631–634. §§2.4, 6.10.

————, and B. de Loer [1924]. Intuitionistischer Beweis des Fundamentalsatzes der Algebra. *Amsterdam Konigl. Akad. van Wetenschapen, Proc.* **27**, 186–188. §§2.4, 6.10.

Brugia, O. [1965]. A noniterative method for the partial fraction expansion of a rational function with high order poles. *SIAM Review* **7**, 381–387. §7.1.

Buckholtz, J. D. [1967]. Sums of powers of complex numbers. *J. Math. Anal. Appl.* **17**, 269–279. §6.4.

Busemann, A. [1928]. Das Förderhöhenverhältnis radialer Kreiselpumpen mit logarithmischspiraligen Schaufeln. *Z. angew. Math. Mech.* **8**, 372–384. §5.8.

Cannon, J., and K. Miller [1965]. Some problems in numerical analytic continuation. *J. Soc. Indust. Appl. Math. Ser. B* **2**, 87–96. §3.6.

BIBLIOGRAPHY

Carrier, G. F., M. Krook, and C. E. Pearson [1966]. *Functions of a complex variable.* McGraw-Hill, New York. §§4.8, 4.9, 5.12, 5.14.

Cartan, H. [1961]. *Théorie élémentaire des fonctions analytiques d'une ou plusieurs variables complexes.* Hermann, Paris. §§1.2, 1.4, 1.6-9, 2.2-4, 3.1.

Cauchy, A. [1825]. Mémoire sur les intégrales définies prises entre des limites imaginaires. §4.3.

―――, [1847]. Méthode générale pour la resolution des systèmes d'equations simultanées. *Comptes Rendus* **25**, 2, 536. §6.14.

Churchill, R. V. [1960]. *Complex variables and applications*, 2nd ed. McGraw-Hill, New York. §§4.8, 5.5, 5.12.

Cockroft, J. D. [1927]. The effect of curved boundaries on the distribution of electrical stress round conductors. *J.I.E.E.* **66**, 385–409. §5.14.

Cohen, K. J. [1962]. Certification of algorithm 30. *Comm. ACM* **4**, 50. §6.14.

Collatz, L. [1945]. *Eigenwertprobleme und ihre numerische Behandlung.* Akad. Verlagsgesellschaft, Leipzig. §2.S.

Cooley, J. W., and J. W. Tukey [1965]. An algorithm for the machine calculation of complex Fourier series. *Math. Comp.* **19**, 297–301. §6.14.

Dahlquist, G. [1959]. Stability and error bounds in the numerical integration of ordinary differential equations. *Trans. Roy. Inst. Technol., Stockholm*, No. 130. §1.3.

David, F. N., M. G. Kendall, and D. E. Barton [1966]. *Symmetric functions and allied tables.* Cambridge University Press, Cambridge. §6.4.

Dejon, B., and K. Nickel [1969]. A never failing, fast converging rootfinding algorithm. *Constructive aspects of the fundamental theorem of algebra*, B. Dejon and P. Henrici, Eds. Wiley-Interscience, London. §6.14.

Delves, L. M., and J. N. Lyness [1967]. A numerical method for locating the zeros of an analytic function. *Math. Comp.* **21**, 543–560. §6.11.

Derr, J. I. [1959]. A unified process for the evaluation of zeros of polynomials over the complex number field. *Math. Tables Aids Comp.* **13**, 29–36. §6.12.

Dienes, P. [1931]. *The Taylor series.* Oxford University Press, Oxford. §§2.5, 3.1, 4.6, 7.5.

Dieudonné, J. [1938]. *La théorie analytique des polynomes d'une variable* (à coefficients quelconques). Mém. des sciences mathématiques, Vol. 93. Herman, Paris. §§6.4, 6.5, 6.7, 6.8.

Dinghas, A. [1968]. *Einführung in die Cauchy-Weierstrass'sche Funktionentheorie.* Bibliographisches Institut, Mannheim. §3.1.

Dirschmied, H. J. [1969]. Bemerkungen zu einer Arbeit von G. Polya zur Bestimmung der Nullstellen ganzer Funktionen. *Numer. Math.* **13**, 344–348. §6.9.

Duncan, J. [1968]. *The elements of complex analysis.* Wiley, New York. §4.3.

Earle, L. J., and R. S. Hamilton [1970]. A fixed point theorem for holomorphic mappings. *Global analysis* (Proc. Sympos. Pure Math. Vol. XVI, Berkeley, Calif., 1968). Amer. Math. Soc., Providence, Rhode Island. §6.12

Edmonds, A. R. [1957]. *Angular momentum in quantum mechanics.* Princeton University Press, Princeton, New Jersey. §1.5.

Effertz, F. H., and F. Kolberg [1963]. *Einführung in die Dynamik selbsttätiger Regelungssysteme.* VDI, Düsseldorf. §4.10.

Ellenberger, K. W. [1960]. Algorithm 30, Numerical solution of polynomial equations. *Comm. ACM* **4**, 643. §6.14.

Erdélyi, A. [1953]. *Higher transcendental functions*, Vol. I. McGraw-Hill, New York. §1.5.

Eve, J. [1964]. The evaluation of polynomials. *Numer. Math.* **6**, 17–21. §6.1.

Feferman, C. [1967]. An easy proof of the fundamental theorem of algebra. *Amer. Math. Monthly* **74**, 854–855. §2.4.

Firth, D. A. [1968]. *A method for generating thick, cambered aerofoils in cascade using a closed mapping function.* Department of Supply, Report ARL/M.E. 121, Melbourne. §5.8.

Friedli, A. [1973]. Optimal covering algorithms in methods of search for solving polynomial equations. *J. Assoc. Comp. Mach.* **20**, 290–300. §6.10.

Gantmacher, F. R. [1959]. *The theory of matrices*, 2 vols. Chelsea, New York. §2.6.

Gargantini, I., and P. Henrici [1972]. Circular arithmetic and the determination of polynomial zeros. *Numer. Math.* **18**, 305–320. §§6.6, 6.13.

Gautschi, W. [1967]. Computational aspects of three-term recurrence relations. *SIAM Review* **9**, 24–82. §4.5.

Gear, C. W. [1971]. *Numerical initial value problems in ordinary differential equations*. Prentice-Hall, Englewood Cliffs, New Jersey. §6.8.

Goldberg, S. [1958]. *Introduction to difference equations*. Wiley, New York. §7.4.

Goldstein, A.A. [1967]. *Constructive real analysis*. Harper and Row, New York. §6.14.

Golusin, G. M. [1957]. *Geometrische Funktionentheorie*. VEB Deutscher Verlag der Wissenschaften, Berlin. §5.10.

Good, I. J. [1960]. Generalizations to several variables of Lagrange's expansion, with applications to stochastic processes. *Proc. Camb. Phil. Soc.* **56**, 367–380. §1.9.

Goursat, E. [1900]. Sur la définition générale des fonctions analytiques d'apres Cauchy. *Trans. Amer. Math. Soc.* **1**, 14–16. §4.3.

Gragg, W. B. [1972]. The Padé table and its relation to certain algorithms of numerical analysis. *SIAM Rev.* **14**, 1–62. §§7.5–7.9.

Hadamard, J. [1892]. Essai sur l'étude des fonctions données par leur développement de Taylor. *J. Math. Pures Appl.* **8**, 101–186. §7.7.

Hardy, G. H., and E. M. Wright [1954]. *An introduction to the theory of numbers*. Clarendon Press, Oxford. §§3.4, 7.3.

Heins, M. [1941]. On the iteration of functions which are analytic and Single-valued in a given multiply-connected domain. *Am. J. of Math.* **63**, 461–480. §6.12.

Henrici, P. [1948]. *Potentialprobleme mit scharfen und abgerundeten Ecken*. Master's thesis, Dept. of Electrical Engineering, ETH Zurich. §§5.12, 5.14.

———— [1958]. The quotient-difference algorithm. *Appl. Math. Series* **49**, 23–46. National Bureau of Standards, Washington, D. C. §§7.6–7.9.

———— [1963a]. Some applications of the quotient-difference algorithm. *Proc. Symp. Appl. Math.* **15**, 159–183. Amer. Math. Soc., Providence, R. I. §§7.6, 7.9.

———— [1963]. Bounds for the eigenvalues of certain tridiagonal matrices. *J. Soc. Appl. Indust. Math.* **11**, 281–290. Erratum, *ibid.*, **12**, 497. §§7.8, 7.9.

———— [1964a]. An algebraic proof of the Lagrange-Bürmann formula. *J. Math. Anal. Appl.* **8**, 218–224. §1.9.

———— [1964b]. *Elements of Numerical Analysis*. Wiley, New York. §7.4.

———— [1966]. An algorithm for analytic continuation. *SIAM J. Numer. Anal.* **3**, 67–78. §3.6.

———— [1967]. Quotient-difference algorithms. *Mathematical Methods for digital computers*, Vol. 2, A. Ralston and H. S. Wilf, eds., Wiley, New York. §§7.6–7.9.

———— [1968]. Uniformly convergent algorithms for the simultaneous determination of all zeros of a polynomial. *Studies in numerical analysis* **2**, 2–8. §6.11.

———— [1970]. Methods of search for solving polynomial equations. *J. Assoc. Comp. Mach.* **17**, 273–283. §6.10.

———— [1971]. An algorithm for the incomplete decomposition of a rational function into partial fractions. *Z. angew. Math. Physik* **22**, 751–755. §7.1.

————, and I. Gargantini [1969]. Uniformly convergent algorithms for the simultaneous determination of all zeros of a polynomial. *Constructive aspects of the fundamental theorem of algebra*, B. Dejon and P. Henrici, Eds., Wiley-Interscience, London. §6.11.

Hervé, M. [1963]. *Several complex variables: Local theory*. Oxford University Press, Oxford. §6.12.

BIBLIOGRAPHY

Hildebrand, F. B. [1956]. *Introduction to numerical analysis*. McGraw-Hill, New York. §§6.9, 6.12, 7.4.
───── [1968]. *Finite-Difference equations and simulations*. Prentice-Hall, Englewood Cliffs, New Jersey. §7.4.
Hille, E. [1959]. *Analytic function theory*, Vol. I. Ginn, Boston. §§2.4, 4.3, 4.6, 4.7, 4.9, and 5.4.
───── [1962]. *Analytic function theory*, Vol. II. Ginn, Boston. §§5.10, 5.12, 5.14.
─────, and R. S. Phillips [1957]. *Functional analysis and semigroups*, rev. ed. Amer. Math. Soc., Providence. §2.1, 2.2.
Householder, A. S. [1964]. *The theory of matrices in numerical analysis*. Blaisdell, New York. §§2.1, 7.8.
───── [1968]. Bigradients and the problem of Routh and Hurwitz. *SIAM Rev.* **10**, 56–66. §6.7.
───── [1970a]. *The numerical treatment of a single non-linear equation*. McGraw-Hill, New York. §§2.5, 6.8, 6.12, 7.6–7.9.
───── [1970b]. *KWIC index for the numerical treatment of non-linear equations*. Oak Ridge National Laboratory ORNL-4595, UC-32. §2.4, Chapters 6 and 7.
Hurwitz, A. [1929]. *Allgemeine Funktionentheorie und Elliptische Funktionen*. Mit einem Anhang von R. Courant. 3. Aufl. Springer, Berlin (4. Aufl. [1964]). §§2.2, 2.4, 2.7, 3.1, 3.3–3.5, 4.4, 4.5, 5.10–5.13.
Isaacson, E., and H. B. Keller [1966]. *Analysis of numerical methods*. Wiley, New York. §6.12.
Jabotinsky, E. [1953]. Representation of functions by matrices. Application to Faber polynomials. *Proc. Amer. Math. Soc.* **4**, 546–553. §1.7.
Jackson, J. D. [1962]. *Classical electrodynamics*. Wiley, New York. §5.7.
Julia, G. [1918]. Sur l'itération des fonctions rationelles. *J. de Math. sér.* 7, **4**, 47–245. §6.12.
Kahan, W. [1967]. Laguerre's method and a circle which contains at least one zero of a polynomial. *SIAM J. Num. Anal.* **4**, 474–482. §§6.4, 6.12.
Kaluza, T. [1928]. Ueber die Koeffizienten reziproker Potenzreihen. *Math. Z.* **28**, 161–170. §1.3.
Kellenberger, W. [1970]. Ein konvergentes Iterationsverfahren zur Bestimmung der Nullstellen eines Polynoms. *Z. angew. Math. Physik* **21**, 647–651. §6.14.
───── [1971]. *Ein konvergentes Iterationsverfahren zur Bestimmung der Wurzeln eines Polynoms*. Diss. Nr. 4653, Eidgen. Techn. Hochschule. Juris-Verlag, Zürich. §6.14
Kerner, I. O. [1966]. Ein Gesamtschrittverfahren zur Berechnung der Nullstellen eines Polynoms. *Numer. Math.* **8**, 290–294. §6.9.
───── [1967]. Bemerkungen zur Berechnung der Nullstellen eines Polynoms durch Iterationsformeln in Verbindung mit ihrer Darstellung durch Lie-Reihen. *Z. angew. Math. Mech.* **47**, 549–550. §6.9.
Kienast, H. [1921]. Untersuchungen über die Lösungen der Differentialgleichung $xy'' + (\alpha - x)y' - \beta y = 0$. *Mitt. Naturf. Ges. Bern* **57**, 247–325. §7.S.
Kirchner, R. B. [1967]. An explicit formula for e^{At}. *Amer. Math. Monthly* **74**, 1200–1204. §2.6.
Knuth, D. E. [1962]. Evaluation of polynomials by computer. *Comm. ACM* **5**, 595. §6.1.
───── [1969]. *The art of computer programming*. Vol. 2: Seminumerical algorithms. Addison-Wesley, Reading, Massachusetts. §§1.3, 1.7, 1.9, 6.1.
Kober, H. [1952]. *Dictionary of conformal representations*. Dover, New York. §5.12.
Koch, E. [1968]. *Berechnung der ebenen, inkompressiblen Potentialströmung durch gerade Schaufelgitter*. VDI Forschungsheft 528, Düsseldorf. §5.8.
Kohfeld, J. J. [1962]. Certification of algorithm 30. *Comm. ACM* **5**, 293. §6.14.
König, J. [1884]. Ueber eine Eigenschaft der Potenzreihen. *Math. Ann.* **23**, 447–449. §7.6.
Kublanowskaya, V. N. [1959]. Application of analytic continuation in numerical analysis by means of change of variables. *Trudy Mat. Inst. Steklov* **53**, 145–185. §3.6.

Kuo, F. F. [1962]. *Network analysis and synthesis*. Wiley, New York. §7.1.

Lavrentiev, M. A., and B. V. Schabat [1967]. *Methoden der komplexen Funktionentheorie*. VEB Deutscher Verlag der Wissenschaften, Berlin. §§4.3, 4.8, 5.8, 5.12, 5.14.

Lehmer, D. H. [1961]. A machine method for solving polynomial equations. *J. Assoc. Comp. Mach.* **8**, 151–162. §§6.8, 6.10.

——— [1969]. Search procedures for polynomial equation solving. *Constructive aspects of the fundamental theorem of algebra*, B. Dejon and P. Henrici, Eds. Wiley-Interscience, London. §§6.10, 6.14.

Levinson, N., and R. Redheffer [1970]. *Complex variables*. Holden-Day, San Francisco. §4.3.

Lewis, G. [1960]. Two methods using power series for solving initial value problems. *NYU Inst. of Math. Sciences* NYO-2881. §3.6.

Lindelöf, E. [1905]. *Le calcul des résidus*. Gauthier-Villars, Paris. §§4.3, 4.7–4.9.

Maehly, H. J. [1954]. Zur iterativen Auflösung algebraischer Gleichungen. *Z. angew. Math. Physik* **5**, 260–263. §6.12.

Marden, M. [1966]. *Geometry of polynomials*. Amer. Math. Soc., Providence, Rhode Island. §§4.10, 6.4, 6.5, 6.7, and 6.8.

Medley, H. J., and R. S. Varga [1968]. On smallest isolated Gerschgorin disks for eigenvalues. *Numer. Math.* **11**, 320–323. §7.9.

Meinardus, G. [1964]. *Approximation von Funktionen und ihre numerische Behandlung*. Springer Tracts in Natural Philosophy, Vol. 4. Springer, Berlin. §7.2.

———, and D. Schwedt [1964]. Nicht-lineare Approximationen. *Arch. Rat. Mech. Anal.* **17**, 297–326. §7.2.

Miller, K. [1970]. Stabilized numerical analytic prolongation with poles. *SIAM J. Appl. Math.* **18**, 346–363. §3.6.

Milne-Thomson, L. M. [1952]. *Theoretical aerodynamics*, 2nd ed. Macmillan, London. §5.8.

Mirsky, L. [1955]. *An introduction to linear algebra*. Clarendon Press, Oxford. §§5.3 and 7.5.

Nasitta, K. [1964]. Ein immer konvergentes Nullstellenverfahren für analytische Funktionen. *Z. angew. Math. Mech.* **44**, 57–63. §6.14.

Nehari, Z. [1952]. *Conformal mapping*. McGraw-Hill, New York. §§5.10, 5.12.

Nering, E. D. [1963]. *Linear algebra and matrix theory*. Wiley, New York. §7.5.

Netto, E. [1927]. *Lehrbuch der Kombinatorik*, 2nd ed. Leipzig. §7.3.

Neumann, J. von [1929]. Ueber die analytischen Eigenschaften von Gruppen linearer Transformationen und ihrer Darstellungen. *Math. Z.* **30**, 3–42. §2.6.

Nickel, K. [1966]. Die numerische Berechnung der Wurzeln eines Polynoms. *Numer. Math.* **9**, 80–98. §6.14.

——— [1970]. Fehlerschranken zu Näherungswerten von Polynomwurzeln. *Computing* **6**, 9–27. §6.4.

Niven, I. [1969]. Formal power series. *Amer. Math. Monthly* **76**, 871–889. §§1.2–1.4, 1.6–1.9.

Nörlund, N. E. [1924]. *Vorlesungen über Differenzenrechnung*. Springer, Berlin. §7.4.

Ostrowski, A. [1940]. Recherches sur la méthode de Graeffe et les zéros de polynomes et les séries de Laurent. *Acta Math.* **72**, 99–257. §6.9.

——— [1966]. *Solution of equations and systems of equations*, 2nd ed. Academic, New York. §§6.12, 6.14, 7.8.

——— [1969]. Eine Methode zur automatischen Auflösung algebraischer Gleichungen. *Constructive aspects of the fundamental theorem of algebra*, B. Dejon and P. Henrici, Eds. Wiley-Interscience, London. §6.14.

Paatero, V. [1931]. Ueber die konforme Abbildung von Gebieten, deren Ränder von beschränkter Drehung sind. *Ann. Acad. Sci. Fenn.* **A33**, 1–77. §5.S.

BIBLIOGRAPHY

Painlevé, P. [1899]. Sur le développement d'une branche uniforme de fonction analytique. *C. R. Acad. Sci. Paris* **128**, 1277–1280. §3.6.
Pan, V. Ya. [1959]. Schemes for the calculation of polynomials with real coefficients. *Dokl. Akad. Nauk SSSR* **127**, 266–269. §6.1.
Parlett, B. [1964]. Laguerre's method applied to the matrix eigenvalue problem. *Math. Comp.* **18**, 464–485. §6.12.
Pederson, R. N. [1969]. The Jordan curve theorem for piecewise smooth curves. *Amer. Math. Monthly* **76**, 605–610. §4.6.
Polya, G. [1968]. Graeffe's method for eigenvalues. *Num. Math.* **11**, 315–319. §6.9.
_____, and G. Szegö [1925]. Aufgaben und Lehrsätze der Analysis, 2 vols. Springer, Berlin. §§1.9, 2.5, 6.2–6.5, 7.3.
_____ [1951]. Isoperimetric inequalities in mathematical physics. Princeton University Press, Princeton, New Jersey. §§5.7, 5.9.
Pomentale, T. [1971]. A class of iterative methods for holomorphic functions. *Numer. Math.* **18**, 193–203. §6.12.
Pottle, C. [1964]. On the partial fraction expansion of a rational function with multiple poles by digital computer. *IEEE Trans. Circuit Theory* **CT-11**, 161. §7.1.
Putzer, E. J. [1966]. Avoiding the Jordan canonical form in the discussion of linear systems with constant coefficients. *Amer. Math. Monthly* **73**, 2–7. §2.6.
Raney, G. [1960]. Functional composition patterns and power series reversion. *Trans. Amer. Math. Soc.* **94**, 441–451. §1.9.
Rasmussen, O. L. [1964]. Solution of polynomial equations by the method of D. Lehmer. *BIT* **4**, 250–260. §6.10.
Redheffer, R. M. [1957]. The fundamental theorem of algebra. *Amer. Math. Monthly* **64**, 582–585. §2.4.
_____ [1964]. What! Another note just on the fundamental theorem of algebra? *Amer. Math. Monthly* **71**, 180–185. §2.4.
_____ [1969]. The homotopy theorems of function theory. *Amer. Math. Monthly* **76**, 778–787. §3.5.
Reiffen, H. J. [1965]. Die Carathéodorische Distanz und ihre zugehörige Differentialmetrik. *Math. Ann.* **161**, 315–324. §6.12.
Rice, J. R. [1962]. Chebyshev approximation by exponentials. *J. Soc. Indust. Appl. Math.* **10**, 149–161. §7.2.
Richmond, H. W. [1923]. On the electrostatic field of a plane or circular grating formed of thick rounded bars. *Proc. London Math. Soc. Ser.* 2, **22**, 483–494.
Riemann, B. [1859]. Ueber die Anzahl der Primzahlen unter einer gegebenen Grösse. *Collected works*, 145–153.
Riordan, J. [1958]. *An introduction to combinatorial analysis*. Wiley, New York. §7.3.
Ritt, J. F. [1920]. On the conformal mapping of a region into a part of itself. *Ann. Math.* **22**, 157–160. §6.12.
Rosenbloom, P. C. [1945]. An elementary constructive proof of the fundamental theorem of algebra. *Amer. Math. Monthly* **52**, 562–570. §§2.4, 6.10.
Rutishauser, H. [1954]. Der Quotienten-Differenzen-Algorithmus. *Z. angew. Math. Physik* **5**, 233–251. §§7.6–7.9.
_____ [1956]. Eine Formel von Wronski und ihre Bedeutung für den Quotienten-Differenzen-Algorithmus. *Z. angew. Math. Physik* **7**, 164–169. §§1.3, 7.6.
_____ [1957]. *Der Quotienten-Differenzen-Algorithmus*. Birkhäuser, Basel. §§7.6–7.9.
_____ [1962]. On a modification of the QD-Algorithm with Graeffe-type convergence. *Z. angew. Math. Physik* **13**, 493–496. §7.6.

———— [1963]. Stabile Sonderfälle des Quotenten-Differenzen-Algorithmus. *Numer. Math.* **5**, 95–112. §7.6.

———— [1969]. Zur Problematik der Nullstellenbestimmung bei Polynomen. *Constructive aspects of the fundamental theorem of algebra*, B. Dejon & P. Henrici, Eds., Wiley-Interscience, London. §§6.1, 6.9, 6.12, 7.1.

Samelson, K. [1959]. Faktorisierung von Polynomen durch funktionale Iteration. *Abh. Bayer. Akad. Wiss. math.-naturw. Klasse, n. F.* **95**, 26 p. §7.4.

Schmidt, J. W., & H. Dressel [1967]. Fehlerabschätzungen bei Polynomgleichungen mit dem Fixpunktsatz von Brouwer. *Numer. Math.* **10**, 42–50. §6.4.

Scholz, N. [1965]. *Aerodynamik der Schaufelgitter*. Braun, Karlsruhe. §5.8.

Schrack, G. F. [1967]. *Lower bounds to the abscissa of stability of stable polynomials*. Dissertation ETH Zurich No. 4065. Elly Huth, Tübingen. §6.7.

Schröder, E. [1870]. Ueber unendlich viele Algorithmen zur Auflösung der Gleichungen. *Math. Ann.* **2**, 317–365. §6.12.

Schur, I. [1947]. Identities in the theory of power series. *Amer. J. Math.* **69**, 14–26. §§1.7, 1.9.

Sebastiao e Silva, J. [1941]. Sur une méthode d'approximation semblable à celle de Graeffe. *Portugal. Math.* **2**, 271–279. §7.4.

Sedow, L. I. [1965]. *Two-dimensional problems in hydrodynamics and aerodynamics*. Wiley-Interscience, New York. §5.8.

Siegel, C. L. [1942]. Iteration of analytic functions. *Ann. Math.* **43**, 607–612. §6.12.

———— [1969]. *Topics in complex function theory, Vol. I: Elliptic functions and uniformization theory*. Wiley-Interscience, New York. §5.13.

Sluis, A. van der [1970]. Upperbounds for roots of polynomials. *Numer. Math.* **17**, 250–262. §6.4.

Smythe, W. R. [1939]. *Static and dynamic electricity*. McGraw-Hill, New York. §5.7.

Soukup, J. [1969]. A method for finding the roots of a polynomial. *Numer. Math.* **13**, 349–353. §6.14.

Specht, W. [1958]. Algebraische Gleichungen mit reellen oder komplexen Koeffizienten. *Enz. Math. Wiss.* Bd. I/1, H. 3.II, 2. Aufl. Teubner, Stuttgart. §§6.1–6.5, 6.7, 6.8.

Spira, R. [1967]. Zeros of approximate functional representations. *Math. Comp.* **21**, 41–48. §6.11.

Stewart, G. W. III [1968]. *Some topics in numerical analysis*. Oak Ridge National Laboratory Report ORNL-4303. Oak Ridge,Tennessee. §6.8.

———— [1971]. On a companion operator for analytic functions. *Num. Math.* **18**, 26–43. §§7.6, 7.7.

———— [1971]. Error analysis of the algorithm for shifting the zeros of a polynomial by synthetic division. *Math. Comp.* **25**, 135–139. §6.1.

Stiefel, E., and G. Scheifele [1971]. *Linear and regular celestial mechanics*. Springer, Berlin. §§2.5, 4.5.

Svejgaard, B. [1967]. Zeros of polynomials. *BIT* **7**, 240–246. §6.14.

Szegö, G. [1959]. *Orthogonal polynomials*, rev. ed. Amer. Math. Soc., New York. §§2.5, 3.4, 7.7.

Takahashi, S. [1929]. Einige Sätze über die Lagen der Wurzeln algebraischer Gleichungen. *Tôhoku Math. J.* **31**, 274–282. §6.5.

Talbot, A. [1960]. The number of zeros of a polynomial in a half-plane. *Proc. Camb. Phil. Soc.* **52** (2), 132–147. §6.7.

Todd, J. [1965], On smallest isolated Gerschgorin disks for eigenvalues. *Num. Math.* **7**, 171–175. §7.9.

Traub, J. F. [1966]. A class of globally convergent iteration functions for the solution of polynomial equations. *Math. Comp.* **20**, 113–138. §§6.9, 7.4.

BIBLIOGRAPHY

Traupel, W. [1944]. Die Berechnung der Potentialströmung durch Schaufelgitter. *Schweizer Archiv* **10**, 363–378. §5.8.

――― [1966]. *Thermische Turbomaschinen*, 1. Band. 2. Aufl. Springer, Berlin. §5.8.

Varga, R. S. [1964]. On smallest isolated Gerschgorin disks for eigenvalues. *Num. Math.* **6**, 366–376. §7.9.

Ward, J. A. [1957]. The downhill method for solving $f(z)=0$. *J. Assoc. Comp. Mach.* **4**, 148–150. §6.14.

Watson, G. N. [1914]. *Complex integration and Cauchy's theorem*. Cambridge University Press, Cambridge. §§4.3, 4.6–4.9.

――― [1944]. *A treatise on the theory of Bessel functions*, 2nd ed. Cambridge University Press, Cambridge. §§1.9, 2.5, 3.4, 4.5, 7.9, 7.S.

Weierstrass, K. [1903]. Neuer Beweis des Satzes, dass jede ganze rationale Funktion einer Veränderlichen dargestellt werden kann als Product aus linearen Funktionen derselben Veränderlichen. *Ges. Werke* **3**, 251–269. §§2.4, 6.9.

Weinig, F. [1935]. *Die Strömung um Schaufeln von Turbomaschinen*. Barth. Leipzig. §5.8.

Weyl, H. [1924]. Randbemerkungen zu Hauptproblemen der Mathematik, II. Fundamentalsatz der Algebra und Grundlagen der Mathematik. *Math. Z.* **20**, 131–150. §§2.4, 6.11.

Wijngaarden, A. van [1953]. A transformation of formal series. *Indagationes Math.* **15**, 522–543. §3.6.

Wilkinson, J. H. [1963]. *Rounding errors in algebraic processes*. Her Majesty's Stationary Office, London. §§4.10, 6.9.

Whittaker, E. T., and G. N. Watson [1927]. *A course of modern analysis*, 4th ed. Cambridge University Press, Cambridge. §§1.9, 2.2–2.5, 2.7, 3.1–3.6, 4.3–4.5.

Whitworth, W. A. [1942]. *Choice and change*. Stechert, New York. §7.3.

Wolfenstein, S. [1967]. Proof of the fundamental theorem of algebra. *Amer. Math. Monthly* **74**, 853–854. §2.4.

Wronski, H. [1811]. *Introduction à la philosophie des mathématiques*, Paris. §1.3.

Added in proof:

Gould, H. W. [1974]. Coefficient identities for powers of Taylor and Dirichlet series. *Amer. Math. Monthly* **81**, 3–14. §1.7.

INDEX

Principal references (referring to definitions) are printed in bold face.

Abel's limit theorem, **82**, 145, 565
   applications of, 83
absolute value, **5**
accumulation, point of, **73**, 88, 146, 147, 152, 655
addition theorem, for Bessel function, **223**, 228
   for exponential function, **107**, 110
   for exponential series, **21**, 122
algebra, **70**
   complete normed, see Banach algebra
   normed, **70**
algebraic complement, 16, 17, **593**
algorithm, for computing poles of meromorphic function, 608–651
   for computing winding number, 239
   for evaluating derivatives of polynomial, 433–437
   for partial fraction decomposition, 553, 557, 562, 656
   for starting Bernoulli's method, 589
   for zeros of polynomials, 497–551
      conditionally convergent, **489**
      convergent, **498**, 549
      globally convergent, **498**
      locally convergent, **498**
      unconditionally convergent, **498**
      uniform convergence of, **498**, 506, 513, 517, 522, 549
   see also quotient-difference algorithm
almost unit, **45**, 47, 49
amount of work, for partial fraction decomposition, 559, 561
   to construct Hadamard polynomials, 634
   in methods of search, 505
   in Weyl's exclusion algorithm, 521
anomaly, eccentric, **119**
   mean, **119**

apolar polynomial, **472**, 473, 474
arc, **165**, 187, 230, 231
   analytic, **391**
   differentiable, **166**
   geometrically equivalent, **166**, 192
   homotopic, **167**
   initial point of, **165**
   piecewise differentiable, **166**, 190, 191
   piecewise regular, **166**, 191, 232, 233
   opposite, **192**
   regular, **166**
   terminal point of, **165**
argument, **5**
   continuous, **230**, 485
   increase of, 231
   principal value of, **294**, 521, 563
   principle of, 234, 276, **278**, 383, 485, 515
area, 339
arithmetic, circular, **475**, 479, 483, 500, 534, 551
arrangement, combinatorial, 578–580
associative law, **2**, 3, 11, 45, 46, 303
astroid, 430
asymptotic expansion, 81

Bairstow's method, 500, 559
Banach algebra, **71**, 72, 77, 81, 87, 89, 90, 90, 92, 93, 95, 99, 101, 103, 133, 143
Banach space, 67, **69**, 74, 75
basis, of natural logarithms, 107
Bernoulli numbers, **13**, 64, 111, 112, 269, 661
Bernoulli polynomials, **276**
Bernoulli's law (in hydrodynamics), 365, 368
Bernoulli's method, 500, **588**, 589, 653, 663
Bessel function, 28, 164, 185, 220,

673

INDEX

221, 607, 619
  addition theorem for, **223,** 228
  asymptotic behavior of, 228
  complex zeros of, 651, 662
  Poisson's integral for, **228**
  zeros of, 138, 284
Bessel's integral, **221,** 227
binary representation, 581
binomial coefficient, **21,** 461
  generalized, **22**
binomial series, **23,** 41, 50, 92, 93, 102, 569
binomial theorem, **22,** 23
Birkhoff's theorem, on polynomial zeros, **456,** 512
Blasius, formula of, **367**
boundary, free, 368
  mapping of, 382
  of set, 203, **235**
boundary layer theory, 425
boundary point, accessible, 385
  of set, 235
breakthrough, electrostatic, 422

cable, excentric, 346
    suspended, 354
  capacity, of condenser, **350**
    per unit length, 350
Cardano, formulas of, 121
cardoid, 194
Casorati-Weierstrass theorem, **215,** 216, 218, 378
Cauchy's estimate, **84,** 86, 90, 91, 96, 98, 134, 156, 159, 211, 332, 530, 597, 657
Cauchy-Hadamard formula, **77,** 142, 155, 157, 571, 656
Cauchy inclusion radius, **458,** 459, 462
Cauchy index, **443**
  for interval, **444,** 485, 486, 487
Cauchy integral formula, **211,** 230, **245,** 388, 389, 600, 629
  for holomorphic functions, 331, 332
Cauchy kernel, 207
Cauchy product, **10,** 11, 41, 46, 90, 219, 226
Cauchy-Riemann equations, **326,** 328, 336, 544
Cauchy sequence, **69,** 134
  convergent, 69

Cauchy's integral theorem, 232
  for disks, **198**
  extended form of, **204**
  for holomorphic functions, 329, 333
  for homotopic curves, 200, 207
  for simply connected regions, 199, **201,** 202, 203, 204
center of mass, 341
circle, generalized, **309,** 317, 318, 389
circulation, **361,** 370
  necessity of, 365
chain rule, for formal power series, **40,** 50, 58
channel, flow through, 357
  with corner, 412
  suddenly narrowing, 412
  turnabout, 413
characteristic, of field, 8
closed curve, homotopic, **199**
closure, **73**
cluster, of polynomial zeros, 499, 514
coefficient, of formal power series, **10**
combinatorial analysis, **578,** 662
commutative law. **2,** 8, 11, 50, 70, 303
complex plane, extended, **301**
component, of open set, **145,** 146, 234
  of set, 637, 638
composition, of analytic functions, 95, 116, 404
  of formal power series, **36,** 45, 46, 50, 55, 132
  of functions, **94**
condenser, 349
  cylindrical, 351
  plate, 344, 355
  semi-infinite, 409, 426
conductor, 342
confluence, method of, **599,** 603, 608, 624
conjugate, of complex number, **7**
continuation, analytic, 146, 148
  along arc, 166
  by continuous continuation, 387
  by functional relation relationships, 152
  by patching, 149
  by power series, 149, 150, **170**
  by rearranging power series, 150, 170
  by reflection, 389
  by removing singularities, 148
  fundamental lemma on, **150**
continued fraction, 490, 614, 663

INDEX 675

continuity, modulus of, 190
  of zeros of polynomial, 281
control theory, 487
convergence, cubic, 538
  locally uniform, **160,** 161, 184
  order of, **498**
  quadratic, 535, 537
  of series, 75
  speed of, 526
  uniform, **74,** 133, 134
  of power series, 81
convergence function, 501, 503
  of exclusion test, 518
  linear, 506
convex hull, closed, 464
corner, concave, **423,** 425
  convex, **429**
  re-entrant, **423**
  rounding of, 422
cotangens function, **110,** 111, 137, 660
covering, of set, **160,** 503
  optimal, 512
cosine function, **110**
cross, mapping of, 413
crosscut, 385
curvature, **186**
curve, closed, **195,** 233
  rectifiable, 188
  simple, 235
  starlike, 205
cycloid, 186

deflation, 438
derivative, complex, 327
  of analytic function, 142
  of complex function of real variable, **180,** 181
  of formal power series, **18,** 135, 141, 142
  of formal Laurent series, **53**
  of function, 129
  logarithmic, 276, 464, 478
  of polynomial, with respect to point, **466**
Descartes, rule of signs of, 442
descent, steepest, 544, 546, 552
descent function, **542,** 546
  continuous, 542, 543, 545
  Kellenberger's, 548, 549
  Nickel's, 547
  Ward's, 547
determinant, 16, 17, 496, 591, **592,** 663

  Laplacian expansion of, 592, **593,** 601
  minor of, **592**
  complement of, **592**
  algebraic complement of, **593**
  of product of matrices, 594
  of sum of matrices, 593, 597
  Vandermondian, 648
diffeomorphism, piecewise, **204,** 246
difference, divided, 247, 248
  of polynomial, 439
difference equation, 584, 588, 663
  basis of solution space of, 586
  of infinite order, 589
  solution of, 585
difference operator, 154
difference scheme, 608
differential, 324, 333
differential equation, formal, **20**
differentiation rules, 181
Dirichlet integral, **340**
Dirichlet problem, 337, 342, 343, 350, 373
distance, 310
  chordal, 311
  Euclidean, 523
distributive law, **2,** 3, 11, 46
  right, for composition, 37, **38**
divisor function, 162
divisor of zero, 2, 11
Dixon's formula, 33, **43,** 44
domain, of definition, **72**
  of values, **72,** 286
drawer problem, 578–580

eigenvalue, of matrix, 636, 638–640
  continuity of, 636, 663
eigenvector, 637
electrostatics, 342
element, regular, **72**
ellipse, 194, 206, 263, 295
  circumference of, 417
emergency step, 547
Eneström's theorem, 284, **462,** 497
epicycloid, 194
equation, analytic, 101
  biquadratic, 121
  cubic, 118
  Diophantine, **582,** 584
  trinomial, 118, 121
equivalence, 299
estimate, a posteriori, 498

Euclidean algorithm, 487, 489, 490
    for rational functions, 447
Euler, first identity of, **32, 44**
    formulas of, **110**
    result of (on representation of integers), 582
    second identity of, **40,** 150
Euler's constant, **275**
evolute, **186**
exclusion algorithm, 500
exclusion test, **517,** 518–522
    convergent, **518**
    one-sided, 520
exponential, of matrix, **122**
    differential equation satisfied by, 123
exponential function, 87, 106, **107,** 284, 293
exponential series, **20,** 50
extension, analytic, **144,** 145–147

factorial, generalized, 24
fast Fourier transform, 549, 552
Fibonacci sequence, **13,** 587, 590
field, **3**
field of complex numbers, **3**
field strength, complex, 346
fixed point, of analytic function, 523–528
    of Moebius transformation, 304, 306
flow, against dam, 407
    against rounded dam, 424
    around airfoil, 365
    around tilted blade, 367
    irrotational, 356
    past obstacle, 358
    through circular grid, 371
    through cascade of airfoils, 371
flow potential, 356
flux, 330, **340**
force, complex, 365
    moment exerted by, 369
    exerted on obstacle, 365, 367
form factor, 346
Fourier-Budan, theorem of, 443
Fourier series, 224, 225, 226, 229, 251, 374, 377, 577
    absolutely convergent, 71, 72
    with coefficients that are rational functions, 269
    considered as Laurent series, 223, 224
Fourier transforms, of rational functions, 254

Fresnel's integral, 205
function, 72
    analytic, **139**
    abstract, 138
    composition of, 95
    defined by definite integral, 183
    defined by limit process, 161
    at infinity, **216**
    inversion of, 97
    at point, 87
    scalar, 89, 139
    on set, 144
    complex, of real variable, 180
    complex-differentiable, **325,** 328
    complex-valued, 286
    continuous, 73
    differentiable, 143
    domain of definition of, 72
    domain of values of, 72
    doubly periodic, 421
    elliptic, 422
    even, 93
    generating, *see* generating function
    harmonic, 332, **336**
    homogeneous, 646
    holomorphic, 143, 328
    inverse, 95
        of exponential function, 113, 291, 293
        of Joukowski function, 296
        of Moebius transformation, 302
    linear, 288
    matrix-valued, 122, 138
    meromorphic, **214,** 215, 249, 312, 596, 602, 607, 612, 641, 655
    odd, 93
    one-to-one, **95,** 291
    periodic, **109,** 223, 224, 227, 269, 659
    quadratic, 289
    range of, 72
    rational, 139, **217,** 553, 625, 626
    scalar complex, **139**
    summatory, **265**
    trivial, **95**
    univalent, 291
    value of, 72
functions, analytic, sequence of, *see* Weierstrass double series theorem
fundamental theorem of algebra, **105,** 159, 378
    proofs of, 138, 280, 552

Galois, theorem of, 497
gamma function, 153, 186, 412, 655
Gauss-Lucas theorem, 463, **465**, 474, 518
Gaussian elimination, 554, 627
Gegenbauer polynomials, **43**, 121, 137
generating function, 60, 61, 138, 578–585, 587, 590
   of Bernoulli numbers, 13, 111
   of Bessel functions, 221
   of Jacobi polynomials, 119
   of Laguerre polynomials, 119
   of Legendre polynomials, 60, 247
geometric series, 207, 210, 225, 460, 580, 597
Gershgorin bounds, 663
Gershgorin disk theorem, **636**, 639, 640
Goursat's theorem, 285, 330
Grace, theorem of, **472**, 473, 474
gradient, complex, **334**, 346, 349, 544
Graeffe's method, 500, 552
graph, of real function, 286
grid, rectangular or hexagonal, 503, 505
group, **1**, 2, 8, 45
   abelian, **2**, 608
   non-abelian, 8
   of Moebius transformations, 302, 307

Hadamard, theorem of, on product of power series, 576
Hadamard polynomials, **625**, 626, 627, 630, 633, 634, 642, 663
Hadamard product, **13**, 576
half-plane, right, **145**
Hankel determinants, 574, 591, **594**, 596, 600, 615, 616, 662, 663
   associated with rational function, 603
   asymptotic behavior of, 602, 612, 646
   recursive computation of, 596
Hankel polynomials, **622**, 625
Heine-Borel lemma, **160**, 161, 168
Helmholtz equation, **342**
Hermite polynomials, **37**, 43
hodograph plane, 369
homeomorphism, 382
homotopy, **167**, 179
Horner algorithm, **435**, 437, 508, 513, 551, 561, 567, 569
Horner scheme, **437**, 438, 556
Hurwitz, theorem of, **283**, 284
hyperbola, 295, 343

mapping of interior of branch of, 395
hypergeometric series, classical, **27**, 34, 39, 40, 42, 44, 48, 62, 118, 150, 226, 229, 417, 421
   confluent, **27**, 29
   zeros of, 662
   generalized, **27**, 32–34, 37, 39, 40, 43, 44, 47, 62, 65, 220, 221
   differential equation satisfied by, 29
   radius of convergence of, 80
hypothesis (H), 642, **649**

identity element, **1**, 8, 11, 45, 303
identity principle, for analytic functions, 88, 147
imaginary part, 7
imaginary unit, 4
inclusion disk, 484, 540
inclusion radius, **457**, 473, 500
   proper, **457**
inclusion region, for poles of rational function, 640
inclusion set, 514
index, critical, **642**, 645, 646, 647, 650, 653
influence, 352
instability, numerical, 613
interpolation, by polynomials, 572
   Lagrangian formula for, 601
   by sums of exponentials, 571, 577, 622, 662
   algorithmic solution of, 631–632
integral, along arc, **188**, 189–194
   along closed curve, 195, 200–206, 233, 243, 245, 246, 247, 249, 252, 257, 265, 272, 277, 350, 351, 353, 366, 628
   elliptic, 417
   improper, 251
   of rational functions, 253–254, 263, 264
   with fractional powers, 257, 264
   with logarithms, 257, 264
   with respect to arc length, **193**
   with respect to parameter, **183**, 184–186
   of trigonometric functions, 183
   improper, 254, 255, 262–264
integral domain, **2**, 11, 12, 51, 89
integration, numerical, 234, 517, 590, 621
   of ordinary differential equations, 662
interior, of Jordan curve, **235**, 243, 245–249,

252, 278, 280, 282, 382, 383
inverse, of element in group, **1**, 4, 8, 12, 46
  in Banach algebra, **72**
  of Moebius transformation, 303
inversion $z \to z^{-1}$, **303**, 381, 591
involution, **306**
isomorphism, 15, 46, 302
iterates, of Moebius transformation, 306
iteration, 523–552
iteration function, 533, 542
  of arbitrary order, 529
  Laguerre's, 500, **532**, 534
  Newton's, **529**, 531, 532
  of order three, 531
  order of, **526**
iteration sequence, **523**, 527, 546

J. C. P. Miller formula, **42**, 43, 64, 65, 579
  extension of, 55
Jacobi polynomial, 119
Jacobian matrix, of conformal map, 339
Jacobi's identity, for Hankel determinants, **595**, 611
  for Hankel polynomials, **622**
Jacobi's transformation, **63**, 65, 228
Jordan block, 128, 130
Jordan canonical form, 127, 130, 305, 306
Jordan curve, **235**, 243, 245, 246, 266, 269, 272, 278, 279, 282, 330, 331, 366, 383, 485, 515, 520, 628
Jordan curve, positively oriented, **235**
Jordan curve theorem, **235**, 285
Jordan region, **382**, 383, 384, 391
Jordan's lemma, **255**, 256
Joukowski flow, 367
Joukowski hypothesis, 365, 370
Joukowski map, **294**, 299, 315, 345, 353, 359, 367, 394, 395, 415

Kahan's theorem, 534
Kakeya-Eneström theorem, 284, **462**, 497
Kapteyn series, **165**
Kepler's equation, **119**, 120, 130, **226**, 230
Kummer's first identity, 29

Lagrange-Bürmann expansion, **58**, 65, 97, 100, 101, 104, 138, 226, 249, 370, 529
  convergent, **100**, 102, 104, 119, 120, 138
Laguerre polynomial, **119**, 618

generating function of, 119
Laguerre's theorem, on polynomial zeros, 466, 473, 480
  generalization of, 478, 534
Laplace's expansion, 592, 593, 607
Laurent series, **209**, 224, 226, 369, 370, 381, 559
  formal, 51, **52**, 54–56, 59, 627, 628, 630
  multiplication of, 219
  uniqueness of, 209, 219
Laurent's theorem, 209
Legendre polynomial, **42**, **60**, 164, 284
  generating function of, 60, 247
Legendre function, 226
Lehmer covering, 512
Lehmer's method, 492
Leibniz rule, for differentiation of product, 144
lemniscate, 299
length, of rectifiable curve, **188**
limes inferior, **76**
limes superior, **76**
limit, of function, 73
Liouville's theorem, **159**, 218, 400, 525, 659
logarithm, **112**, 147, 179, 203, 232, 233, 257, 294
  analytic branch of, 203
  of matrix, **131**, 138
  principal value of, **114**
  simple branch of, **114**
logarithmic series, **49**, 50, 115
L-R algorithm, 641

map, 377
  conformal, **329**
  topological, **382**
mapping, of interior of ellipse, 394
  of interior of parabola, 392
  of regular n-gon, 411
  of triangle, 407
  of two-gon, 406
mapping theorem, for disks, **321**
  for simply connected regions, **378**
matrix, diagonal, 306, 630
  equivalent, **299**
  exponential, **122**
  non-singular, 16, 72, 303
  partitioned, 653
  semi-circulant, **14**
  similar, 123

unitary, 313, 314
upper triangular, **14,** 45, 47, 314
triangular, 171
tridiagonal, 635, 653
maximum, principle of, **86,** 332, 355, 379, 541
  for scalar analytic functions, **104**
method of descent, 541
metric, chordal, 311
Mittag-Leffler theorem, 655
model problem, 344, 346, 352, 357, 358, 364, 368, 407, 409
modulus, **5**
  of doubly connected region, 386, **391**
Moebius transformation, **300,** 348, 376, 384, 389, 418, 419, 463, 465, 474, 491, 524, 662
  composition of Schwarz-Christoffel map with, 404
  entire, 316
  group of, 302
  standard representation of, 304, 314
Moivre's formula, **9**
moment of inertia, polar, **341**
monodromy theorem, **167**
  for simply connected regions, **168**
Montel's theorem, on polynomial zeros, 459

neighborhood, **73**
Neumann series (of Bessel functions), **164**
Newton correction, 454, **467,** 468, 477, 478, 544
Newton's formula, 479
Newton's method, 121, **528,** 532, 533, 547, 651
  for polynomials, 534–541
  with wrong sign, 534
nilpotent, **79**
non-unit, in integral domain of formal power series, **12,** 35, 38, 40, 41, 43, 45, 51
norm, **68,** 173
  column sum, **71**
  Euclidean, **68**
  induced, **71**
  row sum, **71**
  spectral, **71**
  of vector, 173
normality, of formal power series, 605
null function, **464**

Nyquist diagram, 279, 280

obstacle, 380
orthogonality, of Hadamard polynomials, 628
Osgood-Caratheodory theorem, **383,** 384, 397

Padé table, 663
parabola, 290, 392
partial fractions, 159, 553–661
partial fraction decomposition, **218,** 565, 631, 662
  of rational function, 553, 557
  real, 556
partial fraction expansion, 658
  of cotagens function, 660
  of meromorphic function, 655
path, **165**
period, **109**
permutation, 591, 598, 645
Picard's theorem, 216
pivotal points, system of, 187
Plana summation formula, 271, **274,** 285
plane, cut along negative real axis, **145**
  model, 337
  physical, 337
Pochhammer symbol, **24,** 25–30, 32–34, 39, 40, 43, 47, 48, 51, 53, 54, 57, 62, 94, 117, 121, 148, 151, 155–157, 163–165, 174, 254, 417, 633
point at infinity, **300**
Poisson equation, 371
polar representation, of curve, **236**
pole, **214**
  order of, **214**
poles, of equal modulus, 641
polygon, mapping of exterior of, 406, 413
  mapping of interior of, 396
  unbounded, 401
polynomial, **10,** 85, 105, 139, 159, 198, 264, 281, 283
  characteristic, 635
  convergence-generating, 655, 656
  deflated, 513
  reciprocal, **492**
  self-reciprocal, **496**
  stable, **488**
  Taylor coefficients of, 437
  Taylor expansion of, 433
  trigonometric, 284
polynomial zeros, determination of, 137,

497–550, 617–651
polynomials, algebraic theory of, 433
  analytic theory of, 551
power function, general, 115
  principal value of, 116
  simple branch of, 116, 117
power series, formal, 9, 576, 578, 585, 604, 609
  differentiation of, 18
  hypernormal, 606, 607, 615, 617, 620, 621
  normal, 605, 610, 611, 613, 628–630, 638, 642
  radius of convergence of, 77, 151, 156, 656
  reciprocal of, 12, 615
  reversion of, 47, 48, 49, 55–61
  scalar, 84
  ultimately normal, 605
power sum, 454, 481
powers, with matrix exponent, 133
prime end, 385
principal part, 212, 655, 656, 657, 662
  at infinity, 216
  of Laurent series, 555
  of meromorphic function, 596
  of rational function, 217, 574, 583
principal value, of argument, 294
  of logarithm, 294
principal value integral, 260, 262, 265
principle, of boundary map, 401, 424
  of maximum, 86, 104, 332, 355, 541
product, of elements of group, 1, 10, 45
product rule, of differentiation, 19, 53
profiles, cascade of, 367
projection, stereographic, 308
proximity test, 501, 514–519
  equivalent, 503
  with linear convergence function, 506–508, 510, 512, 522
  sharp, 508
  similar, 503

quadratic factor, of denominator of rational function, 559
  multiple, 561
quotient field, 51
quotient-difference algorithm, 17, 500, 588, 591, 613, 640, 641, 663
  extended, 662

progressive form of, 614, 619
quotient-difference scheme, 649
  associated with formal power series, 610
  construction by diagonals, 619, 620
  existence of, 610
  extended, 615
  of rational function, 650, 653

range, of arc, 165
  of function, 72
rational function, 252, 253, 256, 264, 267, 553–584, 603, 613, 626, 638, 639, 653
  integration of, 562
real part, 7
rearrangement, of power series, 140, 142, 144, 151, 152, 170, 171
reciprocal, of formal power series, 12, 13
  of formal Laurent series, 53
rectangle, mapping of, 416
reduction factor, 174
reflection principle, 389, 390, 391, 398, 399
region, 145
  circular, 405, 406, 464, 465–468, 472, 474, 475, 477, 480, 535, 537, 538
  convex, 168
  homeomorphic, 169
  simply connected, 168, 170, 199
  starlike, 285
remainder, in polynomial division, 446
representation, of function by power series, 87
residue, at isolated singularity, 242, 243, 244, 253, 256, 257, 259, 261, 267, 270, 661
  determination of, 243–245, 559, 577, 630–632
  of formal Laurent series, 53, 54–59, 61, 63, 101, 104, 119, 370
  calculation of, 55, 57, 118
  of rational function, 630–632
resolvent, of eigenvalue problem, 607
reversion, of almost unit, 47, 48, 49, 55–61
rhombus rules, 609, 616, 617
Riemann integral, 181, 184
Riemann mapping theorem, 380, 525
Riemann number sphere, 308
  north pole of, 307
  rigid motions of, 312
Riemann sum, 181, 182, 188, 189,

INDEX

190, 193, 659
Riemann zeta function, 279
Riemannian theory, of analytic functions, 143
Riemann's theorem, on isolated singularities, 148, **213**
rotation, 288, 303
Rouché's theorem, **280**, 284, 494, 495, 524, 535
rounding errors, 613, 651
rounding factor, 429
rule, of composition, **1**
Rutishauser's rule, **642**, 645, 651, 652, 653

Saalschütz formula, **33**, 34, 42
Saint-Venant's theorem, **375**
Schur-Cohn algorithm, **494**, 508, 520, 535
  implementation of, 552
Schur-Cohn test, **508**, 509, 510, 519
Schur-Jabotinski theorem, **55**
Schur transform, of polynomial, **493**, 496, 497
Schwarz, lemma of, **379**, 386, 387, 525, 526
Schwarz-Christoffel map, 396–416
  rounding corners in, 422–431
search, methods of, 499, 513
section, of matrix, **14**
semigroup, **2**, 3, 12, 13
sequence, geometric, 509
  of analytic functions, 159
  of entire functions, 284
series, of Bessel functions, 164, 165, 227, 229
  of rational functions, 267, 564, 569
  uniform convergence of, 75
  see also Fourier series; Laurent series; and power series
set, bounded, **160**
  closed, **160**
  compact, **160**
  convex, **168**
  open, **139**, 144
    connected, **145**
sign change, **439**, 487, 489
similarity, **288**
sine function, **110**, 266, 660, 661
singular part, *see* principal part
singularity, essential, **215**
  isolated, 148, **212**, 656, 657

at infinity, **215**
removable, 148, **213**, 379
skew field, **8**
slit, region with, 385
sources and sinks, of flow, 356
space, complete, **69**
  linear, **67**
  metric, **310**
  normed linear, **68**
square, pregnant, **516**
  suspect, 517
square root, under composition, **50**
stability, 279
  abscissa of, **490**, 491
  numerical, **499**, 662
  of polynomial, **551**
stagnation point, 359
star, mapping exterior of, 395, 415
Stieltjes integral, 605
Stirling's formula, 176, 222
Stodola's condition, for stability of polynomial, 488
Stokes, theorem of, 330, 360
streamline, 340, **359**
stretching, **288**
Sturm sequence, **444**, 516
subdivision, **187**, 638
  norm of, **187**
summation, by Lagrange-Bürmann formula, 59–61, 119
  of Fourier series, 270
  of series of rational functions, 267, 264–569
symmetry principle, 320, 392, 396
  see also relection principle
symmetry, with respect to circle, 317
  with respect to generalized circle, 318, 31

Tangens function, **110**, 111, 112, 266
tangent vector, 166, 328
Taylor expansion, **142**, 248, 547, 548, 577
  of Hadamard polynomials, 634
  of meromorphic function, 612
  of rational function, 569, 586, 605, 613, 638, 639, 662
Taylor's formula, 545
Taylor's theorem, **142**, 268, 528
thunderstorm, 352
torsional rigidity, **372**, 377
transfer function, 279

transformation, bilinear, **300**
   fractional linear, **300**
   quadratic, of hypergeometric series, 42, 43, 229, 421
translation, **303**
transplant, of harmonic function, 337
transplantation, conformal, **334,** 335, 343, 372
transposition, **591**
triangle inequality, for complex numbers, 7
   for vectors, 68
turbulence, 422

undetermined coefficients, method of, 553, 562
unit, of integral domain, **12,** 91
unit disk, punctured, **145**

Van Vleck's theorem, on polynomial zeros, **459**
Vandermonde's theorem, **24,** 28, 29, 48, 92, 172, 176, 461
variables, separation of, 342
velocity, complex, **364**
Vieta's formulas, 453, 493

Watson's formula, **62**

Weierstrass double series theorem, **133,** 161, 184, 186, 222, 661
Weierstrass M-test, **75**
weight function, 630
Weyl's algorithm, **517,** 522, 550, 551
Whipple's formula, **43**
winding number, **233,** 277, 383, 485, 515, 516, 517, 520
   numerical determination of, 234, **239**
work function, **506,** 509
Wronski formula, **17,** 616

zeros, clusters of, 514
   in half-plane, 485
   in interval, 448
   of analytic function, 89
   of entire function, by qd algorithm, 619
   of higher multiplicity, of polynomial, 438
   location of, 514
   of meromorphic function, by qd algorithm, 617
   multiplicity of, 89, 514
   number of, in disk, 491
   order of, **89**
   refining of, 484
   simultaneous determination of, 537
   specified number of, 450